W9-CBG-858

RIVERS OF THE U.S.

PENOBSCOT
CONNECTICUT
HUDSON
SUPERIOR
HURON
ONTARIO
WISCONSIN
MICHIGAN
ERIE
SUSQUEHANNA
ILLINOIS
WABASH
MISSOURI
OHIO
POTOMAC
ROANOKE
CUMBERLAND
TENNESSEE
PEE DEE
SAVANNAH
MISSISSIPPI
ALABAMA
APALACHICOLA
ST. JOHNS
MOBILE BAY

PETERSON FIELD GUIDE

TO

FRESHWATER FISHES

of North America North of Mexico

PETERSON FIELD GUIDE TO

FRESHWATER FISHES

of North America North of Mexico

SECOND EDITION

LAWRENCE M. PAGE
Illinois Natural History Survey and
Florida Museum of Natural History

BROOKS M. BURR
Southern Illinois University
at Carbondale

HOUGHTON MIFFLIN HARCOURT
BOSTON NEW YORK

ILLUSTRATIONS BY
Eugene C. Beckham III • John Parker Sherrod
Justin T. Sipiorski • Joseph R. Tomelleri

MAPS BY
Griffin E. Sheehy

SPONSORED BY
National Audubon Society
National Wildlife Federation
Roger Tory Peterson Institute

Text copyright © 2011 by Lawrence M. Page and Brooks M. Burr
Illustrations copyright © 2011 by Eugene C. Beckham, III; John Parker Sherrod; Justin T. Sipiorksi; and Joseph R. Tomelleri

Peterson Field Guides, Peterson Field Guides Series, and the Peterson Field Guides logo are registered trademarks of Houghton Mifflin Harcourt Publishing Company

All rights reserved

For information about permission to reproduce selections from this book, write to Permissions, Houghton Mifflin Harcourt Publishing Company, 215 Park Avenue South, New York, New York 10003.

www.hmhbooks.com

Library of Congress Cataloging-in-Publication Data
Page, Lawrence M.
A field guide to freshwater fishes of North America north of Mexico / Lawrence M. Page, Brooks M. Burr ; illustrations by Eugene C. Beckham III ... [et al.].
p. cm.
Rev. ed. of: A field guide to freshwater fishes : North America north of Mexico. 1991.
Includes bibliographical references and index.

ISBN 978-0-547-24206-4

1. Freshwater fishes—United States—Identification. 2. Freshwater fishes—Canada—Identification. I. Burr, Brooks M. II. Page, Lawrence M. Field guide to freshwater fishes. III. Title.

QL627.P34 2011

597.176097—dc22 2010049219

Book design by Anne Chalmers
Printed in China

SCP 10 9 8 7 6 5 4

The legacy of America's greatest naturalist and creator of this field guide series, Roger Tory Peterson, is kept alive through the dedicated work of the Roger Tory Peterson Institute of Natural History (RTPI). Established in 1985, RTPI is located in Peterson's hometown of Jamestown, New York, near Chautauqua Institution in the southwestern part of the state.

Today RTPI is a national center for nature education that maintains, shares, and interprets Peterson's extraordinary archive of writings, art, and photography. The Institute, housed in a landmark building by world-class architect Robert A. M. Stern, continues to transmit Peterson's zest for teaching about the natural world through leadership programs in teacher development as well as outstanding exhibits of contemporary nature art, natural history, and the Peterson Collection.

Your participation as a steward of the Peterson Collection and supporter of the Peterson legacy is needed. Please consider joining RTPI at an introductory rate of 50 percent of the regular membership fee for the first year. Simply call RTPI's membership department at (800) 758-6841 ext. 226, or email membership@rtpi.org to take advantage of this special membership offered to purchasers of this book. For more information, please visit the Peterson Institute in person or virtually at www.rtpi.org.

PREFACE

The first edition of this guide was completed in 1990 and published in 1991. Since then it has been a primary source of information on identification of North American freshwater fishes. This second edition increases the number of species in the guide from 768 to 909, incorporates new maps and several new and revised plates, and corrects errors. The increase in number of species is the result of adding 114 newly recognized species native to the U.S. and Canada, 19 marine invaders commonly found in freshwater, and 16 newly established non-native (exotic) species. Eight species recognized in the first edition were deleted as names were synonymized or as exotic species thought to be established disappeared. The ichthyofauna of the twenty-first century is not that of the twentieth century, and a revision of this guide was badly needed. We hope we have succeeded in making it current as well as more user-friendly. Suggestions for improvements and notifications of errors are welcome.—LMP and BMB

CONTENTS

How to Use This Guide

Naturalists, anglers, and aquarists derive pleasure and knowledge from observing and catching fishes. Ichthyologists and other scientists study fishes to learn more about the evolution of life, the history of our continent, and how natural resources can be better managed. For these interests and related endeavors, accurate identification of fishes is essential. This guide includes all fishes in fresh waters of North America north of Mexico.

Fishes are aquatic vertebrates with fins and gills throughout life. Currently recognized as valid are about 31,000 species, of which 831 species (3 percent of the total) are native to fresh waters of the United States and Canada. Another 58 species from elsewhere in the world have been established in our area, and 20 marine species are encountered often enough in fresh water to be included in this guide, bringing the total number of species to 909.

Of the 537 families of fishes, 34 (6 percent) are represented by 1 or more species native to freshwater lakes and streams of the United States and Canada, and another 11 families have marine species that occasionally enter our rivers. Eight other families are represented by introduced (exotic) species. Although our fish fauna represents a fraction of the world's total, it is Earth's most diverse temperate freshwater fish fauna.

All freshwater fishes known from North America north of Mexico are included in this guide. The *Peterson Field Guide to Atlantic Coast Fishes* and the *Peterson Field Guide to Pacific Coast Fishes* provide additional information on marine and brackish-water fishes likely to be encountered in fresh water.

Names

Most names of fishes used in this guide are those in *Common and Scientific Names of Fishes from the United States, Canada, and Mexico*, published in 2004 by a joint committee of the American Fisheries Society and the American Society of Ichthyologists and Herpetologists. In a few instances in which the committee changed a common name, we chose to keep the name used in the first edition of this field guide.

Scientific names of species consist of two Latinized and italicized words, e.g., *Lepomis punctatus*. The first is the genus, which begins with a capital letter. The second is the "specific epithet" and is not capitalized. A subspecies has a third descriptor, e.g., *Lepomis punctatus miniatus*. Genera are grouped into families (with names that end in *idae*), families into orders (ending in *iformes*), and orders into classes.

Illustrations

Color plates were painted from live fishes or, more often, from color photographs of live or freshly preserved fishes. Black-and-white plates depict fishes that lack bright colors or show little variation in color among closely related species. Fishes are not drawn to scale, but much larger species usually are shown larger than smaller species. The 57 plates (42 in color, 15 in black and white) show 824 individuals representing 677 species. Additional species are illustrated in text figures.

Measurements

Although ichthyologists use the metric system, guide users remain familiar with inches, feet, and pounds. Measurements are given in both systems. A short rule comparing metric and U.S. units appears below and on the back cover. The maximum total length known (tip of snout, lip, or chin—whichever is farthest forward—to end of longer caudal fin lobe) is given for each species. For small fishes, this number is given in quarter-inches and tenths of centimeters, for intermediate fishes in inches and centimeters, and for large fishes in feet and meters.

If the maximum length recorded was given originally in centimeters, it was converted to inches; if in inches, it was converted to centimeters. Rounding from centimeters to quarter-inches can give various results; for example, 7.4 through 7.9 cm are all given as equivalent to 3 in.

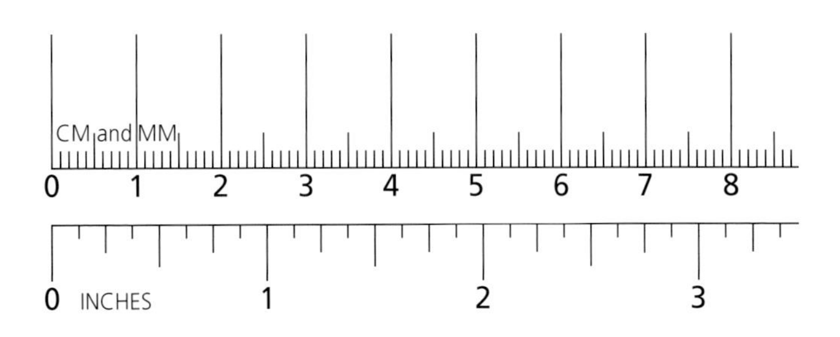

Accounts

Family accounts provide information on distinguishing characters (often anatomical) and distribution. Numbers in parentheses following family names are numbers of native species in the United States and Canada; if introduced species are in our area, number of natives is followed by number of exotics.

Generic accounts are given for large genera and for small genera in which all species share characters useful in identification. *If a character is described in a family or generic account, it usually is not repeated in a species account.*

Species accounts begin with common and scientific names. In the upper right-hand corner of each account is the number of the plate or figure where the species is illustrated, or "Not shown" if not illustrated. A species is not illustrated if it is similar to another species.

Most species accounts contain the following four sections. A similar Species section is omitted if a species is easily identified, and a Remarks section is added if the species has subspecies or other noteworthy characters.

Identification: This section describes the most useful characters for identification. Usually these are color descriptions such as "black stripe along body," shape descriptions such as "dorsal fin origin behind pelvic fin origin," and unusual features such as "barbel at corner of mouth." The most prominent field characters are *italicized* and usually appear early in the account. Accurate field identifications sometimes require consideration of locality and habitat. Large specimens, especially colorful males, are easiest to identify. Positive identification of small or single individuals may require close examination; for that reason, we give some characters useful in identification of preserved fishes (numbers of scales, fin rays, and pharyngeal teeth, etc.).

A color description is included unless a species is noted to be similar or nearly identical to another species. Unless stated otherwise, the description is of an adult fish, and the fish is white below (breast and belly) and has clear fins, conditions that pertain in most species. In many fishes, females retain colors similar to those of young, but males become notably brighter or darker with age. During the spawning season, males often become much brighter in color than at any other time. When known to differ, both "average" and "breeding male" descriptions are given. In some fishes (e.g., darters), large males retain bright colors through much of the year; in others (e.g., most minnows), bright colors are present only during the spawning season.

Counts provided are those considered to be most important for identification and are total ranges unless they are preceded by "usually" or identified as modes (i.e., number[s] occurring most frequently). Counts of bilateral characters are given for one side only, e.g., six branchiostegal rays means six on each side. Pectoral and pelvic fins come one to a side and are referred to collectively (i.e., all four of them) as paired fins. We often discuss these fins and other paired structures (e.g., eyes) in the singular (e.g., pectoral fin, eye) to simplify comparisons between species. Dorsal, caudal, and anal fins are referred to collectively as median fins.

Range: A description of each species' geographic distribution is followed by a comment on abundance (e.g., "Rare"). All species vary in abundance with locality, and the statement on abundance is meant to apply over the species' range or, if introduced, over its range in the U.S. and Canada. The statement is not a relative comparison among species. For example, the Fountain Darter, *Etheostoma fonticola*, is common in its area but is considered an endangered species because it occurs in only one small area. *Abundant* means a species is almost certain to be found in its preferred habitat within its range (see "How to Observe Fishes"); *common* indicates a species is likely to be found; *fairly common*, may be found; *uncommon*, unlikely to be found; *rare*, very unlikely to be found. Species and subspecies described as *threatened* or *endangered* are those appearing on official lists of Canada (Species at Risk Act – SARA) and the United States (USFWS) as of 1 July 2010. Many species also are legally protected by states or provinces.

Habitat: Fishes vary widely in their restriction to particular habitats. Some are extremely limited (e.g., to springs); others can occupy habitats as different from one another as gravel riffles and swamps. For a stream-inhabiting species, a habitat description includes a statement on the size of stream the species generally occupies. Terms used are *streams* (any body of running water), *headwater* (a stream less than 3 ft. [1 m] wide during average condition), *creek* (3–15 ft. [1–5 m]), *small river* (15–80 ft. [5–25 m]), *medium river* (80–165 ft. [25–50 m]), and *large river* (more than 165 ft. [50 m]). A *basin* is a major drainage unit (e.g., Arctic, Hudson Bay, Great Lakes-St. Lawrence, Atlantic, Gulf, Pacific, Mississippi R., Ohio R., and Missouri R. basins) or an independent endorheic drainage unit (e.g., Bonneville basin). Component drainages may be referred to collectively as, for example, Atlantic drainages. A *drainage* is an interconnected group of streams entering an ocean or main river of a basin (e.g., Wabash R. drainage of the Ohio R. basin). A *system* is a subdivision of a drainage (e.g., Embarras R. system of the Wabash R. drainage).

For convenience, we make a distinction between Atlantic and Gulf slope drainages even though the Gulf of Mexico is part of the Atlantic Ocean. Atlantic Slope drainages are those entering the Atlantic Ocean from the Arctic Ocean to the southern tip of Florida. Gulf Slope drainages are those entering the Gulf of Mexico.

Composition of the stream or lake bottom (substrate) is of major importance in distributions of fishes, and habitat descriptions usually include statements on the type(s) of bottom material most often associated with the species. *Mud* refers to a soft bottom (clay or silt); *rock* refers to a hard bottom (gravel, rubble, boulders, or mixtures thereof). More precise terms, in increasing order of particle size, are *clay*, *silt*, *sand*, *gravel*, *rubble*, *boulders*, and *bedrock*.

Similar species: Comparisons are made with species that appear most similar. These species usually, but not always, are closely related forms. When there are many similar species, we compare those closest to the range of the species being identified.

Maps

Range maps are provided for all extant (and some extinct) *freshwater* fishes *native* to North America north of Mexico (except a few restricted to single localities). Range maps are not provided for introduced species or marine invaders. Production of range maps relied heavily on state, provincial, and regional "fish books."

A map shows the total range of a species based on historical and recent records; that is, a map includes an area or drainage even if that population is believed to be extinct. Within these ranges, large gaps in distribution occur in ecologically unsuitable areas. For example, the Rainbow Darter, *Etheostoma caeruleum*, ranges over much of the eastern U.S. but lives in rocky riffles and is absent from many areas within its range. Ranges in Mexico are shown for U.S. species that narrowly extend into Mexico.

Maps for native species with transplanted populations include areas where populations are known to be established. However, species that are continuously being stocked—notably some basses, sunfishes, and trouts—may be found almost anywhere in the U.S. and southern Canada. The notation "Introduced elsewhere" appears on maps for species that are likely to be found outside the range shown. The reader should consult a species account for additional information on geographic distribution.

Hybridization

Crosses between species occur occasionally in nature and are especially common in sunfishes. Identifications of hybrids (as species A x species B) usually are difficult. In making identifications, keep in mind that hybridization occurs most often between closely related species, and hybrids usually have characters intermediate to those of parental species. As aquatic environments degrade, hybridization increases, presumably because of difficulty fishes have in recognizing spawning partners in turbid and polluted water.

Intergrade zones are areas where individuals (known as intergrades) are intermediate in characters used in the recognition of two subspecies. Intergrades may be intermediate because of the mixing of genes ("gene flow") of two subspecies or because of variable environmental conditions leading to selection for characters intermediate to those of two subspecies. Intergrades are named as hybrids between two subspecies (e.g., *Percina caprodes caprodes* x *P. caprodes fulvitaenia*).

How to Observe Fishes

You can watch fishes in clear water from stream banks and lakeshores, and although at first they may all look the same or at best as "minnows" or "sunfishes," you can identify them by knowing what species occur in the area and noting their distinguishing morphological and behavioral traits. Binoculars and polarized sunglasses that eliminate surface glare greatly facilitate fish watching from above water.

Serious fish watchers enter water and join their subjects. With a snorkel and mask, you can view fishes at amazingly short distances and observe their feeding, spawning, and other behaviors. Although fishes tend to swim away from humans on stream banks, they remain close to a person underwater. Often, fishes are curious and readily approach underwater observers.

In areas where many similar species occur, removing fishes from the water may be the only positive way to identify them. Many species can be obtained readily by seining, dipnetting, or angling, and examined while on shore or transferred to aquariums for long-term observation.

Making Identifications

For most identifications, it is best to begin with the plates. Locate the plate with fishes that look most like the one you wish to identify (see "How to Use the Plates" on page 1), and read the short descriptions of distinguishing features on the legend opposite the plate. Arrows on plates pinpoint these features. When you believe you have located the correct species, go to the longer text description (page number given on legend page) and compare characters of the fish with those given in the species account and, if necessary, in generic and family accounts. "Similar species" sections near the end of each account identify the most likely alternative(s) to the species you selected and should be consulted before you decide you have made the correct identification.

At some point, you will need to refer to the distribution map. If you know from past experience that the fish you are working with is one of a few similar species (e.g., a sand darter), you can start with the maps. Eliminating species outside your area will facilitate identification.

Fish Morphology

Fig. 1 illustrates various structures, counts, and measurements used to identify fishes. Most are self-explanatory. The following comments and the Glossary explain others.

Fishes have median fins (dorsal, caudal, and anal) and paired fins (pectoral and pelvic). The dorsal fin in more ancient fishes is supported by flexible, segmented "soft" rays. In more recently evolved fishes, the front section of the dorsal fin contains only inflexible spiny rays ("spines") and may be contiguous with or separated from the soft-rayed part; when the front section is separated (or nearly separated) from the soft-rayed part, the fish is said to have two dorsal fins. Likewise, the anal fin may be spineless or have spines (usually only one to three) preceding rays.

Throughout the evolution of fishes, pelvic fins have tended to move forward on the body, and their position is a quick way to

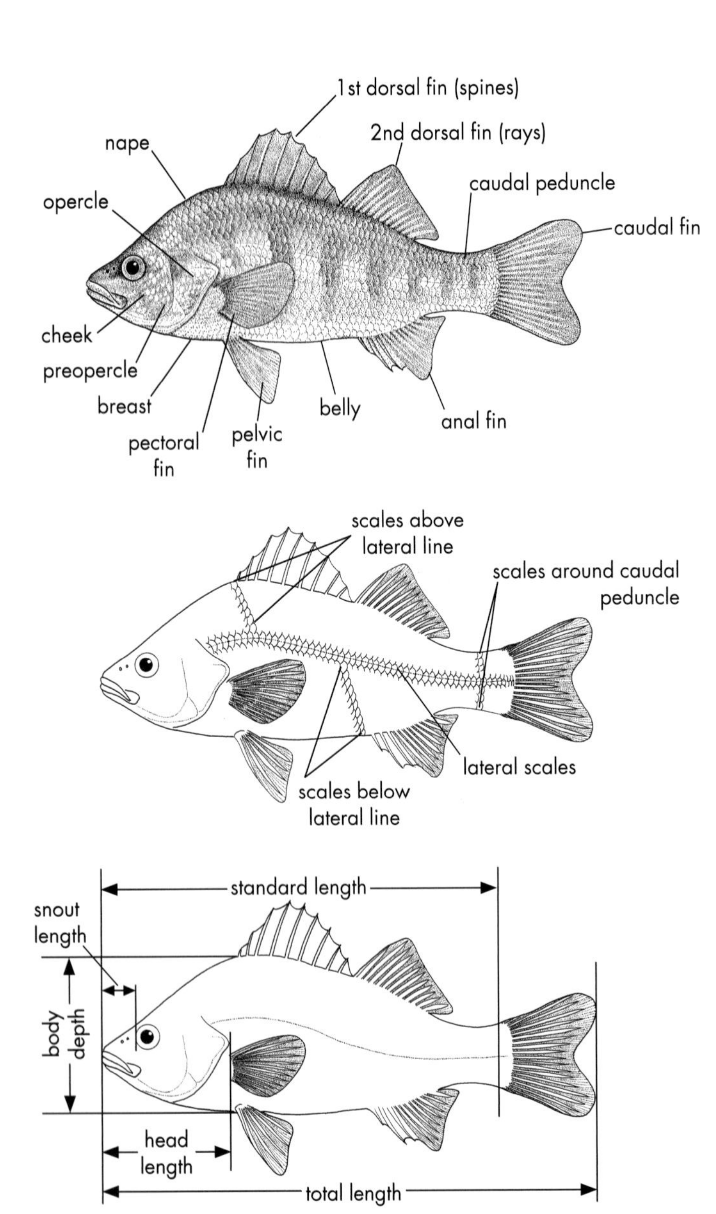

Fig. 1. Structures, counts, and measurements useful in fish identification. See Glossary for definitions of terms.

judge whether a fish belongs to a more ancient or a more recent group. If pelvic fins are abdominal, the fish is a member of an ancient group (e.g., sturgeon), and you can expect to find it near the front of this guide. If the pelvic fins are thoracic (on the breast) or jugular (on the throat), you will find the fish (e.g., a sunfish) closer to the rear of the guide.

The mouth is described as *terminal* if it opens at the front end of the head with the upper and lower jaws being equally far forward; *upturned* if it opens above that point; and *subterminal* if it opens on the underside of the head. You can see the rakers on the first gill arch by lifting the gill cover (Fig. 2); a gill raker count is the total for the entire first arch unless upper or lower limb only is specified. The largest bone in the gill cover is the opercle (Fig. 1).

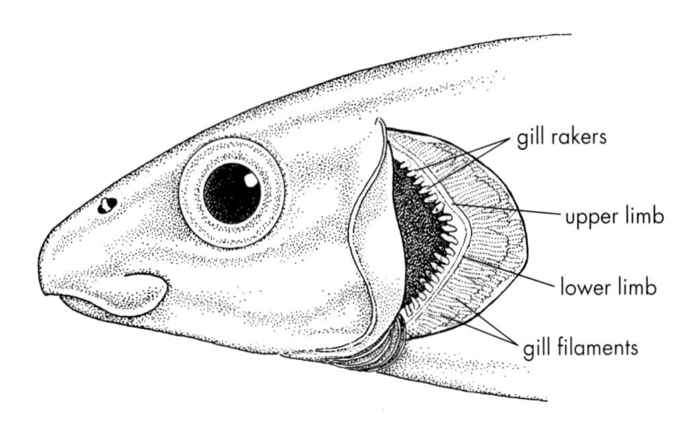

Fig. 2. How to count gill rakers. Count rakers on the first arch exposed as the gill cover is lifted. All rakers, including rudiments, are counted unless "upper limb" or "lower limb" of arch is specified.

Measurements and counts used in this guide are shown in Fig. 1. Measurements of body parts (e.g., snout length) or fins occasionally are used to separate similar species. These measurements, which are always made in a straight line (not along a body contour), usually are compared with another measurement (e.g., total length). A comparison such as snout length into total length is made by dividing one measurement into the other, or by physically "stepping off" one measurement into the other using dividers.

Lateral-scale count (also known as lateral-line scale count) begins just behind the head and continues along the lateral line (or along the midside if the lateral line is absent) to the origin of the caudal fin (which is located by bending the caudal fin to either side and noting the crease between the body and caudal fin). Scales on the caudal fin are not included in lateral-scale counts even if they are pored. Scales above and below the lateral line begin at the origin of the dorsal or anal fin, respectively, and continue diagonally to the lateral line (but do not include the lateral-line scale). A transverse-scale count is a continuation of the count of scales below the lateral line diagonally upward (including the lateral-line scale) to the dorsal fin. Scales around the caudal peduncle are those around the narrowest part. Predorsal scales are those along the nape from the rear of the head to the dorsal fin origin.

Fig. 3 shows how dorsal and anal rays are counted. In pectoral and pelvic fins, all rays are counted. Branchiostegal rays are long slender bones supporting branchiostegal membranes; all (short and long) rays are counted.

To examine pharyngeal teeth, it is necessary to remove the first pharyngeal arch by placing the fish on its side and lifting the gill cover (if necessary, slit the skin along the bottom to loosen the gill cover from the body). The arch lies just to the rear of the gills. Insert a scalpel or strong forceps between the arch and shoulder girdle, beginning at the upper angle of the gill opening and cutting down along the shoulder girdle. Carefully sever the fleshy tendons that hold the upper and lower ends of the arch in position, then lift out the arch and remove the attached flesh to expose the teeth.

Conservation

Pets, bait, and other fishes should never be released into a stream, lake, or pond other than from where they were originally taken. Non-native fishes and their offspring may outcompete or feed on fishes or other organisms and do tremendous harm to native populations.

No objective of this guide is more important than that of increasing humanity's appreciation of fishes and their environments. We often fail to give adequate consideration to the vast and varied forces over millions of years that have forged our present-day biodiversity. Each species on Earth is the product of millions of years of evolution and is fine-tuned to its environment. To conserve the diversity of life, we must reduce our own population, reduce our consumption, and set aside large ecosystems as preserves. We will be able to accomplish those changes only through education and an awareness of the value of diversity. It is our sincere hope that this guide to the rich diversity of North American fishes will contribute to that goal.

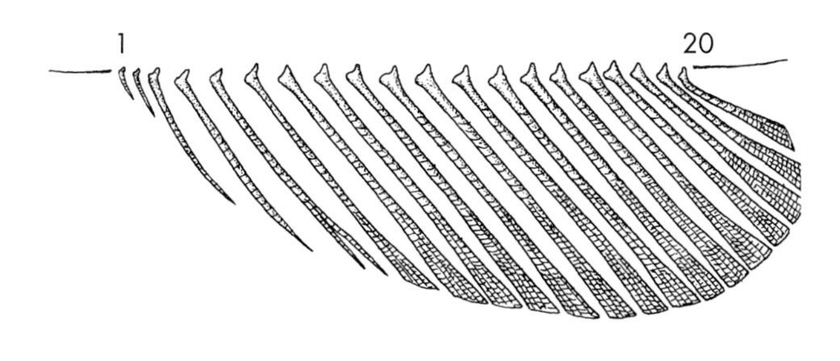

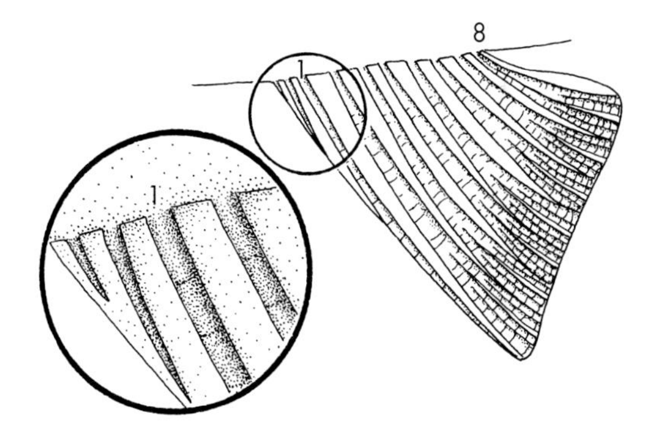

Fig. 3. How to count dorsal and anal rays. In catfishes, trouts, and other fishes in which the rudimentary rays at the beginning of the fin grade gradually in size into branched rays, count all rays. In other fishes, including minnows and suckers, there is a sharp break in size between rudimentary and long rays; count only the long rays. Count the last ray as one even if its branches are separated to the fin base; that is, if the branches of the ray look as if they will join just below the surface, they probably do and are counted as one ray.

PLATES

Plates

The plates portray fishes as they look when alive, and the fish shown usually is an "average" adult. However, the plates emphasize traits that are most useful in separating species, and often these traits are found only on certain individuals (often large males). For example, juvenile and adult female sunfishes typically have similar color patterns, whereas large males develop unique color patterns. For some strongly sexually dimorphic species, a male and female are shown. Black-and-white plates depict fishes that lack bright colors or that show little variation in color among the species on the plate.

How to Use the Plates: (See also How to Use This Guide, p. ix.) Locate the plate with fishes that look most like the one you wish to identify and read the short descriptions of distinguishing features on the legend page opposite the plate. Arrows or other marks highlight these features. (*Be sure to note characters that apply to more than one species—those in italics at top of page.*) When you have found what you believe to be the correct species, go to the species account—the longer text description (page number given on legend page)—and compare the characters of the fish with those in the species account and, if necessary, in preceding genus and family accounts. A quick look at a distribution map will tell you if your identification is reasonable.

Lampreys, American Eel

Lampreys have oral disc, no jaws, no paired fins. See Fig. 6 for more species.

SEA LAMPREY *Petromyzon marinus* **P. 121**
Two dorsal fins. Oral disc as wide or wider than head. Prominent black mottling on back, side, and fins. To 47 in. (120 cm).

CHESTNUT LAMPREY *Ichthyomyzon castaneus* **P. 123**
One slightly notched dorsal fin. Oral disc as wide or wider than head. Usually 51–56 trunk myomeres. Black on lateral-line pores. To 15 in. (38 cm). See also Silver Lamprey, *I. unicuspis*, and Northern Brook Lamprey, *I. fossor*.

SOUTHERN BROOK LAMPREY *Ichthyomyzon gagei* **P. 123**
One slightly notched dorsal fin. Oral disc narrower than head. Blunt, poorly developed teeth. Usually 52–56 trunk myomeres. Black on lateral-line pores. To 6¾ in. (17 cm).

OHIO LAMPREY *Ichthyomyzon bdellium* **P. 124**
One slightly notched dorsal fin. Oral disc as wide or wider than head. Sharp, well-developed teeth. Usually 56–62 trunk myomeres. Black on lateral-line pores. To 12 in. (30 cm).

MOUNTAIN BROOK LAMPREY *Ichthyomyzon greeleyi* **P. 124**
One slightly notched dorsal fin. Oral disc narrower than head. Moderately developed teeth. Usually 57–60 trunk myomeres. Black lateral-line pores on upper side, *no black on pores below gills*. Gray-brown above, small dark flecks on side, cream or yellow fins. To 7¾ in. (20 cm).

PACIFIC LAMPREY *Entosphenus tridentatus* **P. 125**
Two dorsal fins. Oral disc as wide or wider than head. Large, sharp teeth. Usually 64–71 trunk myomeres. Dark blue or brown above, light or silver below. To 30 in. (76 cm). See also Vancouver Lamprey, *E. macrostomus*, Klamath Lamprey, *E. similis*, and Miller Lake Lamprey, *E. minimus*.

KERN BROOK LAMPREY *Entosphenus hubbsi* **P. 129**
Two dorsal fins. Oral disc narrower than head. Usually 52–56 trunk myomeres. Gray to brown above; black specks on dorsal and caudal fins. To 5½ in. (14 cm). See also Pit-Klamath Brook Lamprey, *E. lethophagus*.

AMERICAN BROOK LAMPREY *Lethenteron appendix* **P. 129**
Two dorsal fins connected at base. Oral disc narrower than head. Usually blunt teeth. Usually 67–73 trunk myomeres. Lead gray to slate blue above, yellow fins, black blotch on caudal fin. To 13¾ in. (35 cm). See also Arctic Lamprey, *L. camtschaticum*.

WESTERN BROOK LAMPREY *Lampetra richardsoni* **P. 131**
Two dorsal fins. Oral disc narrower than head. Small, blunt teeth. Usually 58–67 trunk myomeres. Brown to gray above; dark spot on caudal fin. To 7 in. (17 cm). See also River Lamprey, *L. ayresii*.

LEAST BROOK LAMPREY *Lampetra aepyptera* **P. 132**
Two dorsal fins. Oral disc narrower than head. Blunt, extremely degenerate teeth. Usually 52–59 trunk myomeres. Light tan to silver gray above; yellow or white below; yellow or gray fins. To 7 in. (18 cm).

AMERICAN EEL *Anguilla rostrata* **P. 146**
Snakelike body and head. Long dorsal fin continuous with caudal and anal fins. No pelvic fins. To 60 in. (152 cm).

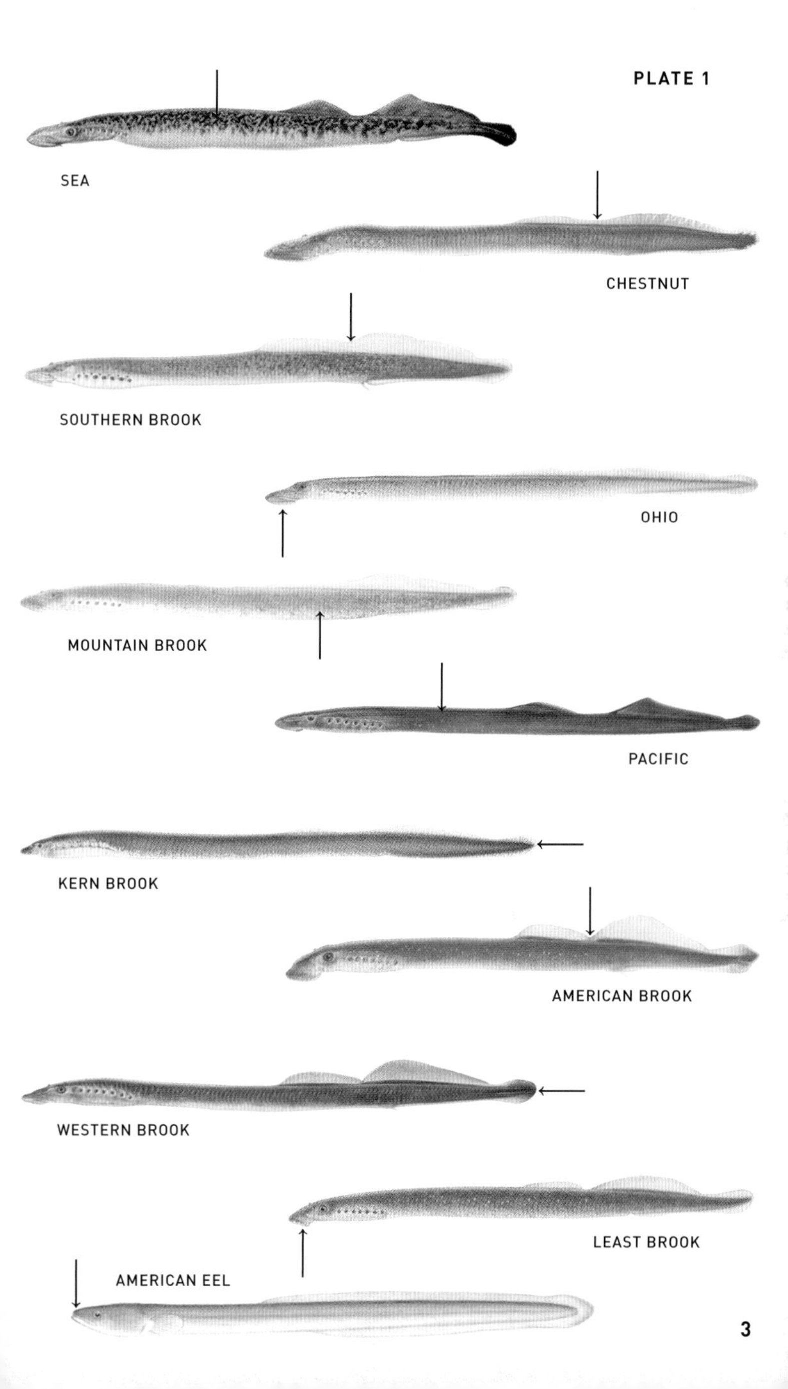
PLATE 1
SEA
CHESTNUT
SOUTHERN BROOK
OHIO
MOUNTAIN BROOK
PACIFIC
KERN BROOK
AMERICAN BROOK
WESTERN BROOK
LEAST BROOK
AMERICAN EEL

Sturgeons, Paddlefish

Sturgeons have shovel-shaped snout, barbels, large bony scutes, heterocercal caudal fin.

SHORTNOSE STURGEON *Acipenser brevirostrum* **P. 134**
Short snout. Anal fin origin beneath dorsal fin origin. To 43 in. (109 cm).

LAKE STURGEON *Acipenser fulvescens* **P. 135**
Anal fin origin behind dorsal fin origin. Scutes on back and along side same color as skin. To 9 ft. (2.7 m).

GREEN STURGEON *Acipenser medirostris* **P. 135**
Barbels usually closer to mouth than to snout tip. Scutes along side paler than skin. To 7 ft. (2.1 m).

WHITE STURGEON *Acipenser transmontanus* **P. 136**
Barbels closer to snout tip than to mouth. No obvious scutes behind dorsal and anal fins. To 20 ft. (6.1 m).

ATLANTIC STURGEON *Acipenser oxyrinchus* **P. 137**
Long, sharply V-shaped snout; 4 small scutes, usually as 2 pairs, between anal fin and caudal fulcrum. To 14 ft. (4.3 m).

Next 2 species have long, slender caudal peduncle.

SHOVELNOSE STURGEON *Scaphirhynchus platorynchus* **P. 138**
Bases of outer barbels in line with or ahead of inner barbels (Fig. 7). To 43 in. (108 cm). See also Alabama Sturgeon, *S. suttkusi*.

PALLID STURGEON *Scaphirhynchus albus* **P. 139**
No scalelike scutes on belly. Bases of outer barbels usually behind bases of inner barbels (Fig. 7). To 73 in. (185 cm).

PADDLEFISH *Polyodon spathula* **P. 140**
Long, canoe-paddle-shaped snout. Large, fleshy, pointed flap on gill cover. To 87 in. (221 cm).

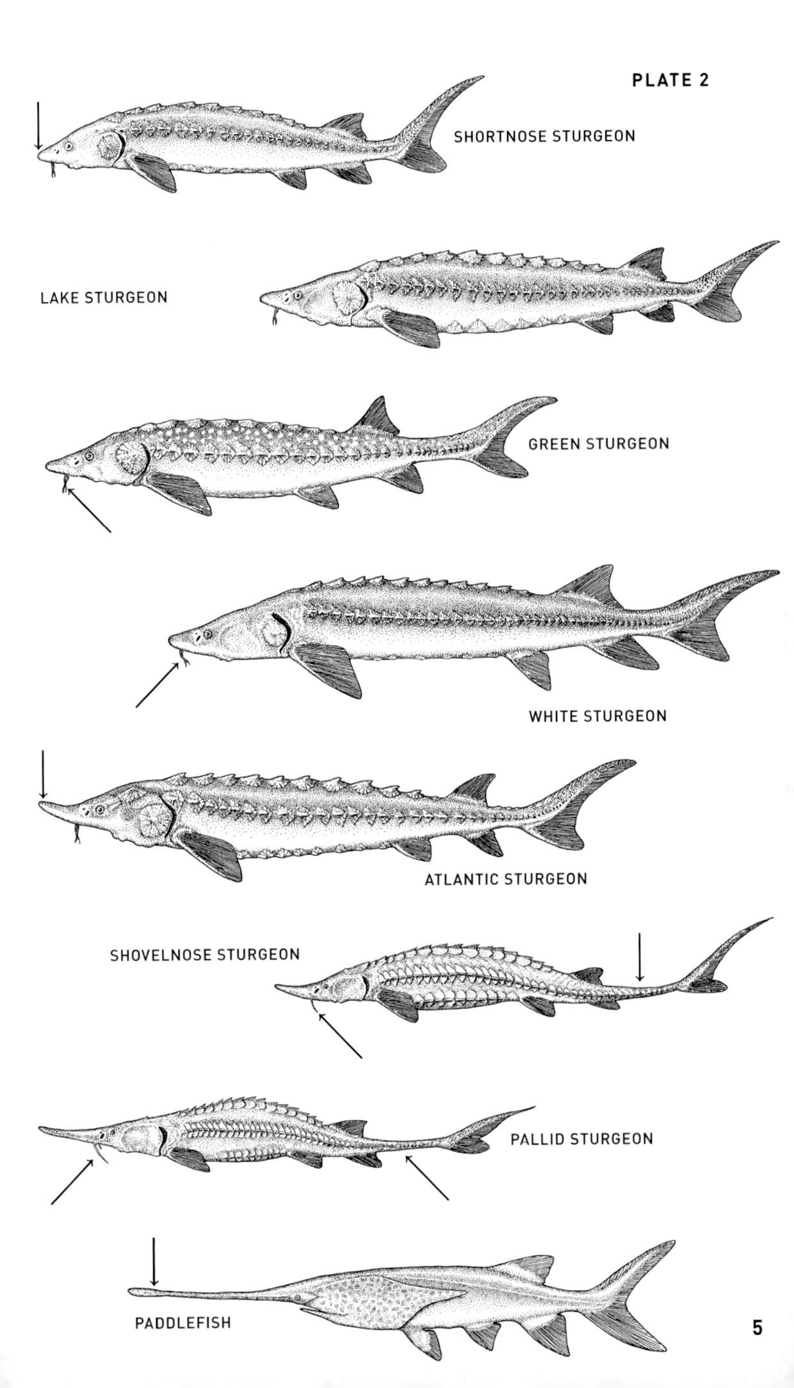
PLATE 2
SHORTNOSE STURGEON
LAKE STURGEON
GREEN STURGEON
WHITE STURGEON
ATLANTIC STURGEON
SHOVELNOSE STURGEON
PALLID STURGEON
PADDLEFISH

Gars, Bowfin

Gars have long, sharply toothed jaws, diamond-shaped scales, dorsal and anal fins far back on body.

ALLIGATOR GAR *Atractosteus spatula* **P. 141**
Giant of the gars. Short, broad snout; 2 rows of teeth on upper jaw. To 12 ft. (3.7 m).

SHORTNOSE GAR *Lepisosteus platostomus* **P. 141**
Short, broad snout. Paired fins usually lack spots. To 33 in. (83 cm).

LONGNOSE GAR *Lepisosteus osseus* **P. 142**
Long, narrow snout. To 72 in. (183 cm).

SPOTTED GAR *Lepisosteus oculatus* **P. 142**
Many dark spots on body, head, and all fins. Bony plates on underside of isthmus. To 44 in. (112 cm).

FLORIDA GAR *Lepisosteus platyrhincus* **P. 143**
Similar to Spotted Gar but lacks bony plates on underside of isthmus. To 52 in. (132 cm).

BOWFIN *Amia calva* **P. 143**
Long, nearly cylindrical body. Long dorsal fin. Tubular nostrils. Large, bony gular plate. To 43 in. (109 cm).

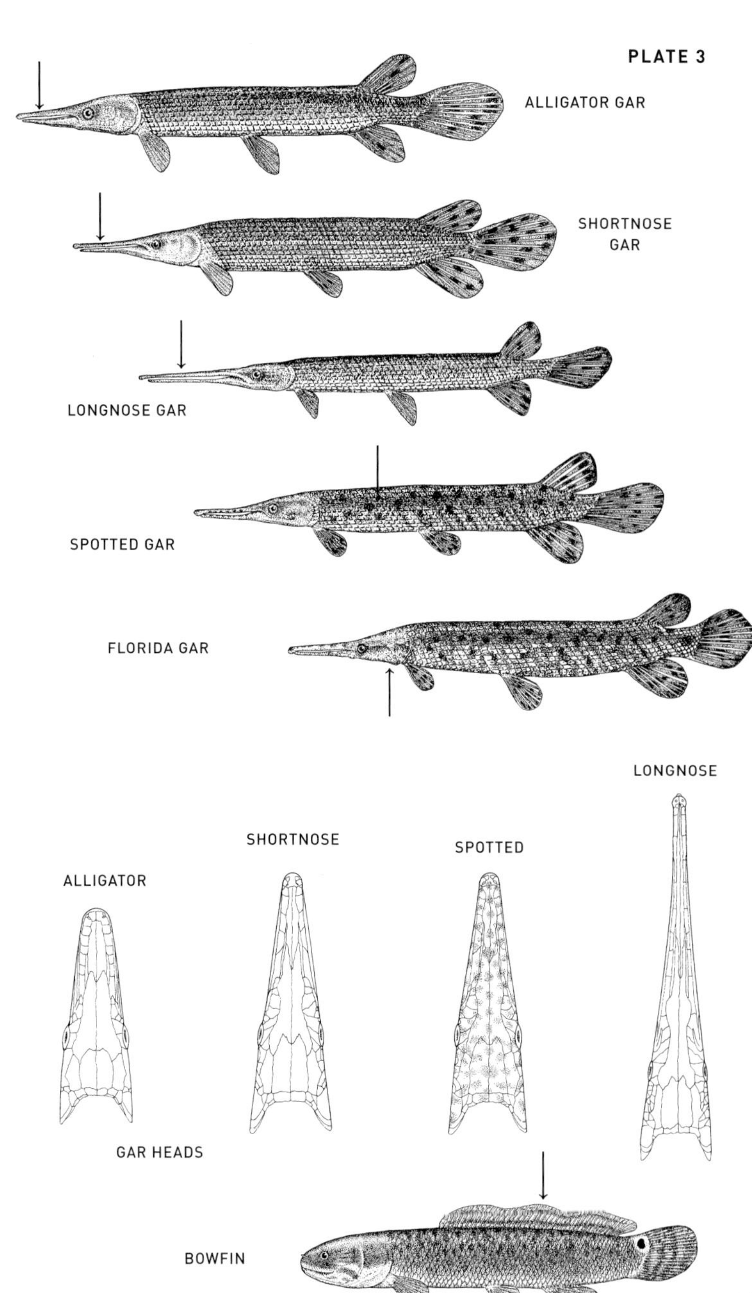
PLATE 3
ALLIGATOR GAR
SHORTNOSE
GAR
LONGNOSE GAR
SPOTTED GAR
FLORIDA GAR
LONGNOSE
SHORTNOSE
SPOTTED
ALLIGATOR
GAR HEADS
BOWFIN

Mooneye, Goldeye, Herrings, Shads, Smelt

All are silver, strongly compressed fishes.

MOONEYE *Hiodon tergisus* **P. 144**

Large eye. Dorsal fin origin in front of anal fin origin. Keel along belly extends from pelvic fin base to anal fin. To 19 in. (47 cm).

GOLDEYE *Hiodon alosoides* **P. 144**

Large eye. Dorsal fin origin opposite or behind anal fin origin. Keel along belly extends from pectoral fin base to anal fin. To 20 in. (51 cm).

Next 7 species (herrings and shads) lack lateral line; have jagged (sawtooth) belly.

BLUEBACK HERRING *Alosa aestivalis* **P. 147**

Strongly oblique mouth. Blue above. Thin dark stripes on upper side. Usually 1 small dark spot behind opercle. Usually 44–50 rakers on lower limb of 1st gill arch. To 16 in. (40 cm).

ALEWIFE *Alosa pseudoharengus* **P. 148**

Similar to Blueback Herring but blue-green above; has larger eye, usually 39–41 rakers on lower limb of 1st gill arch. To 15 in. (38 cm).

SKIPJACK HERRING *Alosa chrysochloris* **P. 149**

No dark spot behind opercle. Blue-green above ends abruptly on silver side. To 21 in. (53 cm). See also Hickory Shad, *A. mediocris*.

AMERICAN SHAD *Alosa sapidissima* **P. 150**

Adult lacks jaw teeth. Cheek deeper than long (Fig. 9). Dark spot behind opercle, usually followed by smaller spots; 59–73 rakers on lower limb of 1st gill arch of adult. To 30 in. (75 cm).

ALABAMA SHAD *Alosa alabamae* **P. 150**

Similar to American Shad, but adult has 42–48 rakers on lower limb of 1st gill arch. To 20¼ in. (51 cm).

GIZZARD SHAD *Dorosoma cepedianum* **P. 151**

Long, whiplike last dorsal ray. Blunt snout. Subterminal mouth. Purple-blue shoulder spot in young and small adult. Has 52–70 lateral scales. To 20½ in. (52 cm).

THREADFIN SHAD *Dorosoma petenense* **P. 151**

Similar to Gizzard Shad but has projecting lower jaw, yellow fins, 40–48 lateral scales. To 9 in. (23 cm).

RAINBOW SMELT *Osmerus mordax* **P. 378**

Adipose fin. Large mouth; 2 large canine teeth. Usually a conspicuous silver stripe along side. To 13 in. (33 cm).

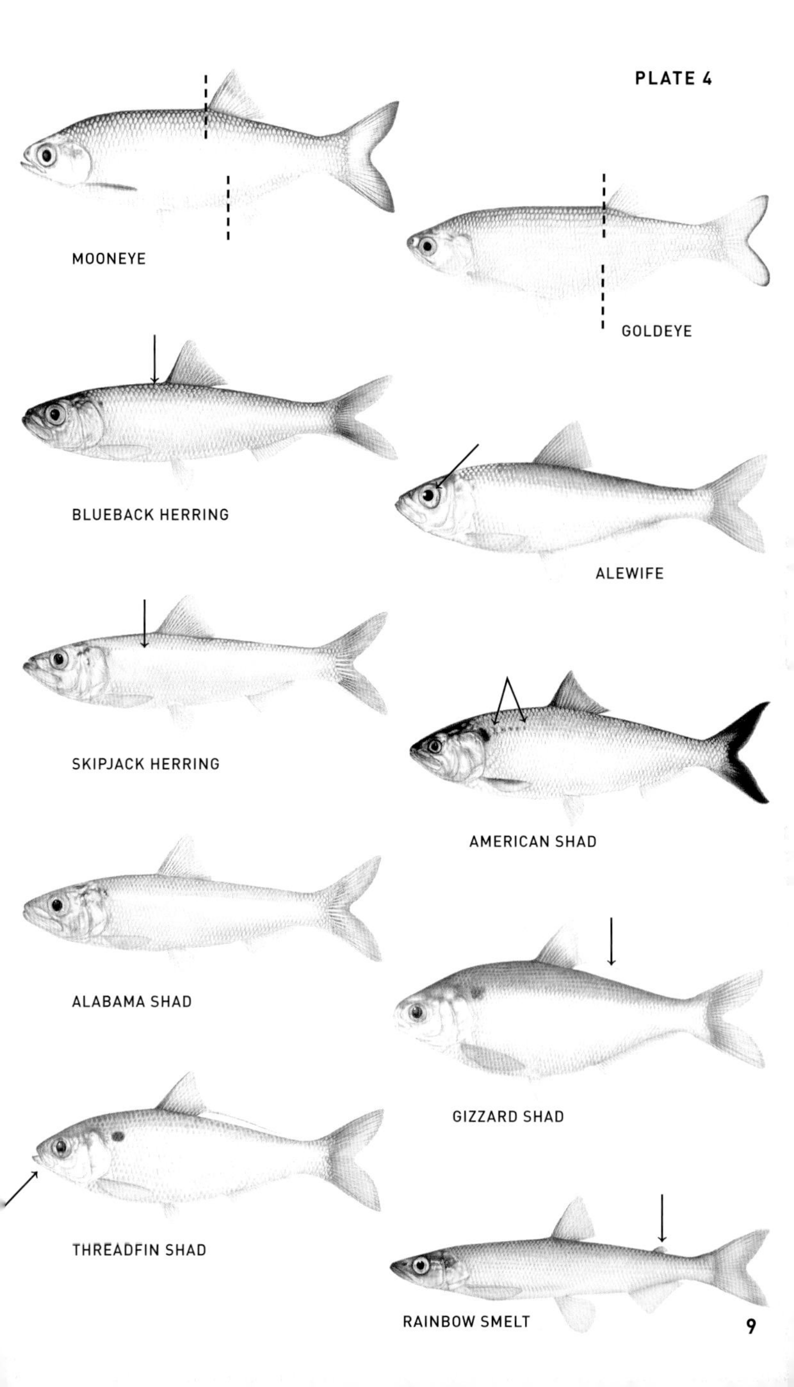
MOONEYE
GOLDEYE
BLUEBACK HERRING
ALEWIFE
SKIPJACK HERRING
AMERICAN SHAD
ALABAMA SHAD
GIZZARD SHAD
THREADFIN SHAD
RAINBOW SMELT

PLATE 5

Native Minnows (1)

All minnows have 1 dorsal fin, abdominal pelvic fins.

GOLDEN SHINER *Notemigonus crysoleucas* **P. 157**
Extremely compressed body. Strongly decurved lateral line. Scaleless keel along belly (Fig. 11). Small, upturned mouth. Herringbone lines on young. To 12½ in. (32 cm).

HITCH *Lavinia exilicauda* **P. 158**
Deep, compressed body tapering to narrow caudal peduncle. Large caudal fin. Strongly decurved lateral line. To 14 in. (36 cm).

SPLITTAIL *Pogonichthys macrolepidotus* **P. 159**
Upper lobe of large caudal fin longer than lower lobe. Barbel at corner of slightly subterminal mouth. To 17½ in. (44 cm). See also Clear Lake Splittail, *P. ciscoides*.

Next 3 species (pikeminnows) have large terminal mouth extending to or past front of eye.

COLORADO PIKEMINNOW *Ptychocheilus lucius* **P. 160**
Has 76–97 lateral scales. Usually 9 dorsal rays, 9 anal rays. To 6 ft. (1.8 m).

SACRAMENTO PIKEMINNOW *Ptychocheilus grandis* **P. 161**
Has 65–78 lateral scales. Usually 8 dorsal rays, 8 anal rays. To 4½ ft. (1.4 m).

NORTHERN PIKEMINNOW *Ptychocheilus oregonensis* **P. 161**
Has 64–79 lateral scales. Usually 9 dorsal rays, 8 anal rays. To 25 in. (63 cm). See also Umpqua Pikeminnow, *P. umpquae*.

HARDHEAD *Mylopharodon conocephalus* **P. 161**
Premaxillary frenum. Long snout. Large terminal mouth reaches front of eye. To 3 ft. (1 m).

SACRAMENTO BLACKFISH *Orthodon microlepidotus* **P. 162**
Small (90–105 lateral) scales. Wide head, flat above. Narrow caudal peduncle. To 21½ in. (55 cm).

ARROYO CHUB *Gila orcuttii* **P. 171**
Deep caudal peduncle. Short, rounded snout. Large eye. To 16 in. (40 cm). See also Yaqui Chub, *G. purpurea*.

CHISELMOUTH *Acrocheilus alutaceus* **P. 163**
Large, forked caudal fin. Wide head. Subterminal mouth; hard plate on lower jaw (Fig. 12). To 12 in. (30 cm).

ALVORD CHUB *Siphatales alvordensis* **P. 173**
High nape. Short, pointed head. Deep caudal peduncle. To 5¼ in. (14 cm). See also Borax Lake Chub, *S. boraxobius*.

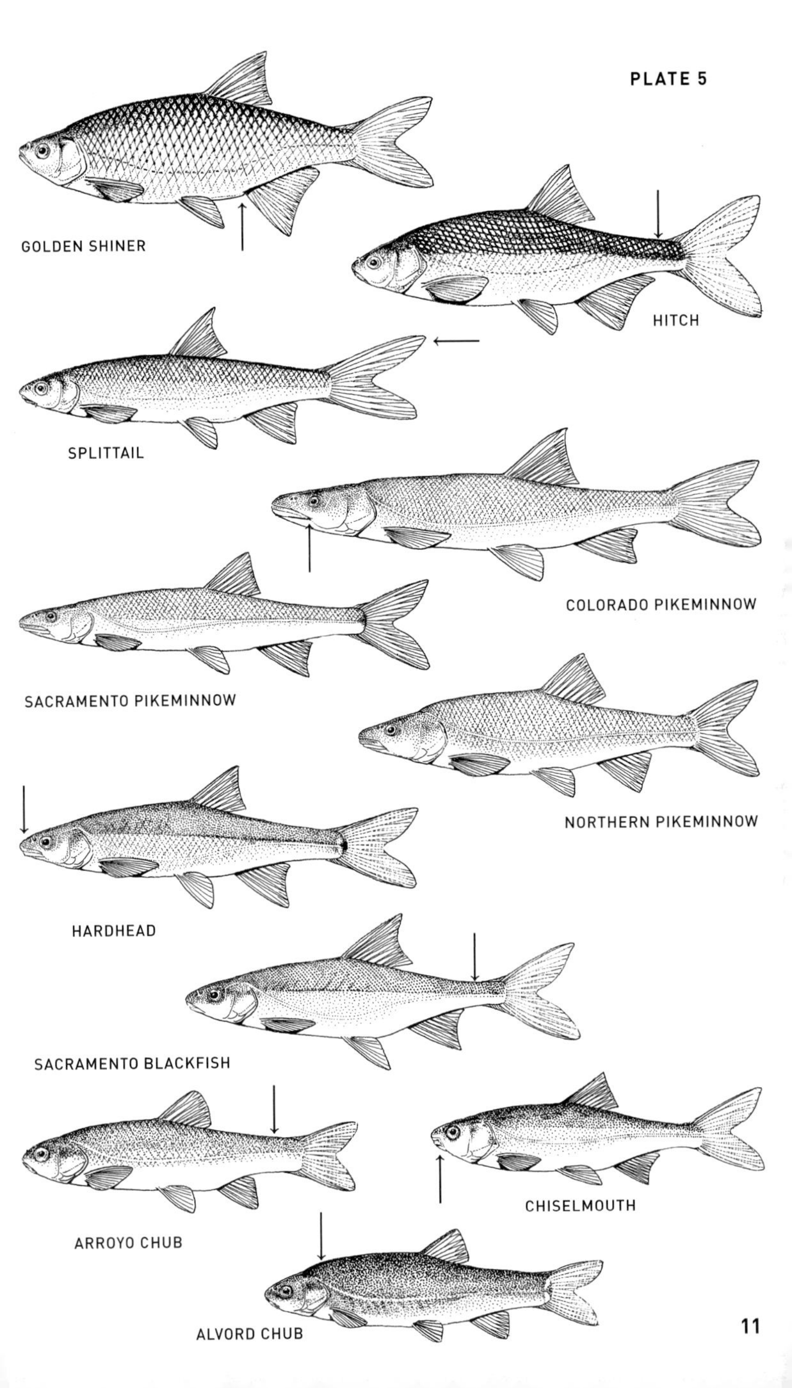
PLATE 5
GOLDEN SHINER
HITCH
SPLITTAIL
COLORADO PIKEMINNOW
SACRAMENTO PIKEMINNOW
NORTHERN PIKEMINNOW
HARDHEAD
SACRAMENTO BLACKFISH
ARROYO CHUB
CHISELMOUTH
ALVORD CHUB

Minnows (2)

MOAPA DACE *Moapa coriacea* **P. 175**
Leatherlike skin (resulting from many small embedded scales). Large black spot on caudal fin base. To 3½ in. (9 cm).

RELICT DACE *Relictus solitarius* **P. 176**
Chubby, soft-bodied. Incomplete lateral line, rarely reaching below dorsal fin. Large terminal mouth. To 5 in. (13 cm).

DESERT DACE *Eremichthys acros* **P. 176**
Deep, chubby body. Hard sheath on jaws. Small mouth. To 3 in. (7.7 cm).

LEAST CHUB *Iotichthys phlegethontis* **P. 177**
Upturned mouth. Large eye. Large scales. Short snout. To 2½ in. (6.4 cm).

OREGON CHUB *Oregonichthys crameri* **P. 177**
Narrow caudal peduncle. Clusters of brown-black spots on back and silver side. To 2¾ in. (7 cm). See also Umpqua Chub, *O. kalawatseti.*

Next 5 species (*Phenacobius*) have long cylindrical body, large fleshy lips on subterminal mouth (Fig. 19).

SUCKERMOUTH MINNOW *Phenacobius mirabilis* **P. 201**
Bicolored body. Intense black spot on caudal fin base. (See also Fig. 19). To 4¾ in. (12 cm).

KANAWHA MINNOW *Phenacobius teretulus* **P. 201**
Many small bumps on very fleshy lips (Fig. 19). Small black blotches on upper half of body. To 4 in. (10 cm).

FATLIPS MINNOW *Phenacobius crassilabrum* **P. 202**
Pelvic fins reach to or past anus. To 4¼ in. (11 cm).

RIFFLE MINNOW *Phenacobius catostomus* **P. 202**
Very long cylindrical body. Eyes high on head, directed upwardly. To 4½ in. (12 cm).

STARGAZING MINNOW *Phenacobius uranops* **P. 203**
Similar to Riffle Minnow but has more elongated body, longer snout, more elliptical eye. To 4½ in. (12 cm).

CUTLIP MINNOW *Exoglossum maxillingua* **P. 204**
Fleshy lobe on either side of central bony plate on lower jaw (Fig. 20). Chubby body. Deep caudal peduncle. To 6¼ in. (16 cm). See also Tonguetied Minnow, *E. laurae.*

Next 4 species have black lower lobe on caudal fin, barbel in corner of mouth.

FLATHEAD CHUB *Platygobio gracilis* **P. 209**
Broad flat head tapering to pointed snout. Large pointed dorsal and pectoral fins. To 12½ in. (32 cm).

SICKLEFIN CHUB *Macrhybopsis meeki* **P. 209**
Large, sharply pointed, sickle-shaped fins. Deep head. Rounded snout. To 4¼ in. (11 cm).

STURGEON CHUB *Macrhybopsis gelida* **P. 210**
Similar to Sicklefin Chub but has straight-edged fins. Keeled scales on back and side. Large papillae on underside of head. To 3¼ in. (8.4 cm).

SILVER CHUB *Macrhybopsis storeriana* **P. 210**
Large eye on upper half of head. Bright silver white side. Short, rounded snout. To 9 in. (23 cm).

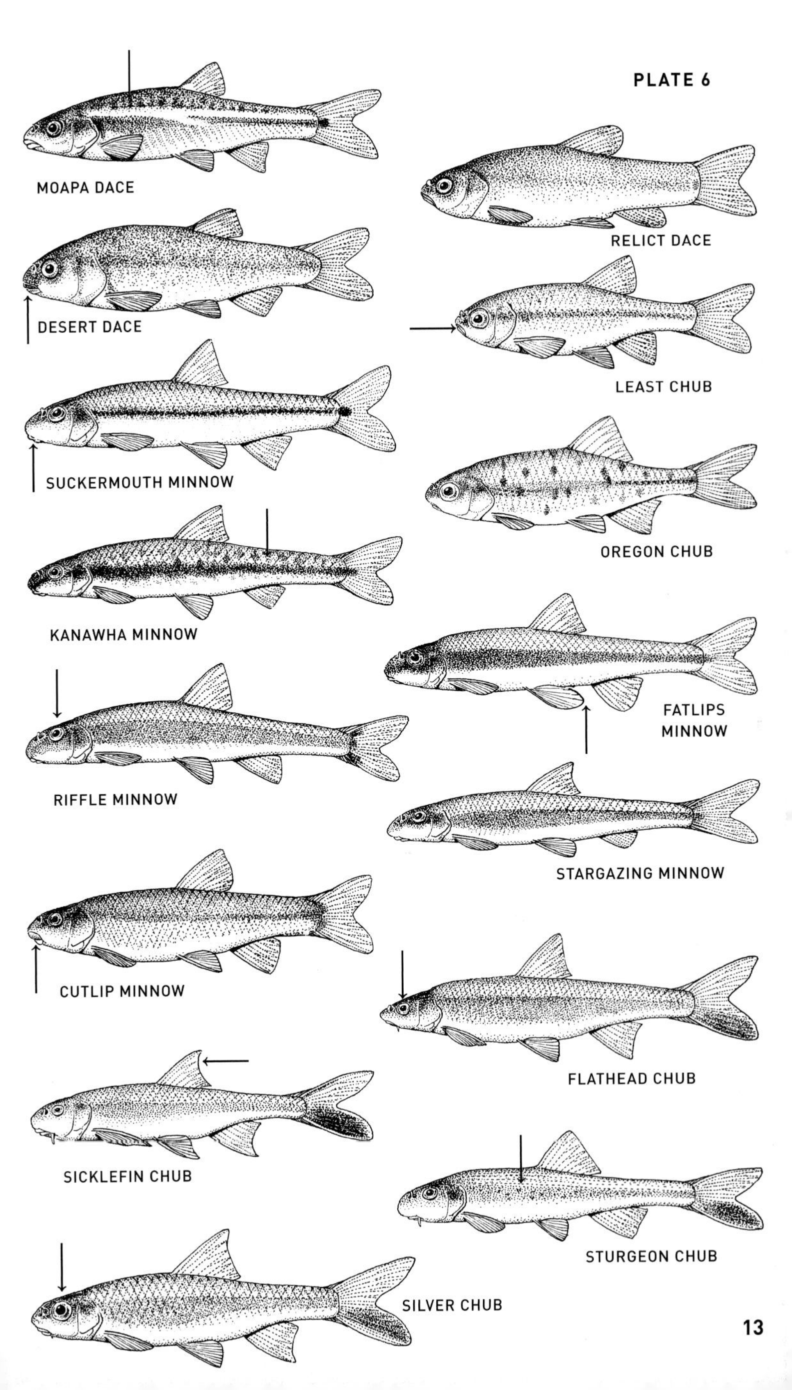
PLATE 6
MOAPA DACE
RELICT DACE
DESERT DACE
LEAST CHUB
SUCKERMOUTH MINNOW
OREGON CHUB
KANAWHA MINNOW
FATLIPS MINNOW
RIFFLE MINNOW
STARGAZING MINNOW
CUTLIP MINNOW
FLATHEAD CHUB
SICKLEFIN CHUB
STURGEON CHUB
SILVER CHUB

Minnows (3)

Subterminal mouth. All but Pallid Shiner have barbel at corner of mouth (Fig. 16).

First 4 species (*Erimystax*) have long slender body deepest at nape, are flattened below.

STREAMLINE CHUB *Erimystax dissimilis* **P. 214**
Has 7–15 horizontally oblong or round dark blotches along side. To 5½ in. (14 cm). See also Ozark Chub, *E. harryi.*

BLOTCHED CHUB *Erimystax insignis* **P. 214**
Has 7–9 large vertical dark blotches and row of black specks along lower edge of dark pigment on side. To 4 in. (10 cm).

GRAVEL CHUB *Erimystax x-punctatus* **P. 215**
Dark Xs on back and upper side. To 4¼ in. (11 cm).

SLENDER CHUB *Erimystax cahni* **P. 216**
Large dark chevrons along rear half of side, darkest and largest on caudal peduncle. To 3½ in. (9 cm).

SHOAL CHUB *Macrhybopsis hyostoma* **P. 211**
Long, bulbous snout. Upwardly directed, elliptical eye. Black spots on back and side. To 3 in. (7.6 cm). See also Remarks for Shoal Chub.

PEPPERED CHUB *Macrhybopsis tetranema* **P. 212**
Similar to Shoal Chub but has small round eye, 2 barbels at corner of mouth, rear barbel longer than eye. To 3 in. (7.7 cm). See also Prairie Chub, *M. australis.*

SPECKLED CHUB *Macrhybopsis aestivalis* **P. 212**
Similar to Shoal Chub but has round eye, more black spots on back and side. To 3½ in. (9 cm).

BURRHEAD CHUB *Macrhybopsis marconis* **P. 213**
Large, round eye. Dark silver stripe along side. Breeding male has yellow pectoral fins. To 2¾ in. (7.3 cm).

Next 5 species have upwardly directed, horizontally elliptical eyes.

THICKLIP CHUB *Cyprinella labrosa* **P. 248**
Small dark brown blotches and crosshatching on back and side. Long snout. Breeding male has yellow-black fins. To 2¾ in. (6.7 cm).

SANTEE CHUB *Cyprinella zanema* **P. 248**
Dark crosshatching on back and side. Long snout. Breeding male has black streaks on dark yellow dorsal and caudal fins. To 3 in. (7.5 cm).

BIGEYE CHUB *Hybopsis amblops* **P. 296**
Black stripe (faded in turbid water) along side and onto snout. Large eye, about equal to snout length. To 3½ in. (9 cm). See also Rosyface Chub, *H. rubrifrons.*

LINED CHUB *Hybopsis lineapunctata* **P. 298**
Clear stripe above black stripe (broad and diffuse at midbody) along side and around snout. Black spot on caudal fin base. To 3 in. (7.9 cm).

PALLID SHINER *Hybopsis amnis* **P. 297**
Back arched at dorsal fin origin. Large eye, about equal to snout length. To 3¼ in. (8.4 cm). See also Clear Chub, *H. winchelli.*

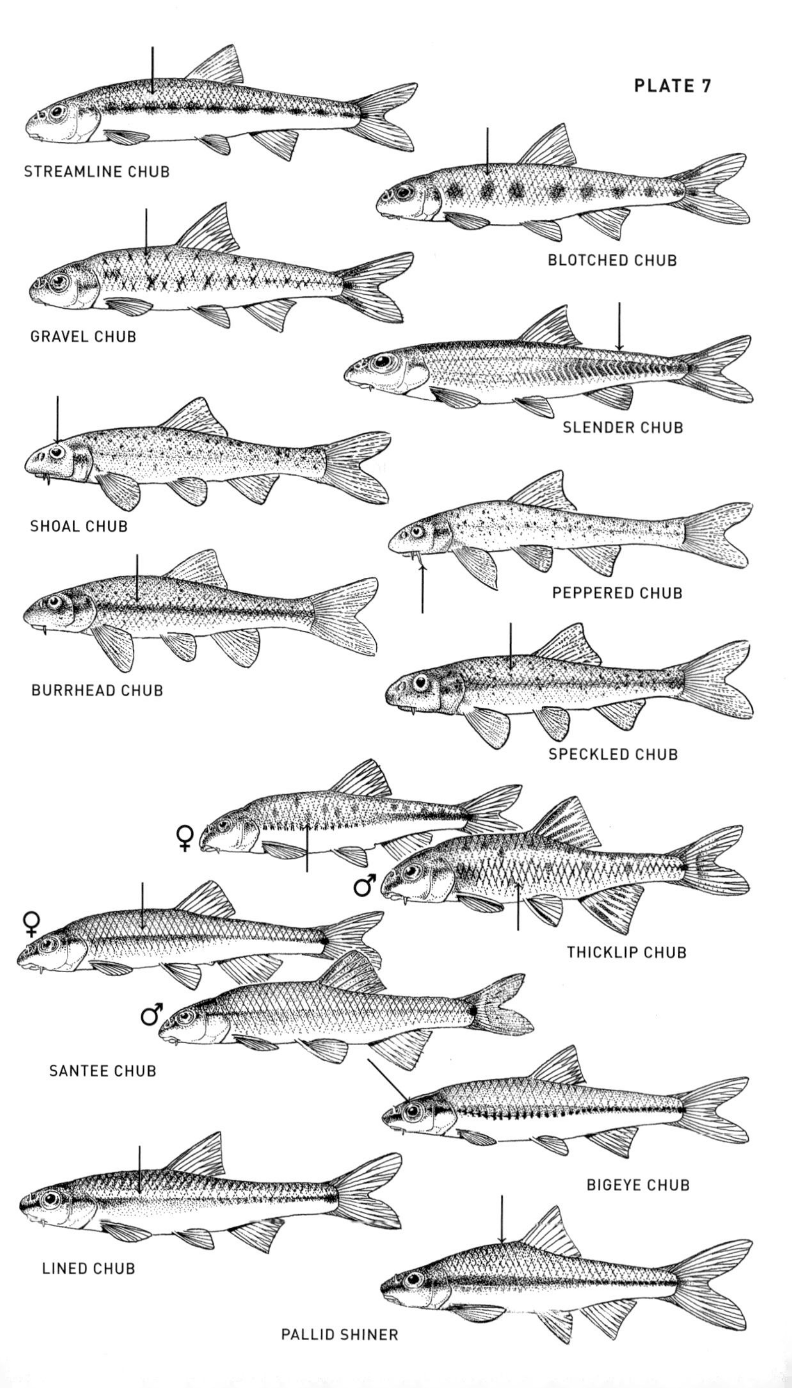
PLATE 7
STREAMLINE CHUB
BLOTCHED CHUB
GRAVEL CHUB
SLENDER CHUB
SHOAL CHUB
PEPPERED CHUB
BURRHEAD CHUB
SPECKLED CHUB
♀
♂
THICKLIP CHUB
♀
♂
SANTEE CHUB
BIGEYE CHUB
LINED CHUB
PALLID SHINER

Minnows (4)

First 4 species (*Hybognathus*) have long, coiled intestine; black peritoneum; small, slightly subterminal mouth (not reaching to eye). See also Fig. 21.

MISSISSIPPI SILVERY MINNOW *Hybognathus nuchalis* **P. 217**
Pointed dorsal fin. Silver side. To 7 in. (18 cm). See also Eastern Silvery Minnow, *H. regius*, and Western Silvery Minnow, *H. argyritis*.

PLAINS MINNOW *Hybognathus placitus* **P. 218**
Small eye. Flat underside of head. Falcate fins in western populations. To 5 in. (13 cm).

BRASSY MINNOW *Hybognathus hankinsoni* **P. 219**
Rounded dorsal fin. Stout, brassy yellow body. To 3¾ in. (9.7 cm).

CYPRESS MINNOW *Hybognathus hayi* **P. 219**
Scales on back and upper side darkly outlined, appear diamond-shaped. Pointed dorsal fin. To 4½ in. (12 cm).

PUGNOSE MINNOW *Opsopoeodus emiliae* **P. 249**
Crosshatched pattern on back and side. Small, strongly upturned mouth. Small scales on front half of nape. To 2½ in. (6.4 cm).

Next 3 species (*Pimephales*) have scales on nape much smaller than elsewhere on body; short, stout 2d dorsal ray.

FATHEAD MINNOW *Pimephales promelas* **P. 251**
Deep body. Short head flat on top. Herringbone lines on upper side. Blunt snout. Breeding male has black head, 2 broad white or gold bars on side. To 4 in. (10 cm).

BLUNTNOSE MINNOW *Pimephales notatus* **P. 251**
Blunt snout. Small, subterminal mouth. Slender body. Black spot on caudal fin base. Breeding male is black with silver bar behind opercle, has about 16 tubercles in 3 rows on snout. To 4¼ in. (11 cm).

BULLHEAD MINNOW *Pimephales vigilax* **P. 252**
Similar to Bluntnose Minnow but has larger eye directed more upwardly, body less slender, bluish sheen on side of body. Breeding male has 5–9 tubercles in 1–2 rows on snout. To 3½ in. (8.9 cm). See also Slim Minnow, *P. tenellus*.

RIVER SHINER *Notropis blennius* **P. 264**
Slender, fairly compressed body. Mouth extends to beneath front of eye. Dorsal fin origin over or slightly behind pelvic fin origin. To 5¼ in. (13 cm).

FLUVIAL SHINER *Notropis edwardraneyi* **P. 265**
Large, round eye, somewhat directed upwardly. Pallid. Dorsal fin origin in front of pelvic fin origin. To 3¼ in. (8 cm).

CHUB SHINER *Notropis potteri* **P. 265**
Head flat above and below. Eye high on head, somewhat directed upwardly. To 4¼ in. (11 cm).

RED RIVER SHINER *Notropis bairdi* **P. 266**
Broad, flat head. Black specks concentrated in large patch on side of body. To 3¼ in. (8 cm). See also Smalleye Shiner, *N. buccula*, and Arkansas River Shiner, *N. girardi*.

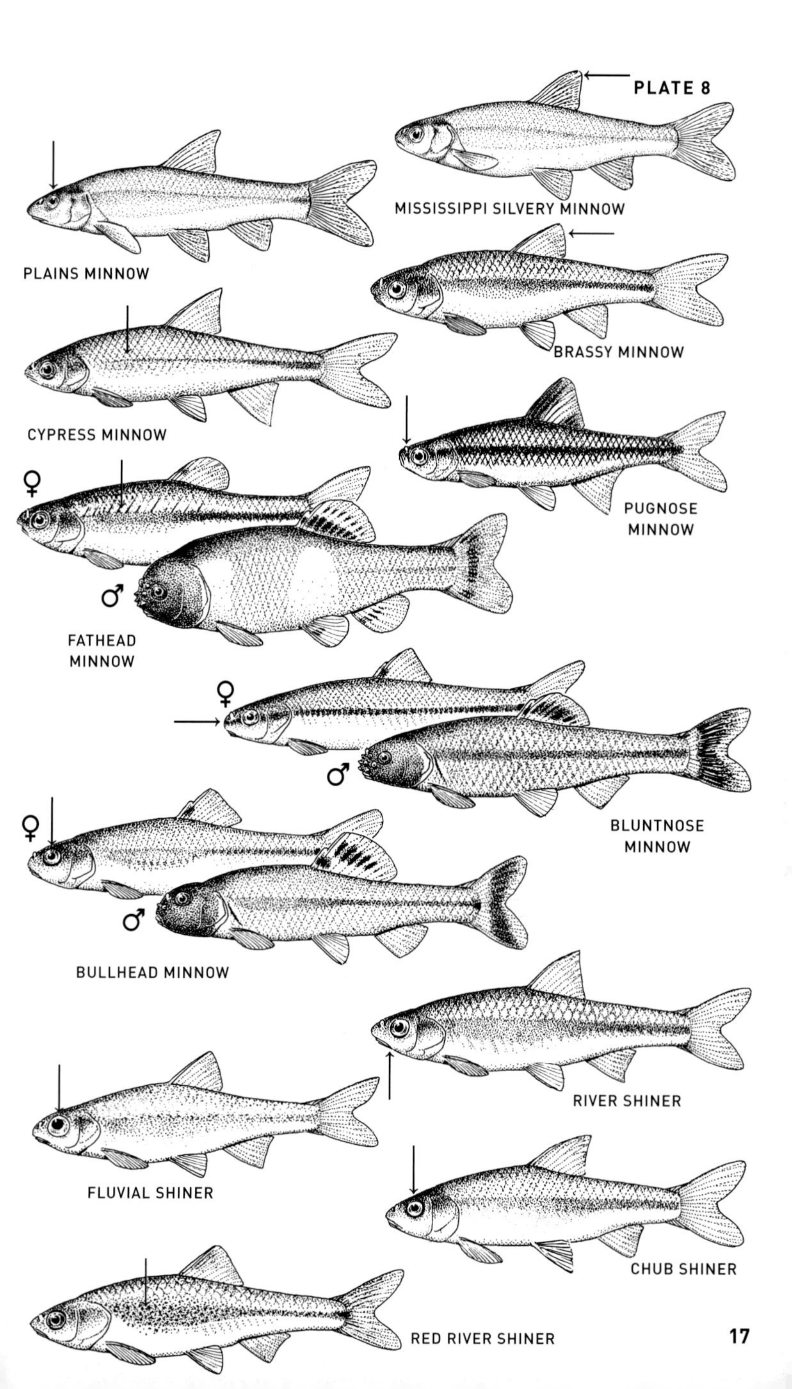
PLATE 8
MISSISSIPPI SILVERY MINNOW
PLAINS MINNOW
BRASSY MINNOW
CYPRESS MINNOW
PUGNOSE
MINNOW
♀
♂
FATHEAD
MINNOW
♀
♂
BLUNTNOSE
MINNOW
♀
♂
BULLHEAD MINNOW
RIVER SHINER
FLUVIAL SHINER
CHUB SHINER
RED RIVER SHINER

MINNOWS (5)

EMERALD SHINER *Notropis atherinoides* **P. 253**
Has 10–12 anal rays. Slender, compressed body. Dorsal fin origin behind pelvic fin origin. To 5 in. (13 cm).

COMELY SHINER *Notropis amoenus* **P. 254**
Nearly identical to Emerald Shiner but has smaller eye, rarely has T- or Y-shaped gill rakers (present in about half of Emerald Shiners). To 4¼ in. (11 cm).

RIO GRANDE SHINER *Notropis jemezanus* **P. 255**
Similar to Emerald Shiner but has larger, less oblique mouth, deeper snout. To 3 in. (7.5 cm).

SHARPNOSE SHINER *Notropis oxyrhynchus* **P. 255**
Sharply pointed snout. To 2½ in. (6.5 cm).

SILVER SHINER *Notropis photogenis* **P. 255**
Slender, compressed body; 2 black crescents (Fig. 28) between nostrils. To 5½ in. (14 cm).

SILVERBAND SHINER *Notropis shumardi* **P. 259**
Tall, pointed dorsal fin; origin slightly in front of pelvic fin origin. To 4 in. (10 cm). See also Silverside Shiner, *N. candidus*.

SILVERSTRIPE SHINER *Notropis stilbius* **P. 258**
Horizontally oval black spot on caudal fin base. Large eye. To 3½ in. (9 cm).

SANDBAR SHINER *Notropis scepticus* **P. 260**
Large, round eye. Darkly outlined scales on back and side. Punctate lateral line. To 3½ in. (9 cm).

POPEYE SHINER *Notropis ariommus* **P. 261**
Huge eye. Darkly outlined scales on back and upper side. To 3¾ in. (9.5 cm).

TELESCOPE SHINER *Notropis telescopus* **P. 262**
Dark wavy lines on back and upper side meet those of other side on caudal peduncle. Large eye. Punctate lateral line. To 3¾ in. (9.4 cm).

TEXAS SHINER *Notropis amabilis* **P. 262**
Large eye. Black lips. Clear stripe above dark stripe (darkest at rear) along side. To 2½ in. (6.2 cm).

ROUGHHEAD SHINER *Notropis semperasper* **P. 263**
Black stripe (darkest at rear) along side. Slender body. Large eye. To 3½ in. (9 cm).

KIAMICHI SHINER *Notropis ortenburgeri* **P. 279**
Pale stripe above silver black stripe along side. Dark-edged scales on back. Strongly upturned mouth. To 2¼ in. (5.5 cm). See also Blackmouth Shiner, *N. melanostomus*.

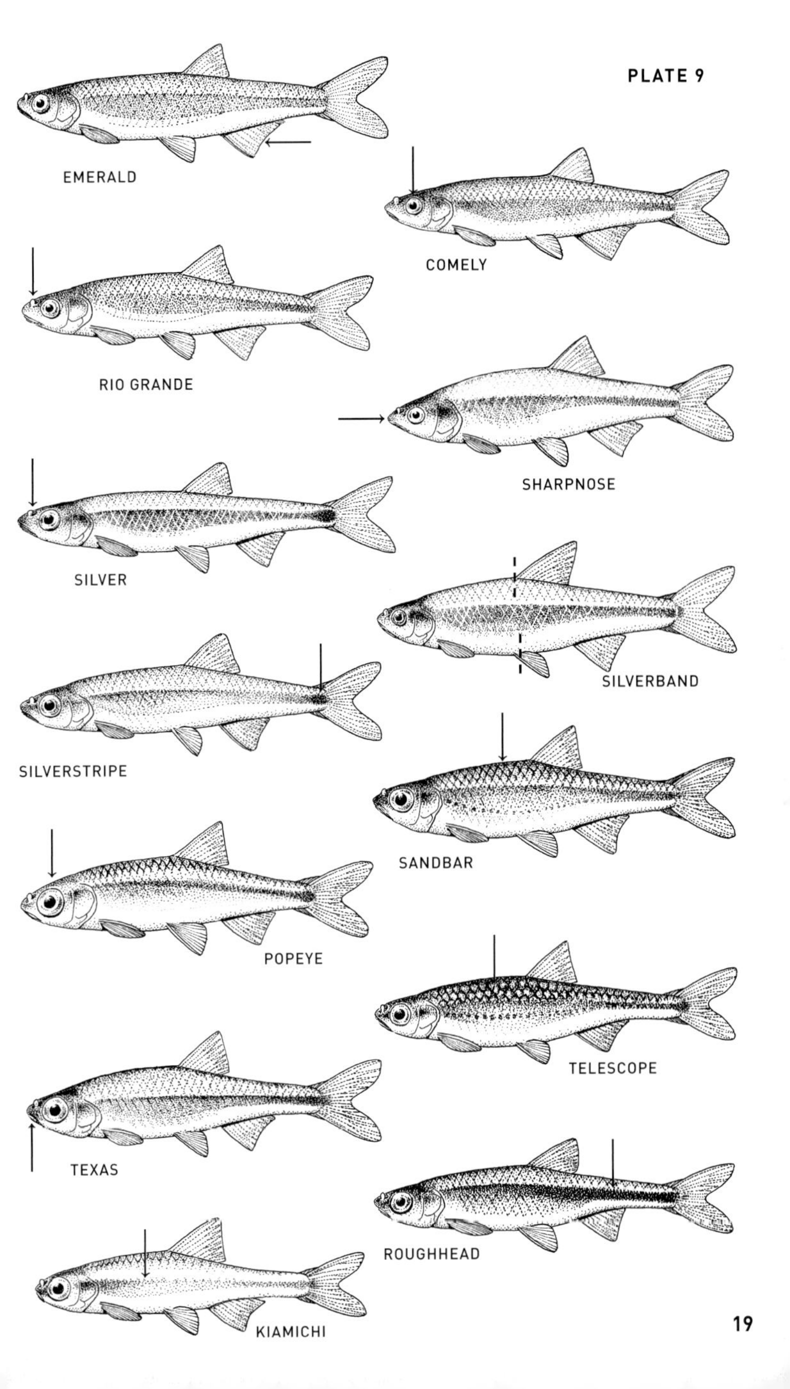
PLATE 9
EMERALD
COMELY
RIO GRANDE
SHARPNOSE
SILVER
SILVERBAND
SILVERSTRIPE
SANDBAR
POPEYE
TELESCOPE
TEXAS
ROUGHHEAD
KIAMICHI

PLATE 10

Minnows (6)

PEPPERED SHINER *Notropis perpallidus* **P. 280**
Many black spots. Slender, compressed body. To 2 in. (5 cm).

CHIHUAHUA SHINER *Notropis chihuahua* **P. 285**
Many black spots. Black wedge on caudal fin base. Stout body. To 3¼ in. (8 cm).

BLUNTNOSE SHINER *Notropis simus* **P. 263**
Blunt snout. Deep, wide head. Small eye. Small black specks on pallid body. To 4 in. (10 cm). See also Phantom Shiner, *N. orca*.

TAMAULIPAS SHINER *Notropis braytoni* **P. 285**
Dusky stripe along side followed by clear area, then small black wedge on caudal fin base. To 2¾ in. (6.9 cm).

SPOTTAIL SHINER *Notropis hudsonius* **P. 293**
Large eye. Short, rounded snout. Nearly horizontal subterminal mouth. Large black caudal spot (inconspicuous in s. Atlantic drainages and often on large individual elsewhere). To 5¾ in. (15 cm).

BLACKSPOT SHINER *Notropis atrocaudalis* **P. 294**
Narrow black stripe along side and around snout. Black rectangle on caudal fin base. Stocky body. Subterminal mouth. To 3 in. (7.6 cm).

BLACKNOSE SHINER *Notropis heterolepis* **P. 290**
Black crescents within black stripe along side and around snout (absent on chin). Slender body. Dorsal fin origin behind pelvic fin origin. (See also Fig. 30.) To 3¾ in. (9.8 cm).

BEDROCK SHINER *Notropis rupestris* **P. 291**
Similar to Blacknose Shiner but has more arched back, complete lateral line, dorsal fin origin over or slightly behind pelvic fin origin. To 2½ in. (6.2 cm).

WHITEMOUTH SHINER *Notropis alborus* **P. 292**
Jagged-edged stripe along side. Black wedge on caudal fin base. No dark stripe along back. Black bridle around snout; no black on lips (Fig. 30). To 2¼ in. (6 cm).

PUGNOSE SHINER *Notropis anogenus* **P. 291**
Small, upturned mouth (Fig. 30). Black peritoneum. Black stripe along side. Black wedge on caudal fin base. To 2¼ in. (5.8 cm).

BRIDLE SHINER *Notropis bifrenatus* **P. 292**
Black spot on caudal fin base usually joined to black stripe along side and around snout (where mostly confined to upper lip; Fig. 30). To 2½ in. (6.5 cm).

NEW RIVER SHINER *Notropis scabriceps* **P. 260**
Broad snout. Large, upwardly directed eye. Darkly outlined scales. Punctate lateral line. To 3¼ in. (8.4 cm).

WEDGESPOT SHINER *Notropis greenei* **P. 261**
Black wedge on caudal fin base. Large, upwardly directed eye. To 3 in. (7.5 cm).

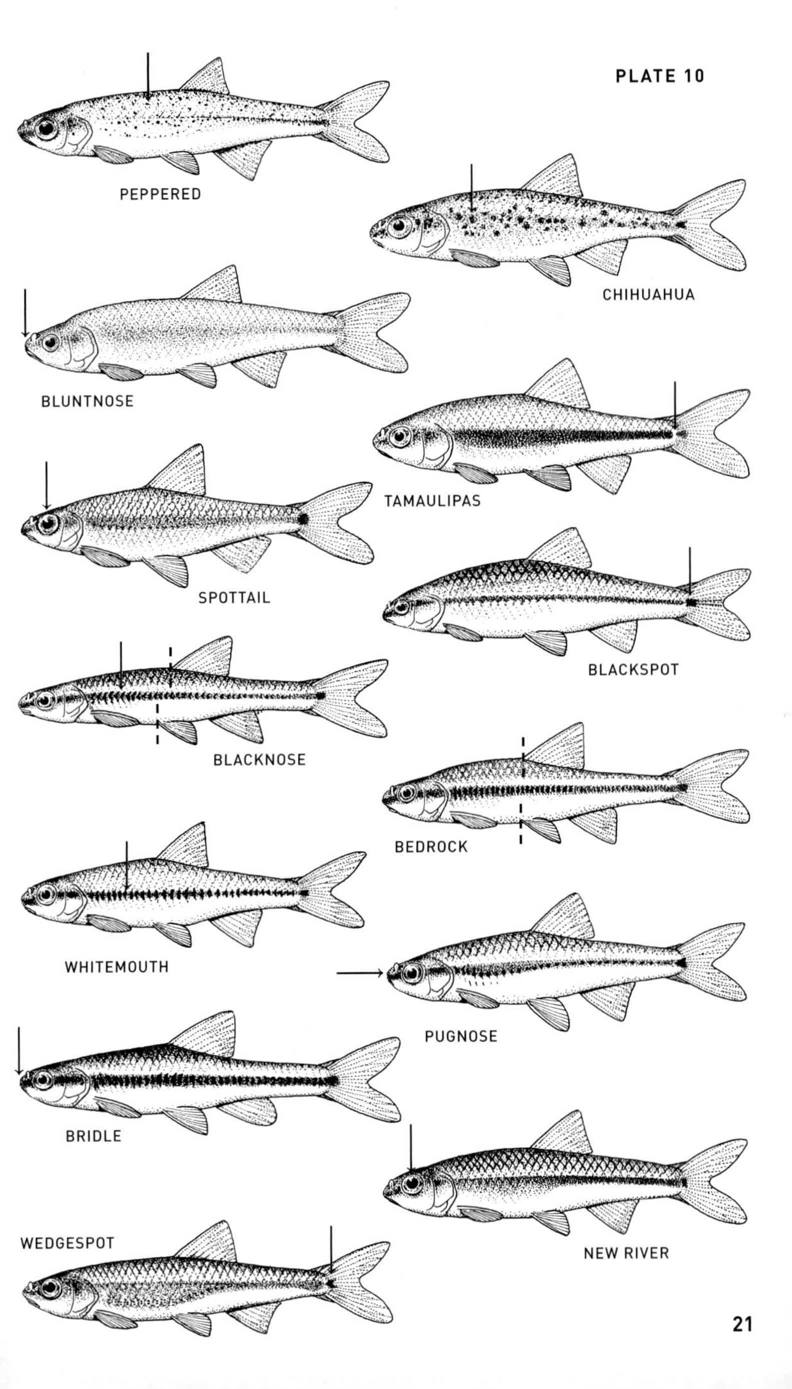
PLATE 10
PEPPERED
CHIHUAHUA
BLUNTNOSE
TAMAULIPAS
SPOTTAIL
BLACKSPOT
BLACKNOSE
BEDROCK
WHITEMOUTH
PUGNOSE
BRIDLE
NEW RIVER
WEDGESPOT

Minnows (7)

First 4 species have clear to yellow stripe above black stripe along side.

BLACKCHIN SHINER *Notropis heterodon* **P. 274**
Black stripe extends around short pointed snout (Fig. 30.) Black stripe often has zigzag appearance on side. To 2¾ in. (7.1 cm).

BIGEYE SHINER *Notropis boops* **P. 276**
Large eye. Punctate lateral line. Black peritoneum. To 3½ in. (9 cm).

COOSA SHINER *Notropis xaenocephalus* **P. 276**
Large black spot on caudal fin base. Darkly outlined scales. Large eye. To 3 in. (7.9 cm).

BURRHEAD SHINER *Notropis asperifrons* **P. 276**
Long, rounded snout. Black spot on caudal fin base. To 3 in. (7.5 cm).

BIGMOUTH SHINER *Notropis dorsalis* **P. 274**
Upwardly directed eye. Head flattened below. Arched body. To 3¼ in. (8 cm).

SILVERJAW MINNOW *Ericymba buccata* **P. 295**
Large silver white chambers on cheek and flattened underside of head. To 3¾ in. (9.8 cm). See also Longjaw Minnow, *E. amplamala*.

SKYGAZER SHINER *Notropis uranoscopus* **P. 278**
Large elliptical eye high on head. Small black wedge on caudal fin base. To 2¾ in. (7.1 cm).

HIGHSCALE SHINER *Notropis hypsilepis* **P. 277**
Blunt snout. Small black wedge on caudal fin base. Eye high on head. To 2½ in. (6.4 cm).

WEED SHINER *Notropis texanus* **P. 278**
Black stripe along side and around snout; some black-edged scales below stripe. Last 3–4 anal rays lined with black in Gulf Coast drainages. To 3½ in. (8.6 cm).

COASTAL SHINER *Notropis petersoni* **P. 279**
Similar to Weed Shiner but lacks black-edged scales below stripe; has all anal rays lined with black, black wedge on caudal fin base. To 3¼ in. (8.2 cm).

GHOST SHINER *Notropis buchanani* **P. 286**
Translucent milky white (sometimes black specks—see text). Body deep at dorsal fin origin, tapering to thin caudal peduncle. To 2½ in. (6.4 cm).

MIMIC SHINER *Notropis volucellus* **P. 287**
Broad, rounded snout. Scales along back in front of dorsal fin wider than adjacent scales. Scales along side on front half of body much deeper than wide (Fig. 31). To 3 in. (7.6 cm). See also Cahaba Shiner, *N. cahabae*.

SAND SHINER *Notropis stramineus* **P. 283**
Punctate lateral line. Dusky stripe along back expanded into dark wedge at dorsal fin origin. To 3¼ in. (8.1 cm).

SWALLOWTAIL SHINER *Notropis procne* **P. 284**
Similar to Sand Shiner but has longer snout, darker stripe (often black) along side, yellow body and fins on breeding male. To 2¾ in. (7.2 cm). See also Palezone Shiner, *N. albizonatus*.

CAPE FEAR SHINER *Notropis mekistocholas* **P. 285**
Long coiled dark gut visible through belly wall. Black stripe along side of body and snout. To 3 in. (7.7 cm).

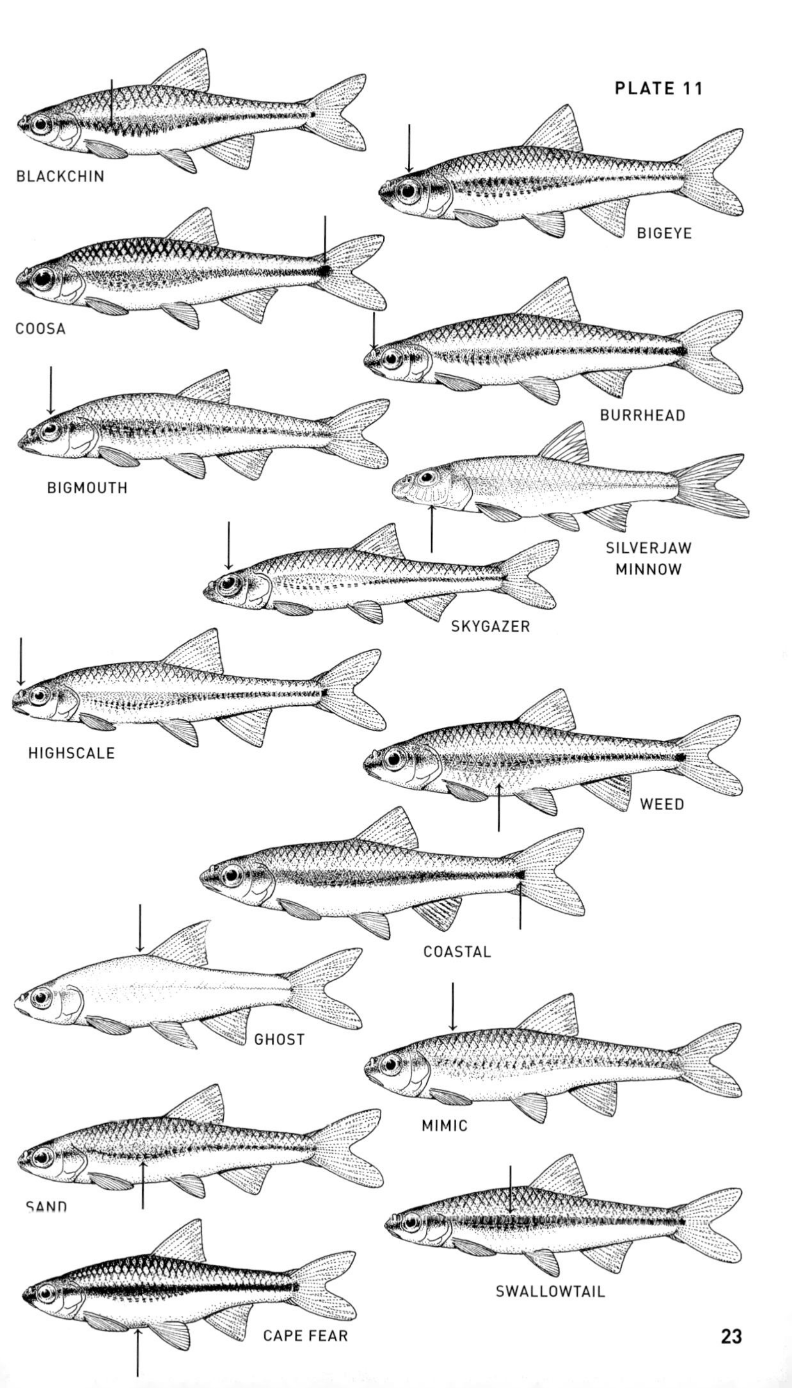
BLACKCHIN
BIGEYE
COOSA
BURRHEAD
BIGMOUTH
SILVERJAW MINNOW
SKYGAZER
HIGHSCALE
WEED
COASTAL
GHOST
MIMIC
SAND
SWALLOWTAIL
CAPE FEAR

PLATE 12

MINNOWS (8)

PEAMOUTH *Mylocheilus caurinus* **P. 162**
Two dark stripes on side; lower 1 ends before anal fin. Large male (shown) has red on side and head. To 14 in. (36 cm).

REDSIDE SHINER *Richardsonius balteatus* **P. 164**
Narrow caudal peduncle. Dorsal fin origin far behind pelvic fin origin. On large individual, red above pectoral fin. To 7 in. (18 cm).

LAHONTAN REDSIDE *Richardsonius egregius* **P. 164**
Similar to Redside Shiner but has longer snout, more slender body. To 6¾ in. (17 cm).

CALIFORNIA ROACH *Hesperoleucus symmetricus* **P. 158**
Slightly subterminal mouth. Deep body tapering to narrow caudal peduncle. To 4¼ in. (11 cm).

ROUNDTAIL CHUB *Gila robusta* **P. 165**
Slender caudal peduncle. Angle along anal fin base continues into middle of caudal fin (Fig. 13). To 17 in. (43 cm). See also Gila Chub, *G. intermedia*, Headwater Chub, *G. nigra*, and Virgin Chub, *G. seminuda.*

HUMPBACK CHUB *Gila cypha* **P. 168**
Long, extremely slender caudal peduncle. Large individual has hump behind small depressed head. Angle along anal fin base continues along upper edge of caudal fin (Fig. 13). To 15 in. (38 cm).

BONYTAIL *Gila elegans* **P. 168**
Similar to Humpback Chub but has terminal mouth, angle along anal fin base continues well above caudal fin (Fig. 13). To 24½ in. (62 cm).

CHIHUAHUA CHUB *Gila nigrescens* **P. 170**
Terminal mouth on rounded snout. Large individual has red-orange mouth and anal and paired fin bases. To 9½ in. (24 cm).

RIO GRANDE CHUB *Gila pandora* **P. 170**
Two dusky stripes along side (darkest on large individual). To 7 in. (18 cm). See also Sonora Chub, *G. ditaenia*.

BLUE CHUB *Gila coerulea* **P. 169**
Pointed snout. Terminal mouth extends to front of eye. Breeding male has blue snout, orange side and fins. To 16 in. (41 cm).

NORTHERN LEATHERSIDE CHUB *Lepidomeda copei* **P. 180**
Leatherlike appearance (created by small scales). Large male has red anal and paired fin bases. To 6 in. (15 cm). See also Southern Leatherside Chub, *L. aliciae*.

UTAH CHUB *Gila atraria* **P. 169**
Yellow to brassy side. Short snout. To 22 in. (56 cm).

TUI CHUB *Siphatales bicolor* **P. 172**
Small, rounded fins. Small mouth does not extend to eye. To 17¾ in. (45 cm).

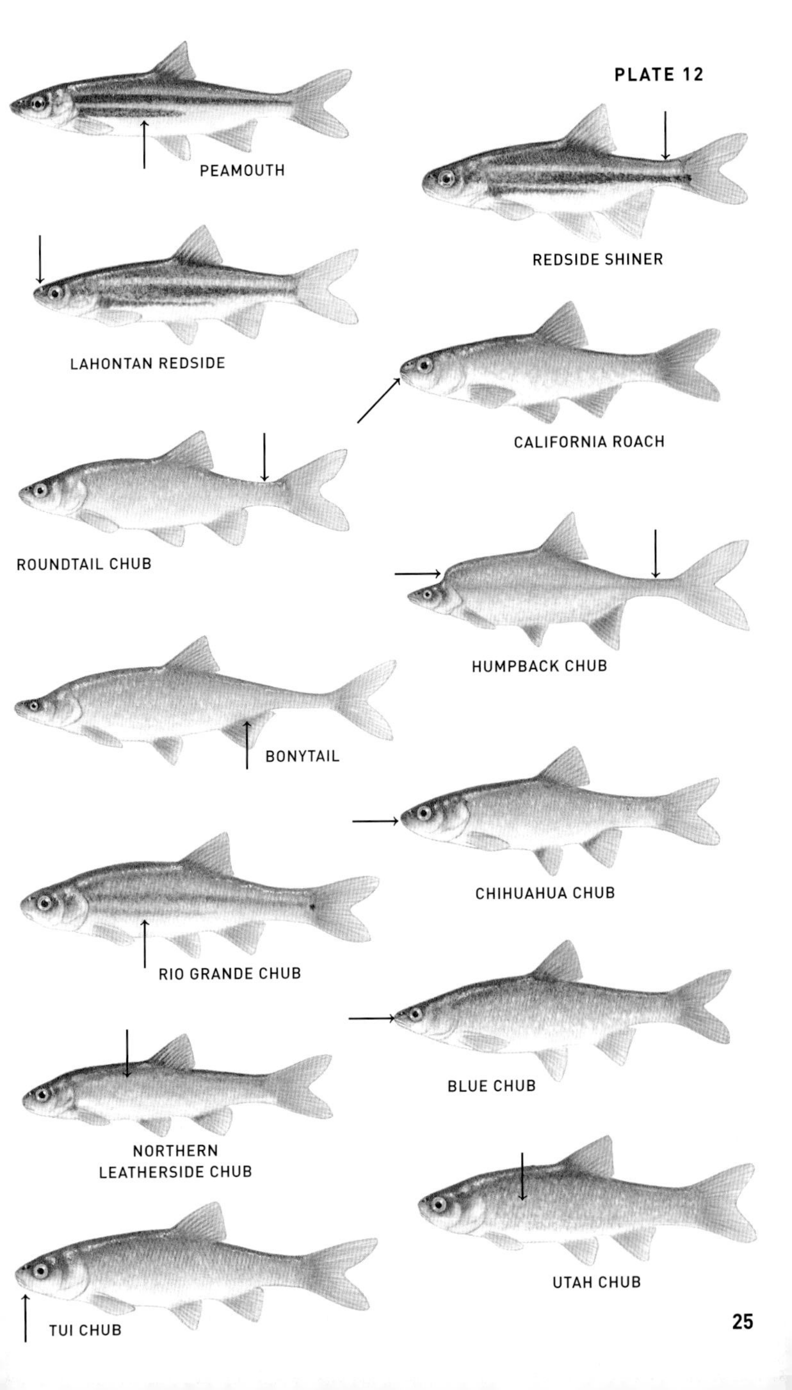
PLATE 12
PEAMOUTH
REDSIDE SHINER
LAHONTAN REDSIDE
CALIFORNIA ROACH
ROUNDTAIL CHUB
HUMPBACK CHUB
BONYTAIL
CHIHUAHUA CHUB
RIO GRANDE CHUB
BLUE CHUB
NORTHERN
LEATHERSIDE CHUB
UTAH CHUB
TUI CHUB

Minnows (9)

ROSYSIDE DACE *Clinostomus funduloides* **P. 173**
Large oblique mouth; long pointed snout. Breeding male (shown) has bright red lower side. Has 43–57 lateral scales. To 4½ in. (11 cm).

REDSIDE DACE *Clinostomus elongatus* **P. 174**
Similar to Rosyside Dace but has 59–75 lateral scales. To 4½ in. (12 cm).

LONGFIN DACE *Agosia chrysogaster* **P. 175**
Small barbel at corner of mouth. Small scales. Large female (shown) has elongated lower lobe on anal fin. To 4 in. (10 cm).

Next 3 species have 2 large spines in dorsal fin, bright silver side.

VIRGIN SPINEDACE *Lepidomeda mollispinis* **P. 179**
Large, gray-black blotches on side, compressed body. To 5¾ in. (15 cm). See also Little Colorado Spinedace, *L. vittata*, White River Spinedace, *L. albivallis*, and Pahranagat Spinedace, *L. altivelis*.

SPIKEDACE *Meda fulgida* **P. 180**
No scales. Slender body; subterminal mouth. To 3½ in. (9.1 cm).

WOUNDFIN *Plagopterus argentissimus* **P. 181**
No scales. Long snout; barbel at corner of mouth. To 3½ in. (9 cm).

FINESCALE DACE *Chrosomus neogaeus* **P. 182**
Dark "cape" on back and upper side. To 4¼ in. (11 cm).

SOUTHERN REDBELLY DACE *Chrosomus erythrogaster* **P. 182**
Two black stripes along side. Small black spots on upper side. To 3½ in. (9.1 cm). See also Northern Redbelly Dace, *C. eos*, and Laurel Dace, *C. saylori*.

MOUNTAIN REDBELLY DACE *Chrosomus oreas* **P. 185**
Black stripe along side broken under dorsal fin. Large black spots on back and upper side. To 2¾ in. (7.2 cm).

TENNESSEE DACE *Chrosomus tennesseensis* **P. 185**
Similar to Mountain Redbelly Dace but has smaller spots, thin black stripe along side. To 2¾ in. (7.2 cm).

BLACKSIDE DACE *Chrosomus cumberlandensis* **P. 184**
Two stripes on side coalesce into wide stripe on large male (shown). To 2¾ in. (7.2 cm).

FLAME CHUB *Hemitremia flammea* **P. 185**
Deep caudal peduncle. Scarlet red below (brightest on male). To 2¾ in. (7.2 cm).

CREEK CHUB *Semotilus atromaculatus* **P. 187**
Large black spot at dorsal fin origin. Breeding male has pink body, orange fins. To 12 in. (30 cm). See also Sandhills Chub, *S. lumbee*.

DIXIE CHUB *Semotilus thoreauianus* **P. 187**
Similar to Creek Chub but has larger scales, yellow fins. To 6 in. (15 cm).

FALLFISH *Semotilus corporalis* **P. 188**
Large (43–50 lateral), darkly outlined scales. To 20¼ in. (51 cm).

NORTHERN PEARL DACE *Margariscus nachtriebi* **P. 188**
Small scales. Barbel as in Creek Chub (Fig. 16). Breeding male (shown) is red-orange along lower side. To 6½ in. (16 cm). See also Allegheny Pearl Dace, *M. margarita*.

LAKE CHUB *Couesius plumbeus* **P. 189**
Large eye. Barbel at corner of mouth. To 9 in. (23 cm).

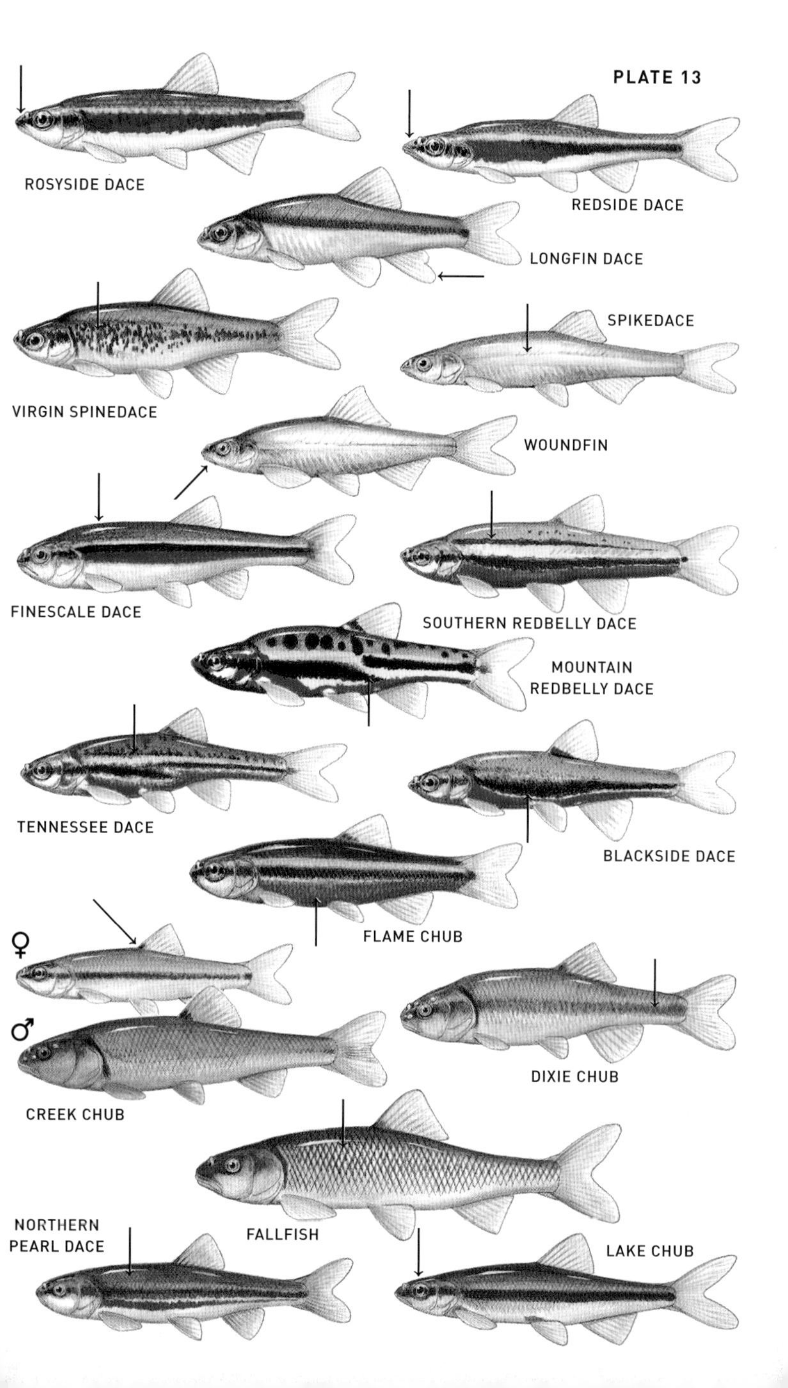
PLATE 13
ROSYSIDE DACE
REDSIDE DACE
LONGFIN DACE
SPIKEDACE
VIRGIN SPINEDACE
WOUNDFIN
FINESCALE DACE
SOUTHERN REDBELLY DACE
MOUNTAIN
REDBELLY DACE
TENNESSEE DACE
BLACKSIDE DACE
FLAME CHUB
♀
♂
DIXIE CHUB
CREEK CHUB
NORTHERN
PEARL DACE
FALLFISH
LAKE CHUB

Minnows (10)

First 4 species (*Nocomis*) have stout body, large scales.

HORNYHEAD CHUB *Nocomis biguttatus* **P. 190**
Bright red (on male) or brassy (female) spot behind eye. To 10¼ in. (26 cm). See also Redspot Chub, *N. asper*.

REDTAIL CHUB *Nocomis effusus* **P. 192**
Bright red-orange fins (especially on young—shown). To 9 in. (23 cm).

RIVER CHUB *Nocomis micropogon* **P. 192**
Long snout. Small eye high on head. Breeding male (shown) has hump on head. To 12½ in. (32 cm). See also Bigmouth Chub, *N. platyrhynchus*, and Bull Chub, *N. raneyi*.

BLUEHEAD CHUB *Nocomis leptocephalus* **P. 193**
Large loop on intestine (visible through body wall on young). Short snout. Breeding male (shown) has hump on blue head. To 10 in. (26 cm).

Next 4 species (*Campostoma*) have cartilaginous ridge on lower jaw (Fig. 17).

CENTRAL STONEROLLER *Campostoma anomalum* **P. 195**
Usually 46–55 lateral scales, 36–46 scales around body at dorsal fin origin. To 6¾ in. (17 cm). See also Highland Stoneroller, *C. spadiceum*.

BLUEFIN STONEROLLER *Campostoma pauciradii* **P. 197**
Has 11–17 (vs. 21–33) rakers on 1st gill arch, usually 33–38 scales around body. To 6¼ in. (16 cm).

LARGESCALE STONEROLLER *Campostoma oligolepis* **P. 197**
Usually has 31–36 scales around body, 43–47 lateral scales. To 8½ in. (22 cm).

MEXICAN STONEROLLER *Campostoma ornatum* **P. 198**
Similar to Central Stoneroller but usually has 58–77 lateral scales, no tubercles on nape of breeding male (shown). To 6¼ in. (16 cm).

ROUNDNOSE MINNOW *Dionda episcopa* **P. 199**
Bicolored. Black stripe along side, zigzagged at front. To 3 in. (7.7 cm).

NUECES ROUNDNOSE MINNOW *Dionda serena* **P. 199**
Bicolored. Body deepest under nape. To 3 in. (7.7 cm). See also Guadalupe Roundnose Minnow, *D. nigrotaeniata,* and Manantial Roundnose Minnow, *D. argentosa*.

DEVILS RIVER MINNOW *Dionda diaboli* **P. 198**
Black wedge on caudal fin base. Black stripe on side. To 2½ in. (6.4 cm).

BLACKNOSE DACE *Rhinichthys atratulus* **P. 204**
Many dark specks. Deep caudal peduncle. To 4 in. (10 cm). Breeding male (shown) of *R. a. atratulus* has gold yellow below red-black stripe, of *R. a. obtusus* (shown) has red below black stripe.

SPECKLED DACE *Rhinichthys osculus* **P. 205**
Usually many dark specks. To 4¼ in. (11 cm). See also Las Vegas Dace, *R. deaconi*.

LEOPARD DACE *Rhinichthys falcatus* **P. 206**
Falcate dorsal fin. Larger dark blotches on side. To 5 in. (12 cm). See also Umatilla Dace, *R. umatilla*.

LONGNOSE DACE *Rhinichthys cataractae* **P. 207**
Long fleshy snout. To 6¼ in. (16 cm). See also Umpqua Dace, *R. evermanni*.

LOACH MINNOW *Rhinichthys cobitis* **P. 208**
Upwardly directed eye. White bar on caudal fin base. To 2¼ in. (6 cm).

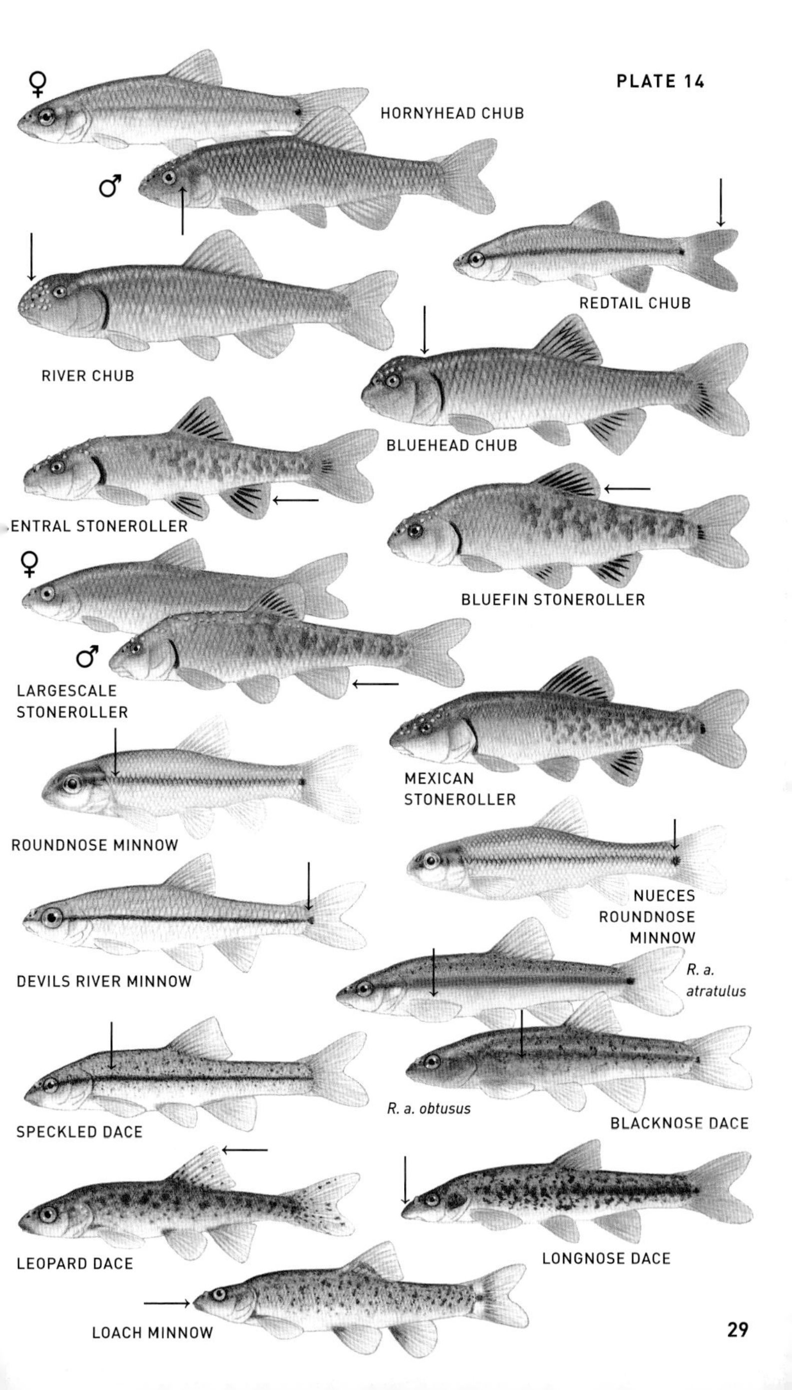
PLATE 14
HORNYHEAD CHUB
REDTAIL CHUB
RIVER CHUB
BLUEHEAD CHUB
ENTRAL STONEROLLER
BLUEFIN STONEROLLER
LARGESCALE
STONEROLLER
MEXICAN
STONEROLLER
ROUNDNOSE MINNOW
NUECES
ROUNDNOSE
MINNOW
DEVILS RIVER MINNOW
R. a.
atratulus
R. a. obtusus
SPECKLED DACE
BLACKNOSE DACE
LEOPARD DACE
LONGNOSE DACE
LOACH MINNOW

Minnows (11)

First 7 species (*Lythrurus*) have very small scales on nape; dorsal fin origin behind pelvic fin origin; usually 10–12 anal rays.

REDFIN SHINER *Lythrurus umbratilis* **P. 225**
Black blotch at dorsal fin origin. Deep body. Breeding male (shown) has red fins, blue head and body. To 3½ in. (8.6 cm).

ROSEFIN SHINER *Lythrurus ardens* **P. 225**
Black blotch at dorsal fin origin. Usually 11 anal rays. Breeding male (shown) has red fins and top of head, faint blue-gray bars on back. To 3½ in. (8.6 cm).

SCARLET SHINER *Lythrurus fasciolaris* **P. 226**
Black blotch at dorsal fin origin. Dusky bars over back (darkest on male). Usually 10 anal rays. To 3½ in. (8.6 cm).

PINEWOODS SHINER *Lythrurus matutinus* **P. 226**
Similar to Rosefin Shiner but more slender. Breeding male (shown) has bright red head, silver body. To 3½ in. (8.6 cm).

RIBBON SHINER *Lythrurus fumeus* **P. 227**
Scales on nape outlined in black. Fairly slender body. Dusky lips and chin (Fig. 23). To 2¾ in. (7 cm). See also Ouachita Mountain Shiner, *L. snelsoni*.

MOUNTAIN SHINER *Lythrurus lirus* **P. 228**
Silver black stripe along side. Black lips, white chin (Fig. 23). To 3 in. (7.5 cm).

CHERRYFIN SHINER *Lythrurus roseipinnis* **P. 228**
Black spots on tips of dorsal and anal fins (Fig. 24). See also Pretty Shiner, *L. bellus*, and Blacktip Shiner, *L. atrapiculus*.

Next 9 species (*Luxilus*) have scales deeper than wide, usually 9 anal rays.

STRIPED SHINER *Luxilus chrysocephalus* **P. 220**
Three dark stripes on upper side meet those of other side to form large Vs (Fig. 22). Large male has pink or red body and fins. To 7¼ in. (18 cm).

COMMON SHINER *Luxilus cornutus* **P. 221**
Stripes on upper side are parallel to stripe along back (Fig. 22). To 7 in. (18 cm).

WHITE SHINER *Luxilus albeolus* **P. 222**
Nearly identical to Common Shiner but more silvery; usually 26–30 scales around body (Common Shiner has 30–35). To 5¼ in. (13 cm).

CRESCENT SHINER *Luxilus cerasinus* **P. 222**
Large black crescents on side. Breeding male (shown) has red on head, body, and fins. To 4½ in. (11 cm).

BLEEDING SHINER *Luxilus zonatus* **P. 222**
Large black bar behind gill cover. Narrow black stripe along side. To 5 in. (13 cm).

CARDINAL SHINER *Luxilus cardinalis* **P. 228**
Broad black stripe along side extends below lateral line. To 4¼ in. (11 cm).

DUSKYSTRIPE SHINER *Luxilus pilsbryi* **P. 223**
Black stripe along body. To 5 in. (13 cm).

BANDFIN SHINER *Luxilus zonistius* **P. 224**
Black band on dorsal fin. Large black spot on caudal fin base. Breeding male (shown) has red bar on caudal fin. To 4 in. (10 cm).

WARPAINT SHINER *Luxilus coccogenis* **P. 224**
Black band on dorsal and caudal fins. To 5½ in. (14 cm).

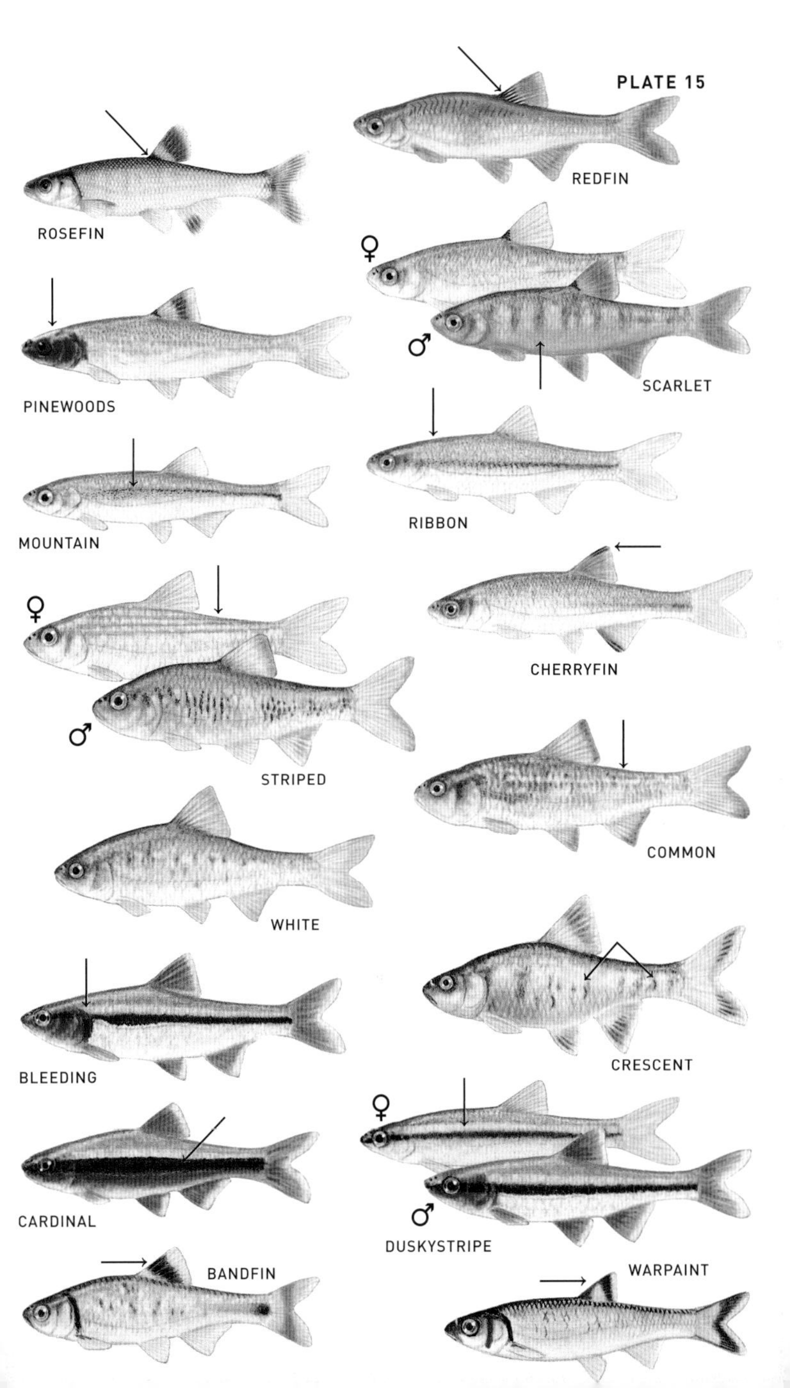
PLATE 15
ROSEFIN
REDFIN
PINEWOODS
♀
♂
SCARLET
MOUNTAIN
RIBBON
♀
♂
STRIPED
CHERRYFIN
WHITE
COMMON
BLEEDING
CRESCENT
CARDINAL
♀
♂
DUSKYSTRIPE
BANDFIN
WARPAINT

PLATE 16

Minnows (12)

Diamond-shaped scales, dusky bar on chin.

STEELCOLOR SHINER *Cyprinella whipplei* **P. 236**
Black blotch on rear half of dorsal fin, black specks on all membranes (Fig. 26). Breeding male (shown) is blue; has large dorsal fin. To 6¼ in. (16 cm). See also Satinfin Shiner, *C. analostana*, and Greenfin Shiner, *C. chloristia*.

SPOTFIN SHINER *Cyprinella spiloptera* **P. 236**
Black blotch on rear half of dorsal fin, little or no black on membranes (Fig. 26). Breeding male is blue; has yellow fins. To 4¾ in. (12 cm).

BLUNTFACE SHINER *Cyprinella camura* **P. 239**
Clear to white bar (sometimes absent) on caudal fin base. Breeding male (shown) has pale orange or red fins and snout, enlarged dorsal fin. To 4¼ in. (11 cm).

BLACKTAIL SHINER *Cyprinella venusta* **P. 238**
Large black spot on caudal fin base. Breeding male is blue; has yellow or (in TX) red-orange fins, dusky dorsal fin. To 7½ in. (19 cm).

RED SHINER *Cyprinella lutrensis* **P. 240**
Blue bar behind head. Breeding male has blue body, red fins. To 3½ in. (9 cm). See also Beautiful Shiner, *C. formosa*, and Plateau Shiner, *C. lepida*.

PROSERPINE SHINER *Cyprinella proserpina* **P. 241**
Subterminal mouth. Black stripe on chin and throat. To 3 in. (7.5 cm).

WHITETAIL SHINER *Cyprinella galactura* **P. 242**
Two large white areas on caudal fin base. Slender body. To 6 in. (15 cm).

ALABAMA SHINER *Cyprinella callistia* **P. 242**
Large black spot on caudal fin base. Pink to red dorsal and caudal fins. To 3¾ in. (9.5 cm).

BLUE SHINER *Cyprinella caerulea* **P. 244**
Blue-black stripe along side, expanded on caudal fin base. Fairly slender body. Pointed snout. To 3½ in. (9 cm).

TRICOLOR SHINER *Cyprinella trichroistia* **P. 243**
Large black spot on caudal fin base fusing into black stripe on side. Yellow to red-orange fins. To 4 in. (10 cm). See also Tallapoosa Shiner, *C. gibbsi*.

FIERYBLACK SHINER *Cyprinella pyrrhomelas* **P. 245**
Black edge on caudal fin (of adult). Black bar behind head. Breeding male has bright red snout, red after white band on caudal fin. To 4¼ in. (11 cm).

OCMULGEE SHINER *Cyprinella callisema* **P. 246**
Deep blue stripe along side. Small black blotch at front of dorsal fin. Subterminal mouth. To 3½ in. (9 cm). See also Bluestripe Shiner, *C. callitaenia*.

ALTAMAHA SHINER *Cyprinella xaenura* **P. 245**
Pointed snout. Black stripe along rear half of side expanded into spot on caudal fin base. To 4½ in. (11 cm).

WHITEFIN SHINER *Cyprinella nivea* **P. 247**
Dark blue to black stripe along side. Black blotch on rear half of dorsal fin. Subterminal mouth. To 3½ in. (8.5 cm).

BANNERFIN SHINER *Cyprinella leedsi* **P. 247**
Small black blotch at front of dorsal fin. Subterminal mouth. Breeding male (shown) has greatly enlarged black dorsal fin. To 4 in. (10 cm).

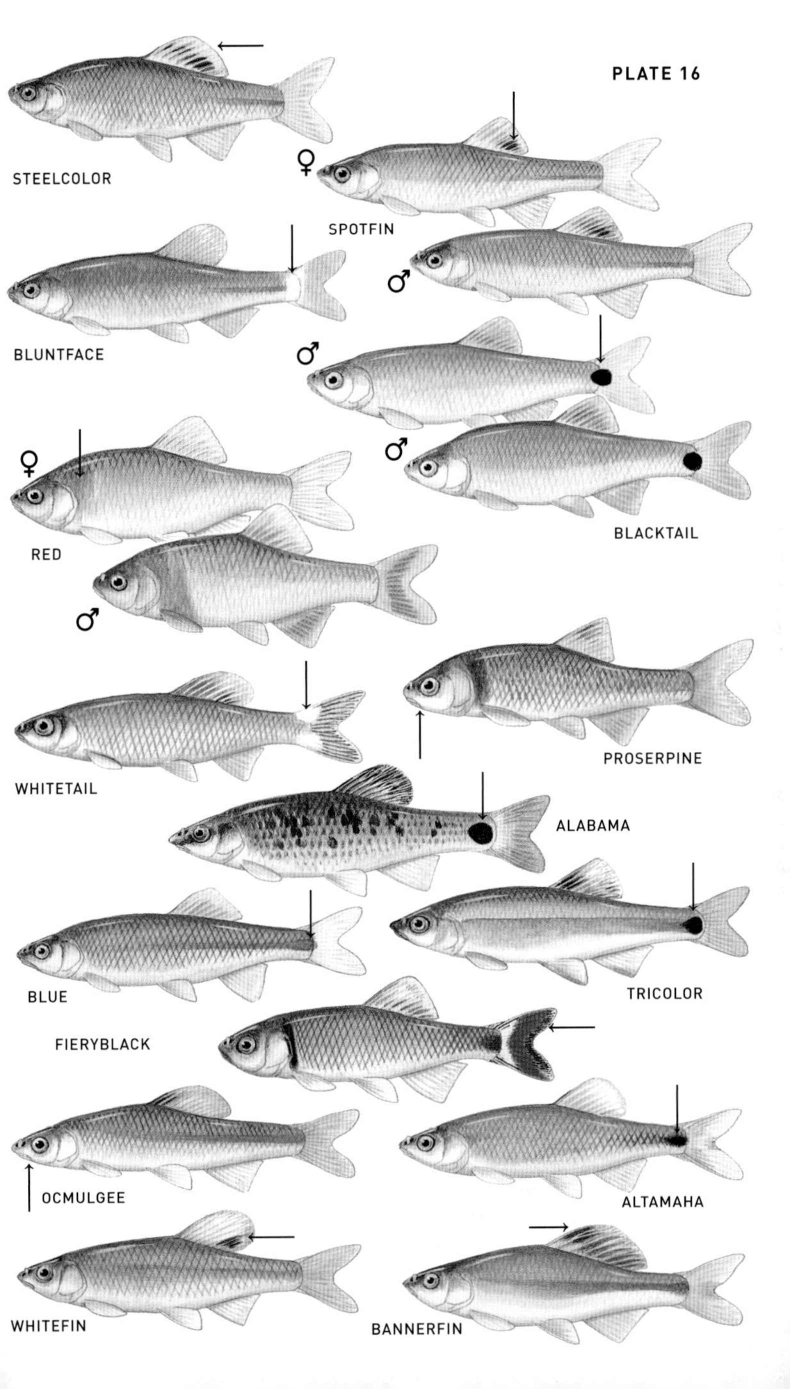
PLATE 16
STEELCOLOR
♀
SPOTFIN
♂
BLUNTFACE
♂
♂
BLACKTAIL
♀
RED
♂
PROSERPINE
WHITETAIL
ALABAMA
TRICOLOR
BLUE
FIERYBLACK
OCMULGEE
ALTAMAHA
WHITEFIN
BANNERFIN

Minnows (13)

LONGNOSE SHINER *Notropis longirostris* **P. 272**
Upwardly directed eye. Long, rounded snout. Subterminal mouth. Yellow fins. To 2½ in. (6.5 cm). See also Yazoo Shiner, *N. rafinesquei.*

ORANGEFIN SHINER *Notropis ammophilus* **P. 272**
Similar to Longnose Shiner but has dorsal fin origin over pelvic fin origin; orange snout and fins on large male. To 2¼ in. (6 cm).

SABINE SHINER *Notropis sabinae* **P. 273**
Strongly arched body, flattened below. Subterminal mouth. Small eye directed upwardly. To 2¼ in. (5.7 cm).

SPOTFIN CHUB *Cyprinella monacha* **P. 249**
Slender, arched body; flattened below. Long snout overhangs mouth. Breeding male (shown) has 2 white bars on blue side. To 4¼ in. (11 cm).

HIGHBACK CHUB *Hybopsis hypsinotus* **P. 299**
Large dorsal fin on arched back. Red fins. Barbel at corner of subterminal mouth. To 2¾ in. (7.2 cm).

ROUGH SHINER *Notropis baileyi* **P. 267**
Red-brown above; brown-black stripe along side uniformly dark from snout to caudal fin base. Black caudal spot. Yellow (sometimes red) fins. To 3½ in. (9 cm).

YELLOWFIN SHINER *Notropis lutipinnis* **P. 268**
Lacks caudal spot, has dorsal fin origin behind pelvic fin origin. Breeding male has yellow to red fins. To 3 in. (7.5 cm).

GREENHEAD SHINER *Notropis chlorocephalus* **P. 268**
Breeding male (shown) is red, has bright white fins. To 2¾ in. (7.2 cm).

REDLIP SHINER *Notropis chiliticus* **P. 269**
Bright red lips. Scattered black blotches on side. Breeding male has scarlet red body and eye, yellow head and fins. To 2¾ in. (7.2 cm).

SAFFRON SHINER *Notropis rubricroceus* **P. 269**
Darkly outlined scales extend below lateral line. Black caudal spot. Breeding male is red-purple above, has bright yellow fins. To 3¼ in. (8.4 cm).

RAINBOW SHINER *Notropis chrosomus* **P. 270**
Clear to red-purple stripe above silver black stripe along side. Iridescent blue and pink head and body (most vivid on large male). To 3¼ in. (8.1 cm).

TENNESSEE SHINER *Notropis leuciodus* **P. 270**
Dark wavy stripes on back and upper side. Black rectangle on caudal fin base. Punctate lateral line. Breeding male is red. To 3¼ in. (8.2 cm).

ROSYFACE SHINER *Notropis rubellus* **P. 256**
Sharply pointed snout longer than eye diameter. Dorsal fin origin well behind pelvic fin origin. Breeding male (shown) has orange to bright red head, front half of body, and fin bases. To 3½ in. (9 cm). See also Carmine Shiner, *N. percobromus*, Highland Shiner, *N. micropteryx*, and Rocky Shiner, *N. suttkusi.*

OZARK MINNOW *Notropis nubilus* **P. 271**
Olive-brown above, white to orange below. Black stripe along side and around snout. Black peritoneum often visible through belly. To 3¾ in. (9.3 cm).

TOPEKA SHINER *Notropis topeka* **P. 294**
Stocky, compressed body. Small eye. Black wedge on caudal fin base. Breeding male (shown) has red-orange fins. To 3 in. (7.6 cm).

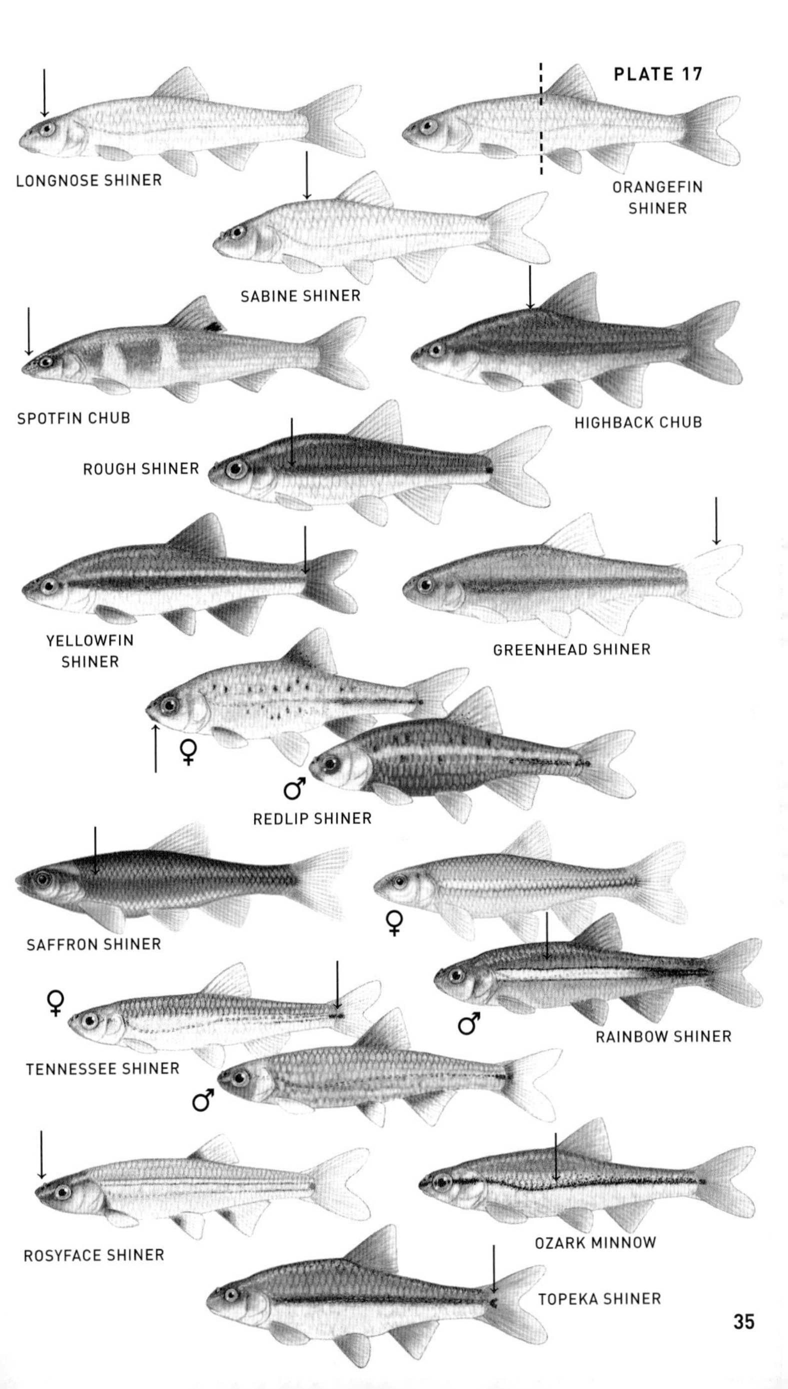
PLATE 17
LONGNOSE SHINER
ORANGEFIN SHINER
SABINE SHINER
SPOTFIN CHUB
HIGHBACK CHUB
ROUGH SHINER
YELLOWFIN SHINER
GREENHEAD SHINER
♀
♂
REDLIP SHINER
SAFFRON SHINER
♀
♂
RAINBOW SHINER
♀
TENNESSEE SHINER
♂
ROSYFACE SHINER
OZARK MINNOW
TOPEKA SHINER

Minnows (14)

IRONCOLOR SHINER *Notropis chalybaeus* **P. 281**
Well-defined black stripe from spot on caudal fin base along side and around snout covering both lips and chin (Fig. 30). Black inside mouth. To 2½ in. (6.5 cm).

DUSKY SHINER *Notropis cummingsae* **P. 281**
Wide black stripe along side, streaking onto caudal fin. To 2¾ in. (7.2 cm).

REDEYE CHUB *Notropis harperi* **P. 282**
Pink-tan above. Red eye. Yellow arc on snout. To 2¼ in. (6 cm).

HIGHFIN SHINER *Notropis altipinnis* **P. 282**
Boldly outlined scales on back separated from silver black stripe along side by clear yellow stripe. To 2½ in. (6.1 cm).

Next 3 species have very broad blue-black stripe along side of deep body, 10–11 anal rays.

FLAGFIN SHINER *Pteronotropis signipinnis* **P. 231**
Red-orange edge on yellow dorsal, caudal, anal, and pelvic fins. No dark predorsal stripe. To 2¾ in. (7 cm).

SAILFIN SHINER *Pteronotropis hypselopterus* **P. 231**
Dark brown predorsal stripe. Large, nearly triangular dorsal and anal fins. To 2¾ in. (7 cm). See Fig. 25 for related species.

BROADSTRIPE SHINER *Pteronotropis euryzonus* **P. 232**
Nearly identical to Sailfin Shiner but has larger dorsal fin; tips of rays at front reach to or beyond those at rear in depressed fin. To 2½ in. (6.5 cm). See Fig. 25 for related species.

BLUEHEAD SHINER *Pteronotropis hubbsi* **P. 235**
Black stripe along side. Deep body. Breeding male has bright blue on top of head, caudal fin, and huge dorsal fin. To 2¼ in. (6 cm).

BLUENOSE SHINER *Pteronotropis welaka* **P. 234**
Black stripe along slender body. Breeding male has bright blue snout, huge dorsal fin. To 2½ in. (6.5 cm).

TAILLIGHT SHINER *Notropis maculatus* **P. 293**
Red above and below large black spot on caudal fin base. Crosshatching on back and side. Large black blotch along front of dorsal fin. Breeding male (shown) has bright red body, red-black edge on fins. To 3 in. (7.6 cm).

Next 3 species have slender, usually arched, body; small subterminal mouth.

MIRROR SHINER *Notropis spectrunculus* **P. 289**
No, or small and crowded, scales on front half of nape. Upwardly directed eye. Large male has orange fins. To 3 in. (7.5 cm).

SAWFIN SHINER *Notropis* species **P. 290**
Similar to Mirror Shiner but has black specks (and orange on large male) confined to front half (not all) of dorsal fin. To 2½ in. (6.6 cm).

OZARK SHINER *Notropis ozarcanus* **P. 288**
Black spot at dorsal fin origin. To 3 in. (7.5 cm).

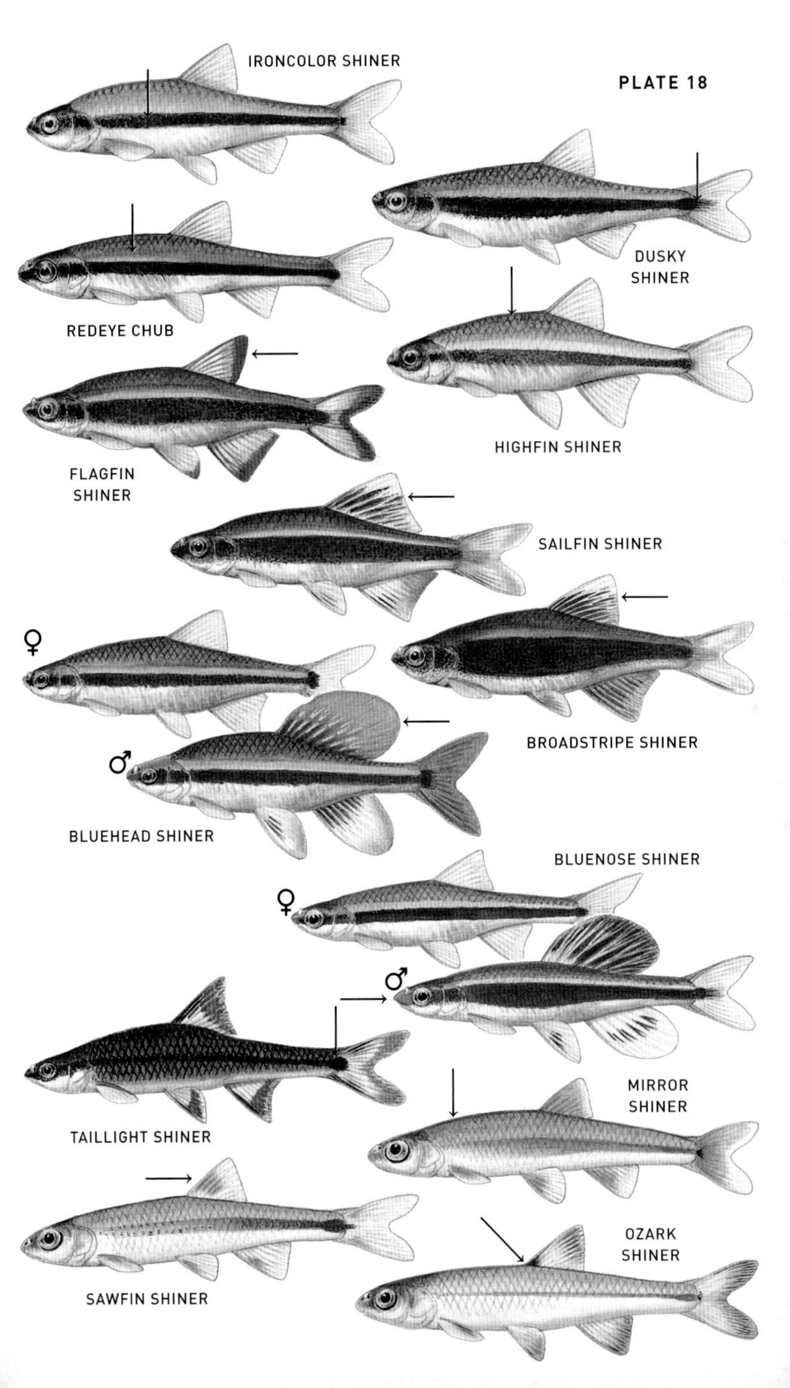
IRONCOLOR SHINER
PLATE 18
DUSKY SHINER
REDEYE CHUB
HIGHFIN SHINER
FLAGFIN SHINER
SAILFIN SHINER
♀
BROADSTRIPE SHINER
♂
BLUEHEAD SHINER
BLUENOSE SHINER
♀
♂
TAILLIGHT SHINER
MIRROR SHINER
SAWFIN SHINER
OZARK SHINER

Eurasian Carps, Suckers (1)

GRASS CARP *Ctenopharyngodon idella* **P. 153**
Wide head; terminal mouth. Large, dark-edged scales. Slender, compressed body. To 5 ft. (1.5 m).

BLACK CARP *Mylopharyngodon piceus* **P. 153**
Dark, often blue-gray or black above; fins dark, almost black; pharyngeal teeth molarlike. To 6½ ft. (2 m).

SILVER CARP *Hypophthalmichthys molitrix* **P. 153**
Deep body; eye below middle of head. Keel from anus to junction of branchiostegal membranes. Large, terminal mouth. Dorsal fin origin about even with pelvic fin origin. To 4 ft. (1.2 m).

BIGHEAD CARP *Hypophthalmichthys nobilis* **P. 154**
Similar to Silver Carp but keel extends from anus to base of pelvic fins. Irregular gray-black blotches on body (except on small young). To 5 ft. (1.5 m).

GOLDFISH *Carassius auratus* **P. 154**
Large scales. Long dorsal fin, 15–21 rays. Terminal mouth; no barbels. To 16 in. (41 cm).

COMMON CARP *Cyprinus carpio* **P. 154**
Two barbels on each side of upper jaw. Long dorsal fin, 17–21 rays. To 4 ft. (1.2 m).

Suckers: One dorsal fin, 9 or more rays; large, thick lips.

First 3 species (*Ictiobus*) have dark body and fins, semicircular subopercle (Fig. 34).

BIGMOUTH BUFFALO *Ictiobus cyprinellus* **P. 299**
Terminal, sharply oblique mouth. Large ovoid head. To 40 in. (100 cm).

SMALLMOUTH BUFFALO *Ictiobus bubalus* **P. 300**
Horizontal, subterminal mouth. Small conical head. Moderately keeled nape (on adult). To 31 in. (78 cm).

BLACK BUFFALO *Ictiobus niger* **P. 301**
Oblique, nearly terminal mouth. Large conical head. Rounded or weakly keeled nape (on adult). To 37 in. (93 cm).

Next 3 species (*Carpiodes*) have silver body, subtriangular subopercle (Fig. 34).

QUILLBACK *Carpiodes cyprinus* **P. 301**
Long 1st dorsal ray usually not reaching rear of dorsal fin base. No nipple on lower lip (Fig. 35). To 26 in. (66 cm).

RIVER CARPSUCKER *Carpiodes carpio* **P. 302**
First dorsal ray usually not reaching beyond middle of dorsal fin. Rounded snout. Nipple at middle of lower lip (Fig. 35). To 25 in. (64 cm).

HIGHFIN CARPSUCKER *Carpiodes velifer* **P. 303**
Long 1st dorsal ray reaches to or beyond rear of dorsal fin base. Blunt snout. Nipple at middle of lower lip (Fig. 35). To 19½ in. (50 cm).

PLATE 19

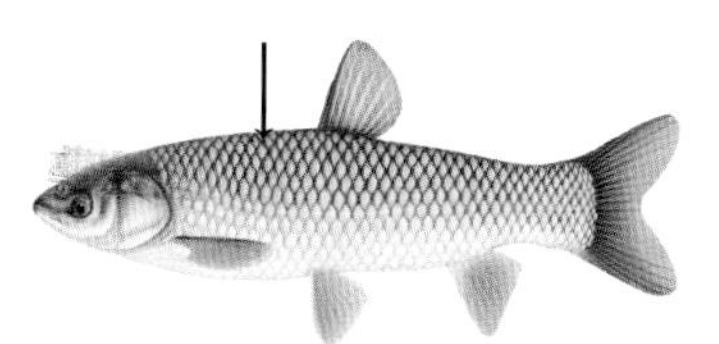
GRASS CARP

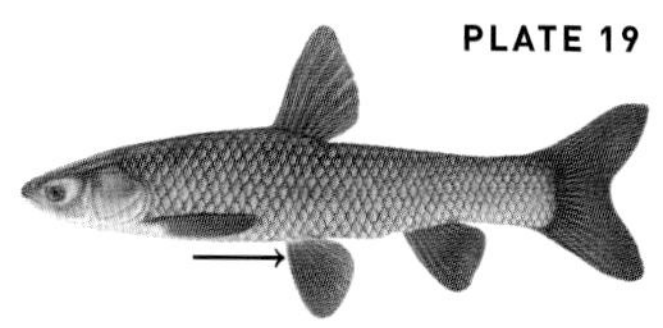
BLACK CARP

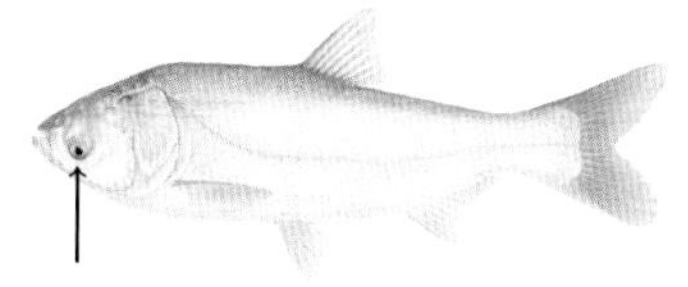
SILVER CARP

BIGHEAD CARP

GOLDFISH

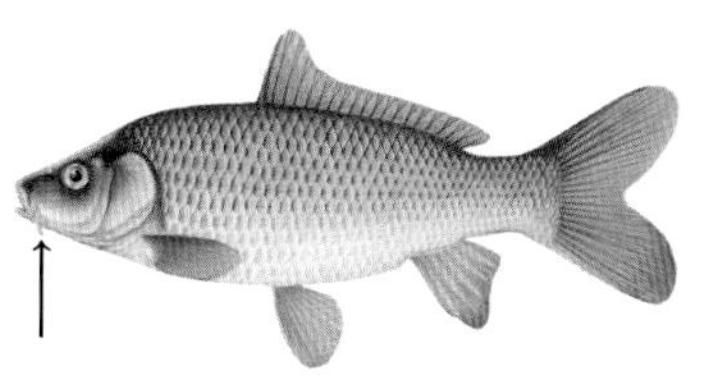
COMMON CARP

BIGMOUTH BUFFALO

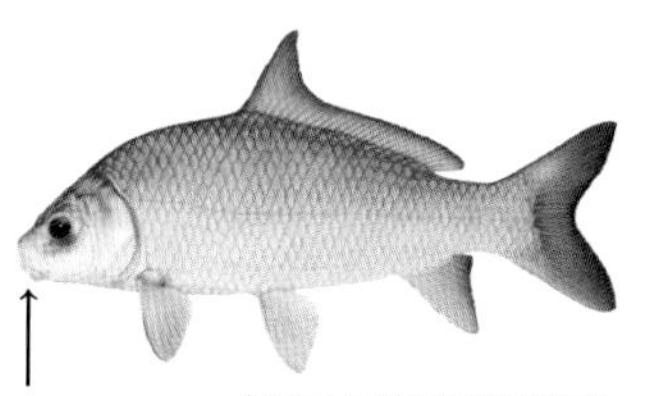
SMALLMOUTH BUFFALO

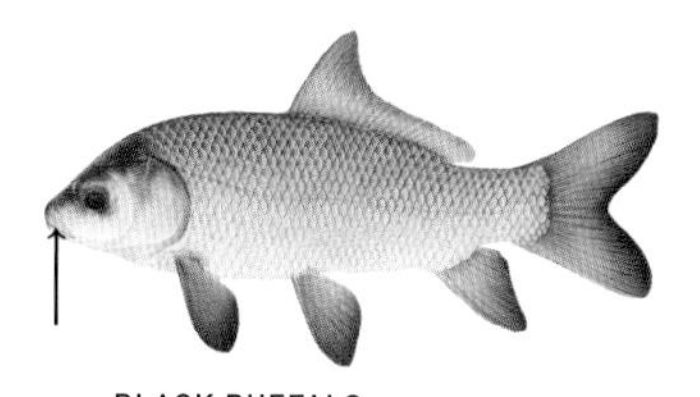
BLACK BUFFALO

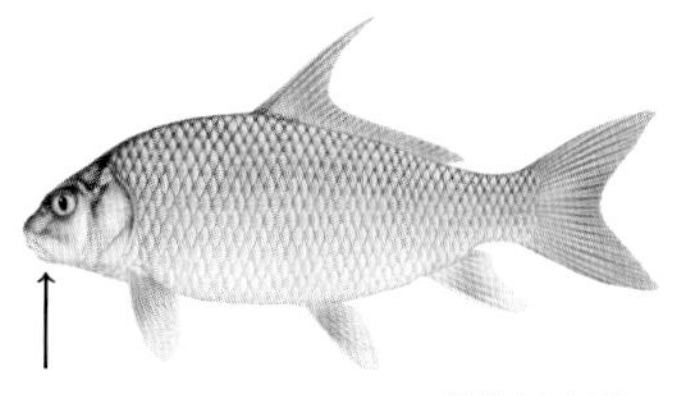
QUILLBACK

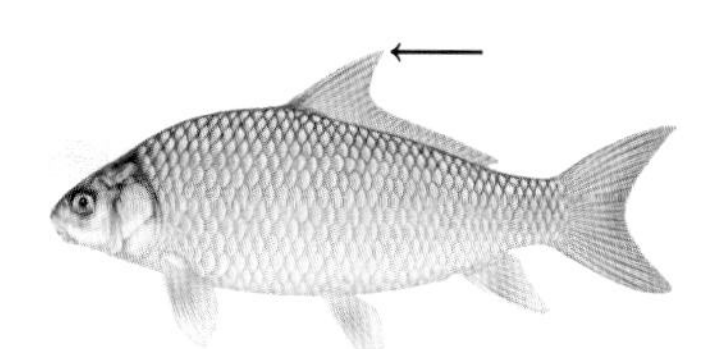
RIVER CARPSUCKER

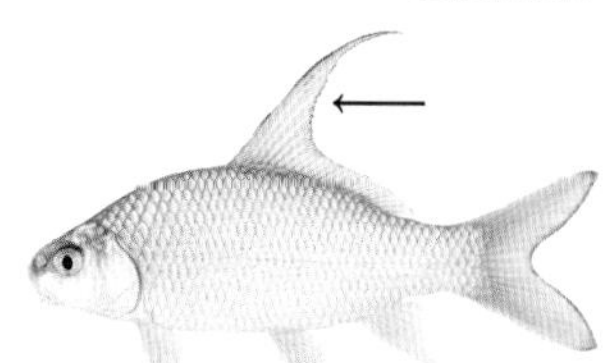
HIGHFIN CARPSUCKER

Suckers (2)

Next 3 species (*Chasmistes*) have thin (relative to other suckers), usually plicate lips (Fig. 36); branched gill rakers.

JUNE SUCKER *Chasmistes liorus* **P. 305**
Large, terminal mouth; 19–20 scales around caudal peduncle. To 20 in. (52 cm). See also Snake River Sucker, *C. muriei*.

CUI-UI *Chasmistes cujus* **P. 306**
Similar to June Sucker but has larger, broader head; 22–26 scales around caudal peduncle. To 26½ in. (67 cm).

SHORTNOSE SUCKER *Chasmistes brevirostris* **P. 306**
Similar to June Sucker but has shorter head, smaller eye, 21–25 scales around caudal peduncle. To 25 in. (64 cm).

RAZORBACK SUCKER *Xyrauchen texanus* **P. 307**
Sharp keel on nape. Long head and body. To 36 in. (91 cm).

LOST RIVER SUCKER *Deltistes luxatus* **P. 307**
Distinct hump on snout. Thin, moderately papillose lips. To 34 in. (86 cm).

***Catostomus* species have thick papillose lips (Fig. 37).**

WHITE SUCKER *Catostomus commersonii* **P. 308**
Lower lip about twice as thick as upper lip (Fig. 37). Has 53–74 lateral scales. To 25 in. (64 cm). See also Summer Sucker, *C. utawana*.

SONORA SUCKER *Catostomus insignis* **P. 309**
Dark-edged scales on upper side. To 31½ in. (80 cm). See also Yaqui Sucker, *C. bernardini*.

UTAH SUCKER *Catostomus ardens* **P. 311**
Dorsal fin membranes densely speckled to edge. (See also Fig. 37.) To 25½ in. (65 cm).

LARGESCALE SUCKER *Catostomus macrocheilus* **P. 311**
Similar to Utah Sucker but has dorsal fin membranes less densely speckled to edge, membrane connecting pelvic fin to body. To 24 in. (61 cm).

FLANNELMOUTH SUCKER *Catostomus latipinnis* **P. 312**
Narrow caudal peduncle. Large fleshy lobes on lower lip (Fig. 37). To 22 in. (56 cm). See also Little Colorado River Sucker, *C.* species.

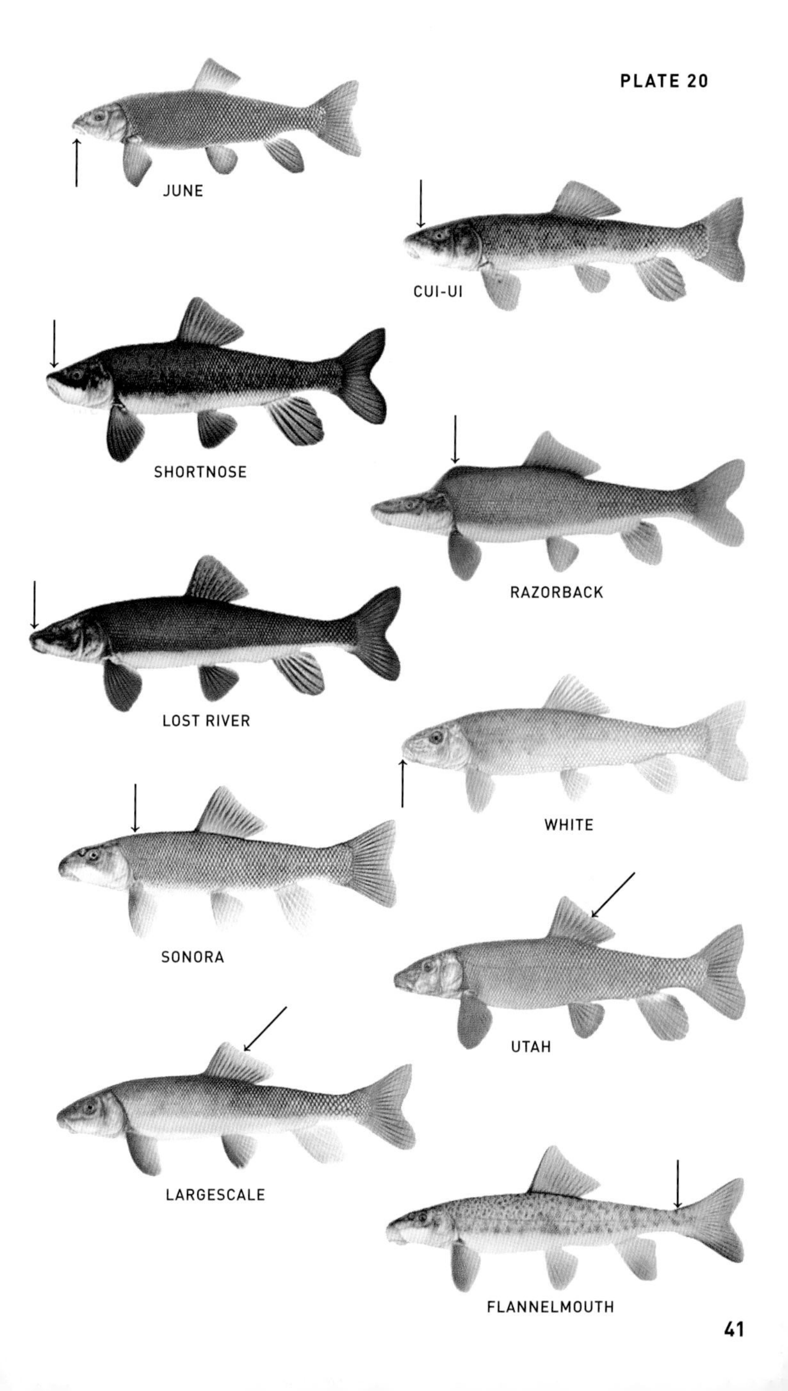
PLATE 20
JUNE
CUI-UI
SHORTNOSE
RAZORBACK
LOST RIVER
WHITE
SONORA
UTAH
LARGESCALE
FLANNELMOUTH

PLATE 21

Suckers (3)

Catostomus species have thick papillose lips (Fig. 37).

SACRAMENTO SUCKER *Catostomus occidentalis* **P. 313**
Yellow-gold below; 12–15 dorsal rays. (See also Fig. 37.) To 23½ in. (60 cm).

KLAMATH LARGESCALE SUCKER *Catostomus snyderi* **P. 313**
Similar to Sacramento Sucker but usually has 11 dorsal rays, distance from pelvic fin origin to caudal fin base equal to or greater than distance from eye to pelvic fin origin. To 21¾ in. (55 cm).

LONGNOSE SUCKER *Catostomus catostomus* **P. 314**
Long snout; 95–120 lateral scales. (See also Fig. 37.) To 25 in. (64 cm). See also Salish Sucker, *C.* species.

WARNER SUCKER *Catostomus warnerensis* **P. 315**
Large head; long snout. Breeding male (shown) has bright red stripe along brassy side. Has 14–16 scale rows below lateral line. To 13¾ in. (35 cm). See also Tahoe Sucker, *C. tahoensis*, Owens Sucker, *C. fumeiventris*, and Klamath Smallscale Sucker, *C. rimiculus*.

MODOC SUCKER *Catostomus microps* **P. 316**
Short head, smaller eye, 9–13 scale rows below lateral line. Breeding male (shown) has red stripe along side. To 13¼ in. (34 cm).

BRIDGELIP SUCKER *Catostomus columbianus* **P. 316**
Mottled brown or blue-black above. Weak or no indentations separate upper and lower lips (Fig. 37). To 12 in. (30 cm).

RIO GRANDE SUCKER *Catostomus plebeius* **P. 318**
Small papillose lips (Fig. 37). Often sharply bicolored. Breeding male (shown) has red stripe along side. To 7¾ in. (20 cm).

MOUNTAIN SUCKER *Catostomus platyrhynchus* **P. 318**
Gray above, blotches on back. Large papillae on lower lip with bare areas on margins of median notch (Fig. 37). To 9¾ in. (25 cm).

DESERT SUCKER *Catostomus clarkii* **P. 320**
Deep indentations separate upper and lower lips. No papillae on front of upper lip; 4–7 rows of papillae at middle of lower lip. To 13 in. (33 cm). See also Santa Ana Sucker, *C. santaanae*.

BLUEHEAD SUCKER *Catostomus discobolus* **P. 320**
Blue head (darkest on adult); 50 or more predorsal scales. To 16 in. (41 cm).

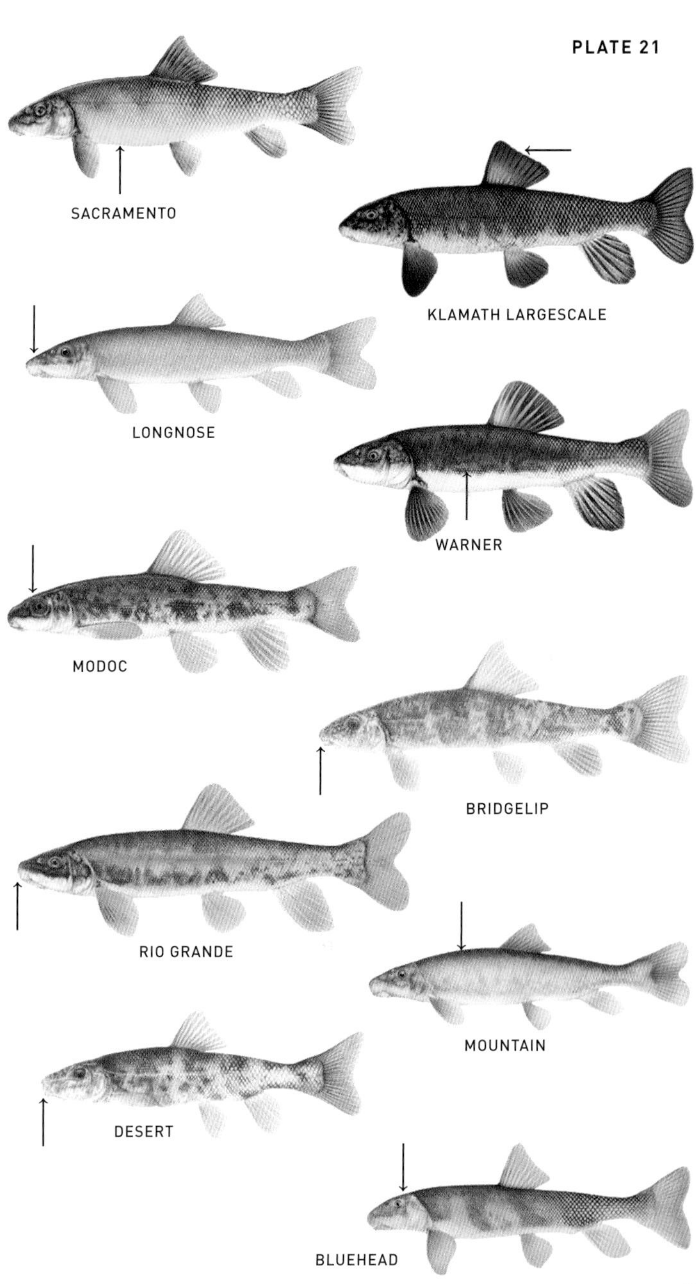
SACRAMENTO
KLAMATH LARGESCALE
LONGNOSE
WARNER
MODOC
BRIDGELIP
RIO GRANDE
MOUNTAIN
DESERT
BLUEHEAD

PLATE 22

Suckers (4)

BLUE SUCKER *Cycleptus elongatus* **P. 304**
Blue (brightest on large individual). Small head. Long, falcate dorsal fin. To 39 in. (99 cm). See also Southeastern Blue Sucker, *C. meridionalis*.

SPOTTED SUCKER *Minytrema melanops* **P. 320**
Parallel rows of dark spots on back and side. Black edge on dorsal fin, lower lobe of caudal fin. To 19½ in. (50 cm).

Next 3 species (*Erimyzon*) have small, oblique, nearly terminal mouth; no lateral line. Young has black stripe along side.

WESTERN CREEK CHUBSUCKER *Erimyzon claviformis* **P. 322**
Has 5–8 dark blotches along side; 37–45 lateral scales. Rounded dorsal fin. To 9 in. (23 cm). See also Eastern Creek Chubsucker, *E. oblongus*.

LAKE CHUBSUCKER *Erimyzon sucetta* **P. 322**
Often dark stripe along side; 34–39 lateral scales. Rounded dorsal fin. To 16 in. (41 cm).

SHARPFIN CHUBSUCKER *Erimyzon tenuis* **P. 323**
Sharply pointed dorsal fin. To 13 in. (33 cm).

NORTHERN HOG SUCKER *Hypentelium nigricans* **P. 323**
Large rectangular head, broadly concave between eyes; 3–6 dark saddles (1 on nape). Papillose lips (Fig. 38). To 24 in. (61 cm). See also Alabama Hog Sucker, *H. etowanum*.

ROANOKE HOG SUCKER *Hypentelium roanokense* **P. 325**
Similar to Northern Hog Sucker but has no or vague saddle on nape, plicate lips (Fig. 38), 39–44 lateral scales. To 6½ in. (16 cm).

STRIPED JUMPROCK *Moxostoma rupiscartes* **P. 336**
Head wider than deep, dark stripes on back and side wider than or equal to pale interspaces, dusky edge on dorsal and caudal fins (in some populations). To 11 in. (28 cm).

BLACKTIP JUMPROCK *Moxostoma cervinum* **P. 336**
Black tips on dorsal and caudal fins. Stripes on back and upper side. Plicate lips; straight lower lip edge (Fig. 40). To 7½ in. (19 cm).

GREATER JUMPROCK *Moxostoma lachneri* **P. 337**
Long slender head and body; slender caudal peduncle. White lower ray on gray caudal fin. To 17½ in. (44 cm).

BIGEYE JUMPROCK *Moxostoma ariommum* **P. 338**
Large eye. Flat, flaring papillose lips (Fig. 40). To 8½ in. (22 cm).

TORRENT SUCKER *Thoburnia rhothoeca* **P. 338**
Two large pale (or dusky) areas on caudal fin base. Red stripe along side of breeding male (shown). Small mouth; each half of lower lip edge nearly triangular (Fig. 40). To 7¼ in. (18 cm). See also Rustyside Sucker, *T. hamiltoni*.

BLACKFIN SUCKER *Thoburnia atripinnis* **P. 339**
Large black blotch on tip of dorsal fin. Black stripes on back and upper side. To 6¾ in. (17 cm).

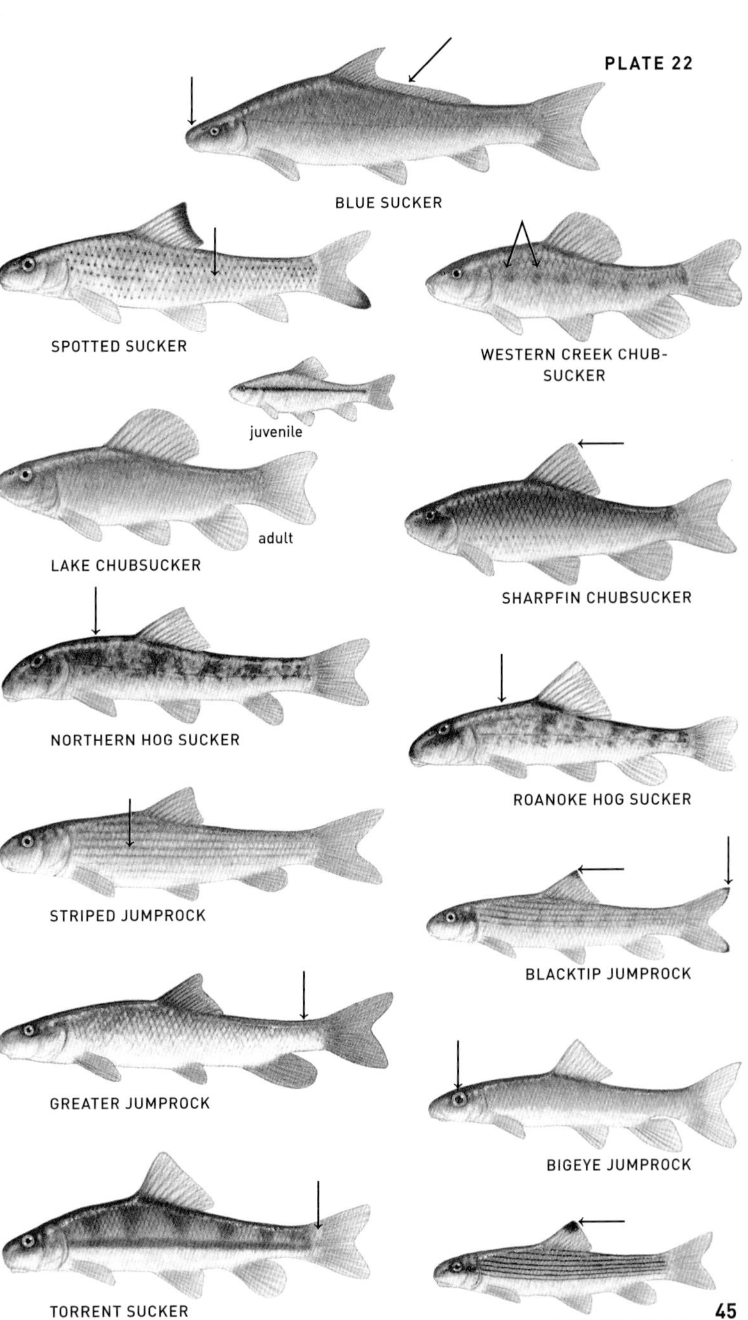
PLATE 22
BLUE SUCKER
SPOTTED SUCKER
WESTERN CREEK CHUB-
SUCKER
juvenile
adult
LAKE CHUBSUCKER
SHARPFIN CHUBSUCKER
NORTHERN HOG SUCKER
ROANOKE HOG SUCKER
STRIPED JUMPROCK
BLACKTIP JUMPROCK
GREATER JUMPROCK
BIGEYE JUMPROCK
TORRENT SUCKER
BLACKFIN SUCKER

Suckers (5)

Redhorses (Moxostoma) *have thick papillose or plicate lips. See also Fig. 40.*

GREATER REDHORSE *Moxostoma valenciennesi* **P. 326**
Large head. Thick plicate lips; V-shaped rear edge on lower lip (Fig. 40). Usually 16 scales around caudal peduncle. To 31½ in. (80 cm).

COPPER REDHORSE *Moxostoma hubbsi* **P. 326**
Deep body. Highly arched back. Short head. To 28 in. (72 cm).

RIVER REDHORSE *Moxostoma carinatum* **P. 326**
Large head. Thick plicate lips. Usually 12–13 scales around caudal peduncle. To 30 in. (77 cm).

ROBUST REDHORSE *Moxostoma robustum* **P. 328**
Stout, wide body. Dusky to dark irregular stripes on side. Plicate lips. To 28½ in. (72 cm).

SHORTHEAD REDHORSE *Moxostoma macrolepidotum* **P. 329**
Short head. Straight rear edge on lower lip (Fig. 40). *M. m. breviceps* has longer upper caudal lobe, more concave dorsal fin. To 29½ in. (75 cm).

BLACKTAIL REDHORSE *Moxostoma poecilurum* **P. 331**
Black stripe on lower caudal fin lobe. Cylindrical body. To 20 in. (51 cm). See also Apalachicola Redhorse, *M.* species.

SICKLEFIN REDHORSE *Moxostoma* species **P. 330**
Similar to Shorthead Redhorse but has falcate dorsal fin—rays 1–3 extend beyond tip of last ray when depressed. To 21½ in. (55 cm).

Next 6 species have gray to light orange (never bright red) fins.

BLACK REDHORSE *Moxostoma duquesnii* **P. 331**
Long, slender caudal peduncle. Plicate lips; broadly V-shaped rear edge on lower lip (Fig. 40). Has 44–47 lateral scales. To 20 in. (51 cm).

GOLDEN REDHORSE *Moxostoma erythrurum* **P. 332**
Stout caudal peduncle. V- or U-shaped rear edge on lower lip (Fig. 40). Has 40–42 lateral scales. To 30½ in. (78 cm). See also Carolina Redhorse, *M.* species.

SILVER REDHORSE *Moxostoma anisurum* **P. 333**
Straight or convex dorsal fin, 14–16 dorsal rays. V-shaped rear edge on deeply divided lower lip (Fig. 40). To 28 in. (71 cm). See also Notchlip Redhorse, *M. collapsum.*

V-LIP REDHORSE *Moxostoma pappillosum* **P. 334**
Long slender body. Concave to falcate dorsal fin. V-shaped rear edge on deeply divided lower lip (Fig. 40). To 17¾ in. (45 cm).

GRAY REDHORSE *Moxostoma congestum* **P. 334**
Broad, U-shaped head (viewed from above). Yellow to light orange fins. To 25½ in. (65 cm). See also Mexican Redhorse, *M. austrinum.*

BRASSY JUMPROCK *Moxostoma* species **P. 335**
Rounded snout (viewed from above). Deep head, distinctly convex between eyes. To 16½ in. (42 cm).

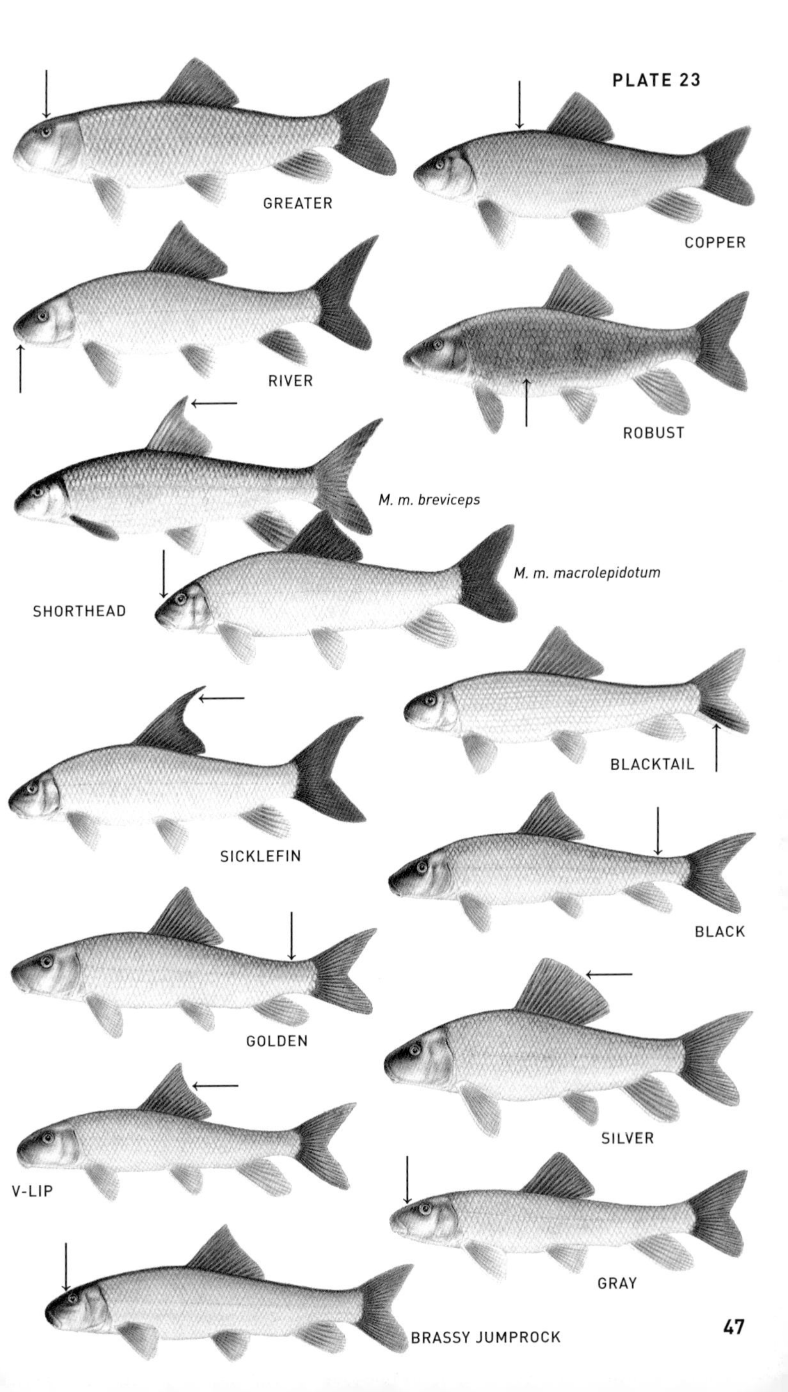
PLATE 23
GREATER
COPPER
RIVER
ROBUST
M. m. breviceps
M. m. macrolepidotum
SHORTHEAD
BLACKTAIL
SICKLEFIN
BLACK
GOLDEN
SILVER
V-LIP
GRAY
BRASSY JUMPROCK

North American Catfishes

Four pairs of barbels; no scales; short base on small adipose fin, its rear edge free from body (except in blindcats); stout spines in dorsal and pectoral fins.

First 3 species have forked caudal fin.

CHANNEL CATFISH *Ictalurus punctatus* **P. 342**

Scattered dark spots on light back and side (except in small young and large adult); rounded anal fin, 24–32 rays. To 50 in. (127 cm). See also Headwater Catfish, *I. lupus*, and Yaqui Catfish, *I. pricei*.

BLUE CATFISH *Ictalurus furcatus* **P. 343**

Long, straight-edged anal fin, tapered like a barber's comb, 30–35 rays. No dark spots on body (except in Rio Grande). To 65 in. (165 cm).

WHITE CATFISH *Ameiurus catus* **P. 344**

Relatively short anal fin base, fin rounded in outline, 22–25 rays. Large, sawlike teeth on rear of pectoral spine. To 24¼ in. (62 cm).

Next 9 species have rounded, straight, or slightly notched caudal fin.

YELLOW BULLHEAD *Ameiurus natalis* **P. 345**

White or yellow chin barbels. Moderately long anal fin, 24–27 rays. Large sawlike teeth on rear of pectoral spine. To 19 in. (47 cm).

BLACK BULLHEAD *Ameiurus melas* **P. 345**

Dusky or black chin barbels. Relatively short anal fin, rounded in outline, 19–23 rays. Usually no large sawlike teeth on rear of pectoral spine (Fig. 41). To 24¼ in. (62 cm).

BROWN BULLHEAD *Ameiurus nebulosus* **P. 346**

Brown or black mottling or spots on body, large sawlike teeth on rear of pectoral spine (Fig. 41). To 21 in. (50 cm).

Next 3 species have large dark blotch at dorsal fin base.

SPOTTED BULLHEAD *Ameiurus serracanthus* **P. 346**

Many small round gray-white spots on dark body. Narrow black edge on fins. To 13¼ in. (34 cm).

SNAIL BULLHEAD *Ameiurus brunneus* **P. 347**

Flat head, rounded snout profile. Black edge on fins (except pectoral). Has 17–20 anal rays. To 11½ in. (29 cm).

FLAT BULLHEAD *Ameiurus platycephalus* **P. 347**

Dark mottling on side. Flat head; relatively straight snout profile. Has 21–24 anal rays. To 11½ in. (29 cm).

FLATHEAD CATFISH *Pylodictis olivaris* **P. 348**

White tip on upper lobe of caudal fin (except on large individual). Wide, flat head; projecting lower jaw. Black or brown mottling on back and side. To 61 in. (155 cm).

WIDEMOUTH BLINDCAT *Satan eurystomus* **P. 348**

No eyes. Jaw teeth well developed. Lower jaw normal in shape, slightly shorter than upper jaw. Separate gill membranes with strong fold between them. To 5¼ in. (13.7 cm).

TOOTHLESS BLINDCAT *Trogloglanis pattersoni* **P. 349**

No eyes. No jaw teeth. Short lower jaw curved upward and into mouth. Fused gill membranes. To 4 in. (10.4 cm).

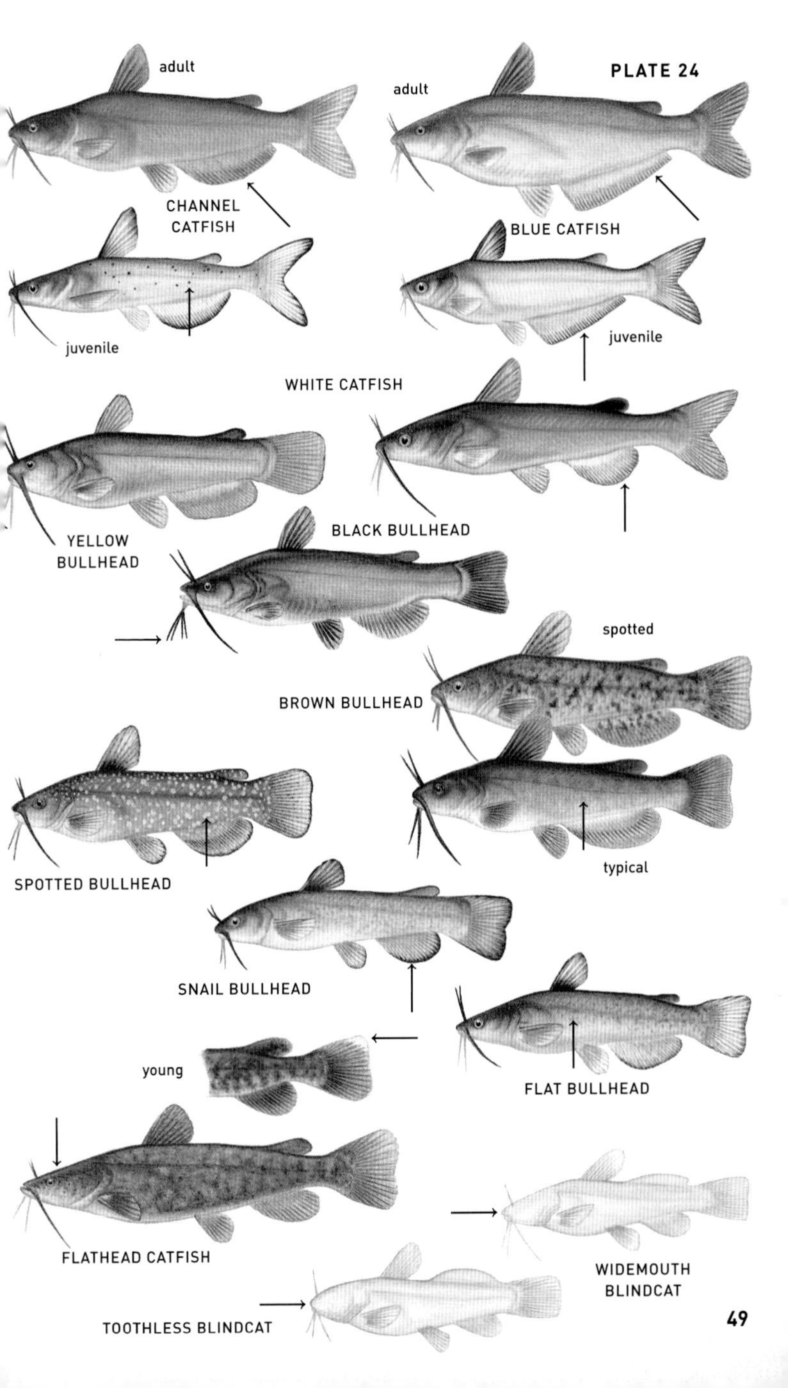
PLATE 24
adult
CHANNEL
CATFISH
juvenile
adult
BLUE CATFISH
juvenile
WHITE CATFISH
YELLOW
BULLHEAD
BLACK BULLHEAD
spotted
BROWN BULLHEAD
typical
SPOTTED BULLHEAD
SNAIL BULLHEAD
young
FLAT BULLHEAD
FLATHEAD CATFISH
WIDEMOUTH
BLINDCAT
TOOTHLESS BLINDCAT

Madtoms (1)

Madtoms have 4 pairs of barbels; no scales; long, low adipose fin joined to, or slightly separated from, caudal fin; stout spines in dorsal and pectoral fins.

STONECAT *Noturus flavus* **P. 350**

Light blotch on nape. Cream white spot at rear of dorsal fin base and on upper edge of gray caudal fin. Backward projections from premaxillary tooth patch (Fig. 42). To 12 in. (31 cm).

Species below have somber color; no dark blotches or saddles; no large teeth on front edge, usually straight teeth on rear edge, of pectoral spine (Fig. 44).

TADPOLE MADTOM *Noturus gyrinus* **P. 350**

Chubby body. Terminal mouth with equal jaws. Dark veinlike line along side. To 5 in. (13 cm).

OUACHITA MADTOM *Noturus lachneri* **P. 351**

Similar to Tadpole Madtom but has shorter, flatter head, more slender body. To 4 in. (10 cm).

SPECKLED MADTOM *Noturus leptacanthus* **P. 351**

Black specks on upper body and fins. No sawlike teeth on rear edge of short pectoral spine (Fig. 44). To 3½ in. (9.4 cm).

BROADTAIL MADTOM *Noturus* species **P. 354**

Chubby body. Dark blotch on caudal fin base. Rounded rear edge on caudal fin. To 2¼ in. (6 cm).

BROWN MADTOM *Noturus phaeus* **P. 354**

Robust. Many brown specks on underside of head and belly. Long anal fin, 20–22 rays. About 6 sawlike teeth on rear of pectoral spine (Fig. 44). To 5¾ in. (15 cm).

BLACK MADTOM *Noturus funebris* **P. 355**

Similar to Brown Madtom but has no or few weak sawlike teeth on rear of pectoral spine (Fig. 44), longer anal fin with 21–27 rays. To 5¾ in. (15 cm).

FRECKLED MADTOM *Noturus nocturnus* **P. 355**

No dark specks on mostly white belly. Dusky black edge on anal fin. Usually 16–18 anal rays, 2–3 weak sawlike teeth on rear of pectoral spine. To 5¾ in. (15 cm).

SLENDER MADTOM *Noturus exilis* **P. 356**

Black border on median fins. Terminal mouth with equal jaws. To 5¾ in. (15 cm).

MARGINED MADTOM *Noturus insignis* **P. 356**

Similar to Slender Madtom but upper jaw projects beyond lower jaw. To 6 in. (15 cm).

ORANGEFIN MADTOM *Noturus gilberti* **P. 357**

White to orange triangle on upper edge of caudal fin. Short anal fin, 14–16 rays. To 3¾ in. (10 cm).

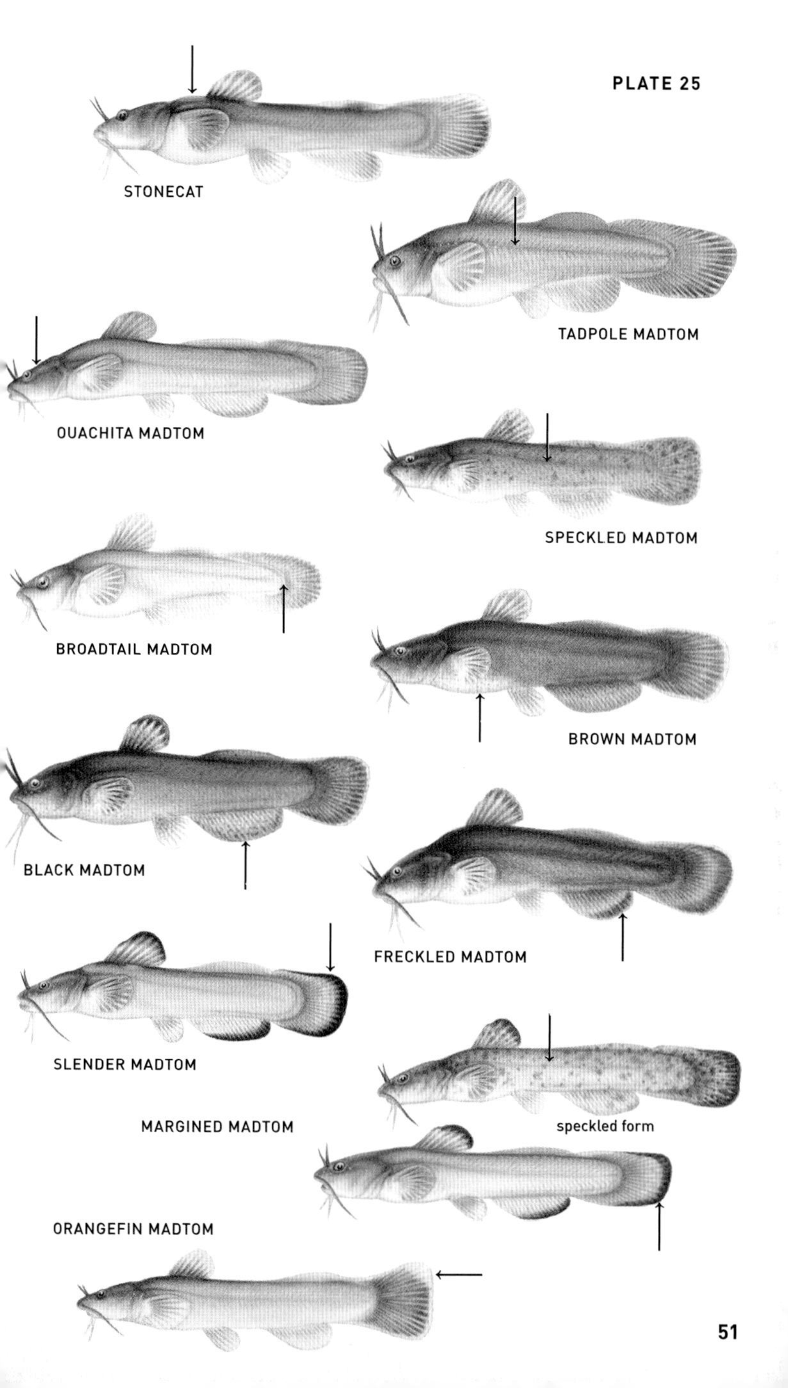
PLATE 25
STONECAT
TADPOLE MADTOM
OUACHITA MADTOM
SPECKLED MADTOM
BROADTAIL MADTOM
BROWN MADTOM
BLACK MADTOM
FRECKLED MADTOM
SLENDER MADTOM
MARGINED MADTOM
speckled form
ORANGEFIN MADTOM

Madtoms (2)

Dark saddles. Sawlike teeth on front and rear edges of curved pectoral spine (Fig. 44).

LEAST MADTOM *Noturus hildebrandi* **P. 358**
White or clear adipose fin. No or small teeth on front edge of short pectoral spine. *N. h. lautus* is strongly bicolored. *N. h. hildebrandi* is blotched above and on lower side. To 2½ in. (6.9 cm). See also Pygmy Madtom, *N. stanauli.*

SMOKY MADTOM *Noturus baileyi* **P. 357**
Four yellow saddles. Dusky band on adipose fin extends nearly to fin edge. To 2¾ in. (7.3 cm).

ELEGANT MADTOM *Noturus elegans* **P. 360**
Brown blotch extends up front of dorsal fin. No black blotch at top of dorsal fin. See also Scioto Madtom, *N. trautmani.*

OZARK MADTOM *Noturus albater* **P. 359**
White upper edge on caudal fin. To 4¾ in. (12 cm). See also Black River Madtom, *N. maydeni.*

SADDLED MADTOM *Noturus fasciatus* **P. 360**
3–4 ivory to yellow saddles alternating with dark saddles. To 3¼ in. (8.5 cm).

CHUCKY MADTOM *Noturus crypticus* **P. 361**
Adipose fin reaches caudal fin. Large black specks on cheek. To 3 in. (7.4 cm).

CADDO MADTOM *Noturus taylori* **P. 362**
Black blotch on upper edge of dorsal fin. Many small teeth on front edge of short pectoral spine (as in Elegant Madtom, Fig. 44). To 3 in. (7.7 cm).

NEOSHO MADTOM *Noturus placidus* **P. 362**
White lower caudal rays. Deep caudal peduncle. To 3¼ in. (8.7 cm).

FRECKLEBELLY MADTOM *Noturus munitus* **P. 364**
Dark brown band to adipose fin edge. Dark specks on belly. To 3¾ in. (9.5 cm).

NORTHERN MADTOM *Noturus stigmosus* **P. 363**
Usually 2 large light spots in front of dorsal fin. Dark band into upper half of adipose fin. To 5 in. (13 cm).

PIEBALD MADTOM *Noturus gladiator* **P. 363**
Dark bar into upper half of adipose fin and across body into anal fin. Dark band on caudal fin joins dark pigment on caudal peduncle. To 5 in. (13 cm).

CAROLINA MADTOM *Noturus furiosus* **P. 365**
No dark specks on belly. Dark band nearly to adipose fin edge. To 4¾ in. (12 cm).

MOUNTAIN MADTOM *Noturus eleutherus* **P. 366**
Dark brown bar on caudal fin base. Dark band on adipose fin usually confined to lower half of fin. To 5 in. (13 cm).

CHECKERED MADTOM *Noturus flavater* **P. 366**
Broad black bar on caudal fin base; black border on caudal fin; black blotch on dorsal fin. To 8 in. (20 cm).

BRINDLED MADTOM *Noturus miurus* **P. 367**
Black blotch on outer ⅓ of dorsal fin extends across first 3–5 rays. Dark saddle to edge of adipose fin. Rounded rear edge on caudal fin. To 5 in. (13 cm).

YELLOWFIN MADTOM *Noturus flavipinnis* **P. 368**
Pale edge on caudal fin; black bar on caudal fin base; 2 light spots in front of dorsal fin. To 6 in. (15 cm).

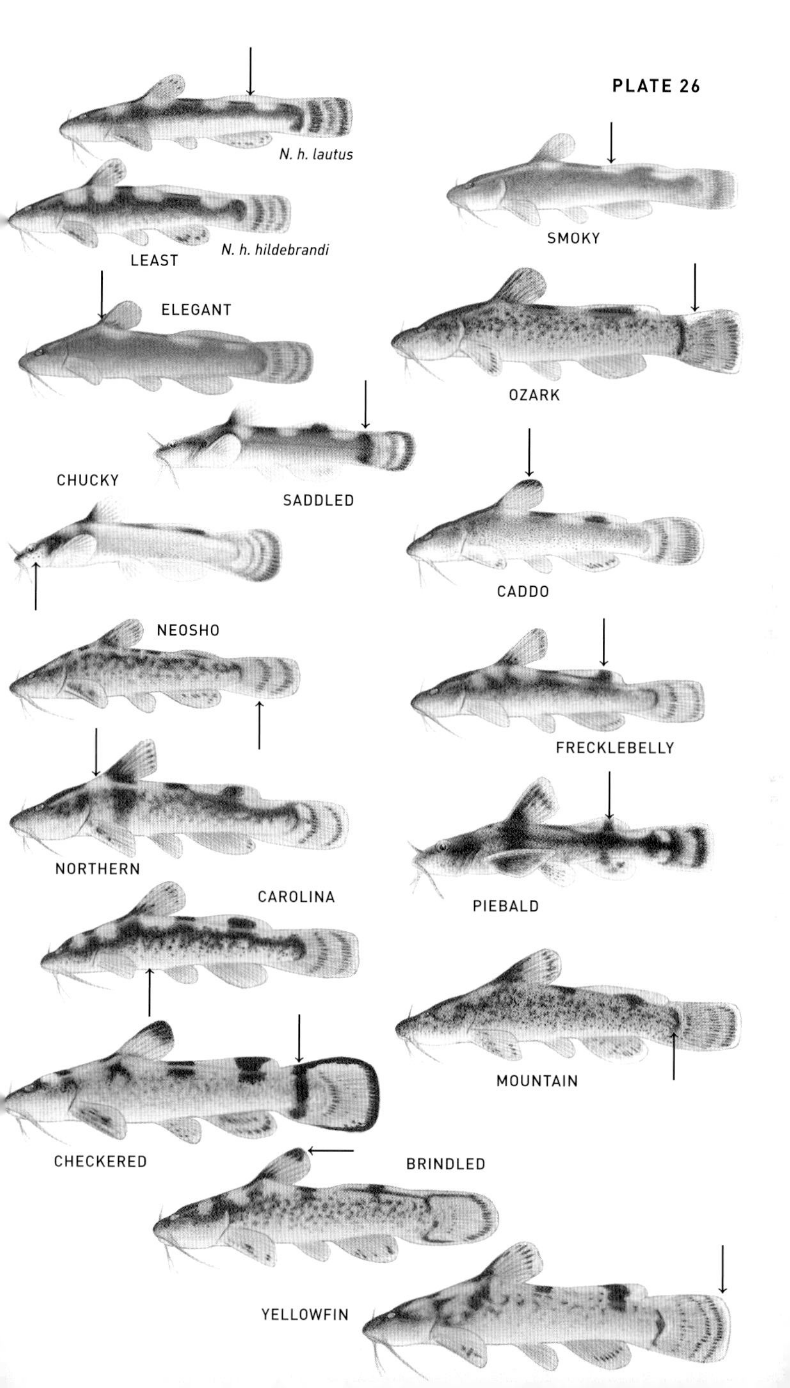

PLATE 26
N. h. lautus
N. h. hildebrandi
LEAST
SMOKY
ELEGANT
OZARK
CHUCKY
SADDLED
CADDO
NEOSHO
FRECKLEBELLY
NORTHERN
CAROLINA
PIEBALD
MOUNTAIN
CHECKERED
BRINDLED
YELLOWFIN

Whitefishes, Chars, Trouts

Coregonus species: Two small flaps of skin between nostrils (Fig. 47).

LAKE WHITEFISH *Coregonus clupeaformis* **P. 383**

Concavity between snout and humped nape. Has 24–33 long rakers on 1st gill arch. To 31 in. (80 cm). See also Humpback Whitefish, *C. pidschian*, Atlantic Whitefish, *C. huntsmani*, and Broad Whitefish, *C. nasus*.

CISCO *Coregonus artedi* **P. 386**

Terminal mouth; 36–64 long rakers on 1st gill arch. To 22½ in. (57 cm). See also Arctic Cisco, *C. autumnalis*, Bering Cisco, *C. laurettae*, Least Cisco, *C. sardinella*, Nipigon Cisco, *C. nipigon*, Shortjaw Cisco, *C. zenithicus*, and Blackfin Cisco, *C. nigripinnis*.

BLOATER *Coregonus hoyi* **P. 389**

Similar to Kiyi but has smaller eye, less dusky upper lip, pelvic fin seldom reaching anus, usually 40–47 long rakers on 1st gill arch. To 14½ in. (37 cm).

KIYI *Coregonus kiyi* **P. 389**

Long paired fins; pelvic fin usually reaches to anus or beyond. Large eye nearly equal to snout length. Black upper lip. Usually 36–41 rakers on 1st gill arch. To 13¾ in. (35 cm).

Prosopium species: One flap of skin between nostrils (Fig. 47).

MOUNTAIN WHITEFISH *Prosopium williamsoni* **P. 393**

Short snout. To 22½ in. (57 cm). See related species in Fig. 50.

BONNEVILLE CISCO *Prosopium gemmifer* **P. 391**

Long, sharply pointed snout. To 8½ in. (22 cm).

ARCTIC GRAYLING *Thymallus arcticus* **P. 394**

Huge purple to black dorsal fin. Small mouth. To 30 in. (76 cm).

Next 5 species (_Salvelinus_) have light (pink, red, or cream) spots on body, minute scales, white leading edge on lower fins.

LAKE TROUT *Salvelinus namaycush* **P. 395**

Deeply forked caudal fin. Many cream or yellow spots on dark head, body, and dorsal and caudal fins. To 49½ in. (126 cm). Fat form in Lake Superior is called Siscowet.

BROOK TROUT *Salvelinus fontinalis* **P. 396**

Bright white leading edge on lower fins. Blue halos around pink or red spots on side. Light wavy lines or blotches on back and dorsal fin. To 28 in. (70 cm).

ARCTIC CHAR *Salvelinus alpinus* **P. 396**

Pink to red spots (largest usually larger than pupil of eye) on back and side. Slightly forked caudal fin. Has usually 20–30 rakers on 1st gill arch; 35–50 pyloric caeca. To 38 in. (96 cm).

DOLLY VARDEN *Salvelinus malma* **P. 397**

Similar to Arctic Char but usually has 14–21 rakers on 1st gill arch, 25–30 pyloric caeca. Anadromous individual dark blue above, pale spots on side; breeding male (shown) green-black above, bright red lower side. Stream-resident gray with pale spots. To about 25 in. (63 cm).

BULL TROUT *Salvelinus confluentus* **P. 398**

Similar to Dolly Varden but has flatter and longer head, eye higher on head. To about 3 ft. (1 m).

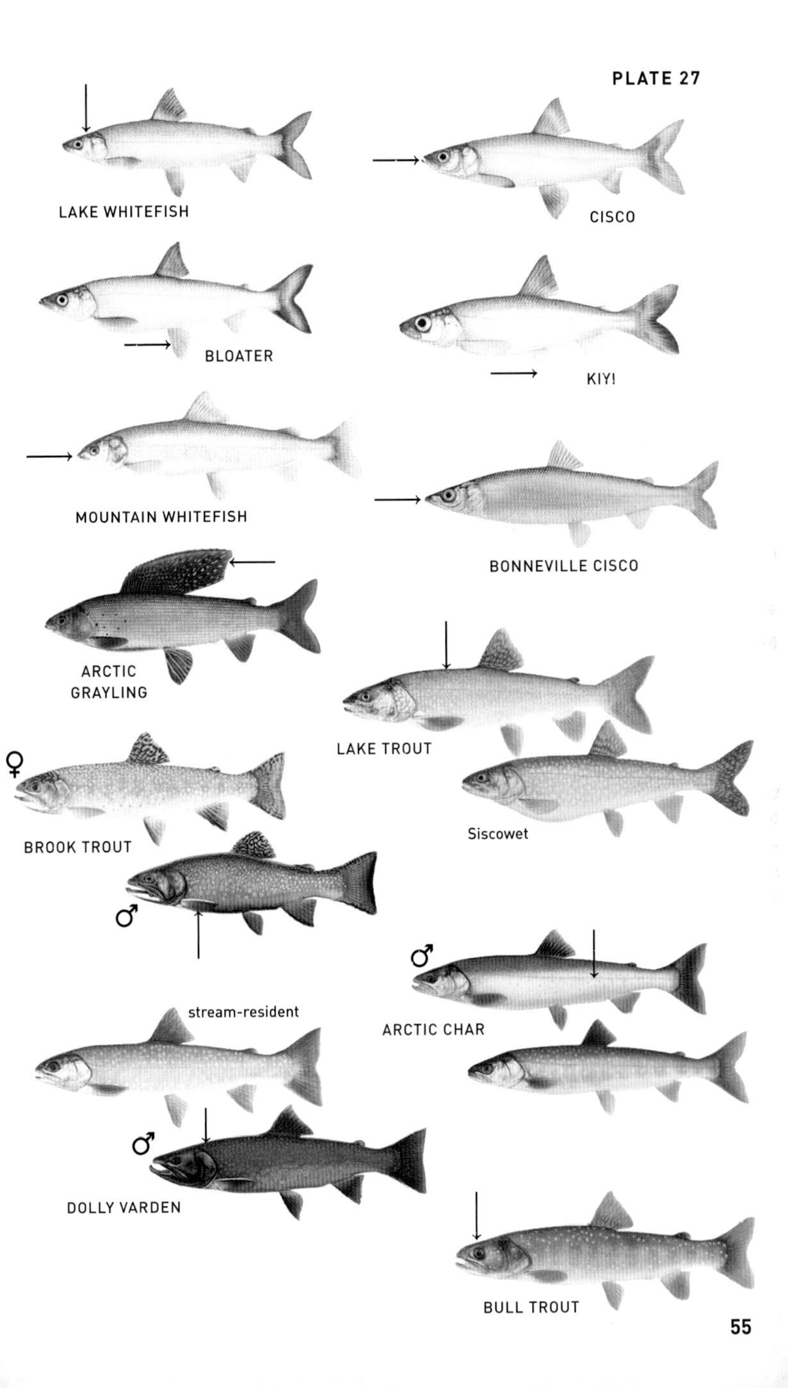
LAKE WHITEFISH
CISCO
BLOATER
KIY!
MOUNTAIN WHITEFISH
BONNEVILLE CISCO
ARCTIC GRAYLING
LAKE TROUT
Siscowet
♀
BROOK TROUT
♂
♂
stream-resident
ARCTIC CHAR
♂
DOLLY VARDEN
BULL TROUT

Salmons, Brown Trout

One dorsal fin; adipose fin; abdominal pelvic fins. Young are shown in Fig. 46.

ATLANTIC SALMON *Salmo salar* **P. 399**

Black spots on head and body; 2–6 large spots on gill cover. X-shaped spots on body of adult. Usually no large spots on caudal fin. Silver at sea (oceanic form); bronze brown in fresh water. Breeding male (shown) is bronze and dark brown, often with red spots. To 55 in. (140 cm).

BROWN TROUT *Salmo trutta* **P. 400**

Red and black spots on body; large black spots on gill cover. Upper jaw reaches to or beyond eye. Breeding male has red lower side. To 40½ in. (103 cm).

Next 5 species (*Oncorhynchus*): Metallic blue-green above, silver below at sea. No white leading edge on lower fins.

SOCKEYE SALMON *Oncorhynchus nerka* **P. 400**

No large black spots on back or caudal fin. Breeding individual has green head, brilliant red body. Male develops hooked upper jaw. Has 28–40 rakers on 1st gill arch. To 33 in. (84 cm).

CHUM SALMON *Oncorhynchus keta* **P. 401**

No large black spots on back or caudal fin. Breeding individual has red, brown, and black bars and blotches on dull green side. To 40 in. (102 cm).

CHINOOK SALMON *Oncorhynchus tshawytscha* **P. 401**

Black spots on back, adipose, and both lobes of caudal fin. Gums black at base of teeth. Large male may have dull red side. To 58 in. (147 cm). Largest salmon; salmon over 30 lb. (14 kg) are almost always this species.

COHO SALMON *Oncorhynchus kisutch* **P. 402**

Black spots on back and upper lobe of caudal fin. Gums white at base of teeth. Breeding individual has pink (female) to red side (male). To 38½ in. (98 cm).

PINK SALMON *Oncorhynchus gorbuscha* **P. 402**

Large black spots on back and both lobes of caudal fin. Breeding individual has brown to pink stripe along side. Male develops humped back. To 30 in. (76 cm).

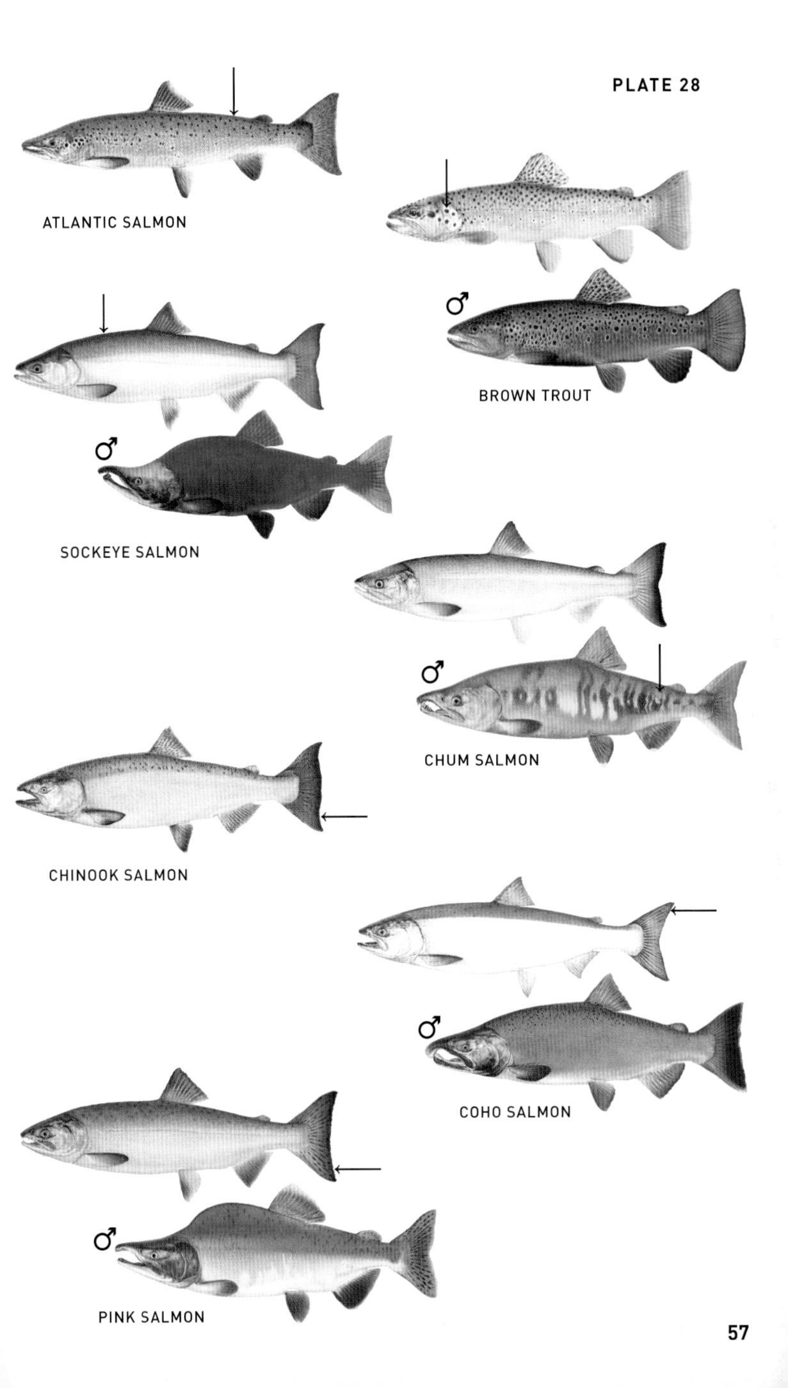

ATLANTIC SALMON
BROWN TROUT
♂
SOCKEYE SALMON
♂
CHUM SALMON
♂
CHINOOK SALMON
COHO SALMON
♂
PINK SALMON
♂

Cutthroat Trouts

Cutthroat Trout (Oncorhynchus clarkii)*: Red "cutthroat" mark under lower jaw; many black spots on body; no or faint red stripe on side.*

COASTAL CUTTHROAT TROUT *O. c. clarkii* **P. 404**
Many black spots over entire body. Sea-run form silver with pink to red-orange anal and pelvic fins. Stream resident with red "cutthroat" mark, white tips on anal and paired fins. Roundish parr marks. To 22 in. (56 cm).

WESTSLOPE CUTTHROAT TROUT *O. c. lewisi* **P. 405**
Black spots on back and upper side; more toward caudal fin. No black spots on lower side between pectoral and anal fins. Breeding male (shown) is green, gold, and red-orange. To 24 in. (61 cm).

YELLOWSTONE CUTTHROAT TROUT *O. c. bouvieri* **P. 405**
Black spots concentrated on caudal peduncle and fin; few spots on lower side. Yellow to bronze body. To 24 in. (61 cm).

FINESPOTTED CUTTHROAT TROUT *O. c. behnkei* **P. 406**
Black specks on body and dorsal, adipose, and caudal fins. Silver green or bronze sheen to yellow-brown body. To 28 in. (71 cm).

BONNEVILLE CUTTHROAT TROUT *O. c. utah* **P. 406**
Light blue parr marks on yellow-green to silver gray body, large black spots on body and fins. To 18 in. (46 cm). Bear Lake form has small, irregularly shaped spots on side and dorsal and caudal fins; to 24 in. (61 cm).

COLORADO RIVER CUTTHROAT TROUT *O. c. pleuriticus* **P. 407**
Medium to large black spots concentrated on caudal peduncle and upper side. Bright yellow-gold body. To 20 in. (51 cm).

GREENBACK CUTTHROAT TROUT *O. c. stomias* **P. 407**
White tip on dorsal fin. Breeding male has red underside. To 18 in. (46 cm).

RIO GRANDE CUTTHROAT TROUT *O. c. virginalis* **P. 408**
Similar to Colorado River Cutthroat Trout but has black spots concentrated on upper side, caudal peduncle, and fin. Pecos R. strain has larger spots. To 15 in. (38 cm).

YELLOWFIN CUTTHROAT TROUT *O. c. macdonaldi* **P. 408**
Silver blue body, yellow lower side; bright yellow fins. Small, irregularly shaped black spots on rear half of body. To 28 in. (71 cm).

LAHONTAN CUTTHROAT TROUT *O. c. henshawi* **P. 408**
Black spots on top of head. To 39 in. (99 cm).

PAIUTE CUTTHROAT TROUT *O. c. seleniris* **P. 409**
Similar to Lahontan Cutthroat Trout but has no black spots on body or caudal fin. Pale parr marks on some adults. To 18 in. (46 cm).

WHITEHORSE CUTTHROAT TROUT *O. c.* subspecies **P. 410**
Nearly identical to Humboldt Cutthroat Trout but has 35–50 pyloric caeca, pale red "cutthroat" mark. To 14 in. (36 cm).

ALVORD CUTTHROAT TROUT *O. c. alvordensis* **P. 409**
Similar to Lahontan Cutthroat Trout but has only 25–50 medium-sized black spots, mostly on upper side. To 20 in. (51 cm).

HUMBOLDT CUTTHROAT TROUT *O. c.* subspecies **P. 410**
Nearly identical to Lahontan Cutthroat Trout but usually has 20–22 rakers on 1st gill arch, 130–160 lateral scales. To 18 in. (46 cm).

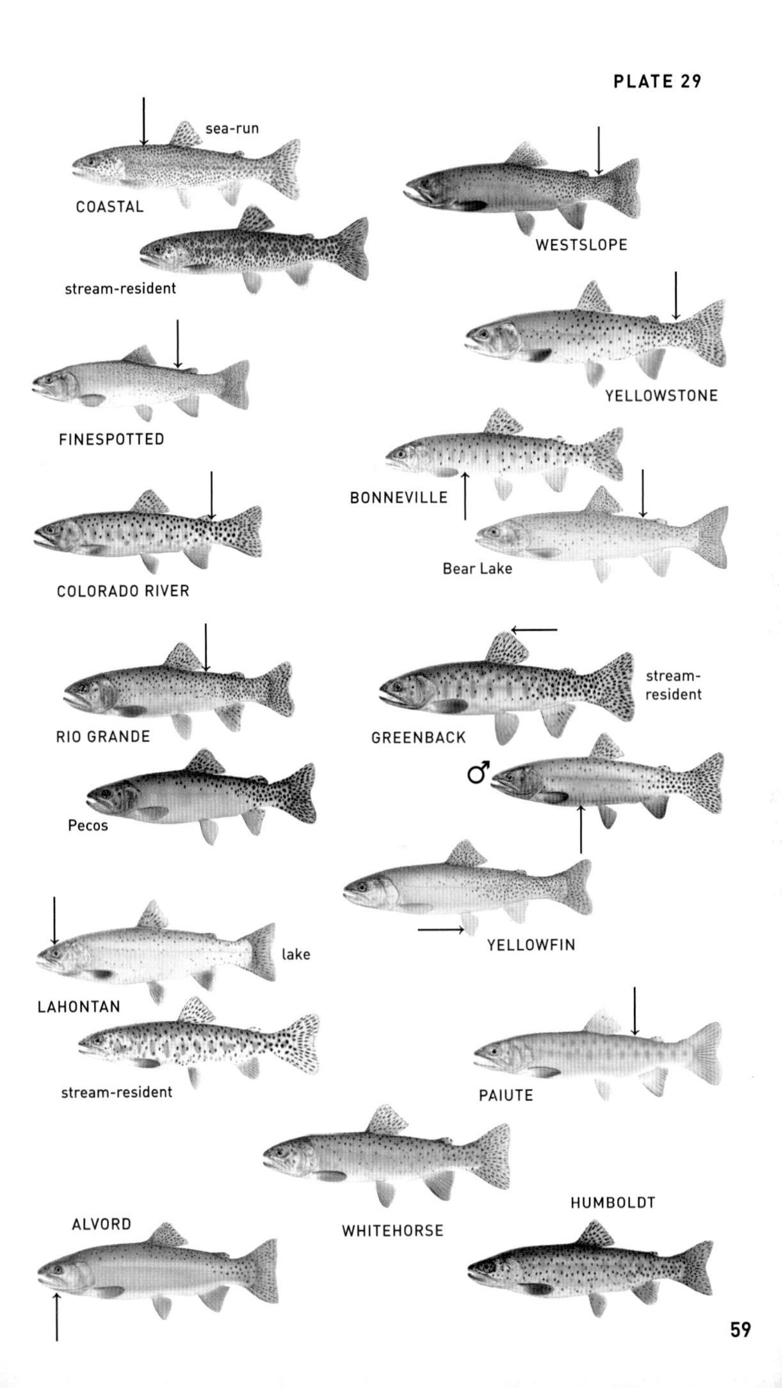
sea-run
COASTAL
stream-resident
WESTSLOPE
YELLOWSTONE
FINESPOTTED
BONNEVILLE
Bear Lake
COLORADO RIVER
RIO GRANDE
GREENBACK
stream-
resident
♂
Pecos
YELLOWFIN
lake
LAHONTAN
stream-resident
PAIUTE
HUMBOLDT
ALVORD
WHITEHORSE

PLATE 30

Rainbow and Related Trouts

Rainbow Trout (Oncorhynchus mykiss): *Usually small irregular black spots on back and fins, radiating rows of black spots on caudal fin. Pink to red stripe on side (except in sea-run form). Often black edge on adipose fin.*

COASTAL RAINBOW TROUT *O. m. irideus* **P. 411**

Stream resident has red-pink stripe along side, most vivid in adult male. Many black spots on top of head, body, and dorsal and caudal fins. To 16 in. (41 cm). Sea-run form ("Steelhead") has smaller black spots, silver or pink side. To 43 in. (110 cm).

COLUMBIA RAINBOW TROUT *O. m. gairdnerii* **P. 412**

Stream resident has many black spots on body and fins, including anal and pelvic fins. Pink to red stripe along side. To 18 in. (46 cm). Sea-run "Red-band Steelhead" has silvery side; adult male (shown) is bronze with bright red lower side, hooked lower jaw. To 40 in. (100 cm). Lake form "Kamloops Trout" has subdued colors.

GREAT BASIN RAINBOW TROUT *O. m. newberrii* **P. 412**

White tip on dorsal, pelvic, and anal fins; purple parr marks. Goose Lake form has fewer spots, subdued color. Williamson R. and upper Klamath Lake form has rounded snout. To 36 in. (91 cm).

SACRAMENTO RAINBOW TROUT *O. m. stonei* **P. 413**

Many black spots on back and upper side, few below. Orange "cutthroat" mark. To 20 in. (51 cm).

EAGLE LAKE RAINBOW TROUT *O. m. aquilarum* **P. 413**

Many black, irregularly shaped spots on pink side, dorsal and caudal fins. To 30 in. (76 cm).

KERN RAINBOW TROUT *O. m. gilberti* **P. 413**

Similar to Sacramento Rainbow Trout but has parr marks, duller colors. To 28 in. (71 cm).

GOLDEN TROUT *Oncorhynchus aguabonita* **P. 414**

Has 10–12 dark parr marks. Red stripe along bright yellow-gold side. Trout Creek Golden Trout, *O. a. aguabonita*, has bright colors, few black spots, usually 170–200 lateral scales. Little Kern River Golden Trout, *O. a. whitei*, has more subdued colors, more black spots, usually 155–160 lateral scales. To 12 in. (30 cm).

GILA TROUT *Oncorhynchus gilae* **P. 414**

Many small black spots (mostly above lateral line) on yellow-gold side, head, dorsal and caudal fins; large spots on adipose fin. To 9 in. (23 cm).

APACHE TROUT *Oncorhynchus apache* **P. 415**

Similar to Gila Trout but has brighter yellow body and fins; larger black spots, black stripe through eye. To 9 in. (23 cm).

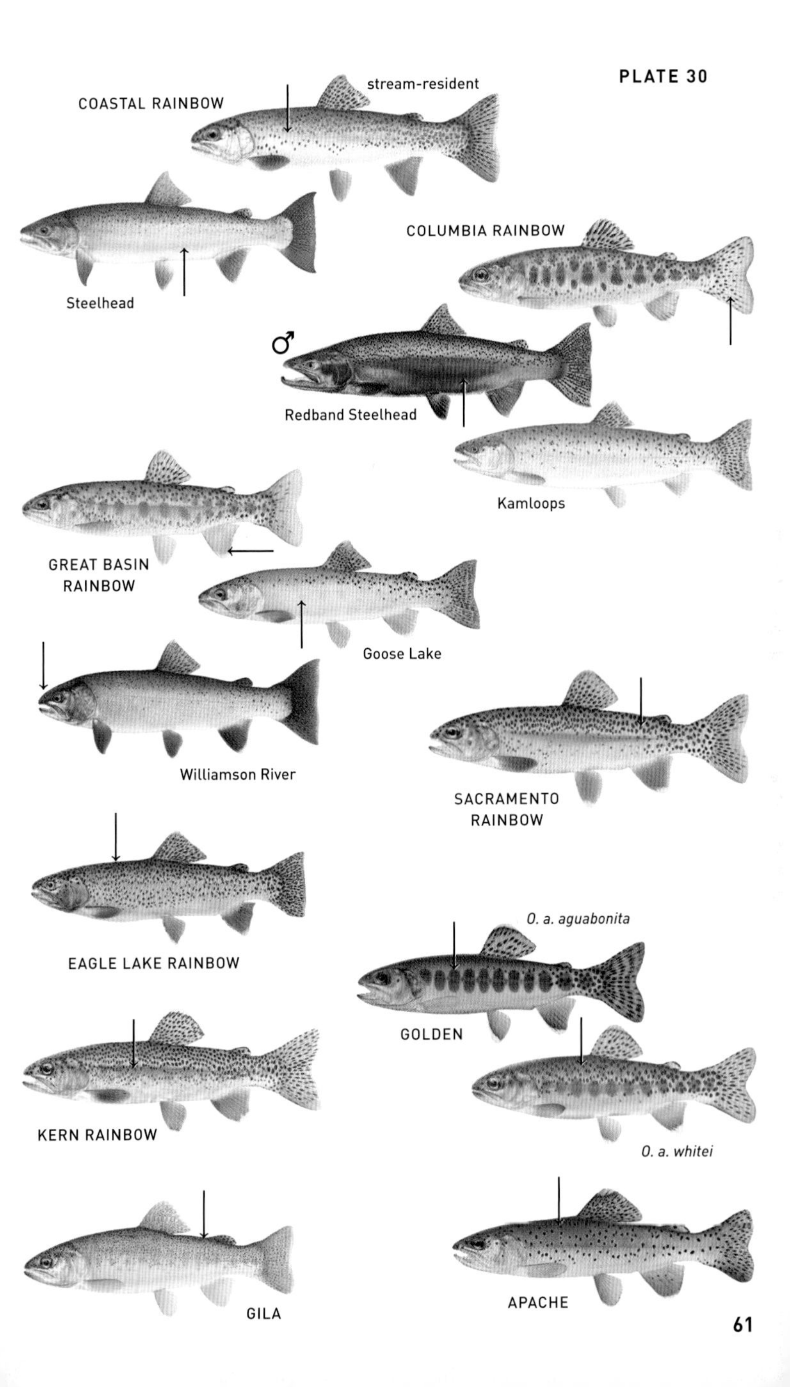
COASTAL RAINBOW
stream-resident
Steelhead
COLUMBIA RAINBOW
♂
Redband Steelhead
Kamloops
GREAT BASIN
RAINBOW
Goose Lake
Williamson River
SACRAMENTO
RAINBOW
EAGLE LAKE RAINBOW
O. a. aguabonita
GOLDEN
KERN RAINBOW
O. a. whitei
GILA
APACHE

PLATE 31

Pirate Perch, Trout-Perch, Sand Roller, Mudminnows, Pikes

One dorsal fin.

PIRATE PERCH *Aphredoderus sayanus* **P. 417**
Large head. Anus on throat. Large individual has purple sheen. To 5½ in. (14 cm).

TROUT-PERCH *Percopsis omiscomaycus* **P. 416**
Rows of 7-12 black spots along back and side. Adipose fin present. Large head. Large silver white chambers on lower jaw and cheek. To 7¾ in. (20 cm).

SAND ROLLER *Percopsis transmontana* **P. 416**
Similar to Trout-Perch but is dark blue-green above; has more arched back, incomplete lateral line. To 3¾ in. (9.6 cm).

Next 4 species (mudminnows) have dorsal fin far back on body, rounded or straight-edged caudal fin.

ALASKA BLACKFISH *Dallia pectoralis* **P. 376**
Black mottling and blotches on body and fins. Tiny pelvic fins, with 2–3 rays. To 13 in. (33 cm).

CENTRAL MUDMINNOW *Umbra limi* **P. 376**
Black bar on caudal fin base. Dorsal fin origin far in front of anal fin origin. To 6 in. (15 cm).

EASTERN MUDMINNOW *Umbra pygmaea* **P. 376**
Similar to Central Mudminnow but has 10–14 dark brown stripes on back and side. To 4½ in. (11 cm).

OLYMPIC MUDMINNOW *Novumbra hubbsi* **P. 375**
Has 10–15 cream to yellow (blue on breeding male), narrow interrupted bars on side. Dorsal fin origin above or slightly in front of anal fin origin. To 3 in. (8 cm).

Next 4 species (pikes) have dorsal fin far back on body, duckbill-like snout, forked caudal fin.

GRASS PICKEREL *Esox americanus vermiculatus* **P. 372**
Dark green to brown wavy bars along side of adult. Black suborbital bar slanted toward rear. Fully scaled cheek and opercle (Fig. 45); 11–13 branchiostegal rays. To 15 in. (38 cm).

REDFIN PICKEREL *Esox americanus americanus* **P. 372**
Similar to Grass Pickerel but with red fins.

CHAIN PICKEREL *Esox niger* **P. 372**
Chainlike pattern on side. Vertical black suborbital bar. Fully scaled cheek and opercle (Fig. 45); 14–17 branchiostegal rays. To 39 in. (99 cm).

NORTHERN PIKE *Esox lucius* **P. 374**
Rows of yellow bean-shaped spots (on adult). Partly scaled opercle; fully scaled cheek (Fig. 45). To 56 in. (142 cm).

MUSKELLUNGE *Esox masquinongy* **P. 374**
Dark spots, blotches, or bars on light yellow-green back and side. Partly scaled cheek and opercle (Fig. 45). To 6 ft. (2 m).

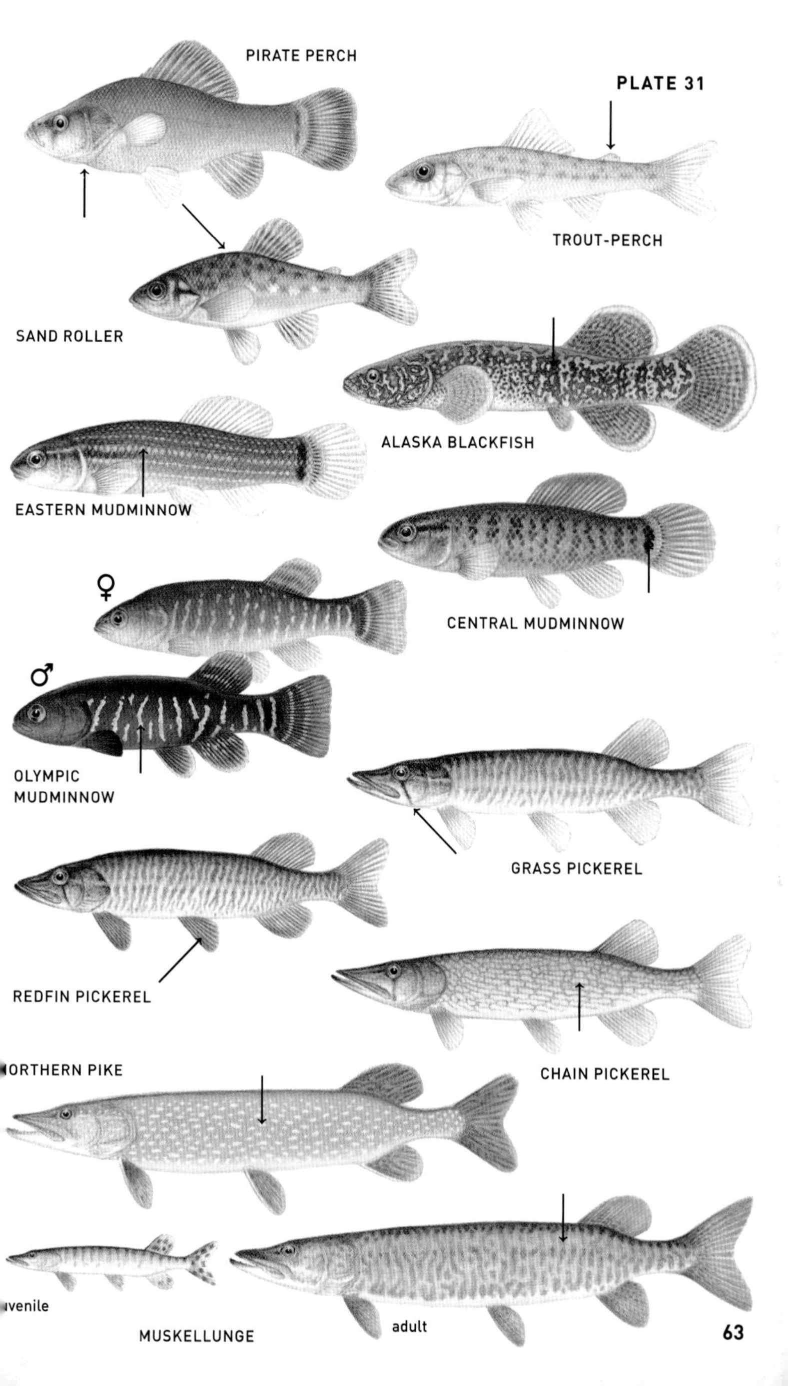
PIRATE PERCH
PLATE 31
TROUT-PERCH
SAND ROLLER
ALASKA BLACKFISH
EASTERN MUDMINNOW
CENTRAL MUDMINNOW
♀
♂
OLYMPIC
MUDMINNOW
GRASS PICKEREL
REDFIN PICKEREL
CHAIN PICKEREL
ORTHERN PIKE
venile
adult
MUSKELLUNGE

PLATE 32

Topminnows, Killifishes (1)

One dorsal fin far back on body; flattened head and back; upturned mouth; no lateral line.

SEMINOLE KILLIFISH *Fundulus seminolis* **P. 424**

Metallic green side, interrupted rows of many small black spots; 15–20 dark green bars (often faint) on female. To 6½ in. (16 cm).

BANDED KILLIFISH *Fundulus diaphanus* **P. 425**

Has 10–20 green-brown bars along silver side. Long, slender body. To 5 in. (13 cm). See also Waccamaw Killifish, *F. waccamensis.*

NORTHERN STUDFISH *Fundulus catenatus* **P. 426**

Rows of small brown (female and young) or red-brown spots (male, shown) on side. Large male has bright blue side. To 7 in. (18 cm).

STIPPLED STUDFISH *Fundulus bifax* **P. 427**

Similar to Northern Studfish but has short interrupted rows of red or brown spots on side. To 4¾ in. (12 cm).

SOUTHERN STUDFISH *Fundulus stellifer* **P. 427**

Many brown or red spots scattered over side (rarely in irregular rows on upper side). Black edge on dorsal and caudal fins of some large males. To 4¾ in. (12 cm).

BARRENS TOPMINNOW *Fundulus julisia* **P. 428**

Many scattered brown (female and young) or red-orange spots (male, shown) on head and body. Yellow-orange fins. Iridescent white-gold stripe along back to dorsal fin. To 3¾ in. (9.4 cm).

WHITELINE TOPMINNOW *Fundulus albolineatus* **P. 428**

Nearly identical to Barrens Topminnow but has interrupted white streaks on rear half of side of large male. To 3¾ in. (8.4 cm).

SPECKLED KILLIFISH *Fundulus rathbuni* **P. 428**

Black line (absent on juvenile) from mouth to eye. Many dark brown spots on back and side of juvenile and female. To 3¾ in. (9.6 cm).

PLAINS TOPMINNOW *Fundulus sciadicus* **P. 429**

Bronze flecks and dark crosshatching on blue-green back and side. Narrow gold stripe in front of dorsal fin. To 2¾ in. (7 cm).

GOLDEN TOPMINNOW *Fundulus chrysotus* **P. 430**

Gold flecks on side; usually 8–11 green bars (often faint) on side of large male. Breeding male has bright red to red-brown spots on rear half of body. To 3 in. (7.5 cm).

BANDED TOPMINNOW *Fundulus cingulatus* **P. 430**

Rows of small brown to red spots on side; 12–15 green bars along side. Clear to light red fins. To 3 in. (7.8 cm). See also Redface Topminnow, *F. rubrifrons.*

PLATE 32

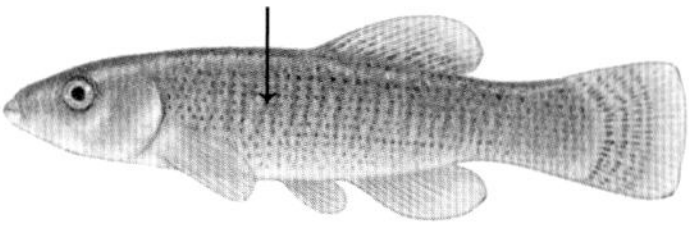
SEMINOLE KILLIFISH

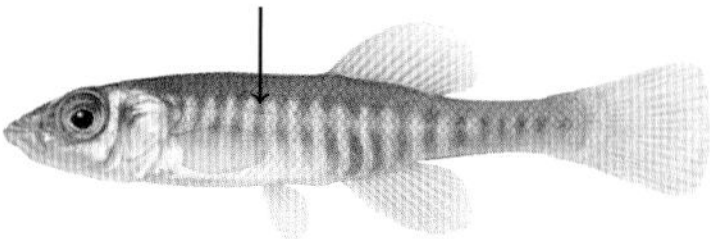
BANDED KILLIFISH

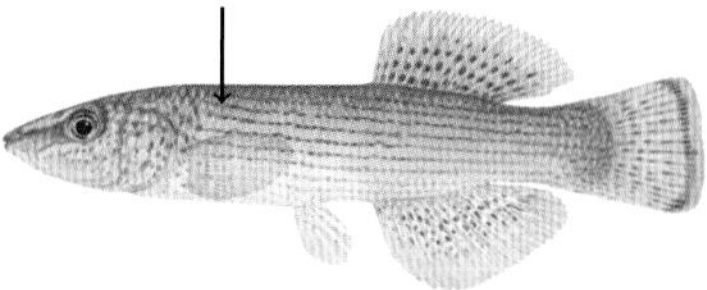
NORTHERN STUDFISH

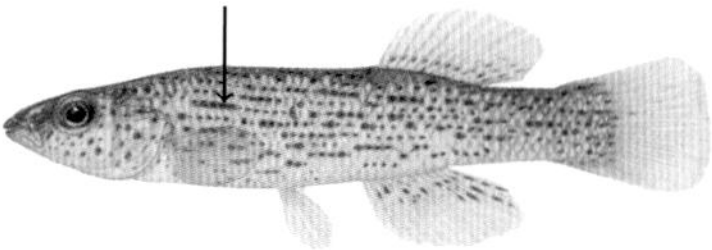
STIPPLED STUDFISH

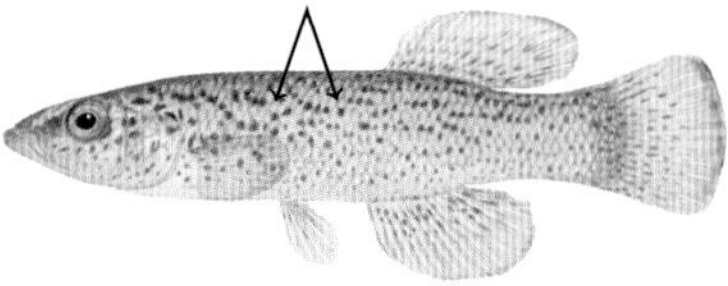
SOUTHERN STUDFISH

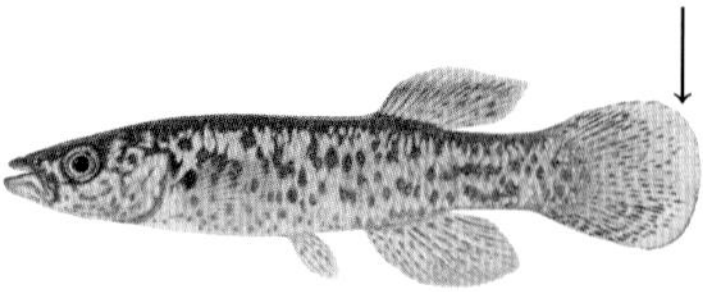
BARRENS TOPMINNOW

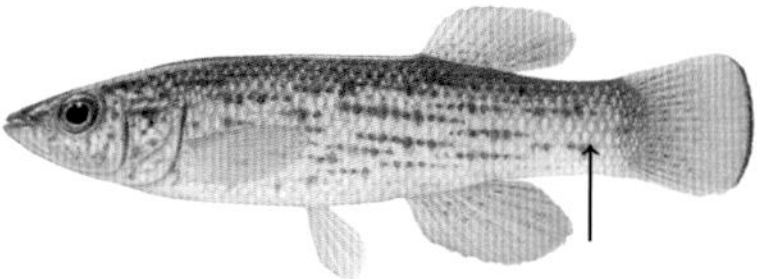
WHITELINE TOPMINNOW

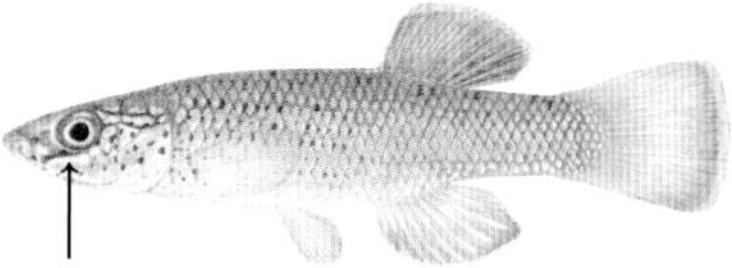
SPECKLED KILLIFISH

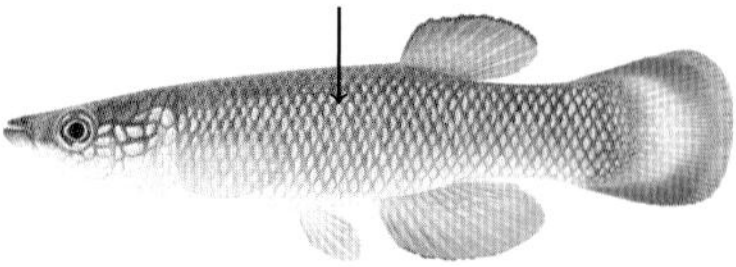
PLAINS TOPMINNOW

BANDED TOPMINNOW

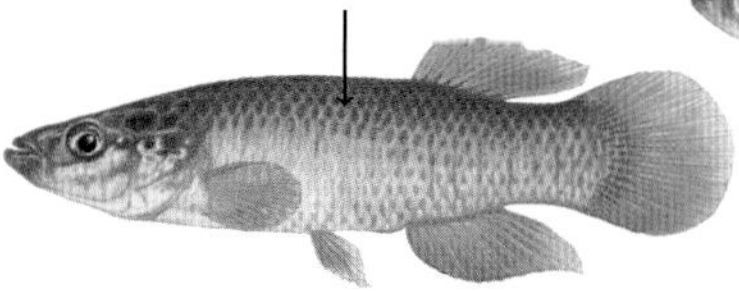

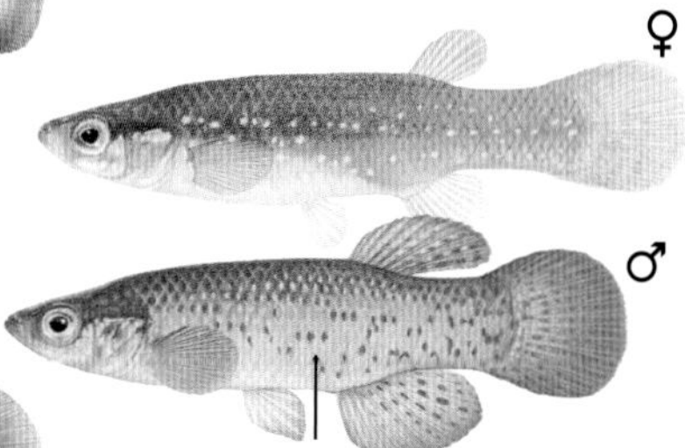
GOLDEN TOPMINNOW

Topminnows, Killifishes (2)

PLAINS KILLIFISH *Fundulus zebrinus* **P. 431**
Has 12–26 gray-green bars (fewer, wider bars on male) on silver white side. Breeding male (shown) has bright orange to red dorsal, anal, and paired fins. To 4 in. (10 cm).

Next 5 species (starhead topminnows) have large, blue-black bar under eye, 6–8 brown to red-brown stripes (young and female) or rows of dots (male) along side, a large gold spot on top of head, small gold spot at dorsal fin origin. Male has dark green bars along side (except Western Starhead Topminnow). See also Fig. 51.

LINED TOPMINNOW *Fundulus lineolatus* **P. 432**
Has 11–15 dark green bars on side of male, thickest at middle; 6–8 black stripes on side of female. To 3¼ in. (8.4 cm).

BAYOU TOPMINNOW *Fundulus nottii* **P. 433**
Has 9–15 dark bars on side of male that extend forward to pectoral fin base. Dark specks between stripes on female. To 3 in. (7.8 cm).

RUSSETFIN TOPMINNOW *Fundulus escambiae* **P. 433**
Nearly identical to Bayou Topminnow but lacks dark specks between stripes on female; has 0–14 dark bars on side of male extending forward only to between paired fins. To 3 in. (7.8 cm).

WESTERN STARHEAD TOPMINNOW *Fundulus blairae* **P. 435**
No dark bars on side of body. Many dark specks between 7–9 dark stripes on side of female. To 3 in. (7.8 cm).

STARHEAD TOPMINNOW *Fundulus dispar* **P. 434**
No or few dark specks between 6–8 thin dark stripes on side of female; 3–13 dark bars on side of male. To 3 in. (7.8 cm).

BLACKSTRIPE TOPMINNOW *Fundulus notatus* **P. 435**
Wide blue-black stripe along side, around snout, and onto caudal fin. Silver white spot on top of head. To 3 in. (7.4 cm).

BLACKSPOTTED TOPMINNOW *Fundulus olivaceus* **P. 436**
Similar to Blackstripe Topminnow but has few to many (male has more) discrete black spots on light tan upper side. To 3¾ in. (9.7 cm).

BROADSTRIPE TOPMINNOW *Fundulus euryzonus* **P. 437**
Similar to Blackstripe and Blackspotted topminnows but has extremely wide purple-brown stripe along side. No crossbars along side of male. To 3¼ in. (8.3 cm).

RAINWATER KILLIFISH *Lucania parva* **P. 438**
Large, dark-edged scales on back and side. Large male has black spot at front of dusky orange dorsal fin. To 2¾ in. (7 cm).

BLUEFIN KILLIFISH *Lucania goodei* **P. 439**
Wide, zigzag, black stripe from tip of snout to black spot on caudal fin base. Bright iridescent blue on front of dorsal and anal fins on large male. To 2 in. (5 cm).

PYGMY KILLIFISH *Leptolucania ommata* **P. 439**
Cream yellow halo around large black spot on caudal peduncle. Male has 5–7 faint bars on rear half of side. Female has dusky stripe along side, black spot on midside. To 1¼ in. (2.9 cm).

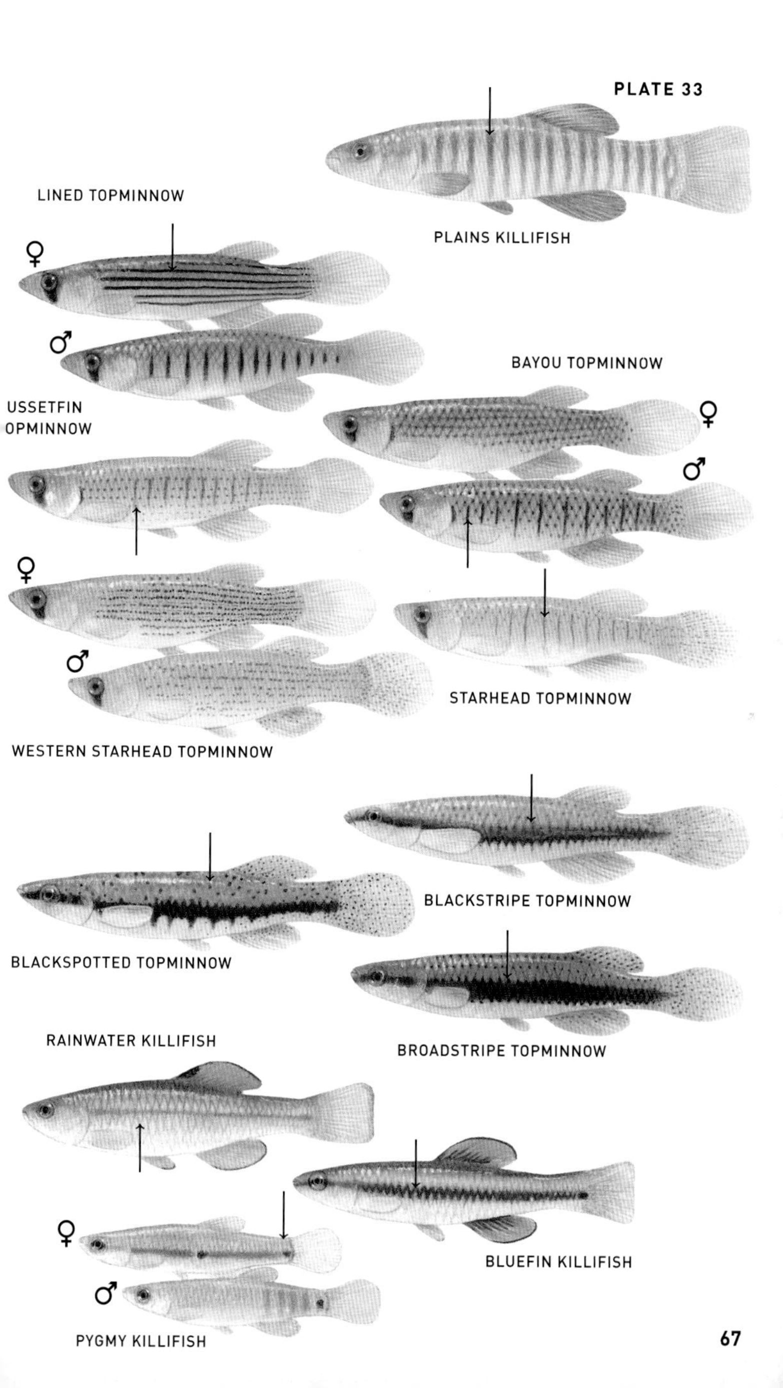
PLAINS KILLIFISH
LINED TOPMINNOW
♀
♂
BAYOU TOPMINNOW
♀
♂
USSETFIN
OPMINNOW
♀
♂
STARHEAD TOPMINNOW
WESTERN STARHEAD TOPMINNOW
BLACKSTRIPE TOPMINNOW
BLACKSPOTTED TOPMINNOW
RAINWATER KILLIFISH
BROADSTRIPE TOPMINNOW
♀
BLUEFIN KILLIFISH
♂
PYGMY KILLIFISH

Pupfishes, Springfishes

Upturned mouth; deep body; no lateral line.

SHEEPSHEAD MINNOW *Cyprinodon variegatus* **P. 454**
Has 5–8 gray-brown bars (wide at top) along silver olive side. To 3 in. (7.5 cm).

RED RIVER PUPFISH *Cyprinodon rubrofluviatilis* **P. 455**
Unscaled belly; 5–8 large triangular brown blotches along silver side. No dark blotches on lower side. To 2¼ in. (5.8 cm).

PECOS PUPFISH *Cyprinodon pecosensis* **P. 455**
Similar to Red River Pupfish but has partly scaled belly, less blue on nape. To 2¼ in. (6 cm).

LEON SPRINGS PUPFISH *Cyprinodon bovinus* **P. 455**
Dark brown blotches along silver side; many small brown blotches on lower side of female (rarely on male). To 2¼ in. (5.6 cm).

WHITE SANDS PUPFISH *Cyprinodon tularosa* **P. 456**
Similar to Leon Springs Pupfish but has dark bars on female connected at bottom, yellow to orange dorsal fin on large male. To 2 in. (5 cm).

COMANCHE SPRINGS PUPFISH *Cyprinodon elegans* **P. 456**
Slender caudal peduncle. Black specks on silver side. Brown-black blotches form "stripe" (often faint on male) along silver side. To 2½ in. (6.2 cm).

DESERT PUPFISH *Cyprinodon macularius* **P. 457**
Breeding male (shown) has blue body, lemon yellow to orange caudal peduncle and fin. To 2¾ in. (7.2 cm). See also Sonoyta Pupfish, *C. eremus*.

OWENS PUPFISH *Cyprinodon radiosus* **P. 458**
Breeding male (shown) has deep blue body, orange edge on blue dorsal and anal fins. To 2¾ in. (7.2 cm).

AMARGOSA PUPFISH *Cyprinodon nevadensis* **P. 458**
Dorsal fin origin nearer to caudal fin base than to tip of snout. To 3 in. (7.8 cm).

SALT CREEK PUPFISH *Cyprinodon salinus* **P. 459**
Slender body. Dorsal fin far back on body. Scales on nape small and crowded. To 3 in. (7.8 cm).

DEVILS HOLE PUPFISH *Cyprinodon diabolis* **P. 460**
No pelvic fins. To 1¼ in. (3.4 cm).

CONCHOS PUPFISH *Cyprinodon eximius* **P. 460**
Faint brown blotches on silver side; rows of small brown spots on upper side. Breeding male (shown) has yellow-orange dorsal fin. To 2 in. (5 cm).

FLAGFISH *Jordanella floridae* **P. 460**
Large black spot on midside. Alternating thin black and red-orange lines, gold flecks, on side. To 2½ in. (6.5 cm).

Next 3 species lack pelvic fins; have dorsal and anal fins far back on body.

WHITE RIVER SPRINGFISH *Crenichthys baileyi* **P. 452**
Row of black spots (or black stripe) along side; 2d row of black spots along lower side from midbody to caudal fin. To 3½ in. (9 cm).

RAILROAD VALLEY SPRINGFISH *Crenichthys nevadae* **P. 453**
Has 1 row of dark spots along side. To 2¼ in. (6 cm).

PAHRUMP POOLFISH *Empetrichthys latos* **P. 453**
Black mottling on silver side. Wide mouth. To 2¼ in. (6 cm). See also Ash Meadows Poolfish, *E. merriami*.

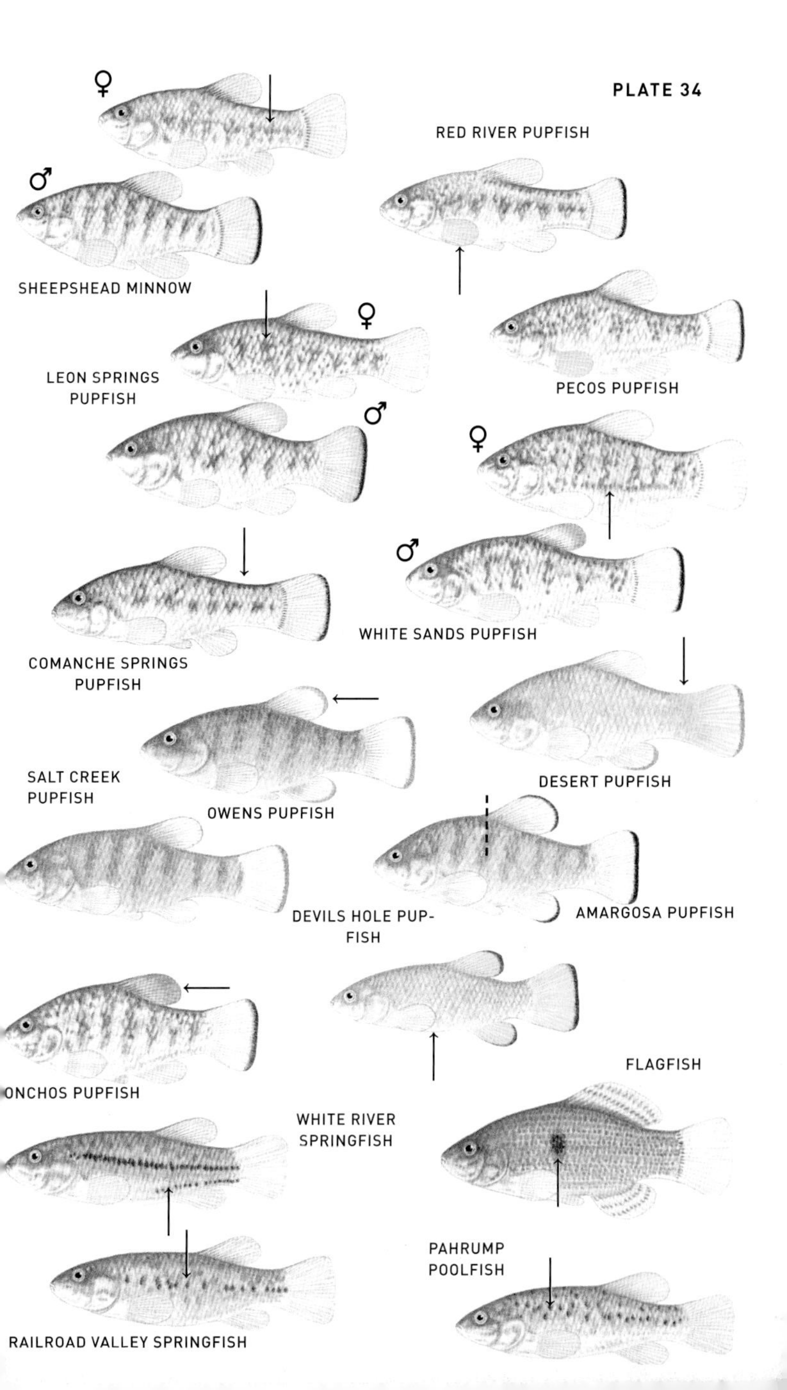
PLATE 34
RED RIVER PUPFISH
SHEEPSHEAD MINNOW
LEON SPRINGS PUPFISH
PECOS PUPFISH
WHITE SANDS PUPFISH
COMANCHE SPRINGS PUPFISH
SALT CREEK PUPFISH
OWENS PUPFISH
DESERT PUPFISH
DEVILS HOLE PUP-FISH
AMARGOSA PUPFISH
FLAGFISH
ONCHOS PUPFISH
WHITE RIVER SPRINGFISH
PAHRUMP POOLFISH
RAILROAD VALLEY SPRINGFISH

Livebearers

Front rays of anal fin of male elongated and modified into a gonopodium to accomplish internal fertilization (Fig. 52). Flattened head; upturned mouth; no lateral line. See p. 440 for introduced livebearers.

AMAZON MOLLY *Poecilia formosa* **P. 441**
All-female species. Similar to Sailfin Molly (female) but lacks rows of brown spots on side. Has 10–12 dorsal rays. To 3¾ in. (9.6 cm).

SAILFIN MOLLY *Poecilia latipinna* **P. 440**
Large male has huge, sail-like dorsal fin. About 5 rows of dark brown spots, iridescent yellow flecks on olive side; 13–16 dorsal rays. To 6 in. (15 cm). See also Shortfin Molly, *P. mexicana*, and Mexican Molly, *P. sphenops*.

WESTERN MOSQUITOFISH *Gambusia affinis* **P. 445**
Has 1–3 rows of black spots on dorsal and caudal fins. Large dusky to black teardrop. To 2½ in. (6.5 cm). See also Eastern Mosquitofish, *G. holbrooki*.

PECOS GAMBUSIA *Gambusia nobilis* **P. 447**
Dusky edges on dorsal and caudal fins, anal fin of female. Darkly outlined scales on back and upper ⅔ of side. To 2 in. (4.8 cm).

BLOTCHED GAMBUSIA *Gambusia senilis* **P. 447**
Dusky stripe (about 1 scale deep) along side. Black spots (often poorly developed on male) on lower side. To 2¼ in. (5.5 cm).

BIG BEND GAMBUSIA *Gambusia gaigei* **P. 448**
Prominent black spots, crescents on upper side (absent on belly). Dark stripe along side. To 2¼ in. (5.4 cm).

AMISTAD GAMBUSIA *Gambusia amistadensis* **P. 449**
Extinct. Similar to Big Bend Gambusia but was more slender, had more terminal mouth. To 2¼ in. (5.8 cm).

LARGESPRING GAMBUSIA *Gambusia geiseri* **P. 449**
Distinct row of black spots on middle of dorsal and caudal fins (indistinct on juvenile). Scattered black spots on side. To 1¾ in. (4.4 cm).

CLEAR CREEK GAMBUSIA *Gambusia heterochir* **P. 449**
Distinct notch at top of pectoral fin of male. No dark stripe along back. To 2¼ in. (5.4 cm).

SAN MARCOS GAMBUSIA *Gambusia georgei* **P. 448**
Presumed extinct. Lemon yellow median fins. Dark edges (best developed on large individual) on dorsal and caudal fins. To 2 in. (4.8 cm).

LEAST KILLIFISH *Heterandria formosa* **P. 450**
Red around black spot on front of dorsal fin. Series of black bars along side; black spot on caudal fin base. To 1½ in. (3.6 cm).

GILA TOPMINNOW *Poeciliopsis occidentalis* **P. 451**
Dark to dusky stripe along side. Large male is black. Extremely long gonopodium, more than ⅓ body length. To 2¼ in. (6 cm). See also Sonora Topminnow, *P. sonoriensis*, and Porthole Livebearer, *P. gracilis*.

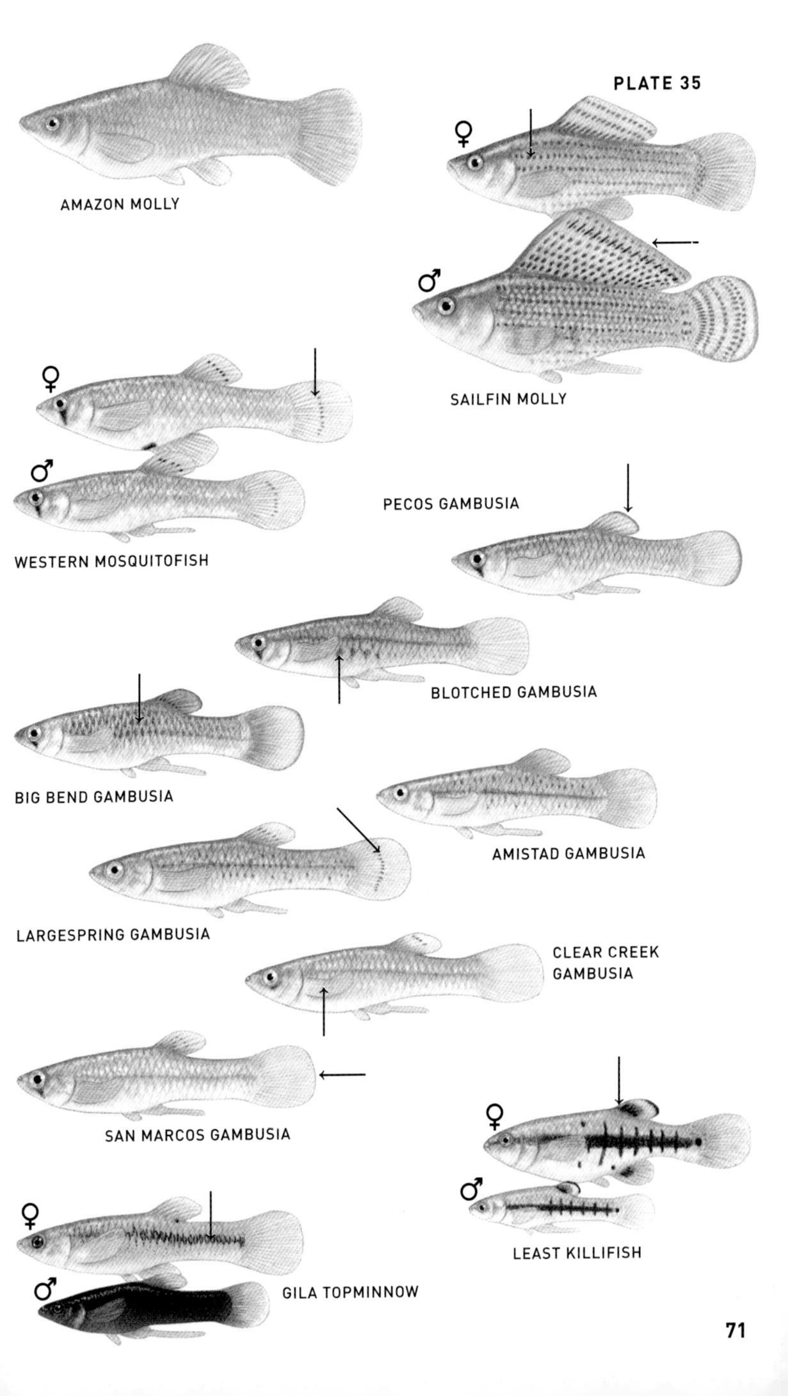
PLATE 35
AMAZON MOLLY
♀
♂
SAILFIN MOLLY
♀
♂
WESTERN MOSQUITOFISH
PECOS GAMBUSIA
BLOTCHED GAMBUSIA
BIG BEND GAMBUSIA
AMISTAD GAMBUSIA
LARGESPRING GAMBUSIA
CLEAR CREEK GAMBUSIA
SAN MARCOS GAMBUSIA
♀
♂
LEAST KILLIFISH
♀
♂
GILA TOPMINNOW

PLATE 36

Tetra, Cavefishes, Burbot, Silversides, Sticklebacks, Drum

MEXICAN TETRA *Astyanax mexicanus* **P. 340**
Adipose fin. Jaw teeth. Black stripe on caudal peduncle and fin. Large individual (shown) has yellow fins, red on caudal and anal fins. To 4¾ in. (12 cm).

Next 5 species (cavefishes) have no pelvic fins (Northern Cavefish has very small pelvic fins), small or rudimentary eyes.

SWAMPFISH *Chologaster cornuta* **P. 418**
Strongly bicolored. Stripe on lower side wide at front, narrow at rear. To 2¾ in. (6.8 cm).

SPRING CAVEFISH *Forbesichthys agassizii* **P. 418**
Long, slender salamanderlike body; 3 axial stripes along side. To 3¼ in. (8.4 cm).

SOUTHERN CAVEFISH *Typhlichthys subterraneus* **P. 419**
Pink-white. No eyes. To 3½ in. (9 cm).

NORTHERN CAVEFISH *Amblyopsis spelaea* **P. 419**
Pink-white. No eyes. Very small pelvic fins. To 4¼ in. (11 cm). See also Ozark Cavefish, *Troglichthys rosae*.

ALABAMA CAVEFISH *Speoplatyrhinus poulsoni* **P. 420**
Pink-white. No eyes. Long, flat head constricted behind snout. To 3 in. (7.4 cm).

BURBOT *Lota lota* **P. 420**
Long slender body. Long barbel at tip of chin. Two dorsal fins; 1st short, 2d very long. To 33 in. (84 cm).

BROOK SILVERSIDE *Labidesthes sicculus* **P. 422**
Long beaklike snout. Two widely separated dorsal fins; origin of 1st above anal fin origin. To 5 in. (13 cm).

MISSISSIPPI SILVERSIDE *Menidia audens* **P. 423**
Two widely separated dorsal fins; origin of 1st in front of anal fin origin. To 6 in. (15 cm). See also Waccamaw Silverside, *M. extensa*.

THREESPINE STICKLEBACK *Gasterosteus aculeatus* **P. 464**
Three dorsal spines, last very short. Keel on caudal peduncle. To 4 in. (10 cm). See also Fourspine Stickleback, *Apeltes quadracus*.

BROOK STICKLEBACK *Culaea inconstans* **P. 463**
Has 4–6 short dorsal spines. No keel on short caudal peduncle. To 3½ in. (8.7 cm).

NINESPINE STICKLEBACK *Pungitius pungitius* **P. 462**
Has 7–12, usually 9, short dorsal spines. Usually a keel on caudal peduncle. To 3½ in. (9 cm).

FRESHWATER DRUM *Aplodinotus grunniens* **P. 603**
Strongly arched, silver body. Short 1st dorsal fin; long 2d dorsal fin. Long pelvic fin ray. Lateral line to end of caudal fin. To 35 in. (89 cm).

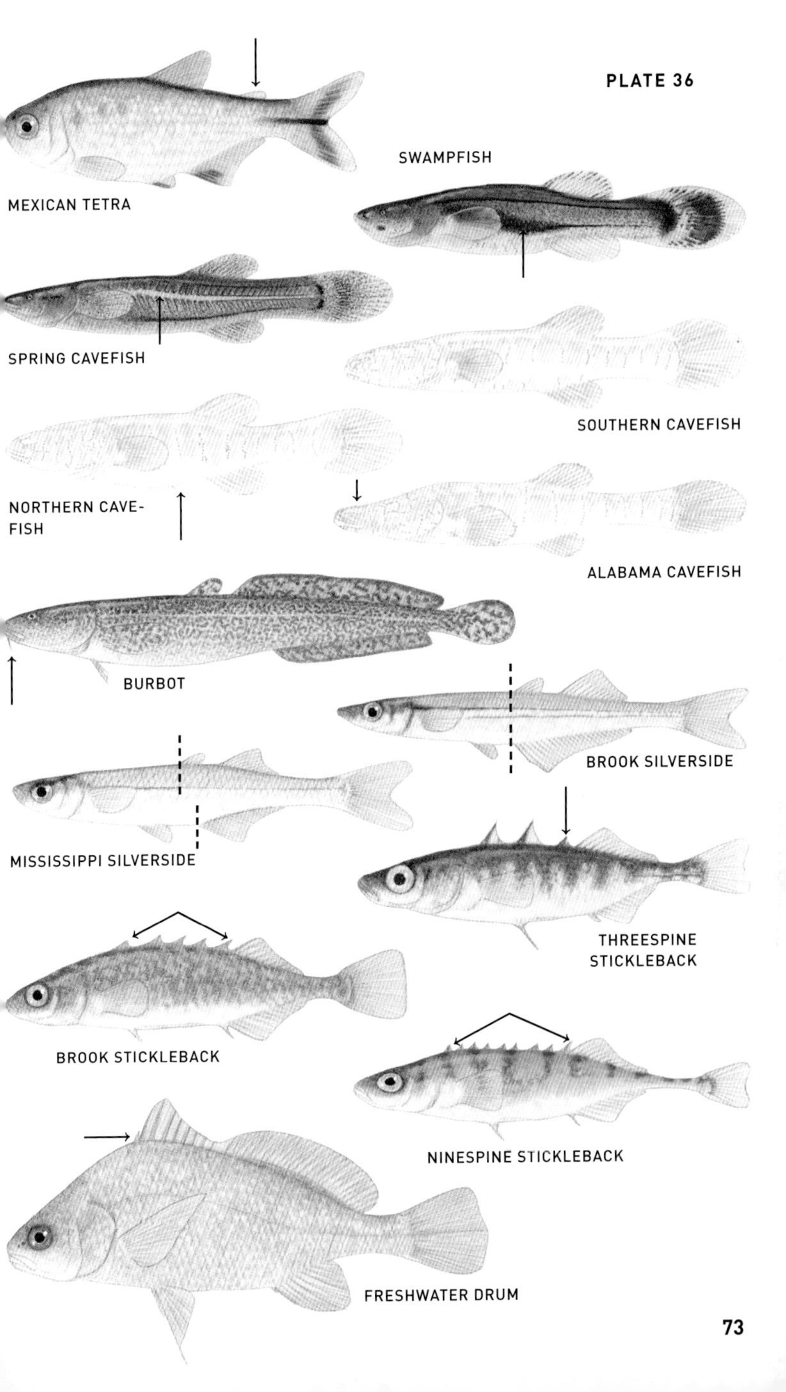
PLATE 36
MEXICAN TETRA
SWAMPFISH
SPRING CAVEFISH
SOUTHERN CAVEFISH
NORTHERN CAVE-
FISH
ALABAMA CAVEFISH
BURBOT
BROOK SILVERSIDE
MISSISSIPPI SILVERSIDE
THREESPINE
STICKLEBACK
BROOK STICKLEBACK
NINESPINE STICKLEBACK
FRESHWATER DRUM

PLATE 37

Sculpins (1)

Sculpins have large, fanlike pectoral fins, large mouth, no or few scales.

DEEPWATER SCULPIN *Myoxocephalus thompsonii* **P. 467**
Large gap between dorsal fins. Extremely wide, flat head. To 9 in. (23 cm). See also Fourhorn Sculpin, *M. quadricornis.*

SPOONHEAD SCULPIN *Cottus ricei* **P. 468**
Wide, flat head. Extremely slender caudal peduncle. Prickles on most of body. To 5 in. (13 cm).

TORRENT SCULPIN *Cottus rhotheus* **P. 469**
"Pinched" caudal peduncle. Two broad dark bars under 2d dorsal fin. To 6 in. (15.5 cm).

BANDED SCULPIN *Cottus carolinae* **P. 470**
Has 4–5 brown-black saddles, last 3 extending down side as dark bars. Usually complete lateral line. To 7¼ in. (18 cm). See also Kanawha Sculpin, *C. kanawhae.*

POTOMAC SCULPIN *Cottus girardi* **P. 471**
Similar to Banded Sculpin but has incomplete lateral line, less regularly bordered bars on side. To 5¼ in. (14 cm).

COASTRANGE SCULPIN *Cottus aleuticus* **P. 471**
Long pelvic fin reaches to anus. Long, tubular posterior nostril. To 6¾ in. (17 cm).

PRICKLY SCULPIN *Cottus asper* **P. 472**
Long anal (usually 16-19 rays) and dorsal (19–23 rays) fins. Large dark brown blotches on upper side. Usually many prickles. To 12 in. (30 cm).

KLAMATH LAKE SCULPIN *Cottus princeps* **P. 472**
Broadly joined dorsal fins. Large pores on flat head. Long fins. Many prickles. To 2¾ in. (7 cm).

SHOSHONE SCULPIN *Cottus greenei* **P. 473**
Deep body. Deep caudal peduncle. To 3½ in. (9 cm).

SLENDER SCULPIN *Cottus tenuis* **P. 473**
Strongly bicolored; brown above, white to brassy below. One or more branched pelvic rays. To 3½ in. (9 cm). See also Rough Sculpin, *C. asperrimus.*

UTAH LAKE SCULPIN *Cottus echinatus* **P. 474**
Presumed extinct. Many prickles on long slender body. No bold saddles or bars. To 4¼ in. (11 cm).

BEAR LAKE SCULPIN *Cottus extensus* **P. 474**
Similar to Utah Lake Sculpin but lacks prickles on breast and belly, has smaller head. To 5¼ in. (13 cm).

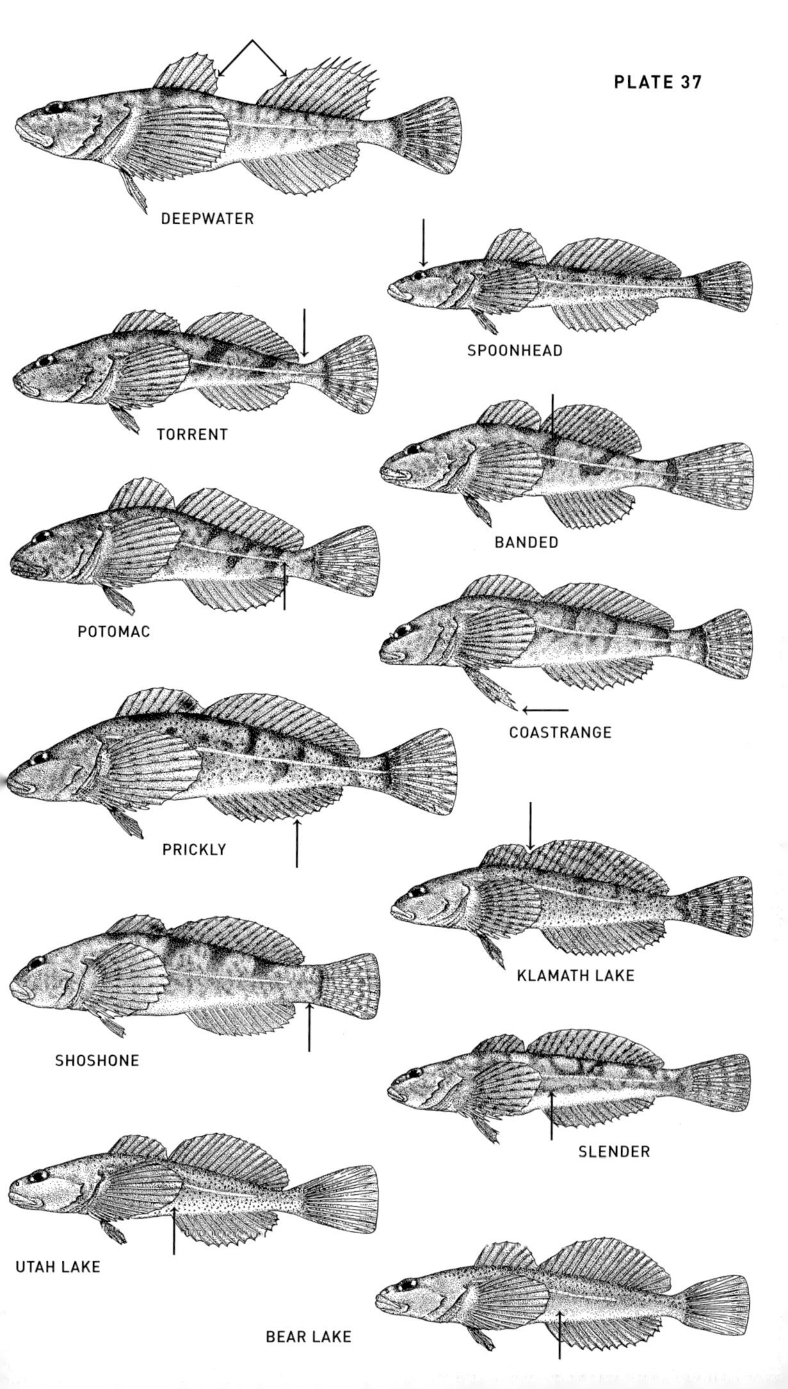
PLATE 37
DEEPWATER
SPOONHEAD
TORRENT
BANDED
POTOMAC
COASTRANGE
PRICKLY
KLAMATH LAKE
SHOSHONE
SLENDER
UTAH LAKE
BEAR LAKE

Sculpins (2)

SLIMY SCULPIN *Cottus cognatus* **P. 475**
Usually 3 pelvic rays. Long slender body. To 4½ in. (12 cm).

Next 4 species have 2 large black spots in 1st dorsal fin, incomplete lateral line.

MOTTLED SCULPIN *Cottus bairdii* **P. 475**
Robust body. Dorsal fins joined at base. To 6 in. (15 cm). See also Blue Ridge Sculpin, *C. caeruleomentum*, Tallapoosa Sculpin, *C. tallapoosae*, Chattahoochee Sculpin, *C. chattahoochee*, Columbia Sculpin, *C. hubbsi*, and Malheur Sculpin, *C. bendirei*.

SHORTHEAD SCULPIN *Cottus confusus* **P. 480**
Slender body. Dorsal fins separate to base. Prickles on body behind pectoral fin. To 5¾ in. (15 cm).

OZARK SCULPIN *Cottus hypselurus* **P. 478**
Wide, wavy black bands on dorsal and caudal fins. To 5½ in. (14 cm).

BLACK SCULPIN *Cottus baileyi* **P. 476**
Nearly identical to Mottled Sculpin but smaller; usually lacks palatine teeth. To 3¼ in. (8.4 cm).

PYGMY SCULPIN *Cottus paulus* **P. 476**
Boldly patterned: black head, white nape, black saddles. To 1¾ in. (4.5 cm).

PAIUTE SCULPIN *Cottus beldingii* **P. 479**
Dorsal fins separate to base. No prickles. Two black spots on dorsal fin. To 5¼ in. (13 cm).

MARGINED SCULPIN *Cottus marginatus* **P. 480**
Similar to Paiute Sculpin but has dorsal fins joined at base, usually 3 pelvic rays. To 5 in. (13 cm).

WOOD RIVER SCULPIN *Cottus leiopomus* **P. 483**
Short head, about 3 times into slender body. Incomplete lateral line. To 4¼ in. (11 cm).

Next 3 species have large black spot at rear (only) of 1st dorsal fin.

RIFFLE SCULPIN *Cottus gulosus* **P. 481**
Deep compressed caudal peduncle. To 4¼ in. (11 cm). See also Reticulate Sculpin, *C. perplexus*.

PIT SCULPIN *Cottus pitensis* **P. 482**
Dark vermiculations and small blotches on side. To 5 in. (13 cm).

MARBLED SCULPIN *Cottus klamathensis* **P. 482**
Deep body. Marbled pattern on fins. To 3½ in. (9 cm).

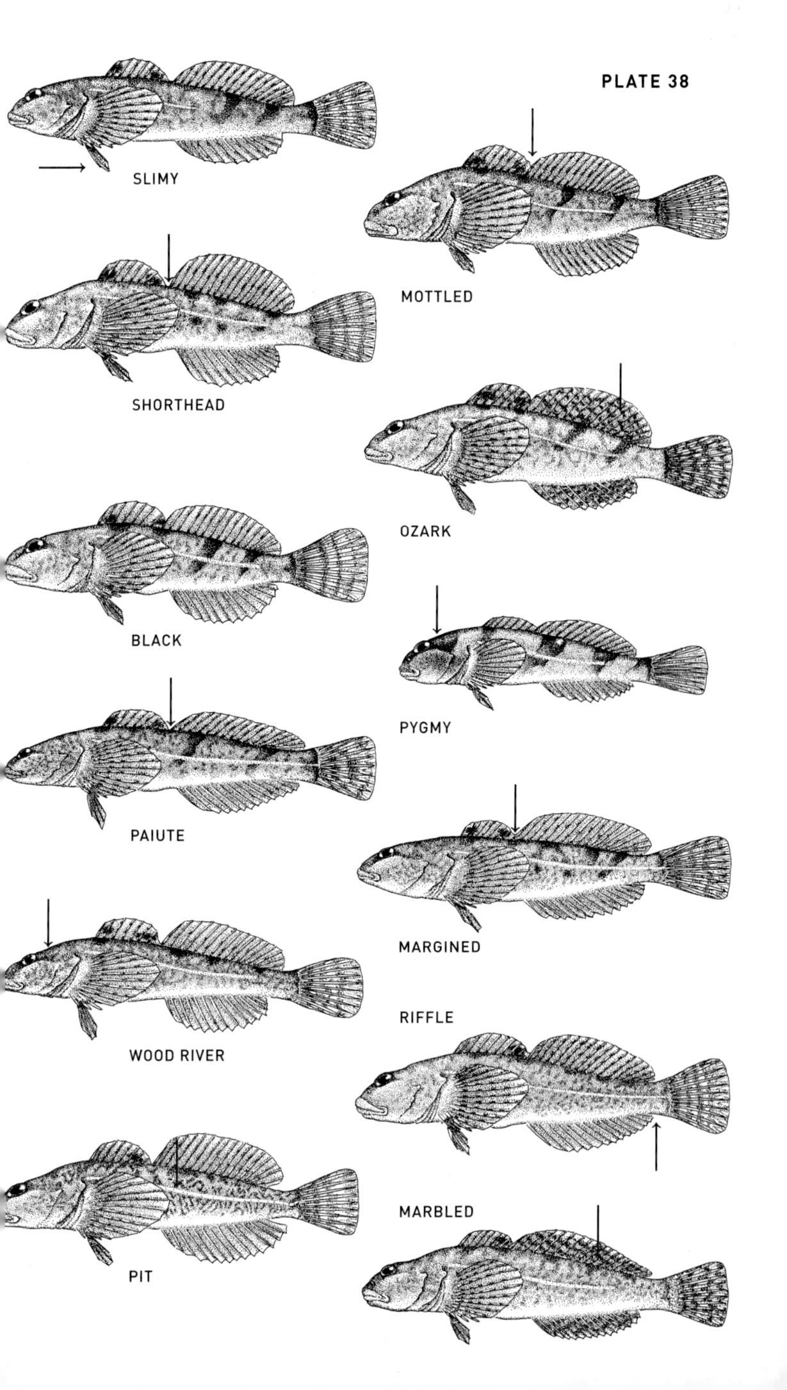
PLATE 38
SLIMY
MOTTLED
SHORTHEAD
OZARK
BLACK
PYGMY
PAIUTE
MARGINED
WOOD RIVER
RIFFLE
PIT
MARBLED

Basses (1)

First 4 species (Morone) *have 2 dorsal fins, 1st tall, usually 9 spines; spine on gill cover flap; small gill on underside of gill cover.*

WHITE PERCH *Morone americana* **P. 485**
No dark stripes along side (of adult). Body deepest under 1st dorsal fin. To 22¾ in. (58 cm).

WHITE BASS *Morone chrysops* **P. 484**
Has 4–7 dark gray-brown stripes on silver white side. Body deepest between dorsal fins. To 17¾ in. (45 cm).

YELLOW BASS *Morone mississippiensis* **P. 485**
Has 5–7 black stripes on silver yellow side broken and offset on lower side. To 18 in. (46 cm).

STRIPED BASS *Morone saxatilis* **P. 483**
Has 6–9 dark gray stripes (on adult) on silver white side. To 6½ ft. (2 m).

Next 7 species (*Micropterus*) have large mouth reaching to or beyond eye, elongate body, black spot at rear of gill cover, 3 anal spines. See juveniles of *Micropterus* on Pl. 42.

Next 6 species have confluent dorsal fins.

REDEYE BASS *Micropterus coosae* **P. 496**
White upper and lower outer edges on orange caudal fin. Rows of dark spots on lower side. Second dorsal, caudal, and front of anal fin brick red on young. To 19 in. (47 cm).

SPOTTED BASS *Micropterus punctulatus* **P. 494**
Rows of small black spots on lower side. Young has 3-colored (yellow, black, white edge) caudal fin. To 24 in. (61 cm). See also Alabama Bass, *M. henshalli.*

GUADALUPE BASS *Micropterus treculii* **P. 496**
Similar to Spotted Bass but has 10–12 dark bars along side (darkest in young). To 16 in. (40 cm).

SHOAL BASS *Micropterus cataractae* **P. 497**
Similar to Redeye Bass but lacks white outer edges on orange caudal fin, patch of teeth on tongue. To 25¼ in. (64 cm).

SUWANNEE BASS *Micropterus notius* **P. 494**
Color as in Largemouth Bass except brown overall; black wavy lines, rows of black spots in dorsal, anal, and caudal fins; black spots on lower side; large male (shown) has bright turquoise cheek, breast, and belly. To 14¼ in. (36 cm).

SMALLMOUTH BASS *Micropterus dolomieu* **P. 497**
Dark brown; bronze specks, often coalesced into 8–16 bars, on yellow-green side. Young has 3-colored (yellow, black, white edge) caudal fin. To 27¼ in. (69 cm).

LARGEMOUTH BASS *Micropterus salmoides* **P. 493**
First dorsal fin highest at middle, low at rear; 1st (spinous) and 2d (soft) dorsal fins nearly separate. Large mouth extends past eye. Broad black stripe (often broken into series of blotches) along side. To 38 in. (97 cm).

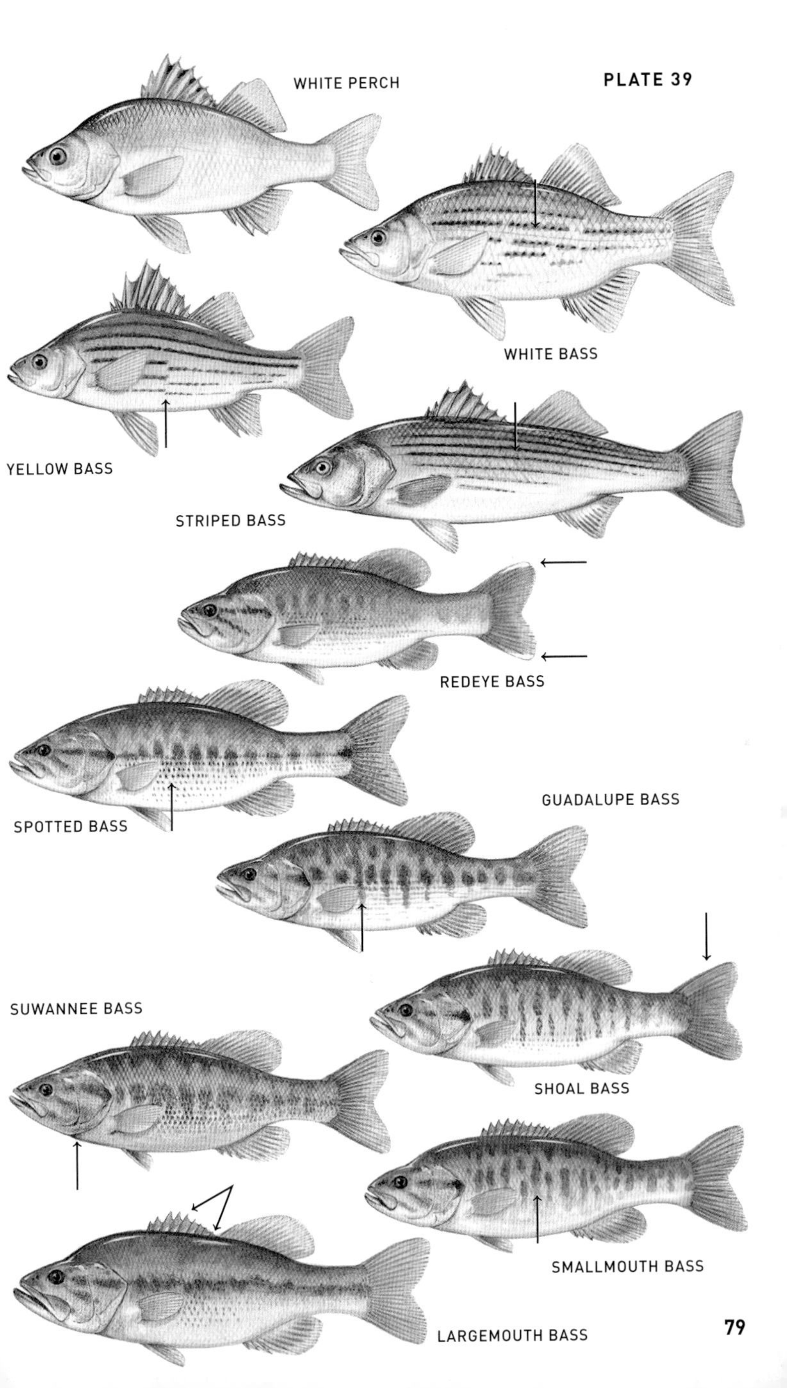
WHITE PERCH
WHITE BASS
YELLOW BASS
STRIPED BASS
REDEYE BASS
GUADALUPE BASS
SPOTTED BASS
SUWANNEE BASS
SHOAL BASS
SMALLMOUTH BASS
LARGEMOUTH BASS

Basses (2), Crappies

SACRAMENTO PERCH *Archoplites interruptus* **P. 486**
Has 6–7 anal spines, 12–13 dorsal spines. Dorsal fin base about twice as long as anal fin base. To 24 in. (61 cm).

Next 4 species (*Ambloplites*) usually have 6 anal spines, red eye.

SHADOW BASS *Ambloplites ariommus* **P. 490**
Irregular marbling of brown or gray on light green or brown side. Large eye. To 8¾ in. (22 cm).

ROANOKE BASS *Ambloplites cavifrons* **P. 490**
Unscaled or partly scaled cheek. Many iridescent gold to white spots on upper side and head. To 14½ in. (36 cm).

OZARK BASS *Ambloplites constellatus* **P. 489**
Similar to Rock Bass but with freckled pattern (scattered dark brown spots) on side of body and head. To 7½ in. (19 cm).

ROCK BASS *Ambloplites rupestris* **P. 489**
Adult has rows of brown-black spots along side, largest and darkest below lateral line. Young has brown marbling on gray side. To 17 in. (43 cm).

WARMOUTH *Lepomis gulosus* **P. 498**
Dark red-brown lines radiating from back of red eye. Large mouth. Thick body. Teeth on tongue. To 12 in. (31 cm). See juvenile on Pl. 42.

FLIER *Centrarchus macropterus* **P. 487**
Large black teardrop. Interrupted rows of black spots along side. Has 7–8 anal spines. Red-orange around black spot near rear of 2d dorsal fin on young. To 7½ in. (19 cm).

WHITE CRAPPIE *Pomoxis annularis* **P. 488**
Very long predorsal region arched with sharp dip over eye; dorsal fin base shorter than distance from eye to dorsal fin origin. Has 6 dorsal spines, 1st much shorter than last. To 21 in. (53 cm).

BLACK CRAPPIE *Pomoxis nigromaculatus* **P. 487**
Long predorsal region arched with sharp dip over eye; dorsal fin base about as long as distance from eye to dorsal fin origin. Has 7–8 dorsal spines, 1st much shorter than last. To 19¼ in. (49 cm).

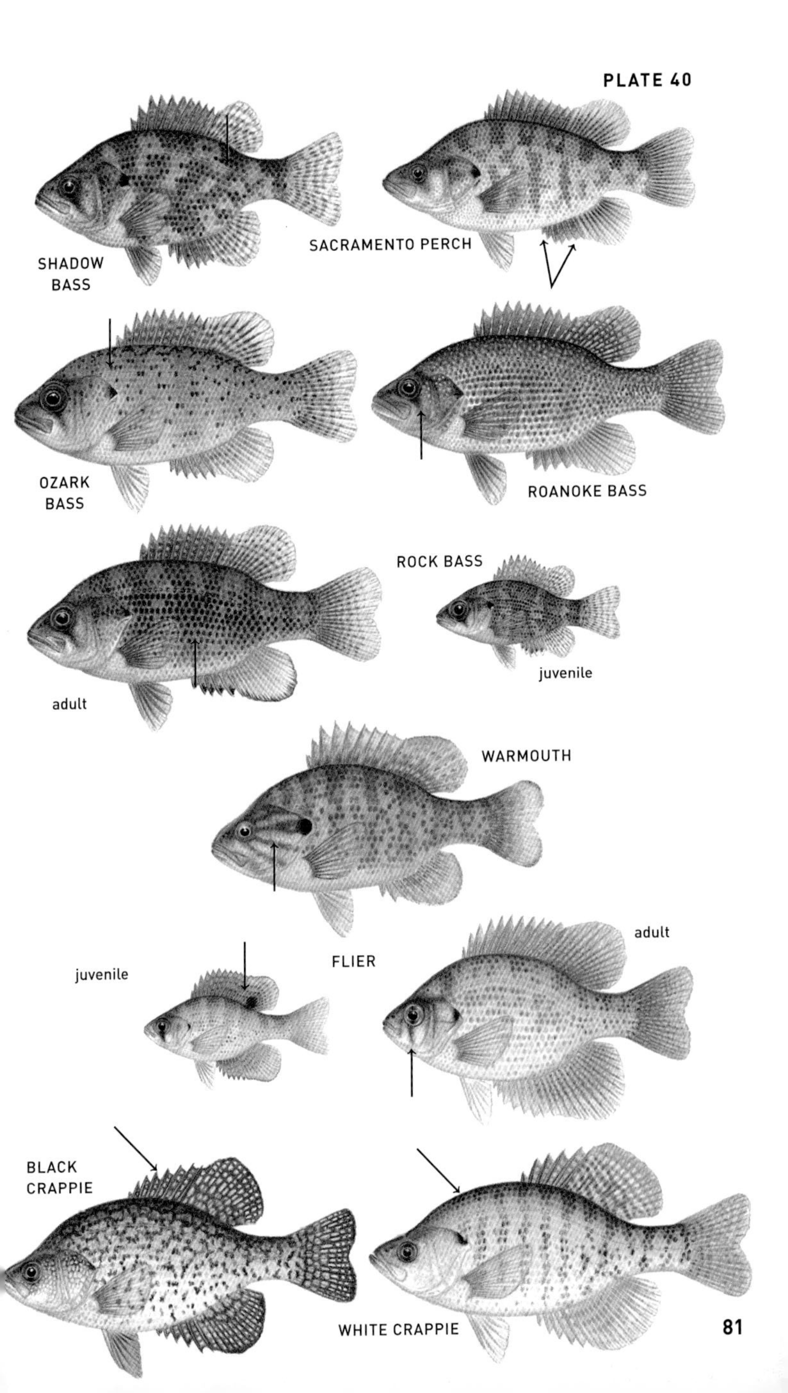
SHADOW BASS
SACRAMENTO PERCH
OZARK BASS
ROANOKE BASS
ROCK BASS
adult
juvenile
WARMOUTH
FLIER
juvenile
adult
BLACK CRAPPIE
WHITE CRAPPIE

Sunfishes (1)

Deep, strongly compressed body; 3 anal spines. See juveniles on Pl. 42.

BANTAM SUNFISH *Lepomis symmetricus* **P. 501**

Chubby. Lacks bright colors of other sunfishes. Usually interrupted, incomplete lateral line. Black spot at rear of dorsal fin on young. To 3½ in. (9 cm).

GREEN SUNFISH *Lepomis cyanellus* **P. 500**

Large mouth; upper jaw extends beneath eye pupil. Large black spot at rear of 2d dorsal and anal fins on adult; yellow or orange margins on 2d dorsal, caudal, and anal fins. To 12 in. (31 cm).

SPOTTED SUNFISH *Lepomis punctatus* **P. 501**

L. p. miniatus has rows of red (on male) or yellow-brown spots (female) on side. *L. p. punctatus* has black specks on side. To 8 in. (20 cm).

BLUEGILL *Lepomis macrochirus* **P. 502**

Large black spot at rear of dorsal fin (faint on young). Dark bars (absent in turbid water; thin and chainlike on young) on deep, extremely compressed body. Long pectoral fin extends far past eye when bent forward. To 16¼ in. (41 cm).

REDEAR SUNFISH *Lepomis microlophus* **P. 503**

Bright red-orange spot, white edge on black ear flap (best developed on large adult). Long pectoral fin usually extends far past eye when bent forward. To 10 in. (25 cm).

PUMPKINSEED *Lepomis gibbosus* **P. 504**

Bright red or orange spot, light-colored margin on black ear flap. Many bold dark brown wavy lines or orange spots on 2d dorsal, caudal, and anal fins. Wavy blue lines on cheek and opercle of adult. To 16 in. (40 cm).

NORTHERN SUNFISH *Lepomis peltastes* **P. 505**

Similar to Longear Sunfish but reaches only 5 in. (13 cm), has large red spot on upwardly slanted ear flap, usually 12 pectoral rays, 40 or fewer lateral scales.

LONGEAR SUNFISH *Lepomis megalotis* **P. 504**

Long ear flap (especially on adult male), horizontal to slanted downward on adult, slanted upward on young. Wavy blue lines on cheek and opercle. Adult is dark red above, bright orange below, marbled and spotted with blue. Usually 13–14 pectoral rays. To 9½ in. (24 cm).

DOLLAR SUNFISH *Lepomis marginatus* **P. 506**

Similar to Longear Sunfish but has shorter upwardly slanted ear flap, red streak along lateral line, usually 12 pectoral rays. To 4¾ in. (12 cm).

ORANGESPOTTED SUNFISH *Lepomis humilis* **P. 507**

Bright orange (on large male) or red-brown (female) spots on silver green side. Wide white margin on long black ear flap. Greatly elongated pores along preopercle margin. To 6 in. (15 cm).

REDBREAST SUNFISH *Lepomis auritus* **P. 506**

Very long, narrow (no wider than eye) ear flap, black to edge. Wavy blue lines on cheek and opercle. Large male has bright orange breast and belly. To 9½ in. (24 cm).

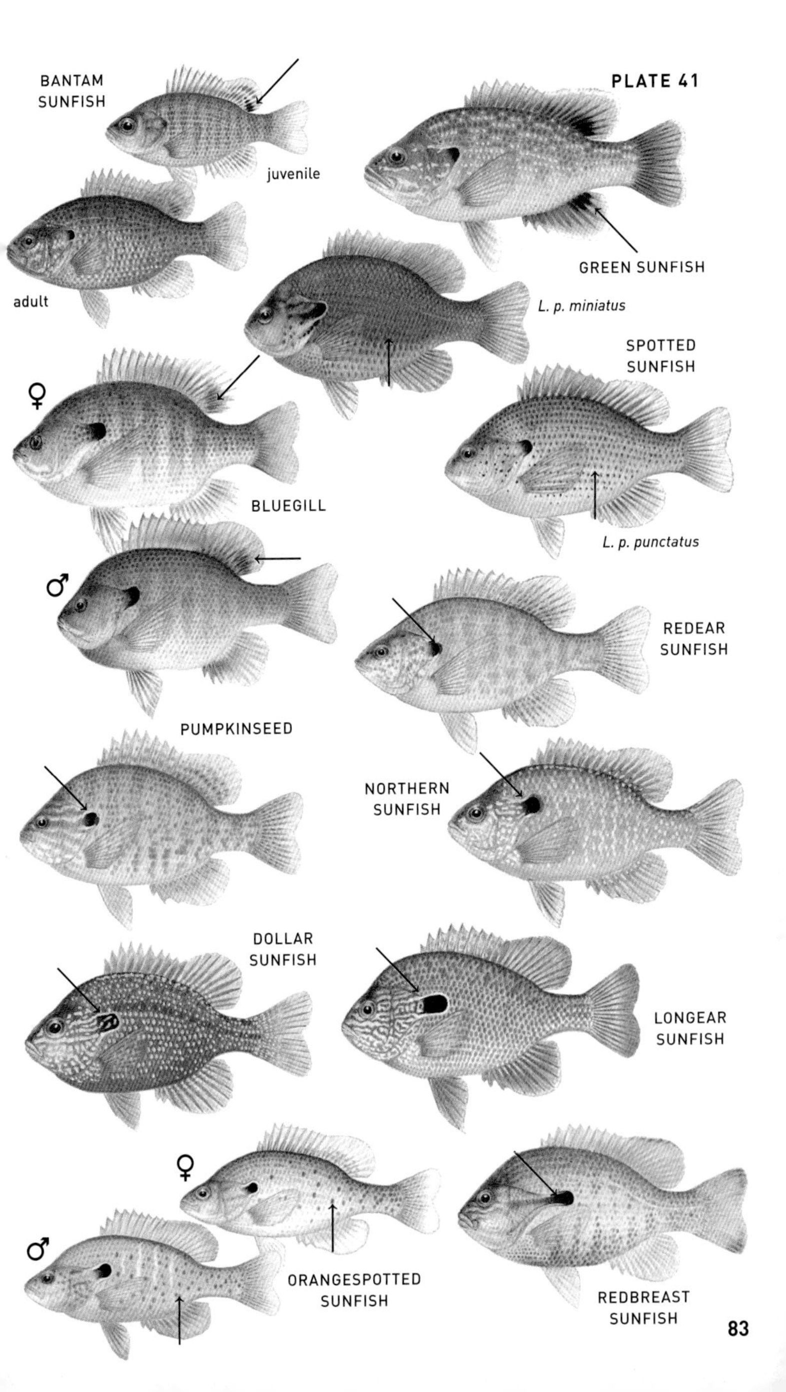
BANTAM
SUNFISH
juvenile
adult
PLATE 41
GREEN SUNFISH
L. p. miniatus
SPOTTED
SUNFISH
♀
BLUEGILL
L. p. punctatus
♂
REDEAR
SUNFISH
PUMPKINSEED
NORTHERN
SUNFISH
DOLLAR
SUNFISH
LONGEAR
SUNFISH
♀
♂
ORANGESPOTTED
SUNFISH
REDBREAST
SUNFISH

Juvenile Sunfishes And Basses

Sunfishes (Lepomis) *1–1½ in. (2.5–4 cm) and basses* (Micropterus) *1½–3 in. (3.5–7.5 cm). See adults on Pls. 39–41.*

Next 3 species have long, pointed pectoral fin; extends past pupil when pressed forward.

BLUEGILL *Lepomis macrochirus* **P. 502**
Dark bars on side. Black spot at rear of dorsal fin.

REDEAR SUNFISH *Lepomis microlophus* **P. 503**
Dark bars on side. White edge on black ear flap.

PUMPKINSEED *Lepomis gibbosus* **P. 504**
Dark spots between chainlike bars on side. Dark wavy lines on dorsal fin.

Next 6 species have short, rounded pectoral fin.

ORANGESPOTTED SUNFISH *Lepomis humilis* **P. 507**
Dark spots, irregularly spaced bars of unequal width on side.

REDBREAST SUNFISH *Lepomis auritis* **P. 506**
Two distinct black lines leading to black spot on ear flap.

LONGEAR SUNFISH *Lepomis megalotis* **P. 504**
Dusky 2–3 wavy dark lines on cheek and opercle.

NORTHERN SUNFISH *Lepomis peltastes* **P. 505**
Has 12 pectoral rays, 40 or fewer lateral scales.

SPOTTED SUNFISH *Lepomis punctatus* **P. 501**
Body and opercle covered with dark brown or black spots.

BANTAM SUNFISH *Lepomis symmetricus* **P. 501**
Chainlike dark bars. Reticulations around black spot at rear of dorsal fin.

Next 2 species have short, rounded pectoral fin extending no farther than pupil when pressed forward. Large mouth; upper jaw extends under pupil.

GREEN SUNFISH *Lepomis cyanellus* **P. 500**
Usually without dark bars. Black spot at rear of dorsal, and often anal, fins.

WARMOUTH *Lepomis gulosus* **P. 498**
Dark blotches between wide dark bars on side; bars with pale centers.

Next 3 species have no or faint dark stripe along side.

SHOAL BASS *Micropterus cataractae* **P. 497**
Dusky caudal fin with light edge. Dusky bars on side.

REDEYE BASS *Micropterus coosae* **P. 496**
Red 2d dorsal, caudal, and anal fins; white edge on caudal fin.

SMALLMOUTH BASS *Micropterus dolomieu* **P. 497**
Dark brown bars along side; 3-colored (yellow, black, white edge) caudal fin.

Next 4 species have dark stripe or series of connected black blotches along side, bold black spot on caudal fin base.

LARGEMOUTH BASS *Micropterus salmoides* **P. 493**
Dorsal fins nearly separate. Caudal fin pale at base, dusky at rear.

SUWANEE BASS *Micropterus notius* **P. 494**
Black wavy lines, rows of black spots in dorsal, anal, and caudal fins.

SPOTTED BASS *Micropterus punctulatus* **P. 494**
Rows of small black spots on lower side; 3-colored caudal fin.

GUADALUPE BASS *Micropterus treculii* **P. 496**
Dusky bars below 10-12 dark blotches along side; 3-colored caudal fin.

PLATE 42
BLUEGILL
PUMPKINSEED
REDEAR SUNFISH
LONGEAR SUNFISH
ORANGESPOTTED SUNFISH
REDBREAST SUNFISH
SPOTTED SUNFISH
BANTAM SUNFISH
NORTHERN SUNFISH
GREEN SUNFISH
WARMOUTH
LARGEMOUTH BASS
SHOAL BASS
SUWANNEE BASS
REDEYE BASS
SPOTTED BASS
SMALLMOUTH BASS
GUADALUPE BASS

Sunfishes (2), Pygmy Sunfishes, Yellow Perch, Sauger, Walleye

First 10 species have rounded caudal fin.

MUD SUNFISH *Acantharchus pomotis* **P. 491**
Has 3–4 parallel dark stripes on face and side of body. To 8¼ in. (21 cm).

BLUESPOTTED SUNFISH *Enneacanthus gloriosus* **P. 492**
Rows of blue or silver spots along side of large young and adult (bars on side of small young). Slender caudal peduncle. To 3¾ in. (9.5 cm).

BANDED SUNFISH *Enneacanthus obesus* **P. 491**
Dark bars on side (darkest on large individual). Rows of purple-gold spots along side. To 3¾ in. (9.5 cm).

BLACKBANDED SUNFISH *Enneacanthus chaetodon* **P. 492**
Six black bars on side, 1st through eye, 6th (often faint) on caudal peduncle. First 2–3 membranes of dorsal fin black. Pink to red and black on pelvic fin. To 3¼ in. (8 cm).

Next 6 species (*Elassoma*) have no lateral line, no notch in dorsal fin.

BANDED PYGMY SUNFISH *Elassoma zonatum* **P. 604**
Has 1–2 large black spots (rarely absent) on upper side; 7–12 dark green to black bars on side. To 1¾ in. (4.7 cm).

SPRING PYGMY SUNFISH *Elassoma alabamae* **P. 604**
Has 6–8 thin gold or blue bars along side. Clear window at rear of 2d dorsal and anal fin bases. Breeding male shown. To 1¼ in. (3 cm).

EVERGLADES PYGMY SUNFISH *Elassoma evergladei* **P. 604**
Scales on top of head. Dark-colored lips. Breeding male shown. To 1¼ in. (3.4 cm).

CAROLINA PYGMY SUNFISH *Elassoma boehlkei* **P. 606**
Has 10–16 narrow black bars along side, about as wide as interspaces. Breeding male shown. To 1¼ in. (3.2 cm).

OKEFENOKEE PYGMY SUNFISH *Elassoma okefenokee* **P. 605**
Brown bars (darkest at rear, often broken into vertically aligned blotches) on side of female. Front of lips light (dark at sides) except in large male. Breeding male (shown) is black with iridescent blue bars. To 1¼ in. (3.4 cm). See also Gulf Coast Pygmy Sunfish, *E. gilberti*.

BLUEBARRED PYGMY SUNFISH *Elassoma okatie* **P. 605**
Has 8–14 wide black bars along side, about 3 times as wide as interspaces. Breeding male shown. To 1¼ in. (3.4 cm).

YELLOW PERCH *Perca flavescens* **P. 511**
Deep, compressed body; dark saddles extend down yellow side. Black blotch at rear of dorsal fin. To 16 in. (40 cm). See also Ruffe, *Gymnocephalus cernua*.

SAUGER *Sander canadensis* **P. 510**
Similar to Walleye but has many black half-moons on 1st dorsal fin; 3–4 dusky brown saddles extending down side as broad bars; no white tips on fins. To 30 in. (76 cm).

WALLEYE *Sander vitreus* **P. 508**
Slender body. Opaque, eye. Large black spot (absent on young) on rear of 1st dorsal fin. White tips on anal fin, lower lobe of caudal fin. To 36 in. (91 cm).

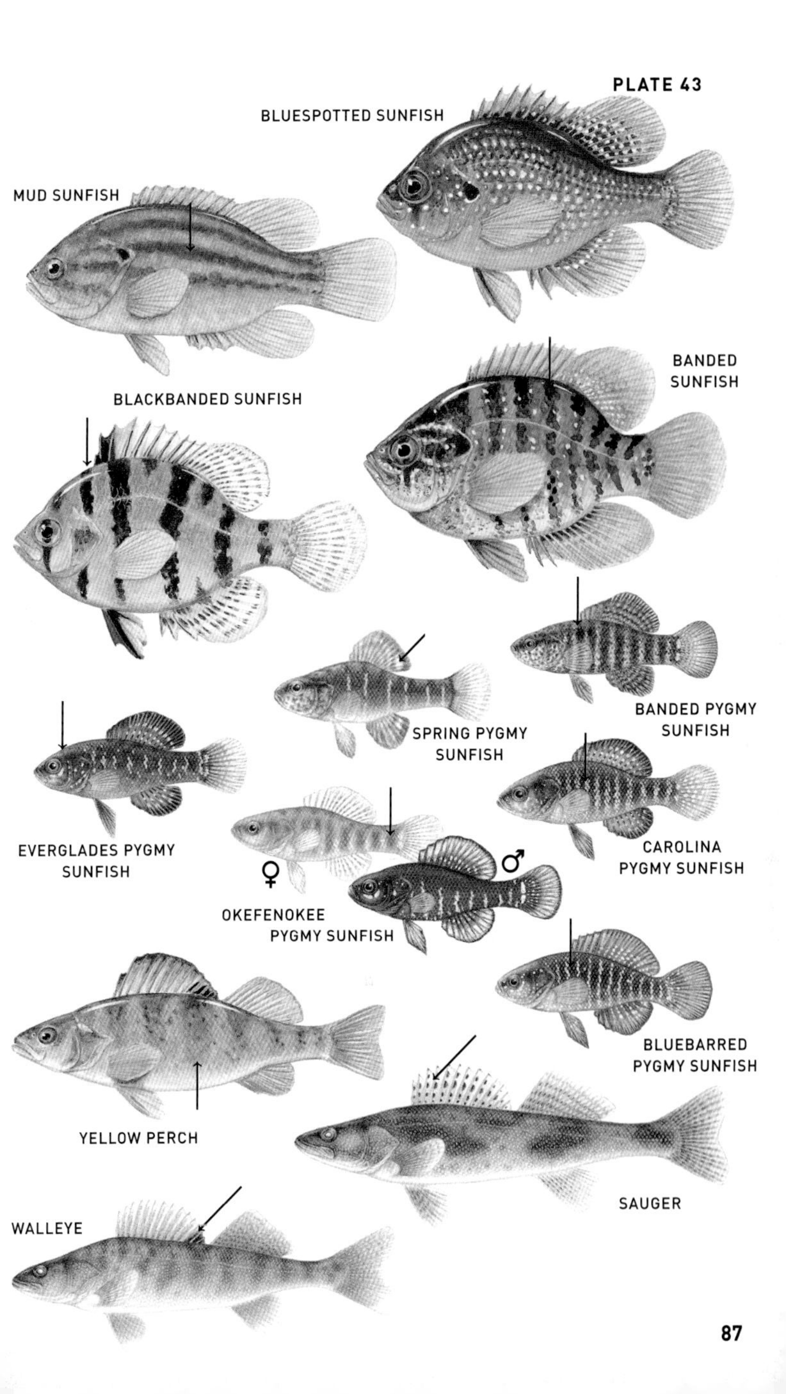
PLATE 43
BLUESPOTTED SUNFISH
MUD SUNFISH
BANDED SUNFISH
BLACKBANDED SUNFISH
BANDED PYGMY SUNFISH
SPRING PYGMY SUNFISH
EVERGLADES PYGMY SUNFISH
CAROLINA PYGMY SUNFISH
♀
♂
OKEFENOKEE PYGMY SUNFISH
BLUEBARRED PYGMY SUNFISH
YELLOW PERCH
SAUGER
WALLEYE

Darters (1)

Scutes on breast and, except on Bluestripe Darter, along belly of male (Fig. 58).

LEOPARD DARTER *Percina pantherina* **P. 515**
Has 10–14 round black spots along side. To 3½ in. (9.2 cm).

LONGHEAD DARTER *Percina macrocephala* **P. 513**
Long snout. Sickle-shaped teardrop. Dark bar below medial black caudal spot. To 4¾ in. (12 cm). See also Sickle Darter, *P. williamsi*.

BRIDLED DARTER *Percina kusha* **P. 514**
Black blotch on caudal fin base centered below black stripe. To 3 in. (7.8 cm). See also Muscadine Darter, *P. smithvanizi*, and Bankhead Darter, *P. sipsi*.

BLACKSIDE DARTER *Percina maculata* **P. 515**
Medial black caudal spot; 6–9 large black ovals along side. To 4¼ in. (11 cm).

SHIELD DARTER *Percina peltata* **P. 516**
Black bar on chin, row of black crescents on 1st dorsal fin (Fig. 60). To 3½ in. (9 cm). See also Chainback Darter, *P. nevisense*, and Appalachia Darter, *P. gymnocephala*.

STRIPEBACK DARTER *Percina notogramma* **P. 516**
Pale yellow stripe along upper side. First dorsal fin dusky, darkest at front. To 3½ in. (8.4 cm).

BLUESTRIPE DARTER *Percina cymatotaenia* **P. 512**
Broad, scallop-edged black stripe along side. To 3½ in. (9 cm). See also Frecklebelly Darter, *P. stictogaster*.

Next 5 species have no or narrow premaxillary frenum (Fig. 57).

CHANNEL DARTER *Percina copelandi* **P. 521**
Blunt snout. Black Xs and Ws on back and upper side. Black caudal spot. To 2½ in. (6.2 cm). See also Pearl Darter, *P. aurora*, and Coal Darter, *P. brevicauda*.

RIVER DARTER *Percina shumardi* **P. 521**
Small black spot at front, large black spot near rear, of 1st dorsal fin; 8–15 black bars along side. To 3 in. (7.8 cm).

SNAIL DARTER *Percina tanasi* **P. 523**
Similar to Stargazing Darter but has gray edge and base on 1st dorsal fin. To 3½ in. (9 cm).

STARGAZING DARTER *Percina uranidea* **P. 523**
Red-brown above; 4 dark brown saddles extend down to lateral line. To 3 in. (7.8 cm).

SADDLEBACK DARTER *Percina vigil* **P. 524**
Five dark brown saddles; 1st under 1st dorsal fin. To 3 in. (7.8 cm).

Next 4 species have 3 dark brown spots on caudal fin base.

BLACKBANDED DARTER *Percina nigrofasciata* **P. 520**
Has 12–15 dark bars along side. To 4½ in. (11 cm).

DUSKY DARTER *Percina sciera* **P. 519**
Has 8–12 oval dark blotches along side. To 5 in. (13 cm). See also Guadalupe Darter, *P. apristis*.

GOLDLINE DARTER *Percina aurolineata* **P. 520**
Similar to Dusky Darter but has amber stripe on upper side. To 3½ in. (9 cm).

FRECKLED DARTER *Percina lenticula* **P. 521**
Large black spot at front of 2d dorsal fin. To 8 in. (20 cm).

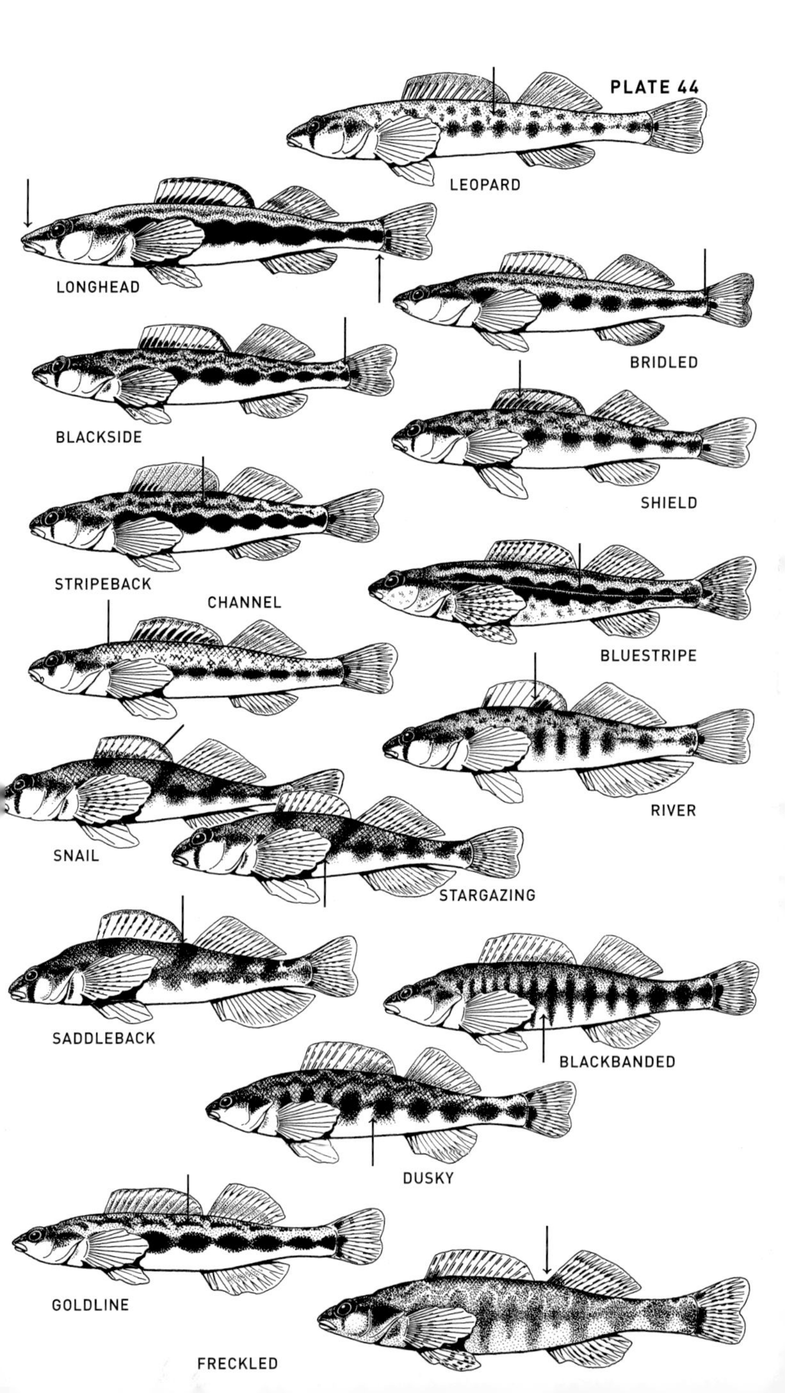
PLATE 44
LEOPARD
LONGHEAD
BRIDLED
BLACKSIDE
SHIELD
STRIPEBACK
CHANNEL
BLUESTRIPE
RIVER
SNAIL
STARGAZING
SADDLEBACK
BLACKBANDED
DUSKY
GOLDLINE
FRECKLED

Darters (2)

No scutes on breast or belly (Fig. 58).

CRYSTAL DARTER *Crystallaria asprella* **P. 533**
Very slender body. Wide, flat head. Four dark saddles on back and upper side. To 6¼ in. (16 cm). See also Diamond Darter, *C. cincotta*.

NAKED SAND DARTER *Ammocrypta beanii* **P. 534**
Middle black band, most prominent near front, on each dorsal fin (Fig. 61). To 2¾ in. (7.2 cm). See also Florida Sand Darter, *A. bifascia*.

EASTERN SAND DARTER *Ammocrypta pellucida* **P. 535**
Has 10–19 horizontal dark green blotches along side. To 3¼ in. (8.4 cm). See also Southern Sand Darter, *A. meridiana*.

WESTERN SAND DARTER *Ammocrypta clara* **P. 534**
Spine on opercle. To 2¾ in. (7.1 cm).

SCALY SAND DARTER *Ammocrypta vivax* **P. 536**
Vertical dark green blotches along side. Black edge and middle black band on each dorsal fin. To 2¾ in. (7.3 cm).

GLASSY DARTER *Etheostoma vitreum* **P. 537**
Translucent body. Many brown specks and black spots on back and side. Dark dashes along lateral line. To 2½ in. (6.6 cm).

JOHNNY DARTER *Etheostoma nigrum* **P. 537**
Dark brown Xs and Ws on side. Blunt snout. No premaxillary frenum. Black preorbital bar extends onto upper lip (Fig. 62). Breeding male has black head and lower fins, black spot on 1st dorsal fin. To 2¾ in. (7.2 cm).

TESSELLATED DARTER *Etheostoma olmstedi* **P. 539**
Similar to Johnny Darter but has uninterrupted infraorbital and supratemporal canals (Fig. 55), enlarged 2d dorsal fin on breeding male (shown). To 4½ in. (11 cm).

WACCAMAW DARTER *Etheostoma perlongum* **P. 540**
Nearly identical to Tessellated Darter but is more slender. To 3½ in. (9 cm).

BLUNTNOSE DARTER *Etheostoma chlorosoma* **P. 540**
Black bridle around extremely blunt snout (Fig. 62). Horizontal dark blotches, Xs, and Ws along side. To 2¼ in. (6 cm). See also Choctawhatchee Darter, *E. davisoni*.

OKALOOSA DARTER *Etheostoma okaloosae* **P. 570**
Has 5–8 rows of small dark brown spots on side. Black spots on lower half of head and breast. To 2 in. (5.3 cm).

TUSCUMBIA DARTER *Etheostoma tuscumbia* **P. 573**
Scales on top of head and often on branchiostegal membranes. Gold specks on back, head, and upper side. To 2½ in. (6.1 cm).

MARYLAND DARTER *Etheostoma sellare* **P. 543**
Asymmetrical caudal fin base (upper half extends more to rear). Wide, flat head; 4 large dark brown saddles. To 3¼ in. (8.4 cm).

GOLDSTRIPE DARTER *Etheostoma parvipinne* **P. 573**
Short blunt snout. Small upturned mouth. Yellow lateral line. Often short wide bars on side. To 2¾ in. (7 cm). See also Rush Darter, *E. phytophilum*

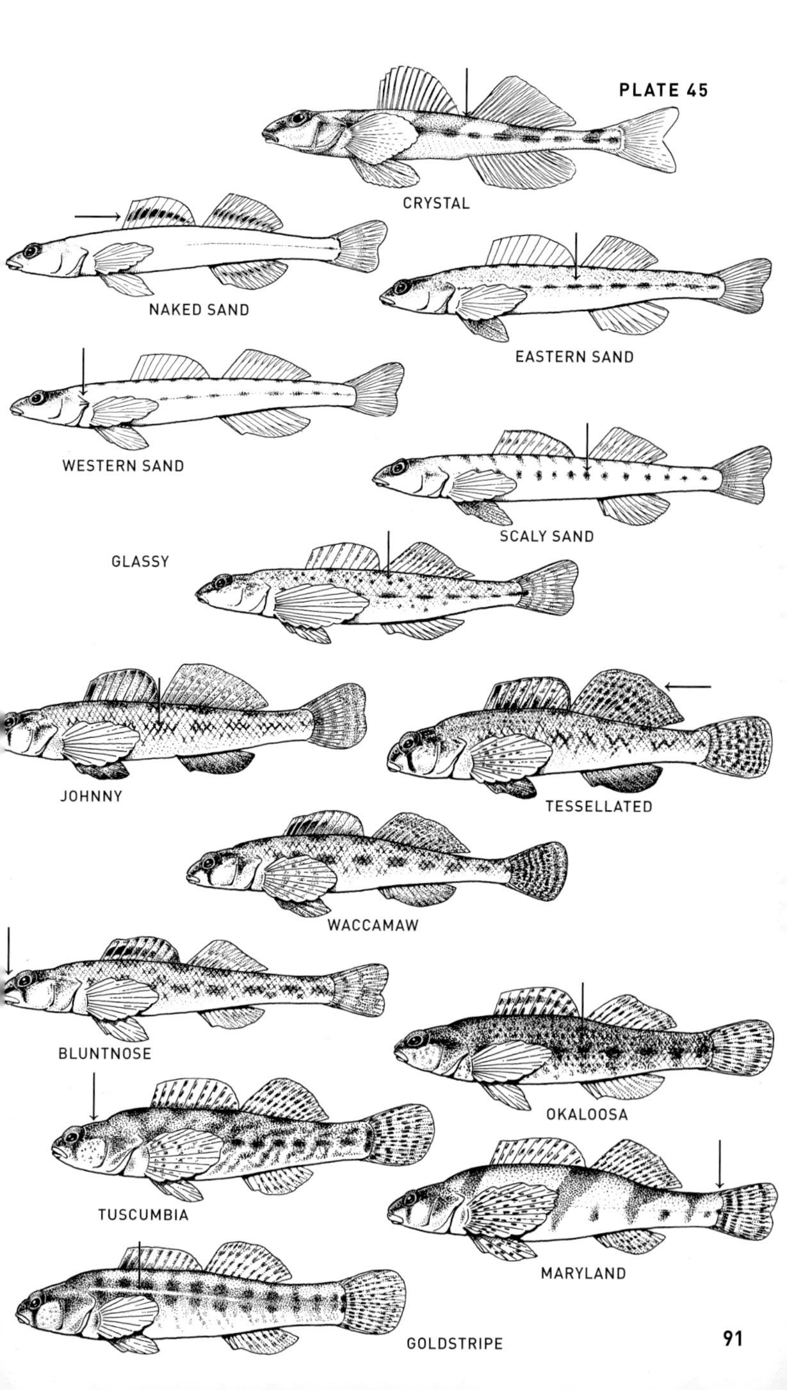
PLATE 45
CRYSTAL
NAKED SAND
EASTERN SAND
WESTERN SAND
SCALY SAND
GLASSY
JOHNNY
TESSELLATED
WACCAMAW
BLUNTNOSE
OKALOOSA
TUSCUMBIA
MARYLAND
GOLDSTRIPE

Darters (3)

Scutes on breast and along belly of male (Fig. 58).

TANGERINE DARTER *Percina aurantiaca* **P. 525**
Small dark brown spots on upper side. Black stripe along back breaks into spots at rear. Breeding male is bright red-orange below. To 7¼ in. (18 cm).

AMBER DARTER *Percina antesella* **P. 524**
Four dark brown saddles (1st in front of 1st dorsal fin). Pointed snout; premaxillary frenum absent or narrow. Breeding male (shown) has anal fin extending to caudal fin. To 2¾ in. (7.2 cm).

GILT DARTER *Percina evides* **P. 525**
Wide dusky green bars (darkest on adult). Yellow to bright orange below. Usually orange 1st dorsal fin; upper Tennessee R. subspecies has orange edge on 1st dorsal fin. To 3¾ in. (9.6 cm).

BRONZE DARTER *Percina palmaris* **P. 524**
Has 8–10 brown saddles; 8–11 brown blotches along side. On large individual wide bars join blotches to saddles. To 3½ in. (9 cm).

PIEDMONT DARTER *Percina crassa* **P. 518**
Has 7–9 oval black blotches along side. Black bar on chin. Row of black crescents (see Fig. 60) on 1st dorsal fin. To 3½ in. (9 cm).

ROANOKE DARTER *Percina roanoka* **P. 518**
Has 8–14 black bars (on adult) or oval blotches (on young) along side. Bright blue side, orange belly, orange band on 1st dorsal fin of large male (shown). To 3 in. (7.8 cm).

Next 3 species have orange band on 1st dorsal fin, black spot on caudal fin base.

LONGNOSE DARTER *Percina nasuta* **P. 527**
Long pointed snout (extreme in some populations); 12–15 dark bars along side. Unscaled or partly scaled breast. To 4½ in. (11 cm).

SLENDERHEAD DARTER *Percina phoxocephala* **P. 527**
Moderately long pointed snout. Has 10–16 round dark blotches along side. Unscaled or partly scaled breast. To 3¾ in. (9.6 cm).

OLIVE DARTER *Percina squamata* **P. 526**
Long, pointed snout. Has 10–12 dark rectangles along side. Fully scaled breast. To 5¼ in. (13 cm). See also Sharpnose Darter, *P. oxyrhynchus.*

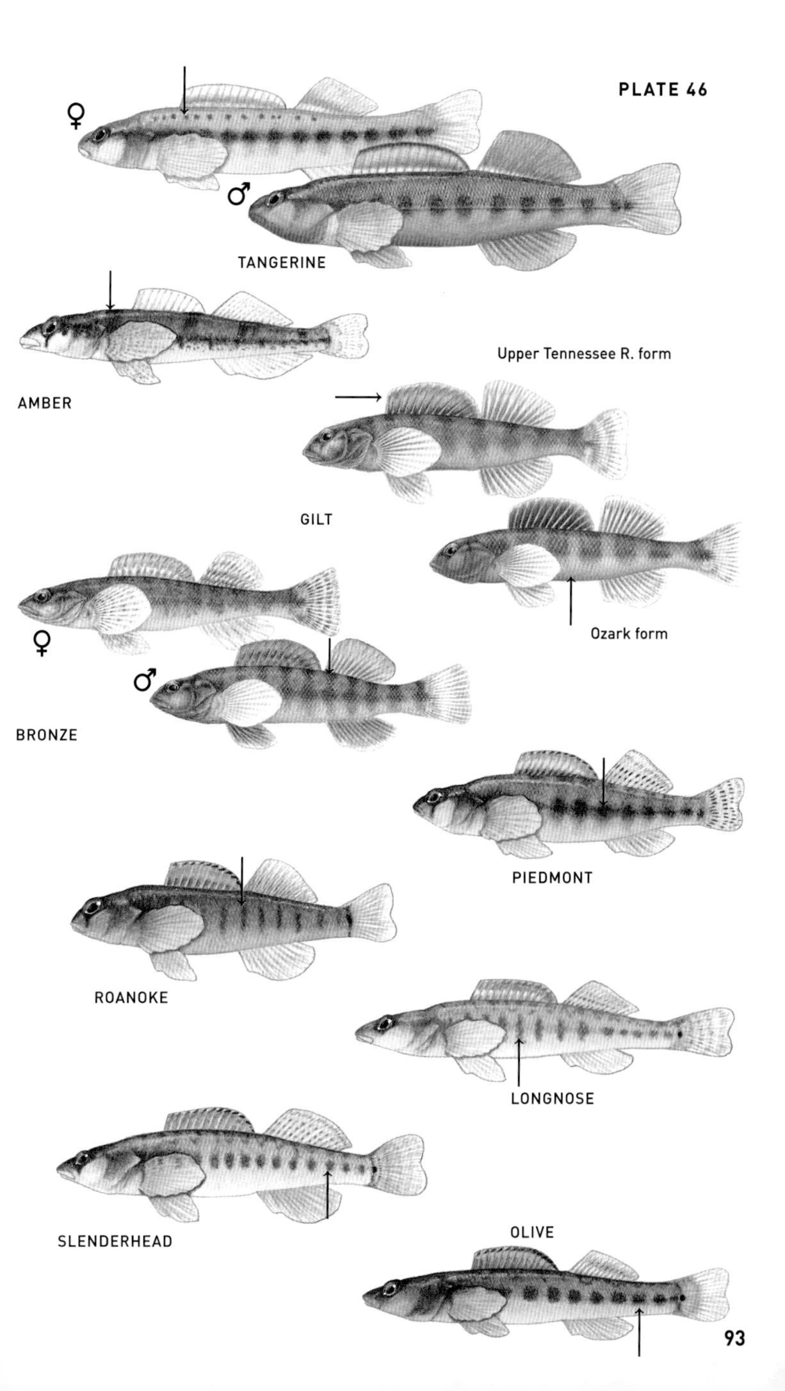
PLATE 46
♀
♂
TANGERINE
AMBER
Upper Tennessee R. form
GILT
Ozark form
♀
♂
BRONZE
PIEDMONT
ROANOKE
LONGNOSE
SLENDERHEAD
OLIVE

PLATE 47

Darters (4)

Logperches have long bulbous snout, wide flat head, scutes on breast and along belly of male (Fig. 58).

BLOTCHSIDE LOGPERCH *Percina burtoni* **P. 528**

Has 8–10 round blotches along side. Large black blotch at rear, orange edge on 1st dorsal fin. No scales on nape. To 6½ in. (16 cm).

LOGPERCH *Percina caprodes* **P. 528**

Many alternating long and short bars along side extend over back and join those of other side. *P. c. caprodes* has fully scaled nape, no middle orange band on 1st dorsal fin. *P. c. semifasciata* has unscaled nape. *P. c. fulvitaenia* has orange band on 1st dorsal fin. *P. c. manitou* has interrupted bars on rear half of side forming light stripe along upper side. To 7¼ in. (18 cm).

CHESAPEAKE LOGPERCH *Percina bimaculata* **P. 530**

Wavy bars along side of body broken into blotches. To 6 in. (15 cm).

ROANOKE LOGPERCH *Percina rex* **P. 529**

Has 10–12 short black bars along side, not joined over back with those of other side. Orange band on 1st dorsal fin. To 6 in. (15 cm).

CONASAUGA LOGPERCH *Percina jenkinsi* **P. 530**

Short bars on side broken into spots and short wavy lines. No red-orange band on 1st dorsal fin. To 5½ in. (14 cm).

MOBILE LOGPERCH *Percina kathae* **P. 531**

First dorsal fin has wide red-orange band below black edge. Long and half bars on side; long bars expanded into blotches. To 6¾ in. (17 cm).

SOUTHERN LOGPERCH *Percina austroperca* **P. 531**

First dorsal fin has narrow red-orange band below black edge. Long, half, and quarter bars; long bars expanded into blotches. To 6¾ in. (17 cm).

TEXAS LOGPERCH *Percina carbonaria* **P. 532**

Long dark bars constricted near middle, producing a series of round blotches along lower side. Large male (shown) has black breast; orange band on 1st dorsal fin. To 5¼ in. (13 cm).

GULF LOGPERCH *Percina suttkusi* **P. 532**

Slender. First dorsal fin has narrow red-orange band below black edge. Long bars along side not expanded into blotches. To 6¼ in. (16 cm).

BIGSCALE LOGPERCH *Percina macrolepida* **P. 533**

Small head. Scales on top of head and in front of pectoral fin. To 4½ in. (11 cm).

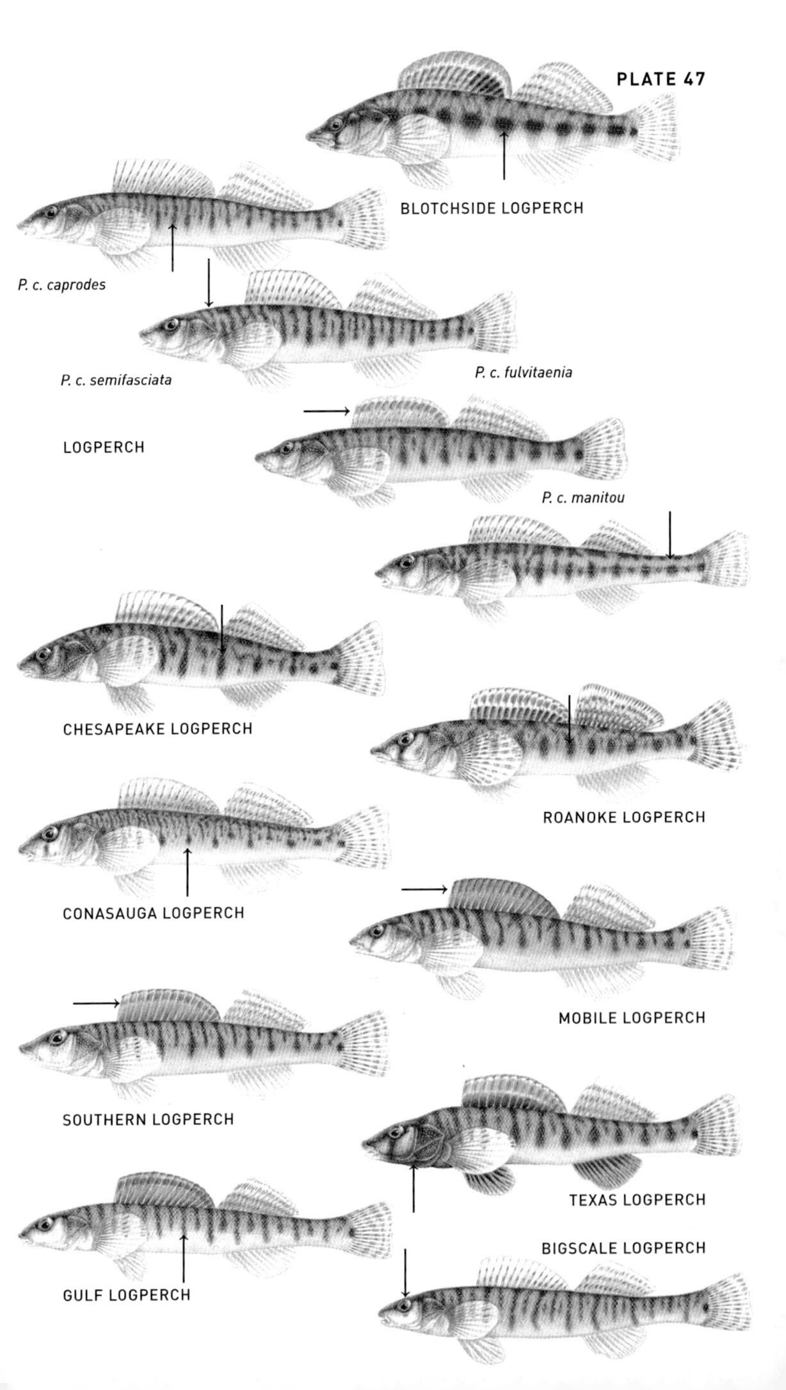
PLATE 47
BLOTCHSIDE LOGPERCH
P. c. caprodes
P. c. semifasciata
P. c. fulvitaenia
LOGPERCH
P. c. manitou
CHESAPEAKE LOGPERCH
ROANOKE LOGPERCH
CONASAUGA LOGPERCH
MOBILE LOGPERCH
SOUTHERN LOGPERCH
TEXAS LOGPERCH
BIGSCALE LOGPERCH
GULF LOGPERCH

Darters (5)

Blunt snout; no scutes on breast or belly (Fig. 58).

GREENSIDE DARTER *Etheostoma blennioides* **P. 543**
Has 5–8 green Ws, Us, or bars (on large male) on side. Brown to dark red spots on upper side. To 6¾ in. (17 cm). See also Tuckasegee Darter, *E. gutselli.*

ROCK DARTER *Etheostoma rupestre* **P. 544**
Similar to Greenside Darter but lacks scales on cheek. To 3¼ in. (8.3 cm).

HOLIDAY DARTER *Etheostoma brevirostrum* **P. 560**
First dorsal fin has red spot at front, blue edge. Breeding male (shown) has red between green bars along side. To 2½ in. (6.4 cm).

EMERALD DARTER *Etheostoma baileyi* **P. 551**
Extremely blunt snout. Red edge on 1st dorsal fin. Has 7–11 small emerald green squares along side (bars on breeding male—shown). To 2¼ in. (5.6 cm).

BANDED DARTER *Etheostoma zonale* **P. 549**
Has 9–13 large dark green bars on side extending onto belly and under caudal peduncle to join those of other side. To 3 in. (7.8 cm).

BRIGHTEYE DARTER *Etheostoma lynceum* **P. 550**
Similar to Banded Darter but interspaces as wide as bars. To 2½ in. (6.4 cm).

LONGFIN DARTER *Etheostoma longimanum* **P. 541**
Has 9–14 dark squares (or Ws) along side. Broadly joined branchiostegal membranes. To 3½ in. (8.9 cm). See also Riverweed Darter, *E. podostomone.*

HARLEQUIN DARTER *Etheostoma histrio* **P. 545**
Two large dark brown to green caudal spots. Many dark brown or black specks on yellow belly and underside of head. To 3 in. (7.7 cm).

SEAGREEN DARTER *Etheostoma thalassinum* **P. 545**
Small dark brown blotches (often Ws) along side. Red edge on 1st dorsal fin. To 3¼ in. (8 cm). See Turquoise Darter, *E. inscriptum.*

SWANNANOA DARTER *Etheostoma swannanoa* **P. 546**
Rows of small red spots on side; 8–9 black blotches on side—vertically oval at front, round at rear. To 3½ in. (9 cm).

BLENNY DARTER *Etheostoma blennius* **P. 546**
Body thick at front, strongly tapering to narrow caudal peduncle. Four large dark saddles extend down side. To 3¼ in. (8.3 cm).

CANDY DARTER *Etheostoma osburni* **P. 547**
Five large dark saddles; 9–11 green bars alternate with orange interspaces (much brighter on male) on side. Has 58–70 lateral scales. To 4 in. (10 cm).

KANAWHA DARTER *Etheostoma kanawhae* **P. 547**
Similar to Candy Darter but has 48–58 lateral scales. To 3½ in. (8.6 cm).

VARIEGATE DARTER *Etheostoma variatum* **P. 548**
Four large saddles angle down and forward to lateral line. Green and orange bars on side. Breeding male is blue and orange. To 4½ in. (11 cm).

MISSOURI SADDLED DARTER *Etheostoma tetrazonum* **P. 548**
Nearly identical to Variegate Darter but has smaller eye, darker saddles. To 3½ in. (9 cm). See also Meramec Saddled Darter, *E. erythrozonum.*

ARKANSAS SADDLED DARTER *Etheostoma euzonum* **P. 549**
Four large saddles angle down and forward to lateral line. Large head and eye. Green and orange spots on upper side. To 4¾ in. (12 cm).

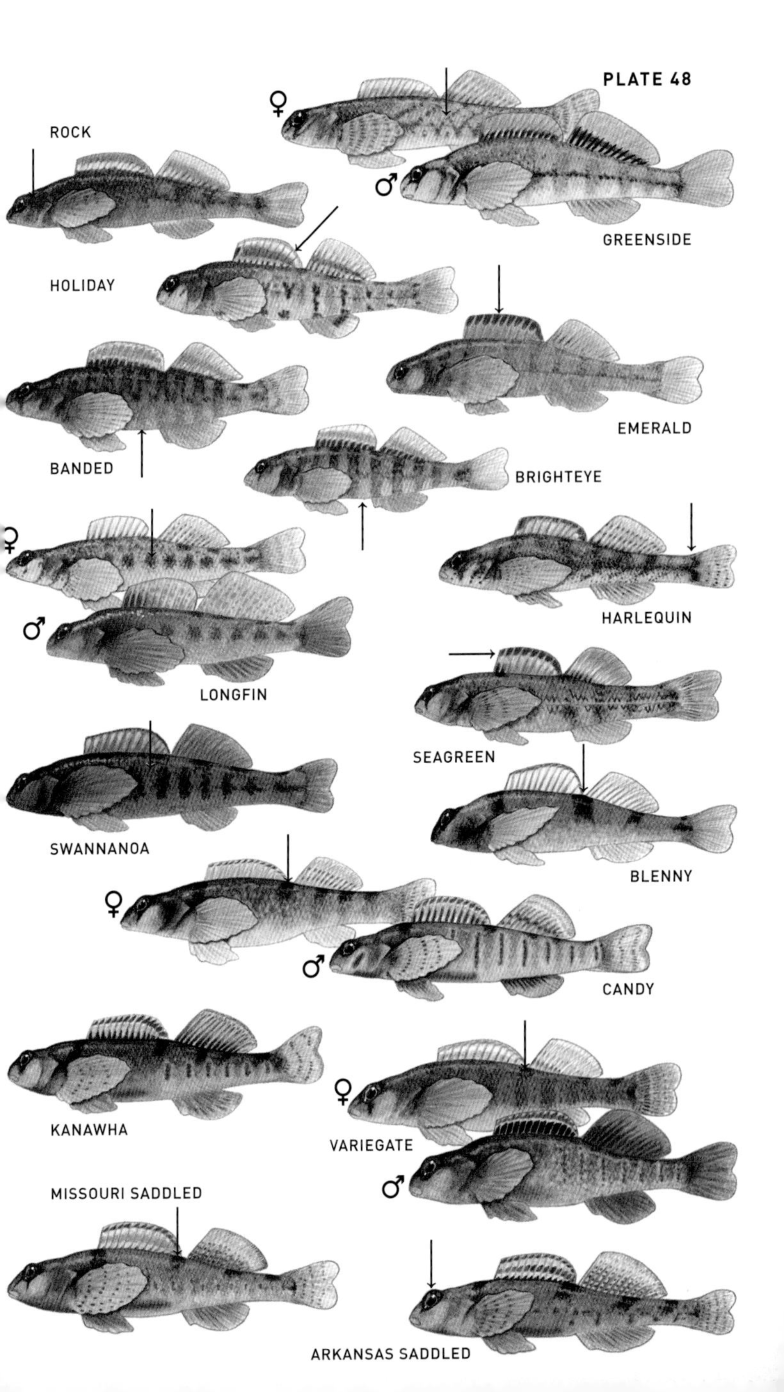
PLATE 48
ROCK
GREENSIDE
HOLIDAY
EMERALD
BANDED
BRIGHTEYE
HARLEQUIN
LONGFIN
SEAGREEN
SWANNANOA
BLENNY
CANDY
KANAWHA
VARIEGATE
MISSOURI SADDLED
ARKANSAS SADDLED

Darters (6)

Blunt snout; broadly joined branchiostegal membranes (Fig. 56); no or narrow premaxillary frenum (Fig. 57); no scutes on breast or belly (Fig. 58).

TENNESSEE SNUBNOSE DARTER *Etheostoma simoterum* **P. 552**
Many small red spots on upper side; 1st dorsal fin has red spot at front, red edge. To 2¾ in. (7.3 cm). See also Tennessee Snubnose Darter lookalikes.

BLACKSIDE SNUBNOSE DARTER *Etheostoma duryi* **P. 554**
Dark blotches along side fused into stripe. First dorsal fin has red spot at front, red edge, wavy red and black lines. To 2¾ in. (7.2 cm). See also Saffron Darter, *E. flavum*.

CHERRY DARTER *Etheostoma etnieri* **P. 555**
Dark lines along upper side. First dorsal fin has small red spot at front, thin black bands. To 3 in. (7.7 cm).

KENTUCKY SNUBNOSE DARTER *Etheostoma rafinesquei* **P. 551**
Crosshatching (created by dark scale edges) and 7–10 short dark bars on side. To 2½ in. (6.5 cm).

SPLENDID DARTER *Etheostoma barrenense* **P. 551**
Has 7–10 black blotches on side fused into stripe. Breeding male is red above and below stripe. Narrow premaxillary frenum. To 2½ in. (6.4 cm).

Next 7 species have thin brown stripe above lateral line interrupted by 7–10 black or brown blotches (front blotches often fused into stripe).

BANDFIN DARTER *Etheostoma zonistium* **P. 556**
First dorsal fin has red spot at front, 2 red bands through middle. 556 has bright red body. To 2¾ in. (7.1 cm).

YAZOO DARTER *Etheostoma raneyi* **P. 558**
Similar to Coastal Darter but has elongated brown blotches along side barely extending below lateral line except on caudal peduncle. To 2¼ in. (6 cm).

FIREBELLY DARTER *Etheostoma pyrrhogaster* **P. 555**
First dorsal fin has red spot at front, red band (wide on male). Breeding male has bright red body. To 2¾ in. (7 cm). See also Chickasaw Darter, *E. cervus*.

WARRIOR DARTER *Etheostoma bellator* **P. 559**
First dorsal fin has red spot at front, red edge. Red lower side on breeding male. To 2¾ in. (7 cm). See also Vermilion Darter, *E. chermocki*.

COOSA DARTER *Etheostoma coosae* **P. 560**
First dorsal fin has bright red spot at front, 1–2 red bands, blue-green edge (faint on female). Has 8–10 blue-brown bars along side of large male. To 2¾ in. (7.2 cm). See also Cherokee Darter, *E. scotti*.

TALLAPOOSA DARTER *Etheostoma tallapoosae* **P. 557**
Similar to Coastal Darter but has larger, blacker blotches on side; large male has low 1st dorsal fin with blue edge, red throughout. To 3 in. (7.7 cm). See also Alabama Darter, *E. ramseyi*.

COASTAL DARTER *Etheostoma colorosum* **P. 556**
First dorsal fin has blue edge, no red spot at front. Breeding male has high1st dorsal fin with blue edge, red band; orange side. To 2¾ in. (7 cm). See also Tombigbee Darter, *E. lachneri*.

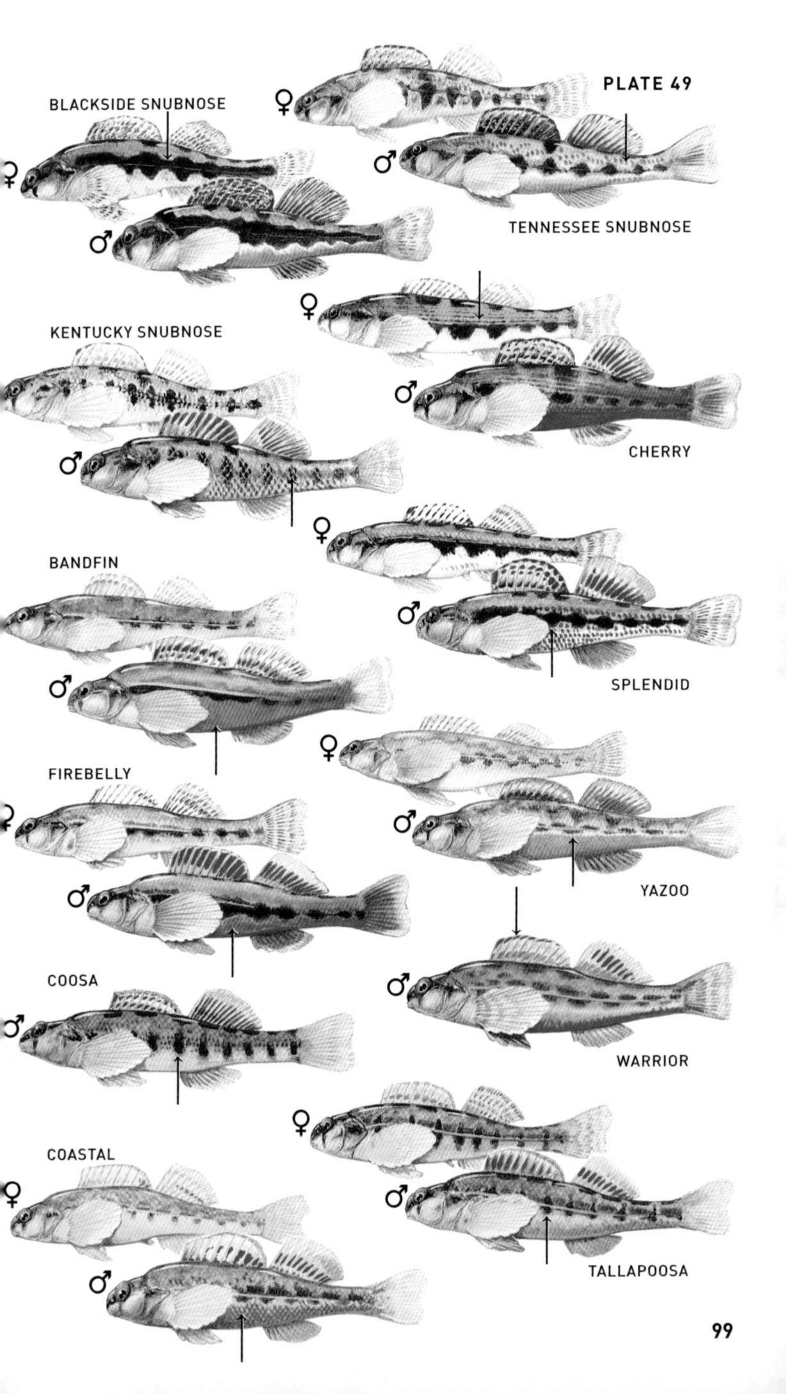
PLATE 49
BLACKSIDE SNUBNOSE
TENNESSEE SNUBNOSE
KENTUCKY SNUBNOSE
CHERRY
BANDFIN
SPLENDID
FIREBELLY
YAZOO
COOSA
WARRIOR
COASTAL
TALLAPOOSA

PLATE 50

Darters (7)

Deep body; no scutes on breast or belly (Fig. 58). All live in fast, rocky riffles.

GREENBREAST DARTER *Etheostoma jordani* **P. 561**
No teardrop. Small dark blotches just below lateral line. Male is blue, has red spots on side. To 3 in. (7.9 cm). See also Tuskaloosa Darter, *E. douglasi*, Etowah Darter, *E. etowahae*, and Lipstick Darter, *E. chuckwachatte*.

Next 7 species have many black spots on fins of females.

REDLINE DARTER *Etheostoma rufilineatum* **P. 562**
Black dashes on cheek and opercle. Red spots on side of male. Cream-colored caudal fin base. To 3½ in. (8.4 cm).

YELLOWCHEEK DARTER *Etheostoma moorei* **P. 563**
Green base, middle red band on 1st dorsal fin of male. Black teardrop on dusky cheek. To 2¾ in. (7.2 cm).

BAYOU DARTER *Etheostoma rubrum* **P. 563**
Cream-colored caudal fin base followed by 2 large black spots. Red spots on side of male. Large black teardrop on white cheek. To 2¼ in. (5.5 cm).

Next 4 species have black halos around red spots on side.

SMALLSCALE DARTER *Etheostoma microlepidum* **P. 564**
Black edge on 2d dorsal, caudal, and anal fins. Large male has bright green and orange median fins. To 2¾ in. (7.2 cm).

SPOTTED DARTER *Etheostoma maculatum* **P. 564**
Pointed snout. No red on fins. To 3½ in. (9 cm). See also Bloodfin Darter, *E. sanguifluum*.

WOUNDED DARTER *Etheostoma vulneratum* **P. 566**
Black edge on median fins. Two red spots at front, 1 at rear, of 1st dorsal fin. To 3¼ in. (8 cm). See also Boulder Darter, *E. wapiti*.

COPPERCHEEK DARTER *Etheostoma aquali* **P. 565**
Wavy copper lines on cheek and opercle. No teardrop. Two red spots at front, 1 at rear, of 1st dorsal fin. Red fins. To 3¼ in. (8 cm).

SHARPHEAD DARTER *Etheostoma acuticeps* **P. 568**
Compressed body. Extremely pointed snout. Dusky bars on side. Female is yellow-brown. Male is olive to blue, has turquoise fins. To 3¼ in. (8.4 cm).

TIPPECANOE DARTER *Etheostoma tippecanoe* **P. 568**
Small—to 1¾ in. (4.3 cm). Blue-black bars on side, darkest at rear; last bar large, encircling caudal peduncle, followed by 2 white (female) or orange (male) caudal spots. See also Golden Darter, *E. denoncourti*.

BLUEBREAST DARTER *Etheostoma camurum* **P. 566**
Male has bright red spots (not surrounded by black halos) on side; red fins, blue breast. Female has brown spots on side. To 3¼ in. (8.4 cm).

ORANGEFIN DARTER *Etheostoma bellum* **P. 567**
Large black teardrop (obscure on large male). Male has blue breast, red spots on side, orange fins. To 3½ in. (9 cm).

GREENFIN DARTER *Etheostoma chlorobranchium* **P. 567**
Large male is deep green with green fins (except sometimes pink pectoral fins). Female similar to Bluebreast Darter. To 4 in. (10 cm).

YOKE DARTER *Etheostoma juliae* **P. 568**
Large black saddle ("yoke") on nape. Orange fins. To 3 in. (7.8 cm).

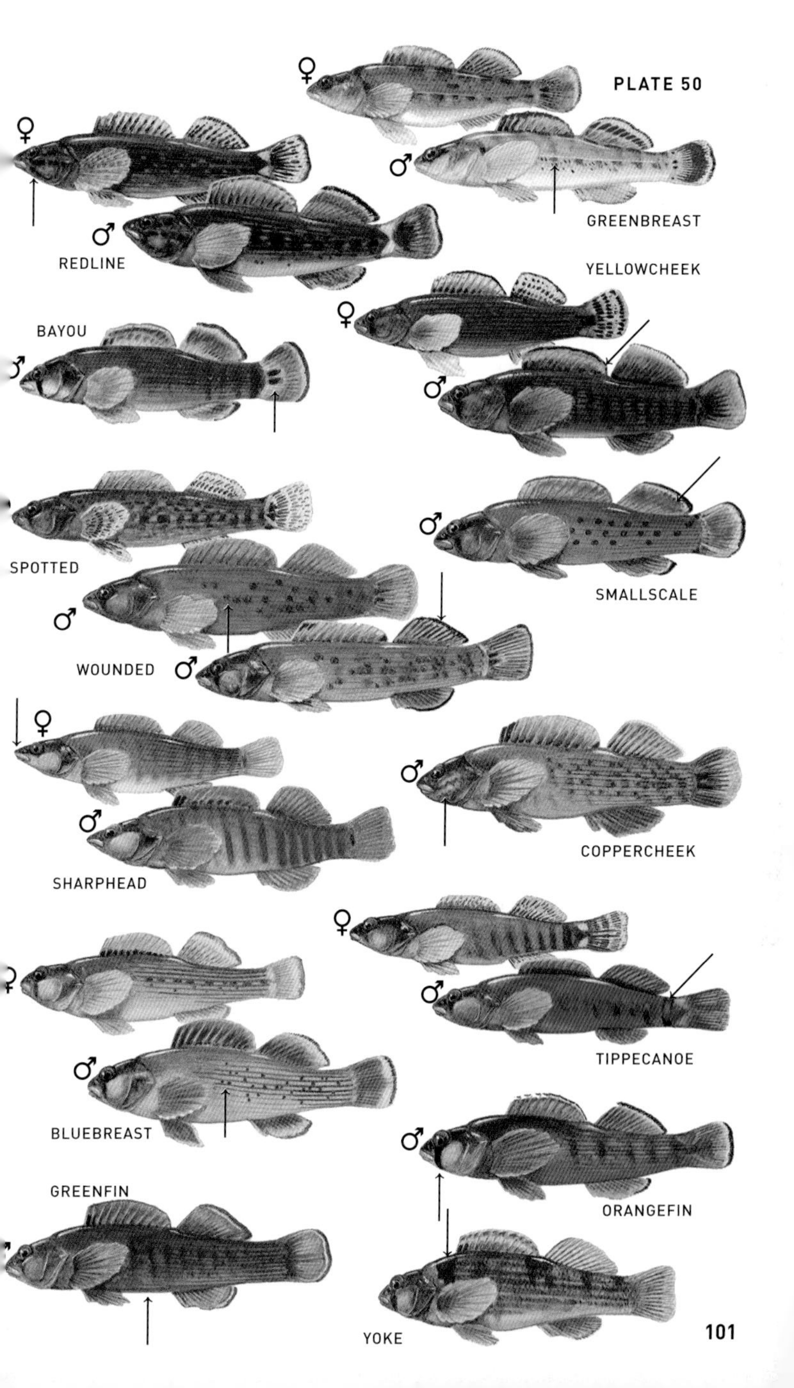
PLATE 50
GREENBREAST
REDLINE
YELLOWCHEEK
BAYOU
SPOTTED
SMALLSCALE
WOUNDED
COPPERCHEEK
SHARPHEAD
TIPPECANOE
BLUEBREAST
GREENFIN
ORANGEFIN
YOKE

Darters (8)

No scutes on breast or belly (Fig. 58).

ARROW DARTER *Etheostoma sagitta* **P. 569**
Two black caudal spots fused into short bar. Long pointed snout; 7–9 green Us alternate with orange bars (brightest on breeding male—shown) along side. Has 55–69 lateral scales. To 4¾ in. (12 cm). See also Cumberland Plateau Darter, *E. spilotum*.

NIANGUA DARTER *Etheostoma nianguae* **P. 569**
Similar to Arrow Darter but has 2 (unfused) jet-black caudal spots, 72–82 lateral scales. To 5¼ in. (13 cm).

PINEWOODS DARTER *Etheostoma mariae* **P. 570**
First dorsal fin with bright red edge, black spot at front. Black spots on lower half of head and breast. To 3 in. (7.6 cm).

SAVANNAH DARTER *Etheostoma fricksium* **P. 570**
Similar to Pinewoods Darter but is green below, with bright orange bars on belly of male (shown). To 3 in. (7.4 cm).

ORANGEBELLY DARTER *Etheostoma radiosum* **P. 575**
Short dark bars (rear ones long) along side, cut into upper and lower halves by yellow lateral line. Orange branchiostegal membranes, blue pelvic fins on large male (shown). To 3½ in. (8.6 cm).

REDFIN DARTER *Etheostoma whipplei* **P. 574**
Many small bright red (on male) or yellow (female) spots on side. Slender; relatively uniform body depth from head to caudal peduncle. To 3½ in. (9 cm). See also Redspot Darter, *E. artesiae*.

RIO GRANDE DARTER *Etheostoma grahami* **P. 584**
Deep bodied. Many small red (on male) or black (female) spots on side. Red 1st dorsal fin (faint on female). To 2¼ in. (6 cm).

Next 5 species have bright orange on head and body of breeding male.

STIPPLED DARTER *Etheostoma punctulatum* **P. 571**
Dark brown mottling, specks on head and body. White to orange below. Large black teardrop. Pointed snout. To 4 in. (10 cm). See also Sunburst Darter, *E.* species.

SLACKWATER DARTER *Etheostoma boschungi* **P. 572**
Many black specks on back and side. Large black teardrop. Blunt snout. To 3 in. (7.8 cm).

ARKANSAS DARTER *Etheostoma cragini* **P. 572**
Strongly bicolored body: upper half dark brown, lower half white to orange. Black specks on body and fins. To 2¼ in. (6 cm).

PALEBACK DARTER *Etheostoma pallididorsum* **P. 572**
Nearly identical to Arkansas Darter but is more slender; has wide, pale olive stripe along back. To 2¼ in. (6 cm).

TRISPOT DARTER *Etheostoma trisella* **P. 573**
Three dark brown saddles. Complete lateral line; 1 anal spine. To 2¼ in. (5.9 cm).

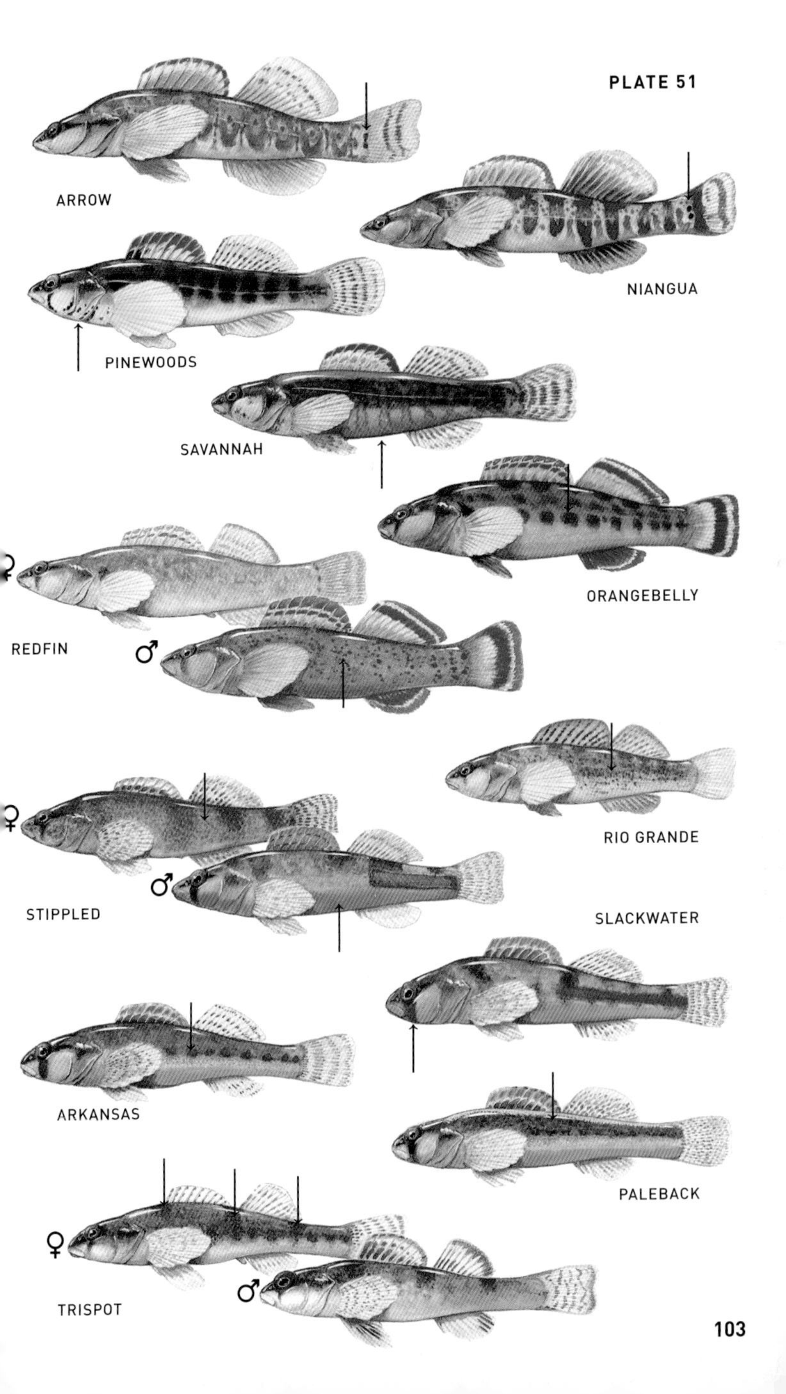
PLATE 51
ARROW
NIANGUA
PINEWOODS
SAVANNAH
ORANGEBELLY
♀
REDFIN
♂
RIO GRANDE
♀
♂
STIPPLED
SLACKWATER
ARKANSAS
PALEBACK
♀
♂
TRISPOT

DARTERS (9)

Arched body, deepest at nape or front of 1st dorsal fin. Interrupted infraorbital canal, usually 3 posterior infraorbital pores (Fig. 55). No scutes on breast or belly (Fig. 58). Breeding males have blue to blue-green anal fin, 6–9 blue bars along side, red or orange between bars.

ORANGETHROAT DARTER *Etheostoma spectabile* **P. 578**
E. s. spectabile has thin dark stripes on side, orange belly. *E. s. pulchellum* has mostly orange 1st dorsal fin, blue-gray breast. *E. s. squamosum* has mostly orange 1st dorsal fin, orange breast, bars darkest on lower side. To 2¾ in. (7.2 cm).

OZARK DARTER *Etheostoma* species **P. 578**
Orange belly; red dots (instead of dark stripes) on upper side. To 2¾ in. (7.2 cm).

IHIYO DARTER *Etheostoma* species **P. 578**
Breast light orange, with few scales; white belly. To 2¾ in. (6.8 cm).

SHELTOWEE DARTER *Etheostoma* species **P. 578**
Orange breast, belly; orange and blue rectangles on side. To 2¾ in. (6.7 cm).

MAMEQUIT DARTER *Etheostoma* species **P. 578**
Fully scaled blue-gray breast; white belly. To 2¼ in. (5.7 cm).

BUFFALO DARTER *Etheostoma bison* **P. 579**
Blue belly. Dark dashes on upper side. To 2¾ in. (6.7 cm).

HIGHLAND RIM DARTER *Etheostoma kantuckeense* **P. 579**
Blue-gray breast (no scales), belly; faint lines (series of dots) on side. To 2½ in. (6.4 cm).

BROOK DARTER *Etheostoma burri* **P. 580**
Red breast, belly; diamond-shaped blue bars on side. To 2½ in. (6.1 cm).

STRAWBERRY DARTER *Etheostoma fragi* **P. 580**
Has 10–12 turquoise bars on side, which meet those of opposite side on belly; fully scaled cheek. To 2½ in. (6.1 cm).

CURRENT DARTER *Etheostoma uniporum* **P. 580**
Has 8–10 oblique turquoise bars on side, which meet those of opposite side on belly, blue-gray belly, few scales on cheek. To 2½ in. (6.1 cm).

SHAWNEE DARTER *Etheostoma tecumsehi* **P. 581**
Similar to Orangethroat Darter, but breeding male has orange bars on front half of body. To 2½ in. (6.4 cm).

HEADWATER DARTER *Etheostoma lawrencei* **P. 581**
Similar to Shawnee Darter but usually has 13 dorsal rays (vs. 12), usually 31 or fewer (vs. 32 or more) pored lateral-line scales. To 3 in. (7.4 cm).

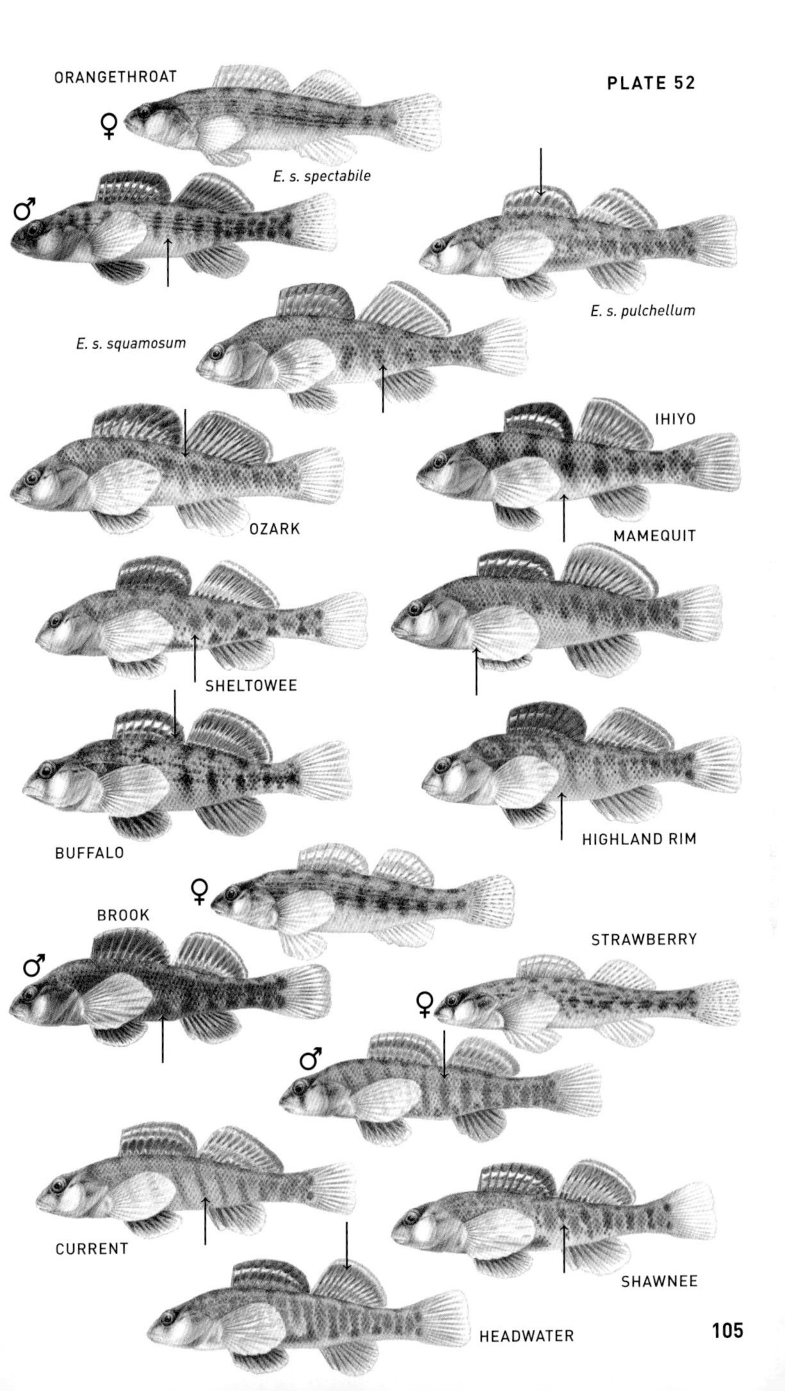
ORANGETHROAT
♀
E. s. spectabile
♂
E. s. pulchellum
E. s. squamosum
IHIYO
OZARK
MAMEQUIT
SHELTOWEE
BUFFALO
HIGHLAND RIM
♀
BROOK
STRAWBERRY
♂
♀
♂
CURRENT
SHAWNEE
HEADWATER

Darters (10)

No scutes on breast or belly (Fig. 58).

RAINBOW DARTER *Etheostoma caeruleum* **P. 576**

Deep bodied. Dark bars on side (blue between red on male; dark brown between yellow-white on female). Anal fin red with blue edge (faint on female). In some populations, males have red spots on side; in others, they do not (both forms shown). To 3 in. (7.7 cm).

MUD DARTER *Etheostoma asprigene* **P. 577**

Large black blotch at rear of 1st dorsal fin (faint on female). Dark bars on side, darkest at rear. Fully scaled cheek. To 2¾ in. (7.1 cm).

REDBAND DARTER *Etheostoma luteovinctum* **P. 582**

Body deepest at front of 1st dorsal fin, strongly tapering to narrow caudal peduncle; 7–9 dark squares just below lateral line. To 2¾ in. (6.8 cm).

GULF DARTER *Etheostoma swaini* **P. 582**

Dark bars on side (often obscured by dark mottling on female). Thin dark lines on upper side, white to orange (large male) below. Has 35–50, usually 38–45, lateral scales. To 3½ in. (9 cm).

CREOLE DARTER *Etheostoma collettei* **P. 583**

Similar to Gulf Darter but has 44–60, usually 46–55, lateral scales; large male is more blue. To 3 in. (7.4 cm).

WATERCRESS DARTER *Etheostoma nuchale* **P. 583**

Similar to Gulf Darter but is smaller, more compressed; has shorter (12–24 pored scales) lateral line. To 2¼ in. (5.4 cm).

COLDWATER DARTER *Etheostoma ditrema* **P. 584**

Dark mottling on back and side, orange belly on male. Three black caudal spots. To 2¼ in. (5.4 cm).

GREENTHROAT DARTER *Etheostoma lepidum* **P. 584**

Red-orange specks or spots between long green bars on side of male; yellow between short brown-black bars on female. To 2½ in. (6.6 cm).

CHRISTMAS DARTER *Etheostoma hopkinsi* **P. 585**

Has 10–12 dark green bars on side, separated by brick red on male, yellow on female. To 2½ in. (6.6 cm).

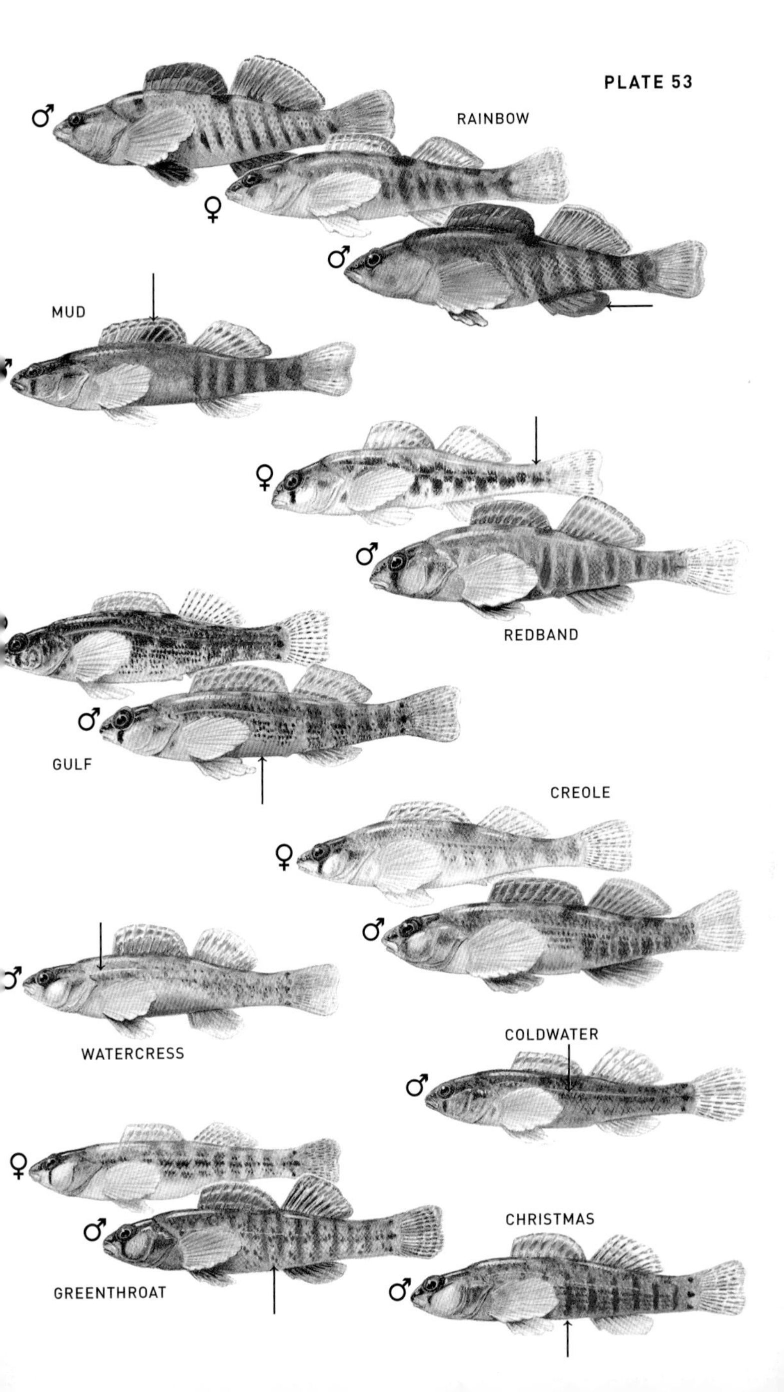
PLATE 53
RAINBOW
MUD
REDBAND
GULF
CREOLE
WATERCRESS
COLDWATER
CHRISTMAS
GREENTHROAT

Darters (11)

No scutes on breast or belly (Fig. 58). Male shown unless female indicated.
First 5 species have iridescent bar on cheek. Breeding male has red dorsal, caudal, and anal fins.

STRIPED DARTER *Etheostoma virgatum* **P. 596**
Dark brown stripes on side. To 3 in. (7.8 cm).

BARCHEEK DARTER *Etheostoma obeyense* **P. 595**
Dark brown blotches, no stripes or rows of dark spots, on side. To 3¼ in. (8.4 cm).

TEARDROP DARTER *Etheostoma barbouri* **P. 597**
Large black teardrop. Rows of small dark brown spots and blotches on side. To 2¼ in. (6 cm).

SLABROCK DARTER *Etheostoma smithi* **P. 595**
Similar to Barcheek Darter but is smaller; has more darkly outlined scales. To 2½ in. (6.2 cm).

STRIATED DARTER *Etheostoma striatulum* **P. 597**
Similar to Teardrop Darter but has narrow bar on cheek, darker rows of spots on side. To 2¼ in. (5.6 cm).

Next 5 species have gold knobs on tips of dorsal spines—large only on male.

STRIPETAIL DARTER *Etheostoma kennicotti* **P. 591**
Black bands on 2d dorsal, caudal fins. Dark brown blotches on upper side, larger dark blotches along side. To 3¼ in. (8.3 cm).

CAROLINA FANTAIL DARTER *Etheostoma brevispinum* **P. 593**
Dark bars on side; wedge-shaped on large male. To 3 in. (8.4 cm).

FANTAIL DARTER *Etheostoma flabellare* **P. 592**
Protruding lower jaw. Broadly joined branchiostegal membranes. To 3¼ in. (8.4 cm).

DUSKYTAIL DARTER *Etheostoma percnurum* **P. 593**
Black specks (largest on juvenile) on side of head. Black edge on pectoral, anal, 2nd dorsal, and caudal fins of breeding male (shown). To 2½ in. (6.4 cm). See also Marbled Darter, *E. marmorpinnum*, and Citico Darter, *E. sitikuense*.

TUXEDO DARTER *Etheostoma lemniscatum* **P. 594**
Slender; anal fin origin behind dorsal fin origin, blacker edge on 2nd dorsal and caudal fins of breeding male (shown). To 2½ in. (6.5 cm).

Next 3 species (and 7 similar species; see Fig. 63) have black bands on 2nd dorsal and caudal fins, 3 black spots on caudal fin base.

DIRTY DARTER *Etheostoma olivaceum* **P. 585**
Thin stripes, black mottling, sometimes black bars on side. Long, sharp snout. Large male is black. To 3¼ in. (8 cm).

LOLLYPOP DARTER *Etheostoma neopterum* **P. 589**
Breeding male has large yellow knobs on tips of 2nd dorsal fin rays (Fig. 63). To 3 in. (7.6 cm). See also Guardian Darter, *E. oophylax*, Egg-mimic Darter, *E. pseudovulatum*, and Relict Darter, *E. chienense*.

FRINGED DARTER *Etheostoma crossopterum* **P. 588**
Breeding male has white edge on 2d dorsal fin (Fig. 63). To 4 in. (10 cm). See also Spottail Darter, *E. squamiceps*, Blackfin Darter, *E. nigripinne*, Barrens Darter, *E. forbesi*, and Crown Darter, *E. corona*.

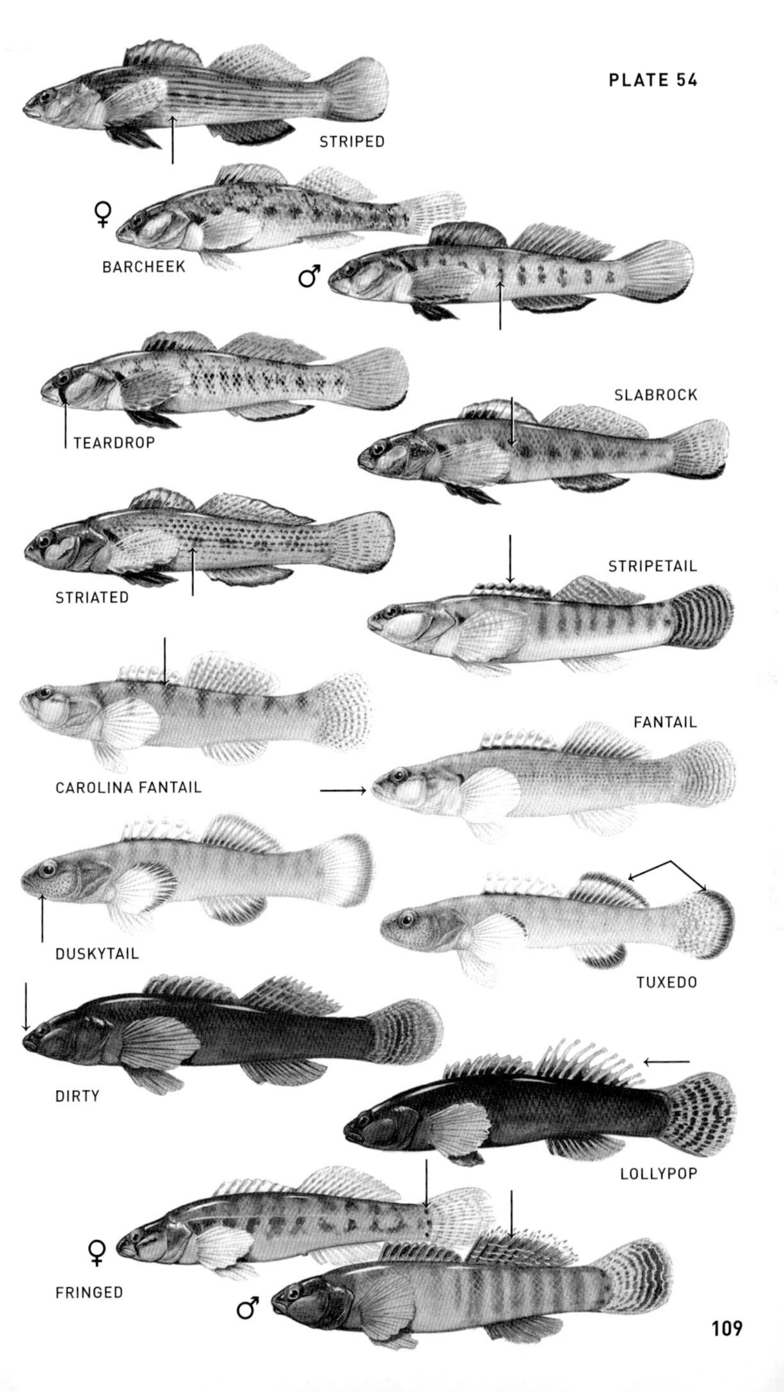
PLATE 54
STRIPED
♀
BARCHEEK
♂
TEARDROP
SLABROCK
STRIATED
STRIPETAIL
CAROLINA FANTAIL
FANTAIL
DUSKYTAIL
TUXEDO
DIRTY
LOLLYPOP
♀
FRINGED
♂

Darters (12)

No scutes on breast or belly (Fig. 58).

ASHY DARTER *Etheostoma cinereum* **P. 561**
Longitudinal rows of small brown spots on upper side. Row of small black rectangles along side. Breeding male has huge 2d dorsal fin. To 4¾ in. (12 cm).

SPECKLED DARTER *Etheostoma stigmaeum* **P. 541**
Has 7–11 dark brown squares (Ws if examined closely) along side just below lateral line on female and juvenile; 7–11 turquoise bars on large male. To 2½ in. (6.1 cm).

BLUESIDE DARTER *Etheostoma jessiae* **P. 542**
Similar to Speckled Darter but has pointed snout, narrow premaxillary frenum, deep blue bars on body and fins of male. To 2¾ in. (6.8 cm).

Next 3 species are small (to 2 in. [5 cm]): have very short lateral line (fewer than 10 pores), large flap on pelvic fin of male.

LEAST DARTER *Etheostoma microperca* **P. 597**
Deep, compressed body. Extremely short (0–3 pores) lateral line. Large teardrop. Orange or red anal and pelvic fins on large male. To 1¾ in. (4.4 cm).

CYPRESS DARTER *Etheostoma proeliare* **P. 598**
Short (0–9 pores) lateral line. Black or brown dashes along side, spots on upper and lower side. To 2 in. (4.8 cm).

FOUNTAIN DARTER *Etheostoma fonticola* **P. 598**
Short (0–6 pores) lateral line; 1 anal spine. Crosshatching on upper and lower sides. To 1¾ in. (4.3 cm).

Next 7 species have incomplete lateral line strongly arched near front (often only slightly arched in Iowa Darter).

SWAMP DARTER *Etheostoma fusiforme* **P. 599**
Slender, compressed body. Many dark specks below. To 2¼ in. (5.9 cm).

SAWCHEEK DARTER *Etheostoma serrifer* **P. 599**
Red around 2 black caudal spots. Yellow lateral line, usually 28–38 pores. To 2¾ in. (6.8 cm).

BACKWATER DARTER *Etheostoma zonifer* **P. 600**
Nearly identical to Slough Darter but has interrupted infraorbital canal, usually with 6 pores (Fig. 55). To 1¾ in. (4.4 cm).

SLOUGH DARTER *Etheostoma gracile* **P. 600**
Bright green bars on side of male, green squares or mottling on female. Middle red band on 1st dorsal fin (faint on female). To 2¼ in. (6 cm).

IOWA DARTER *Etheostoma exile* **P. 601**
Slender body; long, narrow caudal peduncle. Incomplete lateral line, often arched near front. Alternating blue and brick red bars on side of large male. To 2¾ in. (7.2 cm).

CAROLINA DARTER *Etheostoma collis* **P. 601**
Many small dark brown spots on side; brown blotches along side. To 2¼ in. (6 cm).

BROWN DARTER *Etheostoma edwini* **P. 602**
Incomplete, yellow lateral line, sometimes arched at front. Bright red spots on body and fins of male (sometimes of female). To 2 in. (5.3 cm).

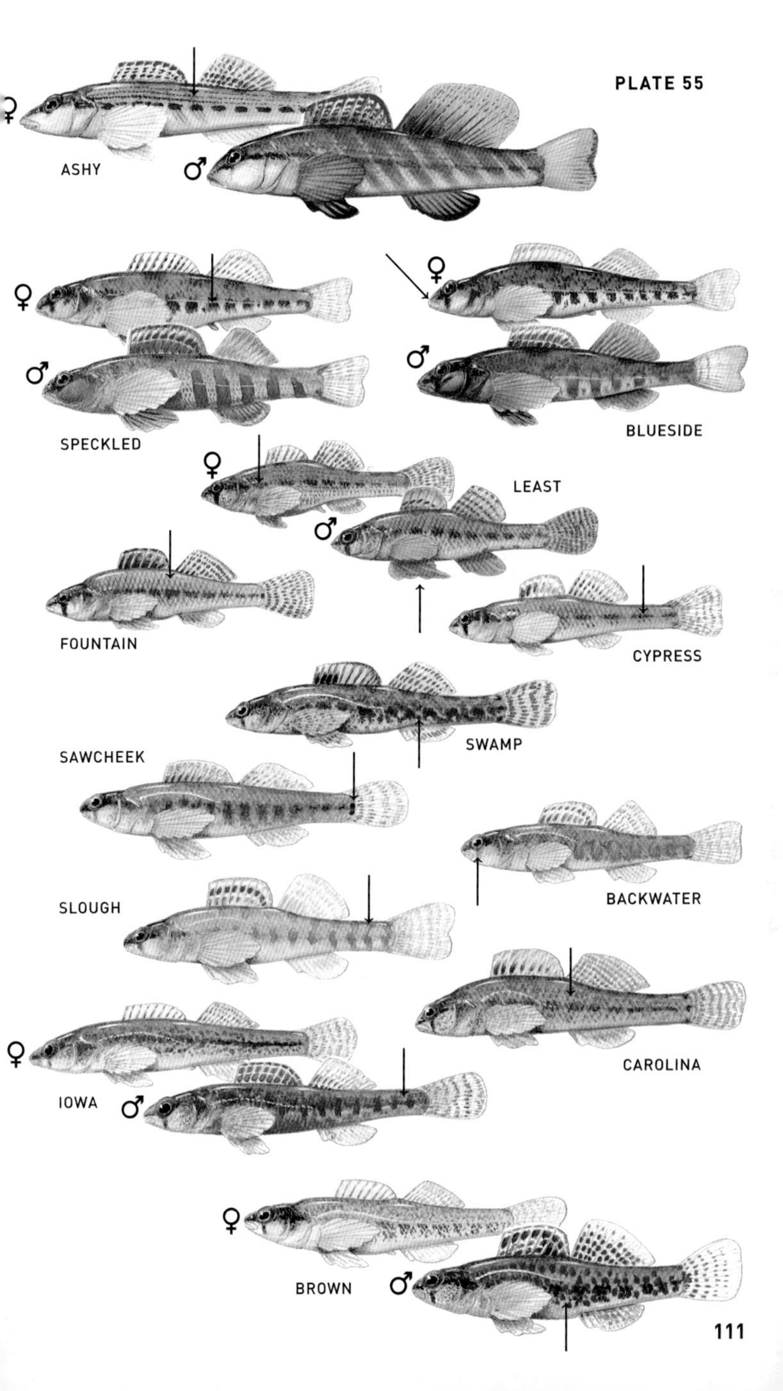
PLATE 55
♀
ASHY
♂
♀
♀
SPECKLED
♂
♂
BLUESIDE
♀
LEAST
♂
FOUNTAIN
CYPRESS
SWAMP
SAWCHEEK
BACKWATER
SLOUGH
CAROLINA
♀
IOWA
♂
♀
BROWN
♂

Cichlids

One nostril on each side; 2-part lateral line, front part higher on body than rear part.

OSCAR *Astronotus ocellatus* **P. 607**
Large rounded 2d dorsal, caudal, and anal fins. Large black spot on upper caudal fin base. To 16 in. (40 cm).

PEACOCK CICHLID *Cichla ocellaris* **P. 608**
Elongate body. Large mouth; projecting lower jaw. Silver halo around large black spot on caudal fin. To 26 in. (66 cm).

REDSTRIPED EARTHEATER *Geophagus surinamensis* **P. 608**
Long snout; eye high on head. Black blotch on side. Three anal spines. To 12 in. (30 cm).

AFRICAN JEWELFISH *Hemichromis letourneuxi* **P. 608**
Rounded caudal fin. Large black blotch on side. To 8 in. (20 cm).

Next 7 species (through Banded Cichlid) have more than 3 anal spines, usually a black blotch on caudal fin base.

RIO GRANDE CICHLID *Herichthys cyanoguttatus* **P. 608**
Has 4–6 dark blotches along rear half of side. Many small white to blue spots on side. To 12 in. (30 cm).

JACK DEMPSEY *Rocio octofasciata* **P. 609**
Two gray to black lines between eyes. Adult is deep blue, has many iridescent spots on head and body. To 10 in. (25 cm).

CONVICT CICHLID *Amatitlania nigrofasciatus* **P. 610**
Usually 7 bars on side extend onto dorsal and anal fins; 1st bar Y-shaped. To 4¾ in. (12 cm).

MIDAS CICHLID *Amphilophus citrinellus* **P. 610**
Color highly variable. Usually 6 dark bars, black blotch on side, and smaller black spot on caudal fin base. Nuchal hump on breeding male (shown). To 12 in. (31 cm). See also Mayan Cichlid, *Cichlasoma urophthalmus*.

BLACK ACARA *Cichlasoma bimaculatum* **P. 611**
Dark blotches along side extend as dark stripe onto opercle. Adult is dark green and yellow, has blue-gray fins. To 8 in. (20 cm). See also Yellowbelly Cichlid, *C. salvini*.

FIREMOUTH *Thorichthys meeki* **P. 612**
Large black blotch on lower half of gill cover. To 6¾ in. (17 cm).

BANDED CICHLID *Heros severus* **P. 611**
Deep, compressed body. Usually 5–7 dusky to black bars on side. Bar on caudal peduncle extends onto dorsal and anal fins. To 8 in. (20 cm).

Next 3 species have black spot on opercle, 3 anal spines.

MOZAMBIQUE TILAPIA *Oreochromis mossambicus* **P. 612**
Mouth reaches under front of eye or beyond. To 15 in. (39 cm). See also Wami Tilapia, *O. urolepis*, Blue Tilapia, *O. aureus*, and Nile Tilapia, *O. niloticus*.

BLACKCHIN TILAPIA *Sarotherodon melanotheron* **P. 613**
Small mouth. Large male has black underside of head. To 10 in. (26 cm).

SPOTTED TILAPIA *Tilapia mariae* **P. 614**
Has 6–9 black blotches or bars on side that continue onto dorsal fin. To 13 in. (33 cm). See also Redbelly Tilapia, *T. zillii*.

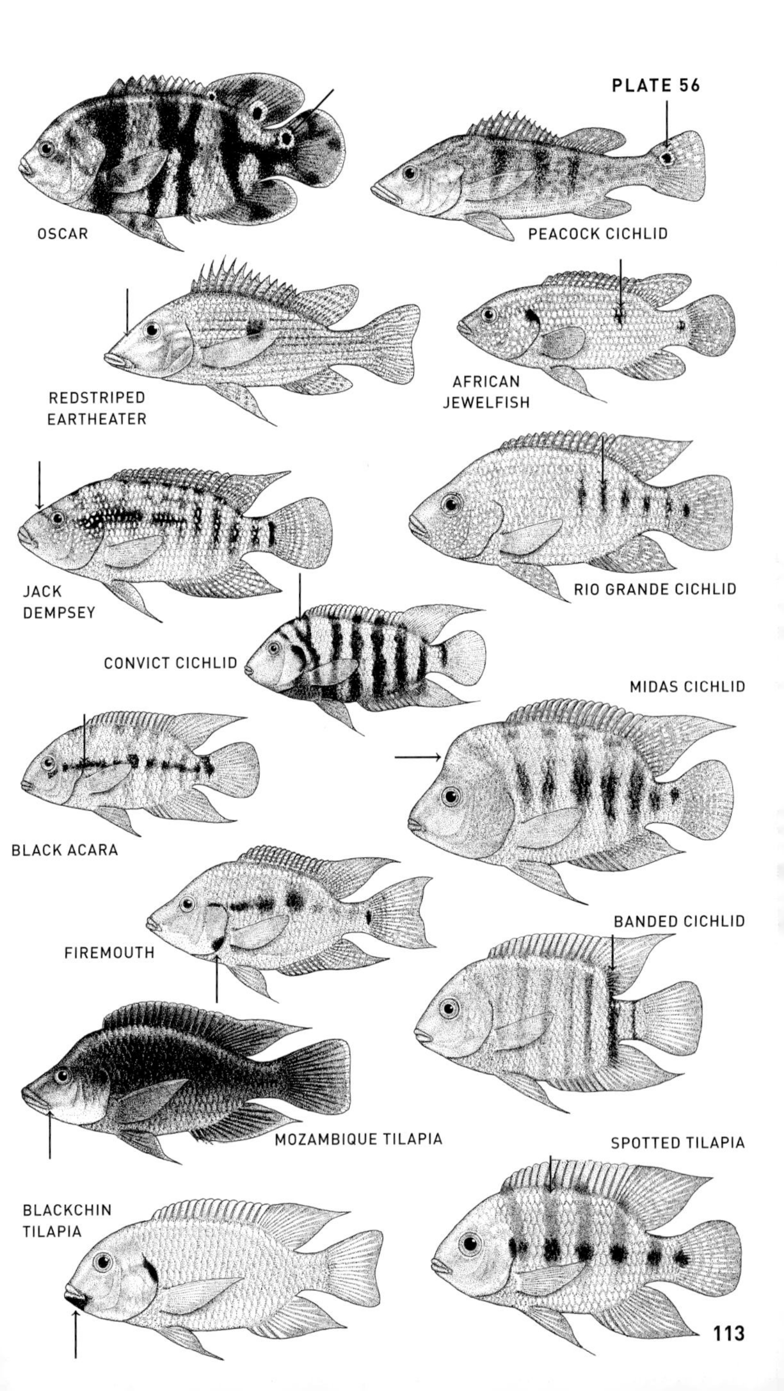

OSCAR
PEACOCK CICHLID
REDSTRIPED EARTHEATER
AFRICAN JEWELFISH
JACK DEMPSEY
RIO GRANDE CICHLID
CONVICT CICHLID
MIDAS CICHLID
BLACK ACARA
FIREMOUTH
BANDED CICHLID
MOZAMBIQUE TILAPIA
SPOTTED TILAPIA
BLACKCHIN TILAPIA

PLATE 57

Marine Invaders

These are among the most likely marine and brackish-water fishes to be found in fresh water in our area.

BAY ANCHOVY *Anchoa mitchilli* **P. 146**

Upper jaw extends well past eye. No fin spines or lateral line. Dorsal fin origin above or slightly in front of anal fin origin. To 4 in. (10 cm).

STRIPED MULLET *Mugil cephalus* **P. 607**

Two widely spaced dorsal fins; 2d fin with 1 spine and 8 rays. Small spots on scales form dusky lateral stripes along blue-green to silver body. To 36 in. (91 cm); rarely more than 20 in. (50 cm) in fresh water.

GULF KILLIFISH *Fundulus grandis* **P. 438**

Short snout, robust body, deep caudal peduncle. Dorsal fin origin in front of anal fin origin. Silver gray to blue-green; many light blue to yellow spots on side and fins, brighter on male. To 7 in. (18 cm). See also Mummichog, *F. heteroclitus.*

ATLANTIC NEEDLEFISH *Strongylura marina* **P. 461**

Extremely slender body; long jaws with needlelike teeth. Dorsal and anal fins far back on body. Tiny scales. To 24 in. (61 cm).

GULF PIPEFISH *Syngnathus scovelli* **P. 465**

Long, slender body encased in bony rings, long tubular snout, 1 dorsal fin, tiny anal fin, no pelvic fins. Ridge along side not continuous (at anal fin) with ridge along underside, fewer than 16 pectoral rays. To 7 in. (18 cm). See also Opossum Pipefish, *Microphis brachyurus.*

DARTER GOBY *Ctenogobius boleosoma* **P. 617**

Pelvic fins fused; no lateral line. Long, slender body; long, pointed caudal fin. No or few scales on nape. Has 4 or 5 brown spots or bars on tan side, large black spot on caudal fin base. Has 11 dorsal rays, 12 anal rays, 16 pectoral rays. To 3 in. (7.5 cm).

FRESHWATER GOBY *Ctenogobius shufeldti* **P. 617**

Similar to Darter Goby but has scales on nape, 12 dorsal rays, 13 anal rays, 17 pectoral fin rays. To 3¼ in. (8 cm).

SOUTHERN FLOUNDER *Paralichthys lethostigma* **P. 620**

Flat body with long dorsal and anal fins covering nearly all of upper and lower edges, origin of dorsal fin over or slightly in front of eyes; eyes on left side of head. Lateral line strongly arched over pectoral fin. To 30 in. (76 cm). See also Starry Flounder, *Platichthys stellatus.*

HOGCHOKER *Trinectes maculatus* **P. 620**

Flat body with long dorsal and anal fins covering nearly all of upper and lower edges, origin of dorsal fin near mouth; eyes on right side of head. Lateral line more or less straight. To 8 in. (20 cm).

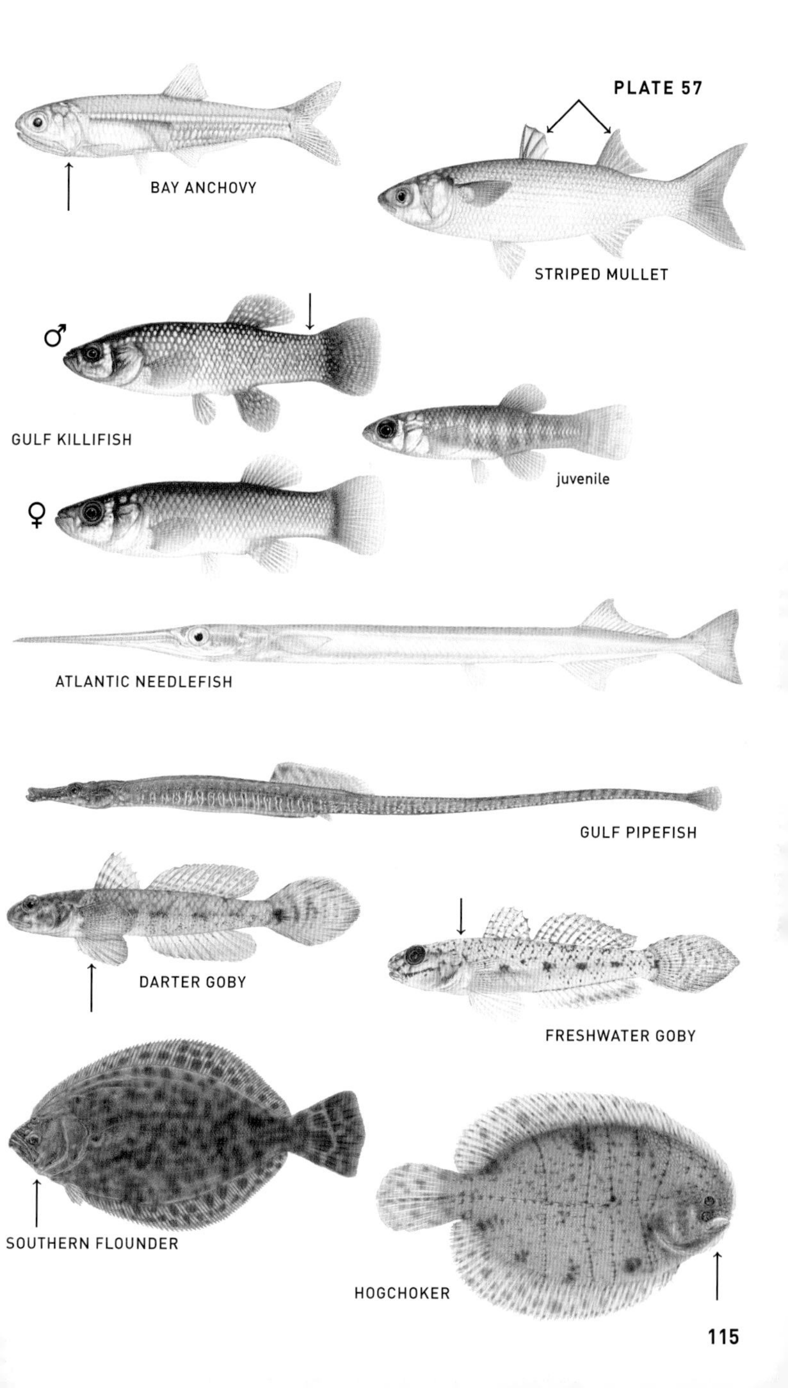
PLATE 57
BAY ANCHOVY
STRIPED MULLET
♂
GULF KILLIFISH
juvenile
♀
ATLANTIC NEEDLEFISH
GULF PIPEFISH
DARTER GOBY
FRESHWATER GOBY
SOUTHERN FLOUNDER
HOGCHOKER

SPECIES ACCOUNTS

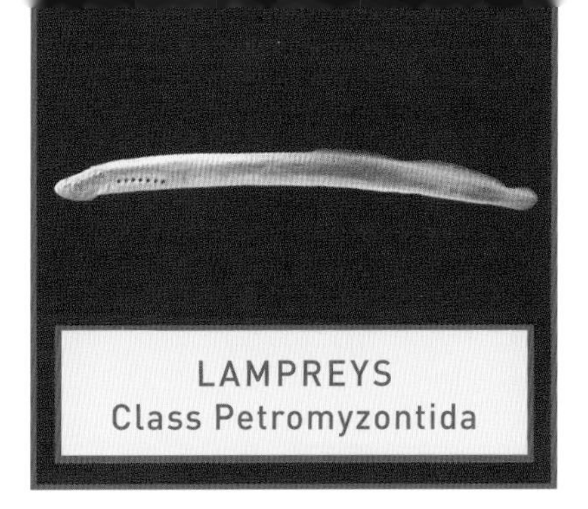

LAMPREYS: Family Petromyzontidae (19)

Lampreys are *primitive, eel-like* fishes that *lack jaws, scales, paired fins,* and *bone*. Forty-one species are known, of which 19 are in fresh waters of North America north of Mexico.

Lampreys excavate pits in stream riffles (rarely in wave-swept areas of lakes) to be used as spawning sites by removing stones with their suction-disc mouths and by fanning out fine particles with vibrations of the body. Eggs hatch into blind larvae called ammocoetes (Fig. 4) which later metamorphose into adults. Larvae may last 3–8 (or, rarely, more) years, living in mud- or sand-bottomed pools and feeding by filtering microorganisms from water. Some species, the so-called brook or nonparasitic lampreys, do not feed as adults and spawn the spring following metamorphosis. Other species are parasitic and feed by attaching to and rasping a hole in the side of a large fish. Adults of several parasitic species migrate to the ocean but must return to fresh water to spawn.

Adults have a cartilaginous skeleton, *1 median nostril*, 7 *pairs of porelike gill openings, 1 or 2 dorsal fins* (Fig. 5) continuous with the caudal fin, and an *oral disc* with rasping teeth on the tongue (Fig. 4). The kind and arrangement of teeth are the best characters to use in identifying lampreys. In species descriptions, a "2" in a tooth formula (e.g., 2-2-2-2) refers to a bicuspid tooth; a "1" refers to a unicuspid tooth (Fig. 4). The count given is the usual condition. Variation in dentition occurs within species, and tooth development may be incomplete in newly transformed individuals. Ammocoetes lack most characters used to identify adults, and seldom can be identified to species with confidence; however, myomere counts are constant throughout life, and in some geographic areas the counts given below can be used to identify ammocoetes as well as adults.

Black pigment is present on the irregularly placed pores of the lateral-line system in species of *Ichthyomyzon*; other lampreys lack

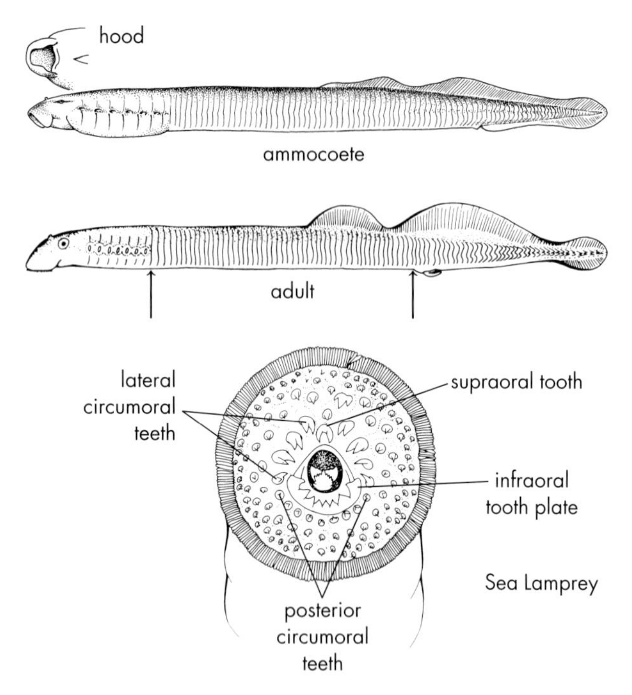

Fig. 4. Lamprey ammocoete and hood surrounding toothless mouth. Adult with arrows indicating 1st and last trunk myomeres. Disc of Sea Lamprey.

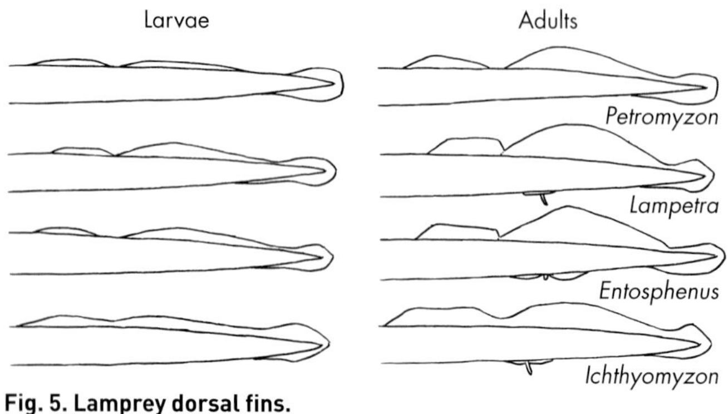

Fig. 5. Lamprey dorsal fins.

black on pores. Within *Ichthyomyzon*, black may be limited to pores on the upper body (Mountain Brook Lamprey, *I. greeleyi*), may also be present on pores on the underside of the body below the gills (most species), or may be absent (Northern Brook Lamprey, *I. fossor*). Pigment is darkest on large individuals and often is absent on ammocoetes and young adults.

SEA LAMPREY *Petromyzon marinus* Pl. 1

IDENTIFICATION: Two dorsal fins. Expanded oral disc *as wide or wider* than head. *Large, sharp disc teeth*; 2 supraoral teeth; usually 2-2-2-2 lateral circumoral teeth, 8–10 posterior circumoral teeth, 7–8 infraoral teeth (Fig. 4). Usually 66–75 trunk myomeres. No black on lateral-line pores. *Round* or *spatulate caudal fin* in adult. *Prominent black mottling* on blue-gray to olive-brown back, side, and fins; cream or yellow-white below. Breeding male has prominent ropelike ridge on nape. Parasitic. To 47 in. (120 cm); landlocked individuals rarely exceed 25 in. (64 cm). **RANGE:** Atlantic Coast from Gulf of St. Lawrence to St. Johns R., FL; St. Lawrence-Great Lakes basin; 1 record for FL panhandle. Also along Atlantic Coast of Europe and Mediterranean Sea. Locally common. **HABITAT:** Individuals with access to ocean are anadromous. Spawning adults in gravel riffles and runs of streams; feeding adults in ocean and lakes. Ammocoetes usually in flowing areas of streams. **SIMILAR SPECIES:** (1) Silver Lamprey, *Ichthyomyzon unicuspis* (Fig. 6), and (2) Chestnut Lamprey, *I. castaneus* (Pl. 1; Fig. 6), have 1 slightly notched dorsal fin, black on lateral-line pores, usually fewer than 56 trunk myomeres, *no* prominent black mottling on body.

SILVER LAMPREY *Ichthyomyzon unicuspis* Not shown

IDENTIFICATION: One slightly notched dorsal fin. Expanded oral disc *as wide or wider* than head. *Large, sharp disc teeth*; usually 2 supraoral

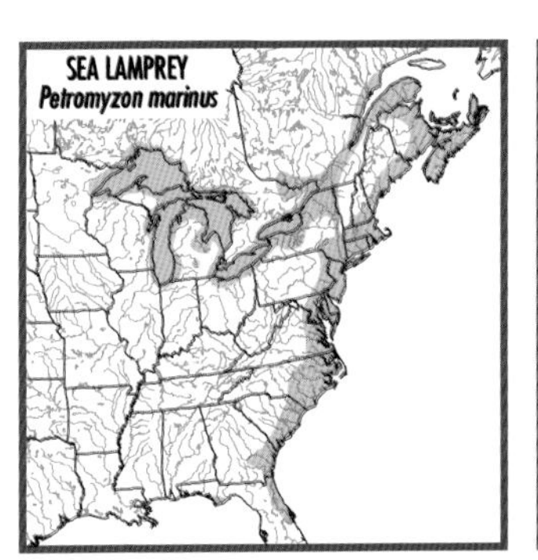

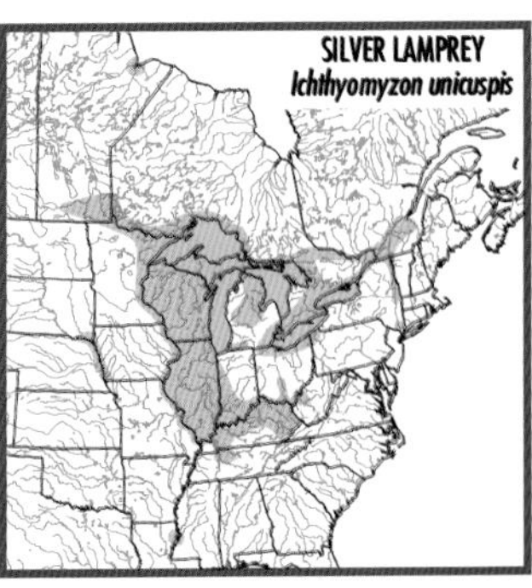

teeth, *1-1-1 or 1-1-1-1 lateral circumoral teeth* (Fig. 6). Usually 49–52 trunk myomeres. Black on lateral-line pores. Gray-brown to yellow-tan above; light yellow or tan below; yellow fins. Parasitic. To 15½ in. (39 cm). **RANGE:** St. Lawrence-Great Lakes basin from QC to sw. ON and south through upper Mississippi and Ohio river basins to TN; Hudson Bay basin, MB; Missouri R., NE, and Mississippi R., MS. Uncommon. **HABITAT:** Feeding adults usually found attached to other fishes in large rivers, lakes, and impoundments; as many as 61 Silver Lampreys have been found on 1 Lake Sturgeon, *Acipenser fulvescens*. Adults migrate upriver to spawn in gravel riffles and runs. Ammocoetes inhabit sandy or muddy pools and backwaters. **SIMILAR SPECIES:** See Fig. 6. (1) Chestnut Lamprey, *I. castaneus*, and (2) Ohio Lamprey, *I. bdellium* (both Pl. 1), have *2-2-2-2 or 2-2-2-2-2 lateral circumoral teeth*, usually more than 51 trunk myomeres. (3) Northern Brook Lamprey, *I. fossor*, has *blunt disc teeth*, pale line along back, no black on lateral-line pores.

NORTHERN BROOK LAMPREY *Ichthyomyzon fossor* Not shown

IDENTIFICATION: One slightly notched dorsal fin. Expanded oral disc *narrower* than head. *Small, blunt disc teeth*; usually 2 supraoral teeth, *1-1 or 1-1-1 lateral circumoral teeth* (Fig. 6). Usually 50–52 trunk myomeres. No black on lateral-line pores. Dark gray or brown above; often a *pale line* along back; pale gray, yellow, or silver white below; gray, yellow, or tan fins. Nonparasitic. To 6¾ in. (17 cm). **RANGE:** St. Lawrence R., QC, west through Great Lakes and n. Mississippi R. basins to Red R. (Hudson Bay basin), s. MB; localized in Ohio R. basin from nw. PA to e. KY; Missouri R. basin, Ozark Uplands, MO. Locally common. **HABITAT:** Clean, clear gravel riffles and runs of small rivers. Ammocoetes occupy quiet water over sand, silt, and debris. **SIMILAR SPECIES:** (1) Other nonparasitic species of *Ichthyomyzon* have *2-2-2 or 2-2-2-2 lateral circumoral teeth* (Fig. 6), black on lateral-line pores, *no*

pale line on back. (2) Silver Lamprey, *I. unicuspis* (Fig. 6), has *large, sharp disc teeth*, black on lateral-line pores, *no* pale line on back.

CHESTNUT LAMPREY *Ichthyomyzon castaneus* Pl. 1

IDENTIFICATION: One slightly notched dorsal fin. Expanded oral disc *as wide or wider* than head. Large, sharp disc teeth; usually 2–3 supraoral teeth, *2-2-2-2 or 2-2-2-2-2 lateral circumoral teeth* (Fig. 6). Usually *51–56 trunk myomeres*. Black on lateral-line pores. Yellow or tan above; white to light olive-yellow below; olive-yellow fins. Parasitic. To 15 in. (38 cm). **RANGE:** St. Lawrence-Great Lakes and Mississippi R. basins from ON and WI south to LA, and from e. TN to e. KS and OK; Red R. (Hudson Bay basin), MB, MN, and SD; Gulf drainages from Mobile Bay, GA and AL, to Sabine Lake, TX. Locally common. **HABITAT:** Lakes and streams. Adults ascend streams to spawn. Ammocoetes occupy sand- and silt-bottomed pools and backwaters. **SIMILAR SPECIES:** See Fig. 6. (1) Silver Lamprey, *I. unicuspis*, has *1-1-1 or 1-1-1-1- lateral circumoral teeth*; usually *49–52 trunk myomeres*. (2) Ohio Lamprey, *I. bdellium* (Pl. 1), usually has *56–62 trunk myomeres*; is gray above and on fins. (3) Southern Brook Lamprey, *I. gagei* (Pl. 1), has blunt disc teeth, expanded oral disc *narrower* than head, nonfunctional digestive tract in adult; reaches only 6¾ in. (17 cm).

SOUTHERN BROOK LAMPREY *Ichthyomyzon gagei* Pl. 1

IDENTIFICATION: One slightly notched dorsal fin. Expanded oral disc *narrower* than head. *Blunt, poorly developed disc teeth*; usually 2–3 supraoral teeth, *2-2-2 or 2-2-2-2 lateral circumoral teeth* (Fig. 6). Usually *52–56 trunk myomeres*. Black on lateral-line pores. Gray or tan above, white to cream below; cream or yellow fins. Nonparasitic. To 6¾ in. (17 cm). **RANGE:** Mississippi R. basin of s. MO, e. OK, AR, MS, and LA; Tennessee R. drainage, w. KY and n. AL; Gulf drainages from Ochlockonee R., FL, to Galveston Bay, TX. Disjunct populations in Mississippi

R. tributaries, WI and MN. Locally common. **HABITAT:** Gravel and sand riffles and runs of creeks and small rivers. Ammocoetes in flowing water near sandbars and debris. **SIMILAR SPECIES:** (1) Chestnut Lamprey, *I. castaneus* (Pl. 1; Fig. 6), has *sharp, well-developed disc teeth,* expanded oral disc *as wide or wider* than head, functional digestive tract in adult; reaches 15 in. (38 cm). (2) Mountain Brook Lamprey, *I. greeleyi* (Pl. 1; Fig. 6), usually has *57–60 trunk myomeres*, larger teeth, no black on pores below gills.

OHIO LAMPREY *Ichthyomyzon bdellium* Pl. 1

IDENTIFICATION: One slightly notched dorsal fin. Expanded oral disc as *wide or wider than head.* Sharp, well-developed disc teeth; usually 2–3 supraoral teeth, *2-2-2-2 or 2-2-2-2-2 lateral circumoral teeth* (Fig. 6). Usually *56–62 trunk myomeres*. Black on lateral-line pores. Blue to lead gray above; white to slightly mottled below; gray fins. Parasitic. To 12 in. (30 cm). **RANGE:** Ohio R. basin from NY to IL, and south to n. GA. Uncommon. **HABITAT:** Large rivers; sometimes creeks and small rivers. Ammocoetes live near debris in muddy pools and backwaters. **SIMILAR SPECIES:** (1) Chestnut Lamprey, *I. castaneus* (Pl. 1; Fig. 6), usually has *51–56 trunk myomeres*; is yellow or tan in life. (2) Mountain Brook Lamprey, *I. greeleyi* (Pl. 1; Fig. 6), has smaller teeth, expanded oral disc *narrower* than head, no black on pores below gills; reaches only 7¾ in. (20 cm).

MOUNTAIN BROOK LAMPREY *Ichthyomyzon greeleyi* Pl. 1

IDENTIFICATION: One slightly notched dorsal fin. Expanded oral disc *narrower* than head. Moderately developed disc teeth; usually 2–3 supraoral teeth, *2-2-2 to 2-2-2-2-2 lateral circumoral teeth* (Fig. 6). Usually *57–60 trunk myomeres*. Black lateral-line pores on upper side, *no black on pores below gills*. Gray-brown above with small dark flecks on side, white or cream below; cream or yellow fins. Nonparasitic. To 7¾

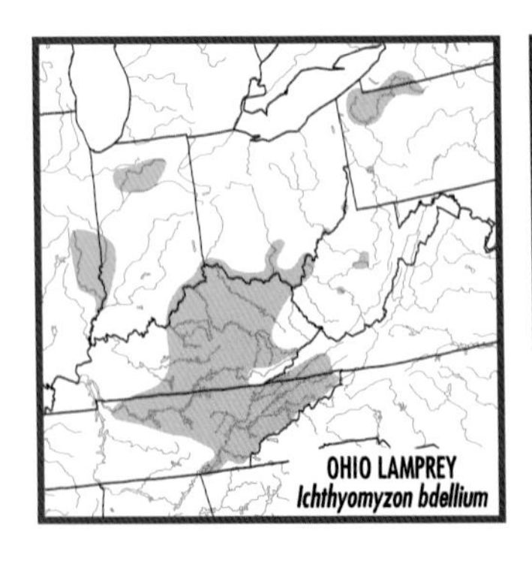

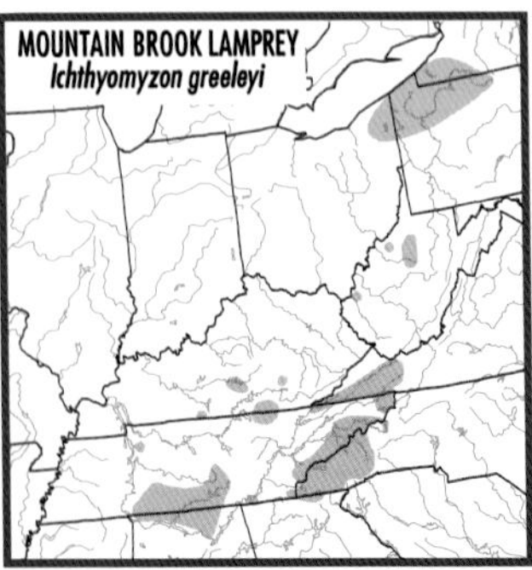

in. (20 cm). **RANGE:** Ohio R. basin from sw. NY to n. AL and GA. Highly localized. **HABITAT:** Gravel riffles and sandy runs of clean, clear high-gradient streams. Ammocoetes in sand, mud, and debris in pools and backwaters. **SIMILAR SPECIES:** See Fig. 6. (1) Ohio Lamprey, *I. bdellium*, and (2) Chestnut Lamprey, *I. castaneus* (both Pl. 1), have large sharp disc teeth, *black on pores below gills*; reach 11 in. (28 cm) or more, have functional digestive tract in adult. (3) Southern Brook Lamprey, *I. gagei* (Pl. 1), usually has *52–56 trunk myomeres*, smaller disc teeth, *black on pores below gills*.

ACIFIC LAMPREY *Entosphenus tridentatus* Pl. 1

IDENTIFICATION: Two dorsal fins. Expanded oral disc *as wide or wider* than head. *Medium-sized eye*. Large, sharp disc teeth, 3 supraoral teeth; usually 2-3-3-2 lateral circumoral teeth; usually 17–20 posterior circumoral teeth, 0–15 bicuspid; usually 5–6 infraoral teeth (Fig. 6). Usually *64–71 trunk myomeres*. Short disc averages 6.7 percent of total length. No black on lateral-line pores. *Dark blue or brown above, light or silver below,* dusky dorsal and caudal fins. Parasitic, anadromous. To 30 in. (76 cm). Dwarf, nonfeeding populations to 10¾ in. (27 cm). **RANGE:** Pacific drainages from AK south to Baja California; also along Pacific Coast of Asia. Common. **HABITAT:** Spawning adults in gravel riffles and runs of clear coastal streams; feeding adults usually in ocean. Ammocoetes in silt, mud, and sand of shallow eddies and backwaters of streams. **REMARKS:** Dwarf, nonanadromous, and nonparasitic populations landlocked in OR and n. CA require further taxonomic study. **SIMILAR SPECIES:** See (1) Vancouver Lamprey, *E. macrostomus*, (2) Klamath Lamprey, *E. similis,* and Miller Lake Lamprey, *E. minimus.*

ANCOUVER LAMPREY *Entosphenus macrostomus* Not shown

IDENTIFICATION: Nearly identical to Pacific Lamprey, *E. tridentatus,* but has *larger eye*; expanded oral disc *much wider* than head; *uniformly*

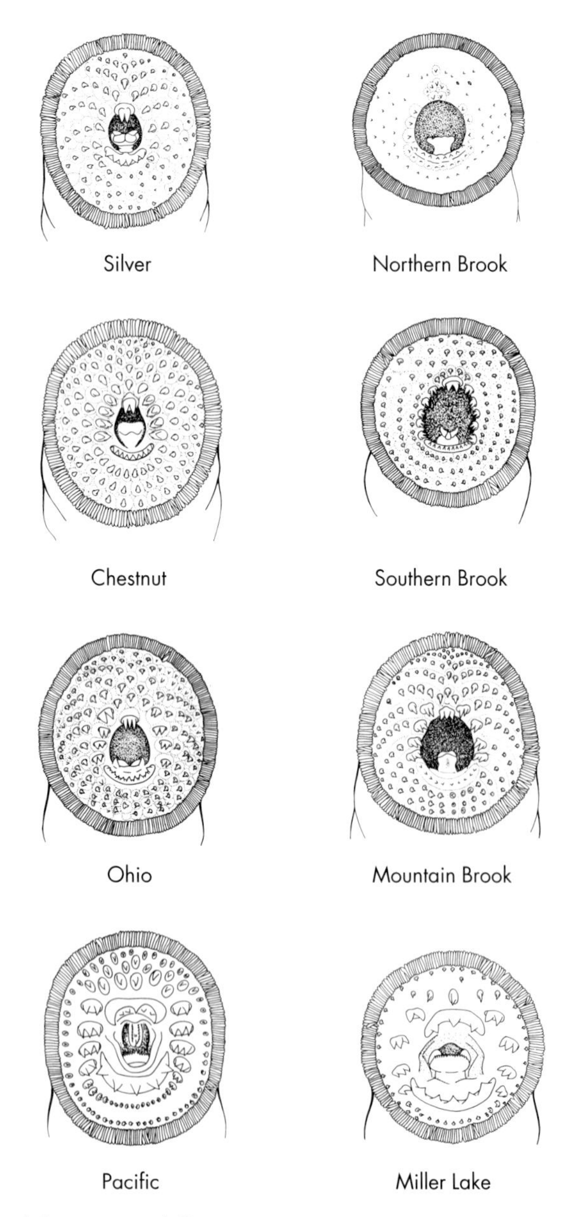

Fig. 6. Lamprey oral discs.

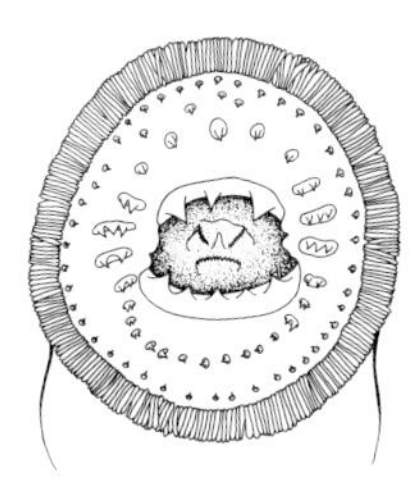
Pit-Klamath Brook

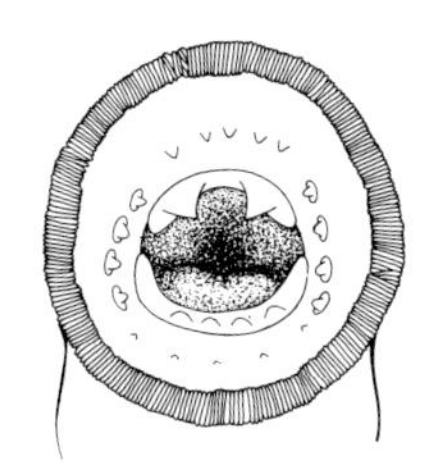
Kern Brook

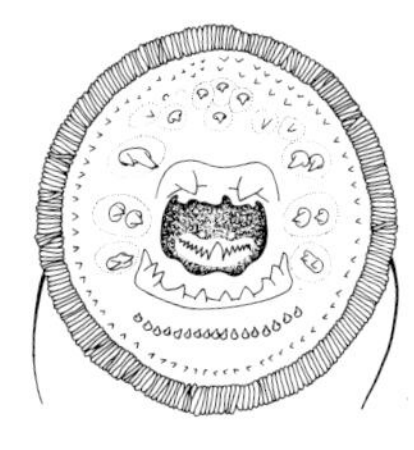
Arctic

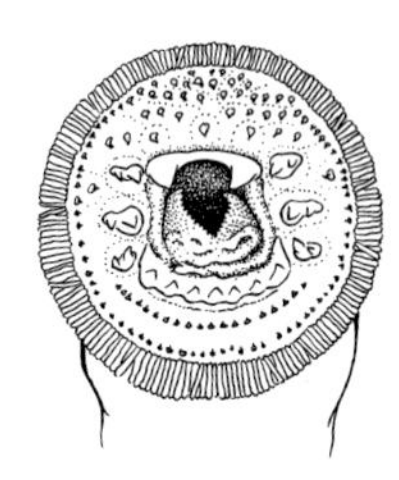
Alaskan Brook

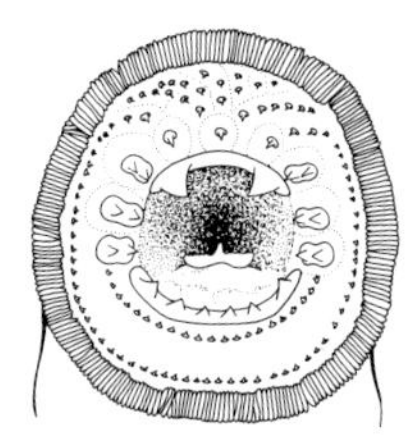
American Brook

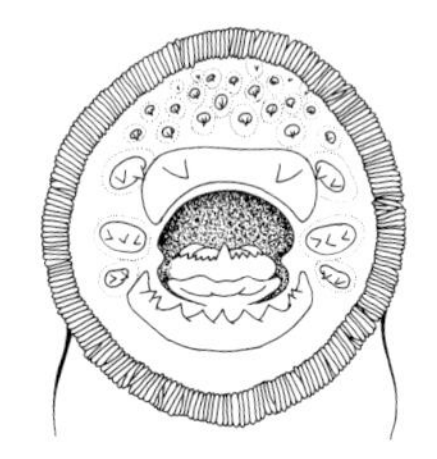
River

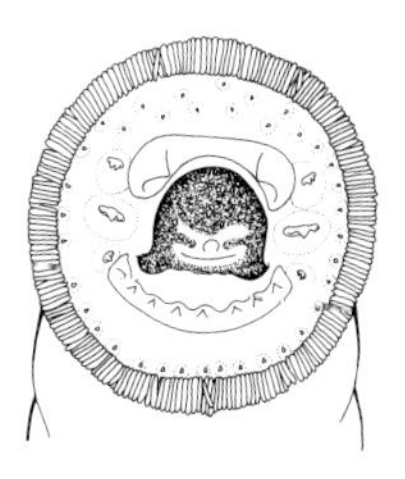
Western Brook

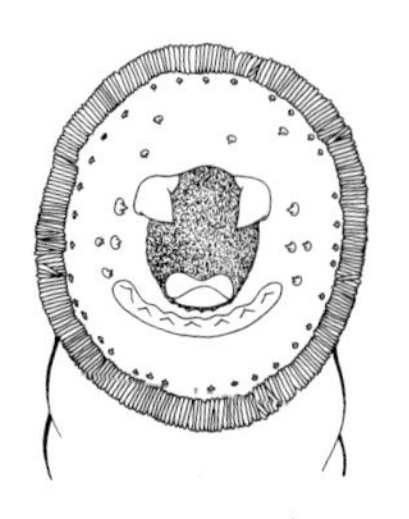
Least Brook

dark, almost black, body and fins. Parasitic. To 10 in. (25 cm). **RANGE:** Mesachie Lake and Lake Cowichan on s. Vancouver I., BC. Uncommon; protected as a *threatened species*. **HABITAT:** Spawns on shallow gravel bars in freshwater lakes or mouths of creeks flowing into lakes. Ammocoetes in silt, mud, or sand in quiet water. **SIMILAR SPECIES:** (1) See Pacific Lamprey, *E. tridentatus* (Pl 1; Fig. 6).

KLAMATH LAMPREY *Entosphenus similis* Not shown

IDENTIFICATION: Similar to Pacific Lamprey, *E. tridentatus*, but usually has *60–63 trunk myomeres*, longer disc averaging 9.2 percent of total length, smaller eye. Parasitic. To 10¾ in. (27 cm). **RANGE:** Klamath and Trinity river drainages, s. OR and n. CA. Rare. **HABITAT:** Large rivers, impoundments, and lakes. **SIMILAR SPECIES:** (1) See Pacific Lamprey, *E. tridentatus* (Pl. 1; Fig. 6). (2) Miller Lake Lamprey, *E. minimus* (Fig. 6), has shorter disc averaging 6.3 percent of total length; reaches only 5 in. (13 cm).

PIT-KLAMATH BROOK LAMPREY *Entosphenus lethophagus* Not shown

IDENTIFICATION: Two dorsal fins. Expanded oral disc *narrower* than head. Disc teeth reduced in size and number; 2–3 supraoral teeth; usually *1-2-2-1 or 2-3-3-2 lateral circumoral teeth;* 9–15 posterior circumoral teeth (10–15 usually unicuspid); usually 5 infraoral teeth (Fig. 6). Usually *60–71 trunk myomeres*. No black on lateral-line pores. Slate gray to brown above; brass or silver below; yellow fins. Nonparasitic. To 8¾ in. (22 cm). **RANGE:** Klamath R., s.-cen. OR, and Pit R., ne. CA. **HABITAT:** Riffles and runs of clear streams. Ammocoetes near weed beds and sandbars. Common. **SIMILAR SPECIES:** See Fig. 6. (1) See Miller Lake Lamprey, *E. minimus*. (2) Kern Brook Lamprey, *E. hubbsi* (Pl. 1), usually has *52–56 trunk myomeres*, usually *1-1-1-1 lateral circumoral teeth*. (3) Pacific Lamprey, *E. tridentatus* (Pl. 1), has large, sharp disc teeth, functional digestive tract in adult; reaches 30 in. (76 cm).

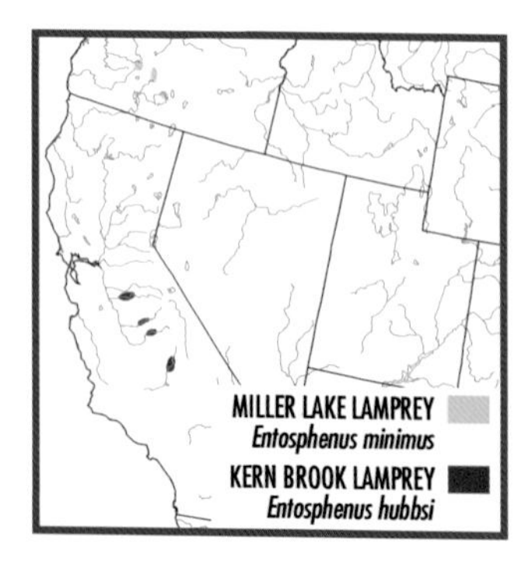

MILLER LAKE LAMPREY *Entosphenus minimus* Not shown

IDENTIFICATION: Similar to Pit-Klamath Brook Lamprey, *E. lethophagus*, but darker overall, has larger eye (ca. 2.6 vs. 1.8 percent total length), and reaches only 5 in. (13 cm). Has 13–17 posterior circumoral teeth (Fig. 6), usually 60–66 trunk myomeres. Parasitic. **RANGE:** Upper Klamath R. drainage in Miller Lake, Miller Creek, Jack Creek, Williamson R. above Klamath Marsh, Sycan R. above Sycan Marsh, Klamath and Lake counties, OR. Rare. **HABITAT:** Clear rocky streams, gravelly areas in Miller Lake. Ammocoetes in silt, mud, or sand. **SIMILAR SPECIES:** (1) See Pit-Klamath Brook Lamprey, *E. lethophagus* (Fig. 6).

KERN BROOK LAMPREY *Entosphenus hubbsi* Pl. 1

IDENTIFICATION: Two dorsal fins. Expanded oral disc *narrower* than head. Usually blunt disc teeth; 2 supraoral teeth; usually *1-1-1-1 lateral circumoral teeth*; 9–12 unicuspid posterior circumoral teeth; 5 infraoral teeth (Fig. 6). Usually *52–56 trunk myomeres.* No black on lateral-line pores. Gray to brown above; white below; black specks on dorsal and caudal fins. Nonparasitic. To 5½ in. (14 cm). **RANGE:** East side of San Joaquin Valley in lower Merced, Kaweah, Kings, and San Joaquin rivers, CA. Uncommon. **HABITAT:** Silty backwaters of rivers; spawns in gravel riffles. **SIMILAR SPECIES:** (1) Pit-Klamath Brook Lamprey, *E. lethophagus* (Fig. 6), usually has *60–71 trunk myomeres, 1-2-2-1 or 2-3-3-2 lateral circumoral teeth.* (2) Pacific Lamprey, *E. tridentatus* (Pl. 1; Fig. 6), has large, sharp disc teeth, usually *64–71 trunk myomeres*, functional digestive tract in adult, reaches 30 in. (76 cm).

AMERICAN BROOK LAMPREY *Lethenteron appendix* Pl. 1

IDENTIFICATION: Two dorsal fins connected at base. Expanded oral disc *narrower* than head. Usually *blunt disc teeth*; 2 supraoral teeth; usually 2-2-2 lateral circumoral teeth; usually *20–24 posterior circumoral teeth*; usually 7–8 infraoral teeth (Fig. 6). Usually *67–73 trunk myomeres.* No black on lateral-line pores. Lead gray to slate blue above, white or silver white below; yellow fins; dark gray to black blotch on caudal fin. Breeding adult is olive green or pink-purple to shiny black above, black stripe at base of dorsal fins. Nonparasitic. To 13¾ in. (35 cm). **RANGE:** Atlantic, Great Lakes, and Mississippi R. basins from St. Lawrence R., QC, west to MN, and south to Chowan R. system, VA, Tennessee R. system, AL, and St. Francis and White river systems, MO and AR. Uncommon. **HABITAT:** Adults in gravel-sand riffles and runs of creeks and small to medium rivers with strong flow; usually in clear water. Ammocoetes in sandy or silty pools. **REMARKS:** Two subspecies, *L. a. appendix* and *L. a. wilderi*, sometimes recognized, but ranges poorly understood. **SIMILAR SPECIES:** See Fig. 6. (1) See Alaskan Brook Lamprey, *L. alaskense*. (2) Arctic Lamprey, *L. camtschaticum*, has *large, sharp disc teeth*; functional digestive tract in adult; reaches 25 in. (63 cm). (3) Least Brook Lamprey, *Lampetra aepyptera* (Pl. 1), has extremely degenerate disc teeth; *usually* 52–59 *trunk myomeres*; usually *lacks* posterior circumoral teeth. (4) River Lamprey, *Lampetra ayresii*, *lacks* posterior circumoral teeth; usually has 2-3-2 lateral circumoral teeth.

ALASKAN BROOK LAMPREY *Lethenteron alaskense* Not shown

IDENTIFICATION: Nearly identical to American Brook Lamprey, *L. appendix*, but has *small supplementary teeth* (absent in American Brook Lamprey) in lateral areas of disc (Fig. 6). Nonparasitic. To 7½ in. (19 cm). **RANGE:** Alaska and Kenai peninsulas to Chatanika and Chena rivers, AK; Martin R. (Arctic basin), NT. Rare. **HABITAT:** Adults in rocky riffles and runs of creeks and small to medium rivers. **SIMILAR SPECIES:** (1) See American Brook Lamprey, *L. appendix* (Pl. 1; Fig. 6).

ARCTIC LAMPREY *Lethenteron camtschaticum* Not shown

IDENTIFICATION: Two dorsal fins. Expanded oral disc *as wide or wider* than head. *Large, sharp, disc teeth*; 2 supraoral teeth; usually *2-2-2 lateral circumoral teeth*; 15–24 posterior circumoral teeth; 5–10 infraoral teeth (Fig. 6). Usually 67–72 trunk myomeres. No black on lateral-line pores. Dark brown to blue-black above, yellow to light brown below; light tan to gray dorsal fins; dark blotch on caudal fin. Parasitic. To 25 in. (63 cm). **RANGE:** Arctic and Pacific drainages from Anderson R. and Mackenzie R. drainage, NT and n. AB, west to Kenai Peninsula, AK. Also in n. Europe and Asia. Common. **HABITAT:** Spawning adults in gravel riffles and runs of clear streams; feeding adults usually in ocean or lakes. Ammocoetes in muddy margins and backwaters of rivers and lakes. **SIMILAR SPECIES:** (1) American Brook Lamprey, *L. appendix* (Pl. 1; Fig. 6), has usually *blunt disc teeth*, nonfunctional digestive tract in adult; reaches only 13¾ in. (35 cm). (2) Pacific Lamprey,

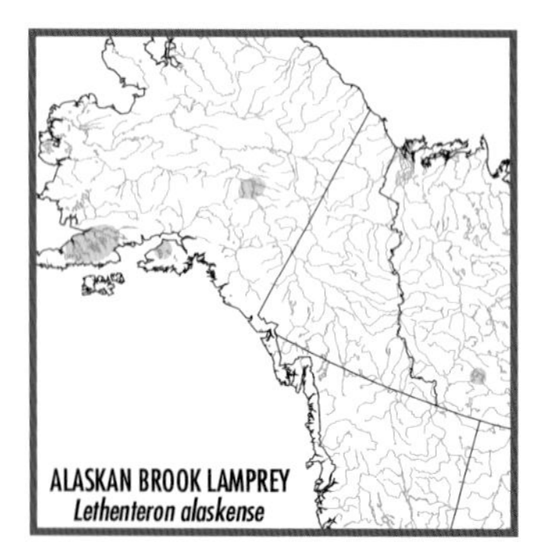

Entosphenus tridentatus (Pl. 1; Fig. 6), has 3 supraoral teeth, usually *2-3-3-2 lateral circumoral teeth.*

RIVER LAMPREY *Lampetra ayresii* **Not shown**

IDENTIFICATION: Two dorsal fins. Expanded oral disc *as wide or wider* than head. *Large, sharp disc teeth*; 2 widely separated supraoral teeth; usually *2-3-2 lateral circumoral teeth*; no posterior circumoral teeth; 7–10 infraoral teeth (Fig. 6). Usually *65–70 trunk myomeres.* No black on lateral-line pores. Yellow to silver gray above, white below; yellow fins; dark gray blotch on caudal fin. Parasitic. To 12¼ in. (31 cm). **RANGE:** Pacific Slope from Tee Harbor, AK, south to Sacramento-San Joaquin drainage, CA. Uncommon. **HABITAT:** Spawning adults in clear gravel riffles of streams; feeding adults in estuaries and ocean. Ammocoetes in sandy and muddy pools of spawning streams. **SIMILAR SPECIES:** (1) Western Brook Lamprey, *L. richardsoni* (Pl. 1; Fig. 6), has *small, blunt disc teeth*, usually *58–67 trunk myomeres*, nonfunctional digestive tract in adults.

WESTERN BROOK LAMPREY *Lampetra richardsoni* **Pl. 1**

IDENTIFICATION: Two dorsal fins. Expanded oral disc *narrower* than head. *Small, blunt disc teeth*; 2 widely separated supraoral teeth; usually *1-2-1 or 2-3-2 lateral circumoral teeth*; *no* posterior circumoral teeth; 6–10 infraoral teeth (Fig. 6). Usually *58–67 trunk myomeres.* No black on lateral-line pores. Brown to gray above; white below; gray fins; *dark spot on caudal fin.* Nonparasitic. To 7 in. (17 cm). **RANGE:** Pacific Slope from Taku R., s. AK, to Sacramento-San Joaquin R. drainage, CA. Uncommon. Morrison Creek Lamprey, an unnamed form on Vancouver I., BC, is protected in Canada as *endangered.* **HABITAT:** Gravel riffles and runs of clear, cool streams. Ammocoetes in muddy and sandy backwaters and pools. **SIMILAR SPECIES:** (1) River Lamprey, *L. ayresii* (Fig. 6), has expanded oral disc *as wide or wider* than head;

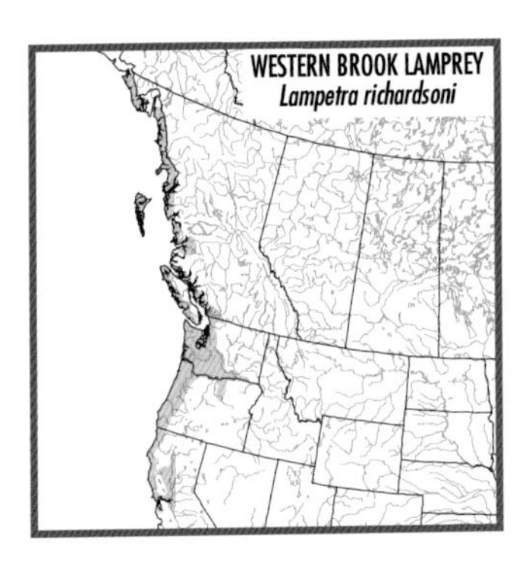

usually has *65–70 trunk myomeres*; *large, sharp disc teeth*; functional digestive tract in adult; reaches 12¼ in. (31 cm).

LEAST BROOK LAMPREY *Lampetra aepyptera* Pl. 1

IDENTIFICATION: Two dorsal fins. Expanded oral disc *narrower* than head. *Blunt, extremely degenerate disc teeth*; 2 widely separated supraoral teeth; usually 1-2-1 or 1-1-1 lateral circumoral teeth; usually no posterior circumoral teeth—if present, usually 4–9 (range 1–22) and unicuspid; usually 7–12 infraoral teeth (Fig. 6). Usually *52–59 trunk myomeres*. No black on lateral-line pores. Light tan to silver gray above; yellow or white below; yellow or gray fins. Breeding adult has mottled gray-brown back, black stripe on side of body through eye and at base of 1st dorsal fin; dusky black edges on dorsal fins, gold stripe from caudal fin through middle of dorsal fins; dark-tipped caudal fin. Nonparasitic. To 7 in. (18 cm); on Atlantic Slope to only 4¾ in. (12 cm). **RANGE:** Atlantic Slope from Susquehanna R. drainage, se. PA, to James R., VA; Tar and Neuse river drainages, NC; Mississippi R. basin from w. PA to s. MO and n. AR, south to s. MS; Gulf Slope drainages from Mobile Bay basin, GA, to Pearl R., MS. Locally common. **HABITAT:** Clean, clear gravel riffles and runs of creeks and small rivers. Ammocoetes in spring-fed wetlands and quiet pools and backwaters of small, sand- or mud-bottomed streams. **REMARKS:** Coastal Plain population of w. KY and TN exhibits characteristics of paedomorphism. Spawning adult from Atlantic Slope lacks black stripe through eye, gold stripe on dorsal fin, and dark-tipped caudal fin. **SIMILAR SPECIES:** (1) American Brook Lamprey, *Lethenteron appendix* (Pl. 1), has *67–73 trunk myomeres*.

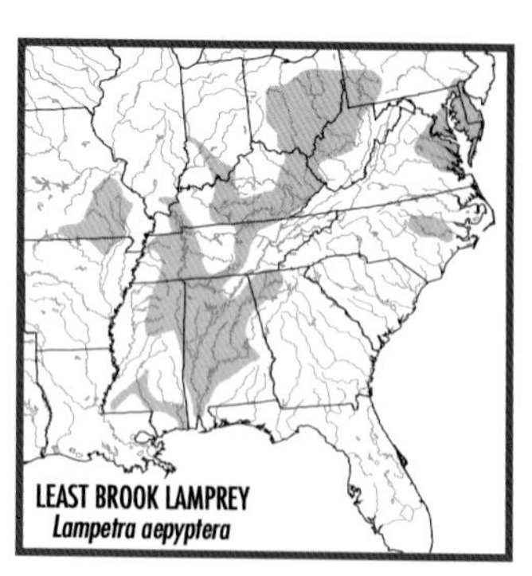

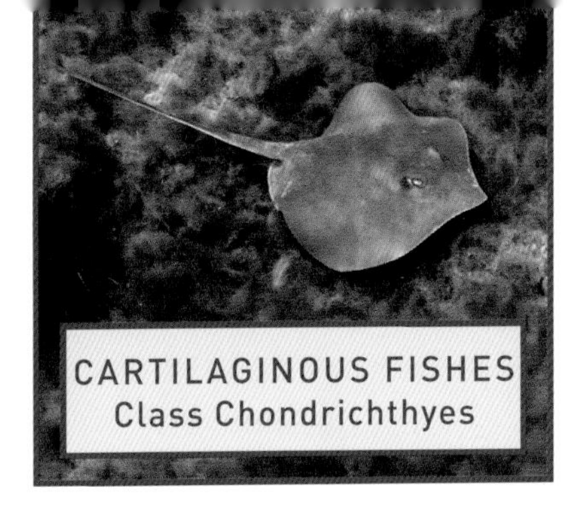

Cartilaginous fishes have jaws, a skeleton of cartilage, and placoid scales. There are about 1,000 species worldwide.

WHIPTAIL STINGRAYS: Family Dasyatidae (1)

Whiptail stingrays have a strongly depressed disc-shaped body with greatly expanded pectoral fins. Eyes and spiracles are on top of the head, and 5 pairs of gill slits are on the underside. A "tail" (long, slender caudal peduncle without a caudal fin) has 1 or more serrated spines that carry a painful venom and can inflict serious wounds. Most of the 81 species live in tropical and temperate oceans. A few live in fresh water, and a few marine species enter brackish and fresh water, including 1 in our area.

ATLANTIC STINGRAY *Dasyatis sabina* Not shown

IDENTIFICATION: *Disc broadly rounded on sides, pointed at front,* width about equal to length. Tail longer than disc, with large serrated spine. Snout longer than interspiracle distance. Brown above, white below. To 24 in. (61 cm) wide. **RANGE:** Atlantic Coast from Chesapeake Bay to Yucatan, Mexico. Resident (reproducing) population in St. Johns R., FL. Inland records for Mississippi R., LA, Tombigbee R., AL, and Chattahoochee R., FL and AL. Common in fresh water only in FL. **HABITAT:** Shallow bays, estuaries, and fresh water over sand or mud. Often partly buried with eyes, spiracles, and tail exposed.

Ray-finned fishes have jaws and, except in some with reduced ossification, a bony skeleton. There are about 30,000 species worldwide, of which about 13,200 (44 percent) occur in fresh water.

STURGEONS: Family Acipenseridae (8)

Sturgeons are large, ancient fishes with a cartilaginous skeleton, *heterocercal caudal fin*, and spiral valve. Twenty-five species survive today. All of them enter fresh water, at least to spawn, but some species spend most of their lives in the ocean. The greatest diversity occurs in Eurasia, with fewer species in North America.

ACIPENSER

GENUS CHARACTERISTICS: Cone-shaped snout; *spiracle*; upper lobe of caudal fin *lacks* long filament; fairly short caudal peduncle, round in cross section and only partly covered by scutes; smooth barbels; 2 fleshy lobes on lower lip.

SHORTNOSE STURGEON *Acipenser brevirostrum* Pl. 2

IDENTIFICATION: *Short snout*, bluntly V-shaped, not upturned at tip. *Anal fin origin beneath dorsal fin origin*. Caudal peduncle relatively short, tip of anal fin reaching caudal fin origin. One row of preanal scutes; 2 scutes between dorsal fin and caudal fulcrum; 1 large scute between anal fin and caudal fulcrum. Dark brown to black above and on upper side, light brown to yellow on lower side; scutes on back and side *paler* than skin; white below; gray or brown fins; white edges on paired fins. Black viscera. Young (under 24 in. [61 cm]) have black or dusky blotches on snout and body. Has 19–22 anal rays; 22–29 rakers on 1st gill arch; 8–13 scutes on back; 22–33 scutes along side. To 43 in.

(109 cm). **RANGE:** Atlantic Coast from St. John R., NB, to St. Johns R., FL. Uncommon; protected as an *endangered species* in U.S. **HABITAT:** River mouths, lakes, estuaries, and bays; occasionally enters open sea. **SIMILAR SPECIES:** See Pl. 2. (1) See Lake Sturgeon, *A. fulvescens*. (2) Green Sturgeon, *A. medirostris,* is green; has 1 large scute behind dorsal fin, 22–28 anal rays, 18–20 rakers on 1st gill arch; white viscera. (3) Atlantic Sturgeon, *A. oxyrinchus,* has *long,* sharply V-shaped *snout*; 2 rows of preanal scutes; white viscera.

LAKE STURGEON *Acipenser fulvescens* Pl. 2

IDENTIFICATION: Similar to Shortnose Sturgeon, *A. brevirostrum,* but *anal fin origin behind dorsal fin origin;* caudal peduncle longer, tip of anal fin reaching only to anterior edge of fulcral caudal scute; olive-brown to gray above and on side; scutes on back and along side *same color as skin*; white below; dark gray or brown fins. Has 25–30 anal rays; 25–40, usually 32–35, rakers on 1st gill arch; 9–17 scutes on back; 29–42 scutes along side. To 9 ft. (2.7 m). **RANGE:** St. Lawrence-Great Lakes, Hudson Bay, and Mississippi R. basins from QC to AB and south to AL and LA. Formerly in Coosa R. system (Mobile Bay drainage), AL. Rare and nearing extinction in Missouri, Ohio, and middle Mississippi river drainages; more common in north although declining. **HABITAT:** Bottom of lakes and large rivers, usually those 16–30 ft. (5–9 m) in depth, over mud, sand, and gravel. Occasionally enters brackish water. **SIMILAR SPECIES:** See Pl. 2. (1) See Shortnose Sturgeon, *A. brevirostrum*. (2) Green Sturgeon, *A. medirostris,* is green; has scutes along side *paler* than skin, 18–20 rakers on 1st gill arch, 23–30 scutes along side, white viscera.

GREEN STURGEON *Acipenser medirostris* Pl. 2

IDENTIFICATION: Moderately blunt snout in adult; sharper and more shovel-like in young; *barbels usually closer to mouth than to snout tip.*

One row of preanal scutes; 1 large scute behind dorsal fin; 1 large scute behind anal fin. *Olive to dark green above and on side*; white-green below; gray or green fins; scutes along side *paler* than skin. Some northern fish have stripes on belly and lower side. White viscera. Has 22–28 anal rays; 18–20 rakers on 1st gill arch; 9–11 scutes on back; 23–30 scutes along side. To 7 ft. (2.1 m). **RANGE:** Pacific Coast from Aleutian Is. to Mexico. Generally uncommon; protected as an *endangered species* in Canada. **HABITAT:** Estuaries, lower reaches of large rivers, salt or brackish water off river mouths. Ascends far up Columbia R., WA, and Klamath and Trinity rivers, CA. **SIMILAR SPECIES:** See Pl. 2. (1) Shortnose Sturgeon, *A. brevirostrum*, is *brown or black*; has 2 scutes between dorsal fin and caudal fulcrum, 19–22 anal rays, 22–29 rakers on 1st gill arch, black viscera. (2) Lake Sturgeon, *A. fulvescens*, is *olive gray or brown* above and on side; has 2 scutes between dorsal fin and caudal fulcrum, 25–40 rakers on 1st gill arch, 25-30 anal rays, black viscera. (3) White Sturgeon, *A. transmontanus*, is *gray, pale olive, or gray-brown*; has barbels *closer to snout tip than to mouth*, no obvious scutes behind dorsal and anal fins, 2 rows of preanal scutes, 38–48 scutes along side, black viscera.

WHITE STURGEON *Acipenser transmontanus* Pl. 2

IDENTIFICATION: Moderately blunt snout in adult, sharper in young; *barbels closer to snout tip than to mouth. No obvious scutes behind dorsal and anal fins* (other than caudal fulcra). Two rows (6–9 scutes) of preanal scutes. Gray, pale olive, or gray-brown above and on upper side; white to pale gray on lower side; all scutes light in color; white below; gray fins. Black viscera. Has 28–30 anal rays; 34–36 rakers on 1st gill arch; 11–14 scutes on back; 38–48 scutes along side. To 20 ft. (6.1 m). **RANGE:** Pacific Coast and drainages from Alaska Bay, AK, to Mexico. Introduced into lower Colorado R., AZ. Rare; protected as an *endangered species* in Canada and U.S. **HABITAT:** Estuaries of large rivers; moves far inland to spawn in Fraser (BC), Columbia (WA), and

Sacramento rivers (CA). Landlocked in Columbia R. drainage, MT. **REMARKS:** Largest freshwater fish in N. America. **SIMILAR SPECIES:** (1) All other *Acipenser* species (Pl. 2) have obvious scutes behind dorsal or anal fins.

ATLANTIC STURGEON *Acipenser oxyrinchus* Pl. 2

IDENTIFICATION: *Long, sharply V-shaped snout.* Snout tip upturned in young. Has 2 rows of preanal scutes; *4 small scutes, usually as 2 pairs, between anal fin and caudal fulcrum* (1st pair may overlap anal fin base, 2d pair may look like 1 scute); 6–9 scutes, mostly in pairs, behind dorsal fin. Blue-black above, paler on side; white below; gray to blue-black fins; white leading edge on paired fins, lower lobe of caudal fin, anal fin. White spines on scutes on back and side contrast with dark skin. White viscera. Has 26–28 anal rays; 15–27 rakers on 1st gill arch; 7–16 scutes on back; 24–35 scutes along side. To 14 ft. (4.3 m). **RANGE:** Atlantic Coast from Hamilton R., Labrador, to FL; Gulf of Mexico from Tampa Bay, FL, to Lake Pontchartrain, LA. One record in Mississippi R. basin, MS. Also in n. Europe. Uncommon; greatly depleted throughout most of range. **HABITAT:** Shallow waters of continental shelf; ascends coastal rivers to spawn. **REMARKS:** Two subspecies. *A. o. oxyrinchus*, along Atlantic Coast, has relatively short head, pectoral fin, and spleen; scutes on back longer than wide. *A. o. desotoi* (Gulf Sturgeon), along Gulf Coast, has longer head, pectoral fin, and spleen; scutes on back much shorter than wide; protected as a *threatened subspecies.* **SIMILAR SPECIES:** See Pl. 2. (1) Shortnose Sturgeon, *A. brevirostrum*, has *shorter snout, 1 large scute* between anal fin and caudal fulcrum, 2 scutes between dorsal fin and caudal fulcrum, 1 row of preanal scutes. (2) White Sturgeon, *A. transmontanus, lacks* obvious scutes behind anal and dorsal fins. (3) Green Sturgeon, *A. medirostris,* is green; has *1 large scute* between anal fin and caudal fulcrum, 1 row of preanal scutes.

SHOVELNOSE STURGEON *Scaphirhynchus platorynchus* Pl. 2

IDENTIFICATION: *Flat, shovel-shaped snout.* Four fleshy lobes on lower lip; 4 fringed barbels. Long, slender caudal peduncle flat in cross section and covered with bony scutes. Long filament on upper lobe of caudal fin (sometimes broken off). No spiracle. Scalelike scutes on belly (except in small young). Bases of outer barbels in line with, or ahead of, inner barbels (Fig. 7); inner barbels same length as outer barbels. Light brown or buff above and on side, white below. To 43 in. (108 cm). **RANGE:** Mississippi R. basin from w. PA to MT and south to LA; upper Rio Grande, NM. Common in Mississippi basin, extirpated from Rio Grande. **HABITAT:** Bottoms of main channels and embayments of large turbid rivers; frequently in flowing water over sand mixed with gravel and mud. **SIMILAR SPECIES:** See (1) Alabama Sturgeon, *S. suttkusi*, and (2) Pallid Sturgeon, *S. albus* (Pl. 2).

ALABAMA STURGEON *Scaphirhynchus suttkusi* Not shown

IDENTIFICATION: Similar to Shovelnose Sturgeon, *S. platorynchus*, but orange-brown above, *yellow on lower side*; *lacks* large backward-directed spines on snout; no or few (at rear) scalelike scutes on belly. To 40 in. (100 cm). **RANGE:** Below Fall Line in Mobile Bay drainage, AL and MS. Rare; protected as an *endangered species*. **HABITAT:** Bottoms of main channels of large rivers; usually in current over sand or gravel; less often over mud. **SIMILAR SPECIES:** See Pl. 2. (1) See Shovelnose

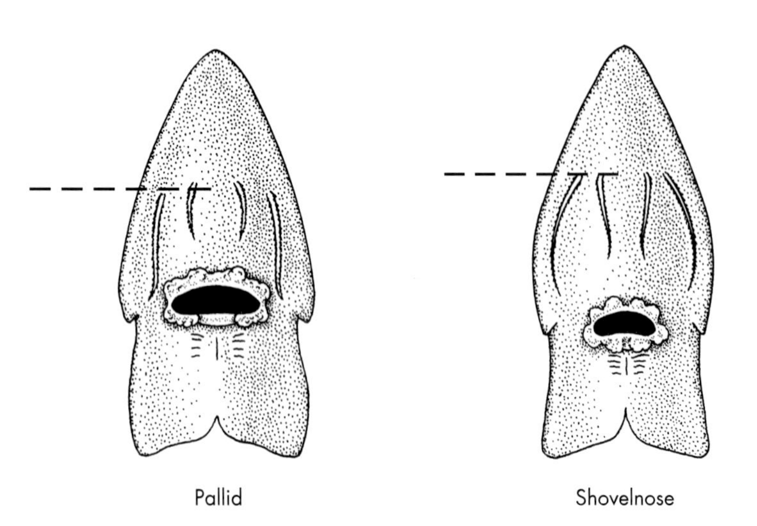

Fig. 7. Pallid and Shovelnose sturgeons—underside of head.

Sturgeon, *S. platorynchus*. (2) Pallid Sturgeon, *S. albus*, is *gray-white above and on side;* bases of outer barbels usually behind bases of inner barbels (Fig. 7), inner barbels shorter than outer barbels.

PALLID STURGEON *Scaphirhynchus albus* **Pl. 2**

IDENTIFICATION: Similar to Shovelnose Sturgeon, *S. platorynchus*, but has *no scalelike scutes* on belly; bases of outer barbels usually *behind* bases of inner barbels (Fig. 7), inner barbels shorter than outer barbels; gray-white above and on side. To 73 in. (185 cm). **RANGE:** Nearly restricted to main channels of Missouri R. and lower Mississippi R. from MT to LA. Rare; protected as an *endangered species*. **HABITAT:** Large, deep, turbid river channels; usually in strong current over firm sand or gravel. **SIMILAR SPECIES:** (1) See Shovelnose Sturgeon, *S. platorynchus* (Pl. 2). (2) Alabama Sturgeon, *S. suttkusi*, is orange-brown above, yellow on lower side; lacks large backward-directed spines on snout; bases of outer barbels *in line with, or ahead of*, inner barbels (Fig. 7); inner barbels same length as outer barbels.

PADDLEFISHES: Family Polyodontidae (1)

Paddlefishes are a family of 2 strange and ancient species that have a skeleton of cartilage, *heterocercal caudal fin, spiral valve,* and *long paddle-shaped snout*. One species, Paddlefish, *Polyodon spathula*, inhabits the Mississippi and adjacent Gulf drainages in North America. The other, Chinese Paddlefish, *Psephurus gladius*, is China's largest freshwater fish (to 21 ft. [6.4 m]) and inhabits the Yangtze River system. The Paddlefish's huge snout is covered with taste buds and may aid in locating plankton and other food organisms.

PADDLEFISH *Polyodon spathula* Pl. 2

IDENTIFICATION: *Long, canoe-paddle-shaped snout* about ⅓ body length. Huge mouth; toothless jaws. Unscaled except for tiny scales on caudal peduncle. Large, *fleshy, pointed flap* on rear edge of gill cover. Tiny eyes. Gray to blue-gray (nearly black), often mottled, above and on upper side; white below. Many long, slender gill rakers. To 87 in. (221 cm). **RANGE:** Mississippi R. basin from sw. NY to MT and south to LA; Gulf Slope drainages from Mobile Bay, AL, to Galveston Bay, TX. Formerly in Great Lakes basin; probably extirpated. Fairly common. **HABITAT:** Slow-flowing water of large rivers, lakes, impoundments; usually in water deeper than 4 ft. (1.2 m).

Gars: Family Lepisosteidae (5)

Gars are a group of 7 North American species of ancient fishes easily recognized by their *long, sharply toothed jaws, diamond-shaped and non-overlapping ganoid scales*, and dorsal and anal fins placed *far back* on the body. The primitiveness of gars is demonstrated by the retention in living species of a spiral valve, a lunglike gas bladder used to assist the gills in respiration, ganoid scales, and an abbreviated (rounded externally) heterocercal caudal fin. Small gars (less than 10 in. [25 cm]) have a rayless fleshy filament extending above the caudal fin.

In addition to the 5 species north of the Rio Grande are the Tropical Gar, *Atractosteus tropicus*, from southern Mexico to Costa Rica, and the Cuban Gar, *A. tristoechus*, native to western Cuba and Isle of Pines.

ALLIGATOR GAR *Atractosteus spatula* Pl. 3

IDENTIFICATION: *Giant of gars*—to 12 ft. (3.7 m). *Short, broad snout*; upper jaw shorter than rest of head. *Two rows of teeth* on upper jaw. Young has light stripe along back from tip of snout to upper base of caudal fin. Dark olive-brown (sometimes black) above and on side, occasionally spotted; white to yellow below; dark brown fins, few dark spots on median fins. Has 58–62 lateral scales; 48–54 predorsal scales; 59–66 rakers on 1st gill arch. **RANGE:** Mississippi R. basin from sw. OH and IL south to Gulf of Mexico; Gulf Coastal Plain from Econfina R., FL, to Veracruz, Mexico. Uncommon, except locally in swamps and bayous in s.-cen. U.S. **HABITAT:** Sluggish pools and backwaters of large rivers, swamps, bayous, and lakes. Rarely in brackish and marine water. **SIMILAR SPECIES:** (1) Shortnose Gar, *Lepisosteus platostomus* (Pl. 3), has upper jaw *longer than rest of head*, *1 row* of teeth on upper jaw, 16–25 rakers on 1st gill arch.

SHORTNOSE GAR *Lepisosteus platostomus* Pl. 3

IDENTIFICATION: *Short* (relative to those of other gars), *broad snout*; upper jaw longer than rest of head; 1 row of teeth on upper jaw. Olive or brown above and on side; white below; black spots on median fins, paired fins usually lack spots (some spots on head and paired fins of individuals from clear water). Young has fairly broad dark brown stripes along back and side. Has 59–65 lateral scales; 50–60 predorsal scales; 16–25 rakers on 1st gill arch. To 33 in. (83 cm). **RANGE:** Lowlands in Mississippi R. basin from s.-cen. OH and WI to MT and south to LA; Lake Michigan drainage, WI; Calcasieu and Mermentau rivers, LA. Common. **HABITAT:** Quiet pools and backwaters of creeks and small to large rivers, swamps, and lakes. Often found near vegetation and submerged logs. **SIMILAR SPECIES:** (1) Florida Gar, *L. platyrhincus*, and (2) Spotted Gar, *L. oculatus* (both Pl. 3), have many dark spots on body, head, and *all* fins, 53–59 lateral scales.

ALLIGATOR GAR
Atractosteus spatula

SHORTNOSE GAR
Lepisosteus platostomus

LONGNOSE GAR *Lepisosteus osseus* Pl. 3

IDENTIFICATION: *Long, narrow snout*; on adult, more than twice as long as rest of head. One row of teeth on upper jaw. Olive-brown above and on side; white below; dark spots on median fins and, in individuals from clear water, on body. Young has narrow brown stripe along back and broad dark brown stripe along side. Has 57–63 lateral scales; 47–55 predorsal scales; 14–31 rakers on 1st gill arch. To 72 in. (183 cm). **RANGE:** Atlantic Slope from Delaware R., NJ, to cen. FL; St. Lawrence R., QC, through Great Lakes and Mississippi R. basin to Red R. (Hudson Bay basin), ND, and south to LA; Gulf Slope drainages from cen. FL to Rio Grande drainage, TX and Mexico. Locally common, especially in clear lakes and impoundments. **HABITAT:** Sluggish pools, backwaters and oxbows of medium to large rivers, lakes. Usually near vegetation. Occasionally in brackish waters. **SIMILAR SPECIES:** (1) Other gars (Pl. 3) have much shorter, wider snout.

SPOTTED GAR *Lepisosteus oculatus* Pl. 3

IDENTIFICATION: Many *olive-brown to black spots on body, head, and all fins*. Moderately long snout; upper jaw longer than rest of head; 1 row of teeth on upper jaw. *Bony plates on underside of isthmus*. Olive-brown to black above and on side; white to yellow below. Young has dark stripes along back and side. Has 53–59 lateral scales; 45–54 predorsal scales; 15–24 rakers on 1st gill arch. To 44 in. (112 cm). **RANGE:** Lake Erie and s. Lake Michigan drainages; Mississippi R. basin from IL to Gulf Coast; Gulf Slope drainages from lower Apalachicola R., FL, to San Antonio R., TX. Locally common; protected in Canada as a *threatened species*. **HABITAT:** Quiet, clear pools and backwaters of lowland creeks; small to large rivers, oxbow lakes, swamps, sloughs; ditches with an abundance of vegetation or debris. Occasionally enters brackish water. **SIMILAR SPECIES:** (1) See Florida Gar, *L. platyrhincus* (Pl. 3). (2) Shortnose Gar, *L. platostomus* (Pl. 3), usually *lacks* dark spots on top of head, body, and paired fins; usually has 60–63 lateral scales.

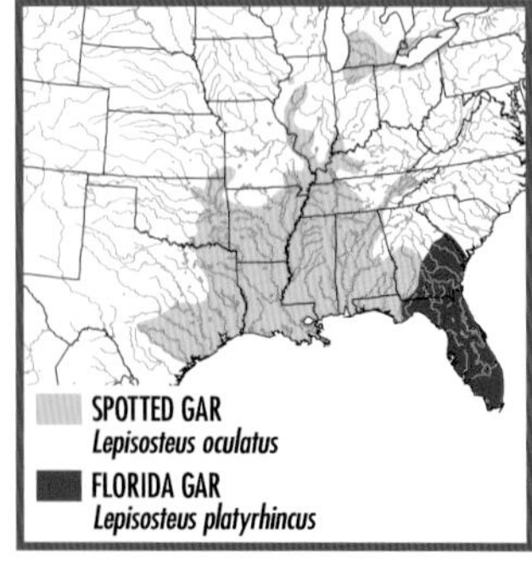

FLORIDA GAR *Lepisosteus platyrhincus* Pl. 3

IDENTIFICATION: Similar to Spotted Gar, *L. oculatus,* but *lacks* bony plates on underside of isthmus; has shorter, broader snout and lower jaw. Has 54–59 lateral scales; 47–51 predorsal scales; 19–33 rakers on 1st gill arch. To 52 in. (132 cm). **RANGE:** Savannah R. drainage, GA, to Ocklockonee R. drainage, FL and GA; throughout peninsular FL. Common. **HABITAT:** Sluggish mud- or sand-bottomed pools of quiet lowland streams and lakes; usually near vegetation. **SIMILAR SPECIES:** (1) See Spotted Gar, *L. oculatus* (Pl. 3). (2) Other gars (Pl. 3) usually lack large dark spots on anterior part of body, head, and paired fins.

BOWFINS: Family Amiidae (1)

Now represented by 1 living species, Amiidae once (back to the Jurassic) was distributed in Europe, Asia, Africa, South America, and North America. An ancient fish, the Bowfin, *Amia calva,* retains an abbreviated (rounded externally) heterocercal caudal fin, lunglike gas bladder, vestiges of a spiral valve, and a large gular plate on the underside of the head.

BOWFIN *Amia calva* Pl. 3

IDENTIFICATION: *Long, nearly cylindrical body*; large head. *Long dorsal fin* extends more than half length of back; 42–53 rays. *Tubular nostrils*. Large mouth; upper jaw extends beyond eye. Rounded pectoral, pelvic, and caudal fins. Cycloid scales. *Large, bony gular plate*. Mottled olive above; cream-yellow to pale green below. Black bands on dark green dorsal and caudal fins. Yellow to orange halo around prominent *black spot* near base of upper caudal fin rays in young; less distinct spot in adult. Brilliant turquoise green lips, throat, belly, and ventral fins on breeding male. To 43 in. (109 cm). **RANGE:** Native to St. Lawrence R.-Great Lakes and Mississippi R. basins from QC to n. MN and south to Gulf, and on Atlantic

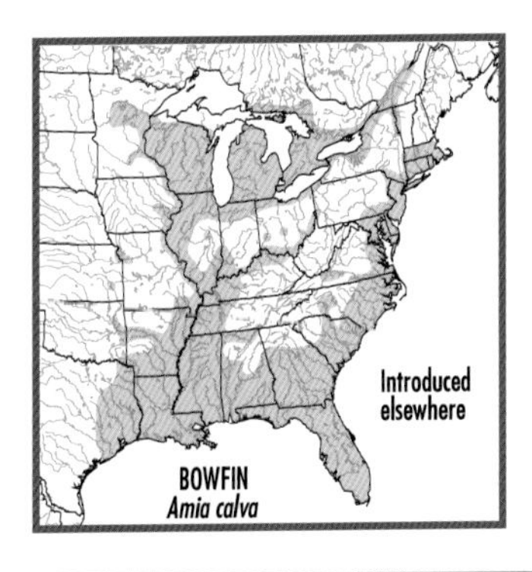

and Gulf coastal plains from Susquehanna R. drainage, se. PA, to Colorado R., TX. Introduced sparingly elsewhere, including on Atlantic Slope north to MA. Locally common. **HABITAT:** Swamps, sloughs, lakes, and pools and backwaters of lowland streams. Usually near vegetation.

MOONEYES: Family Hiodontidae (2)

This family contains only 2 living species, both restricted to N. America. Mooneyes resemble shads but have a *lateral line* and an *untoothed keel* along the belly. *They are strongly compressed, large-eyed fishes* with no scales on the head, cycloid scales on the body, no spines in the fins, 1 dorsal fin, an adipose eyelid, abdominal pelvic fins, and an axillary process at the base of the pelvic fin. Mooneyes feed on aquatic invertebrates and small fishes. Their eggs are semibouyant and drift downstream or into quiet water.

GOLDEYE *Hiodon alosoides* Pl. 4

IDENTIFICATION: *Deep, compressed silver body.* Dorsal fin origin *opposite or behind* anal fin origin. Fleshy keel along belly *from pectoral fin base to anal fin.* Large mouth; maxillary extends behind pupil of eye. Blunt, round snout. Usually 9–10 dorsal rays, 29–34 anal rays, 57–62 lateral scales. Silver blue-green above, silver white below; clear to dusky fins. Gold iris. To 20 in. (51 cm). **RANGE:** Tributaries of James Bay, QC and ON; Arctic and Mississippi R. basins from Mackenzie R. drainage, NT, to w. PA and south to LA. Locally common. **HABITAT:** Deep, open pools and channels of medium to large, often turbid, lowland rivers; lakes and impoundments. **SIMILAR SPECIES:** (1) See Mooneye, *H. tergisus* (Pl. 4).

MOONEYE *Hiodon tergisus* Pl. 4

IDENTIFICATION: Similar to Goldeye, *H. alosoides*, but has dorsal fin origin *in front of* anal fin origin, larger eye, deeper body, fleshy keel

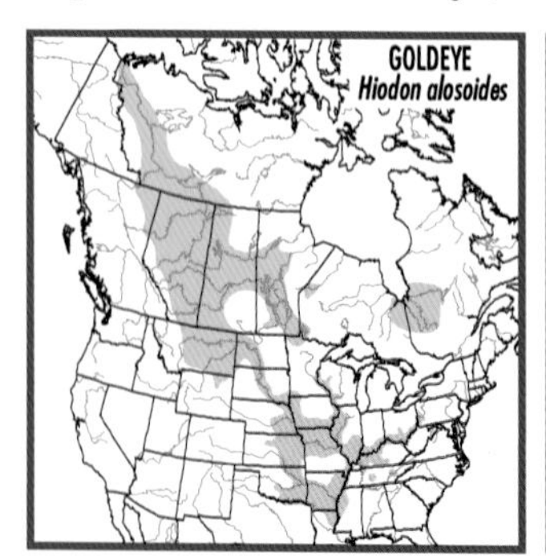

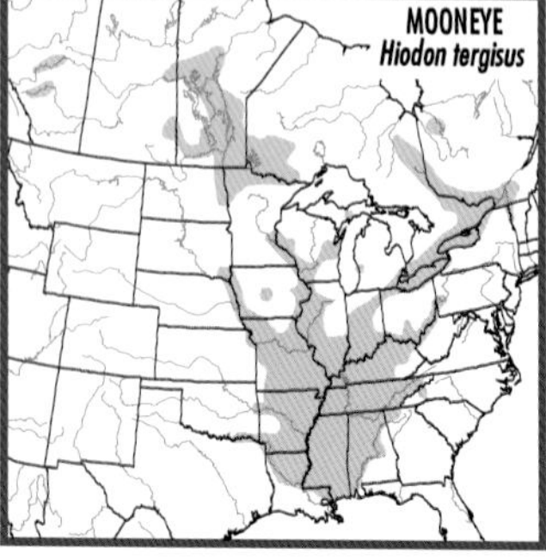

along belly from *pelvic fin base to anal fin*, maxillary extends to front or middle of pupil, more silver body and iris. Usually 11–12 dorsal rays, 26–29 anal rays, 52–57 lateral scales. To 19 in. (47 cm). **RANGE:** St. Lawrence-Great Lakes, Mississippi R., and Hudson Bay basins from QC to AB and south to LA; Gulf Slope drainages from Mobile Bay, AL, to Lake Pontchartrain, LA. Locally common. **HABITAT:** Deep pools and backwaters of medium to large rivers; lakes and impoundments. **SIMILAR SPECIES:** (1) See Goldeye, *H. alosoides* (Pl. 4).

FEATHERFIN KNIFEFISHES: Family Notopteridae (1 introduced)

This is a small family of freshwater fishes with a deep body, an *extremely long anal fin* that is *confluent with the caudal fin, no or tiny dorsal fin*, and no or small pelvic fins. Featherfin knifefishes are generally uniformly gray, tan, or silver, with many small black spots, wavy stripes, or ocelli above the anal fin. They grow to 5 ft. (1.5 m). The 10 species are native to Africa and tropical Asia.

CLOWN KNIFEFISH *Chitala ornata* Not shown

IDENTIFICATION: Deep, compressed body; *concave head-nape profile*; *6–10 large ocelli* on silver gray body above anal fin (on individuals over 4 in. [10 cm]); vertical bars on side of juvenile. Long anal fin (121–126 rays) confluent with caudal fin; small dorsal fin; tiny pelvic fins. To 3 ft. (1 m). **RANGE:** Native to Mekong, Chao Phraya, and Mekong river basins, Indo-China and Thailand. Established in Palm Beach Co., FL, where uncommon. **HABITAT:** Medium to large rivers, lakes; often near vegetation. **REMARKS:** A nocturnal predator. Can breathe air. Male builds nest of branches and leaves and guards eggs and young.

FRESHWATER EELS: Family Anguillidae (1)

Freshwater eels occur on every continent except Antarctica. There are 19 species, 1 of them in our area. They *lack* pelvic fins and have scales so small they appear to be absent. Adults of American Eels, *Anguilla rostrata,* and European Eels, *A. Anguilla,* migrate from fresh water to the Atlantic Ocean (usually assumed to be the Sargasso Sea between Bermuda and the West Indies), where they spawn and die. The leptocephalus (leaflike larva) migrates to shores of North America or Europe. The trip to North America takes 1 year; to Europe, 3 years. After arrival the larva transforms, in sequence, into a "glass-eel" (small and transparent but with adult morphology), an elver (darkly pigmented but still small), and finally an adult. Males remain in brackish water and in streams along coasts; females may migrate far upstream and remain there up to 15 years.

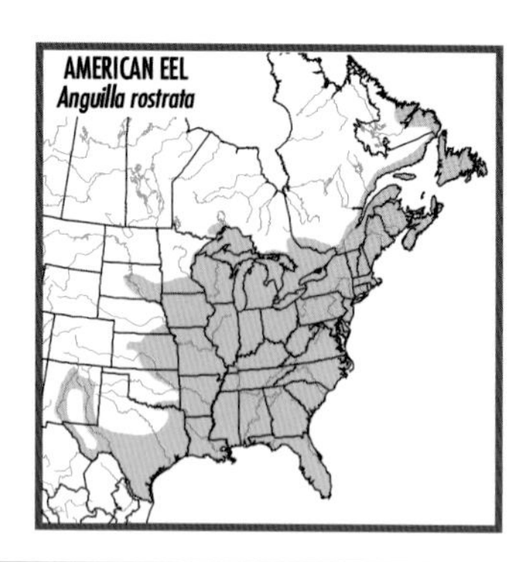

AMERICAN EEL *Anguilla rostrata* **Pl. 1**

IDENTIFICATION: Slender, *snakelike body*; small, pointed head. *Long dorsal fin* extends along more than half of body, continuous with caudal and anal fins. No pelvic fins. Small gill opening just in front of pectoral fin. Lower jaw projects well beyond upper jaw. Yellow to olive-brown above; pale yellow to white below; fins similar in color to adjacent body. To 60 in. (152 cm). **RANGE:** Catadromous species that spawns in Atlantic Ocean and ascends streams in N. and S. America. Found in Atlantic, Great Lakes, Mississippi, and Gulf basins from NL to SD and south to S. America. Common near sea; uncommon in more inland streams and lakes. **HABITAT:** Usually in permanent streams with continuous flow. Hides during daylight in undercut banks and deep pools near logs and boulders.

ANCHOVIES: Family Engraulidae (1)

These small, compressed silver fishes have a *long snout* overhanging a large mouth. *The upper jaw extends well past the eye*. There are no fin spines or lateral line. Anchovies usually school along warm coasts. Some species ascend rivers, and a few live in fresh water. Anchovies are economically important and are used as bait, harvested for food, or processed into fish meal and oil. There are 144 species, 1 of which enters fresh waters of our area.

BAY ANCHOVY *Anchoa mitchilli* **Pl. 57**

IDENTIFICATION: *Dorsal fin origin above or slightly in front of anal fin origin. Short snout* slightly overhanging mouth. Blue-green above; *silver stripe* along side, often faint at front; clear fins. No sawtooth edge on belly. Has 11–14 pectoral rays; 23–31 anal rays. To 4 in. (10 cm). **RANGE:** Atlantic and Gulf coasts from Harraseeket R., ME, to Yucatán,

Mexico. Abundant. **HABITAT:** Marine; enters open water of bays, estuaries, and lower reaches of coastal rivers. **SIMILAR SPECIES:** (1) Herrings and shads (Clupeidae) have upper jaw *not* extending past eye, sharply pointed scales on belly.

HERRINGS AND SHADS: Family Clupeidae (8)

Most herrings and shads are marine, but a few species live only in fresh water, and some marine species frequently enter fresh water. All have a *strongly compressed body, no lateral line*, and no scales on the head. They have cycloid scales on the body, 1 dorsal fin, a conspicuous adipose eyelid (Fig. 8), abdominal pelvic fins, an axillary process just above the base of the pelvic fin, no adipose fin, and no spines in the fins. The belly has sharply pointed scales, creating a *sawtooth edge*. Included among the world's 210 species of clupeids are sardines, menhaden, shads, and other economically important fishes.

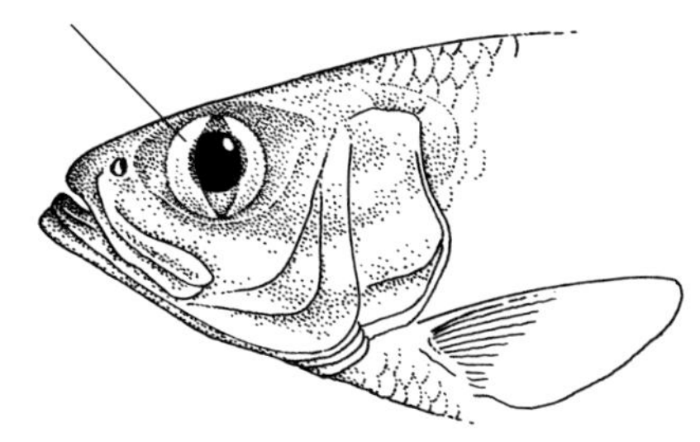

Fig. 8. Clupeid—adipose eyelid.

BLUEBACK HERRING *Alosa aestivalis* Pl. 4

IDENTIFICATION: *Strongly oblique mouth* (about 45° to horizontal); lower jaw *equal to or projecting only slightly beyond* snout. Cheek longer than or about equal to its depth (Fig. 9). *Blue above*; silver side, usually 1 small blue-black spot near upper edge of gill opening. Thin dark stripes on back and upper side, light green or yellow fins on adult. Teeth on lower jaw. Fairly small eye, diameter about equal to snout length. Usually 44–50 rakers on lower limb of 1st gill arch. Black peritoneum. To 16 in. (40 cm). **RANGE:** Atlantic Coast from NS to St. Johns R., FL. Ascends lower reaches of coastal rivers during spring spawning season; young move to ocean at about 1 month of age (1–2 in. [3–5 cm]). Introduced into VA and NC reservoirs and upper Tennessee R.

system, TN. Common; dams have reduced range and abundance. **FRESHWATER HABITAT:** Usually in current over rocky bottom. **SIMILAR SPECIES:** See Pl. 4. (1) See Alewife, *A. pseudoharengus*. (2) Hickory Shad, *A. mediocris,* and (3) Skipjack Herring, *A. chrysochloris,* have lower jaw projecting *far beyond* snout, usually fewer than 24 rakers on lower limb of 1st gill arch; are *gray-green* above. (4) American Shad, *A. sapidissima,* has cheek deeper than long (Fig. 9), *less oblique mouth*; lacks jaw teeth.

ALEWIFE *Alosa pseudoharengus* Pl. 4

IDENTIFICATION: Similar to Blueback Herring, *A. aestivalis,* but *blue-green above*; has *larger eye* (diameter greater than snout length), deeper body, usually *39–41 rakers* on lower limb of 1st gill arch. Gray peritoneum. To 15 in. (38 cm); usually less than 10 in. (25 cm) in land-locked populations. **RANGE:** Atlantic Coast from Red Bay, Labrador,

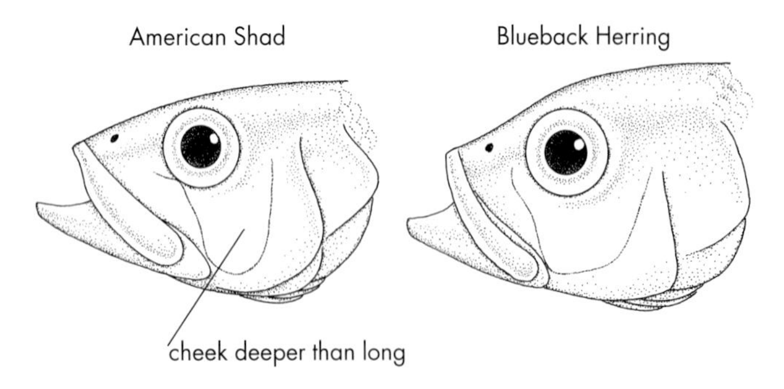

Fig. 9. American Shad and Blueback Herring.

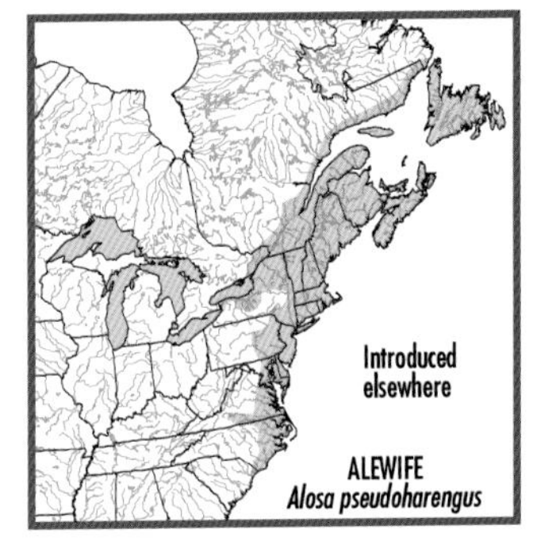

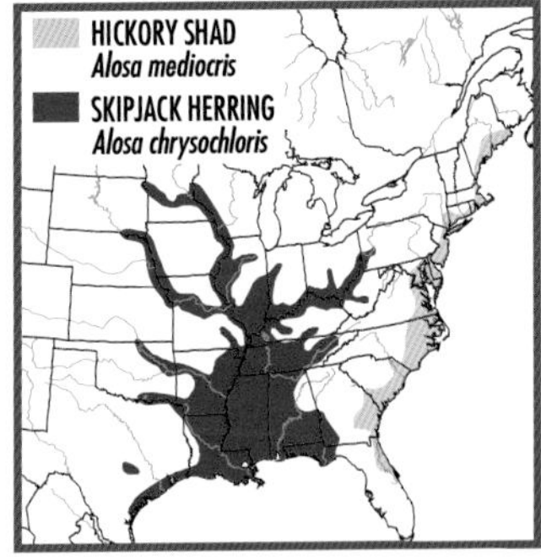

to SC; many landlocked populations. Individuals with access to ocean are anadromous, ascend coastal rivers during spring spawning migrations. Native to Lake Ontario; introduced into other Great Lakes via Welland Canal (first taken in Lake Erie in 1931). Introduced elsewhere, including New R., WV and VA, and upper Tennessee R. system, TN. Often abundant, although dams impede migrations. **HABITAT:** Open water over all bottom types. **SIMILAR SPECIES:** (1) See Blueback Herring, *A. aestivalis* (Pl. 4).

HICKORY SHAD *Alosa mediocris* Not shown

IDENTIFICATION: *Strongly oblique mouth* (about 45° to horizontal); lower jaw *projects beyond* snout. Cheek longer than or about equal to its depth. Blue-black spot near upper edge of gill cover followed by *row of poorly defined dusky spots* reaching to below dorsal fin. Gray-green above, *gradually shading to silver side*; clear fins, except often black edges on dusky dorsal and caudal fins. Teeth on lower jaw. Has 18–23 rakers on lower limb of 1st gill arch. To 24 in. (60 cm). **RANGE:** Atlantic Coast from Kenduskeag R., ME (possibly Campobello I., NB), to St. Johns R., FL; ascends coastal rivers during spring and fall. Least common of Atlantic Coast *Alosa*. **FRESHWATER HABITAT:** Open water of large rivers. **SIMILAR SPECIES:** See Pl. 4. (1) See Skipjack Herring, *A. chrysochloris*. (2) Blueback Herring, *A. aestivalis,* and (3) Alewife, *A. pseudoharengus,* have lower jaw *equal to or projecting only slightly beyond* snout, usually 39 or more rakers on lower limb of 1st gill arch; are blue or blue-green above. (4) American Shad, *A. sapidissima,* has cheek deeper than long (Fig. 9), less oblique mouth.

SKIPJACK HERRING *Alosa chrysochloris* Pl. 4

IDENTIFICATION: Similar to Hickory Shad, *A. mediocris,* but *lacks* blue-black spot behind gill cover; *blue-green above ends abruptly,* not gradually shading into silver side. Has 20–24 rakers on lower limb of 1st gill arch. To 21 in. (53 cm). **RANGE:** Red R. (Hudson Bay basin) and Mississippi R. basin from cen. MN and sw. PA to Gulf; Gulf Slope drainages from Ochlockonee R., FL, to San Antonio Bay, TX. Common; extirpated from upper Mississippi R. following construction of dams. **HABITAT:** Open water of medium to large rivers and large reservoirs. Usually in current over sand and gravel. Occasionally enters brackish and marine waters. **SIMILAR SPECIES:** (1) See Hickory Shad, *A. mediocris*. (2) Alabama Shad, *A. alabamae* (Pl. 4), has cheek deeper than long, less oblique mouth; adult lacks jaw teeth.

AMERICAN SHAD *Alosa sapidissima* Pl. 4

IDENTIFICATION: Adult *lacks* jaw teeth. Cheek decidedly *deeper than long* (Fig. 9); mouth less oblique than in above 4 species. Lower jaw equal to or projecting only slightly beyond snout. Blue-black spot near upper edge of gill opening usually followed by *1 or 2 rows of smaller spots*. Green or blue above; silver side; clear to light green fins; dusky dorsal and caudal fins; black edge on caudal fin. Has 59–73 rakers on lower limb of 1st gill arch of adult. To 30 in. (75 cm). **RANGE:** Atlantic Coast from Sand Hill R., Labrador, to St. Johns R., FL; ascends coastal rivers during spring spawning migrations. Introduced into Sacramento R., CA, in 1870s; has spread along Pacific Coast from Kamchatka, Russia, to Todos Santos Bay, Mexico. Landlocked in Millerton Lake, CA. Common. **FRESHWATER HABITAT:** Open water of large rivers. **SIMILAR SPECIES:** See Pl. 4. (1) See Alabama Shad, *A. alabamae*. (2) Blueback Herring, *A. aestivalis*, (3) Alewife, *A. pseudoharengus*, (4) Hickory Shad, *A. mediocris*, and (5) Skipjack Herring, *A. chrysochloris*, have *teeth* on lower jaw, strongly oblique mouth, *cheek longer than or about equal to its depth.*

ALABAMA SHAD *Alosa alabamae* Pl. 4

IDENTIFICATION: Nearly identical to American Shad, *A. sapidissima*, but adult has *42–48 rakers* on lower limb of 1st gill arch. To 20¼ in. (51

cm). **RANGE:** Gulf Coast from Suwannee R., FL, to Mississippi R., LA; ascends rivers during spawning runs. Sporadic in Mississippi R. basin, but reported as far north as Keokuk, IA, and known in Cumberland, Tennessee, Missouri, Arkansas, Ouachita, and Red river systems. Uncommon. **FRESHWATER HABITAT:** Open water of medium to large rivers. **SIMILAR SPECIES:** (1) See American Shad, *A. sapidissima* (Pl. 4). (2) Skipjack Herring, *A. chrysochloris* (Pl. 4), has teeth on lower jaw, strongly oblique mouth, cheek longer than or about equal to its depth.

GIZZARD SHAD *Dorosoma cepedianum* Pl. 4

IDENTIFICATION: *Long, whiplike last dorsal ray. Blunt snout*; distinctly *subterminal mouth*; deep notch at center of upper jaw. *No scales* on nape. Dorsal fin origin above or behind pelvic fin origin. *Large purple-blue spot* near upper edge of gill cover in young and small adult; faint or absent in large adult. Silver blue above, grading to silver white side, often blue-green sheen on head and body; 6–8 dark stripes on back and upper side; dusky fins, no black specks on chin or floor of mouth. Has 52–70 lateral scales; 10–13 dorsal rays; 25–36 anal rays. To 20½ in. (52 cm). **RANGE:** St. Lawrence-Great Lakes, Mississippi, Atlantic, and Gulf drainages from QC to cen. ND, and south to s. FL and Mexico. Common. **HABITAT:** Open water of medium to large rivers, lakes, and impoundments; ascends creeks and small rivers with well-developed pools. Enters brackish water. **SIMILAR SPECIES:** (1) See Threadfin Shad, *D. petenense* (Pl. 4).

THREADFIN SHAD *Dorosoma petenense* Pl. 4

IDENTIFICATION: Similar to Gizzard Shad, *D. cepedianum,* but has *more pointed snout, projecting lower jaw, black specks* on chin and floor of mouth, *yellow fins*, purple shoulder spot near upper edge of gill cover *persisting in adult, 40–48 lateral scales*. Has 11–14 dorsal rays; 17–27 anal rays. To 9 in. (23 cm). **RANGE:** Native range presumably Gulf Slope drainages from Mississippi R. to Cen. America. Widely introduced as forage; from Ohio R., OH, and Mississippi R., cen. IL, south through Mississippi R. basin to Gulf, and Atlantic and Gulf drainages from Chesapeake Bay to Guatemala. Also introduced to Colorado R. drainage, AZ, NV, and CA, Pacific drainages, CA, and elsewhere. Common. **HABITAT:** Lakes, backwaters and pools of medium to large rivers; usually in open water over sand, mud, and debris. Occasionally enters brackish water. **SIMILAR SPECIES:** (1) See Gizzard Shad, *D. cepedianum* (Pl. 4).

CARPS AND MINNOWS: Family Cyprinidae (261 native; 10 introduced)

Cyprinidae is the largest family of fishes, containing about 2700 species, and is present on all continents except South America, Australia, and Antarctica. The greatest diversity is in Southeast Asia.

"Minnow" is sometimes applied to any small fish but properly refers to species of Cyprinidae. Minnows have *1 dorsal fin, abdominal pelvic fins*, a lateral line (rarely absent), and cycloid scales on the body. They have no true spines in the fins, although some (e.g., carp and spinefins) have soft rays that develop into spinelike structures (referred to below as spines). They lack teeth on the jaws but have 1–3 rows of teeth on pharyngeal bones that grind food against a horny pad on the basioccipital. Although not useful on live minnows, numbers and configuration of pharyngeal teeth are helpful in identification of preserved specimens. A notation of "2,5-4,2," for example, means that teeth are in 2 rows on each side of the head, with 2 teeth in each outer row, 5 in the inner row on the left side, and 4 in the inner row on the right side. In genus and species accounts, *anal ray and pharyngeal tooth counts are modes*; other counts are ranges unless stated otherwise.

Minnows are as varied ecologically as they are morphologically. Species of *Notropis* are mostly small midwater carnivores that feed mainly on small crustaceans and insects. Herbivorous minnows, feeding predominantly on algae, include stonerollers, *Campostoma;* Chiselmouth, *Acrocheilus alutaceus;* some silvery minnows, *Hybognathus;* and a few species of *Notropis* (e.g., Cape Fear Shiner, *N. mekistocholas*). Plants takes longer than animals to digest, and herbivores have long intestines. The long and usually coiled gut is often visible through the belly wall. The body cavity lining (peritoneum) is usually white in carnivorous minnows and black in herbivorous ones. The color of the peritoneum sometimes can be seen through the body wall; unless stated otherwise, a species has a white peritoneum.

GRASS CARP *Ctenopharyngodon idella* Pl. 19

IDENTIFICATION: *Wide head*; terminal mouth. *Large (34–45 lateral) scales, dark-edged* with *black spot* at base. Slender (especially adult), fairly compressed body; dorsal fin origin in front of pelvic fin origin. Gray to brassy green above; white to yellow below; clear to gray-brown fins. Has 7 dorsal rays; 8 anal rays. Pharyngeal teeth 2,5-4,2 or 2,4-4,2, elongate with *prominent parallel grooves* on grinding surfaces, often hooked at tip. Has 12–16 rakers on 1st gill arch. To 5 ft. (1.5 m). **RANGE:** Native to e. Asia. Introduced into AR in 1960s; now known in at least 34 states. Uncommon but increasing in lower Mississippi R. **HABITAT:** Lakes, ponds, pools, and backwaters of large rivers. **REMARKS:** Grass Carp was introduced into U.S. to control aquatic weed problems in lakes and ponds. However, many fishes, waterfowl, and other native species are dependent on aquatic vegetation. In many places, introduction of Grass Carp has proven to be more costly than beneficial. **SIMILAR SPECIES:** (1) See Black Carp, *Mylopharyngodon piceus* (Pl. 19).

BLACK CARP *Mylopharyngodon piceus* Pl. 19

IDENTIFICATION: Similar to Grass Carp, *Ctenopharyngodon idella,* but is dark, often *blue-gray or black above* and on side; fins dark, almost black; *pharyngeal teeth 0,4-5,0* (rarely, 1 or 2 small minor-row teeth), massive, *molarlike, and smooth.* Usually *18–21 rakers* on 1st gill arch. To 6½ ft. (2 m). **RANGE:** Native to e. Asia. Introduced into U.S. in 1970s. Rare, but thought to be reproducing and increasing in lower Mississippi R. basin. **HABITAT:** Floodplain lakes, backwaters, main channels of large rivers. **REMARKS:** Black Carp was introduced into U.S. to control snails and other mollusks in lakes and ponds. Its escape into river basins is likely to exacerbate the already serious decline in N. American mussels and snails. **SIMILAR SPECIES:** (1) See Grass Carp, *Ctenopharyngodon idella* (Pl. 19).

SILVER CARP *Hypophthalmichthys molitrix* Pl. 19

IDENTIFICATION: Deep, laterally compressed body; *eye below middle of head.* Small (91–124 lateral) scales on silvery sides. *Keel from anus to junction of branchiostegal membranes.* Large, terminal mouth. Dorsal fin origin about even with pelvic fin origin. Olive to gray above, silver or white on side and below. Has 7 dorsal rays; 11–14 anal rays. Over 100 rakers fused into *spongelike filtering apparatus* on 1st gill arch. Pharyngeal teeth 0,4-4,0. To 4 ft. (1.2 m). **RANGE:** Native to e. China. Introduced into U.S. in 1973; now occurs in at least 16 states. Established in middle and lower Mississippi R. basin from IA, IL, IN, and KY to LA. Abundant where established. **HABITAT:** Open water of large rivers, lakes, ponds, and reservoirs. **REMARKS:** Feeds on phytoplankton and was introduced to improve water quality in aquaculture. Jumps into boats with outboard motors, sometimes seriously injuring humans. **SIMILAR SPECIES:** (1) See Bighead Carp, *H. nobilis* (Pl. 19).

BIGHEAD CARP *Hypophthalmichthys nobilis* Pl. 19

IDENTIFICATION: Similar to Silver Carp, *H. molitrix, but keel from anus to base of pelvic fins;* dorsal fin origin usually behind pelvic fin origin; dark gray above and on side to off-white below; young are silver until about 2 months of age when *irregular gray-black blotches* develop on body. About *130 unfused rakers* on 1st gill arch. To 5 ft. (1.5 m). **RANGE:** Native to e. Asia. Introduced into AR in 1972. Established in Missouri, Mississippi, and Ohio river basins from cen. KY to SD and south to LA. Common. **HABITAT:** Open water of large rivers, backwaters, floodplain lakes, reservoirs, and ponds; young sometimes in small creeks. **REMARKS:** Feeds on zooplankton and was introduced to improve water quality in aquaculture facilities. **SIMILAR SPECIES:** (1) See Silver Carp, *H. molitrix* (Pl. 19).

GOLDFISH *Carassius auratus* Pl. 19

IDENTIFICATION: *Large scales*, 25–31 along lateral line. *Long dorsal fin*, 15–21 rays. *Stout, saw-toothed spine* (and 2 smaller spines) at front of dorsal and anal fins. No barbels. Deep, thick body; terminal mouth; large caudal fin. Gray-green above; brassy sheen on back and side; white to yellow below; gray to brown dorsal and caudal fins (see Remarks). Has 5–6 anal rays; pharyngeal teeth 0,4-4,0. To 16 in. (41 cm). **RANGE:** Native to Asia; first introduced into U.S. in late 1600s. Now established in much of U.S., s. ON, s. AB, and s. BC. Locally common. **HABITAT:** Shallow, muddy pools and backwaters of sluggish rivers; ponds and lakes. Usually in warm turbid or vegetated water; more tolerant than most fishes of some forms of pollution. **REMARKS:** Recently released goldfish with "pet-store" colors (red, white, blue, black) may be encountered in the wild. However, because of natural selection, reproducing populations quickly revert back to natural cryptic colors. Hybridizes readily with Common Carp, *Cyprinus carpio;* in areas of major pollution, hybrids (which are fertile) outnumber parent species. **SIMILAR SPECIES:** (1) Common Carp, *Cyprinus carpio* (Pl. 19), has barbels, 32–38 lateral scales, pharyngeal teeth 1,1,3-3,1,1; is more flattened below.

COMMON CARP *Cyprinus carpio* Pl. 19

IDENTIFICATION: *Two barbels*, rear one much larger, on each side of upper jaw. *Long dorsal fin*, 17–21 rays. *Stout, saw-toothed spine* (and 2 smaller spines) at front of dorsal and anal fins. Deep, thick body, strongly arched to dorsal fin, flattened below. Mouth terminal on young, subterminal on adult. Gray (young) to brassy green (adult) above; scales on back and upper side dark-edged, with black spot at base; white to yellow below; clear to dusky fins. Large adult has red-orange caudal and anal fins. Large scales (32–38 lateral scales); 5–6 anal rays; pharyngeal teeth 1,1,3-3,1,1. To 4 ft. (1.2 m). **RANGE:** Native to Eurasia; First introduced to N. America in 1831 and now widely distributed in s. Canada and most of U.S. Common. **HABITAT:** Muddy pools of small to large rivers; lakes and ponds. Usually in impoundments and turbid, sluggish streams with organic matter. **REMARKS:** Individuals known as "mirror carp" (with few, enlarged scales) or "leather carp" (scaleless) are fairly common. Common Carp roots in mud, increases turbidity, and can result in decreased populations of native fishes. It is considered, as are most introduced species, a nuisance. **SIMILAR SPECIES:** (1) Goldfish, *Carassius auratus* (Pl. 19), *lacks* barbels and dark-edged scales; is less flattened below; has 25–31 lateral scales, pharyngeal teeth 0,4-4,0.

IDE *Leuciscus idus* Not shown

IDENTIFICATION: Slender body somewhat compressed, *humped* (adult) behind head. *Red fins*, especially anal and paired fins of adult. Rounded snout, terminal mouth. Dorsal fin origin behind pelvic fin origin. Gray to gold-brown above, silver side. Has 53–63 lateral scales; 8 dorsal rays; 9–10 anal rays; pharyngeal teeth 3,5-5,3. To 40 in. (102 cm). **RANGE:** Native to Eurasia. Introduced into U.S. in 1870s and reported from several states (in east and as far west as MN, NE, and TX) but does not appear to be established. **HABITAT:** Large lowland rivers and lakes; moves into smaller streams to spawn.

TENCH *Tinca tinca* Not shown

IDENTIFICATION: *Small scales*, 95–105 in lateral line; thick, *leatherlike skin*; *short, deep caudal peduncle*; truncate caudal fin. *Long barbel* at corner of terminal mouth. Dorsal fin origin over or behind pelvic fin origin. Olive to dark green above; white to bronze below with gold sheen; gray fins. Has 8 dorsal rays; 7–8 anal rays; pharyngeal teeth 0,5-4,0 or 0,5-5,0. To 33 in. (84 cm). **RANGE:** Native to Eurasia. Introduced into U.S. in 1880s; now reproducing in CT, CO, BC, WA, CA, and possibly NY and MD. Localized and uncommon. **HABITAT:** Mud-bottomed, usually vegetated lakes; backwaters of small to large rivers.

BITTERLING *Rhodeus sericeus* Not shown

IDENTIFICATION: *Incomplete lateral line*, confined to 1st 4–10 scales. *Orange to red fins* on adult. Fairly deep, compressed body; dorsal fin origin behind pelvic fin origin. Rounded snout, terminal to slightly subterminal mouth. Gray-green above, metallic blue stripe along rear half of silver side; dusky orange fins. Breeding male has iridescent blue side, bright red dorsal and anal fins. Has 34–40 lateral scales; 9–10 dorsal rays; 8–9 anal rays; pharyngeal teeth 0,5-5,0. To 4¼ in. (11 cm). **RANGE:** Native to Eurasia. Introduced and possibly established in Bronx R. (lower Hudson R. drainage), NY. **HABITAT:** Ponds, lakes, marshes; muddy and sandy pools and backwaters of rivers. **REMARKS:** Female develops extremely long ovipositor to lay eggs in shells of living mussels.

RUDD *Scardinius erythrophthalmus* **Fig. 10**

IDENTIFICATION: *Deep, compressed body; scaled bony keel* along belly from pelvic to anal fin. *Bright red* anal, pelvic, and pectoral fins; red-brown dorsal and caudal fins. Small head; terminal oblique mouth; dorsal fin origin behind pelvic fin origin. Concave margin on dorsal and anal fins (tips of front rays reach beyond fin when depressed). Brown-green above, brassy yellow side; gold eye with red spot at top. Has 36–45 lateral scales; usually 9–11 dorsal rays, 10–11 anal rays; 10–13 rakers on 1st gill arch; pharyngeal teeth 3,5-5,3. To 19 in. (48 cm). **RANGE:** Native to Eurasia. Reproducing populations in lower Hudson R. drainage and Finger Lakes region, NY, and in ME. Uncommon in U.S. **HABITAT:** Lakes; sluggish pools of medium to large rivers. **SIMILAR SPECIES:** (1) Golden Shiner, *Notemigonus crysoleucas* (Pl. 5), usually *lacks* red fins (except in southern populations; young elsewhere may have light red-orange median fins); has scaleless keel, usually 7–9 dorsal rays, 11–14 anal rays, 17–19 rakers on 1st gill arch; pharyngeal teeth 0,5-5,0.

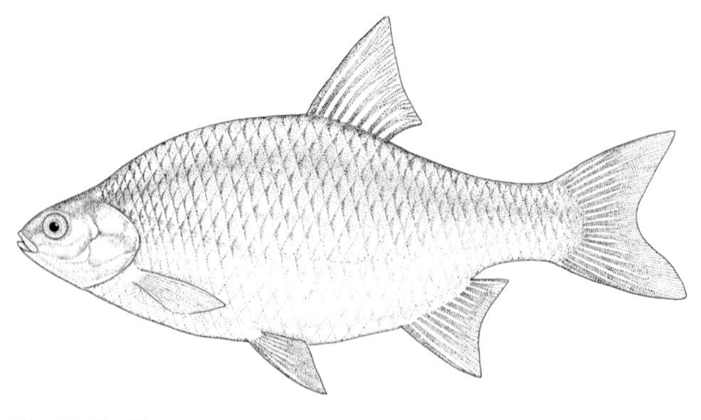

Fig. 10. Rudd.

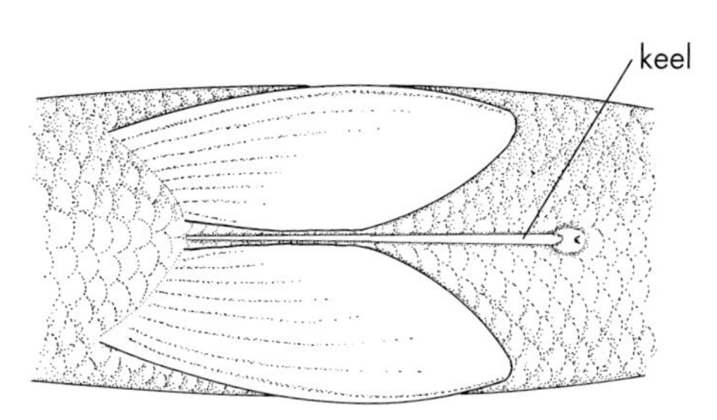

Fig. 11. Golden Shiner—belly.

GOLDEN SHINER *Notemigonus crysoleucas* Pl. 5

IDENTIFICATION: *Extremely compressed body*; strongly *decurved* lateral line; *scaleless keel* along belly from pelvic to anal fin (Fig. 11). Small, *upturned mouth* on pointed snout; dorsal fin origin behind pelvic fin origin. Silver in clear and turbid water; gold side and fins in coffee-colored water; fins red in southern populations. Dusky stripe along side, *herringbone lines* on upper side of young. Has 44–54 lateral scales; 7–9 dorsal rays; 8–19, usually 11–14, anal rays; 17–19 rakers on 1st gill arch; pharyngeal teeth 0,5-5,0. To 12½ in. (32 cm). **RANGE:** Native to Atlantic and Gulf slope drainages from NS to s. TX, Great Lakes, Hudson Bay, and Mississippi R. basins west to SK, MT, and w. OK and TX. Introduced (via bait buckets) elsewhere in U.S. Common except in uplands. **HABITAT:** Vegetated lakes, ponds, swamps, backwaters and pools of creeks and small to medium rivers. **SIMILAR SPECIES:** (1) Rudd, *Scardinius erythrophthalmus* (Fig. 10), has red fins, *scaled keel*, usually 9–11 dorsal rays, 10–11 anal rays, 10–13 rakers on 1st gill arch; 3,5-5,3 pharyngeal teeth.

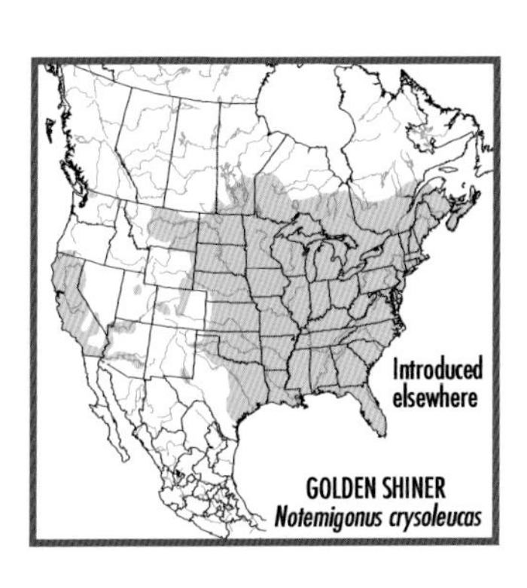

HITCH *Lavinia exilicauda* **Pl. 5**

IDENTIFICATION: *Deep, compressed body,* tapering to narrow caudal peduncle; large caudal fin; strongly *decurved lateral line*. Small compressed head, terminal (slightly upturned) mouth. Dorsal fin origin behind pelvic fin origin. Brown-yellow above, scales on back and silver side darkly outlined (crosshatched pattern on small individuals); dusky fins. Small individuals have black caudal spot. Has 54–62 lateral scales, 10–13 dorsal rays, 11–13 anal rays; pharyngeal teeth 0,5-4,0 or 0,5-5,0. To 14 in. (36 cm). **RANGE:** Sacramento-San Joaquin, Clear Lake, Russian R., and Pajaro-Salinas river drainages, CA. Common. **HABITAT:** Lakes, ponds, sloughs, backwaters and sluggish sandy pools of small to large rivers. **REMARKS:** Three subspecies: *L. e. chi*, lake-adapted form in Clear Lake; *L. e. harengus*, deep-bodied form with low fin-ray counts in Pajaro and Salinas rivers; *L. e. exilicauda* in rest of range. **SIMILAR SPECIES:** (1) Golden Shiner, *Notemigonus crysoleucas* (Pl. 5), has keel on belly, 7–9 dorsal rays, more compressed body.

CALIFORNIA ROACH *Hesperoleucus symmetricus* **Pl. 12**

IDENTIFICATION: *Slightly subterminal mouth* on fairly short snout. Deep compressed body tapering to *narrow caudal peduncle*; dorsal fin origin behind pelvic fin origin; large caudal fin. Dusky gray to steel blue above; dark stripe along side; silver white below. Breeding individual has red-orange chin, gill cover, and anal and paired fin bases. Has 47–63 lateral scales; 7–10 dorsal rays; pharyngeal teeth 0,5-4,0. To 4¼ in. (11 cm). **RANGE:** Sacramento-San Joaquin, Russian, Pajaro-Salinas, and smaller coastal drainages, CA. Introduced into Cuyama R., Soquel Creek, and Eel R. drainages, CA. Common, but decreasing. **HABITAT:** Rocky pools of headwaters, creeks, and small to medium rivers. **REMARKS:** Several subspecies are recognized through combinations of characters not described here because comprehensive

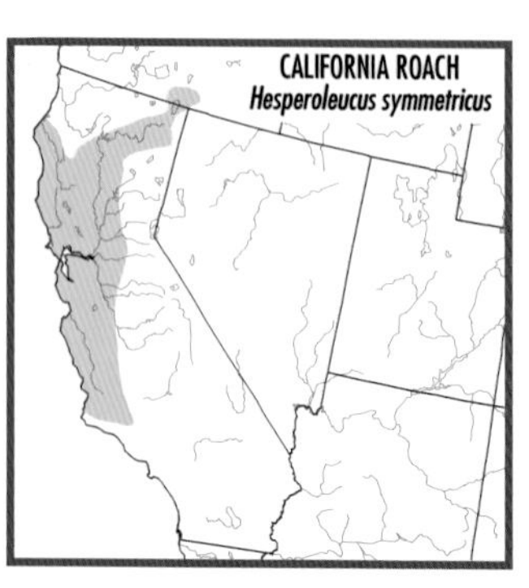

study of intraspecific variation is needed. **SIMILAR SPECIES:** (1) Lahontan Redside, *Richardsonius egregius* (Pl. 12), and (2) Redside Shiner, *R. balteatus* (Pl. 12), have bright red side on adult, *terminal mouth* on pointed snout, axillary process at pelvic fin base; pharyngeal teeth in 2 rows. (3) Hitch, *Lavinia exilicauda* (Pl. 5), has deeper, more compressed body, *terminal mouth*, 10–13 dorsal rays, no red coloration.

SPLITTAIL *Pogonichthys macrolepidotus* Pl. 5

IDENTIFICATION: Upper lobe of *large caudal fin longer* than lower lobe. *Barbel* at corner of slightly subterminal mouth. Long, slender, compressed body; dorsal fin origin in front of pelvic fin origin; small head; large eye. Olive-gray above, silver gold side. Large individual has hump on nape, red-orange caudal and paired fins. Has 9–10 dorsal rays; 57–64 (usually 60–62) lateral scales; 14–18 rakers on 1st gill arch; pharyngeal teeth 2,5-5,2. To 17½ in. (44 cm). **RANGE:** Formerly throughout Sacramento-San Joaquin R. drainage, CA; now mostly in San Francisco Bay Delta and lower Sacramento and San Joaquin rivers. Common in reduced range. **HABITAT:** Backwaters and pools of rivers; lakes. Tolerant of brackish water. **SIMILAR SPECIES:** (1) See Clear Lake Splittail, *P. ciscoides*. Other minnows have symmetrical caudal fin.

CLEAR LAKE SPLITTAIL *Pogonichthys ciscoides* Not shown

IDENTIFICATION: Similar to Splittail, *P. macrolepidotus*, but *no* (or small) *barbel* on *terminal mouth*, less asymmetrical caudal fin; usually 62–65 lateral scales; 18–23 rakers on 1st gill arch. To 14 in. (36 cm). **RANGE:** Clear Lake and tributaries, Lake Co., CA. Extinct. **HABITAT:** Shoreline (as young) and open water of Clear Lake; spawned in lake tributaries. **SIMILAR SPECIES:** (1) See Splittail, *P. macrolepidotus* (Pl. 5).

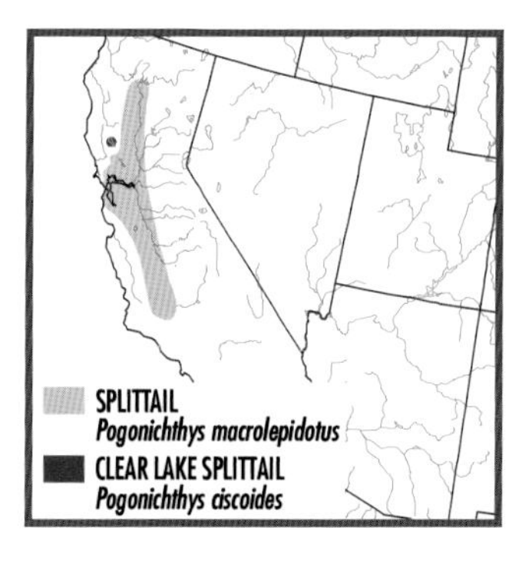

PTYCHOCHEILUS

GENUS CHARACTERISTICS: Long conical head, flattened between eyes; *very large, horizontal, terminal mouth* extending to or past front of eye; slender, barely compressed body; large, *deeply forked caudal fin*; dorsal fin origin slightly behind pelvic fin origin; *narrow caudal peduncle*, appears pinched near caudal fin; moderately decurved, complete lateral line. Gray-green above, silver side; fins clear to slate with yellow to orange margins on large individual. Large male has bright orange anal, pelvic, and pectoral fins. Young has dusky stripe on rear half of side, black caudal spot. Pharyngeal teeth 2,4-4,2 to 2,5-5,2. Pikeminnows are the largest N. American minnows. They differ from one another mainly in numbers of lateral scales and fin rays, which can be difficult to determine on live specimens. However, no river system contains more than 1 species.

COLORADO PIKEMINNOW *Ptychocheilus lucius* Pl. 5

IDENTIFICATION: *Very small scales*: 76–97 (usually 80–87) lateral scales, 18–23 scales above lateral line; usually 9 dorsal rays, 9 anal rays. To 6 ft. (1.8 m). **RANGE:** Colorado R. drainage, WY, CO, UT, NM, AZ, NV, CA, and Mexico. Now mostly restricted to UT and CO; extirpated from southern portion of range by construction of large dams. Protected as an *endangered species*. **HABITAT:** Pools of medium to large rivers. Large individuals usually in deep, flowing rocky or sandy pools. **SIMILAR SPECIES:** (1) Other pikeminnows (Pl. 5) have *larger scales* (usually fewer than 75 lateral scales), *deeper caudal peduncle*, usually 8 anal rays; reach a smaller maximum size (to 4½ ft. [1.4 m]). (2) Roundtail Chub, *G. robusta* (Pl. 12), has mouth extending only to *front* of eye; young lacks black spot on caudal fin base.

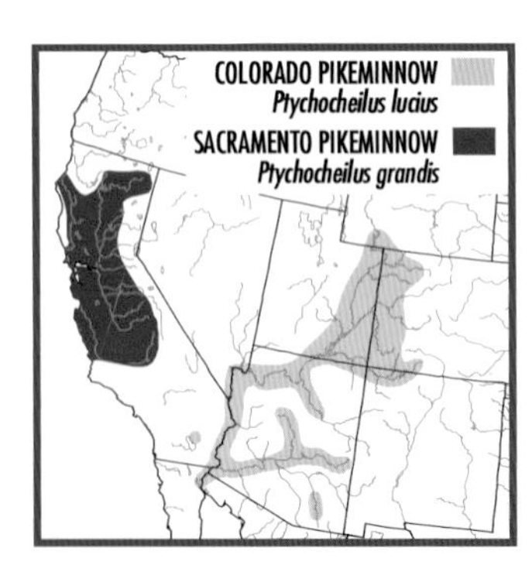

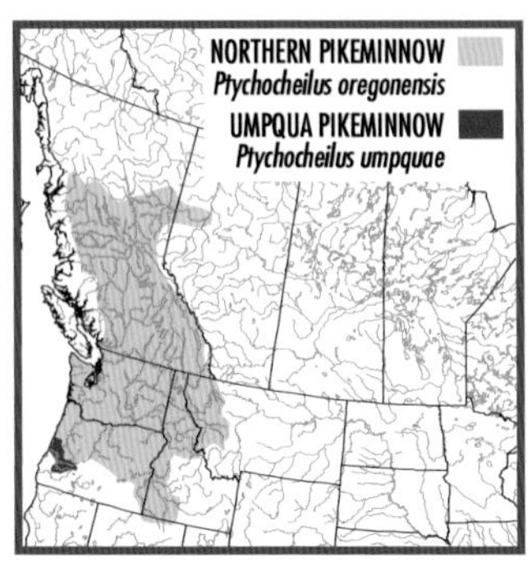

SACRAMENTO PIKEMINNOW *Ptychocheilus grandis* Pl. 5

IDENTIFICATION: 38–44 scales on back from head to dorsal fin, 65–78 (usually 67–75) lateral scales, 12–15 scales above lateral line; usually 8 dorsal rays, 8 anal rays. To 4½ ft. (1.4 m). **RANGE:** Sacramento-San Joaquin, Pajaro-Salinas, Russian, Clear Lake drainages, CA. Introduced elsewhere in CA, including Eel R. and tributaries to Morro Bay. Common in clear, warm streams. **HABITAT:** Rocky and sandy pools and runs of small to large rivers. **SIMILAR SPECIES:** (1) Other pikeminnows, *Ptychocheilus* (Pl. 5), have 48 or more scales on back from head to dorsal fin, 9 dorsal rays.

NORTHERN PIKEMINNOW *Ptychocheilus oregonensis* Pl. 5

IDENTIFICATION: Has 48–72 (usually 51–62) scales on back from head to dorsal fin, 64–79 (usually 66–75) lateral scales, 12–20 scales above lateral line; usually 9 dorsal rays, 8 anal rays. To 25 in. (63 cm). **RANGE:** Pacific drainages from Nass R., BC, to Columbia R., NV; Harney R. basin, OR; Peace R. system (Arctic basin), BC and AB. Common; locally abundant. **HABITAT:** Lakes, pools, and sometimes runs of small to large rivers. Large individuals in deep water. **SIMILAR SPECIES:** (1) Umpqua Pikeminnow, *P. umpquae,* usually has 18–20 scales above lateral line, 60–83 scales on back from head to dorsal fin. (2) Sacramento Pikeminnow, *P. grandis* (Pl. 5), has 38–44 scales on back from head to dorsal fin, usually 12–14 scales above lateral line, 8 dorsal rays.

UMPQUA PIKEMINNOW *Ptychocheilus umpquae* Not shown

IDENTIFICATION: Has 60–83 scales on back from head to dorsal fin, 66–81 (usually 68–75) lateral scales, 16–22 scales above lateral line; usually 9 dorsal rays, 8 anal rays. To 17½ in. (44 cm). **RANGE:** Umpqua and Siuslaw river drainages, OR. Common. **HABITAT:** Pools and sluggish runs of creeks and small rivers; lakes. **SIMILAR SPECIES:** See Pl. 5. (1) Northern Pikeminnow, *P. oregonensis*, usually has 14–18 scales above lateral line, 51–62 scales on back from head to dorsal fin. (2) Sacramento Pikeminnow, *P. grandis*, usually has 38–44 scales on back from head to dorsal fin, 12–15 scales above lateral line, 8 anal rays.

HARDHEAD *Mylopharodon conocephalus* Pl. 5

IDENTIFICATION: *Premaxillary frenum*; long, pointed snout; large terminal mouth reaches front of eye. Long, slender body; dorsal fin origin behind pelvic fin origin. Brown to dusky bronze above (darkest on large individual); silver side. Has 69–81 lateral scales; 8 dorsal rays; pharyngeal teeth 2,5-4,2. To 3 ft. (1 m). **RANGE:** Sacramento-San Joaquin and Russian river drainages, CA. Fairly common, but becoming localized. **HABITAT:** Deep, rock- and sand-bottomed pools of small to large rivers. **SIMILAR SPECIES:** (1) Sacramento Pikeminnow, *Ptychocheilus grandis*, and (2) Sacramento Blackfish, *Orthodon microlepidotus* (both

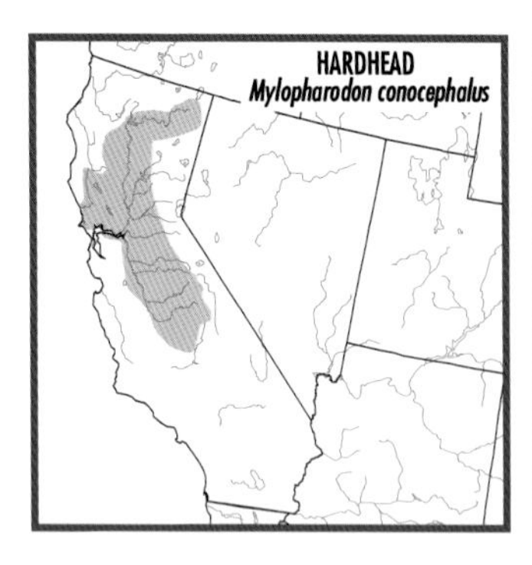

Pl. 5), *lack* premaxillary frenum; Sacramento Pikeminnow has mouth extending under eye; Sacramento Blackfish has dorsal fin origin in front of pelvic fin, 90–105 lateral scales.

SACRAMENTO BLACKFISH *Orthodon microlepidotus* Pl. 5

IDENTIFICATION: *Small* (90–105 lateral) *scales. Wide head*, flat above; slightly upturned mouth. Slender, compressed body; long *narrow caudal peduncle*; dorsal fin origin in front of pelvic fin origin. Light to dark gray above, often with olive sheen; silver side. Has 9–11 dorsal rays; pharyngeal teeth 0,6-6,0 or 0,6-5,0. To 21½ in. (55 cm). **RANGE:** Native to Sacramento-San Joaquin, Pajaro, and Salinas river drainages, and Clear Lake, CA; also Russian R., CA, where possibly introduced. Introduced to Santa Ana R., CA, and Humboldt R. system, NV. Common in native range (abundant in Clear Lake). **HABITAT:** Lakes, backwaters and sluggish pools of small to large rivers; usually in warm, turbid water. **SIMILAR SPECIES:** (1) Hardhead, *Mylopharodon conocephalus* (Pl. 5), has premaxillary frenum, *larger* (69–81 lateral) *scales*, dorsal fin origin behind pelvic fin origin. (2) Sacramento Pikeminnow, *Ptychocheilus grandis* (Pl. 5), has more *compressed head*, longer snout, larger mouth reaching below eye, *larger* (65–78 lateral) *scales*.

PEAMOUTH *Mylocheilus caurinus* Pl. 12

IDENTIFICATION: Dark gray-brown to green above; *2 dark stripes, lower one ending in front of anal fin*, on silver yellow side; yellow to brown fins. Slender body, somewhat compressed; large eye; long rounded snout; *barbel* at corner of slightly subterminal mouth; large forked caudal fin; dorsal fin origin over or in front of pelvic fin origin. *Axillary process* at pelvic fin base. Large male has red on side, belly, mouth, gill cover, and pectoral fin base. Complete lateral line; 66–84 lateral scales; 8 dorsal rays; 8 anal rays; short intestine; molarlike pharyngeal teeth 1,5-5,1. To 14 in. (36 cm). **RANGE:** Nass (Pacific Slope) and Peace R. (Arctic basin) systems, BC, south to Columbia R. drainage, OR and ID;

Vancouver I., BC; Mackenzie R. drainage (Arctic basin), NT. Common; locally abundant. **HABITAT:** Lakes and slow-flowing areas of small to medium rivers; usually in vegetation. **SIMILAR SPECIES:** (1) Species of *Ptychocheilus* (Pl. 5) have larger mouth, no barbel, usually no stripes along side, no red; pharyngeal teeth 2,4-4,2 or 2,5-5,2 (not molarlike).

CHISELMOUTH *Acrocheilus alutaceus* Pl. 5

IDENTIFICATION: *Large, forked caudal fin.* Wide head; subterminal mouth, *hard plate* on lower jaw (Fig. 12); large eye. Moderately compressed body, deepest in front of dorsal fin, strongly tapering to *narrow* caudal peduncle; dorsal fin origin over pelvic fin origin; rounded, protruding snout. Gray above, brassy silver side; yellow to brown fins. Large individual has gray dorsal and caudal fins, orange at pectoral and pelvic fin bases. Decurved, complete lateral line; long intestine (at least twice length of body); black peritoneum. Has 85–93 lateral

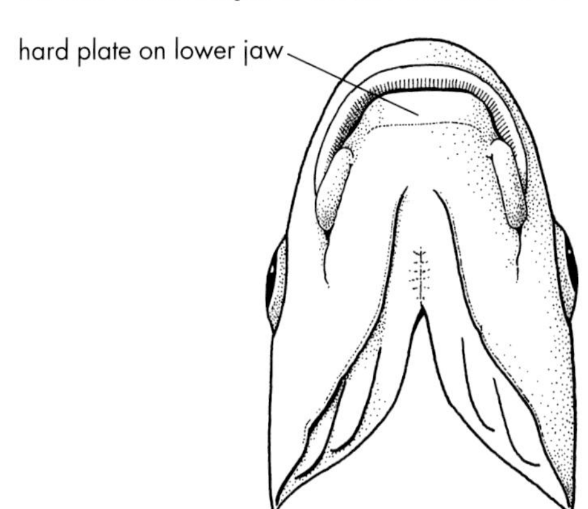

Fig. 12. Chiselmouth—underside of head.

scales; 10–11 dorsal rays; 9–10 anal rays; pharyngeal teeth 0,4-4,0 to 0,5-5,0. To 12 in. (30 cm). **RANGE:** Fraser and Columbia river drainages, BC, WA, ID, OR, and NV; Harney R. basin, OR. Fairly common in Columbia R. drainage. **HABITAT:** Flowing pools and runs over sand and gravel in creeks and small to medium rivers; margins of lakes. **REMARKS:** Sharp, cartilaginous plates on upper (concealed under lip) and lower jaws (Fig. 12) scrape food from rocks. **SIMILAR SPECIES:** (1) Redside Shiner, *Richardsonius balteatus* (Pl. 12), and (2) Lahontan Redside, *R. egregius* (Pl. 12), *lack* protruding snout and hard plates on jaws, have dorsal fin origin well *behind* pelvic fin origin.

REDSIDE SHINER *Richardsonius balteatus* Pl. 12

IDENTIFICATION: Deep, compressed body; *dorsal fin origin far behind pelvic fin origin. Narrow caudal peduncle*; deeply forked caudal fin. Short pointed snout, terminal mouth, large eye. *Axillary process* at pelvic fin base. Olive-gray to brown above, clear to yellow streak above dark stripe along side. Red above pectoral fin base on large individual; clear to yellow-brown fins. Breeding male is brassy yellow, bright red along lower side. Decurved, complete lateral line; 52–67 lateral scales; 8–12 (usually 10) dorsal rays; 10–24 (usually 15) anal rays; long intestine; silver peritoneum; pharyngeal teeth 2,4-4,2 to 2,5-5,2. To 7 in. (18 cm). **RANGE:** Pacific Slope drainages from Nass R., BC, to Rogue, Klamath, and Columbia river drainages, OR, ID, NV, and WY; Bonneville basin, s. ID, w. WY, and UT; Peace R. system (Arctic basin), AB and BC. Introduced into upper Missouri R. basin, MT, and Colorado R. drainage, WY, UT, CO, and AZ. Common; often abundant. **HABITAT:** Runs and flowing to standing pools of headwaters, creeks, and small to medium rivers; lakes and ponds. Usually over mud or sand, often near vegetation. **SIMILAR SPECIES:** (1) See Lahontan Redside, *R. egregius* (Pl. 12).

LAHONTAN REDSIDE *Richardsonius egregius* Pl. 12

IDENTIFICATION: Similar to Redside Shiner, *R. balteatus,* but has longer snout, more slender body, less slender caudal peduncle, 8–9

(rarely 10) anal rays. To 6¾ in. (17 cm). **RANGE:** Lahontan and other interior basins in n. NV and n. CA, including Humboldt, Walker, Carson, Truckee, Susan, Quinn, and Reese river systems, Walker, Tahoe, and Pyramid lakes. Introduced elsewhere, including upper Sacramento R. system, CA. Common; locally abundant. **HABITAT:** All types of stream habitats but usually in pools and slow runs; margins of lakes. **SIMILAR SPECIES:** (1) See Redside Shiner, *R. balteatus* (Pl. 12).

GILA

Some species previously in *Gila* are now in *Siphatales* and *Lepidomeda*. Six of 9 *Gila* species from U.S. are native to Colorado R. system. Morphological variation among these species is confusing, hybrids are common, and individuals often are difficult to identify.

THICKTAIL CHUB *Gila crassicauda* — Not shown

IDENTIFICATION: *Thick, deep caudal peduncle*; deep, compressed body with *high nape* rising steeply from short, pointed head; 49–60 lateral scales, 8–9 dorsal rays, 8–9 anal rays; pharyngeal teeth 2,5-4,2. To 1¼ in. (3.2 cm). **RANGE:** Sacramento-San Joaquin R., Clear Lake (Lake Co.), Pajaro R., Salinas R., and San Francisco Bay drainages, CA. Extinct. **HABITAT:** Marshes and backwaters along rivers and lake margins. **SIMILAR SPECIES:** (1) Tui Chub, *Siphatales bicolor* (Pl. 12), has pharyngeal teeth 0,5-5,0 or 0,4-4,0.

ROUNDTAIL CHUB *Gila robusta* — Pl. 12

IDENTIFICATION: Deep, compressed body; flat head; slender caudal peduncle; large, forked caudal fin; angle along anal fin base continues into *middle* of caudal fin (Fig. 13). Terminal mouth extends to front of eye. Dark olive-gray above; silver side. Breeding male may develop red or orange on lower half of cheek and paired fin bases. Has 71–99, *usually more than 78,* lateral scales, 9 dorsal rays, 9 anal rays, 10 or more rakers on 1st gill arch; pharyngeal teeth 2,5-4,2. To 17 in. (43

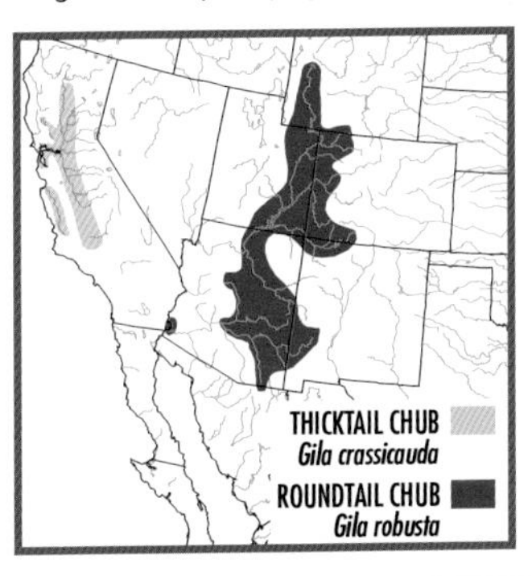

cm). **RANGE:** Colorado R. drainage, WY, CO, UT, NM, AZ, and Mexico. Locally common. **HABITAT:** Rocky runs, sometimes pools, of creeks and small to large rivers; impoundments. **REMARKS:** Roundtail Chub and Humpback Chub, *G. cypha*, are thought to hybridize throughout their shared range except in Yampa R. A taxonomically confusing population in White R., s. NV, is protected as *G. robusta jordani, an endangered subspecies.* **SIMILAR SPECIES:** See (1) Gila Chub, *G. intermedia*, (2) Headwater Chub, *G. nigra*, and (3) Virgin Chub, *G. seminuda.* (4) Humpback Chub, *G. cypha* (Pl. 12), and (5) Bonytail, *G. elegans* (Pl.

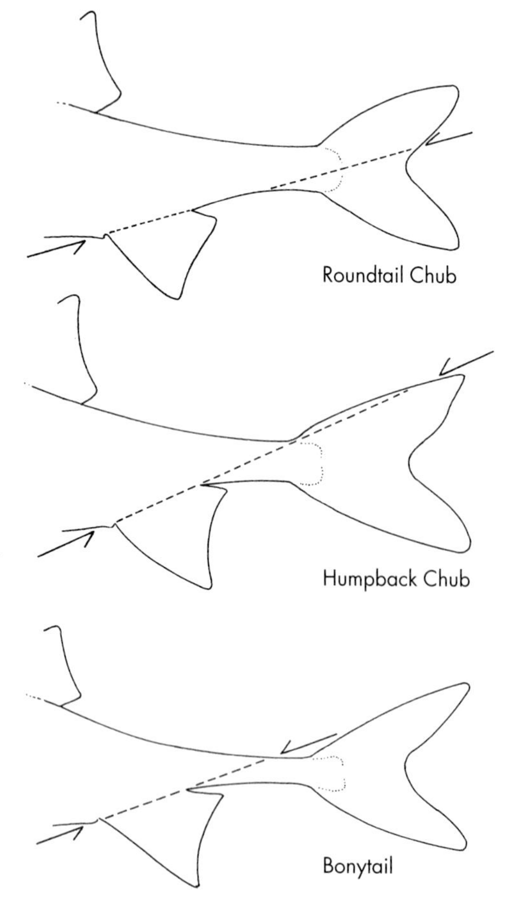

Fig. 13. Roundtail Chub, Humpback Chub, and Bonytail—angle along anal fin base.

12), have extremely slender caudal peduncle, smaller eye, angle along anal fin base continuing *above* or along upper edge of caudal fin (Fig. 13); large individual has hump on nape, depressed head. (6) Colorado Pikeminnow, *Ptychocheilus lucius* (Pl. 5), has mouth extending to rear of eye; young has black spot on caudal fin base.

GILA CHUB *Gila intermedia* Not shown

IDENTIFICATION: Similar to Roundtail Chub, *G. robusta*, but has 51–83, *usually fewer than 70,* lateral scales; usually *8 dorsal rays*, *8 anal rays*, *9 or fewer rakers* on 1st gill arch. Scales often darkly outlined. Presumed to reach about 15 in. (38 cm). **RANGE:** Gila R. system (Colorado R. drainage), NM and AZ; Sonora, Mexico. Uncommon; protected as an *endangered species*. **HABITAT:** Pools and undercut banks of creeks and small rivers, often near woody debris; marshes. **SIMILAR SPECIES:** See (1) Roundtail Chub, *G. robusta* (Pl. 12), and (2) Headwater Chub, *G. nigra*.

HEADWATER CHUB *Gila nigra* Not shown

IDENTIFICATION: Thought to be of hybrid origin and similar to parental Roundtail Chub, *G. robusta,* and Gila Chub, *G. intermedia*. Has 71–90, *usually fewer than 80,* lateral scales, 8 dorsal rays, 8 anal rays; 10 or more rakers on 1st gill arch. Often with diffuse stripes on dark gray or brown side, dark fin membranes. Presumed to reach about 15 in. (38 cm). **RANGE:** Gila R. system (Colorado R. drainage), NM and AZ. Locally common. **HABITAT:** Pools and undercut banks of headwaters, creeks, and small rivers. **SIMILAR SPECIES:** See (1) Roundtail Chub, *G. robusta* (Pl. 12), and (2) Gila Chub, *G. intermedia.*

VIRGIN CHUB *Gila seminuda* Not shown

IDENTIFICATION: Thought to be of hybrid origin and similar to parental Roundtail Chub, *G. robusta,* and Bonytail, *G. elegans*. *Usually more than 85 lateral scales, no or deeply embedded scales* on back, breast, and

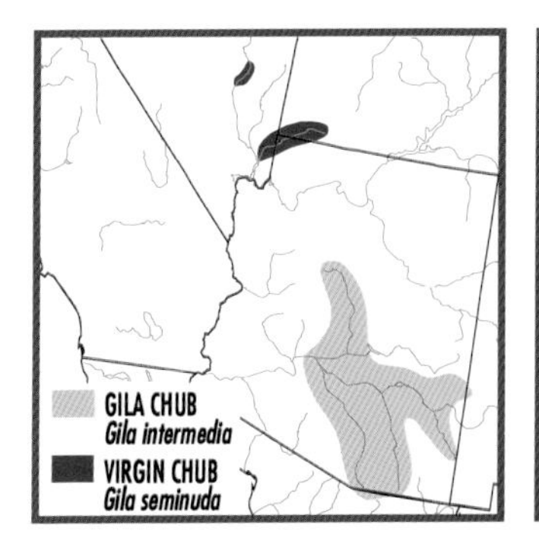

belly. Presumed to reach about 15 in. (38 cm). **RANGE:** Virgin R. system (Colorado R. drainage), sw. UT, s. NV, and nw. AZ; Muddy (Moapa) R. system, NV. Rare. **HABITAT:** Pools and undercut banks of headwaters, creeks, and small rivers. **SIMILAR SPECIES:** See (1) Roundtail Chub, *G. robusta*, and (2) Bonytail, *G. elegans* (both Pl. 12).

HUMPBACK CHUB *Gila cypha* Pl. 12

IDENTIFICATION: Deep, compressed body; *long, extremely slender caudal peduncle*; falcate fins; large, forked caudal fin; angle along anal fin base continues along *upper edge* of caudal fin (Fig. 13). Large individual has large, nearly scaleless *hump* behind *small, depressed head. Subterminal mouth* extends to front of small eye. Dark olive-gray above; silver side. Large individual has orange lower side, pectoral and anal fin bases. Has 73–90 lateral scales; usually 9 dorsal rays, 10 anal rays; pharyngeal teeth 2,5-4,2. To 15 in. (38 cm). **RANGE:** Upper Colorado R. drainage, WY, CO, UT, and AZ. Rare; protected as an *endangered species*. **HABITAT:** Swift, rocky runs and flowing pools. **REMARKS:** Hump at back of head enables fish to maintain position in swift water. Similar humps are on other Colorado R. fishes, Bonytail, *G. elegans*, and Razorback Sucker, *Xyrauchen texanus* (Pl. 20). Large falcate fins and narrow caudal peduncle are also characteristic of fishes in swift water. **SIMILAR SPECIES:** (1) See Bonytail, *G. elegans* (Pl. 12). (2) Roundtail Chub, *G. robusta* (Pl. 12), *lacks* hump on nape; has *deeper head, deeper caudal peduncle*, larger eye, more *terminal mouth*; angle along anal fin base continues into *middle* of caudal fin (Fig. 13).

BONYTAIL *Gila elegans* Pl. 12

IDENTIFICATION: Similar to Humpback Chub, *G. cypha,* but has *terminal mouth*, angle along anal fin base continues *well above* caudal fin (Fig. 13), more slender caudal peduncle, smaller hump on nape, usually 10 dorsal rays. Breeding male has red lower side and pectoral and anal fin bases. Has 75–99 lateral scales. To 24½ in. (62 cm). **RANGE:**

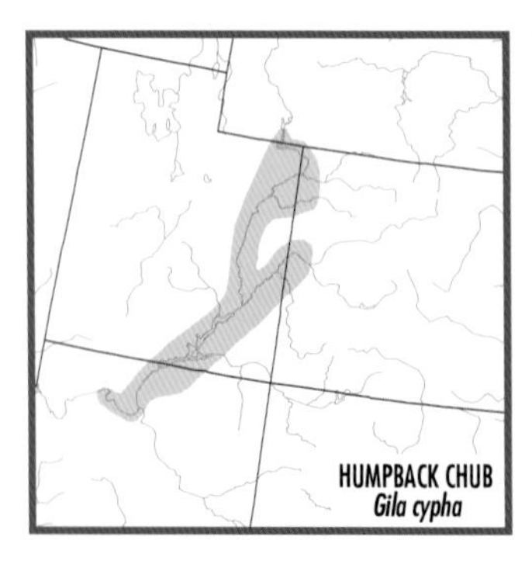

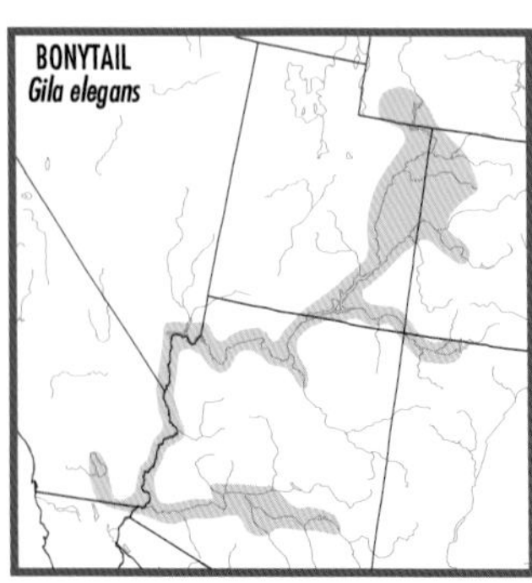

Colorado R. drainage, WY, CO, UT, NM, AZ, CA, and Mexico. Extremely rare; protected as an *endangered species.* **HABITAT:** Flowing pools and backwaters, usually over mud or rocks. Streamlined body suggests occasional occupation of swift runs. **SIMILAR SPECIES:** See (1) Humpback Chub, *G. cypha* (Pl. 12), and (2) Virgin Chub, *G. seminuda.* (3) Roundtail Chub, *G. robusta* (Pl. 12), has larger eye, deeper head, deeper caudal peduncle; *lacks* hump on nape; angle along anal fin base continues into *middle* of caudal fin (Fig. 13).

BLUE CHUB *Gila coerulea* Pl. 12

IDENTIFICATION: Large eye; *pointed snout*; terminal mouth extends to front of eye. Fairly slender, compressed body; *slender caudal peduncle.* Dusky olive above; silver blue side. Breeding male has blue snout, orange side and fins. Has 58–71 lateral scales, 9 dorsal rays, 8–9 anal rays; pharyngeal teeth 2,5-5,2. To 16 in. (41 cm). **RANGE:** Klamath and Lost river systems, OR and CA. Common; abundant in impoundments. **HABITAT:** Rocky pools of creeks and small to large rivers; rocky shores of lakes and impoundments. **SIMILAR SPECIES:** (1) Tui Chub, *Siphatales bicolor* (Pl. 12), has smaller mouth not extending to front of eye, deeper body, *more rounded snout*, larger (41–64 lateral) scales.

UTAH CHUB *Gila atraria* Pl. 12

IDENTIFICATION: Olive-brown to blue-black above; *yellow to brassy side*; clear to olive-yellow fins. Large male has yellow to gold fin bases, mouth, and side of head; may be gold overall. Deep, compressed body; large eye; *short, blunt snout.* Has 45–65 lateral scales; usually 9 dorsal rays, 8 anal rays; pharyngeal teeth 2,5-4,2. To 22 in. (56 cm). **RANGE:** Snake R. system above Shoshone Falls, WY and ID; Lake Bonneville basin (including Great Salt Lake drainage and Sevier R. system), se. ID and UT. Introduced into e. NV, upper Missouri R. basin, MT, and Colorado R. drainage, WY and UT. Locally common. **HABITAT:** Lakes; quiet pools of headwaters, creeks, and small to medium rivers; often

in vegetation over mud or sand. **SIMILAR SPECIES:** (1) Northern Leatherside Chub, *Lepidomeda copei* (Pl. 12), and (2) Southern Leatherside Chub, *L. aliciae*, are *silver blue*; have red on lower body of large male, larger (68–85 lateral) scales, 8 dorsal rays.

CHIHUAHUA CHUB *Gila nigrescens* Pl. 12

IDENTIFICATION: Deep, compressed body; long, fairly slender caudal peduncle. *Terminal mouth on rounded snout*. Olive-gray above; silver side; clear to slate gray fins. Breeding individual has *red-orange mouth* and paired and anal fin bases. Has 67–78 lateral scales, 9 dorsal rays, 8 anal rays, 9–14 rakers on 1st gill arch; pharyngeal teeth 2,5-4,2. To 9½ in. (24 cm). **RANGE:** Mimbres R., NM; also Lagunas Guzman and Bustillos, Chihuahua, Mexico. Rare. Protected in U.S. as a *threatened species*. **HABITAT:** Flowing pools of creeks and small rivers; usually near brush or other cover. **SIMILAR SPECIES:** (1) Rio Grande Chub, *G. pandora* (Pl. 12), and (2) Sonora Chub, *G. ditaenia*, have shorter, deeper caudal peduncle, 2 dusky black stripes on side, blunter snout, larger eye, 8 dorsal rays, 6–10 rakers on 1st gill arch. Rio Grande Chub has 51–67 lateral scales. (3) Yaqui Chub, *G. purpurea*, has deep caudal peduncle, black wedge on caudal fin base, 48–62 lateral scales, 8 dorsal rays.

RIO GRANDE CHUB *Gila pandora* Pl. 12

IDENTIFICATION: Olive-gray above; *2 dusky stripes* along silver side (darkest on large individual), upper one extending to caudal fin, lower one to anal fin; dusky to black caudal spot. Fairly deep, compressed body; deep caudal peduncle. Slightly subterminal mouth extends to front of large eye; rounded, fairly blunt snout. Breeding individual has red-orange (brightest on male) anal, dorsal, and paired fin bases and side of head; orange lower side. Has 51–67 lateral scales, 8 dorsal rays, 8 anal rays, 6–10 rakers on 1st gill arch; pharyngeal teeth 2,5-4,2. To 7 in. (18 cm). **RANGE:** Upper Rio Grande and Pecos river systems, CO and NM; isolated population in Davis Mts. (Pecos R. system),

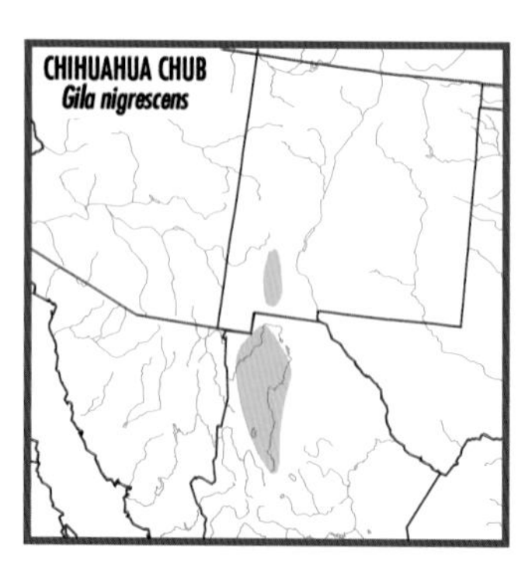

TX. Introduced into headwaters of Canadian R. (Red R. drainage), NM. Common, but declining. **HABITAT:** Flowing pools of headwaters, creeks, and small rivers; usually near brush. **SIMILAR SPECIES:** (1) See Sonora Chub, *G. ditaenia.*

SONORA CHUB *Gila ditaenia* Not shown

IDENTIFICATION: Nearly identical to Rio Grande Chub, *G. pandora,* but has *smaller* (63–75 lateral) *scales*. To 10 in. (25 cm). **RANGE:** Sycamore (Bear) Canyon, Santa Cruz Co., AZ; also in Río de la Concepcion, Sonora, Mexico. Uncommon in extremely small area in U.S.; protected in U.S. as a *threatened species*. **HABITAT:** Rocky and sandy pools (intermittent during dry season) of creeks; springs. **SIMILAR SPECIES:** (1) See Rio Grande Chub, *G. pandora* (Pl. 12).

ARROYO CHUB *Gila orcuttii* Pl. 5

IDENTIFICATION: Deep body; *deep caudal peduncle*. Small, slightly subterminal mouth; short, rounded snout; large eye. Gray-olive above; often a dusky gray stripe along silver side. Has 48–62 lateral scales, 8 dorsal rays, 7 anal rays; pharyngeal teeth 2,5-4,2. To 16 in. (40 cm). **RANGE:** Native to Malibu and San Juan creeks, and Los Angeles, San Gabriel, San Luis Rey, Santa Ana, and Santa Margarita river drainages, s. CA. Introduced north to Santa Ynez R. and in Mojave (Death Valley basin) R. drainage, CA. Extirpated from much of native range, but common in a few streams. **HABITAT:** Sand- and mud-bottomed flowing pools and runs of headwaters, creeks, and small to medium rivers; often in intermittent streams. **SIMILAR SPECIES:** (1) See Yaqui Chub, *G. purpurea*. (2) Thicktail Chub, *G. crassicauda*, has *deeper caudal peduncle*, high nape, 8 anal rays.

YAQUI CHUB *Gila purpurea* Not shown

IDENTIFICATION: Similar to Arroyo Chub, *G. orcuttii,* but has *black wedge* on caudal fin base, 8 anal rays. Yellow-brown to steel blue (large male) above.

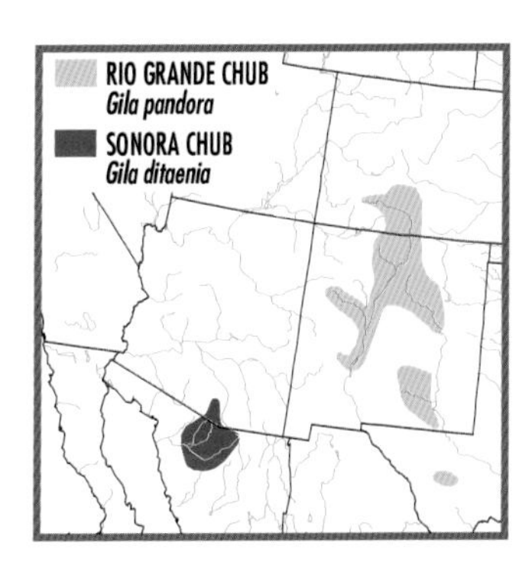

To 5½ in. (14 cm). **RANGE:** Native to Rio Yaqui basin, se. AZ and Mexico. Introduced into Leslie Creek (Whitewater Draw drainage), extreme se. AZ. Extremely rare; possibly extinct. Protected in U.S. as an *endangered species*. **HABITAT:** Quiet pools of headwaters and creeks; usually in vegetation. **SIMILAR SPECIES:** (1) See Arroyo Chub, *G. orcuttii* (Pl. 5).

TUI CHUB *Siphatales bicolor* Pl. 12

IDENTIFICATION: Highly variable (see Remarks). Deep, compressed body; dorsal fin origin over pelvic fin origin; fairly deep caudal peduncle; *small, rounded fins*. Small, terminal to slightly subterminal mouth; does *not* extend to eye. Dusky olive to dark green above; brassy brown side, often mottled in adult; silver white to yellow below; clear to dusky olive fins. Young has dusky stripe along side. Large individual may have yellow to copper fins with pink, red, or orange bases, red-orange lower side. Has complete lateral line, 41–64 lateral scales; usually 8 dorsal rays, 7–8 anal rays; pharyngeal teeth 0,5-5,0 to 0,4-4,0. To 17¾ in. (45 cm). **RANGE:** Columbia R. drainage, WA, OR, and ID, south in Klamath and upper Pit rivers (Sacramento R. drainage), and interior drainages of NV and CA to Mohave R., s. CA. Common; locally abundant but decreasing in some areas because of habitat degradation and introduced species. **HABITAT:** Quiet, vegetated, mud- or sand-bottomed pools of headwaters, creeks, and small to large rivers; lakes. **REMARKS:** Many distinctive forms of Tui Chub in isolated (endorheic) drainages of w. U.S. are recognized through combinations of characters not described here because comprehensive study of intraspecific variation is needed. Among distinctive subspecies are *S. b. snyderi* (protected as an *endangered subspecies*) in Owens R., CA; *S. b. mohavensis* (protected as an *endangered subspecies*), Mohave R., CA; and *S. b. bicolor*, Klamath R. system, OR and CA. Both *S. b. obesa*, a stream- and spring-inhabiting form, and *S. b. pectinifer*, a lake-inhabiting form, are in Lake Lahontan basin, NV. **SIMILAR SPECIES:** (1) Alvord Chub, *S. alvordensis* (Pl. 5), and (2) Borax Lake Chub,

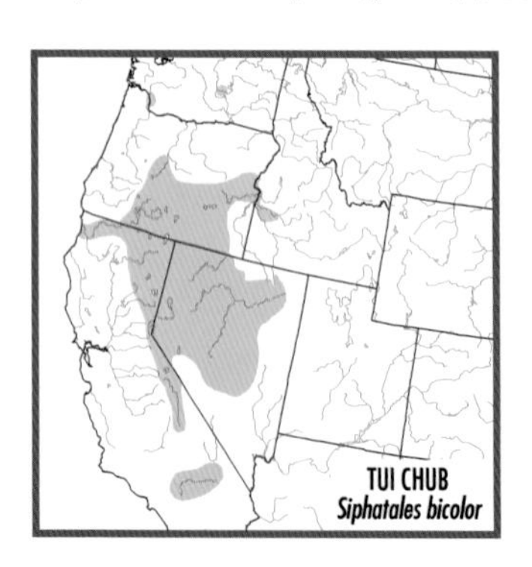

S. boraxobius, have higher nape, usually 7 dorsal rays. (3) Blue Chub, *Gila coerulea* (Pl. 12), has larger mouth extending to *front of eye*, 58–71 lateral scales; is more slender.

ALVORD CHUB *Siphatales alvordensis* Pl. 5

IDENTIFICATION: *High nape* rising steeply from short pointed head; slender, fairly compressed body; deep caudal peduncle. Small eye high on head; rounded snout; terminal mouth. Dusky olive to dark green above, line of black specks along upper side; silver side; clear to dusky olive fins. Has 58–72 lateral scales; usually 7 dorsal rays, 7 anal rays, 14–15 pectoral rays; pharyngeal teeth 0,5-4,0. To 5¼ in. (14 cm). **RANGE:** Alvord basin, se. OR and nw. NV. Common in small area. **HABITAT:** Springs and spring-fed streams; impoundments. **SIMILAR SPECIES:** (1) See Borax Lake Chub, *S. boraxobius*. (2) Tui Chub, *S. bicolor* (Pl. 12), usually has 8 dorsal rays.

BORAX LAKE CHUB *Siphatales boraxobius* Not shown

IDENTIFICATION: Similar to Alvord Chub, *S. alvordensis*, but *larger head is concave between eyes*; has *incomplete lateral line*, more slender caudal peduncle, usually 13 pectoral rays. To 4¼ in. (11 cm). **RANGE:** Borax Lake and outflows (Alvord basin), Harney Co., OR. Common in extremely small range; protected as an *endangered species*. **HABITAT:** Borax Lake is a small, shallow clear lake fed by hot springs. Borax Lake Chub is only fish in the lake. **SIMILAR SPECIES:** (1) See Alvord Chub, *S. alvordensis* (Pl. 5).

ROSYSIDE DACE *Clinostomus funduloides* Pl. 13

IDENTIFICATION: *Large oblique mouth; long pointed snout*. Large, forked caudal fin. Compressed body; small scales. Dorsal fin origin behind pelvic fin origin. Olive above, dark stripe along back, dark-edged scales on upper side; green to yellow-gold stripe (brightest on large individuals) above dusky stripe along silver side; scattered dark blotches on

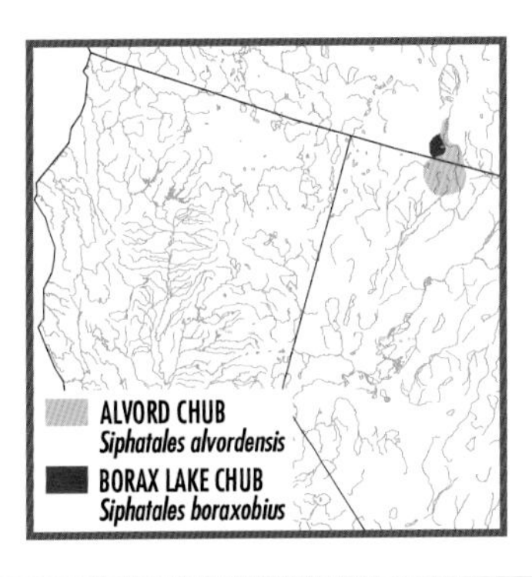

side of large individual; white, orange, or red lower side. Breeding male is dark blue above, has bright brick red lower side. Complete, decurved lateral line; 43–57 lateral scales; 9 anal rays; pharyngeal teeth 2,5-4,2. To 4½ in. (11 cm). **RANGE:** Atlantic Slope (mostly above Fall Line) from lower Delaware R. drainage, PA, to Savannah R. drainage, GA; Ohio R. basin from WV and OH to ne. MS. Common to abundant on Atlantic Slope and parts of Ohio basin; absent in Ohio R. basin between cen. OH and Cumberland R. (including most of KY); rare in Cumberland R. drainage. **HABITAT:** Rocky flowing pools of headwaters, creeks, and small rivers; usually in small clear streams. **REMARKS:** Three subspecies. *C. f. funduloides*, on Atlantic Slope from Delaware R. drainage, PA, to lower Savannah R. drainage, SC and GA, in upper Tennessee R. drainage, TN and VA, and in Ohio R. basin, s. OH, e. KY, and s. WV, usually has 47–54 lateral scales, 34–43 scales around body, slightly pigmented lower side. *C. f. estor*, in lower and middle Tennessee and Cumberland river drainages, KY, TN, and AL, usually has 43–50 lateral scales, 31–37 scales around body, slightly pigmented lower side. An undescribed subspecies, endemic to Little Tennessee R. system, TN and NC, has blunt snout, darkly pigmented lower side, usually 50–53 lateral scales, 36–46 scales around body. Intergrades between *C. f. funduloides* and undescribed subspecies occur in headwaters of Little Tennessee R. system, NC and GA, and headwaters of Savannah R. drainage, GA. Hiwassee R., TN and NC, may contain intergrades between *C. f. estor* and undescribed subspecies. **SIMILAR SPECIES:** (1) See Redside Dace, *C. elongatus* (Pl. 13). (2) Striped Shiner, *Luxilus chrysocephalus*, and (3) Common Shiner, *L. cornutus* (both Pl. 15), have *smaller mouth, shorter rounded snout*, dark stripes on upper side, dorsal fin origin over pelvic fin origin, 36–43 lateral scales, pharyngeal teeth 2,4-4,2.

REDSIDE DACE *Clinostomus elongatus* Pl. 13

IDENTIFICATION: Similar to Rosyside Dace, *C. funduloides*, but has *longer, more pointed snout; more slender body*; brighter (carmine) red

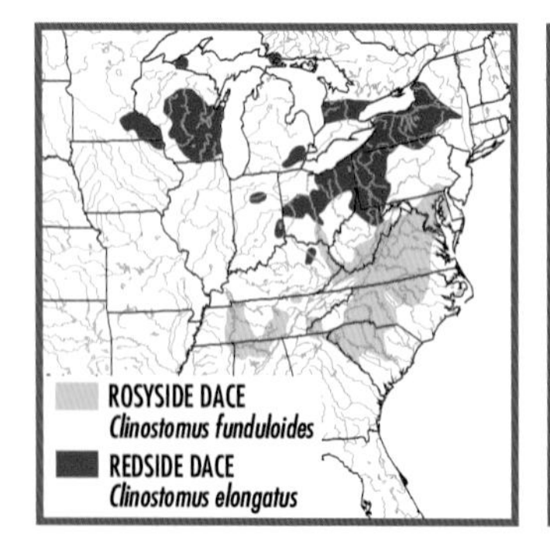

on lower side (of adult); smaller (59–75 lateral) scales. Breeding male is steel blue above, has yellow-gold stripe along side, bright red on lower side. To 4½ in. (12 cm). **RANGE:** Hudson and upper Susquehanna river drainages, NY and PA; Great Lakes (except Lake Superior) and Mississippi R. basins from NY and s. ON to MN and south to WV and KY. Common in eastern part of range (but declining in many areas); localized in west. Protected in Canada as an *endangered species.* **HABITAT:** Rocky and sandy pools of headwaters, creeks, and small rivers; largest populations in clear, spring-fed streams. **SIMILAR SPECIES:** (1) See Rosyside Dace, *C. funduloides* (Pl. 13).

LONGFIN DACE *Agosia chrysogaster* Pl. 13

IDENTIFICATION: Large female has *elongated lower lobe* on anal fin. *Small (70–95 lateral) scales. Small barbel* at corner of mouth; long coiled intestine; black peritoneum. Rounded snout; slightly subterminal mouth. Dorsal fin origin above or slightly in front of pelvic fin origin. Dark gray above; dusky black band along silver side enlarged into black spot at caudal fin base; often gold flecks on side. Large male is light yellow below, has yellow paired fin bases, large dorsal fin. Has 7 anal rays; pharyngeal teeth 0,4-4,0. To 4 in. (10 cm). **RANGE:** Lower Colorado R. drainage (primarily Gila and Bill William river systems), NM and AZ, and south through s. AZ and Pacific drainages of Mexico. Introduced into Mimbres R. and Rio Grande, NM. Common. **HABITAT:** Shallow sandy and rocky runs; flowing pools of creeks and small to medium rivers; often near cover. **SIMILAR SPECIES:** (1) Speckled Dace, *Rhinichthys osculus*, and (2) Longnose Dace, *R. cataractae* (both Pl. 14), *lack* elongated lower lobe on anal fin of female, have dorsal fin origin behind pelvic fin origin, short intestine, pharyngeal teeth 1,4-4,1 or 2,4-4,2. (3) *Gila* species (Pls. 5 & 12) *lack* elongated lower lobe on anal fin of female, barbel; have short intestine.

MOAPA DACE *Moapa coriacea* Pl. 6

IDENTIFICATION: *Leatherlike skin* (resulting from many small embedded scales). *Large black spot* on caudal fin base. Fairly slender body, dorsal fin origin over or slightly behind pelvic fin origin; long rounded snout; slightly subterminal mouth. Olive above, broad dusky stripe along back, cream spot at dorsal fin origin, green-brown blotches along upper side; turquoise stripe above dark gold-brown stripe along side (faint at front). Complete or incomplete lateral line; 69–79 lateral scales; 8 (often 7) anal rays; pharyngeal teeth 0,5-4,0. To 3½ in. (9 cm). **RANGE:** Endemic to warm springs area of Muddy (Moapa) R., Clark Co., se. NV. Common within extremely restricted range. Protected as an *endangered species*. **HABITAT:** Pools of Muddy (Moapa) R. and feeder springs over gravel, sand, and mud. **SIMILAR SPECIES:** (1) Relict Dace, *Relictus solitarius* (Pl. 6), has larger (50–70 lateral) scales, *less* leatherlike skin; terminal mouth. (2) Speckled Dace, *Rhinichthys osculus* (Pl. 14), has barbel at corner of mouth, pointed snout overhanging mouth; pharyngeal teeth 1,4-4,1 to 2,4-4,2.

RELICT DACE *Relictus solitarius* Pl. 6

IDENTIFICATION: *Chubby; soft-bodied. Incomplete lateral line*, rarely reaching below dorsal fin. *Large terminal mouth*, no frenum or barbel. Small fins. Dorsal fin origin slightly in front of to behind pelvic fin origin; anal fin well behind dorsal fin. Color highly variable; dusky violet, yellow, or green above, speckled with brown; yellow fins. Has 50–70 lateral scales; usually 7 anal rays; pharyngeal teeth 0,4-4,0. To 5 in. (13 cm). **RANGE:** Lakes Franklin, Gale, Waring, Steptoe, and Spring basins in e. NV. Common in small range. **HABITAT:** Springs and their effluents. **SIMILAR SPECIES:** (1) Desert Dace, *Eremichthys acros* (Pl. 6), has *small subterminal mouth*, hard sheath on jaws, 68–78 lateral scales. (2) Tui Chub, *Siphatales bicolor* (Pl. 12), has *complete* lateral line, dorsal fin origin over pelvic fin origin.

DESERT DACE *Eremichthys acros* Pl. 6

IDENTIFICATION: *Deep chubby body; hard sheath* on upper and lower jaws; *small*, slightly subterminal mouth. Black lips; short blunt snout; no barbel or frenum. Dorsal fin origin behind pelvic fin origin. Olive above, bright green stripe along upper side; yellow-brassy side; silver below; amber-white fins. Complete or incomplete lateral line; 68–78 lateral lines; 7–8 anal rays; pharyngeal teeth 0,5-4,0. To 3 in. (7.7 cm). **RANGE AND HABITAT:** Warm springs of Soldier Meadows, Lahontan basin, Humboldt Co., NV. Recorded in water as warm as 100°F (38°C). Protected as a *threatened species*. **SIMILAR SPECIES:** (1) Relict Dace, *Relictus solitarius* (Pl. 6), has *large terminal mouth*, 70 or fewer lateral scales, *no* hard sheath on jaws. (2) Speckled Dace, *Rhinichthys osculus* (Pl. 14), has barbel in corner of *larger* mouth, is more slender, *no* hard sheath on jaws; pharyngeal teeth 1,4-4,1 to 2,4-4,2.

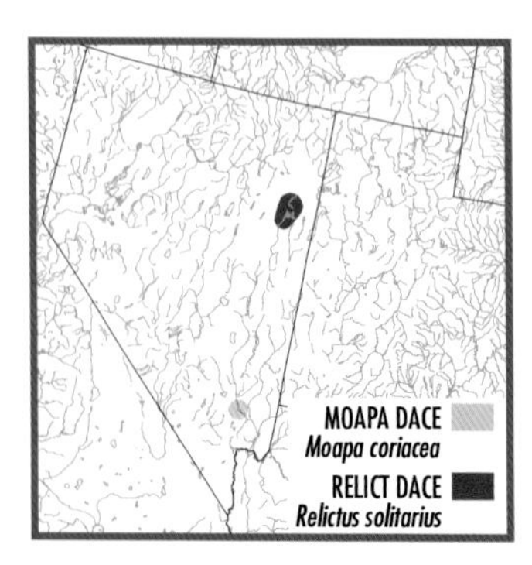

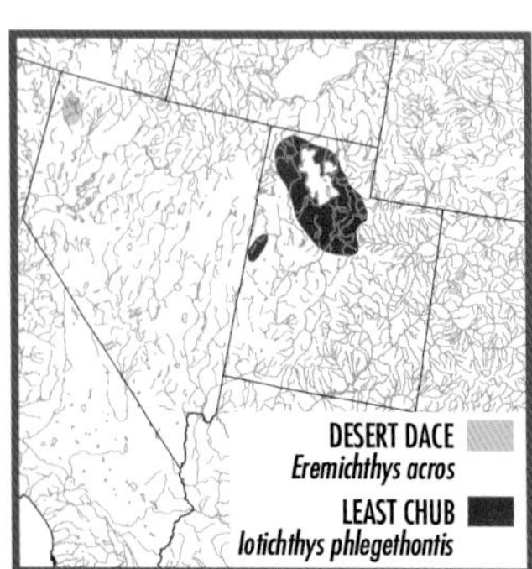

LEAST CHUB *Iotichthys phlegethontis* Pl. 6

IDENTIFICATION: *Upturned mouth*; large eye; short rounded snout. *No lateral line* (rarely 1–2 pores at front). *Large (34–38 lateral) scales.* Fairly deep, compressed body; dorsal fin origin behind pelvic fin origin; slender peduncle. Olive above; black specks on back and side; gold stripe along blue side; white to yellow fins. Breeding male has red-gold lower side, gold eye and fins. Has 8 anal rays; pharyngeal teeth 2,5-4,2. To 2½ in. (6.4 cm). **RANGE:** Bonneville basin, n. UT. Common. **HABITAT:** Marshes, ponds, vegetated areas of streams and lakes; usually over mud. **SIMILAR SPECIES:** (1) Utah Chub, *Gila atraria* (Pl. 12), (2) Northern Leatherside Chub, *Lepidomeda copei* (Pl. 12), (3) Southern Leatherside Chub, *L. aliciae,* and (4) Tui Chub, *Siphatales bicolor* (Pl. 12), have *terminal mouth, complete lateral line, smaller scales*, deeper peduncle; reach much larger size.

OREGON CHUB *Oregonichthys crameri* Pl. 6

IDENTIFICATION: Distinctive body compressed and deepest at dorsal fin origin, strongly tapering to *very narrow caudal peduncle.* Olive-tan above, green-brown stripe along back in front of dorsal fin; *clusters of large brown-black spots* scattered over back and silver side giving salt-and-pepper appearance; darkly outlined scales on back; lines of brown specks on lower side. Small *barbel* (sometimes absent) at corner of terminal mouth; rounded snout. Dorsal fin origin over or slightly behind pelvic fin origin. Fully scaled breast. Complete lateral line; 35–39 lateral scales; 7 anal rays; pharyngeal teeth 1,4-4,1. To 2¾ in. (7 cm). **RANGE:** Willamette R. drainage, OR. Localized and rare because of habitat alteration; protected as an *endangered species.* **HABITAT:** Sluggish sand- and gravel-bottomed pools and backwaters of creeks and small rivers; often near vegetation. Ponds and sloughs. **SIMILAR SPECIES:** (1) See Umpqua Chub, *O. kalawatseti.*

UMPQUA CHUB *Oregonichthys kalawatseti* **Not shown**

IDENTIFICATION: Nearly identical to Oregon Chub, *O. crameri,* but has unscaled or partly scaled breast, slightly subterminal mouth. Has 34–40 lateral scales. To 2¼ in. (5.9 cm). **RANGE:** Umpqua R. drainage, OR. Uncommon. **HABITAT:** Sluggish sand- and gravel-bottomed runs, pools, and backwaters of creeks and small rivers; sloughs. Often near vegetation. **SIMILAR SPECIES:** (1) See Oregon Chub, *O. crameri* (Pl. 6).

LEPIDOMEDA

GENUS CHARACTERISTICS: *Large gray-black blotches and specks* (darkest on large individuals) scattered on side. *Compressed, silver white to brassy olive body, often with blue sheen.* Has *2 large spines* at front of dorsal fin; 2d spine fits in groove on 1st. Dorsal fin origin behind pelvic fin origin; large eye. Complete lateral line; usually 7 pelvic rays (most N. American minnows have 8 pelvic rays).

LITTLE COLORADO SPINEDACE *Lepidomeda vittata* **Not shown**

IDENTIFICATION: *Eight anal rays*; usually more than 90 (89–105) lateral scales; pharyngeal teeth 1,4-4,1 or 2,4-4,2. Rounded snout; large, terminal, fairly oblique mouth. Large male has faint yellow-orange at paired fin bases. To 4 in. (10 cm). **RANGE:** Upper Little Colorado R. system, e. AZ. Uncommon and highly localized. Protected as a *threatened species.* **HABITAT:** Rocky and sandy runs and pools of creeks and small rivers. **SIMILAR SPECIES:** (1) Virgin Spinedace, *L. mollispinis* (Pl. 13), and (2) White River Spinedace, *L. albivallis,* have *9 anal rays,* usually fewer than 90 lateral scales, pharyngeal teeth 2,5-4,2. (3) Pahranagat Spinedace, *L. altivelis,* has pointed snout, large oblique mouth, large pointed dorsal fin (Fig. 14), 9 anal rays, pharyngeal teeth 2,5-4,2.

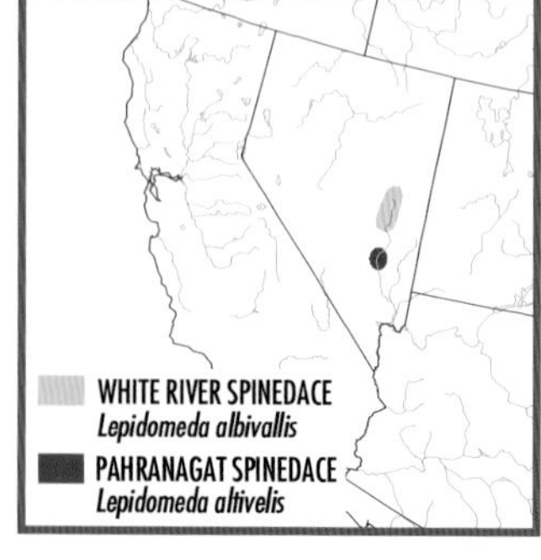

VIRGIN SPINEDACE *Lepidomeda mollispinis* Pl. 13

IDENTIFICATION: Rounded snout; large terminal mouth. Large male has red-orange at paired and anal fin bases, upper edge of gill cover, sometimes on belly. Black specks on opercle confined to upper half. Has 77–91 lateral scales, 9 anal rays; pharyngeal teeth 2,5-4,2. To 5¾ in. (15 cm). **RANGE:** Virgin R. system, UT, NV, and AZ. Generally common but reduced in impounded or channelized streams. **HABITAT:** Gravel- and sand-bottomed flowing pools and runs of fast and usually clear creeks and small rivers. **REMARKS:** Two subspecies: *L. m. mollispinis* and *L. m. pratensis* (protected as a *threatened subspecies*); latter is restricted to Big Spring and surrounding area, Lincoln Co., NV, has higher, more pointed dorsal fin, longer pelvic fins, and smaller, more oblique mouth. **SIMILAR SPECIES:** (1) See White River Spinedace, *L. albivallis.* (2) Little Colorado Spinedace, *L. vittata,* has 8 anal rays, usually more than 90 lateral scales, more oblique mouth, pharyngeal teeth 2,4-4,2. (3) Pahranagat Spinedace, *L. altivelis,* has pointed snout; large, oblique mouth; large, pointed dorsal fin (Fig. 14).

WHITE RIVER SPINEDACE *Lepidomeda albivallis* Not shown

IDENTIFICATION: Similar to Virgin Spinedace, *L. mollispinis,* but has dark specks over *all* of opercle, much larger pharyngeal arch; red-orange bases on olive to pink-brown dorsal and caudal fins, red-orange anal and pelvic fins (colors brightest on large male). Has 79–92 lateral scales. To 5¾ in. (15 cm). **RANGE:** Upper White R. system, NV. Highly localized in small area. Protected as an *endangered species.* **HABITAT:** Springs and outflows of upper White R.; usually in shallow, cool, clear water over sand and gravel. **SIMILAR SPECIES:** (1) See Virgin Spinedace, *L. mollispinis* (Pl. 13).

PAHRANAGAT SPINEDACE *Lepidomeda altivelis* Fig. 14

IDENTIFICATION: *Pointed snout; large, oblique mouth,* front tip of lower jaw above middle of eye; compressed head. Large, pointed dorsal fin. Has 84–95 lateral scales, 9 anal rays; pharyngeal teeth 2,5-4,2. To 3 in. (7.9 cm). **RANGE:** White R., Pahranagat Valley, NV. Rare. **HABITAT:** Spring outflows. **SIMILAR SPECIES:** (1) Other spinedace, *Lepidomeda*

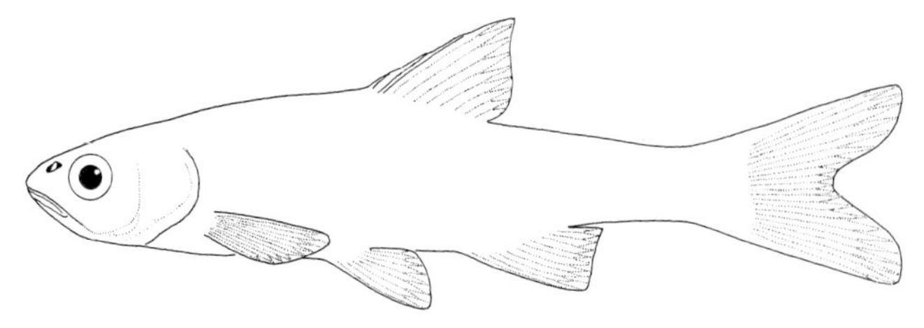

Fig. 14. Pahranagat Spinedace.

species, have *rounded snout*; front tip of mouth not extending above middle of eye; smaller, less pointed dorsal fin; wider head.

NORTHERN LEATHERSIDE CHUB *Lepidomeda copei* Pl. 12

IDENTIFICATION: *Leatherlike appearance* created by small (68–85 lateral) scales, black specks on *silver blue back and side*. Breeding male has red paired and anal fin bases, lower lobe of caudal fin, upper edge of gill cover. Slender, barely compressed body; large eye; terminal mouth on short rounded snout. Has 8 dorsal rays, 8 anal rays; pharyngeal teeth 1,4-4,1 to 2,5-4,2. To 6 in. (15 cm). **RANGE:** Bear and Upper Snake river systems, WY, ID, and ne. UT. Rare. **HABITAT:** Sluggish pools and backwaters, usually over mud or sand, of creeks and small to medium rivers. **SIMILAR SPECIES:** (1) See Southern Leatherside Chub, *L. aliciae*. (2) Utah Chub, *Gila atraria* (Pl. 12), has brassy side, larger (45–64 lateral) scales, 9 dorsal rays; lacks red.

SOUTHERN LEATHERSIDE CHUB *Lepidomeda aliciae* Not shown

IDENTIFICATION: Nearly identical to Northern Leatherside Chub, *L. copei,* but has shallower head, longer snout (and fixed differences in several mitochondrial and nuclear genetic markers). To 6 in. (15 cm). **RANGE:** Utah Lake and Sevier river systems, UT. Historically in Beaver R. system, where it now appears to be extinct. Possibly introduced into Colorado R. drainage, UT. Common. **HABITAT:** Sluggish pools and backwaters, usually over mud or sand, of creeks and small to medium rivers. **SIMILAR SPECIES:** (1) See Northern Leatherside Chub, *L. copei* (Pl. 12).

SPIKEDACE *Meda fulgida* Pl. 13

IDENTIFICATION: *No scales*. Slender body, somewhat compressed at front, strongly compressed at caudal peduncle; fairly pointed snout; slightly subterminal mouth; large eye. Dorsal fin origin behind pelvic fin origin. Olive-gray to light brown above; *brilliant silver side*, often with blue sheen; black specks and blotches on back and upper side.

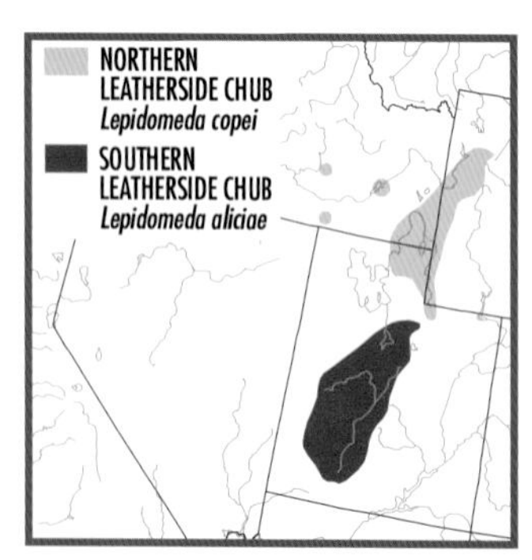

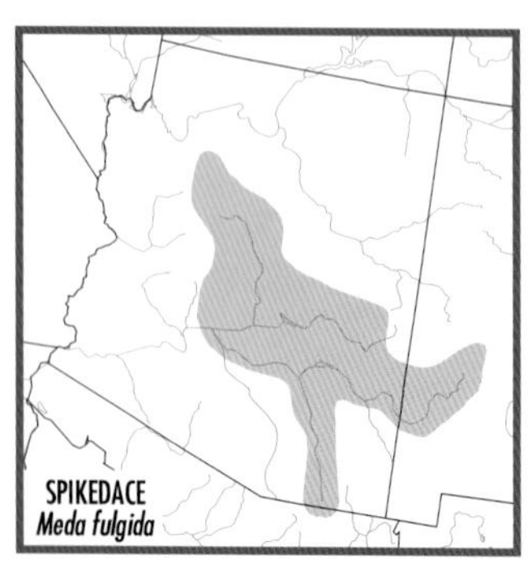

Breeding male has bright brassy yellow head and fin bases, yellow belly and fins. Has 9 anal rays; pharyngeal teeth 1,4-4,1. To 3½ in. (9.1 cm). **RANGE:** Gila R. system, AZ and NM. Rare or extirpated over most of range. Protected as a *threatened species*. **HABITAT:** Sandy and rocky runs and pools; often near riffles of creeks and small rivers. **SIMILAR SPECIES:** (1) Woundfin, *Plagopterus argentissimus* (Pl. 13), has wider, flatter head; barbel at corner of mouth. (2) Virgin Spinedace, *Lepidomeda mollispinis* (Pl. 13), and other *Lepidomeda* species *have scales*, large gray-black blotches on compressed body.

WOUNDFIN *Plagopterus argentissimus* **Pl. 13**

IDENTIFICATION: *No scales. Wide, flat head*; body compressed at rear. Long, rounded snout; subterminal mouth, *barbel* at corner of mouth; fairly small eye high on head. Dorsal fin origin behind pelvic fin origin. Large, forked caudal fin. Dusky gray above; *brilliant silver side*, often with blue sheen; sometimes faint yellow at paired fin bases. Pink lower side on breeding male. Has 9–10 anal rays; pharyngeal teeth 1,5-4,1. To 3½ in. (9 cm). **RANGE:** Known from Virgin and Gila river systems (both lower Colorado R. drainage), UT, NV, and AZ; extant only in Gila R. system. Extremely rare; protected as an *endangered species*. **HABITAT:** Fast sandy runs and pools of small to medium rivers; usually in warm turbid water. **SIMILAR SPECIES:** (1) Spikedace, *Meda fulgida* (Pl. 13), has *more compressed head*, dark specks and blotches on back and upper side, *lacks* barbel. (2) Virgin Spinedace, *Lepidomeda mollispinis* (Pl. 13), and other *Lepidomeda* species *have scales*, large gray-black blotches on *compressed body*; *lack* barbel.

CHROSOMUS

GENUS CHARACTERISTICS: Adult males are spectacularly colored. Lateral line usually ends at about *middle* of body, occasionally is absent. *Scales are so small* they appear to be absent and give fish a metallic look. Males have bright *silver spots* at fin bases. Slender, fairly compressed body; dorsal fin origin above or behind pelvic fin origin. Eight dorsal rays; 7–8 anal rays.

FINESCALE DACE *Chrosomus neogaeus* Pl. 13

IDENTIFICATION: *Dark brown to gray "cape"* on back and upper side. *Body profusely speckled with black.* Dark olive to gold stripe along side, light olive between cape and stripe; silver white below; usually a black caudal spot; clear to yellow fins. Large male has red along side. Large head; *large terminal mouth* extends under eye; rounded snout. Has 63–92 lateral scales; pharyngeal teeth 2,5-4,2. To 4¼ in. (11 cm). **RANGE:** Atlantic, Great Lakes, Hudson Bay, and upper Mississippi, Missouri, and Peace-Mackenzie river drainages from NB to NT and BC; south to NY, WI, and WY. Common in east and north; sporadic in Missouri drainage. **HABITAT:** Lakes, ponds, and sluggish pools of headwaters, creeks, and small rivers; usually over silt and near vegetation. **REMARKS:** Finescale Dace and Northern Redbelly Dace, *C. eos,* commonly hybridize. Hybrids, intermediate in characters, are always females and in some areas are more common than parental species. Apparently hybrids breed with males of parental species and can outnumber and even replace one of the parent species. **SIMILAR SPECIES:** (1) Southern Redbelly Dace, *C. erythrogaster* (Pl. 13), and (2) Northern Redbelly Dace, *C. eos*, have 2 black stripes along side, *smaller head and mouth.* (3) Northern Pearl Dace, *Margariscus nachtriebi* (Pl. 13), *lacks* dark "cape" on back and upper side; has blunter snout, *smaller mouth* (rarely reaching eye), often has herringbone lines on back, usually complete lateral line, usually barbel near corner of mouth.

SOUTHERN REDBELLY DACE *Chrosomus erythrogaster* Pl. 13

IDENTIFICATION: *Two black stripes* along side; upper one thin, broken into spots at rear; lower one wide, becoming thin on caudal peduncle. Olive-brown above, dusky stripe along back; *black spots* (sometimes

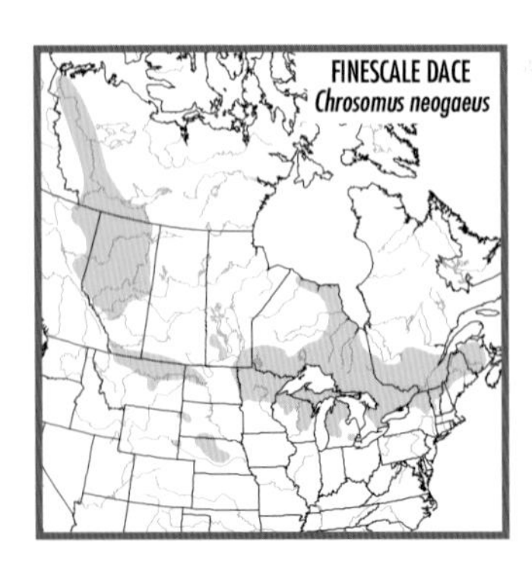

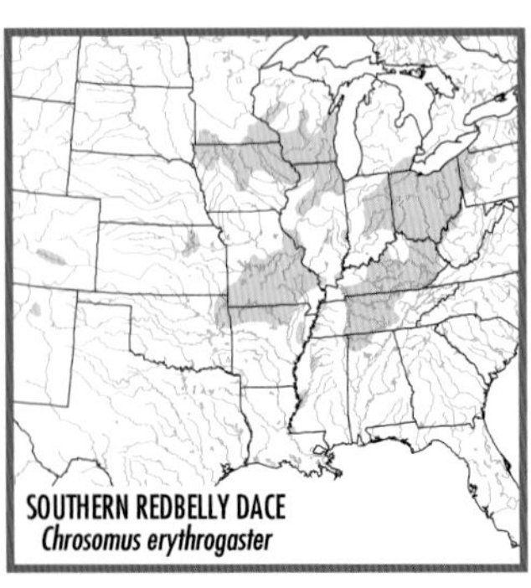

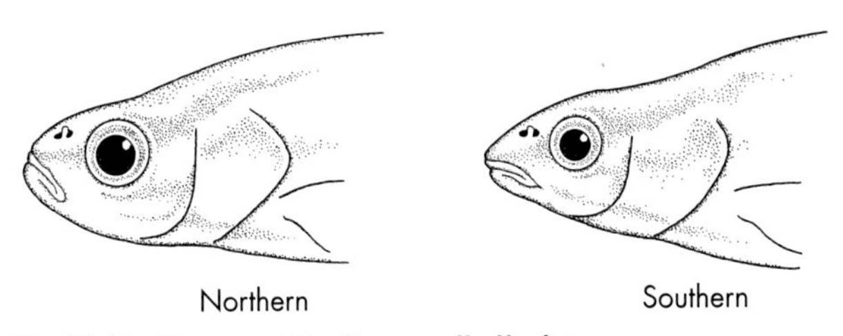

Fig. 15. Southern and Northern redbelly dace.

absent) on upper side, often arranged in row; silver yellow side; black wedge-shaped caudal spot; white, yellow, or red below. Large male is vividly colored, with bright red belly, lower head, and base of dorsal fin; yellow fins. Moderately pointed snout, longer than eye in adult; small, moderately oblique (less than 45°), slightly subterminal mouth ending in front of eye (Fig. 15). Has 67–95 lateral scales; pharyngeal teeth 0,5-5,0. To 3½ in. (9.1 cm). **RANGE:** Basins of Lakes Erie and Michigan and Mississippi R. from NY to s. MN; south to Tennessee R. drainage, AL, and White and Arkansas river drainages, AR and OK. Isolated populations on Former Mississippi Embayment, TN, MS, and AR; Kansas R. system, KS; and upper Arkansas R. drainage, CO and NM. Common in uplands and spring-fed streams; absent in lowlands. **HABITAT:** Rocky, usually spring-fed, pools of headwaters and creeks. **SIMILAR SPECIES:** See (1) Northern Redbelly Dace, *C. eos,* and (2) Laurel Dace, *C. saylori.* (3) Mountain Redbelly Dace, *C. oreas*, and (4) Tennessee Dace, *C. tennesseensis* (both Pl. 13), have 1 broken stripe along side, spots in rows on back and upper side.

NORTHERN REDBELLY DACE *Chrosomus eos* **Not shown**

IDENTIFICATION: Similar to Southern Redbelly Dace, *C. erythrogaster,* but has more rounded, shorter (about equal to eye diameter) snout;

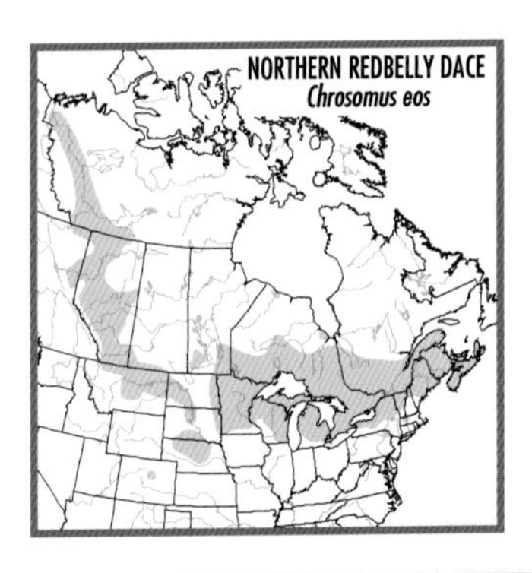

more upturned mouth, with chin *in front* of upper lip (Fig. 15). Large males may have red or yellow belly, head, and fins. Has 70–90 lateral scales. To 3¼ in. (8 cm). **RANGE:** Atlantic, Great Lakes, Hudson Bay, and upper Mississippi, Missouri, and Peace-Mackenzie river drainages, from NS west to NT and BC; south to n. PA, WI, and NE. Isolated population in South Platte R. system, CO. Common. **HABITAT:** Lakes, ponds, bogs, and pools of headwaters and creeks. Usually over silt, often near vegetation. **SIMILAR SPECIES:** (1) See Southern Redbelly Dace, *C. erythrogaster* (Pl. 13). (2) Finescale Dace, *C. neogaeus* (Pl. 13), has larger head and mouth, 1 stripe along side, dark "cape" on back and upper side, many small specks.

LAUREL DACE *Chrosomus saylori* **Not shown**

IDENTIFICATION: Similar to Southern Redbelly Dace, *C. erythrogaster*, but has S-shaped gut (double-looped in Southern Redbelly Dace), black on underside of head of breeding male. Has 72–90 lateral scales. To 3 in. (7.4 cm). **RANGE:** Upper Tennessee R. system, TN. Known only from Walden Ridge of Cumberland Plateau. Common. **HABITAT:** Rocky pools of headwaters and creeks; usually along undercut banks or around large rocks. **SIMILAR SPECIES:** (1) See Southern Redbelly Dace, *C. erythrogaster* (Pl. 13).

BLACKSIDE DACE *Chrosomus cumberlandensis* **Pl. 13**

IDENTIFICATION: Two dusky stripes along side *converge on caudal peduncle*, coalesce into *wide black stripe* on large male. *Many black specks on back and upper side.* Olive to green-gold above, silver white to red below. Large male is bright red below, behind opercle, and at base of dorsal fin; has bright silver pectoral and pelvic fin bases, yellow fins. Moderately compressed body, pointed snout, slightly subterminal mouth. Has 66–81 lateral scales; pharyngeal teeth 0,5-5,0. To 2¾ in. (7.2 cm). **RANGE:** Upper Cumberland R. drainage (Big South

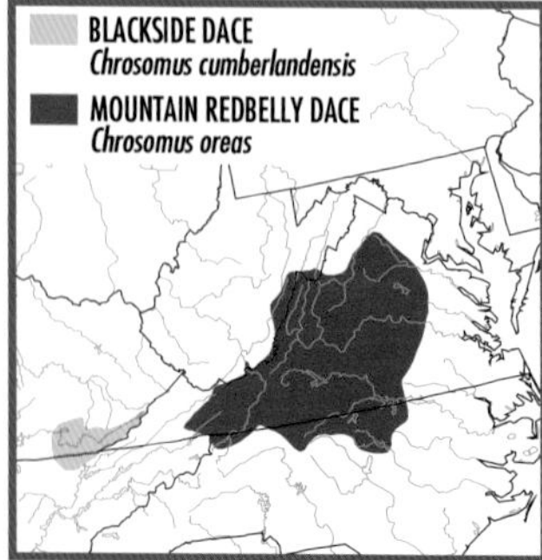

Fork and above), KY and TN. Rare; protected as a *threatened species.* **HABITAT:** Rocky pools of headwaters and creeks; usually along undercut banks, near large rocks or among detritus. **SIMILAR SPECIES:** See Pl. 13. (1) Mountain Redbelly Dace, *C. oreas*, and (2) Tennessee Dace, *C. tennesseensis*, have broken stripe on side, spots in rows on upper side; are less compressed; *lack* wide black stripe on large male. (3) Southern Redbelly Dace, *C. erythrogaster, lacks* many black specks on upper side, has 2 *parallel stripes* on side.

MOUNTAIN REDBELLY DACE *Chrosomus oreas* Pl. 13

IDENTIFICATION: Black stripe along side *broken* under dorsal fin; *large black spots* in row along back and in row along upper side. Olive to green-gold above; silver white to red below. Large male is bright red below (including lower half of opercle), behind opercle, and at base of dorsal fin; has bright silver pectoral and pelvic fin bases, black chin and breast, yellow fins. Moderately pointed snout; slightly subterminal mouth. Has 64–81 lateral scales; pharyngeal teeth 0,5-5,0. To 2¾ in. (7.2 cm). **RANGE:** Montane and Piedmont regions of Atlantic Slope from Shenandoah R. (Potomac R. drainage), VA, to Neuse R. drainage, NC; upper New R. drainage, WV, VA, and NC. Introduced into upper Holston R. system, VA, and Big Sandy R. system, KY. Abundant. **HABITAT:** Rocky pools and runs of headwaters, creeks, and small to medium rivers. **SIMILAR SPECIES:** (1) See Tennessee Dace, *C. tennesseensis* (Pl. 13). (2) Other *Chrosomus* species (Pl. 13) *lack* broken stripe on side.

TENNESSEE DACE *Chrosomus tennesseensis* Pl. 13

IDENTIFICATION: Similar to Mountain Redbelly Dace, *C. oreas,* but has *smaller spots* (smaller than eye pupil) on back and upper side, usually a *thin black stripe* along side (above larger broken stripe), and red on caudal fin and below stripe along side of large male. Has 67–95 lateral scales. To 2¾ in. (7.2 cm). **RANGE:** Upper Tennessee R. drainage (from lower Clinch R. system, TN, and Holston R. system, VA, to near GA border), VA and TN. Rare and localized. **HABITAT:** Gravel-, sand-, and silt-bottomed pools of spring-fed headwaters. **SIMILAR SPECIES:** (1) See Mountain Redbelly Dace, *C. oreas* (Pl. 13).

FLAME CHUB *Hemitremia flammea* Pl. 13

IDENTIFICATION: Chubby, barely compressed body, *deep caudal peduncle*; short head, with extremely *short snout*; small, slightly subterminal mouth; round eye. Dorsal fin origin slightly behind pelvic fin origin. Olive above, dark stripe along back; dark streaks along upper side, then light stripe, then black stripe ending at black caudal spot or wedge; white to red below. Large individual (especially male) has bright *scarlet red* along bottom ⅓ of body and at base of dorsal fin. Silver peritoneum flecked with black. Incomplete lateral line, fewer than

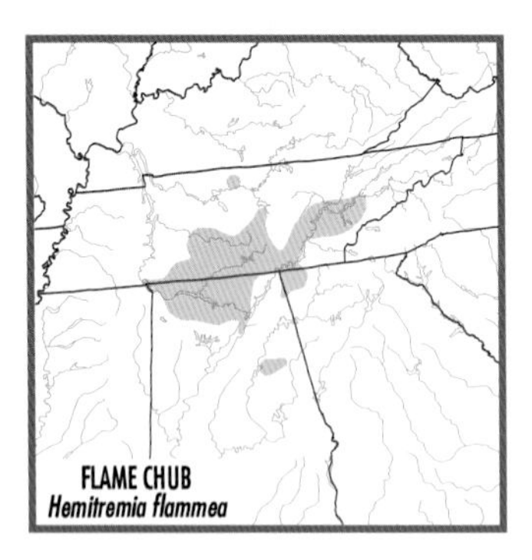

half of scales pored; 38–44 lateral scales; 7–8 anal rays; pharyngeal teeth 2,5-4,2. To 2¾ in. (7.2 cm). **RANGE:** Middle Cumberland (mostly Caney Fork) and Tennessee river drainages, TN, GA, and AL; Kelly Creek (Coosa R. system), ne. AL. Uncommon; extirpated from many areas because of alterations of springs. **HABITAT:** Springs and spring-fed streams; usually over gravel. **SIMILAR SPECIES:** (1) Southern Redbelly Dace, *Chrosomus erythrogaster* (Pl. 13), has 2 black stripes along side, black spots on upper side (sometimes missing), often has yellow on body and fins, much smaller (67–95 lateral) scales.

SEMOTILUS

GENUS CHARACTERISTICS: Large minnows with thick, barely to moderately compressed body, *broad head*, small *flaplike barbel* in groove *above* (but not in) corner of mouth (Fig. 16), usually 8 anal rays, dorsal fin origin slightly behind pelvic fin origin, complete lateral line, short intestine, silver peritoneum with black specks, pharyngeal teeth usually 2,5-4,2. Flaplike barbel is most easily seen when mouth is held open.

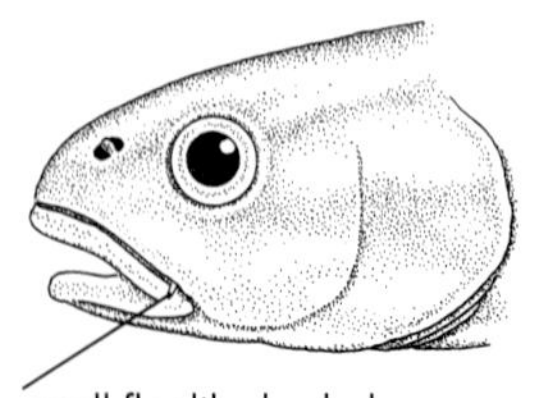

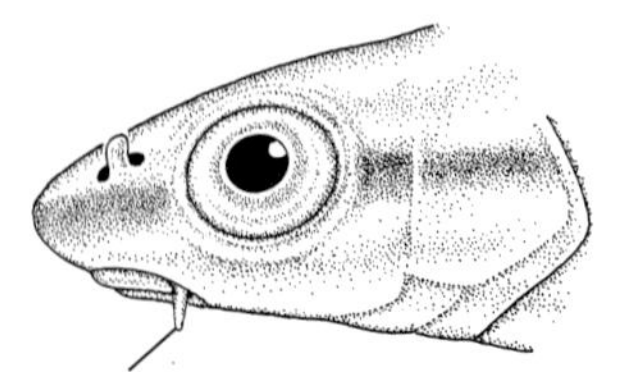

Fig. 16. Minnows—barbels.

CREEK CHUB *Semotilus atromaculatus* Pl. 13

IDENTIFICATION: *Large black spot* at front of dorsal fin base, black caudal spot (indistinct in large individual). *Large terminal mouth* reaching past front of eye. Body barely compressed at front, compressed at caudal peduncle; pointed snout. Gray-brown above, dark stripe along back; herringbone lines on upper side in young; dusky black stripe (darkest on young) along olive-silver side, around snout, and onto upper lip; black bar along back of gill cover. Breeding male has orange at dorsal base, orange lower fins, blue on side of head, pink on lower half of head and body; 6–12 large tubercles on head. Has 47–65 lateral scales; 8 dorsal rays. To 12 in. (30 cm). **RANGE:** Most of e. U.S. and se. Canada in Atlantic, Great Lakes, Hudson Bay, Mississippi, and Gulf basins as far west as SK, WY, and Brazos R., TX, but absent from FL and s. GA; isolated population in upper Pecos and Canadian river systems, NM. Introduced elsewhere in U.S. One of most common fishes in e. N. America. **HABITAT:** Rocky and sandy pools of headwaters, creeks, and small rivers. **SIMILAR SPECIES:** (1) See Dixie Chub, S. *thoreauianus* (Pl. 13), and (2) Sandhills Chub, *S. lumbee*. (3) Fallfish, *S. corporalis* (Pl. 13), *lacks* black spot at dorsal fin base; has darkly outlined, larger (43–50 lateral) scales, larger eye.

DIXIE CHUB *Semotilus thoreauianus* Pl. 13

IDENTIFICATION: Similar to Creek Chub, *S. atromaculatus,* but has *larger* (45–52 lateral) *scales*; *wider, more diffuse stripe* (darker on young) along side; stouter body; less distinct dorsal and caudal spots. Breeding male has orange to pink underside, yellow fins, usually 8 large hooked tubercles on head. To 6 in. (15 cm). **RANGE:** Gulf Slope from Ochlockonee R. system, GA and FL, to Tombigbee R. system, AL. Common below Fall Line. **HABITAT:** Sand- and gravel-bottomed pools of creeks and small rivers. **SIMILAR SPECIES:** (1) See Creek Chub, *S. atromaculatus* (Pl. 13). (2) Sandhills Chub, *S. lumbee*, is strongly bicolored; has black caudal spot; 9, rather than 8, dorsal rays.

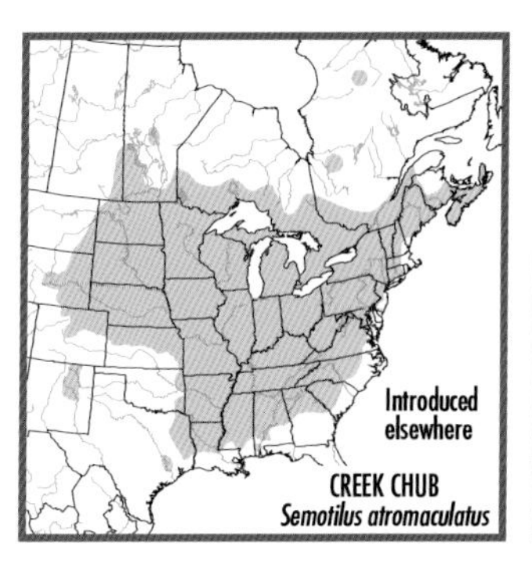

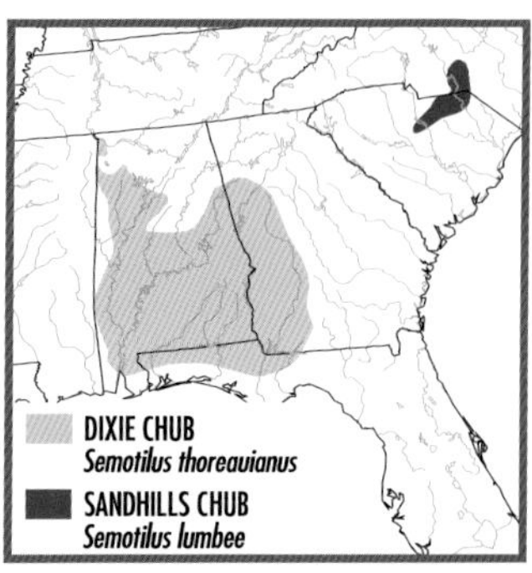

SANDHILLS CHUB *Semotilus lumbee* **Not shown**

IDENTIFICATION: Similar to Creek Chub, *S. atromaculatus,* but *lacks* discrete black spot at dorsal fin base (dusky spot may be present), is more strongly bicolored (dark above, white below), has *9 dorsal rays,* larger scales (usually 18 scales around caudal peduncle; Creek Chub usually has 19–20). Breeding male has *red fins.* Has 42–52 lateral scales. To 9½ in. (24 cm). **RANGE:** "Carolina Sandhills" of s.-cen. NC and ne. SC, encompassing parts of Cape Fear, Peedee, and Santee river drainages. Fairly common in small range. **HABITAT:** Flowing sand- and gravel-bottomed pools and runs of headwaters, creeks, and small rivers. **SIMILAR SPECIES:** (1) See Creek Chub, *S. atromaculatus* (Pl. 13). (2) Dixie Chub, *S. thoreauianus* (Pl. 13), has *8 dorsal rays,* is less strongly bicolored, has less distinct black caudal spot.

FALLFISH *Semotilus corporalis* **Pl. 13**

IDENTIFICATION: *Large* (43–50 lateral) *scales. Large eye.* Moderately compressed body; fairly long, rounded snout slightly overhanging large mouth; dorsal fin origin over or slightly behind pelvic fin origin. Scales on back and upper side *darkly outlined* on adult; young has black stripe along side, dark caudal spot. Olive to gold-brown above, dark stripe along back; bright silver side, sometimes with purple or blue sheen; black bar along back of gill cover. Breeding male has fairly large tubercles on head. Has 8 dorsal rays; 8 anal rays. Largest minnow native to e. N. America; to 20¼ in. (51 cm). **RANGE:** Hudson Bay, Lake Ontario, and Atlantic Slope drainages from QC and ON to James R. drainage, VA. Introduced into Tonawanda R. (Lake Erie drainage), NY. Common. **HABITAT:** Gravel- and rubble-bottomed pools and runs of small to medium rivers; lake margins. **SIMILAR SPECIES:** (1) Creek Chub, *S. atromaculatus* (Pl. 13), has large, black spot at front of dorsal fin base, *smaller scales* (usually more than 50 along lateral line) that are less distinctly outlined, *smaller eye.* (2) River Chub, *Nocomis micropogon,* and (3) Hornyhead Chub, *N. bigutattus* (both Pl. 14), have much *smaller eye,* larger barbel *in corner* of mouth, 7 anal rays, pharyngeal teeth 0,4-4,0 or 1,4-4,1.

NORTHERN PEARL DACE *Margariscus nachtriebi* **Pl. 13**

IDENTIFICATION: *Small* (60–75 lateral) *scales,* short head, fairly deep caudal peduncle. *Flaplike barbel* in groove above mouth (often missing on 1 or both sides). Nearly cylindrical body; rounded snout; small, slightly subterminal mouth (seldom reaching front of eye); dorsal fin origin behind pelvic fin origin. Dark olive to gray above, often with black herringbone lines, dark stripe along back; many small *black and brown specks* (absent in western populations) on silver side; white, yellow, or red below. Black stripe along side, black caudal spot on young;

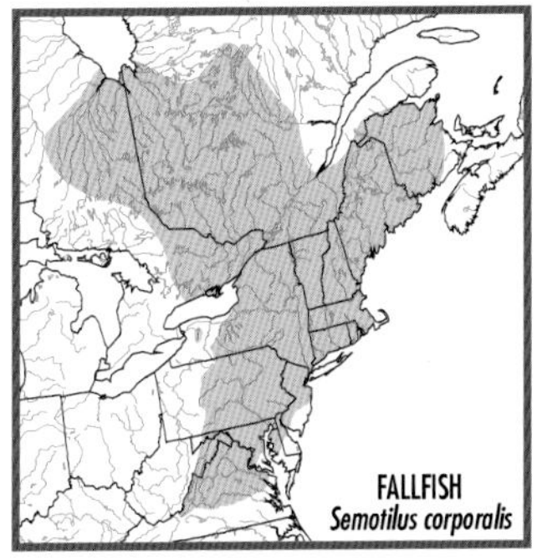

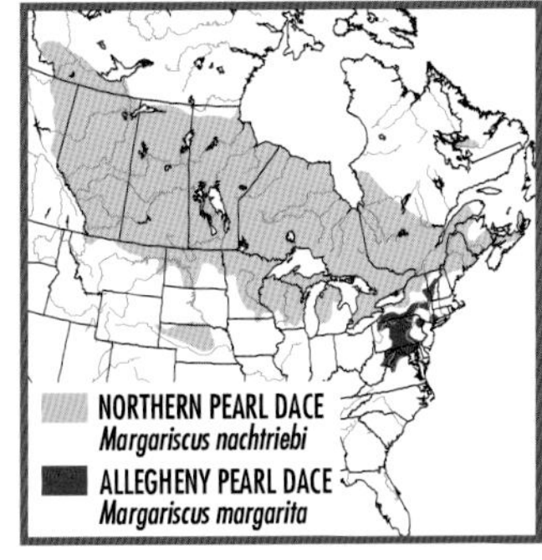

stripe vague on adult. Breeding male is bright orange-red along lower side; has pale yellow stripe along belly, many small tubercles on head. Usually complete lateral line; 8 dorsal rays, 8 anal rays; pharyngeal teeth 2,5-4,2. To 6½ in. (16 cm). **RANGE:** Atlantic, Hudson Bay, Great Lakes, and Mississippi R. basins in s. Canada and n. U.S. from Atlantic Coast to s. NT, e. BC, and MT; south to NY, WI, and IA. Isolated population in upper Missouri R. basin, SD, NE, and WY. Common over most of range. **HABITAT:** Pools of creeks and small rivers; ponds and lakes. Usually over sand or gravel. **SIMILAR SPECIES:** (1) See Allegheny Pearl Dace, *M. margarita*. (2) Creek Chub, *Semotilus atromaculatus* (Pl. 13), has black spot at front of dorsal fin base, large terminal mouth, *lacks* brown and black specks on side. (3) Lake Chub, *Couseius plumbeus* (Pl. 13), has longer, sharper snout, more compressed body; lacks red along lower side of body of large male. (4) Finescale Dace, *Chrosomus neogaeus* (Pl. 13), *lacks* barbel, herringbone lines; has more pointed snout, larger mouth (to below eye), incomplete lateral line.

ALLEGHENY PEARL DACE *Margariscus margarita* Not shown

IDENTIFICATION: Nearly identical to Northern Pearl Dace, *M. nachtriebi*, but has *larger scales; usually 50–62 lateral scales.* To 6½ in. (16 cm). **RANGE:** Atlantic Slope from Hudson R. drainage, VT and NY, south to Potomac R. drainage, VA; upper Ohio R. drainages, s. NY to WV. Common. **HABITAT:** Pools of upland creeks and small rivers; ponds and lakes. Usually over sand or gravel. **SIMILAR SPECIES:** (1) See Northern Pearl Dace, *M. nachtriebi* (Pl. 13).

LAKE CHUB *Couesius plumbeus* Pl. 13

IDENTIFICATION: Barbel at corner of large, barely subterminal mouth. *Large eye*; head flattened above and below; moderately pointed snout. Moderately compressed, slender body; dorsal fin origin over or slightly behind pelvic fin origin. Brown to green above; dark stripe along silver

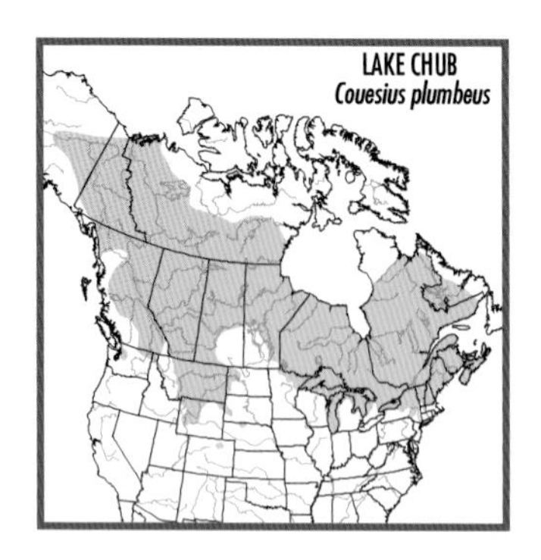

gray side, darkest on young and large male; sometimes black specks on side and belly; dusky caudal spot. Large male may have red at pectoral and pelvic fin origins, corners of mouth. Complete lateral line; 53–70 lateral scales; 8 anal rays; pharyngeal teeth 2,4-4,2. To 9 in. (23 cm). **RANGE:** Most northern minnow in N. America; only minnow in AK (Yukon R. drainage). Throughout much of Canada and extreme n. U.S.; south to Delaware R., NY, Lake Michigan, IL, Platte R. system, WY. Relict populations in upper Missouri R. drainage, SD, NE, CO, and WY, and Twin Springs Creek (Mississippi R. tributary), IA. Common throughout much of range. **HABITAT:** Virtually any body of water, standing or flowing, large or small. Usually in gravel-bottomed pools and runs of streams and along rocky lake margins. **SIMILAR SPECIES:** (1) Northern Pearl Dace, *Margariscus nachtriebi* (Pl. 13), has shorter, blunter snout; less compressed body; bright red along lower side of large male.

NOCOMIS

GENUS CHARACTERISTICS: Large, bronze-colored, stout body; *large* (36–45 lateral) scales; dark-edged scales on back and upper side; *barbel* at corner of *large, slightly subterminal mouth*. Complete lateral line, 7 anal rays. Dorsal fin origin slightly in front of to slightly behind pelvic fin origin.

HORNYHEAD CHUB *Nocomis biguttatus* Pl. 14

IDENTIFICATION: *Red* (on young) to yellow (adult) caudal fin; other fins yellow to orange. *Bright red spot* behind eye on large male (brassy on female). Dark olive to brown above; iridescent green on yellow-brown side; white to light yellow below. Dusky iridescent yellow stripe along back, yellow streak above dusky stripe along side and around snout, black caudal spot (all darkest on young). Breeding male is pink below with pink-orange fins; has many large tubercles on top of head.

Rounded snout. Has 38–45 lateral scales; usually 16–17 scales around caudal peduncle; pharyngeal teeth 1,4-4,1. To 10¼ in. (26 cm). **RANGE:** Mohawk R. system, NY, west through Great Lakes and Mississippi R. basin to Red R. drainage (Hudson Bay basin), MB and ND, and south to Ohio R. drainage; Ozark drainages, MO and AR. Isolated populations in lower Kentucky R. system, KY, Platte and Cheyenne river systems, NE, WY, and CO, and Kansas R., KS. Common throughout much of range. **HABITAT:** Rocky pools and runs of creeks and small to medium rivers. **SIMILAR SPECIES:** See Pl. 14. See (1) Redspot Chub, *N. asper*, and (2) Redtail Chub, *N. effusus*. (3) River Chub, *N. micropogon*, has longer snout (about same as length of head behind eye), smaller eye higher on head, usually no stripe along side, *no* red spot behind eye, *no* bright red caudal fin, large hump on head of breeding male, pharyngeal teeth 0,4-4,0. (4) Bluehead Chub, *N. leptocephalus*, *lacks* bright red caudal fin and red spot behind eye; usually has large loop on intestine; pharyngeal teeth 0,4-4,0; breeding male has large hump on dark blue head.

REDSPOT CHUB *Nocomis asper* Not shown

IDENTIFICATION: Similar to Hornyhead Chub, *N. biguttatus,* but adult and large juvenile have *rows of tubercles* (or tubercle spots) on side of body (often 2–3 tubercles/scale); spot behind eye bright red on large juvenile as well as adult. Breeding male has yellow-pink fins. Has 38–45 lateral scales, usually 17–21 scales around caudal peduncle; pharyngeal teeth 1,4-4,1. To 8½ in. (22 cm). **RANGE:** Arkansas R. drainage, sw. MO, se. KS, ne. OK, and nw. AR; isolated populations in Blue R. (Red R. drainage), OK, and upper Ouachita R. drainage, AR. **HABITAT:** Rocky, usually clear, runs and pools of creeks and small to medium rivers. Locally common. **SIMILAR SPECIES:** (1) See Hornyhead Chub, *N. biguttatus* (Pl. 14). (2) Redtail Chub, *N. effusus* (Pl. 14), has red-orange paired fins, usually 1 tubercle/scale, pharyngeal teeth 0,4-4,0.

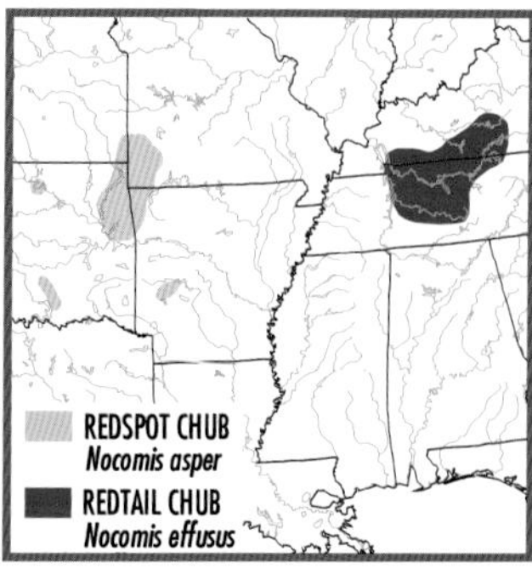

REDTAIL CHUB *Nocomis effusus* **Pl. 14**

IDENTIFICATION: Similar to Hornyhead Chub, *N. biguttatus,* but has *brighter red-orange fins* (especially on young), usually 19–20 scales around caudal peduncle; large juvenile and adult have *rows of tubercles* (or tubercle spots) on side of body (usually 1 tubercle/scale); pharyngeal teeth 0,4-4,0; 39–44 lateral scales. To 9 in. (23 cm). **RANGE:** Upper Green, upper Barren, Cumberland (Little South Fork and downstream), Duck, and lower Tennessee (between Cumberland and Duck rivers) river drainages, KY and TN. **HABITAT:** Clear rocky runs and pools of creeks and small rivers. Uncommon over much of range; locally common. **SIMILAR SPECIES:** (1) See Hornyhead Chub, *N. biguttatus* (Pl. 14). (2) Redspot Chub, *N. asper,* has less brightly colored fins, often 2–3 tubercles/scale, pharyngeal teeth 1,4-4,1.

RIVER CHUB *Nocomis micropogon* **Pl. 14**

IDENTIFICATION: *Long snout* (about same as length of head behind eye); *small eye* high on head. Large mouth; width greater than interpelvic width. Dark olive to brown above; brassy, iridescent olive green side; white to light yellow below; olive to light orange caudal fin; other fins clear to yellow-pink. Dusky iridescent yellow stripe along back, sometimes a dusky stripe along side ending in darker caudal spot. Breeding male has pink-blue head, body, and fins; *large tubercles on snout* (including in front of nostrils); *large hump* on top of head. Has 37–43 (usually 38–41) lateral scales; more than 30 scales around body at dorsal fin origin; pharyngeal teeth 0,4-4,0. To 12½ in. (32 cm). **RANGE:** Atlantic drainages from Susquehanna R., NY, to James R., VA; Great Lakes basin, NY and ON to MI; Ohio R. basin, NY to e. IL and south to n. GA and AL except absent in sw. IN, w. ⅔ of KY, and most of w. TN. Introduced in Ottawa R. system, ON. Also present and possibly introduced in upper Santee R., NC, Savannah R., SC and GA, and Coosa R., GA. Common; locally abundant. **HABITAT:** Rocky runs and

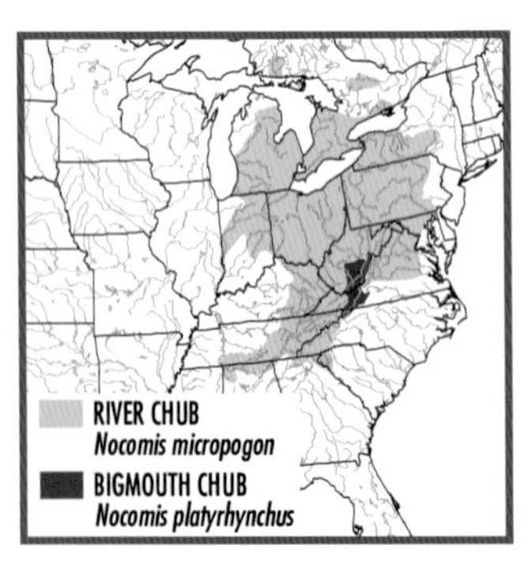

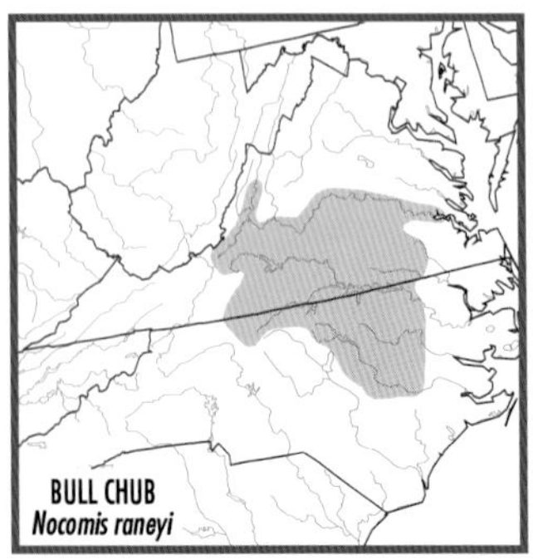

flowing pools of small to medium rivers. **SIMILAR SPECIES:** See Pl. 14. (1) See Bigmouth Chub, *N. platyrhynchus,* and (2) Bull Chub, *N. raneyi.* (3) Bluehead Chub, *N. leptocephalus, lacks* tubercles on snout in front of nostrils; has snout *shorter* than length of head behind eye, eye lower on head, deeper body, usually fewer than 30 scales around body, usually large loop on intestine. (4) Hornyhead Chub, *N. biguttatus,* (5) Redspot Chub, *N. asper,* and (6) Redtail Chub, *N. effusus,* have snout *shorter* than length of head behind eye, *larger eye* lower on head, more distinct stripe along side, red caudal fin on young, red spot behind eye on large male, no hump on head of breeding male.

BIGMOUTH CHUB *Nocomis platyrhynchus* — Not shown

IDENTIFICATION: Nearly identical to River Chub, *N. micropogon,* but has tubercles (on adult; largest on breeding male) extending over top of head from snout to well behind eyes. River Chub has fewer, larger tubercles usually restricted to area in front of eyes. To 9½ in. (24 cm). **RANGE:** New R. drainage, WV, VA, and NC. Generally uncommon. **HABITAT:** Rocky pools and runs of creeks and small to medium rivers. **SIMILAR SPECIES:** (1) See River Chub, *N. micropogon* (Pl. 14).

BULL CHUB *Nocomis raneyi* — Not shown

IDENTIFICATION: Nearly identical to River Chub, *N. micropogon,* but has *smaller mouth* (width equal to or less than interpelvic width), 39–45 (usually 40–43) lateral scales. To 12½ in. (32 cm). **RANGE:** Atlantic Slope drainages from James R., VA, to Neuse R. (1 record in upper Cape Fear R.), NC. Generally common above Fall Line. **HABITAT:** Rocky pools and runs of creeks and small to medium rivers. **SIMILAR SPECIES:** (1) See River Chub, *N. micropogon* (Pl. 14).

BLUEHEAD CHUB *Nocomis leptocephalus* — Pl. 14

IDENTIFICATION: *Large loop* on intestine visible through body wall of young (see Remarks). Fairly short, rounded snout; deep body. Tan to dark olive above, dusky iridescent yellow stripe along back; dusky stripe along brassy iridescent green side; light yellow to red-orange fins. Breeding male has *large hump* on top of *dark blue head,* large tubercles on head *behind nostrils,* orange or blue side, orange fins. Has 36–43 lateral scales, usually fewer than 30 scales around body at dorsal fin origin; pharyngeal teeth 0,4-4,0. To 10 in. (26 cm). **RANGE:** Atlantic and Gulf slope drainages from Shenandoah R., VA, to Mississippi R., MS; tributaries of Mississippi R. north to Yazoo R. system, MS; upper New R. drainage, WV, VA, and NC; Bear Creek (Tennessee R. drainage), AL and MS. Introduced into Little Tennessee R. and French Broad systems, NC and TN. Common. **HABITAT:** Rocky and sandy pools and runs of headwaters, creeks, and small to medium rivers. **REMARKS:** Three subspecies. *N. l. leptocephalus,* in New R.

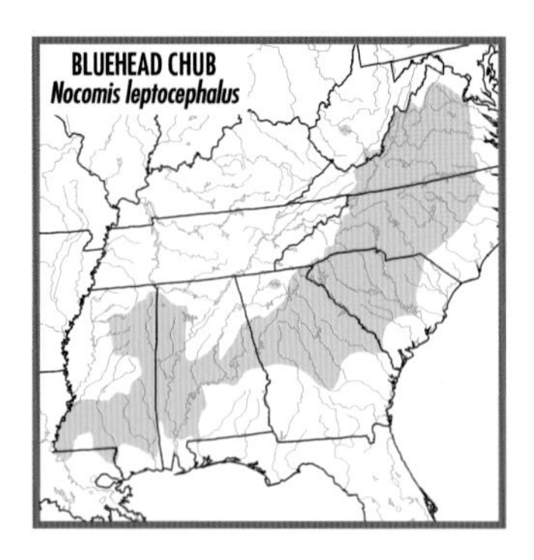

and Atlantic drainages south to Santee R., has 16 or more tubercles on head. *N. l. interocularis*, in Savannah, Altamaha, and Apalachicola drainages, has 7–9 tubercles on head. *N. l. bellicus*, in Gulf Slope and Mississippi R. drainages west of Apalachicola drainage, has only 4–6 tubercles on head. Intestinal loop is absent in some *N. l. interocularis* in Chattahoochee R. system. **SIMILAR SPECIES:** See Pl. 14. (1) River Chub, *N. micropogon*, and (2) Bull Chub, *N. raneyi*, have longer snout, eye higher on head, more slender body, tubercles on snout in front of nostrils, usually more than 30 scales around body, *no* large loop on intestine. (3) Hornyhead Chub, *N. biguttatus, lacks* large loop on intestine, large hump on head of breeding male; has bright red caudal fin, spot behind eye.

CAMPOSTOMA

GENUS CHARACTERISTICS: *Hard cartilaginous ridge on* lower jaw (Fig. 17) of *subterminal mouth. Thick,* barely compressed body; dorsal fin origin over or slightly behind pelvic fin origin. Tan to brown above; often a dark stripe along side, dark caudal spot on young, *irregular dark brown to black blotches* on back and side of large individual. Breeding male has *white lips,* bright red eye. Stonerollers use hard ridge on lower jaw to scrape algae and other food from rocks. Algae are difficult to digest, and stonerollers have a long intestine to aid in digestion. To accommodate its great length (about 18 in. [46 cm] in a 5-in. [13-cm] individual), intestine is coiled around gas bladder (except in Mexican Stoneroller, *C. ornatum*). Distributions and variation of Central Stoneroller, *C. anomalum,* and Largescale Stoneroller, *C. oligolepis,* are unclear.

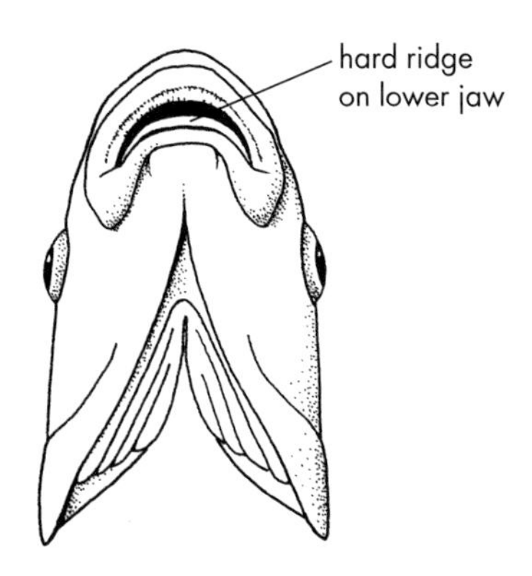

Fig. 17. *Campostoma*—underside of head.

:ENTRAL STONEROLLER *Campostoma anomalum* **Pl. 14**

IDENTIFICATION: Breeding male has black band on orange dorsal and *anal* fins (see Remarks), crescent-shaped row of 1–3 large tubercles on inner edge of nostril (Fig. 18), tubercles on nape and adjacent area of body, body strongly arched at nape. Complete lateral line; usually *46–55 lateral scales, 36–46 scales around body* at dorsal fin origin, 21–33 rakers on 1st gill arch; pharyngeal teeth 0,4-4,0 or rarely 1,4-4,1. To 6¾ in. (17 cm). **RANGE:** Widespread across most of e. and cen. U.S. in Atlantic, Great Lakes, Mississippi R., and Hudson Bay (Red R.) basins from NY and ON to ND and WY, and south to GA, LA, and TX (but absent from most of lower Ohio R. basin); Gulf Slope drainages of TX. Also in Río San Juan basin, Mexico. Common to abundant throughout much of range but generally absent on Piedmont and Coastal Plain, uncommon on Great Plains. **HABITAT:** Rocky riffles, runs, and pools of headwaters, creeks, and small to medium rivers. **REMARKS:** Three subspecies often recognized but in need of study. *C. a. anomalum,* in Ohio R. and upper Atlantic drainages, usually has 15–16 pectoral rays (other populations generally have 16–19). *C. a. michauxi,* in Santee and Savannah river drainages, NC and SC, uniquely lacks black band on anal fin of breeding male. *C. a. pullum,* apparently occupying rest of range, usually has 18–20 scales over body from lateral line to lateral line at dorsal fin origin (including lateral-line scales); other populations have 15–17 scales. **SIMILAR SPECIES:** See Pl. 14. See (1) Highland Stoneroller, *C. spadiceum,* (2) Bluefin Stoneroller, *C. pauciradii,* (3) Largescale Stoneroller, *C. oligolepis*, and (4) Mexican Stoneroller, *C. ornatum.*

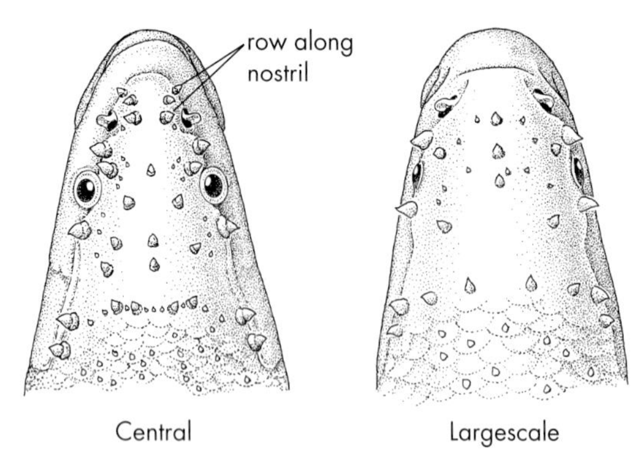

Fig. 18. Central and Largescale stonerollers—head tubercles on breeding male.

HIGHLAND STONEROLLER *Campostoma spadiceum* **Not shown**

IDENTIFICATION: Similar to Central Stoneroller, *C. anomalum,* but has *bright red fins* throughout life; pharyngeal teeth in *2 rows* (usually 1,4-4,1 or 2,4-4,1); breeding male has *black vertical bar* near base of caudal fin (large black spot or wedge in Central Stoneroller), *many small tubercles* in rows on nape (fewer, scattered tubercles on Central Stoneroller). Has 49–63 lateral scales, 36–49 scales around body at dorsal fin origin, 23–30 rakers on 1st gill arch, 15–16 pectoral rays. To 6¾ in. (17 cm). **RANGE:** Red, Ouachita, and lower Arkansas drainages, cen. AR to e. OK. Common. **HABITAT:** Rocky riffles, runs, and sometimes pools of headwaters, creeks, and small rivers. **SIMILAR SPECIES:** (1) See Central Stoneroller, *C. anomalum* (Pl. 14). (2) Largescale Stoneroller, *C. oligolepis* (Pl. 14), has 17–19 pectoral rays; *orange dorsal and anal fins* on breeding male, no crescent-shaped row of 1–3 large tubercles on inner edge of nostril.

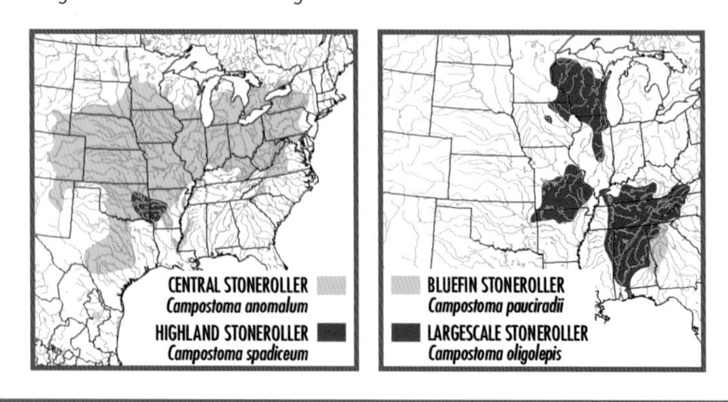

BLUEFIN STONEROLLER *Campostoma pauciradii* Pl. 14

IDENTIFICATION: Similar to Central Stoneroller, *C. anomalum*, but has only *11–17 rakers* on 1st gill arch; *usually 33–38 scales* around body at dorsal fin origin; *blue-green dorsal and anal fins,* blue-green upper side, and brassy yellow lower side on breeding male. Usually 42–49 lateral scales; 15–17 pectoral rays; pharyngeal teeth 0,4-4,0 or 1,4-4,1. To 6¼ in. (16 cm). **RANGE:** Apalachicola and Oconee river drainages (mostly above Fall Line), GA and AL; extreme upper Tallapoosa and Etowah river drainages, GA; Toccoa R. system (Tennessee R. drainage), GA. Fairly common only in upper and middle Chattahoochee R. drainage. **HABITAT:** Rocky riffles, runs, and sometimes pools of headwaters, creeks, and small rivers. **SIMILAR SPECIES:** (1) See Central Stoneroller, *C. anomalum* (Pl. 14). (2) Largescale Stoneroller, *C. oligolepis* (Pl. 14), usually has *19–30 rakers* on 1st gill arch; breeding male with orange dorsal and anal fins, no (or a weak) black band on anal fin, no crescent-shaped row of 1–3 large tubercles on inner edge of nostril.

LARGESCALE STONEROLLER *Campostoma oligolepis* Pl. 14

IDENTIFICATION: Similar to Central Stoneroller, *C. anomalum,* but breeding male has no (or only a weak) black band on orange anal fin, no crescent-shaped row of 1–3 large tubercles on inner edge of nostril (Fig. 18), body deepest near dorsal fin origin. Usually *31–36 scales around body* at dorsal fin origin, *43–47 lateral scales* (see Remarks). Has 19–30 rakers on 1st gill arch, 17–19 pectoral rays; pharyngeal teeth 0,4-4,0 or 1,4-4,1. To 8½ in. (22 cm). **RANGE:** Upper Mississippi R. and Lake Michigan drainages, WI, e. MN, e. IA, and n. IL; Ozarkian streams of MO, n. AR, and ne. OK; Green, Cumberland, and Tennessee river drainages from KY to AL and east to VA and NC; Mobile Bay drainage, GA, AL, and MS. Introduced into Escambia R., AL. Usually common; extirpated from cen. IL. **HABITAT:** Rocky riffles and runs of clear creeks and small to medium rivers. Less tolerant of siltation than is Central Stoneroller. **REMARKS:** Largescale Stonerollers from more eastern drainages have higher scale counts: usually 36–42 scales around body at dorsal fin origin, 48–53 lateral scales. **SIMILAR SPECIES:** (1) See Central Stoneroller, *C. anomalum* (Pl. 14). (2) Highland Stoneroller, *C. spadiceum,* has 15–16 pectoral rays, bright red fins throughout life; crescent-shaped row of 1–3 large tubercles on inner edge of nostril of breeding male. (3) Bluefin Stoneroller, *C. pauciradii* (Pl. 14), has 11–17 rakers on 1st gill arch, 15–17 pectoral rays; blue-green fins; crescent-shaped row of 1–3 large tubercles on inner edge of nostril; black band on anal fin of breeding male.

MEXICAN STONEROLLER *Campostoma ornatum* Pl. 14

IDENTIFICATION: Similar to Central Stoneroller, *C. anomalum,* but usually has *58–77 lateral scales, 47–60 scales around body* at dorsal fin origin, 15–19 rakers on 1st gill arch; breeding male lacks tubercles on nape and adjacent area of body; intestine rarely coiled around gas bladder (see *Campostoma* account). Pharyngeal teeth 0,4-4,0. To 6¼ in. (16 cm). **RANGE:** Rio Grande system of Big Bend region, s. TX; Rucker Canyon and Leslie Creek, extreme se. AZ. Fairly common. Widespread and common in n. Mexico. **HABITAT:** Rocky riffles and adjacent pools of headwaters and creeks. **SIMILAR SPECIES:** (1) See Central Stoneroller, *C. anomalum* (Pl. 14).

DEVILS RIVER MINNOW *Dionda diaboli* Pl. 14

IDENTIFICATION: *Black wedge* on caudal fin base. Complete, often *punctate,* lateral line. Black stripe along side and onto snout. Slender, fairly compressed body. *Large eye about as long as snout:* short blunt snout; small subterminal mouth. *Dark-edged scales* on silver olive back and upper side, dark stripe along back; silver white below, often with dark green belly. Dorsal fin origin over to slightly behind pelvic fin origin. Has 32–36 lateral scales; 8 anal rays; pharyngeal teeth 0,4-4,0. To 2½ in. (6.4 cm). **RANGE:** Devils R. and nearby San Felipe, Sycamore, and Las Moras creeks, Val Verde and Kinney counties, TX. Also in n. Mexico. Common in extremely small range; protected as a *threatened species.* **HABITAT:** Rocky runs and flowing pools. **REMARKS:** *Dionda* species are long-gutted minnows that feed mainly on algae and other vegetation. Plant material in gut tends to make belly dark. **SIMILAR SPECIES:** (1) Other *Dionda* species (Pl. 14) have *round black spot* on caudal fin base; *lack* conspicuous dark-edged scales on back and upper side, punctate lateral line (may be punctate at front).

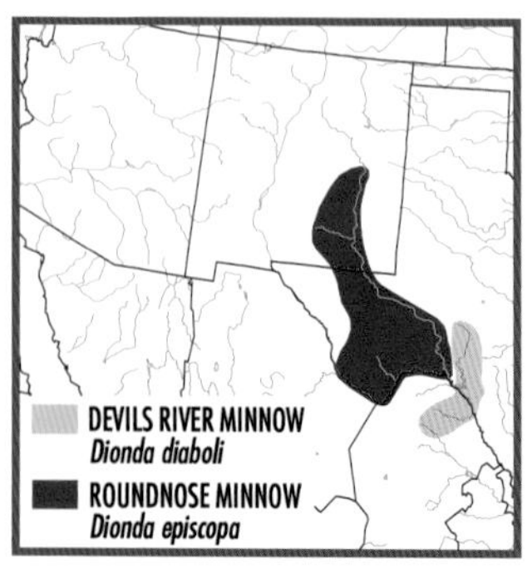

ROUNDNOSE MINNOW *Dionda episcopa* Pl. 14

IDENTIFICATION: *Strongly bicolored;* olive above, silver white below (but often with dark green belly). Light yellow stripe above dark stripe (*zigzagged* at front) along side and around snout, followed by *black spot* on caudal fin base. Dark green stripe along back, widest at dorsal fin origin. Yellow fins on adult. Fairly compressed body; abruptly rising nape, *body deepest just behind head*; dorsal fin origin over to slightly behind pelvic fin origin; rounded snout; small subterminal mouth; eye shorter than snout. Complete lateral line; 34–45 lateral scales; 8 anal rays; pharyngeal teeth 0,4-4,0. To 3 in. (7.7 cm). **RANGE:** Rio Grande drainage above Devils R., TX, NM, and Mexico. Locally common in TX, declining elsewhere; extirpated from Rio Grande, NM. **HABITAT:** Rocky pools, sometimes runs, of headwaters, creeks, and small rivers. Often among filamentous algae. **SIMILAR SPECIES:** (1) Nueces Roundnose Minnow, *D. serena* (Pl. 14), (2) Guadalupe Roundnose Minnow, *D. nigrotaeniata*, and (3) Manantial Roundnose Minnow, *D. argentosa*, have more slender body, *deepest under nape*. Nueces Roundnose Minnow also has eye about as long as snout, 7 anal rays. (4) Devils River Minnow, *D. diaboli* (Pl. 14), has *black wedge* on caudal fin base, eye about as long as snout, dark-edged scales on back and upper side.

NUECES ROUNDNOSE MINNOW *Dionda serena* Pl. 14

IDENTIFICATION: *Strongly bicolored;* olive to light brown above, silver white to green below. Dark stripe (*zigzagged* at front) along side and around snout, followed by *black spot* on caudal fin base. Dark green stripe along back, widest at dorsal fin origin. Yellow fins on adult. Slender, compressed body; *body deepest under nape*; dorsal fin origin over to slightly behind pelvic fin origin; rounded snout; small subterminal mouth; *eye about as long as snout*. Complete lateral line; 34–45 lateral scales; *7 anal rays*; pharyngeal teeth 0,4-4,0. To 3 in. (7.7 cm). **RANGE:** Upper Nueces R. drainage, TX. Locally common.

HABITAT: Rocky pools, sometimes runs, of headwaters, creeks, and small rivers. Often among filamentous algae. **SIMILAR SPECIES:** See (1) Guadalupe Roundnose Minnow, *D. nigrotaeniata*, and (2) Manantial Roundnose Minnow, *D. argentosa*. (3) Roundnose Minnow, *D. episcopa* (Pl. 14), has abruptly rising nape, *body deepest just behind head*; eye *shorter* than snout, *8 anal rays*.

GUADALUPE ROUNDNOSE MINNOW *Dionda nigrotaeniata* **Not shown**

IDENTIFICATION: Similar to Nueces Roundnose Minnow, *D. serena*, but has eye *shorter* than snout, *8 anal rays*. To 3 in. (7.6 cm). **RANGE:** Colorado and San Antonio R. drainages, TX. Locally common. **HABITAT:** Rocky pools, sometimes runs, of headwaters, creeks, and small rivers. Often among filamentous algae. **SIMILAR SPECIES:** (1) See Nueces Roundnose Minnow, *D. serena* (Pl. 14).

MANANTIAL ROUNDNOSE MINNOW *Dionda argentosa* **Not shown**

IDENTIFICATION: Genetically distinct but morphologically indistinguishable from Guadalupe Roundnose Minnow, *D. nigrotaeniata*. **RANGE:** Devils R. and San Felipe Creek, TX. Locally common. **HABITAT:** Rocky pools, sometimes runs, of headwaters, creeks, and small rivers. Often among filamentous algae. **SIMILAR SPECIES:** (1) See Guadalupe Roundnose Minnow, *D. nigrotaeniata*.

PHENACOBIUS

GENUS CHARACTERISTICS: *Long, cylindrical body; large fleshy lips* on *subterminal mouth* (Fig. 19). Round snout; dorsal fin origin in front of pelvic fin origin. Complete, straight lateral line; pharyngeal teeth 0,4-4,0.

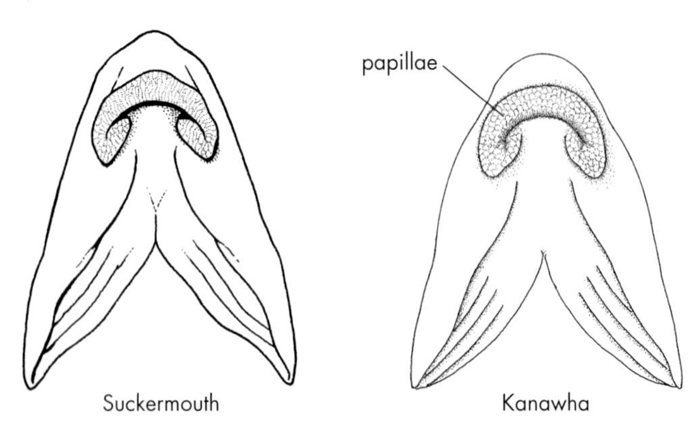

Fig. 19. Suckermouth and Kanawha minnows—underside of head.

SUCKERMOUTH MINNOW *Phenacobius mirabilis* Pl. 6

IDENTIFICATION: *Bicolored* (olive-brown above, silver white below) *body;* large *fleshy lips* (Fig. 19); *intense black spot* on caudal fin base following thin dark stripe along side of body. Thin dark stripe along back; darkly outlined scales on back and upper side. Has 42–51 lateral scales; 15–17 scales around caudal peduncle. To 4¾ in. (12 cm). **RANGE:** Mississippi R. basin from OH and WV to WY, CO, and NM, and from se. MN to n. AL and n. TX; w. Lake Erie drainage, OH and MI; isolated populations in Gulf drainages: Sabine R., LA and TX; Trinity and Colorado rivers, TX; Pecos R., NM. Common throughout most of range; rare in Gulf drainages. **HABITAT:** Gravel and rubble riffles and runs of clear to turbid creeks and small to medium, sometimes large, rivers. **SIMILAR SPECIES:** See Pl. 6. (1) Kanawha Minnow, *P. teretulus, lacks* black spot on caudal fin base; is more slender; has black blotches on upper side, *larger lips.* (2) Fatlips Minnow, *P. crassilabrum,* (3) Riffle Minnow, *P. catostomus,* and (4) Stargazing Minnow, *P. uranops, lack* black spot on caudal fin base, are much more slender, have 52 or more lateral scales.

KANAWHA MINNOW *Phenacobius teretulus* Pl. 6

IDENTIFICATION: Many papillae on *very fleshy lips* (Fig. 19). Gray-brown above, *small black blotches* scattered on upper half of body; dark stripe along back; dusky stripe along silver side. Has 45–49 lateral scales; 16–19 scales around caudal peduncle. To 4 in. (10 cm). **RANGE:** New (upper Kanawha) R. drainage, WV, VA, and NC. Generally uncommon. **HABITAT:** Rubble and gravel riffles and runs of creeks and small to medium rivers. **SIMILAR SPECIES:** See Pl. 6. (1) Suckermouth Minnow, *P. mirabilis,* has black spot on caudal fin base, *thinner lips* with few bumps, deeper head and body, no black blotches on upper side. (2) Fatlips Minnow, *P. crassilabrum,* (3) Riffle Minnow, *P. catostomus,* and (4) Stargazing Minnow, *P. uranops,* are more slender, *lack* black blotches, have 52 or more lateral scales.

FATLIPS MINNOW *Phenacobius crassilabrum* Pl. 6

IDENTIFICATION: *Pelvic fins reach to or past anus.* Lips as in Suckermouth Minnow, *P. mirabilis* (Fig. 19). Dark olive above; dark stripe along back wider in front of dorsal fin; light green streak above black stripe along silver side; 2 yellow spots on caudal fin base. Has 56–68 lateral scales; 19–20 scales around caudal peduncle. To 4¼ in. (11 cm). **RANGE:** Upper Tennessee R. drainage from South Fork Holston R., VA, to Little Tennessee R., GA; extreme upper Savannah R. drainage, NC, SC, and GA. Fairly common. **HABITAT:** Gravel and rubble runs and riffles of creeks and small to medium rivers. **SIMILAR SPECIES:** See Pl. 6. (1) Riffle Minnow, *P. catostomus*, and (2) Stargazing Minnow, *P. uranops*, are more slender; have larger, more upwardly directed eyes; pelvic fins *not* reaching past anus. Riffle Minnow has 15–19, Stargazing Minnow 13–16, scales around caudal peduncle. (3) Suckermouth Minnow, *P. mirabilis,* and (4) Kanawha Minnow, *P. teretulus,* have pelvic fins *not* reaching past anus, fewer than 52 lateral scales; Suckermouth Minnow has black spot on caudal fin base; Kanawha Minnow has black blotches on upper side.

RIFFLE MINNOW *Phenacobius catostomus* Pl. 6

IDENTIFICATION: *Very long, cylindrical body;* lips as in Suckermouth Minnow, *P. mirabilis* (Fig. 19). Eyes high on head, *directed upwardly.* Dark gray-brown back; light brown upper side; silver sheen over dusky stripe along side. Has 56–69 lateral scales; 15–19 scales around caudal peduncle. To 4½ in. (12 cm). **RANGE:** Mobile Bay drainage, se. TN, nw. GA, and AL. Local and uncommon in Tallapoosa and Black Warrior river systems, fairly common in Coosa and Cahaba river systems. **HABITAT:** Gravel and rubble runs and riffles of creeks and small to medium rivers. **SIMILAR SPECIES:** See Pl. 6. (1) See Stargazing Minnow, *P. uranops.* (2) Fatlips Minnow, *P. crassilabrum*, is *less* slender, has pelvic fins reaching past anus. (3) Suckermouth Minnow,

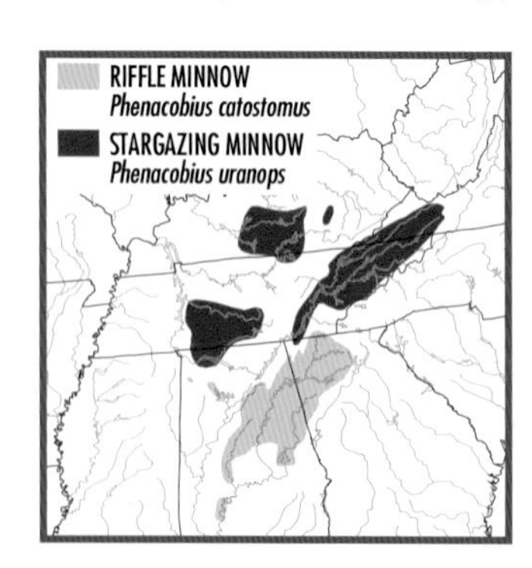

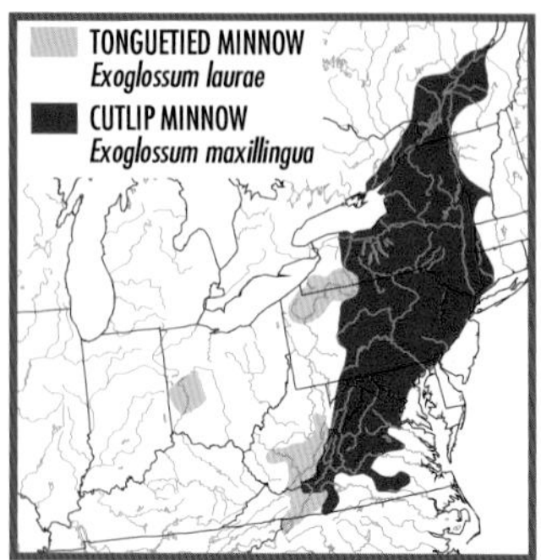

P. mirabilis, is deeper bodied; has black spot on caudal fin base, 42–51 lateral scales.

STARGAZING MINNOW *Phenacobius uranops* **Pl. 6**

IDENTIFICATION: Similar to Riffle Minnow, *P. catostomus,* but has *more elongated body,* longer snout, *more elliptical eye.* Has 52–61 lateral scales; 13–16 scales around caudal peduncle. To 4½ in. (12 cm). **RANGE:** Upper Green, middle Cumberland, and Tennessee river drainages, KY, VA, TN, GA, and AL. Common only in upper Tennessee and Green river drainages. **HABITAT:** Gravel and rubble runs and riffles of clear, fast creeks and small to medium rivers. **SIMILAR SPECIES:** (1) See Riffle Minnow, *P. catostomus* (Pl. 6).

TONGUETIED MINNOW *Exoglossum laurae* **Not shown**

IDENTIFICATION: Central *bony plate* with *fleshy lobe* to either side on lower jaw (Fig. 20). *Chubby body,* barely compressed; deep caudal peduncle; dorsal fin origin over pelvic fin origin. Rounded snout, subterminal mouth; thick upper lip thinner at middle; premaxillary frenum present; usually a small barbel near corner of mouth. Olive-gray above; silver green-purple side; clear to light olive or light red fins; small individual has dusky stripe along side and onto snout, black caudal spot. Complete lateral line; 47–53 lateral scales; 7 anal rays; pharyngeal teeth 1,4-4,1. To 6¼ in. (16 cm). **RANGE:** Three areas of upper Ohio R. basin: (1) upper Allegheny R. drainage, NY and PA, and upper Genesee R. (Lake Ontario drainage), NY and PA; (2) upper New R. drainage, WV, VA, and NC; and (3) Great Miami and Little Miami river systems, OH. Fairly common, but less widespread and abundant than historically. **HABITAT:** Rocky pools and runs of creeks and small to medium rivers; often near vegetation or other cover. **REMARKS:** Breeding

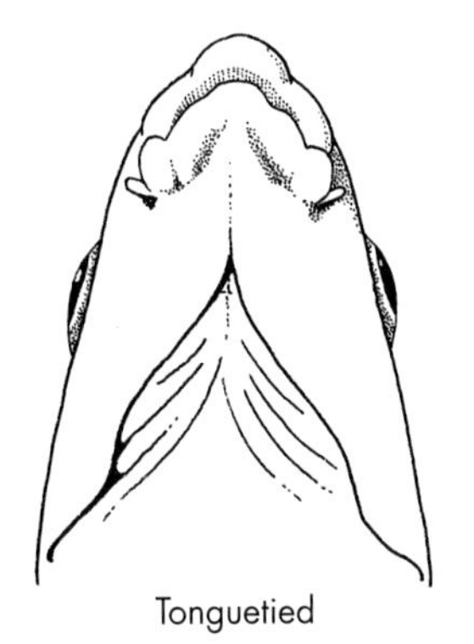
Tonguetied

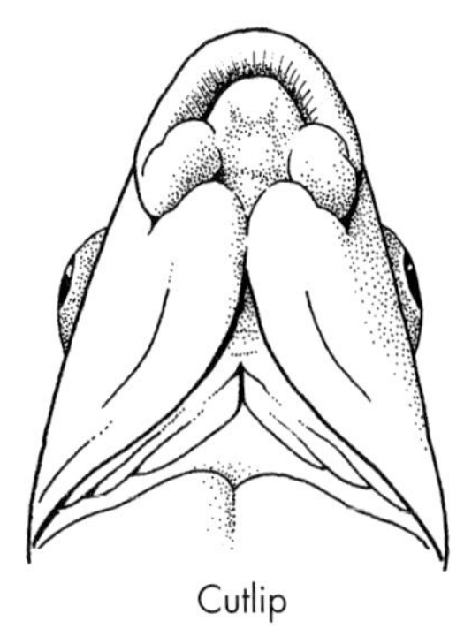
Cutlip

Fig. 20. Tonguetied and Cutlip minnows—underside of head.

male Tonguetied and Cutlip minnows build large circular or rectangular nests by piling pebbles carried in mouth. **SIMILAR SPECIES:** (1) See Cutlip Minnow, *E. maxillingua* (Pl. 6). No other species has similar mouth structure.

CUTLIP MINNOW *Exoglossum maxillingua* Pl. 6

IDENTIFICATION: Similar to Tonguetied Minnow, *E. laurae*, but has *much larger fleshy lobe* on each side of lower jaw, well separated from central bony plate and followed by another fleshy lobe on underside of head (Fig. 20); no barbel near corner of mouth. To 6¼ in. (16 cm). **RANGE:** Atlantic Slope from St. Lawrence R. drainage, QC, to upper Roanoke R., NC (absent in most of New England; 1 record in Connecticut R., VT); Lake Ontario drainage, ON and NY. Also in upper New R. drainage, WV and VA, where may be introduced. Common in clear streams. **HABITAT:** Rocky pools and runs of creeks and small to medium rivers; usually in quiet water near boulders. **REMARKS:** Cutlip Minnow, known to fishermen in some areas as "eye-picker," plucks out eyes of other fishes, a peculiar behavior also found in some African cichlids. **SIMILAR SPECIES:** (1) See Tonguetied Minnow, *E. laurae.*

RHINICHTHYS

GENUS CHARACTERISTICS: *Long, slender, streamlined body,* deepest at nape, flattened below. *Small scales;* more than 40 along complete, straight lateral line. Usually *many black specks* on strongly bicolored (dark above, light below) body. Dorsal fin origin behind pelvic fin origin. Several species are wide ranging and highly variable; additional populations may warrant recognition as species.

BLACKNOSE DACE *Rhinichthys atratulus* Pl. 14

IDENTIFICATION: *Many brown-black specks* on back and side. *Deep caudal peduncle. Barbel* in corner of mouth; no groove separating snout from upper lip; pointed snout slightly overhangs mouth. Light brown above, black spot followed by silver spot on dorsal fin base; *black stripe* along side, through eye and onto snout, continuous in young, as blotches in adult; often a silver stripe above black stripe; silver white below. Breeding male develops pads on upper surface of pectoral fin, has yellow-white pectoral and pelvic fins, white to red stripe (see Remarks) below black body stripe. Has 53–70 lateral scales, 7 anal rays; pharyngeal teeth 2,4-4,2. To 4 in. (10 cm). **RANGE:** Atlantic, Great Lakes, Hudson Bay, Mississippi R., and upper Mobile Bay drainages from NS to MB and south to n. GA and n. AL. Generally common, especially in montane and spring-fed streams. **HABITAT:** Rocky runs and pools of headwaters, creeks, and small rivers. **REMARKS:** Two subspecies (possibly species) appear to be recognizable, although they and their ranges are poorly defined. *R. a. atratulus,* in n. and cen. Atlantic Slope drainages (including James and Roanoke rivers) and e. Great Lakes-St. Lawrence basin (west to e. Lake Ontario), has silver white to gold yellow along and below black-red stripe on side of breeding male. *R. a. obtusus,* over rest of range but apparently also inhabiting James and Roanoke river drainages, has orange to brick red stripe along and below black stripe on side of breeding male. **SIMILAR SPECIES:** See Pl. 14. (1) Speckled Dace, *R. osculus*, and (2) Leopard Dace, *R. falcatus*, usually have groove separating snout from upper lip; large individuals have red lips, snout, and fin bases. (3) Longnose Dace, *R. cataractae*, has long fleshy snout, eyes high on head.

SPECKLED DACE *Rhinichthys osculus* Pl. 14

IDENTIFICATION: Extremely variable; see Remarks. Usually dark olive back and side heavily *speckled with black;* gold specks on back; dusky stripe along side, through eye, and onto snout; black spot on caudal fin base. (Young has black stripe along side, often lacks black specks of adult.) Pointed snout slightly overhangs mouth. *Barbel* (absent in Canada and occasionally elsewhere) in corner of mouth; usually a groove separating snout from upper lip. Deep caudal peduncle (see Remarks). Silver yellow side, white below, yellow fins. Large individual has red-orange lips, snout, and bases of pectoral, pelvic, and anal fins. Has 47–89 lateral scales, 7 (often 8) anal rays; pharyngeal teeth 1,4-4,1 or 2,4-4,2. To 4¼ in. (11 cm). **RANGE:** Western drainages (Pacific and endorheic) from Columbia R., BC, to Colorado R., AZ and NM, and south into Sonora, Mexico. Most ubiquitous fish in w. U.S. Protected in Canada as an *endangered species.* **HABITAT:** Rocky riffles, runs, and pools of headwaters, creeks, and small to medium rivers; rarely in lakes. **REMARKS:** Occupying many isolated western drainages, Speckled Dace has diversified tremendously. Forms in swift water

(e.g., in Colorado R. drainage) are streamlined, have large falcate fins and slender caudal peduncle; those in slower water are more chubby and small-finned. Speckled Dace is treated as a complex of subspecies, but morphological variation and distributions are poorly known. *R. o. nevadensis,* in Ash Meadows, NV, *R. o. oligoporus,* Clover Valley, NV, *R. o. lethoporus,* Independence Valley, NV, and *R. o. thermalis,* Kendall Warm Springs, WY, are protected as *endangered subspecies;* and Foskett Speckled Dace, *R. o.* subspecies, Foskett Spring, Warner Basin, OR, as a *threatened subspecies.* **SIMILAR SPECIES:** See (1) Las Vegas Dace, *R. deaconi,* and (2) Leopard Dace, *R. falcatus* (Pl. 14). (3) Loach Minnow, *R. cobitis* (Pl. 14), has nearly terminal mouth, white spots at front and rear of dorsal fin, white bar on caudal peduncle, no scales on breast, belly, and part of back; *lacks* barbel, groove on snout. (4) Blacknose Dace, *R. atratulus* (Pl. 14), lacks groove separating snout from upper lip; breeding male develops pads on pectoral fin, has white, yellow, or red stripe along lower side, no red-orange lips, snout, or fin bases. (5) Longnose Dace, *R. cataractae* (Pl. 14), and (6) Umpqua Dace, *R. evermanni,* have long, fleshy snout in front of mouth, no groove separating snout from upper lip.

LAS VEGAS DACE *Rhinichthys deaconi* — Not shown

IDENTIFICATION: Similar to Speckled Dace, *R. osculus,* but has *very small pectoral fins* (less than 1/6 total length of fish), anal fin with 1st and last rays about *equal length,* and larger scales (40–52 lateral). To 3 in. (7.4 cm). **RANGE:** Las Vegas Creek, NV. Extinct. **HABITAT:** Inhabited springs and outflows along Las Vegas Creek. **SIMILAR SPECIES:** (1) See Speckled Dace, *R. osculus* (Pl. 14).

LEOPARD DACE *Rhinichthys falcatus* — Pl. 14

IDENTIFICATION: Streamlined version of Speckled Dace, *R. osculus,* with *falcate* (concave upper edge) *dorsal fin,* more forked caudal fin, more slender body, narrower caudal peduncle, longer and more

LEOPARD DACE
Rhinichthys falcatus

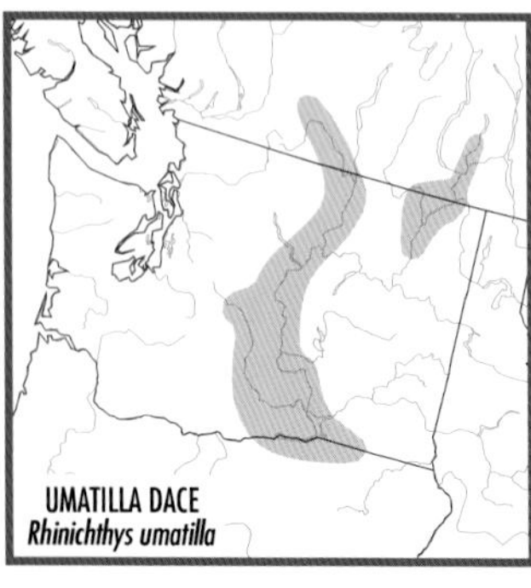
UMATILLA DACE
Rhinichthys umatilla

pointed snout, bigger eye, and larger black blotches on back, side, and *fins. Long barbel* protrudes beyond corner of mouth. Has 51–64 lateral scales, 20–32 scales around caudal peduncle, 7 anal rays; pharyngeal teeth 2,4-4,2. To 5 in. (12 cm). **RANGE:** Fraser and Columbia river drainages, BC, WA, OR, and ID. Generally uncommon. **HABITAT:** Flowing pools and gravel runs of creeks and small to medium rivers; rocky margins of lakes. **SIMILAR SPECIES:** See (1) Speckled Dace, *R. osculus* (Pl. 14), which has straight edge on dorsal fin, is less slender, has smaller black spots on body, and (2) Umatilla Dace, *R. umatilla.*

UMATILLA DACE *Rhinichthys umatilla* **Not shown**

IDENTIFICATION: Similar to Leopard Dace, *R. falcatus,* but has *short barbel not protruding beyond corner of mouth, large dark blotches on side,* stouter body, 56–72 lateral scales, 29–40 scales around caudal peduncle. To 5 in. (12 cm). **RANGE:** Columbia R. drainage, BC, WA, and OR. Locally common. Protected in Canada as a *threatened species.* **HABITAT:** Rubble riffles and runs of large rivers. **REMARKS:** Umatilla Dace may be of hybrid origin between Leopard Dace, *R. osculus,* and Speckled Dace, *R. falcatus.* Distributions of these 3 species in Pacific Northwest are poorly understood. **SIMILAR SPECIES:** (1) See Leopard Dace, *R. falcatus* (Pl. 14).

LONGNOSE DACE *Rhinichthys cataractae* **Pl. 14**

IDENTIFICATION: Deep caudal peduncle; *long fleshy snout* extends in front of *subterminal mouth, barbel* in corner of mouth; no groove separating snout from upper lip. Eyes high on head; caudal fin moderately forked; straight-edged dorsal and anal fins. Olive-brown to dark red-purple above, brown-black spots and mottling on back and side of some individuals; dark stripe along side (darkest on young); dusky spot on caudal fin base; silver to yellow below. Breeding male may have bright red on head and fin bases. Has 48–76 lateral scales, 7–9 (usually 8) anal rays, 8 dorsal rays; pharyngeal teeth 2,4-4,2. To

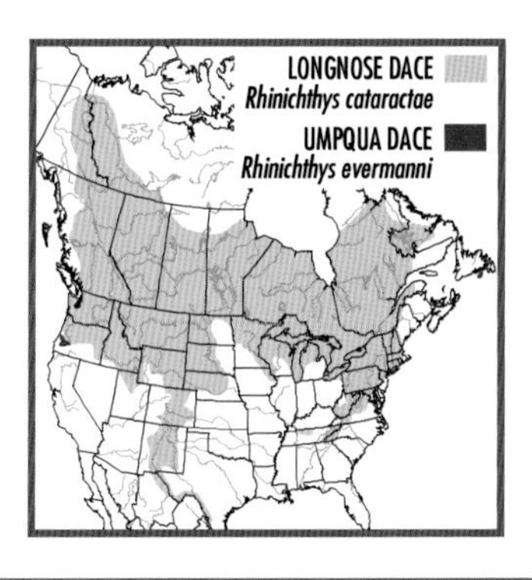

6¼ in. (16 cm). **RANGE:** Widest distribution of any N. American minnow. Generally distributed above 40°N from coast to coast; occurs as far north as Arctic Circle in Mackenzie R. drainage; south in Appalachian Mts. to n. GA and in Rocky Mts. south into Rio Grande drainage of TX and n. Mexico. Common in n. U.S. and along Atlantic Slope to VA; fairly common in west but absent from AK and from western drainages south of Columbia and Coos river drainages. **HABITAT:** Rubble and gravel riffles (sometimes runs and pools) of fast creeks and small to medium rivers; rocky shores of lakes. **REMARKS:** Subspecies sometimes recognized, but they and their ranges are poorly defined. Nooksack Dace, an undescribed form in Nooksack R. drainage in s. BC and nw. WA, is protected in Canada as *endangered*. **SIMILAR SPECIES:** (1) See Umpqua Dace, *R. evermanni*. (2) Blacknose Dace, *R. atratulus*, (3) Speckled Dace, *R. osculus*, and (4) Leopard Dace, *R. falcatus* (all Pl. 14) *lack* long fleshy snout, have eye more on side of head.

UMPQUA DACE *Rhinichthys evermanni* Not shown

IDENTIFICATION: Similar to Longnose Dace, *R. cataractae*, but has *narrow caudal peduncle* bordered above and below by *keel* leading into caudal fin rays, concave edge on dorsal and anal fins, more deeply forked caudal fin, 9–10 dorsal rays. Has 57–61 lateral scales; 7–8 anal rays; pharyngeal teeth 2,4-4,2. To 4¼ in. (11 cm). **RANGE:** Umpqua R. drainage, OR. Common. **HABITAT:** Fast rocky riffles. **SIMILAR SPECIES:** (1) See Longnose Dace, *R. cataractae* (Pl. 14).

LOACH MINNOW *Rhinichthys cobitis* Pl. 14

IDENTIFICATION: *Small, nearly terminal mouth; upwardly directed eyes.* Olive-brown above; many black specks and blotches on back and side; *white spots* at front and rear of dorsal fin; black spot in middle of *white bar* on caudal fin base. Breeding male has bright red fin bases, mouth

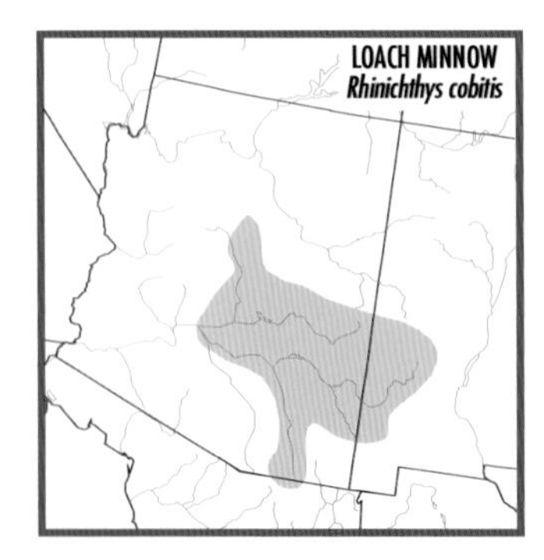

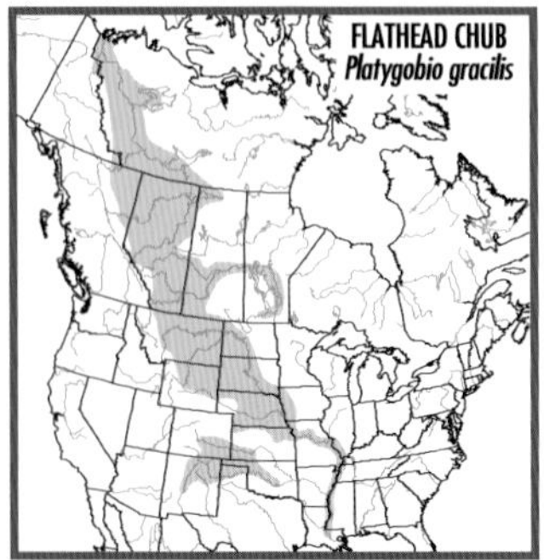

and lower head, sometimes belly. Large female has yellow belly, fins. No scales on breast, belly, and part of back. No barbel in corner of mouth; no groove separating snout from upper lip; 65–70 lateral scales; 7 anal rays; pharyngeal teeth 1,4-4,1. To 2¼ in. (6 cm). **RANGE:** Upper Gila R. system, NM and AZ; San Pedro R., AZ and n. Sonora, Mexico. Locally common in NM; uncommon in AZ. Protected as a *threatened species.* **HABITAT:** Rocky, often vegetated, riffles of creeks and small to medium rivers. **SIMILAR SPECIES:** (1) Speckled Dace (Pl. 14) *lacks* large white spots at front and rear of dorsal fin, white bar on caudal fin base; has *subterminal mouth,* scales on nape and belly, barbel.

FLATHEAD CHUB *Platygobio gracilis* Pl. 6

IDENTIFICATION: *Broad, flat head* tapering to pointed snout; *small barbel* in corner of large subterminal mouth. *Large, pointed, sickle-shaped dorsal and pectoral fins on* large individual; 1st dorsal fin ray extends beyond last ray in depressed fin. Body slightly compressed at front, more compressed along caudal peduncle; dorsal fin origin over or in front of pelvic fin origin. Fairly small eyes. Light dusky brown or olive above; silver side; lower lobe of caudal fin dusky black. Has 42–59 lateral scales; 8 anal rays; pharyngeal teeth 2,4-4,2. To 12½ in. (32 cm). **RANGE:** Mackenzie, Saskatchewan, and Lake Winnipeg drainages in YT, NT, MB, SK, AB, and BC; Missouri-Mississippi river basin from s. AB and MT to LA; Arkansas R. drainage in KS, OK, CO, TX, and NM. Upper Rio Grande (including Pecos) drainage, NM. Common in northern part of range; restricted to Mississippi R. proper in MO, IL, and south. **HABITAT:** Sandy runs of small to large turbid rivers. **REMARKS:** Two subspecies usually recognized. *P. g. gracilis,* in northern and eastern parts of range, usually in large rivers, has pointed head and usually 48 or more lateral scales. *P. g. gulonella,* in southern and western parts of range, inhabits smaller streams; has rounder snout, smaller eye, usually fewer than 48 lateral scales, and is smaller (to 6 in. [15 cm]). Two forms broadly intergrade (most of Missouri R. drainage). **SIMILAR SPECIES:** (1) Sicklefin Chub, *Macrhybopsis meeki*, and (2) Sturgeon Chub, *M. gelida* (both Pl. 6), have less flattened and pointed head, smaller eyes, small papillae on underside of head. On Sicklefin Chub, tip of pectoral fin reaches beyond pelvic fin origin. Dorsal and pectoral fins of Sturgeon Chub are straight-edged.

SICKLEFIN CHUB *Macrhybopsis meeki* Pl. 6

IDENTIFICATION: *Large, sharply pointed, sickle-shaped fins;* 1st dorsal fin ray extends beyond last ray in depressed fin; tip of pectoral fin reaches beyond pelvic fin origin. Lower lobe of caudal fin *black with white edge* in large individual. *Long barbel in* corner of subterminal mouth. Body deepest under nape, strongly tapering to narrow caudal

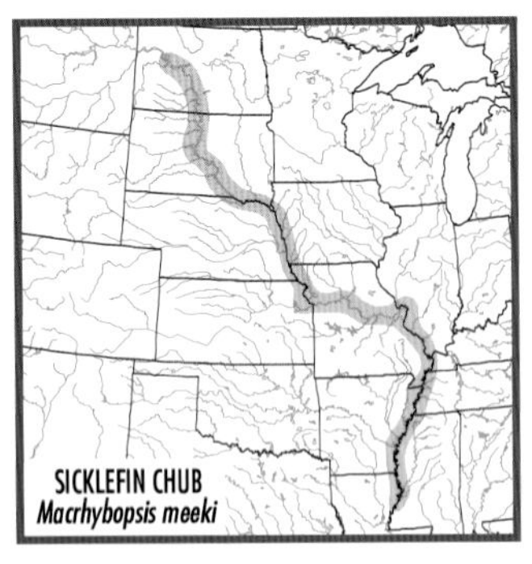

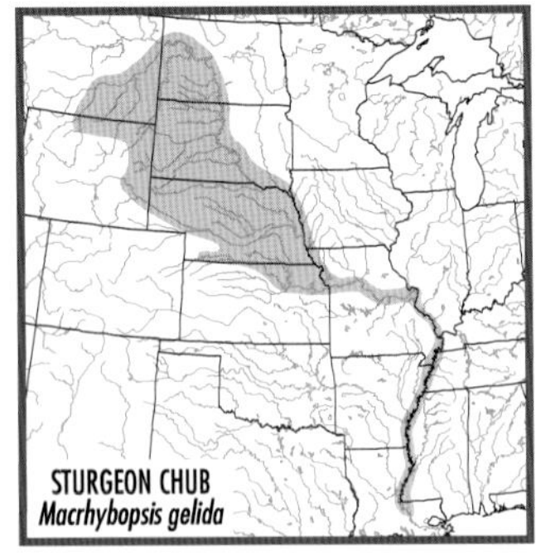

peduncle; barely compressed at front, strongly compressed at peduncle. Dorsal fin origin over or slightly behind pelvic fin origin. Small eye high on deep head; rounded snout. Head broad with many small papillae on underside (gular area). Scales without bony keels. Light green to brown above, often with many dark brown and silver specks; silver side. Has 43–50 lateral scales; 8 anal rays; pharyngeal teeth 0,4-4,0. To 4¼ in. (11 cm). **RANGE:** Missouri R. from ND to mouth; Mississippi R. from mouth of Missouri R. to s. MS; lower Kansas R., e. KS. Fairly common in middle Missouri R.; rare elsewhere. **HABITAT:** Sand and gravel runs of large rivers. **SIMILAR SPECIES:** (1) See Sturgeon Chub, *M. gelida* (Pl. 6). (2) Flathead Chub, *Platygobio gracilis* (Pl. 6) has broad, flat head; larger eye; no papillae on underside of head; pectoral fin not reaching pelvic fin origin.

STURGEON CHUB *Macrhybopsis gelida* Pl. 6

IDENTIFICATION: Similar to Sicklefin Chub, *M. meeki*, but has straight-edged fins; last dorsal fin ray extends beyond 1st ray in depressed fin; keeled scales on back and side; snout longer and projecting well beyond upper lip; large papillae on underside of head; larger brown specks on back. Has 39–45 lateral scales; pharyngeal teeth 1,4-4,1. To 3¼ in. (8.4 cm). **RANGE:** Missouri R. basin, MT and WY to IL; Mississippi R. from mouth of Missouri R. to LA. Fairly common in middle Missouri R.; rare elsewhere. **HABITAT:** Shallow sand and gravel runs of medium to large turbid rivers. **SIMILAR SPECIES:** (1) See Sicklefin Chub, *M. meeki* (Pl. 6).

SILVER CHUB *Macrhybopsis storeriana* Pl. 6

IDENTIFICATION: *Large eye* on upper half of head; short, rounded snout; barbel in corner of *subterminal mouth.* Slender, fairly compressed body, flattened below; dorsal fin origin in front of pelvic fin origin. Light olive above; *bright silver white side;* white edge on dusky black

(absent on juveniles) lower lobe of caudal fin. Complete lateral line; 35–48 lateral scales; 8 anal rays; pharyngeal teeth 1,4-4,1. To 9 in. (23 cm). **RANGE:** Lake Erie drainage; Red R. drainage from MB south to MN; Mississippi R. basin, PA and WV west to MN, NE, KS, and OK, and south to Gulf Coast; Gulf drainages from Mobile Bay drainage, AL, to Lake Pontchartrain drainage, LA; isolated population in Brazos R., TX. Common. **HABITAT:** Sand-, silt-, and sometimes gravel-bottomed pools and backwaters of small to large rivers; lakes. **SIMILAR SPECIES:** (1) Bigeye Chub, *Hybopsis amblops* (Pl. 7), has black stripe along side and onto snout, reaches only 3½ in. (9 cm).

SHOAL CHUB *Macrhybopsis hyostoma* Pl. 7

IDENTIFICATION: *Long, bulbous snout* overhangs subterminal mouth; *long barbel* (sometimes 2) at corner of mouth. *Black spots* on back and side. *Upwardly directed, elliptical eye.* Slender, barely compressed body, deepest under nape, flattened below. Dorsal fin origin over or in front of pelvic fin origin; anus closer to anal fin origin than pelvic fin origin; pectoral fin of male short, rounded, rarely reaching pelvic fin origin. Translucent; light olive to gray above, dark scale margins; silver to iridescent blue stripe along side (darkest on caudal peduncle); silver white below. Complete lateral line; 32–43 lateral scales; usually 8 anal rays; pharyngeal teeth 0,4-4,0. To 3 in. (7.6 cm). **RANGE:** Mississippi R. basin, from e. OH, to s. MN and NE south to LA; Wolf R. (Lake Michigan drainage), WI; Gulf drainages from Mississippi R., LA, to Lavaca R., TX. Common over much of range but declining. **HABITAT:** Sand and gravel runs of small to large rivers. **REMARKS:** Three undescribed species related to Shoal Chub occur in Gulf drainages east of Mississippi R. GULF CHUB, in drainages from Mobile Bay basin (below Fall Line) to Lake Pontchartrain drainage, AL, MS, and LA, has dorsal fin origin *over or slightly in front of* pelvic fin origin, anus closer to anal fin origin than pelvic fin origin, *no exposed scales* just in front

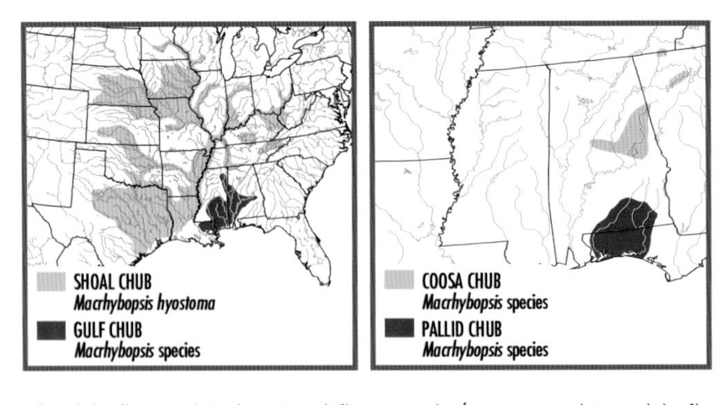

of pelvic fins, *pointed pectoral fin* on male (may extend to pelvic fin origin), 1 barbel at corner of mouth, usually 8 anal rays, pharyngeal teeth 0,4-4,0. COOSA CHUB, above Fall Line in Cahaba, Tallapoosa, and Coosa river systems, GA, AL, and TN, has dorsal fin origin *behind* pelvic fin origin, anus *midway* between origins of pelvic and anal fins, *exposed scales* just in front of pelvic fins, short rounded pectoral fin, 1 barbel at corner of mouth, usually 8 anal rays, pharyngeal teeth *1,4-4,1.* PALLID CHUB, reaching only 2¼ in. (6 cm), in Gulf drainages from Choctawhatchee R. to Escambia R., AL and FL, has *2 barbels* at corner of mouth, *tiny black spots* (barely visible) on back and side, dorsal fin origin *over* pelvic fin origin, anus *midway* between origins of pelvic and anal fins, *long pointed pectoral fin* on male (often extends past pelvic fin origin), usually *7 anal rays,* pharyngeal teeth 0,4-4,0. **SIMILAR SPECIES:** See Pl. 7. See (1) Speckled Chub, *M. aestivalis*, (2) Peppered Chub, *M. tetranema*, (3) Prairie Chub, *M. australis*, and (4) Burrhead Chub, *M. marconis*.

SPECKLED CHUB *Macrhybopsis aestivalis* Pl. 7

IDENTIFICATION: Similar to Shoal Chub, *M. hyostoma*, but has *round eye*, more black spots on back and side; *lacks* dark scale margins, stripe along side. Has 31–42 lateral scales. To 3½ in. (9 cm). **RANGE:** Rio Grande drainage, TX and NM; south in Mexico to San Fernando R. drainage. Common in cen. Rio Grande; absent in upper, rare in lower, drainage. **HABITAT:** Sand and gravel runs of small to large rivers. **SIMILAR SPECIES:** (1) See Shoal Chub, *M. hyostoma* (Pl. 7).

PEPPERED CHUB *Macrhybopsis tetranema* Pl. 7

IDENTIFICATION: Similar to Shoal Chub, *M. hyostoma*, but has *small round eye, 2 barbels* at corner of mouth, rear barbel *longer* than eye; *fleshy lips* greatly expanded at rear. *Pointed pectoral fin* reaches pelvic fin on large male. Has 35–48 lateral scales. To 3 in. (7.7 cm). **RANGE:**

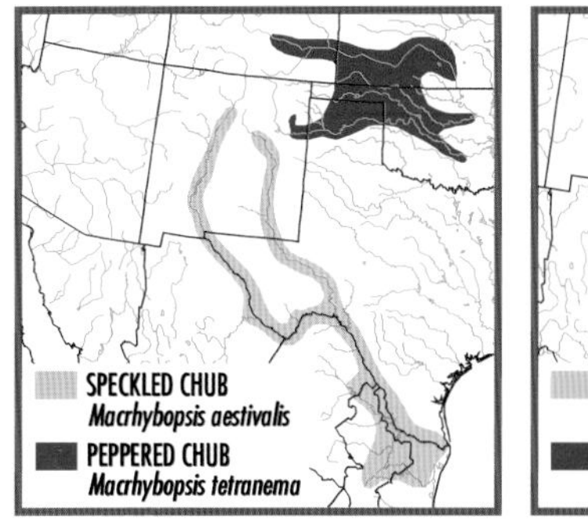

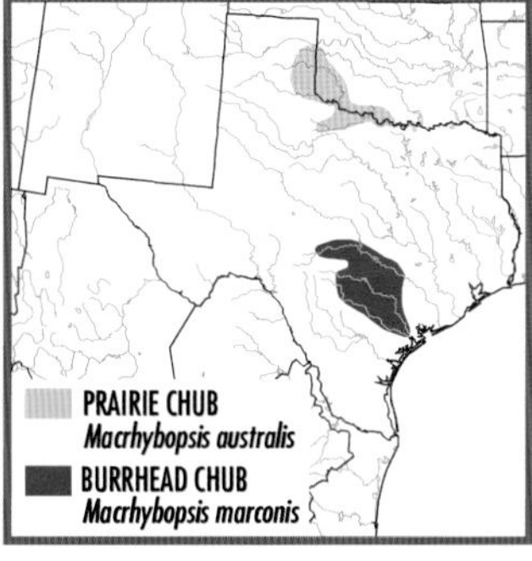

Upper Arkansas R. drainage, KS, OK, TX, NM, and formerly CO. Localized and declining over most of range. **HABITAT:** Sand and gravel runs of creeks and small to large rivers. **SIMILAR SPECIES:** (1) See Shoal Chub, *M. hyostoma* (Pl. 7).

RAIRIE CHUB *Macrhybopsis australis* Not shown

IDENTIFICATION: Nearly identical to Peppered Chub, *M. tetranema*, but usually has 7 (not 8) *anal rays, long pectoral fin* reaches *past* pelvic fin on large male. Has 36–42 lateral scales. To 2¾ in. (7 cm). **RANGE:** Upper Red R. drainage, OK and TX. Common. **HABITAT:** Sand and gravel runs of creeks and small to large rivers. **SIMILAR SPECIES:** (1) See Peppered Chub, *M. tetranema* (Pl. 7).

URRHEAD CHUB *Macrhybopsis marconis* Pl. 7

IDENTIFICATION: Similar to Shoal Chub, *M. hyostoma*, but has *large round eye, dark silver stripe* along side from head to caudal fin; breeding male has *yellow pectoral fins, tubercles on head* (no tubercles on Shoal Chub). Has 35–39 lateral scales. To 2¾ in. (7.3 cm). **RANGE:** Colorado, Guadalupe, and San Antonio river drainages, TX. Generally common; in Colorado R. drainage, restricted to Edwards Plateau. **HABITAT:** Sand and gravel riffles and runs of small to large rivers. **SIMILAR SPECIES:** (1) See Shoal Chub, *M. hyostoma* (Pl. 7).

RIMYSTAX

GENUS CHARACTERISTICS: Fast- and often deep-water-inhabiting, silvery minnows with distinctive pigment patterns. *Long, slender body* deepest at nape, *flattened below;* dorsal fin origin in front of pelvic fin origin. *Barbel* at corner of subterminal mouth; long, bulbous snout; *large eye*; large, horizontal pectoral fins. Complete lateral line; 7 anal rays; pharyngeal teeth 0,4-4,0.

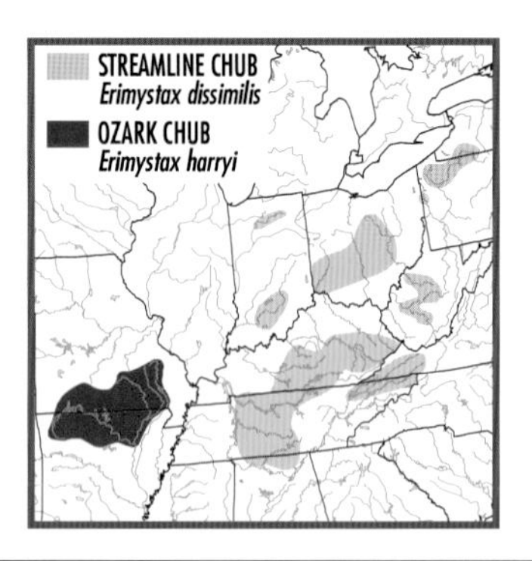

STREAMLINE CHUB *Erimystax dissimilis* Pl. 7

IDENTIFICATION: Has *7–15 horizontally oblong or round dark gray blotches* along side; blotches about same size as black caudal spot. White to gold spot at front and back of dorsal fin. Dark olive above, scales darkly outlined, series of dark dashes along back; often dark specks on back and upper side; often a gray stripe along silver side. Upper lip uniform, not expanded at front; gut S-shaped. Has 38–53 lateral scales. To 5½ in. (14 cm). **RANGE:** Ohio R. basin from w. NY to n. IN, and south to n. AL; uncommon and localized. One record for Lake Erie, OH. **HABITAT:** Riffles, runs, and current-swept pools over gravel and rubble in clear small to large rivers. **SIMILAR SPECIES:** (1) See Ozark Chub, *E. harryi.* (2) Blotched Chub, *E. insignis* (Pl. 7), has 7–9 large dark gray *rectangles*, row of black specks along lower edge of dark pigment along side. (3) Gravel Chub, *E. x-punctatus* (Pl. 7), *lacks* dark blotches, has Xs along side.

OZARK CHUB *Erimystax harryi* Not shown

IDENTIFICATION: Similar to Streamline Chub, *E. dissimilis*, but has upper lip *wider at front* than on sides, *double-looped gut*. Has 42–52 lateral scales. To 4½ in. (11 cm). **RANGE:** St. Francis and White river drainages, MO and AR. Common. **HABITAT:** Riffles, runs, and current-swept pools over gravel and rubble in clear, small to large rivers. **SIMILAR SPECIES:** (1) See Streamline Chub, *E. dissimilis* (Pl. 7).

BLOTCHED CHUB *Erimystax insignis* Pl. 7

IDENTIFICATION: Has *7–9 large, vertical dark gray rectangles* (obscure on young) along side; rectangles *larger* than black caudal spot. *Row of black specks* along lower edge of dark pigment on side; iridescent yellow stripe along light green-silver side. Has 36–49 lateral scales. To 4 in. (10 cm). **RANGE:** Cumberland and Tennessee river drainages, VA, NC, KY, TN, GA, and AL. Locally common. **HABITAT:** Rocky riffles and runs of clear small to medium rivers. **REMARKS:** Two subspecies. *E. i. eristigma,* upper Tennessee R. drainage except Clinch and Powell rivers, has large mouth (upper jaw about ⅓ head length), upper lip enlarged at middle. *E. i. insignis,* Cumberland and lower and middle Tennessee rivers (upstream to Sequatchie R.), has smaller mouth (upper jaw about¼ head length), upper lip barely enlarged. Clinch and Powell river populations are intergrades. **SIMILAR SPECIES:** (1) Streamline Chub, *E. dissimilis* (Pl. 7), has 7–10 *small dark blotches* (not rectangles) along side, *no* row of black specks along bottom of dark pigment. (2) Gravel Chub, *E. x-punctatus* (Pl. 7), *lacks* dark rectangles, has Xs along side.

GRAVEL CHUB *Erimystax x-punctatus* Pl. 7

IDENTIFICATION: Few to many *dark Xs* on back and side. Light olive above with dusky stripe along back midline; scales darkly outlined; blue sheen along silver side; dusky caudal spot. Has 38–45 lateral scales. To 4¼ in. (11 cm). **RANGE:** Thames R. system, s. ON; Ohio R. basin from NY and PA to Wabash R., IL (absent south of Ohio R. except 1 record in upper Green R., KY); Mississippi R. basin from s. WI and s. MN south to n. AR and e. OK; Ouachita R. system, AR. Locally common but declining over much of range. **HABITAT:** Gravel riffles and runs of creeks and small to large rivers. **REMARKS:** Two subspecies. *E. x. trautmani,* Wabash R. drainage and eastward, usually has 12 scales around caudal peduncle, is more slender. *E. x. x-punctatus,* west of Wabash R. drainage, usually has 16 scales around caudal peduncle, is stouter. **SIMILAR SPECIES:** (1) Streamline Chub, *E. dissimilis* (Pl. 7), (2) Ozark Chub, *E. harryi*, and (3) Blotched Chub, *E. insignis* (Pl. 7), have dark blotches or rectangles along side, *lack* dark Xs.

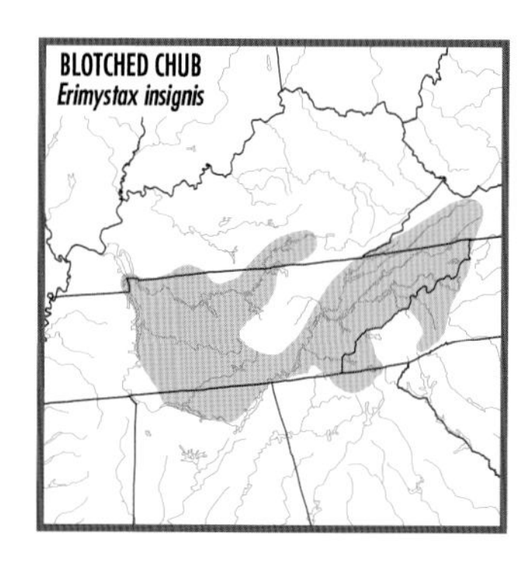

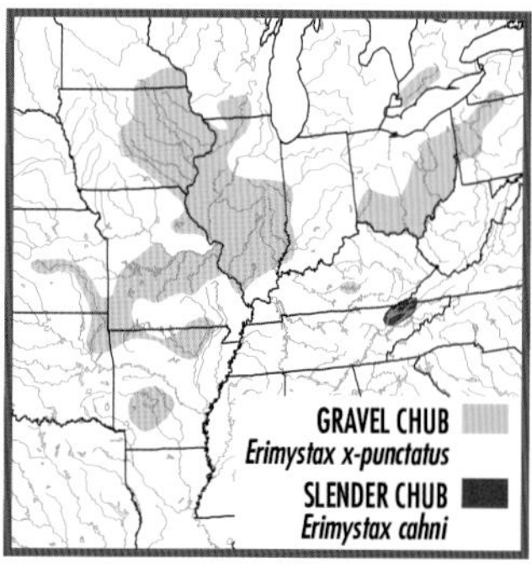

SLENDER CHUB *Erimystax cahni* Pl. 7

IDENTIFICATION: *Large dark* <s along rear half of side, darkest and largest on caudal peduncle. Dusky olive above; silver side; black caudal spot. Has 40–46 lateral scales. To 3½ in. (9 cm). **RANGE:** Upper Tennessee R. drainage (Holston, Clinch, and Powell rivers), TN and VA. Extremely rare in small range; protected as a *threatened species.* **HABITAT:** Gravel runs and riffles of medium-sized rivers. **SIMILAR SPECIES:** (1) Streamline Chub, *E. dissimilis*, and (2) Blotched Chub, *E. insignis* (both Pl. 7), have dark *blotches or rectangles* along side. (3) Stargazing Minnow, *Phenacobius uranops* (Pl. 6), has shorter snout; *lacks* dark <s on side, barbel at corner of mouth.

HYBOGNATHUS

GENUS CHARACTERISTICS: Long, *coiled intestine; black peritoneum. Small, slightly subterminal mouth* (rear edge of mouth in front of eye). Fairly deep caudal peduncle; round snout, dorsal fin origin in front of pelvic fin origin. Large male has light yellow along side and on lower fins. Complete, straight lateral line; 34–41 lateral scales; usually 8

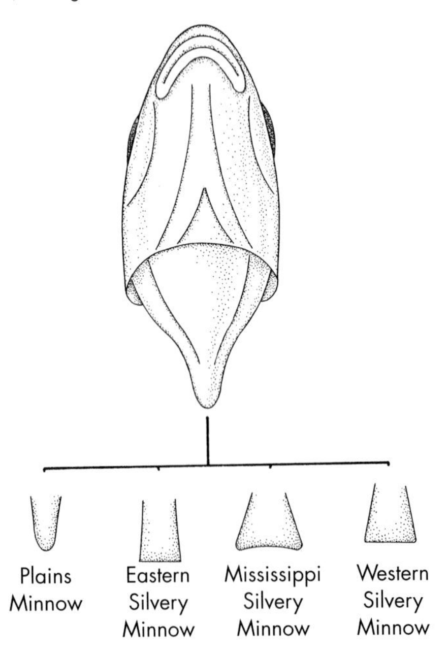

Fig. 21. *Hybognathus* species—basioccipital process.

anal rays; pharyngeal teeth 0,4-4,0. *Hybognathus* species ingest mud and organic matter, and long intestine facilitates digestion of algae and other plant material. First 4 species of *Hybognathus* are extremely similar and can be identified only by examining basioccipital process (Fig. 21).

MISSISSIPPI SILVERY MINNOW *Hybognathus nuchalis* Pl. 8

IDENTIFICATION: *Stout body;* moderately compressed, deepest and widest in front of dorsal fin. Moderately small eye (about¼ head length); pointed dorsal fin. Light brown to yellow-olive above, wide dusky to yellow-green stripe along back; silver side (often brilliant). Has 15–16 pectoral rays. Basioccipital process *broad and distinctly concave at rear* (Fig. 21). To 7 in. (18 cm). **RANGE:** Lowlands of Mississippi R. basin, from OH to MN and south to LA; Gulf drainages from Mobile Bay, AL, to Brazos R., TX. Generally common but less so along periphery of range; now absent in MN and e. TN. **HABITAT:** Pools and backwaters of low-gradient creeks and small to large rivers. **SIMILAR SPECIES:** See (1) Rio Grande Silvery Minnow, *H. amarus,* (2) Eastern Silvery Minnow, *H. regius*, (3) Western Silvery Minnow, *H. argyritis,* and (4) Plains Minnows, *H. placitus* (Pl. 8). (5) River Shiner, *Notropis blennius* (Pl. 8), has larger mouth, 7 anal rays, much shorter intestine.

RIO GRANDE SILVERY MINNOW *Hybognathus amarus* Not shown

IDENTIFICATION: Nearly identical to Mississippi Silvery Minnow, *H. nuchalis*, but has basioccipital process *broader and shallowly concave at rear*, less compressed body, more rounded snout. To 4 in. (10 cm). **RANGE:** Rio Grande drainage (including Pecos R.), TX and NM. Rare; apparently extant only in Rio Grande of NM. Protected as an *endangered species.* **HABITAT:** Pools and backwaters of low-gradient creeks and small to large rivers. **SIMILAR SPECIES:** (1) See Mississippi Silvery Minnow, *H. nuchalis* (Pl. 8).

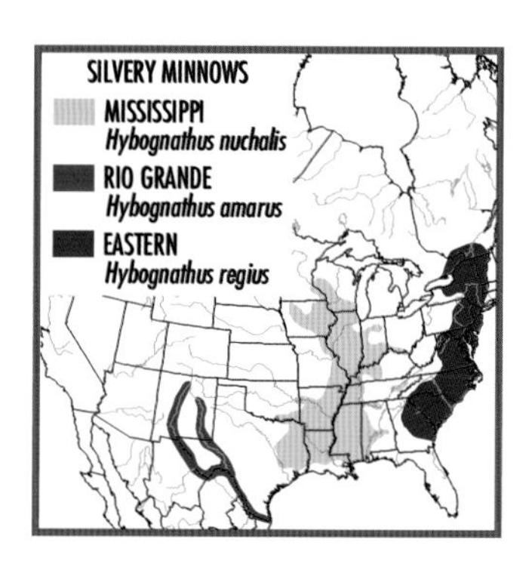

EASTERN SILVERY MINNOW *Hybognathus regius* Not shown

IDENTIFICATION: Nearly identical to Mississippi Silvery Minnow, *H. nuchalis*, but has *narrower, straight-edged basioccipital process* (Fig. 21). To 4¾ in. (12 cm). **RANGE:** Atlantic Slope from St. Lawrence R. drainage, QC, to Altamaha R. drainage, GA; Lake Ontario drainage, ON and NY. Generally common. **HABITAT:** Pools and backwaters of low-gradient creeks and small to large rivers. **SIMILAR SPECIES:** (1) See Mississippi Silvery Minnow, *H. nuchalis* (Pl. 8).

WESTERN SILVERY MINNOW *Hybognathus argyritis* Not shown

IDENTIFICATION: Nearly identical to Mississippi Silvery Minnow, *H. nuchalis*, but has rear edge of basioccipital process *straight* or *barely concave* (Fig. 21), slightly *smaller eye*. To 4¾ in. (12 cm). **RANGE:** Missouri R. basin, AB and MT to MO, Mississippi R. from mouth of Missouri R. to mouth of Ohio R.; South Saskatchewan R. (Hudson Bay basin), extreme s. AB. Locally common; protected in Canada as an *endangered species*. **HABITAT:** Sluggish pools and backwaters, usually over mud or sand, of small to large rivers. **SIMILAR SPECIES:** (1) See Mississippi Silvery Minnow, *H. nuchalis* (P1. 8).

PLAINS MINNOW *Hybognathus placitus* Pl. 8

IDENTIFICATION: Similar to Mississippi Silvery Minnow, *H. nuchalis*, but has *smaller eye* (about ⅕ head length), *peglike* basioccipital process (Fig. 21), underside of head more flattened; more falcate fins in western populations. Has 16–17 pectoral rays. To 5 in. (13 cm). **RANGE:** Missouri, Arkansas, Red, Brazos, and Colorado river drainages, from MT and ND south to NM and TX; Mississippi R. from mouth of Missouri R. to TN. Introduced in Pecos R., NM. One of most characteristic and common (sometimes abundant) fishes of Great Plains. **HABITAT:** Usually in shallow sandy runs and pools of creeks and small to large rivers. **SIMILAR SPECIES:** See (1) Mississippi Silvery Minnow, *H. nuchalis* (Pl. 8), and (2) Western Silvery Minnow, *H. argyritis*.

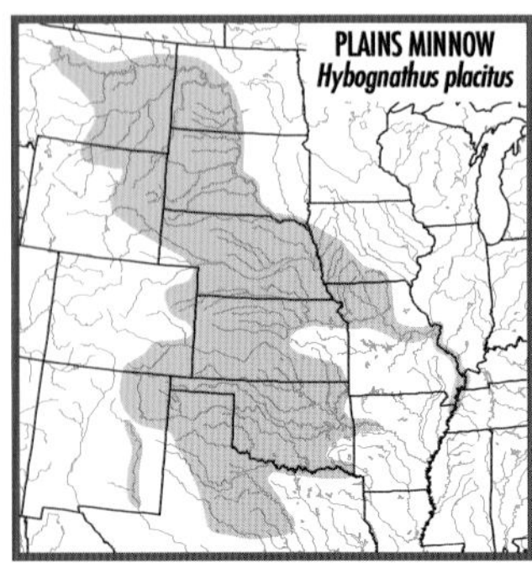

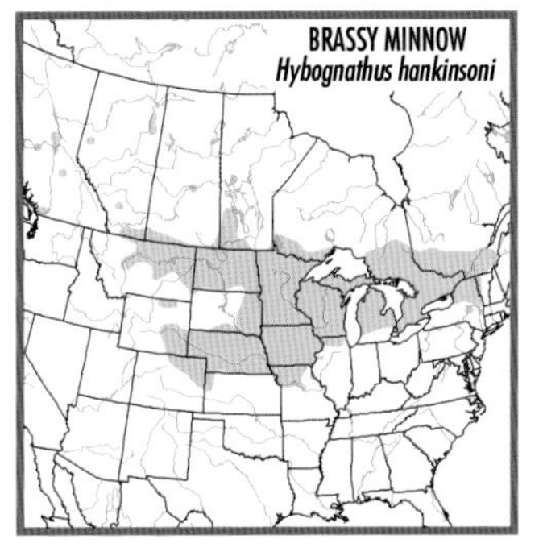

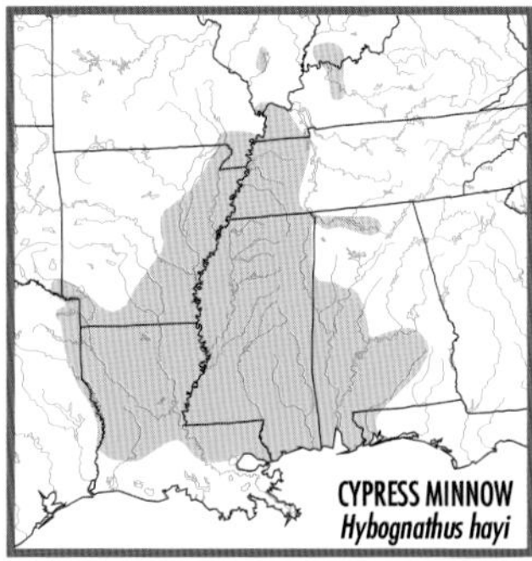

BRASSY MINNOW *Hybognathus hankinsoni* Pl. 8

IDENTIFICATION: *Rounded dorsal fin. Stout, brassy yellow body;* moderately compressed, deepest and widest in front of dorsal fin. Moderately small eye (about¼ head length). Dusky olive above, wide dusky to yellow-green stripe along back; often with thin lines along upper side similar to those on Common and Striped shiners, *Luxilus cornutus* and *L. chrysocephalus* (Fig. 22). Brassy yellow side (best developed on large adult; smaller individual may be dull silver); diffuse dusky stripe best developed on rear half of side. Has 13–15 pectoral rays. Basioccipital process *straight* or *barely concave* posteriorly. To 3¾ in. (9.7 cm). **RANGE:** St. Lawrence R. and Lake Champlain drainages, QC and VT, across Great Lakes, Hudson Bay, and Missouri-upper Mississippi river basins of s. Canada and n. U.S. south to MO and CO; Mackenzie R. system (Arctic basin), AB; Fraser R. system (Pacific Slope), BC. Common in some areas. **HABITAT:** Pools of sluggish, clear creeks and small rivers; usually over sand or gravel. **SIMILAR SPECIES:** (1) Mississippi Silvery Minnow, *H. nuchalis* (Pl. 8), (2) Western Silvery Minnow, *H. argyritis*, and (3) Eastern Silvery Minnow, *H. regius*, *lack* brassy yellow color, have *pointed* dorsal fin, usually lack longitudinal lines along upper side.

CYPRESS MINNOW *Hybognathus hayi* Pl. 8

IDENTIFICATION: Scales on back and upper side thinly outlined with black, appearing *diamond-shaped.* Compressed body, deepest and widest at dorsal fin origin. Pointed dorsal fin. Moderately large eye (⅓ head length). Light to dark olive above, thin dusky to yellow-green stripe along back; silver side, sometimes overlaid by a dusky stripe usually best developed on caudal peduncle. Has 14–16 pectoral rays. Basioccipital process *broad* and *straight* to slightly concave posteriorly. To 4½ in. (12 cm). **RANGE:** Ohio and Mississippi river basins from sw. IN and s. IL to LA; Gulf Slope drainages from Escambia R., FL and AL, to Sabine R., TX. Mostly on Former Mississippi Embayment; rarely above Fall Line (e.g., Tennessee R. in n. AL). Locally common, but disappearing from northern part of range. **HABITAT:** Swamps, oxbows, and backwaters and pools of sluggish streams; usually over mud and near detritus. **SIMILAR SPECIES:** (1) Other *Hybognathus* species (Pl. 8) *lack* diamond-shaped scales; have smaller eye and shallower, less compressed body. (2) *Cyprinella* species (Pl. 16) have diamond-shaped scales but have dorsal fin origin behind pelvic fin origin, much shorter intestine.

LUXILUS

GENUS CHARACTERISTICS: Large scales; those on front half of side *much deeper than wide. Deep, strongly compressed body;* dorsal fin origin over or slightly behind pelvic fin origin. *Dusky to black bar* (often thin dusky line on juvenile) on side behind gill cover. Large, oblique terminal mouth. Large male has hooked tubercles on snout. Usually 9 anal rays; 36–46 lateral scales, pharyngeal teeth 2,4-4,2.

STRIPED SHINER *Luxilus chrysocephalus* Pl. 15

IDENTIFICATION: *Three dark stripes* on upper side meet those of other side behind dorsal fin to form *large Vs* (Fig. 22). Olive above, dark stripe along middle of back; silver bronze side. Dark crescents on side, prominent on large individual. Large male and sometimes female has pink or red body and fins. Usually 24–29 scales around body at dorsal fin origin. To 7¼ in. (18 cm). **RANGE:** Great Lakes and Mississippi R. basins from w. NY and WI, south to AL, LA, and e. TX; Gulf drainages from Mobile Bay, GA and AL, to Sabine R., LA. Common to abundant. Introduced into Escambia R. system, FL and AL. **HABITAT:** Rocky pools near riffles in clear to fairly turbid creeks and small to medium rivers. **REMARKS:** Two subspecies. *L. c. isolepis,* in Mississippi R. basin below confluence of White R., AR, and in Gulf drainages (except Coosa

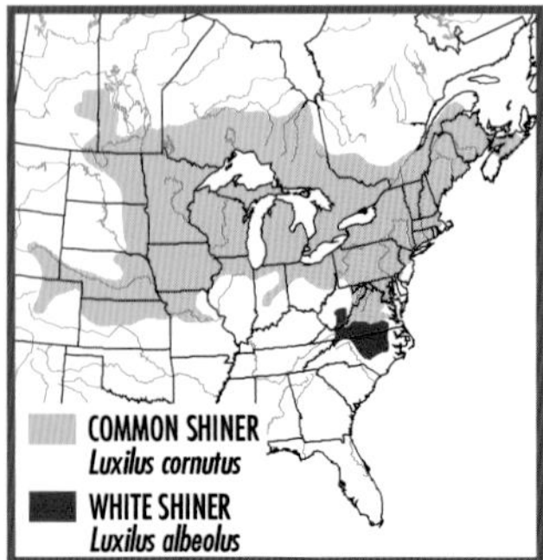

R. system), has straight dark stripes on upper side (Fig. 22), usually 13–14 scales on upper side from head to beneath dorsal fin origin. *L. c. chrysocephalus,* rest of range, has crooked stripes on upper side (Fig. 22), usually 14–17 scales on upper side from head to beneath dorsal fin origin. **SIMILAR SPECIES:** See Pl. 15. (1) Common Shiner, *L. cornutus*, has only *1 or 2 stripes* on upper side parallel to stripe along back, usually 30–35 scales around body. (2) Bleeding Shiner, *L. zonatus*, (3) Duskystripe Shiner, *L. pilsbryi,* and (4) Cardinal Shiner, *L. cardinalis*, have dark stripe along side and around snout, are less deep bodied.

COMMON SHINER *Luxilus cornutus* Pl. 15

IDENTIFICATION: Olive above, dark stripe along middle of back; 1 or 2 dark stripes (often faint) on upper side *parallel to* stripe along back (Fig. 22); silver bronze side. Dark crescents on side, prominent only on large individual. Breeding male (sometimes female) has pink body, pink or red fins. Usually 30–35 scales around body at dorsal fin origin. To 7 in. (18 cm). **RANGE:** Atlantic, Great Lakes, Hudson Bay, and Missis-

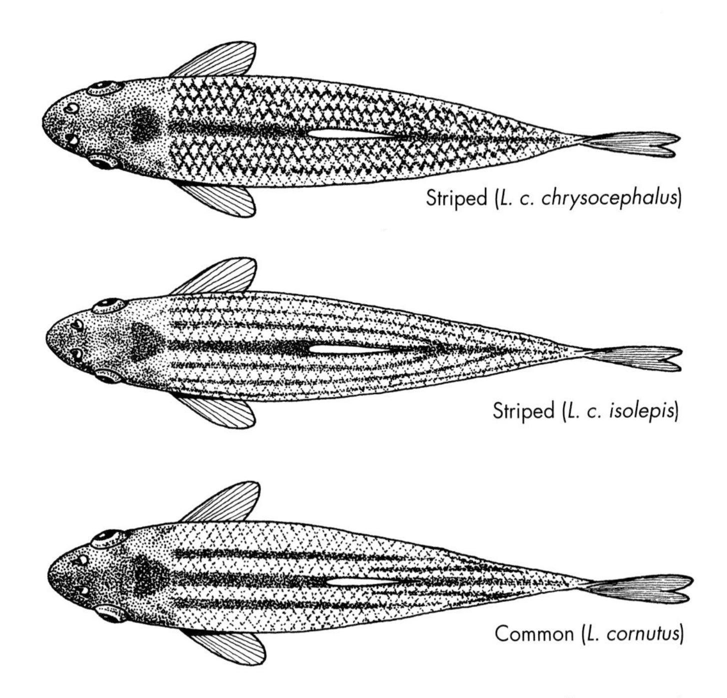

Fig. 22. Striped (*Luxilus chrysocephalus*) and Common (*L. cornutus*) shiners.

sippi R. basins, from NS to se. SK, and south to James R. drainage, VA, n. OH, cen. MO, and CO. Abundant. **HABITAT:** Rocky pools near riffles in clear, cool creeks and small to medium rivers; sometimes in lakes in northern part of range. **REMARKS:** Hybridization with Striped Shiner, *L. chrysocephalus*, occurs frequently where ranges overlap. **SIMILAR SPECIES:** (1) See White Shiner, *L. albeolus* (Pl. 15). (2) Striped Shiner, *L. chrysocephalus* (Pl. 15), has 3 dark stripes on upper side meeting those of other side behind dorsal fin to form Vs, usually 24–29 scales around body.

WHITE SHINER *Luxilus albeolus* Pl. 15

IDENTIFICATION: Nearly identical to Common Shiner, *L. cornutus*, but more silvery; *lacks* dark crescents (rarely present) on side; usually has *26–30 scales* around body at dorsal fin origin. To 5¼ in. (13 cm). **RANGE:** Atlantic Slope from Chowan R. system, VA, to Cape Fear R. drainage, NC; upper New R. drainage (Ohio R. basin), WV, VA, and NC. Common; abundant in upper Roanoke R. drainage. **HABITAT:** Rocky pools near riffles in clear creeks and small to medium rivers. **SIMILAR SPECIES:** (1) See Common Shiner, *L. cornutus* (Pl. 15). (2) Crescent Shiner, *L. cerasinus* (Pl. 15), has dark crescents on side; red on head, body, and fins of large male; strongly contrasting upper and lower halves of head.

CRESCENT SHINER *Luxilus cerasinus* Pl. 15

IDENTIFICATION: *Large black crescents* on side. Dusky olive above, dark stripe along middle of back; silver side. Dark upper half of head strongly contrasts with light lower half. Breeding male has blue back, red on head, body, fins; large female may have red on body, fins. Usually 24–28 scales around body at dorsal fin origin. To 4½ in. (11 cm). **RANGE:** James, Roanoke, Chowan, New, and extreme upper Cape Fear river drainages, VA and NC. Common to abundant in Roanoke drainage; possibly introduced in other drainages. **HABITAT:** Rocky and sandy pools and runs of headwaters, creeks, and small rivers. **SIMILAR SPECIES:** (1) White Shiner, *L. albeolus* (Pl. 15), *lacks* dark crescents on side; has less contrasting upper and lower halves of head. (2) Common Shiner, *L. cornutus* (Pl. 15), has less distinct crescents, less contrasting upper and lower halves of head, 30–35 scales around body, pink on large male.

BLEEDING SHINER *Luxilus zonatus* Pl. 15

IDENTIFICATION: *Large black bar* on side behind gill cover. *Narrow black stripe* along side and around snout, constricted behind head, not extending below lateral line. Olive above, black stripe along back; dusky stripes on upper side meet those of other side behind dorsal fin. Large individual has red head and fins (brightest on breeding male). Usually

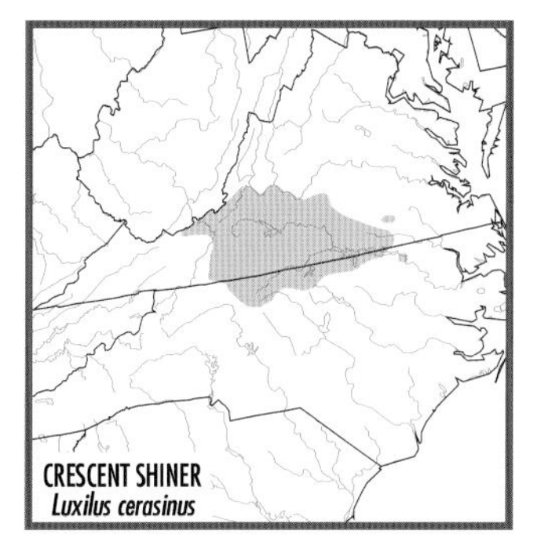

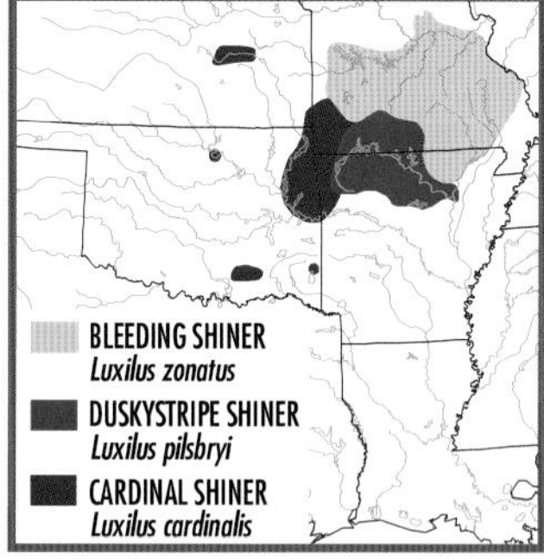

26 scales around body at dorsal fin origin, 7 rakers on 1st gill arch. To 5 in. (13 cm). **RANGE:** Ozark-draining tributaries of Missouri (west to Sac R.), Mississippi (including Meramec R.), Little, St. Francis, and Black rivers, s. MO and ne. AR. Common to abundant. **HABITAT:** Rocky runs, riffles, and deep, flowing pools of clear, fast creeks and small to medium rivers. **SIMILAR SPECIES:** (1) See Duskystripe Shiner, *L. pilsbryi* (Pl. 15). (2) Cardinal Shiner, *L. cardinalis* (Pl. 15), *lacks* black bar on side behind gill cover, dusky stripes on upper side; has *broad black stripe* along side (not constricted behind head) extending below lateral line; blue snout and crimson red head, lower body, and fins on large adult.

DUSKYSTRIPE SHINER *Luxilus pilsbryi* Pl. 15

IDENTIFICATION: Similar to Bleeding Shiner, *L. zonatus*, but *lacks* bold black bar behind gill cover, dusky stripes on upper side; black stripe along side *not* notably constricted behind head, *extends to lateral line* under dorsal fin; blue snout, bright red head and fins on large individual. Usually 24–26 scales around body at dorsal fin origin, 7 rakers on 1st gill arch. To 5 in. (13 cm). **RANGE:** White (excluding Black R. system) and Little Red river systems, s. MO and n. AR. Common to abundant. **HABITAT:** Rocky pools, runs, and deep riffles of clear, fast creeks and small to medium rivers. **SIMILAR SPECIES:** See (1) Bleeding Shiner, *L. zonatus*, and (2) Cardinal Shiner, *L. cardinalis* (both Pl. 15).

CARDINAL SHINER *Luxilus cardinalis* Pl. 15

IDENTIFICATION: Similar to Duskystripe Shiner, *L. pilsbryi*, but has *broad black stripe* along side extending *below* lateral line; *crimson red head, lower side of body, and fins* on large individual; usually 26–27 scales around body at dorsal fin origin, 8–9 rakers on 1st gill arch. To 4¼ in. (11 cm). **RANGE:** Arkansas R. drainage, sw. MO, nw. AR, e. KS and e. OK. Also Red R. drainage, s. OK, where probably introduced.

Common in Arkansas R. drainage; rare in Red R. drainage. **HABITAT:** Rocky runs, riffles, and flowing pools of creeks and small to medium rivers. **SIMILAR SPECIES:** (1) See Duskystripe Shiner, *L. pilsbryi* (Pl. 15). (2) Bleeding Shiner, *L. zonatus* (Pl. 15), has black bar on side behind gill cover; dusky stripes on upper side; *narrow black stripe (constricted* behind head) along side, *not extending below* lateral line; less intense red color on head and fins.

WARPAINT SHINER *Luxilus coccogenis* Pl. 15

IDENTIFICATION: *Black band* (red-orange on young) on yellow dorsal fin; *red bar on* opercle; wide *black edge* (darkest at fork) on caudal fin. Olive above, dark stripe along middle of back; large black bar behind gill cover on silver side. Breeding male has pink or red side, red snout, red on dorsal fin. Usually 25–29 scales around body at dorsal fin origin. To 5½ in. (14 cm). **RANGE:** Upper Tennessee R. drainage, VA, NC, TN, n. GA, and n. AL, and adjacent tributaries of Savannah R., NC and SC, Santee R., NC, and New R., NC. Common; may be introduced in New and Santee river drainages. **HABITAT:** Gravel and rubble riffles and adjacent pools of clear, fast creeks and small to medium rivers. **SIMILAR SPECIES:** (1) Bandfin Shiner, *L. zonistius* (Pl. 15), *lacks* wide black edge on caudal fin, has large black spot on caudal fin base. (2) Striped Shiner, *L. chrysocephalus* (Pl. 15), *lacks* black band on dorsal fin, wide black edge on caudal fin.

BANDFIN SHINER *Luxilus zonistius* Pl. 15

IDENTIFICATION: *Black band* (red-orange in young) on dorsal fin, *large black spot* on caudal fin base. Olive above, dark stripe along middle of back; dusky stripe (darkest at rear) along silver copper side; faint red bar on cheek, black (red in young) bar behind gill cover. Breeding male is blue above, has red bar on caudal fin. Usually 27–30 scales around body at dorsal fin origin, 9–10 anal rays. To 4 in. (10 cm). **RANGE:** Apalachicola R. drainage, GA, AL, FL; adjacent tributaries of Savannah, Altamaha, and Coosa rivers, GA, and Tallapoosa R., GA and AL. Possibly introduced into Hiwassee R. system, GA. Fairly common. **HABITAT:** Rocky pools near riffles in clear creeks and small rivers. **SIMILAR SPECIES:** (1) Warpaint Shiner, *L. coccogenis* (Pl. 15), *lacks* large black spot on caudal fin base, has wide black edge on caudal fin. (2) Striped Shiner, *L. chrysocephalus* (Pl. 15), *lacks* black band on dorsal fin, large black spot on caudal fin base.

LYTHRURUS

GENUS CHARACTERISTICS: *Very small scales* on nape; dorsal fin origin *behind* pelvic fin origin. Fairly large, oblique terminal mouth. Usually 10–12 anal rays (often 9 in Scarlet Shiner, *L. fasciolaris); pharyngeal*

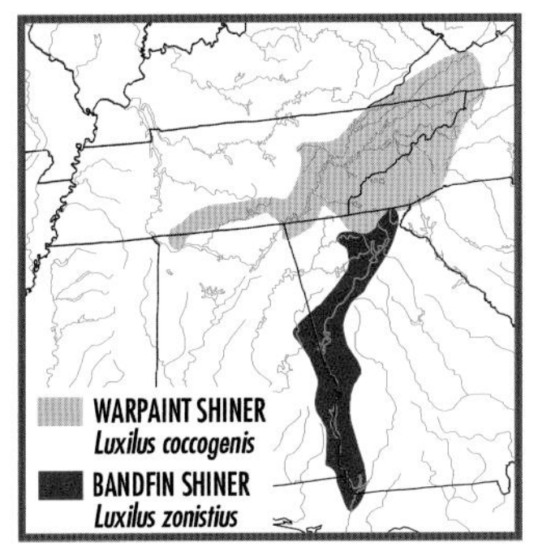

teeth 2,4-4,2. Large males develop bright red or yellow fins and, in some species, metallic blue bodies.

REDFIN SHINER *Lythrurus umbratilis* **Pl. 15**

IDENTIFICATION: *Dark blotch* at dorsal fin origin. *Deep, compressed body;* fairly large eye. Pale olive to steel blue above, dusky stripe along back; black specks on back and upper side; herringbone lines on upper side of large individual. Breeding male has black membranes on *red fins,* blue head and body, often a large dark blotch on side. Has 10–11 anal rays; 37–56 lateral scales. To 3½ in. (8.6 cm). **RANGE:** Great Lakes and Mississippi R. basins, w. NY and s. ON to se. MN and south to LA; Gulf drainages west of Mississippi R. to Trinity and San Jacinto rivers, TX. Common; locally abundant. **HABITAT:** Quiet to flowing pools (often turbid) of headwaters, creeks, and small to medium rivers. **REMARKS:** Two subspecies. *L. u. umbratilis,* Missouri, Arkansas, and upper Salt river drainages, MO, e. KS, e. OK, and nw. AR, has many black specks behind gill cover (coalescing into bar in large male), dusky spot at dorsal fin origin, breeding tubercles on male's opercle. *L. u. cyanocephalus,* rest of range, has few black specks behind gill cover, black spot at dorsal fin origin, few or no tubercles on opercle. Intergrades occupy lower Salt R., MO, and Arkansas R. tributaries in w. AR. **SIMILAR SPECIES:** See Pl. 15. (1) Scarlet Shiner, *L. fasciolaris,* has dusky bars over back, no black on fin membranes; is more slender. (2) Ribbon Shiner, *L. fumeus,* (3) Ouachita Mountain Shiner, *L. snelsoni,* and (4) Mountain Shiner, *L. lirus, lack* dark blotch at dorsal fin origin, herringbone lines, red fins; are more slender.

ROSEFIN SHINER *Lythrurus ardens* **Pl. 15**

IDENTIFICATION: *Dark blotch* at dorsal fin origin. Fairly deep, compressed body; fairly large eye. Olive to steel blue above, dusky stripe along back; black specks on back and upper side; dusky lips and chin.

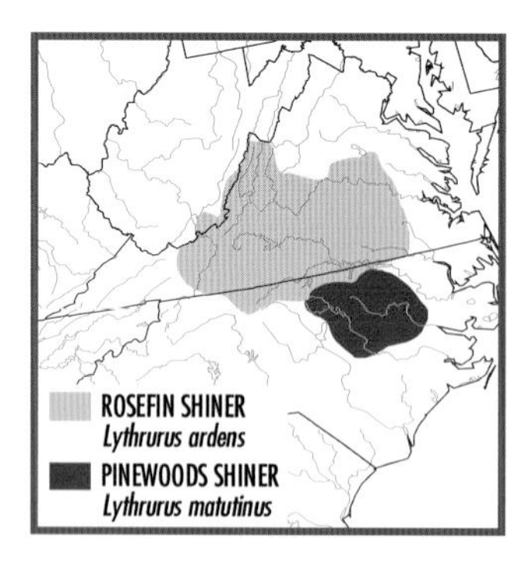

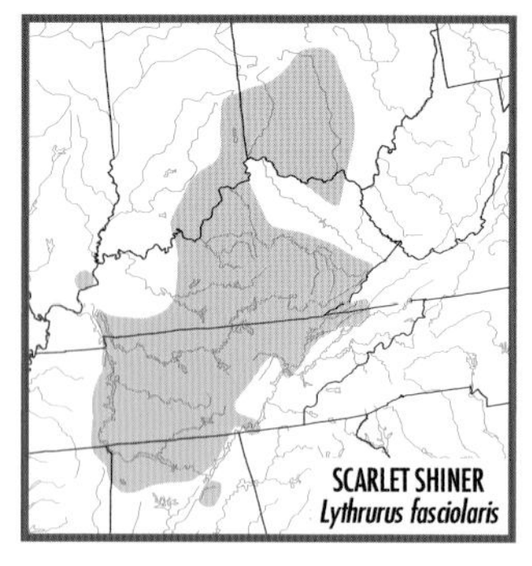

Breeding male has faint *red fins and top of head, faint blue-gray bars on back*. Usually 11 anal rays (range 10–12); 41–51 lateral scales. To 3½ in. (8.6 cm). **RANGE:** Atlantic Slope from York R. drainage (where introduced), VA, to Roanoke R. drainage, NC; upper New R. system (Kanawha-Ohio drainage) above Kanawha Falls, WV and VA. Common. **HABITAT:** Sandy and rocky runs and flowing pools of creeks and small to medium rivers. **SIMILAR SPECIES:** See Pl. 15. (1) See Scarlet Shiner, *L. fasciolaris*, and (2) Pinewoods Shiner, *L. matutinus*. (2) Redfin Shiner, *L. umbratilis*, *lacks* dusky bars over back (rarely has faint bars); has *deeper* body, black on fins of breeding male. (3) Mountain Shiner, *L. lirus*, (4) Ribbon Shiner, *L. fumeus*, and (5) Cherryfin Shiner, *L. roseipinnis, lack* black blotch at dorsal fin origin, dusky bars over back; Mountain Shiner has silver black stripe along side, black lips; Ribbon Shiner has yellow fins; Cherryfin Shiner has large black spots at tips of dorsal and anal fins, dark stripe along rear of side.

PINEWOODS SHINER *Lythrurus matutinus* Pl. 15

IDENTIFICATION: Similar to Rosefin Shiner, *L. ardens*, but is *more slender*; breeding male has *bright red head* (mostly blue on Rosefin Shiner), *little or no blue* on body, no red on paired fins. Usually 11 anal rays; 40–49 lateral scales. To 3½ in. (8.6 cm). **RANGE:** Tar and Neuse river drainages, NC. Common. **HABITAT:** Sandy runs and flowing pools of creeks and small to medium rivers. **SIMILAR SPECIES:** (1) See Rosefin Shiner, *L. ardens* (Pl. 15). (2) Scarlet Shiner, *L. fasciolaris* (Pl. 15), usually has 10 anal rays, *deeper body*, dusky bars over back; breeding male has red or orange fins and lower body, but *lacks* bright red head.

SCARLET SHINER *Lythrurus fasciolaris* Pl. 15

IDENTIFICATION: Similar to Rosefin Shiner, *L. ardens*, but usually has 10 anal rays, deeper body, *dusky bars* over back (often absent, especially on small individual). Breeding male has keeled nape, dark blue-gray bars on back and upper side, intense red or orange on fins and lower side but not on top of head. Has 38–53 lateral scales. To 3½ in. (8.6 cm). **RANGE:** Ohio R. basin from Muskingum R. drainage, OH, to se. IL (extirpated in IL) and south to Tennessee R. drainage, GA, AL, and MS; upper Black Warrior R. system (Gulf basin), AL. Common to abundant. **HABITAT:** Rocky pools and runs of clear, fairly fast headwaters, creeks, and small rivers. **SIMILAR SPECIES:** (1) See Rosefin Shiner, *L. ardens* (Pl. 15). (2) Pinewoods Shiner, *L. matutinus* (Pl. 15), usually has 11 anal rays, more slender body, *no dusky bars* over back; breeding male has bright red head, little or no blue on body.

RIBBON SHINER *Lythrurus fumeus* Pl. 15

IDENTIFICATION: Scales on nape *outlined in black. Fairly slender, compressed body;* fairly large eye. Pale olive above, dusky stripe along back; silver black stripe along side (darkest at rear but often weak) and around snout; dusky lips and chin (Fig. 23). Yellow fins on large individual. Has 11–12 anal rays; 35–45 lateral scales. To 2¾ in. (7 cm). **RANGE:** Mississippi R. basin, cen. IL, to nw. AL, LA, and e. OK; Gulf drainages from Lake Pontchartrain, LA, to Navidad R., TX. Mostly below Fall Line, but also in lowlands of s. IL. Common. **HABITAT:** Quiet, usually turbid, mud- or sand-bottomed pools of headwaters, creeks, and small rivers. **SIMILAR SPECIES:** (1) See Ouachita Mountain Shiner, *L. snelsoni*. (2) Redfin Shiner, *L. umbratilis* (Pl. 15), has dark blotch at dorsal fin origin, herringbone lines on upper side, red fins on large male, *deeper body.* (3) Mountain Shiner, *L. lirus* (Pl. 15), has black lips, white chin, more slender body, blacker stripe on side.

OUACHITA MOUNTAIN SHINER *Lythrurus snelsoni* Not shown

IDENTIFICATION: Similar to Ribbon Shiner, *L. fumeus*, but *lacks* black-outlined scales on nape; has *red on head and throat* (brightest on male) during breeding season (spring), no yellow on fins, 9–11 (usually 10) anal rays; is more slender. Has 27–33 scales around body at dorsal fin origin. To 2 in. (5.3 cm). **RANGE:** Above Fall Line in Little R. system, AR

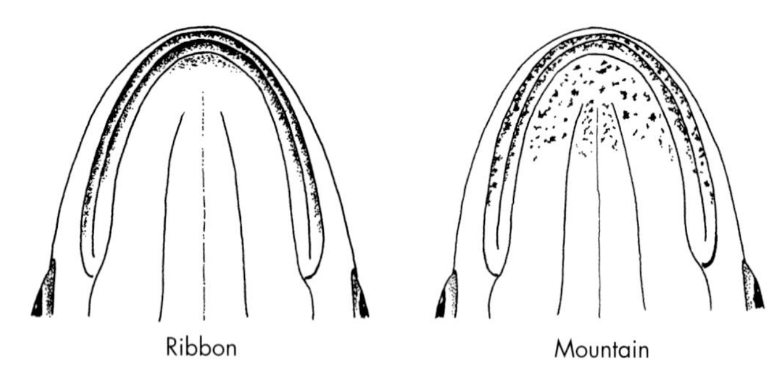

Fig. 23. Ribbon and Mountain shiners—underside of head.

and OK (Ouachita Mts.). Common. **HABITAT:** Rocky pools of small to medium rivers. Often near vegetation. **SIMILAR SPECIES:** See Pl. 15. (1) See Ribbon Shiner, *L. fumeus*. (2) Mountain Shiner, *L. lirus*, has black stripe around snout (across lips), pink or yellow (*no red*) on head and throat. (3) Redfin Shiner, *L. umbratilis*, has dark blotch at dorsal fin origin, deeper body, herringbone lines on upper side, red fins on large male, usually 33 or more scales around body at dorsal fin origin.

MOUNTAIN SHINER *Lythrurus lirus* **Pl. 15**

IDENTIFICATION: *Slender*, compressed body. *Silver black* stripe along side (solid at rear, diffuse at front) and around snout. Fairly large eye. Pale olive above; black specks on back and upper side, often outlining scales on nape; black lips, white chin (Fig. 23). Pale pink or yellow on head and body of large individual. Has 10–11 anal rays; 36–49 lateral scales. To 3 in. (7.5 cm). **RANGE:** Tennessee and Alabama river drainages, VA, TN, nw. GA, and AL. In Alabama R. drainage nearly restricted to Coosa R. system above Fall Line (where common). Uncommon in Tennessee R. drainage. **HABITAT:** Sandy and rocky pools and runs of clear creeks and small rivers. **SIMILAR SPECIES:** See Pl. 15. (1) Ouachita Mountain Shiner, *L. snelsoni*, has dusky lips and chin, red on head and throat of large individual. (2) Ribbon Shiner, *L. fumeus*, has dusky lips and chin, *deeper body*, less distinct stripe along side. (3) Scarlet Shiner, *L. fasciolaris*, has black blotch at dorsal fin origin, dusky chin, *no* black stripe along side. (4) Redfin Shiner, *L. umbratilis*, has *deeper body*, dark blotch at dorsal fin origin, herringbone lines on upper side, *no* black stripe on side.

CHERRYFIN SHINER *Lythrurus roseipinnis* **Pl. 15**

IDENTIFICATION: *Black spots* on tips of dorsal and anal fins (Fig. 24). Deep, compressed body; fairly large eye. Pale olive above, dusky stripe along back; black specks on back and upper side, sometimes

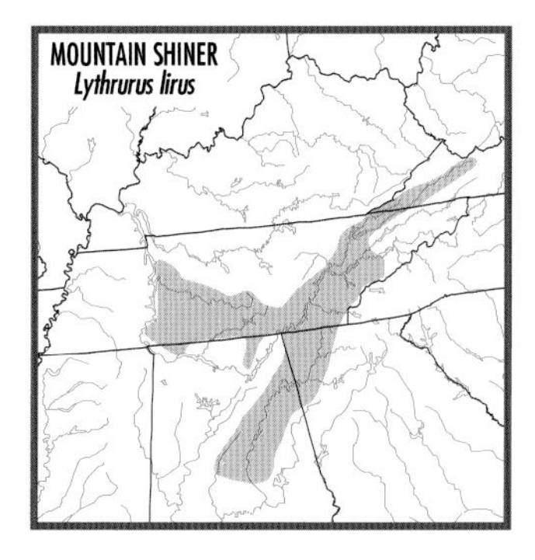

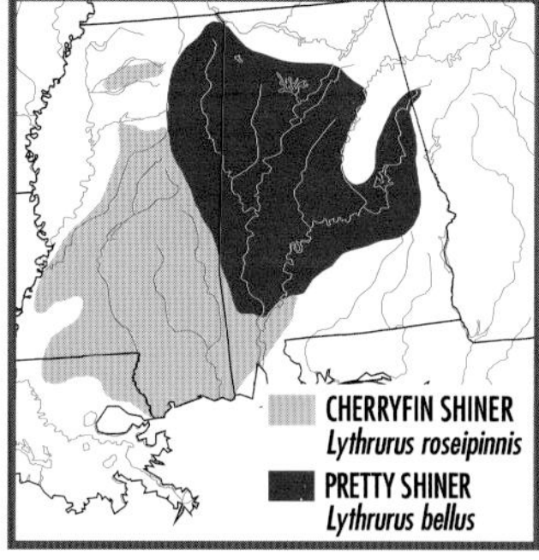

outlining scales on nape; herringbone lines rarely present; dark stripe along rear half of side; dusky lips and chin. Breeding male has pale to bright red fins. Has 11–12 anal rays; 36–49 lateral scales. To 3 in. (7.5 cm). **RANGE:** Gulf drainages from extreme lower Mobile Bay basin, AL, to Lake Pontchartrain, LA; Yazoo R., Big Black R., and Bayou Pierre drainages (Mississippi R. basin), MS. Common. **HABITAT:** Sand- and gravel-bottomed pools of headwaters, creeks, and small rivers. **SIMILAR SPECIES:** See (1) Pretty Shiner, *L. bellus*, and (2) Blacktip Shiner, *L. atrapiculus*. (3) Ribbon Shiner, *L. fumeus* (Pl. 15), *lacks* large black spots in dorsal and anal fins; has yellow fins. (4) Redfin Shiner, *L. umbratilis* (Pl. 15), has dark blotch at dorsal fin origin, *lacks* large black spots in dorsal and anal fins.

PRETTY SHINER *Lythrurus bellus* Not shown

IDENTIFICATION: Nearly identical to Cherryfin Shiner, *L. roseipinnis*, but has *broad black band* on edge of dorsal, anal, and pelvic fins (Fig.

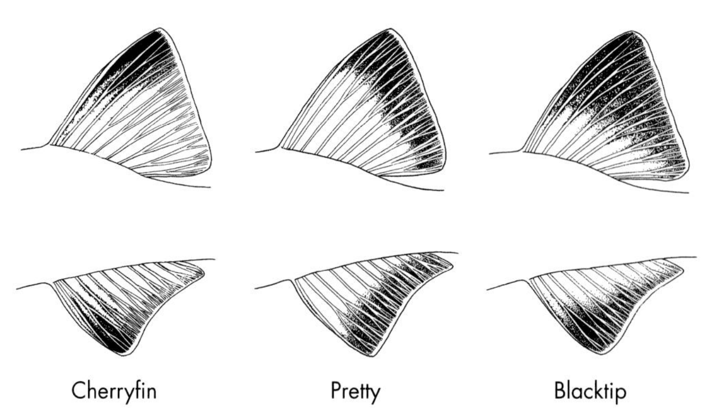

Fig. 24. Cherryfin, Pretty, and Blacktip shiners—dorsal and anal fins.

24). Has 10–11 anal rays; 35–47 lateral scales. To 3 in. (7.5 cm). **RANGE:** Mobile Bay drainage, Bear and Yellow creek systems (Tennessee R. drainage), AL and MS. Common below Fall Line although avoids lower Coastal Plain. **HABITAT:** Sand- and clay-bottomed pools of headwaters, creeks, and small rivers. **REMARKS:** Two subspecies. *L. b. alegnotus,* above Fall Line in Black Warrior R. system, AL, has dark stripe along side, usually 12–13 scales around caudal peduncle. *L. b. bellus,* in rest of range, lacks dark stripe on side, usually has 14–15 scales around caudal peduncle, dark body on breeding male. Intergrades occur in North R. and Hurricane Creek, Tuscaloosa and Warrior counties, AL. **SIMILAR SPECIES:** See Pl. 15. (1) See Cherryfin Shiner, *L. roseipinnis.* (2) Blacktip Shiner, *L. atrapiculus,* has *tapering* black band on dorsal, anal, and pelvic fins (Fig. 24). (3) Mountain Shiner, *L. lirus,* (4) Scarlet Shiner, *L. fasciolaris,* and (5) Ribbon Shiner, *L. fumeus, lack* black band on dorsal and anal fins, are more slender; Mountain Shiner has black stripe along side, black lips, white chin; Scarlet Shiner has black blotch at dorsal fin origin, dusky bars over back; Ribbon Shiner has yellow fins.

BLACKTIP SHINER *Lythrurus atrapiculus* Not shown

IDENTIFICATION: Nearly identical to Cherryfin Shiner, *L. roseipinnis,* but has *tapering* (wide at front, narrow at rear) *black band* on edge of dorsal, anal, and pelvic fins (Fig. 24). Has 10–11 anal rays; 36–45 lateral scales. To 2½ in. (6.5 cm). **RANGE:** Apalachicola (including upper Flint R.), Choctawhatchee, Yellow, and Escambia river drainages, w. GA, se. AL, and FL. Introduced into Old Town Creek (Tallapoosa R. system), Bullock Co., AL. Above Fall Line only in Apalachicola drainage. Fairly common. **HABITAT:** Sand- and gravel-bottomed pools, sometimes runs, of headwaters, creeks, and small rivers. **SIMILAR SPECIES:** (1) See Cherryfin Shiner, *L. roseipinnis* (Pl. 15). (2) Pretty Shiner, *L. bellus,* has *uniformly broad* black band in dorsal, anal, and pelvic fins (Fig. 24).

PTERONOTROPIS

GENUS CHARACTERISTICS: *Broad, blue-black stripe along side; large dorsal and anal fins. Fairly deep to deep, compressed body,* deepest under or near dorsal fin origin; dorsal fin origin behind pelvic fin origin; complete lateral line; pharyngeal teeth 2,4-4,2. Large male develops large dorsal and anal fins (extremely so in Bluenose Shiner, *P. welaka*, and Bluehead Shiner, *P. hubbsi*) and bright colors on head and body.

FLAGFIN SHINER *Pteronotropis signipinnis* Pl. 18

IDENTIFICATION: Olive-gold above; upper side yellow at front, red at rear; *broad blue-black stripe* along side with vertical orange dashes; pale gold lower side; gold snout. *Red-orange edge* on yellow dorsal, caudal, anal, and pelvic fins; yellow pectoral fins. *Deep, compressed body,* strongly tapering to narrow caudal peduncle. Large dorsal and anal fins; tips of rays at front of dorsal fin extend *to or beyond* those at rear in depressed fin. Decurved lateral line. Has 10–11 anal rays; 22–26 (usually 23–24) scales around body; 32–36 lateral scales; pharyngeal teeth 2,4-4,2. To 2¾ in. (7 cm). **RANGE:** Coastal Plain from Apalachicola R. drainage, FL, to Lake Pontchartrain drainage, LA. Common; locally abundant. **HABITAT:** Flowing pools and runs of headwaters, creeks, and small rivers; usually over sand and near vegetation. **SIMILAR SPECIES:** (1) Sailfin Shiner, *P. hypselopterus*, and closely related species (Pl. 18; Fig. 25) have dark predorsal stripe, black dorsal fin membranes, small bright red spots on caudal fin base, usually 27–30 scales around body; *lack* red-orange edge on dorsal and pelvic fins.

SAILFIN SHINER *Pteronotropis hypselopterus* Pl. 18

IDENTIFICATION: Pink-brown to olive above, *dark brown predorsal stripe*; *broad steel blue stripe* along side bordered above by thin pink to red line; white to light pink below; small *red spots* above and below black spot on caudal fin base. Dusky middle band on dorsal fin; other

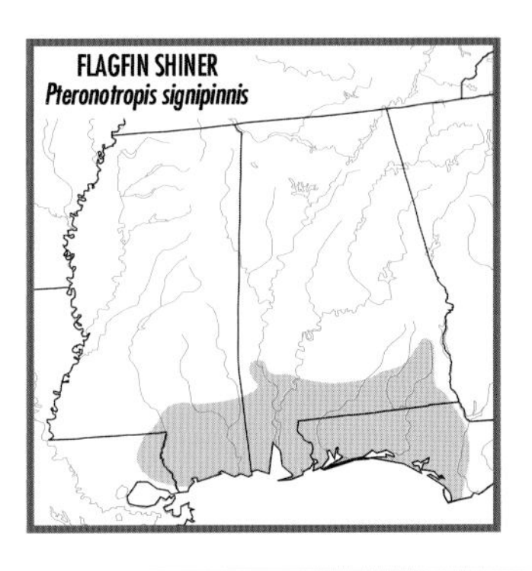

fins yellow to light orange. *Deep, compressed body* tapering to narrow caudal peduncle. *Large, nearly triangular dorsal and anal fins,* rays at front shorter than those at rear in depressed fin. Breeding male with light orange edge on black dorsal fin, orange caudal and anal fins. Decurved lateral line. Has 10–12 (usually 11) anal rays; 34–43 lateral scales. To 2¾ in. (7 cm). **RANGE:** Below Fall Line in Gulf drainages from St. Andrews Bay, FL, to Mobile Bay, AL; absent in Choctawhatchee R. system above mouth of Pea R., AL. Common. **HABITAT:** Sand- and clay-bottomed pools and runs of headwaters, creeks, and small rivers; often among debris and vegetation. **SIMILAR SPECIES:** See Pl. 18. (1) Sailfin Shiner lookalikes (Fig. 25): Next 5 species (through Metallic Shiner, *P. metallicus*) are easily distinguished from one another and Sailfin Shiner only as breeding males. It often is necessary to rely on microscopic examination of specimens and geography to identify them. (2) Flagfin Shiner, *P. signipinnis, lacks* dark predorsal stripe; has yellow to red upper side, red-orange edge on dorsal, caudal, and anal fins, yellow on caudal fin base.

ORANGETAIL SHINER *Pteronotropis merlini* Fig. 25

IDENTIFICATION: Nearly identical to Sailfin Shiner, *P. hypselopterus*, but is deeper bodied, usually has 10 anal rays (range 9–11). Breeding male has *chevron or lunate-shaped black blotch* at caudal fin origin, slightly separated from black stripe along side; orange caudal and anal fins. Has 35–42 lateral scales. To 2½ in. (6.5 cm). **RANGE:** Choctawhatchee R. system above confluence with Pea R., se. AL. Common. **HABITAT:** Sand- and silt-bottomed pools and runs of headwaters, creeks, and small rivers; usually along undercut banks and debris. Common. **SIMILAR SPECIES:** (1) See Sailfin Shiner, *P. hypselopterus* (Pl. 18, Fig. 25).

BROADSTRIPE SHINER *Pteronotropis euryzonus* Pl. 18

IDENTIFICATION: Nearly identical to Sailfin Shiner, *P. hypselopterus*, but is deeper bodied; has *larger dorsal fin* (tips of rays at front extend

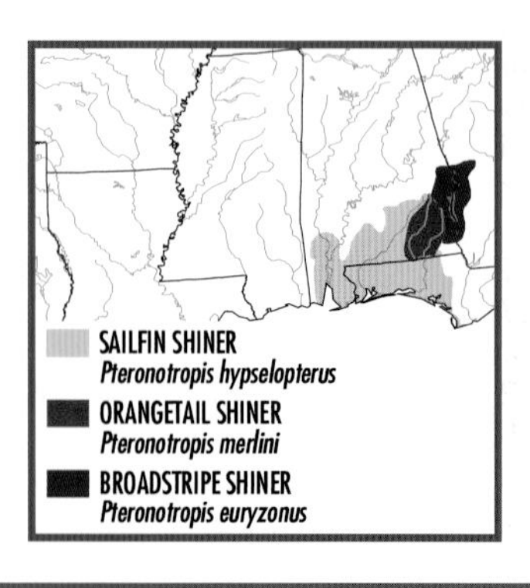

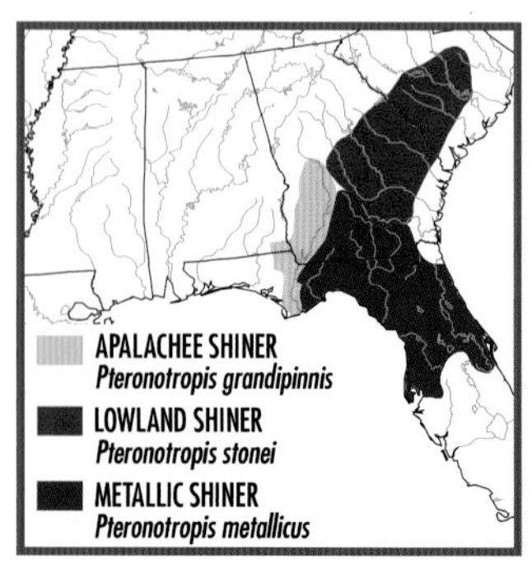

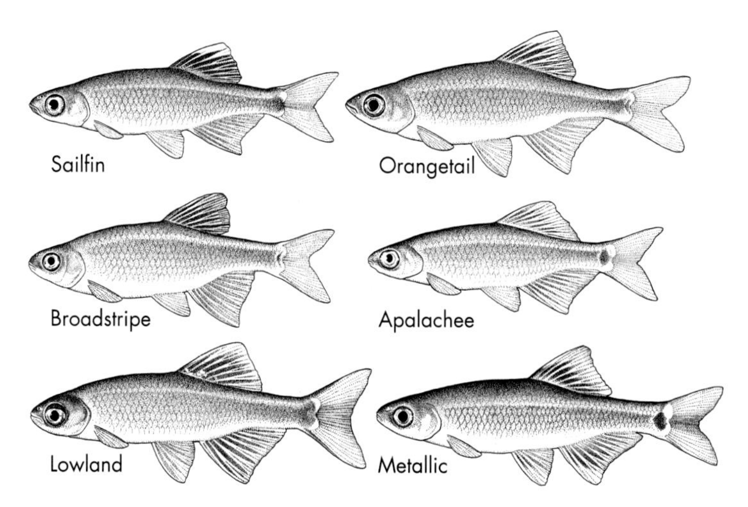

Fig. 25. Sailfin, Orangetail, Broadstripe, Apalachee, Lowland, and Metallic shiners.

to or beyond those at rear in depressed fin), usually 10 anal rays (range 9–11). Breeding male *lacks* light edge on black dorsal fin (fin is black throughout); has *chevron or lunate-shaped black blotch* at caudal fin origin, slightly separated from black stripe along side; bright orange caudal and anal fins. Has 34–42 lateral scales. To 2½ in. (6.5 cm). **RANGE:** Middle Chattahoochee R. drainage, GA and AL (Lee Co., AL, and Talbot Co., GA, south to Houston Co., AL). Locally common. **HABITAT:** Clay, sand, and bedrock pools of headwaters, creeks, and small rivers; often among vegetation and debris. **SIMILAR SPECIES:** (1) See Sailfin Shiner, *P. hypselopterus* (Pl. 18, Fig. 25).

APALACHEE SHINER *Pteronotropis grandipinnis* **Fig. 25**

IDENTIFICATION: Nearly identical to Sailfin Shiner, *P. hypselopterus*, but breeding male has larger dorsal and anal fins—*anal fin reaches caudal fin*. Blue-black stripe along side continues to black oval spot at base of caudal fin; *well-defined* lower edge on stripe (dusky in Sailfin Shiner). *Yellow edge* on dusky black anal fin of breeding male. Usually 10 anal rays (range 9–11). Has 33–42 lateral scales. To 2½ in. (6.5 cm). **RANGE:** Apalachicola R. drainage, GA, AL, and FL (but absent in Chattahoochee R. system north of Cedar Creek, Houston Co., AL, where replaced by Broadstripe Shiner, *P. euryzonus*). Common. **HABITAT:** Sand- and silt-bottomed pools and runs of headwaters, creeks, and small rivers; usually along undercut banks and debris. Common. **SIMILAR SPECIES:** (1) See Sailfin Shiner, *P. hypselopterus* (Pl. 18, Fig. 25).

LOWLAND SHINER *Pteronotropis stonei* Fig. 25

IDENTIFICATION: Nearly identical to Sailfin Shiner, *P. hypselopterus*, but has *dark snout* (light in Sailfin Shiner). Dull blue-black stripe along side continues to caudal fin with no or poorly defined spot on caudal fin base; *well-defined* lower edge on stripe (dusky in Sailfin Shiner); clear (rarely orange) edge on black dorsal fin. *Yellow edge* on dusky black anal fin of breeding male. Usually 9–10 anal rays (range 7–12). Has 34–42 lateral scales. To 2½ in. (6.6 cm). **RANGE:** Below Fall Line from Little Lynches R. system (Peedee R. drainage), SC, to Satilla R. drainage, GA. Common. **HABITAT:** Sand- and silt-bottomed pools and runs of creeks and small rivers; usually near vegetation. **SIMILAR SPECIES:** (1) See Sailfin Shiner, *P. hypselopterus* (Pl. 18, Fig. 25).

METALLIC SHINER *Pteronotropis metallicus* Fig. 25

IDENTIFICATION: Nearly identical to Sailfin Shiner, *P. hypselopterus*, but has *clear to light brown edge* on dusky black anal fin. Black on dorsal, anal, and pelvic fins mostly in middle of fin, creating *dark crescents*. Breeding male has black spot at caudal fin origin *slightly separated* from black stripe along side. Usually 10 anal rays (range 9–11). Has 33–42 lateral scales. To 2½ in. (6.5 cm). **RANGE:** Atlantic and Gulf drainages from St. Marys R., GA and FL, to New R., FL panhandle. South in FL to Alafia R. system (Tampa Bay drainage). Common. **HABITAT:** Sand- and silt-bottomed pools and runs of headwaters, creeks, and small rivers; usually near vegetation. **SIMILAR SPECIES:** (1) See Sailfin Shiner, *P. hypselopterus* (Pl. 18, Fig. 25).

BLUENOSE SHINER *Pteronotropis welaka* Pl. 18

IDENTIFICATION: *Black* stripe along side from chin and snout to caudal fin, where expanded into black spot. Slender body; fairly pointed snout; slightly subterminal mouth; dorsal fin origin slightly behind pelvic fin origin. Breeding male has *bright blue snout; huge, mostly black dorsal fin;* black band on yellow anal (also enlarged) and pelvic fins; silver on side of body. Dusky olive-brown above, scales outlined in black; dark streak along back; light yellow stripe above dark stripe along side; white below; clear to yellow fins. Incomplete lateral line, 5–12 pores. Has 34–37 lateral scales; 8 dorsal rays; 8 anal rays; pharyngeal teeth 1,4-4,1. To 2½ in. (6.5 cm). **RANGE:** Middle St. Johns R. drainage, FL; Gulf drainages (mostly below Fall Line) from Apalachicola R., GA and FL, to Pearl R., MS and LA. Locally common. **HABITAT:** Backwaters and quiet vegetated pools of creeks and small to medium rivers, over mud and sand. Schools in fairly deep (3–6 ft. [1–2 m]) water. **SIMILAR SPECIES:** (1) Bluehead Shiner, *P. hubbsi* (Pl. 18), is deeper bodied, has blunter snout, 9–10 dorsal rays, pharyngeal teeth 0,4-4,0; *lacks* blue on tip of snout. (2) Ironcolor Shiner, *Notropis chalybaeus* (Pl. 18), has

shorter, pointed snout, dorsal fin origin over pelvic fin origin; *lacks* blue snout, enlarged fins on large male; pharyngeal teeth 2,4-4,2.

BLUEHEAD SHINER *Pteronotropis hubbsi* Pl. 18

IDENTIFICATION: *Black stripe* along side from chin (but absent on upper lip and snout) to caudal fin, where expanded into spot. Has 9–10 *dorsal* and anal rays. Breeding male has *bright blue* top of head, dorsal fin, and caudal fin; *huge dorsal and anal fins.* Deep body; short, blunt snout; terminal mouth. Dusky orange-brown above, scales outlined in black; dark stripe along back from head to dorsal fin. Light orange stripe above black stripe along side. Dusky dorsal fin; other fins clear to faintly yellow or orange. Incomplete lateral line, 2–9 pores; 34–38 lateral scales; pharyngeal teeth 0,4-4,0. To 2¼ in. (6 cm). **RANGE:** Ouachita and Red river drainage lowlands of s. AR, LA, and ne. TX; Wolf Lake in sw. IL. Local and uncommon. **HABITAT:** Backwaters, oxbows, and sluggish pools of creeks and small rivers; usually near vegetation over mud or sand. **SIMILAR SPECIES:** (1) Bluenose Shiner, *P. welaka* (Pl. 18), is more slender; has more pointed snout, black stripe along side extending onto upper lip and snout, usually *8 dorsal* and anal rays, pharyngeal teeth 1,4-4,1. (2) Ironcolor Shiner, *Notropis chalybaeus* (Pl. 18), has more pointed snout, black stripe along side extending onto upper lip and snout, dorsal fin origin over pelvic fin origin, *8 dorsal* and anal rays, pharyngeal teeth 2,4-4,2; *lacks* blue color, enlarged fins on large male.

CYPRINELLA

GENUS CHARACTERISTICS: Scales on side appear *diamond-shaped,* taller than wide. *Dusky to black stripe* (faint on young and sometimes on adult) on chin; *white edge* on fins of large male. Strongly compressed body; dorsal fin origin slightly behind pelvic fin origin (or slightly in front or behind in Bannerfin Shiner, *C. leedsi*). Pharyngeal teeth 0,4-4,0 or 1,4-4,1. Often most abundant fishes in streams in e. N. America. Success of *Cyprinella* species may be related to habit of hiding eggs in rock crevices and the ability to produce sounds and communicate with one another. Although *Cyprinella* species can be difficult to separate as juveniles and females, large males develop specific colors and, in some species, enlarged fins, and are easily identified. *Cyprinella* species commonly hybridize with one another, sometimes forming hybrid swarms.

SPOTFIN SHINER *Cyprinella spiloptera* Pl. 16

IDENTIFICATION: *Black blotch* on rear half of dorsal fin; little or no black on membranes of front half of dorsal fin (Fig. 26) (except large males). Fairly deep body; pointed snout; terminal mouth. Dusky olive above, black stripe along back; sometimes a dusky bar on side behind head; diffuse dark stripe along rear half of silver side. Breeding male has blue back and side, yellow-white fins, dusky dorsal fin. Usually 26 scales around body, 14 around caudal peduncle; 8 anal rays. Has 34–41 (usually 35–39) lateral scales; pharyngeal teeth 1,4-4,1. To 4¾ in. (12 cm). **RANGE:** Atlantic Slope from St. Lawrence drainage, QC, to Potomac R. drainage, VA; Great Lakes (except Lake Superior), Hudson Bay (Red R.), and Mississippi R. basins from ON and NY to se. ND and south to AL and e. OK; isolated populations in Ozarks. Generally common. **HABITAT:** Sand and gravel runs and pools of creeks, and small to medium (sometimes large) rivers. **SIMILAR SPECIES:** See Pl. 16. (1) Steelcolor Shiner, *C. whipplei*, and (2) Satinfin Shiner, *C. analostana*, usually have 9 anal rays, black specks on all membranes of dorsal fin (Fig. 26). (3) Red Shiner, *C. lutrensis*, has red fins on male, deeper body, *no* black blotch on rear half of dorsal fin, usually 9 anal rays; pharyngeal teeth 0,4-4,0.

STEELCOLOR SHINER *Cyprinella whipplei* Pl. 16

IDENTIFICATION: Dorsal fin has black specks on *all* membranes, *black blotch* on rear half (Fig. 26). Fairly deep body; pointed snout; terminal

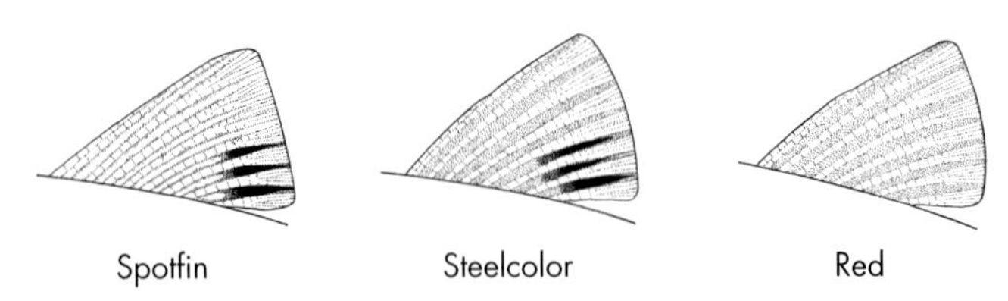

Fig. 26. Spotfin, Steelcolor, and Red shiners—black pigment on dorsal fin of female and nonbreeding male.

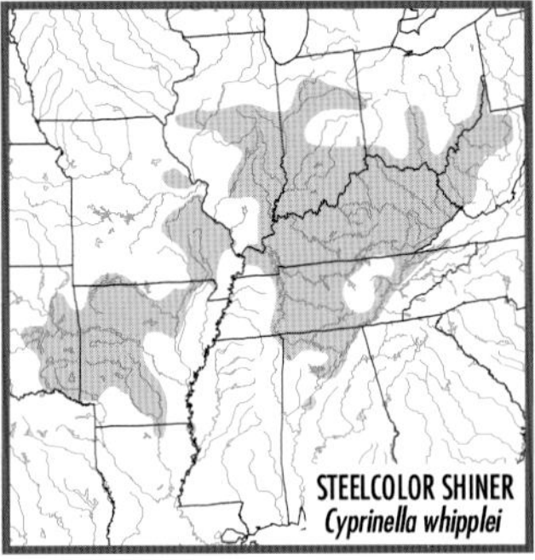

mouth. Dusky olive above, often has blue sheen; dark stripe along back; diffuse dark stripe along rear half of silver side. Breeding male has blue back and side, red snout, white-edged yellow fins, enlarged dusky dorsal fin. Usually 26 scales around body, 14 scales around caudal peduncle, 15 pectoral rays, 9 anal rays. Has 36–40 (usually 37–38) lateral scales; pharyngeal teeth 1,4-4,1. To 6¼ in. (16 cm). **RANGE:** Mississippi R. basin, from OH and WV to MO and e. OK, and south to n. AL and n. LA; Black Warrior R. system (Mobile Bay drainage), AL. Mostly absent on Coastal Plain. Common. **HABITAT:** Rocky and sandy runs, less often pools, of creeks and small to medium rivers. Usually near riffles. **SIMILAR SPECIES:** See Pl. 16. See (1) Satinfin Shiner, *C. analostana*, and (2) Greenfin Shiner, *C. chloristia*. (3) Spotfin Shiner, *C. spiloptera*, usually has 8 anal rays, *little or no black* on membranes of front half of dorsal fin (Fig. 26), lacks enlarged dorsal fin on large male. (4) Bluntface Shiner, *C. camura*, has blunt snout, white bar on caudal fin base, may have pale orange or red dorsal and caudal fins on large male. (5) Red Shiner, *C. lutrensis,* has red fins on male, *no* black blotch on rear half of dorsal fin, blunter snout; pharyngeal teeth 0,4-4,0.

ATINFIN SHINER *Cyprinella analostana* Not shown

IDENTIFICATION: Nearly identical to Steelcolor Shiner, *C. whipplei,* but usually has 13–14 pectoral rays, 33–38 lateral scales. To 4¼ in. (11 cm). **RANGE:** Atlantic Slope from Hudson R. drainage, NY, to Peedee R. drainage, SC; isolated records in Lake Ontario drainage, NY. Common. **HABITAT:** Rocky and sandy pools and runs of creeks and small to medium rivers; occasionally in headwaters and large rivers. **SIMILAR SPECIES:** See Pl. 16. See (1) Steelcolor Shiner, *C. whipplei*, and (2) Greenfin Shiner, *C. chloristia*. (3) Spotfin Shiner, *C. spiloptera*, has 8 anal rays, little or no black on membranes of front half of dorsal fin (Fig. 26).

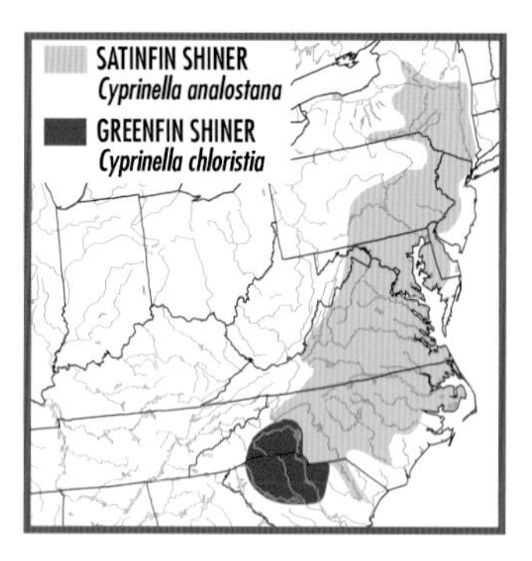

(4) Whitefin Shiner, *C. nivea*, is more slender; has subterminal mouth, darker stripe along side, usually 8 anal rays.

GREENFIN SHINER *Cyprinella chloristia* **Not shown**

IDENTIFICATION: Similar to Satinfin Shiner, *C. analostana*, and Steelcolor Shiner, *C. whipplei*, but has 8 anal rays, usually 24 scales around body, more distinct dark stripe on side. Has 32–36 (usually 34–36) lateral scales; pharyngeal teeth 1,4-4,1. To 3 in. (7.5 cm). **RANGE:** Santee R. drainage, NC and SC, and Peedee R. drainage, SC. Mostly above Fall Line. Common. **HABITAT:** Rocky and sandy pools and runs of creeks and small to medium rivers. **SIMILAR SPECIES:** See Pl. 16. See (1) Satinfin Shiner, *C. analostana*, and (2) Steelcolor Shiner, *C. whipplei*. (3) Whitefin Shiner, *C. nivea*, is more slender; has subterminal mouth, darker stripe along side, usually 37–38 lateral scales.

BLACKTAIL SHINER *Cyprinella venusta* **Pl. 16**

IDENTIFICATION: *Large black spot* on caudal fin base. Dorsal fin dusky, dark blotch on rear half. Fairly deep body; pointed snout, terminal mouth. Dusky olive above, narrow black stripe along back; diffuse dark stripe along rear half of silver side. Breeding male has blue back and side, yellow-white or red-orange (in s. TX) fins, dusky dorsal fin. Usually 28–29 scales around body, 14 around caudal peduncle; 8 anal rays (often 9 in upper Chattahoochee R. system, GA). Has 34–48 (usually 36–43) lateral scales; pharyngeal teeth 1,4-4,1. To 7½ in. (19 cm). **RANGE:** Gulf drainages from Suwannee R., GA and FL, to Rio Grande, TX; Mississippi R. basin (mostly on Former Mississippi Embayment) from s. IL and MO to LA and west in Red R. drainage to w. OK. Introduced into Sac R. (Missouri R. drainage), MO. Abundant over much of range. **HABITAT:** Usually in sandy pools and runs of small to medium rivers; also in creeks and rocky pools and runs. **REMARKS:** Three subspecies. *C. v. venusta*, in Mississippi R. basin and drainages to west, has 34–39 (usually 36–38) lateral scales and relatively deep body. *C. v. cercostigma,* in Gulf drainages (except Mobile Bay) east of Mississippi R., has 37–43 (usually 38–41) lateral scales and slightly more diffuse spot on caudal fin base. *C. v. stigmatura,* in upper Alabama and Tombigbee river systems, has 38–48 (usually 40–44) lateral scales and relatively slender body. Intergrades between *cercostigma* and *stigmatura* occur in Cahaba and Tallapoosa river systems, AL. **SIMILAR SPECIES:** See Pl. 16. (1) Tricolor Shiner, *C. trichroistia,* and (2) Tallapoosa Shiner, *C. gibbsi,* have black spot on caudal fin base fusing with black stripe along side, jet-black blotch on rear half of dorsal fin, orange fins on large male, usually 26 scales around body. (3) Alabama Shiner, *C. callistia*, is less compressed, has red dorsal and caudal fins, more subterminal mouth, usually 24 scales around body. (Blacktail Shiner has yellow fins within range of Tricolor, Tallapoosa, and Alabama shiners.)

LUNTFACE SHINER *Cyprinella camura* Pl. 16

IDENTIFICATION: Usually a *clear to white bar* on caudal fin base. *Black blotch* on rear half of dorsal fin. Deep body; blunt snout; terminal or slightly subterminal mouth. Dusky olive above, narrow dark stripe along back; sometimes a dark bar on side behind head, dusky stripe along rear half of silver blue side; dusky dorsal and caudal fins. Breeding male has pale orange or red fins and snout, enlarged dorsal fin. Usually 26 scales around body, 14 around caudal peduncle; 9 anal rays. Has 35–39 (usually 36–37) lateral scales; pharyngeal teeth 1,4-4,1. To 4¼ in. (11 cm). **RANGE:** Tributaries of Mississippi and Tennessee rivers on Former Mississippi Embayment from KY to LA; Arkansas R. drainage, sw. MO, e. KS, nw. AR, and e. OK. Generally common. **HABITAT:** Sandy and rocky pools and runs of clear to turbid creeks and small to medium rivers. **SIMILAR SPECIES:** See Pl. 16. (1) Steelcolor

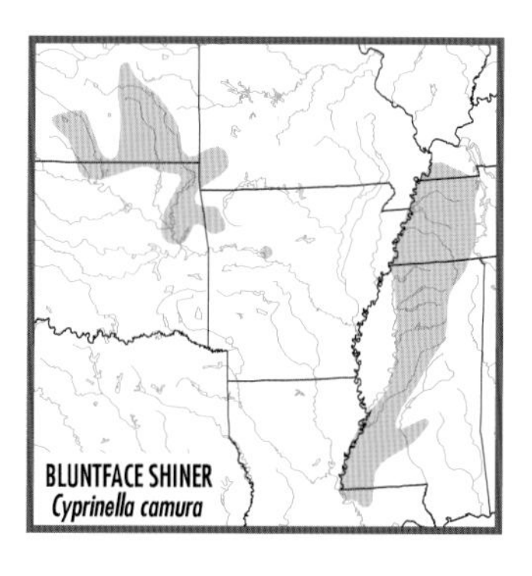

Shiner, *C. whipplei*, and (2) Spotfin Shiner, *C. spiloptera*, *lack* clear or white bar on caudal fin base, have more pointed snout; Spotfin Shiner usually has 8 anal rays. (3) Whitetail Shiner, *C. galactura*, is much more slender, has sharper snout, usually 39–41 lateral scales.

RED SHINER *Cyprinella lutrensis* Pl. 16

IDENTIFICATION: *Deep body;* terminal mouth; rounded snout. Dusky blue triangular bar behind head on side (faint on young). Dusky olive to blue back and upper side, black stripe along back; diffuse dark stripe along rear half of silver side. Dusky dorsal fin, no black blotch on rear half (Fig. 26). Breeding male has *red fins* (except dorsal), blue back and side, dark blue bar before pink bar behind head on side. Usually 26 scales around body, 14 around caudal peduncle, 9 anal rays, 14 pectoral rays. Has 32–36 lateral scales; pharyngeal teeth 0,4-4,0. To 3½ in. (9 cm). **RANGE:** Mississippi R. basin from sw. WI and e. IN to WY and south to LA; Gulf drainages west of Mississippi R. to Rio Grande, TX, NM, and CO. Absent in Ozark and Ouachita uplands. Also in n. Mexico. Common bait fish widely introduced elsewhere in U.S. Abundant. **HABITAT:** Silty, sandy, and rocky pools and runs, sometimes riffles, of creeks and small to medium rivers. Tolerant of siltation and high turbidity. **SIMILAR SPECIES:** (1) See Beautiful Shiner, *C. formosa*, and (2) Plateau Shiner, *C. lepida*. (3) Proserpine Shiner, *C. proserpina* (Pl. 16), has black stripe along side and on chin and throat; subterminal mouth; *yellow to orange fins.*

BEAUTIFUL SHINER *Cyprinella formosa* Not shown

IDENTIFICATION: Nearly identical to Red Shiner, *C. lutrensis*, but has *orange or yellow back* and *silver side, red-orange caudal peduncle;* orange caudal and lower fins on large male. Small, crowded scales on nape. Usually 26 scales around body, 14 around caudal peduncle; 8 anal rays. Has 34–47 lateral scales. To 3½ in. (9 cm). **RANGE:** San

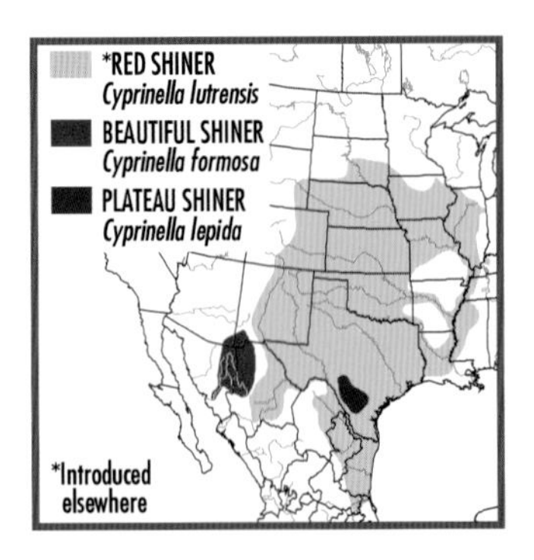

Bernardina Creek, sw. NM and se. AZ; also in Mexico. Recognized as a *threatened species* and possibly extirpated from U.S. **HABITAT:** Sandy and rocky pools of creeks. **SIMILAR SPECIES:** See (1) Red Shiner, *C. lutrensis* (Pl. 16), and (2) Plateau Shiner, *C. lepida*.

PLATEAU SHINER *Cyprinella lepida* Not shown

IDENTIFICATION: Similar to Red Shiner, *C. lutrensis*, and Beautiful Shiner, *C. formosa*, but breeding male has green back, *yellow-purple side, purple bar* on side of *gold-orange head,* yellow-orange fins. To 3 in. (7.5 cm). **RANGE:** Upper Nueces R. drainage, Edwards Plateau, TX. Common. **HABITAT:** Springs and spring-fed creeks; usually in clear water over gravel. **SIMILAR SPECIES:** See (1) Red Shiner, *C. lutrensis* (Pl. 16), and (2) Beautiful Shiner, *C. formosa*.

PROSERPINE SHINER *Cyprinella proserpina* Pl. 16

IDENTIFICATION: *Black stripe* on chin and throat; *black stripe* along side; yellow to orange fins (on male). Fairly deep body; *subterminal mouth.* Dusky olive to blue back and side, dark stripe along back. Breeding male has white-edged orange fins, brassy yellow head, blue-black bar on side behind head. Usually 26 scales around body, 14 around caudal peduncle; 8 anal rays, 13 pectoral rays. Has 34–36 lateral scales; pharyngeal teeth 0,4-4,0. To 3 in. (7.5 cm). **RANGE:** Devils R., lower Pecos R., and nearby tributaries of Rio Grande, TX; also Río San Carlo (Rio Grande drainage), Coahuila, Mexico. Fairly common. **HABITAT:** Rocky runs and pools of creeks and small rivers. **SIMILAR SPECIES:** (1) Red Shiner, *C. lutrensis* (Pl. 16), (2) Plateau Shiner, *C. lepida*, and (3) Beautiful Shiner, *C. formosa*, *lack* black stripe along side and black stripe on chin and throat (may have dusky stripe on chin); have *terminal mouth,* red or orange fins.

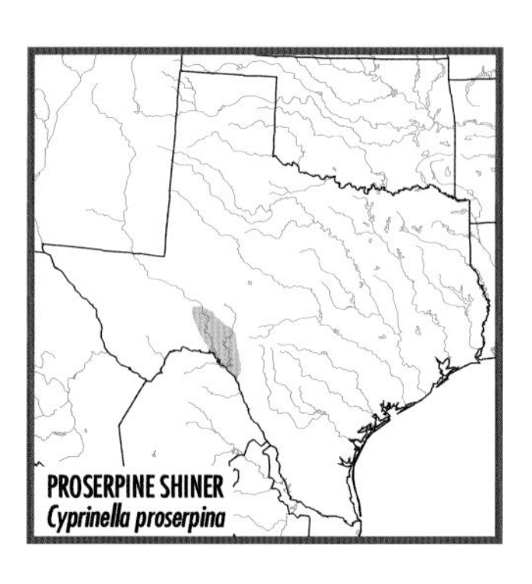

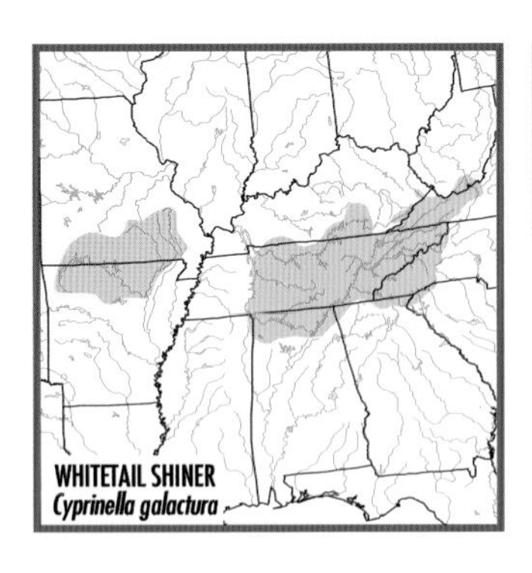

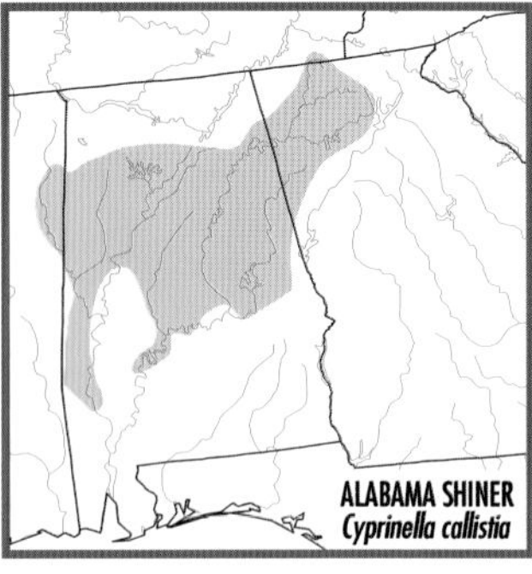

WHITETAIL SHINER *Cyprinella galactura* Pl. 16

IDENTIFICATION: *Two large clear to white areas* on caudal fin base. *Black blotch* on rear half of dorsal fin. *Slender body*; moderate snout; terminal or slightly subterminal mouth. Dusky olive above, dark streak along back; silver side; dusky dorsal and caudal fins. Breeding male has blue back and side, white fins, enlarged dorsal fin; dorsal and anal fins may have red tint. Usually 26 scales around body, 14 around caudal peduncle; 9 anal rays. Has 38–43 (usually 39–41) lateral scales; pharyngeal teeth 1,4-4,1. To 6 in. (15 cm). **RANGE:** Disjunct range east and west of Former Mississippi Embayment. Cumberland and Tennessee river drainages, VA, NC, KY, TN, GA, AL, and MS; upper Savannah and Santee drainages (Atlantic Slope), NC, SC, and GA; and upper New R. drainage, WV and VA. St. Francis and White river drainages, MO and AR. Common; locally abundant. **HABITAT:** Rocky runs, less often pools and riffles, of clear headwaters, creeks, and small rivers. **SIMILAR SPECIES:** See Pl. 16. (1) Bluntface Shiner, *C. camura*, has *deeper body*, blunt snout, usually 36–37 lateral scales. (2) Spotfin Shiner, *C. spiloptera*, *lacks* clear or white areas on caudal fin base and dusky pigment on front half of dorsal fin (except on large male); has deeper body, usually 35–39 lateral scales, 8 anal rays. (3) Steelcolor Shiner, *C. whipplei*, *lacks* clear or white areas on caudal fin base, has deeper body, usually 37–38 lateral scales.

ALABAMA SHINER *Cyprinella callistia* Pl. 16

IDENTIFICATION: *Large black spot on* caudal fin base (fusing on small individual into black stripe on rear half of side). *Pink to red dorsal and caudal fins.* Blunt snout, *subterminal mouth,* large eye. Olive above, dark stripe along back; silver side. Breeding male has black edge on rear half of red dorsal fin, red caudal fin; other fins white. Usually 24 scales around body, 16 around caudal peduncle; 8 anal rays. Has 37–41 lateral scales; pharyngeal teeth 1,4-4,1. To 3¾ in. (9.5 cm). **RANGE:**

Mobile Bay drainage, se. TN, nw. GA, AL, and ne. MS; mostly above Fall Line. Generally common. **HABITAT:** Gravel- and rubble-bottomed pools and runs of creeks and small to medium rivers. **SIMILAR SPECIES:** See Pl. 16. (1) Blacktail Shiner, *C. venusta*, (2) Tricolor Shiner, *C. trichroistia*, and (3) Tallapoosa Shiner, *C. gibbsi*, have more pointed snout, more *terminal mouth*, dark blotch on rear half of dorsal fin, usually 26–29 scales around body, 14 scales around caudal peduncle, yellow or orange fins on large male.

TRICOLOR SHINER *Cyprinella trichroistia* Pl. 16

IDENTIFICATION: *Large black spot* on caudal fin base *fusing* into *black stripe* on rear half of side; *yellow to red-orange fins; jet-black blotch on* rear half of dorsal fin. Fairly deep body; moderate snout; terminal mouth. Dusky olive above, dark stripe along back; silver side. Breeding male has orange and white fins, blue side; tubercles in 1 row on lower jaw, large and in 2 rows on top of head (Fig. 27). Usually 26 scales around body, 14 around caudal peduncle; 9 anal rays. Has 36–44 lateral scales; pharyngeal teeth 1,4-4,1. To 4 in. (10 cm.). **RANGE:** Alabama R. drainage (mostly Coosa and Cahaba river systems), se. TN, nw. GA, and AL; localized in Black Warrior R. system, AL. Common. **HABITAT:** Rocky and sandy runs and pools of creeks and small to medium rivers. **SIMILAR SPECIES:** (1) See Tallapoosa Shiner, *C. gibbsi*. (2) Blue Shiner, *C. caerulea* (Pl. 16), is more slender, has more subterminal mouth, less prominent spot on caudal fin base, darker stripe along side, usually 22 scales around body, 8 anal rays.

TALLAPOOSA SHINER *Cyprinella gibbsi* Not shown

IDENTIFICATION: Similar to Tricolor Shiner, *C. trichroistia*, but has longer, more overhanging snout, subterminal mouth; tubercles in *2 rows* on lower jaw, smaller and more scattered on top of head of breeding male (Fig. 27). Has 37–42 lateral scales. To 3¾ in. (9.5 cm).

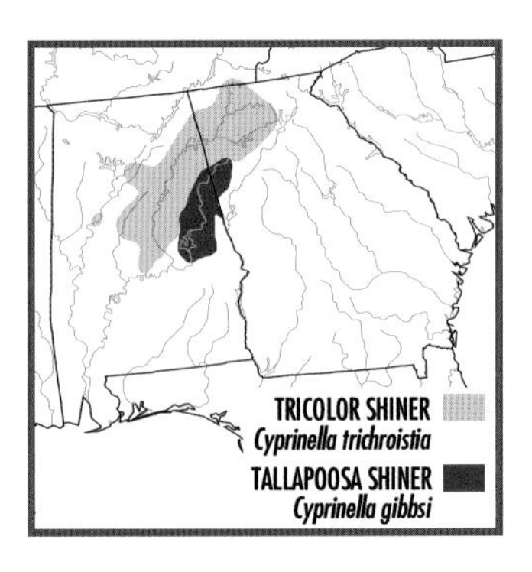

RANGE: Tallapoosa R. system (Alabama R. drainage), AL and GA; 1 record in Chattahoochee R., AL, where probably introduced. Most common minnow in Tallapoosa R. tributaries. **HABITAT:** Sandy and rocky runs of small to medium rivers; less often in flowing pools. **SIMILAR SPECIES:** (1) See Tricolor Shiner, *C. trichroistia* (Pl. 16).

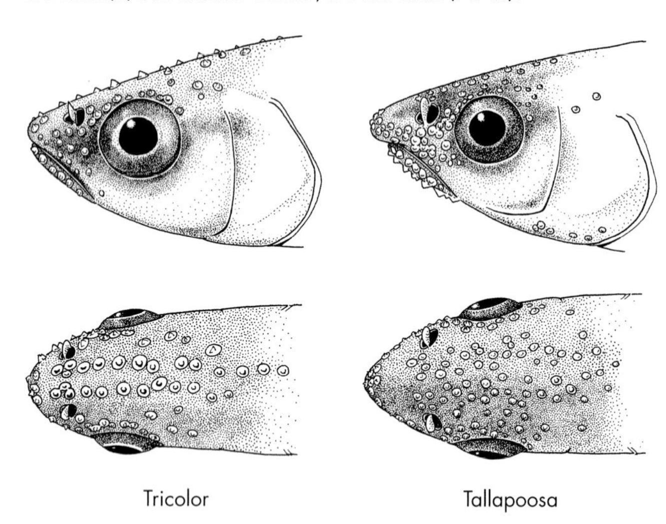

Fig. 27. Tricolor and Tallapoosa shiners—head tubercles on breeding male.

BLUE SHINER *Cyprinella caerulea* **Pl. 16**

IDENTIFICATION: *Blue-black stripe* along side from head to caudal fin, darkest and widest at rear; expanded slightly into spot on caudal fin base. Often a *black blotch* on rear half of dorsal fin. Fairly *slender body*; *pointed snout;* slightly subterminal mouth. Light brown above, dark stripe along back; silver side. Breeding male has yellow side, yellow-white fins. Usually *22 scales* around body, 14 around caudal peduncle; 8 anal rays. Has 37–39 lateral scales; pharyngeal teeth 1,4-4,1. To 3½ in. (9 cm). **RANGE:** Coosa and Cahaba river systems, se. TN, nw. GA, and AL. Local and uncommon. Protected as a *threatened species*; extirpated from Cahaba R. system. **HABITAT:** Rocky runs of small to medium rivers. **SIMILAR SPECIES:** (l) Tricolor Shiner, *C. trichroistia* (Pl. 16), (2) Tallapoosa Shiner, *C. gibbsi*, and (3) Alabama Shiner, *C. callistia* (Pl. 16), have deeper body, less diamond-shaped scales, usually 24–26 scales around body.

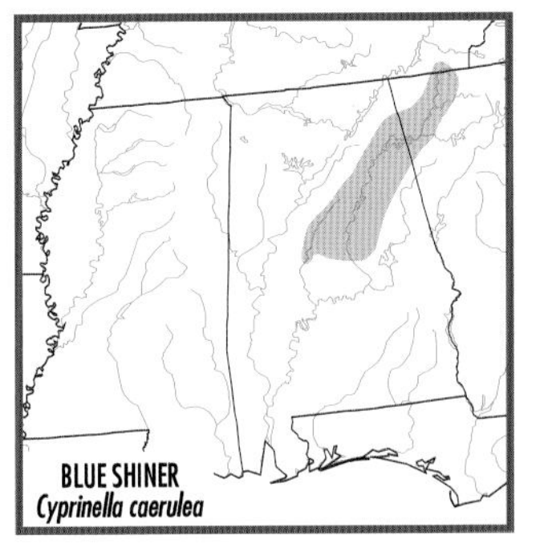

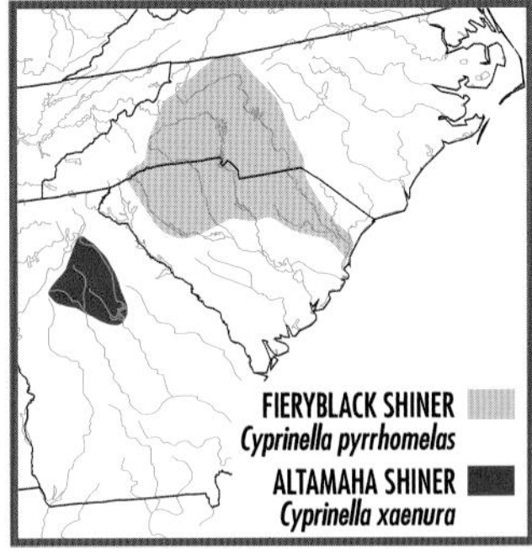

FIERYBLACK SHINER *Cyprinella pyrrhomelas* Pl. 16

IDENTIFICATION: *Black edge* on caudal fin of adult; *black bar* on side behind head. *Large eye.* Dark olive above, narrow black stripe along back; dusky stripe along rear half of side, often expanded into black spot on caudal fin base; silver side. Breeding male has blue back and side, *bright red snout, bright red band* after white band on caudal fin, dorsal fin red at front; other fins white. *Black blotch* on rear half of dorsal fin. Deep body; terminal mouth. Usually 26 scales around body, 14 around caudal peduncle; 10–11 anal rays. Has 34–39 lateral scales; pharyngeal teeth 1,4-4,1. To 4¼ in. (11 cm). **RANGE:** Peedee and Santee river drainages, NC and SC; introduced to Chattooga R. (Savannah R. drainage), GA. Common, especially in montane streams. **HABITAT:** Rocky runs and pools near riffles of creeks and small to medium rivers. **SIMILAR SPECIES:** (1) Whitefin Shiner, *C. nivea* (Pl. 16), (2) Satinfin Shiner, *C. analostana*, (3) Greenfin Shiner, *C. chloristia*, and (4) Altamaha Shiner, *C. xaenura* (Pl. 16), *lack* black edge on caudal fin; have smaller eye; all but Altamaha Shiner have 8–9 anal rays.

ALTAMAHA SHINER *Cyprinella xaenura* Pl. 16

IDENTIFICATION: Pointed snout; terminal or slightly subterminal mouth. Dusky olive above, dark stripe along back; silver black stripe along rear half of side, often expanded into *spot* on caudal fin base. Breeding male has blue side, white fins; *yellow to orange dorsal, caudal, and anal fins.* Usually 26 scales around body, 16 around caudal peduncle; 10–11 anal rays. Has 38–40 lateral scales; pharyngeal teeth 1,4-4,1. To 4½ in. (11 cm). **RANGE:** Upper Altamaha R. drainage, n.-cen. GA. Fairly common. **HABITAT:** Rocky and sandy pools of creeks and small rivers. **SIMILAR SPECIES:** See Pl. 16. (1) Most similar to Fieryblack Shiner, *C. pyrrhomelas,* which has black edge on caudal fin of adult; red on snout, dorsal and caudal fins of male; 14 scales around caudal peduncle; larger eye. (2) Ocmulgee Shiner, *C. callisema,* and

(3) Bannerfin Shiner, *C. leedsi*, both in Altamaha R. drainage, have distinctly subterminal mouth, 8 anal rays, 14 scales around caudal peduncle; pharyngeal teeth 0,4-4,0.

OCMULGEE SHINER *Cyprinella callisema* Pl. 16

IDENTIFICATION: *Deep blue stripe* along side (sometimes faint near front), often ending in darker spot on caudal fin base. *Small black blotch* at front of dorsal fin—near tip of 1st ray, middle of 2d ray. Fairly deep body; long, round snout; *subterminal mouth.* Dusky olive above; dark stripe along back to dorsal fin, dark streak behind fin; silver side. Breeding male has white lower fins, orange dorsal and caudal fins; enlarged, dusky dorsal fin. Usually 26 scales around body, 14 around caudal peduncle; 8 anal rays. Has 37–40 (usually 38–39) lateral scales; pharyngeal teeth, 0,4-4,0. To 3½ in. (9 cm). **RANGE:** Altamaha and Ogeechee river drainages, GA. Locally common in Altamaha, uncommon in Ogeechee R. drainage. **HABITAT:** Sandy and rocky runs of small to medium rivers. **SIMILAR SPECIES:** See Pl. 16. (1) See Bluestripe Shiner, *C. callitaenia*. (2) Bannerfin Shiner, *C. leedsi,* has more diffuse stripe along side, more flattened underside. (3) Altamaha Shiner, *C. xaenura,* is deeper bodied; has 10–11 anal rays, 16 scales around caudal peduncle; pharyngeal teeth 1,4-4,1.

BLUESTRIPE SHINER *Cyprinella callitaenia* Not shown

IDENTIFICATION: Nearly identical to Ocmulgee Shiner, *C. callisema,* but has *crescent-shaped line* of black specks from eye to mouth, *darker spot* on caudal fin base. Has 37–40 lateral scales; pharyngeal teeth 1,4-4,1. To 3½ in. (9 cm). **RANGE:** Apalachicola R. drainage, GA, AL, and FL. Localized and uncommon. **HABITAT:** Sandy and rocky runs of small to medium rivers. **SIMILAR SPECIES:** (1) See Ocmulgee Shiner, *C. callisema* (Pl. 16).

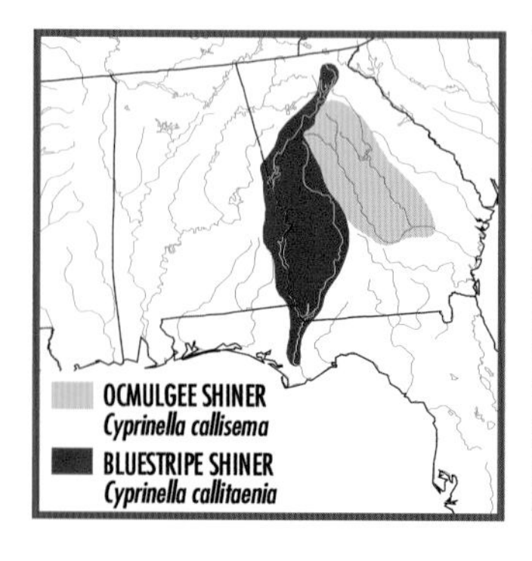

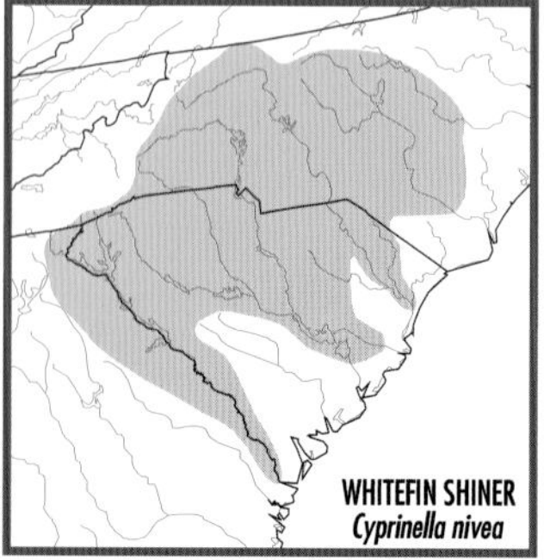

WHITEFIN SHINER *Cyprinella nivea* Pl. 16

IDENTIFICATION: *Dark blue to black stripe* along side, darkest on rear half. *Black blotch* on rear half of dorsal fin. Slender body; rounded snout; *subterminal mouth.* Dusky olive above, dark stripe along back; silver side. Large male has white fins. Usually 26 scales around body, 14 around caudal peduncle; 8 anal rays. Has 35–41 (usually 37–38, see Remarks) lateral scales; pharyngeal teeth 1,4-4,1. To 3½ in. (8.5 cm). **RANGE:** Atlantic Slope from Neuse R. drainage, NC, to Savannah R. drainage, GA. Common. **HABITAT:** Sand and gravel runs and riffles, usually in small to medium rivers, less often in creeks. **REMARKS:** Savannah R. population has smaller (usually 39–40 lateral) scales than other populations of Whitefin Shiner. **SIMILAR SPECIES:** (1) Bluestripe Shiner, *C. callitaenia*, and (2) Ocmulgee Shiner, *C. callisema* (Pl. 16), have black blotch on *front,* not rear, half of dorsal fin. (3) Bannerfin Shiner, *C. leedsi* (Pl. 16), has small black blotch at *front,* not rear, of dorsal fin; flattened underside; pharyngeal teeth 0,4-4,0.

BANNERFIN SHINER *Cyprinella leedsi* Pl. 16

IDENTIFICATION: *Small black blotch* at front of dorsal fin—near tip of 1st ray, middle of 2d ray; outer half of fin dusky. *Subterminal mouth;* long, rounded snout. Blue to black stripe along silver side, darkest on rear half, usually faint at front. Slender body; deepest in adult near dorsal fin origin, strongly tapering to caudal peduncle and snout; *flattened underside.* Dorsal fin origin slightly in front of to slightly behind pelvic fin origin. Dusky olive above, black stripe along back (wider in front of dorsal fin); white below. Breeding male has white lower fins, orange dorsal and caudal fins, black outer half of greatly enlarged dorsal fin. Usually 26 scales around body, 14 around caudal peduncle; 8 anal rays. Has 35–39 lateral scales; pharyngeal teeth 0,4-4,0. To 4 in. (10 cm). **RANGE:** Atlantic Slope from Edisto R. drainage, SC, to Altamaha R. drainage, GA; Gulf Slope in Suwannee and Ochlock-

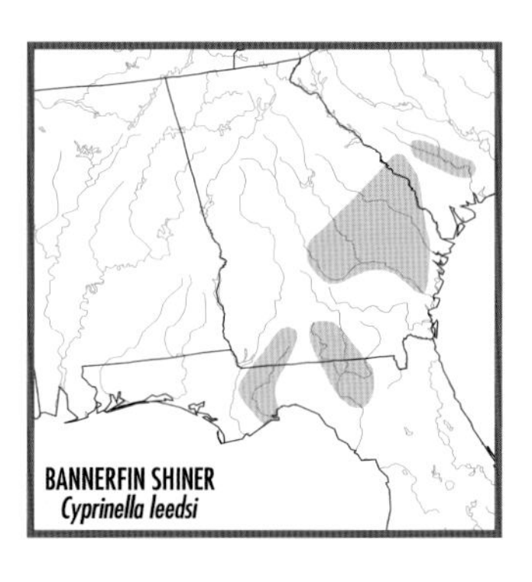

onee drainages, GA and FL. Restricted to Coastal Plain; uncommon. **HABITAT:** Sandy runs of medium to large rivers. **SIMILAR SPECIES:** (1) Ocmulgee Shiner, *C. callisema* (Pl. 16), and (2) Bluestripe Shiner, *C. callitaenia*, have dark blue-black stripe along entire side; *lack* flattened underside. (3) Whitefin Shiner, *C. nivea* (Pl. 16), has black blotch on *rear* half (not front) of dorsal fin; pharyngeal teeth 1,4-4,1; *lacks* flattened underside.

THICKLIP CHUB *Cyprinella labrosa* Pl. 7

IDENTIFICATION: *Small dark brown blotches* and crosshatching on back and side. Compressed body, *flattened below,* deepest under nape; dorsal fin origin over or slightly behind pelvic fin origin. *Long snout* overhanging mouth; large barbels; eyes directed upwardly. Large, horizontal pectoral fins. Straw yellow above; dark stripe (darkest at rear) along silver side, punctate lateral line, black spot on caudal fin base, black stripe from eye to snout. Breeding male is dark gray, has black membranes in yellow fins. Complete lateral line; 37–40 lateral scales; 8 anal rays; pharyngeal teeth l,4-4,l. To 2¾ in. (6.7 cm). **RANGE:** Upper Peedee and Santee river drainages, VA, NC, and SC. Locally common, especially in montane and upper Piedmont streams. **HABITAT:** Rocky riffles and runs of creeks and small rivers. **SIMILAR SPECIES:** (1) Santee Chub, *C. zanema* (Pl. 7), *lacks* dark blotches on back and side, is more slender.

SANTEE CHUB *Cyprinella zanema* Pl. 7

IDENTIFICATION: *Dark crosshatching on back and side.* Slender, compressed body, *flattened below,* deepest under nape; dorsal fin origin behind pelvic fin origin. Long *snout* overhanging mouth; large barbels; eyes directed upwardly. Large, horizontal pectoral fins. Yellow above and on side; dark stripe (darkest at rear) along silver side, black spot on caudal fin base, black stripe from eye to snout. Breeding male is silver with dark yellow fins, black streaks on dorsal and caudal fins.

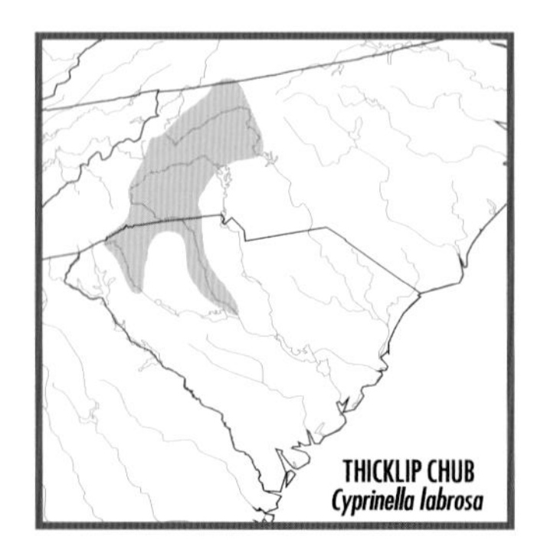

THICKLIP CHUB
Cyprinella labrosa

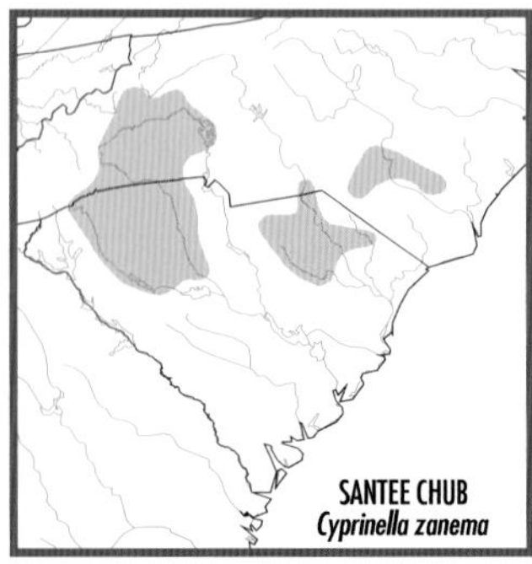

SANTEE CHUB
Cyprinella zanema

Complete lateral line; 38–42 lateral scales; 8 anal rays; pharyngeal teeth 1,4-4,1. To 3 in. (7.5 cm). **RANGE:** Cape Fear, Peedee, and upper Santee river drainages, NC and SC. Locally common, especially in upper Piedmont streams of Santee drainage. **HABITAT:** Sandy and rocky runs and current-swept pools of creeks and small rivers. **SIMILAR SPECIES:** (1) Thicklip Chub, *C. labrosa* (Pl. 7), has dark blotches on back and side, is deeper bodied.

SPOTFIN CHUB *Cyprinella monacha* Pl. 17

IDENTIFICATION: Slender body, *flattened below,* arched above, deepest at dorsal fin origin. *Long snout overhangs mouth; small barbel* at corner of mouth; small eyes slightly directed upwardly. *Black blotch* on rear half of dorsal fin (darkest on large adult). Dorsal fin origin behind pelvic fin origin. Olive to gray above, iridescent green stripe along back, another along upper side; dark stripe along silver side, darkest on rear half, expanded into large black caudal spot. Breeding male has *2 large white bars on blue side* (dark above, light below), white edges on blue fins. Complete lateral line; 52–62 lateral scales; 8 anal rays; 0,4-4,0 pharyngeal teeth. To 4¼ in. (11 cm). **RANGE:** Historically in several tributaries of Tennessee R., VA, NC, TN, GA, and AL; now only in Little Tennessee R., NC, Buffalo and Emory rivers, TN, and Holston R., VA and TN. Rare; protected as a *threatened species.* **HABITAT:** Rocky riffles and runs of clean, small to medium rivers. **REMARKS:** Relationships uncertain; sometimes put in monotypic genus *Erimonax.* **SIMILAR SPECIES:** (1) Slender Chub, *Erimystax cahni* (Pl. 7), has dark <s along side, *no* dark blotch on dorsal fin, dorsal fin origin in front of pelvic fin origin, 7 anal rays.

PUGNOSE MINNOW *Opsopoeodus emiliae* Pl. 8

IDENTIFICATION: *Crosshatched pattern* on back and upper half of side; *small, strongly upturned* mouth; *small crowded scales* on front half of nape; 9 dorsal rays; 2 dark areas (front and rear—clear area between)

on dorsal fin, most prominent on large male (dark areas absent in parts of FL). Fairly slender body; dorsal fin origin over to slightly behind pelvic fin origin. Dusky olive-yellow above; dark stripe along silver side of head and body, sometimes ending in small black spot on caudal fin base; large individuals in FL have pink, red, or orange fins. Breeding male has bright white lower half of anal and pelvic fins. Usually complete lateral line, 36–40 lateral scales, 8 anal rays; pharyngeal teeth usually 0,5-5,0 (see Remarks). To 2½ in. (6.4 cm). **RANGE:** Edisto R. drainage, SC, to s. FL, and across Gulf Slope to Nueces R. drainage, TX; north in Mississippi R. and Great Lakes basins to se. KS, se. MN, and s. ON. Generally restricted to lowlands. Common but declining in parts of range. **HABITAT:** Clear to turbid vegetated lakes, swamps, oxbows, and sluggish streams of all sizes. **REMARKS:** Two subspecies. *O. e. peninsularis,* in peninsular FL, commonly has 4 pharyngeal teeth on right side (0,4) and 5 on left (0,5), lacks dark spots on dorsal fin; breeding male has tubercles in a cluster to either side of upper lip. *O. e. emiliae,* in rest of range except where intergrades occur (s. GA and ne. FL west to Ochlockonee R.), has pharyngeal tooth count of 0,5-5,0; 2 dark spots on dorsal fin of adult; and breeding tubercles across snout of breeding male. **SIMILAR SPECIES:** (1) Taillight Shiner, *Notropis maculatus* (Pl. 18), also with crosshatched pattern and often with Pugnose Minnow, has large black spot on caudal fin base, red on head and body (especially large male), more *horizontal* and *subterminal* mouth, smaller eye, 8 dorsal rays. (2) Pugnose Shiner, *Notropis anogenus* (Pl. 10), has upturned mouth but *lacks* crosshatched pattern, has black peritoneum, 8 dorsal rays, pharyngeal teeth 0,4-4,0.

PIMEPHALES

GENUS CHARACTERISTICS: Species of *Pimephales* are among most common fishes in e. N. America. All have *much smaller scales on nape* (usually more than 20 in row from head to dorsal fin) than elsewhere on body; 2d ray of dorsal fin is short and stout, distinctly separated from 3d ray by a membrane.

FATHEAD MINNOW *Pimephales promelas* Pl. 8

IDENTIFICATION: Deep, compressed body; *head short, flat on top. Herringbone lines on upper side.* Blunt snout; terminal, oblique mouth; round eye on side of head. Dorsal fin origin over pelvic fin origin. Dark olive above and on side, dusky stripe along back, dusky stripe along side; dull yellow to white below with black peritoneum visible; fins clear except often a dusky to black blotch at front of dorsal fin (about midway). Breeding male has dark black head, 2 broad white to gold bars on side (1 behind head, 1 under dorsal fin); black fins; large gray fleshy pad on nape, about 16 large tubercles in 3 rows on snout. Intestine long, with several loops. Usually incomplete lateral line; 40–54 lateral scales; 7 anal rays; pharyngeal teeth 0,4-4,0. To 4 in. (10 cm). **RANGE:** Over much of N. America from QC to NT and BC, and south to AL, TX, and Mexico. Widely introduced, including in Colorado R. drainage, AZ and NM. Absent on Atlantic and Gulf slopes between Potomac R., VA, and Trinity R., TX, except where introduced. Common over much of range; uncommon in uplands. **HABITAT:** Muddy pools of headwaters, creeks, and small rivers; ponds. Tolerant of conditions (e.g., turbid, hot, poorly oxygenated, intermittent streams) unsuitable for most fishes. **REMARKS:** Frequent use of this species as fishing bait has resulted in many introduced populations. A strain called "rosy-reds" (red-orange body and fins) has been bred for the pet trade and bait. **SIMILAR SPECIES:** (1) Other *Pimephales* species (Pl. 8) are more slender, *lack* herringbone lines, have complete lateral line.

BLUNTNOSE MINNOW *Pimephales notatus* Pl. 8

IDENTIFICATION: *Blunt snout* overhanging small, subterminal, horizontal mouth. Slender body nearly *square* in cross section, with top

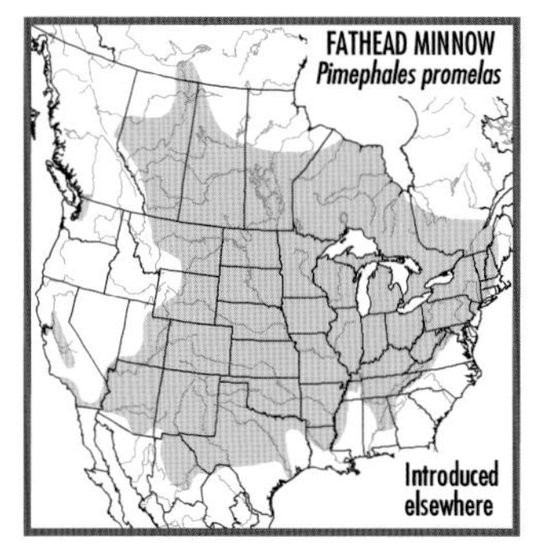

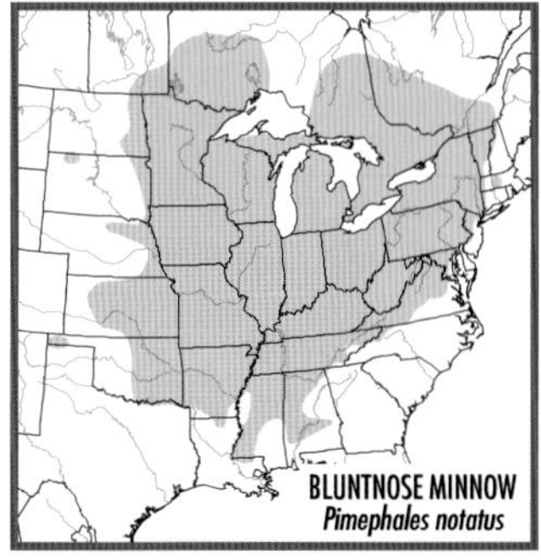

of head and nape *flattened;* round eye on side of head. Dorsal fin origin slightly behind pelvic fin origin. Light olive to tan above, scales darkly outlined (often with crosshatched appearance), black streak along back; dusky to *black stripe* around snout and along silver side to conspicuous *black spot* on caudal fin base; fins clear except for dusky to black blotch at front of dorsal fin (about midway). Breeding male black with silver bar behind opercle, about 16 large tubercles in 3 rows on snout, large gray fleshy pad on nape. Intestine long with several loops; black peritoneum. Complete lateral line; 39–50 lateral scales; 7 anal rays; pharyngeal teeth 0,4-4,0. To 4¼ in. (11 cm). **RANGE:** Great Lakes, Hudson Bay, and Mississippi R. basins from ON and NY to s. MB, and south to LA; Atlantic Slope from St. Lawrence R., QC, to Roanoke R., VA (absent from most of New England); Gulf Slope from Mobile Bay drainage, AL, to Mississippi R. Abundant; probably most common freshwater fish in e. N. America. **HABITAT:** Can be almost anywhere in its range, but usually in clear rocky streams. **SIMILAR SPECIES:** (1) Bullhead Minnow, *P. vigilax* (Pl. 8), and (2) Slim Minnow, *P. tenellus*, have eyes higher on head, directed more upwardly; more terminal mouth; larger head; bluish sheen on side; silver white peritoneum.

BULLHEAD MINNOW *Pimephales vigilax* Pl. 8

IDENTIFICATION: *Large eye,* directed somewhat *upwardly,* on upper half of head. *Large black spot at front of dorsal fin.* Slender body nearly *square* in cross section, with top of head and nape flattened; rounded snout; small, terminal, nearly horizontal mouth; upper lip slightly thickened at middle. Dorsal fin origin over pelvic fin origin. Light to dark olive above, scales darkly outlined (often appearing crosshatched); often a dusky to black stripe along silver blue side, ending just before large black spot at caudal fin base; dusky stripe along underside of caudal peduncle. Breeding male dark with black head, silver bar behind opercle, 5–9 large tubercles in 1–2 rows on snout, large gray fleshy pad on nape. Intestine short. Peritoneum silvery with black specks. Complete lateral line; 37–45 lateral scales; usually 7–8 scales above lateral line, 7 anal rays; pharyngeal teeth 0,4-4,0. To 3½ in. (8.9 cm). **RANGE:** Mississippi R. basin from OH to MN, and south to Gulf; Gulf Slope drainages from Mobile Bay, GA and AL, to Rio Grande, TX, NM, and Mexico. Introduced in Rio Grande of NM and w. TX. Common over much of range; locally abundant. **HABITAT:** Quiet pools and runs over sand, silt, or gravel, in small to large rivers. Most common in medium-sized rivers. **SIMILAR SPECIES:** (1) See Slim Minnow, *P. tenellus*. (2) Bluntnose Minnow, *P. notatus* (Pl. 8), has eyes *lower* on head, directed more to side; black peritoneum; no bluish sheen on side.

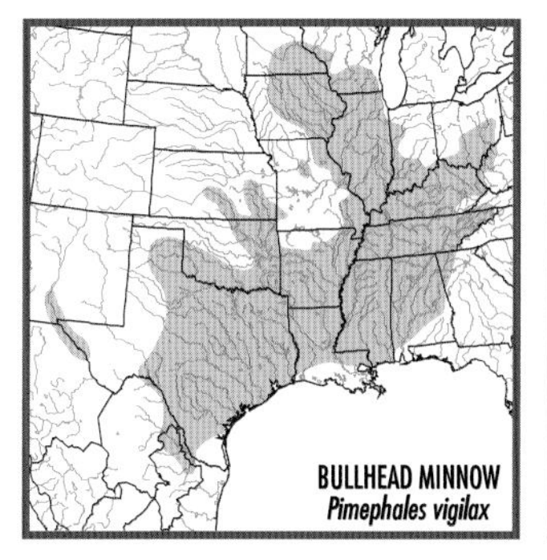

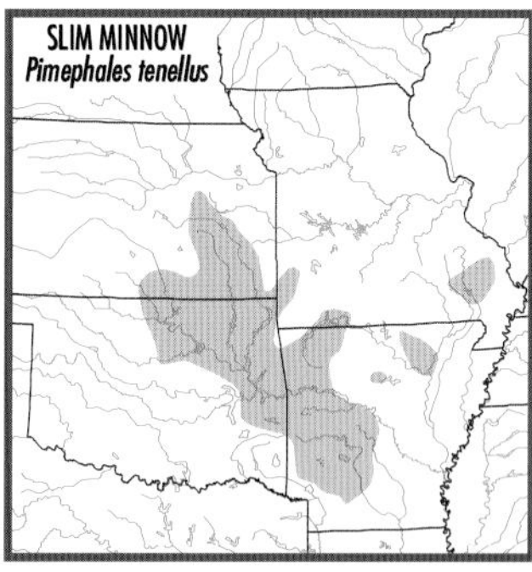

SLIM MINNOW *Pimephales tenellus* **Not shown**

IDENTIFICATION: Similar to Bullhead Minnow, *P. vigilax*, but has upper lip decidedly *thickened* at middle, narrower and darker stripe along underside of caudal peduncle, 6 (rarely 5 or 7) scales above lateral line; breeding male with usually 11–13 tubercles in 3 rows on snout. To 2¾ in. (7 cm). **RANGE:** S. MO, e. KS, AR, and ne. OK. Locally common in northwestern part of range. **HABITAT:** Sand- and gravel-bottomed pools and runs of creeks and small rivers. **REMARKS:** Two subspecies. *P. t. parviceps,* in Castor, St. Francis, Black, White, lower Arkansas, Ouachita, and Little (Red R. drainage) river systems, is more slender and has snout projecting slightly beyond upper lip. *P. t. tenellus,* in Arkansas R. drainage of OK and KS, has a terminal upper lip. **SIMILAR SPECIES:** (1) See Bullhead Minnow, *P. vigilax* (Pl. 8).

NOTROPIS

GENUS CHARACTERISTICS: Eight dorsal rays; no barbel (except in Redeye Chub, *N. harperi*); short intestine with 1 loop at front (except Ozark Minnow, *N. nubilus*, and Cape Fear Shiner, *N. mekistocholas*); scales on front half of side not much taller than wide (except Mimic Shiner, *N. volucellus*, Cahaba Shiner, *N. cahabae*, and Ghost Shiner, *N. buchanani*); scales on nape about same size as those on upper side (except Mirror Shiner, *N. spectrunculus*, and undescribed Sawfin Shiner, *N.* species); scales usually not appearing diamond-shaped; pharyngeal teeth 0,4-4,0 to 2,4-4,2. *Notropis* is 2d largest genus (ca. 75 species) of freshwater fishes in N. America and includes most of our small minnows.

EMERALD SHINER *Notropis atherinoides* **Pl. 9**

IDENTIFICATION: *Slender, compressed body;* large, terminal oblique mouth (reaching to front of eye) on fairly pointed snout; dorsal fin origin *behind* pelvic fin origin. Black (front half) lips. Light olive above, narrow dusky stripe along back; partly dusky, silver stripe with emerald sheen along side. Rakers on 1st gill arch often (ca. half of individuals) T- or Y-shaped. Complete lateral line; 35–40 lateral scales; 10–12 anal rays; 8 pelvic rays; pharyngeal teeth 2,4-4,2. To 5 in. (13 cm). **RANGE:** St. Lawrence R. drainage, QC, and Hudson R. drainage, NY, to Mackenzie R. drainage (Arctic basin), NT and BC, and south through Great Lakes and Mississippi R. basins to Gulf; Gulf Slope drainages from Mobile Bay, AL, to lowermost Trinity R., TX. Common; probably most abundant fish in Mississippi and other large rivers. **HABITAT:** Pools and runs of medium to large rivers, lakes; usually in clear water over sand or gravel. **SIMILAR SPECIES:** See Pl. 9. See (1) Comely Shiner, *N. amoenus*, (2) Rio Grande Shiner, *N. jemezanus*, and (3) Sharpnose Shiner, *N. oxyrhynchus*. (4) Silver Shiner, *N. photogenis*, has 2 dark crescents between nostrils, larger eye, 9 pelvic rays.

COMELY SHINER *Notropis amoenus* Pl. 9

IDENTIFICATION: Nearly identical to Emerald Shiner, *N. atherinoides*, but has *smaller eye,* less oblique mouth, rarely has T- or Y-shaped rakers on 1st gill arch. Within zone of sympatry (cen. and se. NY), Comely Shiner usually has 26–31 scales around body at dorsal fin origin, 13–17 around caudal peduncle; Emerald Shiner usually has 22–27 and 12–14, respectively. Has 35–47 lateral scales; 11 anal rays. To 4¼ in. (11 cm). **RANGE:** Atlantic Slope drainages from Hudson R., NY, to Cape Fear R., NC and SC; 1 record from Seneca Lake (Lake Ontario drainage), NY. Introduced into Yadkin R. system, NC. Common. **HABITAT:** Runs and flowing pools, over sand, gravel, or rubble, of creeks and medium to large rivers. **SIMILAR SPECIES:** (1) See Emerald Shiner, *N. atherinoides* (Pl. 9).

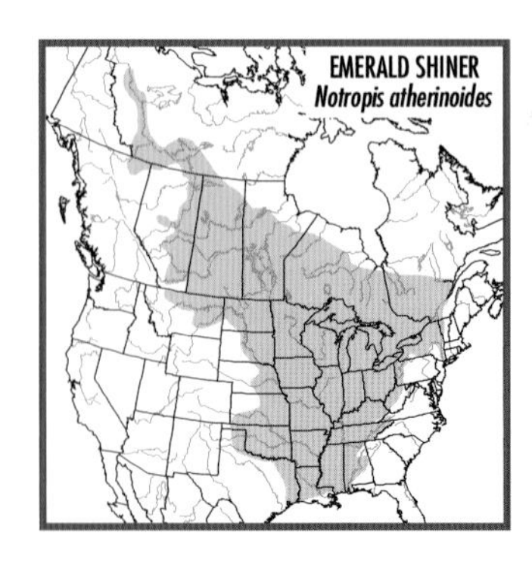

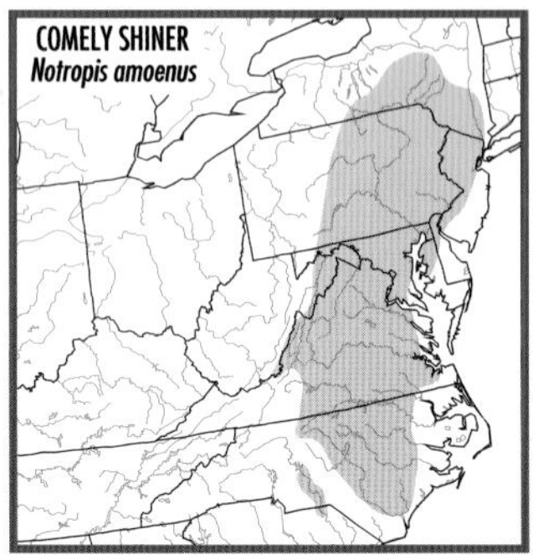

RIO GRANDE SHINER *Notropis jemezanus* Pl. 9

IDENTIFICATION: Similar to Emerald Shiner, *N. atherinoides,* but has *larger, less oblique mouth extending under smaller eye*; deeper snout; *lacks* black lips (may be dusky), black around anal fin base and along underside of caudal peduncle. To 3 in. (7.5 cm). **RANGE:** Rio Grande drainage, TX, NM, and Mexico. Common in lower Rio Grande; less common elsewhere; extirpated from Rio Grande, NM. **HABITAT:** Sandy and rocky runs and flowing pools of small to large rivers. **SIMILAR SPECIES:** (1) See Emerald Shiner, *N. atherinoides* (Pl. 9).

SHARPNOSE SHINER *Notropis oxyrhynchus* Pl. 9

IDENTIFICATION: Similar to Emerald Shiner, *N. atherinoides*, but has *sharply pointed snout;* upper jaw level with upper edge of eye. Has 34–37 lateral scales; 10 anal rays. To 2½ in. (6.5 cm). **RANGE:** Brazos R. basin, TX. Introduced into Colorado R., TX. Common. **HABITAT:** Sand and gravel runs of medium to large rivers; less often in sand- or mud-bottomed pools. **SIMILAR SPECIES:** (1) See Emerald Shiner, *N. atherinoides* (Pl. 9).

SILVER SHINER *Notropis photogenis* Pl. 9

IDENTIFICATION: *Slender, compressed body; 2 black crescents* (Fig. 28) between nostrils. Large, terminal mouth (reaching to front of eye) on long snout; thickened tip of lower jaw projects beyond upper jaw; large eye; dorsal fin origin behind pelvic fin origin. Light olive above, black stripe along back; partly dusky silver stripe with blue sheen along side. Black lips (front half). Complete lateral line; 36–40 lateral scales; 10–12 anal rays; 9 pelvic rays; pharyngeal teeth 2,4-4,2. To 5½ in. (14 cm). **RANGE:** Lakes Erie and Ontario, and Ohio R. drainages from ON and NY to MI, and south to n. GA and AL. Common in east, uncommon in west. **HABITAT:** Rocky runs and riffles of small to large rivers. **SIMILAR SPECIES:** (1) Emerald Shiner, *N. atherinoides* (Pl. 9), *lacks* black crescents; has smaller eye, shorter snout, deeper body, 8 pelvic rays. (2) Rosyface Shiner, *N. rubellus* (Pl. 17), and (3) Carmine Shiner, *N. percobromus*, have sharper snout, black streak above silver stripe along side, red on head and body of large male; *lack* black crescents between nostrils.

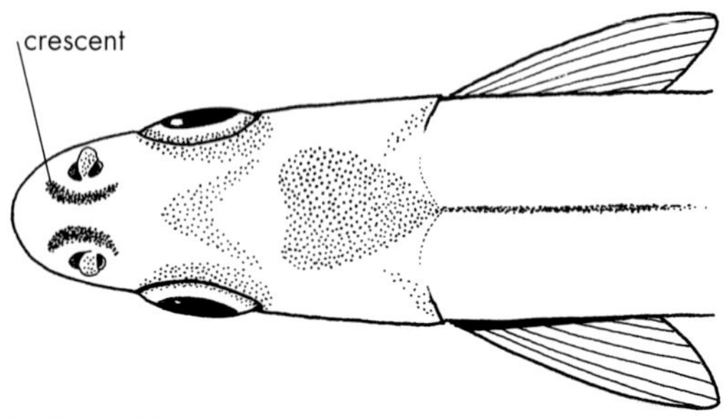

Fig. 28. Silver Shiner.

ROSYFACE SHINER *Notropis rubellus* Pl. 17

IDENTIFICATION: *Sharply pointed snout* longer than eye diameter. Slender, compressed body; dorsal fin origin *well behind* pelvic fin origin (over *middle* of pelvic fin). Olive above, often with narrow dusky stripe along back; faint *red* at base of dorsal fin; *black streak* just above silver stripe along side; blue sheen overall. Scale margins on back readily discernable; lateral-line pores distinctly outlined by black specks. Breeding male is blue above; has *orange to bright rosy red head, front half of body,* and *fin bases.* Some red on large female. Complete lateral line; 36–45 lateral scales; 9–11 anal rays; pharyngeal teeth 2,4-4,2 or 1,4-4,1. Usually 13–14 scales around caudal peduncle, 25–26 scales around body at dorsal fin origin. To 3½ in. (9 cm). **RANGE:** Uplands on Atlantic Slope from St. Lawrence R. drainage, QC, and Hudson R. drainage, VT (absent in rest of New England) to James R. drainage, VA. Great Lakes and upper Ohio R. drainages (above mouth of Green R.) from QC to WI and south to KY; Cumberland R. system above Cumberland Falls, KY. Common in clear gravelly streams with fast current. **HABITAT:** Rocky runs and flowing pools of small to medium rivers. **SIMILAR SPECIES:** See (1) Carmine Shiner, *N. percobromus*, (2) Highland Shiner, *N. micropteryx*, and (3) Rocky Shiner, *N. suttkusi*. (4) Emerald Shiner, *N. atherinoides*, and (5) Comely Shiner, *N. amoenus* (both Pl. 9), have *blunter snout, no* red coloration, *no* black streak above silver stripe along side, dorsal fin origin close to pelvic fin origin.

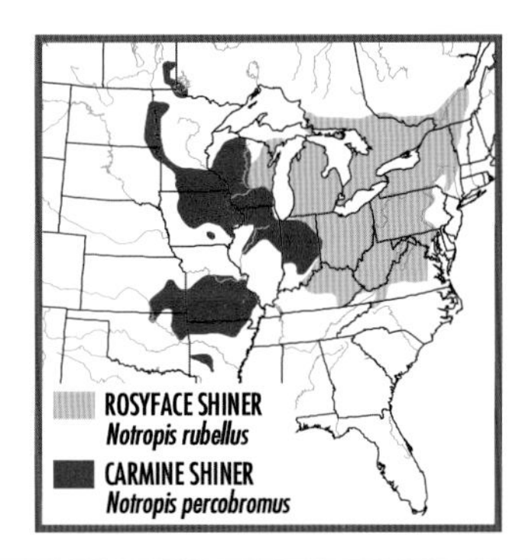

CARMINE SHINER *Notropis percobromus* **Not shown**

IDENTIFICATION: Genetically distinct but morphologically indistinguishable from Rosyface Shiner, *N. rubellus*. To 3 in. (9 cm). **RANGE:** Three disjunct areas: Hudson Bay (Red R.) and upper Mississippi R. basins from MB south to cen. IN and cen. IL; Ozark tributaries of Mississippi, Missouri, White, and Arkansas rivers in s. MO, n. AR, and ne. OK, and west to s.-cen. KS; and Ouachita R. system, s. AR. Common in upland streams with fast current; protected in Canada as a *threatened species*. **HABITAT:** Rocky runs and flowing pools of creeks and small to medium rivers; usually in clear water. Common. **SIMILAR SPECIES:** (1) See Rosyface Shiner, *N. rubellus* (Pl. 17).

HIGHLAND SHINER *Notropis micropteryx* **Not shown**

IDENTIFICATION: Similar to Rosyface Shiner, *N. rubellus*, and Carmine Shiner, *N. percobromus*, but breeding male has diffuse red-orange on lower side, jaws, and fin bases; top of red head barely or not contrasting with nape. Usually 12 scales around caudal peduncle, 23–24 scales around body at dorsal fin origin; pharyngeal teeth 2,4-4,2. To 3 in. (7.4 cm). **RANGE:** Upland regions of Green, Cumberland, and Tennessee river systems, KY, TN, VA, NC, AL, and MS; absent from Cumberland R. system above Cumberland Falls. Common. **HABITAT:** Rocky runs and flowing pools of creeks and small to medium rivers. **SIMILAR SPECIES:** See (1) Rosyface Shiner, *N. rubellus* (Pl. 17), and (2) Carmine Shiner, *N. percobromus*.

ROCKY SHINER *Notropis suttkusi* **Not shown**

IDENTIFICATION: Similar to Rosyface Shiner, *N. rubellus*, and Carmine Shiner, *N. percobromus*, but scale margins not readily discernable on back, lateral-line pores not distinctly outlined by black specks, lateral line deeply decurved. Breeding male has red on jaws and pectoral fin base, and red-orange on side below lateral stripe, not concentrated

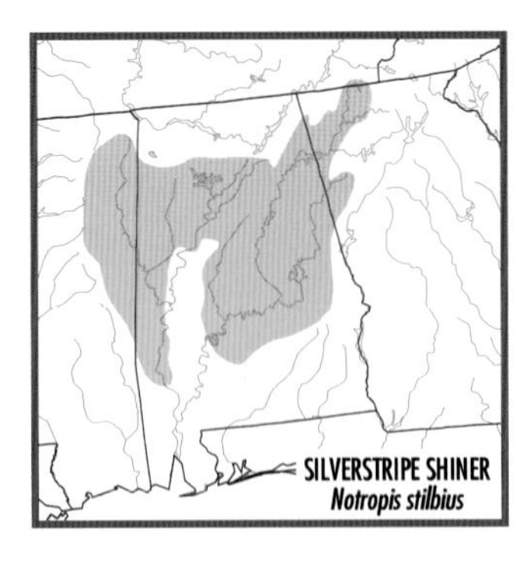

along lateral line. Has 33–39 lateral scales. To 2½ in. (6.6 cm). **RANGE:** Red R. system draining Ouachita Highlands, sw. AR and se. OK (Little R., AR, to Blue R., OK). Common. **HABITAT:** Gravel- and rubble-bottomed runs and flowing pools. **SIMILAR SPECIES:** See (1) Rosyface Shiner, *N. rubellus* (Pl. 17), and (2) Carmine Shiner, *N. percobromus*.

SILVERSTRIPE SHINER *Notropis stilbius* Pl. 9

IDENTIFICATION: *Horizontally oval black caudal spot. Large terminal mouth* on *pointed snout; large eye.* Slender, compressed body; dorsal fin origin behind pelvic fin origin. Olive above; dusky stripe along back; darkly outlined scales on back and upper side; broad silver black stripe along side; narrow dark stripe around snout (on both lips); punctate lateral line (Fig. 29). Complete lateral line; 35–37 lateral scales; 10–11 anal rays; 8 pelvic rays; pharyngeal teeth 2,4-4,2. To 3½ in. (9 cm). **RANGE:** Mobile Bay drainage, GA, AL, MS, and se. TN; Bear and Indian creeks (Tennessee R. drainage), AL and MS. Common above Fall Line; uncommon below. **HABITAT:** Gravel- and sand-bottomed runs and

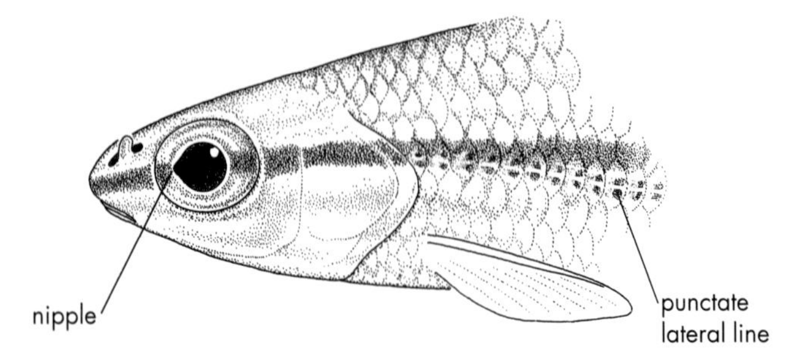

Fig. 29. Minnow—nipple on eye and punctate lateral line.

flowing pools of small to large rivers; often near vegetation. **SIMILAR SPECIES:** (1) Emerald Shiner, *N. atherinoides* (Pl. 9), has smaller eye, no caudal spot, no punctations along lateral line.

SILVERBAND SHINER *Notropis shumardi* Pl. 9

IDENTIFICATION: *Tall, pointed dorsal fin;* front rays extend well beyond rear rays when fin depressed. Compressed body; deep caudal peduncle; dorsal fin origin slightly in front of pelvic fin origin, about *midway* between tip of snout and caudal fin base. Terminal, slightly upturned mouth on short pointed snout. Light olive above, dusky stripe along back; silver stripe along side (often dusky at rear). Complete lateral line; 33–39 lateral scales; 9 (often 8) anal rays; usually 9 pelvic rays (unusual in *Notropis*); pharyngeal teeth 2,4-4,2. To 4 in. (10 cm). **RANGE:** Large rivers of Missouri-Mississippi basin (mainly Missouri, Mississippi, Illinois, lower Ohio, Arkansas, and Red rivers), from SD and cen. IL south to Gulf; Gulf drainages, TX, from Trinity R. to Lavaca R.; 1 record from Pearl R., MS. Fairly common. **HABITAT:** Flowing pools and runs of large, often turbid, rivers; usually over sand and gravel. **SIMILAR SPECIES:** (1) See Silverside Shiner, *N. candidus*. (2) Emerald Shiner, *N. atherinoides* (Pl. 9), has *shorter* dorsal fin with origin behind pelvic fin origin, *closer to* caudal fin base than to tip of snout; 10–12 anal rays; more slender body.

SILVERSIDE SHINER *Notropis candidus* Not shown

IDENTIFICATION: Nearly identical to Silverband Shiner, *N. shumardi*, but has longer, more pointed snout, more slender body, larger eye, usually 8 anal rays. To 4½ in. (11 cm). **RANGE:** Alabama and Tombigbee rivers and lower portions of their major tributaries, AL and MS. Common. **HABITAT:** Sand-gravel runs of medium to large rivers. **SIMILAR SPECIES:** (1) See Silverband Shiner, *N. shumardi* (Pl. 9).

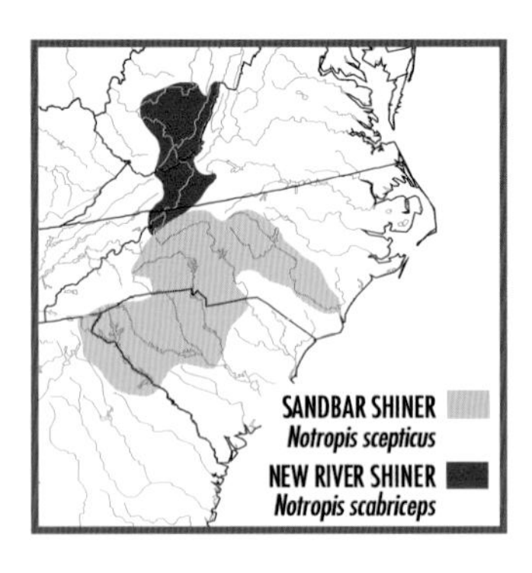

SANDBAR SHINER *Notropis scepticus* Pl. 9

IDENTIFICATION: *Large, round eye,* diameter more than snout length; snout turns down sharply above nostril. *Darkly outlined scales* on back and upper side form lines (often faint) meeting those of other side on caudal peduncle. *Falcate anal fin.* Straw yellow to olive above, dark stripe along back (darker in front of dorsal fin); *punctate,* complete lateral line, dusky stripe (darkest at rear) on snout and along silver side, followed by small black caudal spot. Deep, compressed body, dorsal fin origin slightly behind pelvic fin origin. Large terminal mouth on fairly pointed snout. Has 34–38 lateral scales; 12–13 scales around caudal peduncle; 10–11 anal rays; pharyngeal teeth 2,4-4,2. To 3½ in. (9 cm). **RANGE:** Cape Fear R. drainage, NC, to Savannah R. drainage, GA. Common on Piedmont. **HABITAT:** Flowing sand-bottomed pools, often near riffles, in creeks and small to medium rivers. **SIMILAR SPECIES:** (1) Telescope Shiner, *N. telescopus* (Pl. 9), has slender body, darker wavy stripes along back and upper side, snout not turning down sharply above nostril, *straight-edged* (barely concave) *anal fin.* (2) New River Shiner, *N. scabriceps* (Pl. 10), has broad snout, upwardly directed eyes, subterminal mouth, 8 anal rays.

NEW RIVER SHINER *Notropis scabriceps* Pl. 10

IDENTIFICATION: *Broad snout; large, upwardly directed eye; slightly subterminal mouth. Darkly outlined scales on* back and upper side form lines (often faint) meeting those of other side on caudal peduncle. Olive above, dusky stripe along back in front of dorsal fin; *punctate lateral line,* dusky stripe (darkest at rear) on snout and along silver side. Compressed body, somewhat flattened below; large, horizontal pectoral fins; dorsal fin origin over pelvic fin origin. Complete lateral line; 35–39 lateral scales; 8 anal rays, pharyngeal teeth 2,4-4,2. To 3¼ in. (8.4 cm). **RANGE:** Above Kanawha Falls in New R. drainage, WV, VA, and NC. Common. **HABITAT:** Sandy and rocky runs and flowing pools of creeks and small to medium rivers. **SIMILAR SPECIES:** (1) Sandbar Shiner, *N. scepticus* (Pl. 9), has *narrower snout, eye directed to side,* deeper body, terminal mouth, 10–11 anal rays.

WEDGESPOT SHINER *Notropis greenei* Pl. 10

IDENTIFICATION: *Black wedge* on caudal fin base; *large, upwardly directed eye,* nipple at front of pupil (Fig. 29). Large, horizontal mouth; long, rounded snout. Fairly compressed body, dorsal fin origin over pelvic fin origin. Gray above, scales on back and side outlined in black; wide dusky stripe along back in front of dorsal fin, thin stripe behind dorsal fin. Silver black stripe along side, darkest at rear; punctate lateral line. Complete lateral line, 35–38 lateral scales, 8 (often 9) anal rays; pharyngeal teeth 2,4-4,2. To 3 in. (7.5 cm). **RANGE:** Ozarkian tributaries of Mississippi, Missouri, White, and Arkansas rivers, MO, AR, and OK. Fairly common but declining. **HABITAT:** Flowing pools and runs over sand, gravel, and rubble of creeks and small to medium rivers. **SIMILAR SPECIES:** (1) Sand Shiner, *N. stramineus* (Pl. 11), *lacks* black wedge on caudal fin base (often a dusky edge), has small, outwardly directed eye, dark wedge at dorsal fin origin, 7 anal rays, pharyngeal teeth 0,4-4,0.

POPEYE SHINER *Notropis ariommus* Pl. 9

IDENTIFICATION: *Huge eye* (largest of any species of *Notropis*), more than 1.5 times snout length. Fairly deep, compressed body; dorsal fin origin over to slightly behind pelvic fin origin; broad, moderately pointed snout; terminal mouth. Light brown above, *darkly outlined scales* on back and upper side; gray-black stripe along back; dusky stripe along silver side, darkest at rear and often expanded into black caudal spot. Complete lateral line; 35–39 lateral scales; usually 9 (often 10) anal rays; pharyngeal teeth 2,4-4,2. To 3¾ in. (9.5 cm). **RANGE:** Ohio R. basin from PA to IN and south to Tennessee R. drainage, GA and AL; 1 record from Maumee R. (Lake Erie drainage), OH. Rare

and highly localized. **HABITAT:** Clear, gravel-bottomed, flowing pools and runs of creeks and small to medium rivers. **SIMILAR SPECIES:** (1) Telescope Shiner, *N. telescopus* (Pl. 9) has dark wavy stripes on back and upper side, punctate lateral line, more slender body, *smaller eye,* usually 10–11 anal rays.

TELESCOPE SHINER *Notropis telescopus* Pl. 9

IDENTIFICATION: *Dark wavy stripes* along olive-brown back and upper side *meet* those of other side on caudal peduncle. *Large eye,* longer than moderately pointed snout; terminal mouth. *Punctate* (usually), complete lateral line; dusky stripe (darkest at rear) along silver side; often expanded into black caudal spot. Bold stripe along back gray-black to dorsal fin, then narrower and light gray. Fairly slender, compressed body; dorsal fin origin behind pelvic fin origin. Has 35–39 lateral scales; usually 10–11 anal rays; pharyngeal teeth 2,4-4,2. To 3¾ in. (9.4 cm). **RANGE:** Disjunct. Cumberland and Tennessee river drainages, VA, NC, KY, TN, GA, and AL; Little, St. Francis, and White river drainages, MO and AR; extreme upper Santee drainage (Atlantic Slope), NC and SC. Introduced into New R. drainage, WV and VA; upper Yadkin and Catawba river drainages, NC. Fairly common. **HABITAT:** Rocky runs and flowing pools, often near riffles, of clear creeks and small to medium rivers. **SIMILAR SPECIES:** (1) Popeye Shiner, *N. ariommus* (Pl. 9), *lacks* wavy stripes on upper side and back, black punctations along lateral line; has deeper body, larger eye, usually 9 anal rays. (2) Tennessee Shiner, *N. leuciodus* (Pl. 17), has *smaller eye,* black rectangle on caudal fin base, red on large male, 8–9 anal rays; is more slender.

TEXAS SHINER *Notropis amabilis* Pl. 9

IDENTIFICATION: *Large eye; black lips.* Clear stripe above *dark stripe* along side (darkest at rear) and onto caudal fin; dark stripe along back darkest in front of dorsal fin; darkly outlined scales above clear

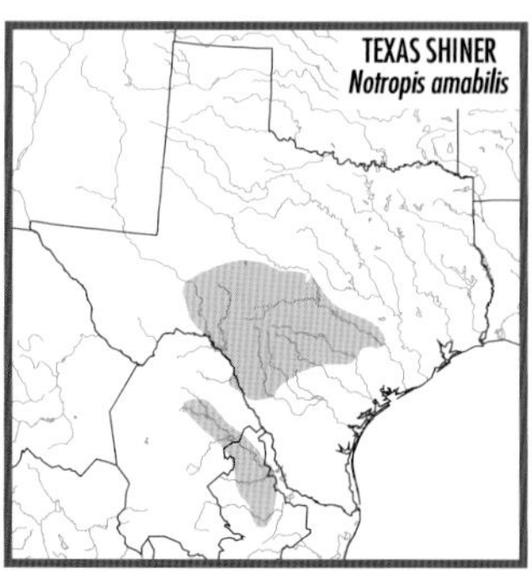

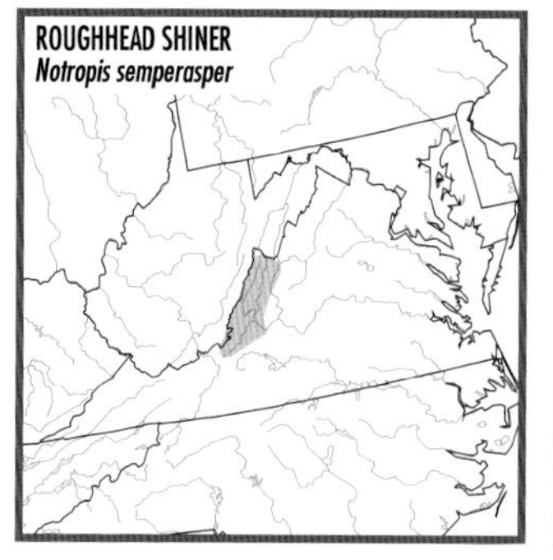

stripe. Oblique, terminal mouth; moderately pointed snout. Deep, compressed body; dorsal fin origin behind pelvic fin origin. Usually complete lateral line; 32–36 lateral scales; 9 anal rays; pharyngeal teeth 2,4-4,2. To 2½ in. (6.2 cm). **RANGE:** Colorado R. to Rio Grande (into lower Pecos R.) drainage, TX and Mexico. Fairly common, especially in springs and spring-fed streams on Edwards Plateau. **HABITAT:** Rocky and sandy runs and pools of headwaters, creeks, and small to medium rivers; usually in clear water. **SIMILAR SPECIES:** No similar species within range.

ROUGHHEAD SHINER *Notropis semperasper* Pl. 9

IDENTIFICATION: *Large eye. Black stripe* along side from head to caudal fin, darkest on rear half of body. Slender, compressed body; fairly pointed snout; slightly subterminal mouth. Dorsal fin origin behind pelvic fin origin. Dark green above, dusky iridescent stripe along back; silver side, thin iridescent green stripe above black stripe. Usually 38–41 lateral scales, 10–11 anal rays; pharyngeal teeth 2,4-4,2. To 3½ in. (9 cm). **RANGE:** Upper James R. drainage, VA. Fairly common. **HABITAT:** Clear rocky pools and backwaters of small to large rivers. **SIMILAR SPECIES:** (1) Comely Shiner, *N. amoenus* (Pl. 9), (2) Rosyface Shiner, *N. rubellus* (Pl. 17), and (3) Rosefin Shiner, *Lythrurus ardens* (Pl. 15), *lack* black stripe along side. Comely Shiner has longer, more pointed snout; Rosyface and Rosefin shiners have smaller eye, red on body and fins (seasonally variable).

BLUNTNOSE SHINER *Notropis simus* Pl. 10

IDENTIFICATION: *Blunt snout,* often overhanging upper lip. *Deep, wide head;* large, slightly subterminal mouth usually ending under pupil. Small eye. Fairly deep, compressed body; deepest under nape; dorsal fin origin behind pelvic fin origin. Generally *pallid* with only small black specks on head and back and along silver side; sometimes slight concentration forming faint dusky stripe along side; clear fins. Incomplete lateral line (ending on caudal peduncle); 33–38 lateral scales; usually 9 (often 10) anal rays; 4–9 (usually 6–8) rakers on 1st gill arch; pharyngeal teeth 2,4-4,2. To 4 in. (10 cm). **RANGE:** Upper Rio Grande (above El Paso, TX), TX and NM; Pecos R., NM. Rare; gone from much of former range, apparently because of reduced water levels in Rio Grande system. **HABITAT:** Main channels of Rio Grande and Pecos R., usually over sand or gravel. **REMARKS:** Two subspecies. *N. s. simus,* Rio Grande above El Paso, TX, possibly extinct, has embedded scales on breast (adults appear unscaled), slender caudal peduncle, usually 9–10 anal rays. *N. s. pecosensis,* recognized as a *threatened subspecies,* smaller (to 3 in. [7.5 cm]), and only in Pecos R., NM, has fully scaled breast, relatively deep caudal peduncle, usually 8–9 anal rays, and more pigment, including a broader dusky stripe and thin black streak along side. **SIMILAR SPECIES:** (1) See Phantom Shiner, *N. orca.* (2) Tamaulipas Shiner, *N. braytoni* (Pl. 10), has larger eye, darker stripe along side, usually 7 anal rays, pharyngeal teeth 1,4-4,1.

PHANTOM SHINER *Notropis orca* — Not shown

IDENTIFICATION: Similar to Bluntnose Shiner, *N. simus*, but has mouth ending *in front* of eye pupil, less arched back, relatively deeper caudal peduncle, usually 8 anal rays (within range of Phantom Shiner, Bluntnose Shiner usually has 9–10 rays), usually 9–10 (range 8–11) rakers on 1st gill arch. To 3½ in. (9 cm). **RANGE:** Rio Grande from mouth to n.-cen. NM (unknown from Pecos R.). Near extinction. **HABITAT:** Main channel of Rio Grande, usually over sand and often in turbid water. **SIMILAR SPECIES:** (1) See Bluntnose Shiner, *N. simus* (Pl. 10).

RIVER SHINER *Notropis blennius* — Pl. 8

IDENTIFICATION: Slender, fairly compressed body; mouth extends to *beneath front of eye*, uniformly *dark stripe* along back and encircling dorsal fin base. Moderately pointed snout, usually overhanging slightly

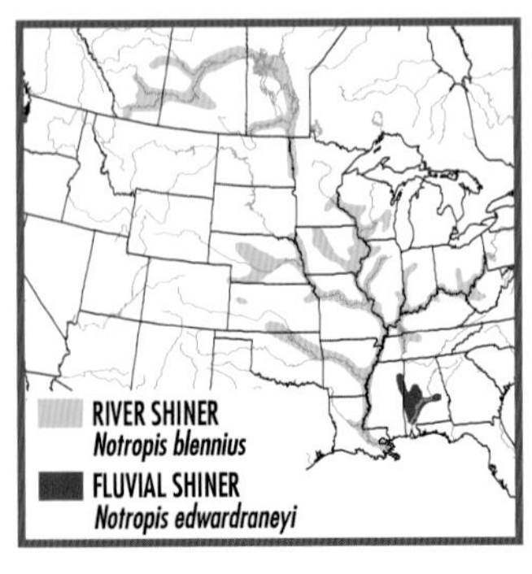

subterminal mouth. Dorsal fin origin over or slightly behind pelvic fin origin. Straw colored above, scales on upper side faintly outlined; dusky stripe along rear half of silver side. Scaled (or mostly scaled) breast. Complete lateral line; 34–41 (usually 35–36) lateral scales, 7 anal rays; pharyngeal teeth 2,4-4,2 (often 1,4-4,1). To 5¼ in. (13 cm). **RANGE:** Hudson Bay basin from MB to AB, and south through Red R., MN and ND; Lake Winnebago system (Lake Michigan drainage), WI; Mississippi R. basin from WI and MN south to Gulf; east to PA and west to CO. Common in central part of range, especially upper Mississippi and lower Ohio rivers. **HABITAT:** Pools and main channels of medium to large rivers, usually over sand and gravel. **SIMILAR SPECIES:** (1) Mississippi Silvery Minnow, *Hybognathus nuchalis* (Pl. 8), and other species of *Hybognathus* have much *smaller mouth,* extending only about half distance from tip of snout to eye; long gut (and soft belly); dorsal fin origin in front of pelvic fin origin; 8 anal rays. (2) Sand Shiner, *N. stramineus* (Pl. 11), has thin stripe along back *expanded at* dorsal fin origin, not encircling dorsal fin; punctate lateral line; smaller mouth. (3) Fluvial Shiner, *N. edwardraneyi* (Pl. 8), has *larger, more upwardly directed eye;* is more pallid.

FLUVIAL SHINER *Notropis edwardraneyi* Pl. 8

IDENTIFICATION: Very large, round eye (eye longer than snout), somewhat directed *upwardly. Pallid;* slightly dusky above with thin dark stripe along back; silver side; clear falcate fins. Fairly slender, compressed body; deepest just in front of dorsal fin. Dorsal fin origin in front of pelvic fin origin, 1st ray reaching beyond rest of fin when pressed against back. Terminal to slightly subterminal mouth; moderately pointed snout. Unscaled breast. Complete lateral line, 32–35 lateral scales; usually 7 anal rays; pharyngeal teeth 2,4-4,2. To 3¼ in. (8 cm). **RANGE:** Mobile Bay drainage, AL and MS; mostly below Fall Line. Locally abundant. **HABITAT:** Main channels, usually with current, of small to large rivers; often over sand or gravel. **SIMILAR SPECIES:** (1) River Shiner, *N. blennius* (Pl. 8), has *smaller, less upwardly directed eye;* is more darkly pigmented. (2) Skygazer Shiner, *N. uranoscopus* (Pl. 11), with upwardly directed eye, is much more slender, has black pigment along lateral line and outlining upper scales, black wedge on caudal fin base, pharyngeal teeth 0,4-4,0.

CHUB SHINER *Notropis potteri* Pl. 8

IDENTIFICATION: *Head flat* above and below, tapering into pronounced snout (very wide viewed from above). Eye *high on head* and somewhat *directed upwardly.* Slightly subterminal mouth. Dorsal fin origin in front of or over pelvic fin origin; long and pointed, with 1st rays reaching beyond rest of fin when depressed against body. Moderately compressed body, deepest just in front of dorsal fin. Dull olive to tan above, thin dusky stripe along back; faintly outlined scales on back and upper side; scattered black specks along silver side, concentrated into black streak on rear half; orange along top of eye. Complete (at least to caudal peduncle) lateral line; 34–37 lateral scales; usually 7 anal rays; pharyngeal teeth 2,4-4,2. To 4¼ in. (11 cm). **RANGE:** Red and Brazos river drainages, LA, AR, OK, and TX; lower Mississippi R., LA (near mouth of Red R.), 1 record in San Jacinto R. system, TX. Fairly common in mainstream of Red R.; localized elsewhere. **HABITAT:** Sand and gravel runs of small to large rivers; tolerant of turbidity. **SIMILAR SPECIES:** (1) Smalleye Shiner, *N. buccula,* and (2) Red River Shiner, *N. bairdi* (Pl. 8), have black specks concentrated in patch on side, eye directed more to side than upwardly, shorter snout, pharyngeal teeth 0,4-4,0. (3) River Shiner, *N. blennius* (Pl. 8), has larger eye located more on side of head, duskier back and side, more pointed snout.

RED RIVER SHINER *Notropis bairdi* Pl. 8

IDENTIFICATION: *Broad, flat head;* large, nearly terminal mouth (ending behind front of eye); rounded snout. Round eye high on head. Black specks scattered over back and upper side, concentrated in *large patch* on front half of side. Tan above, thin dusky stripe along back; silver side, often with light blue iridescence. Compressed body, deepest under nape; arched back; dorsal fin origin in front of pelvic fin origin. Breast and nape partly unscaled. Lateral line usually with several interruptions; 33–37 lateral scales; usually 7 anal rays, 15 pectoral rays; pharyngeal teeth 0,4-4,0. To 3¼ in. (8 cm). **RANGE:** Red R. drainage from sw. AR to w. OK and nw. TX. Introduced into Arkansas R. system and possibly Canadian R., OK. Common to abundant. Introduced into Cimarron R. (Arkansas R. drainage), s. KS and OK, and now more common there than related, native Arkansas River Shiner, *N. girardi*. **HABITAT:** Sandy, turbid channels of small to large rivers. **SIMILAR SPECIES:** See (1) Smalleye Shiner, N. *buccula*, and (2) Arkansas River Shiner, *N. girardi.*

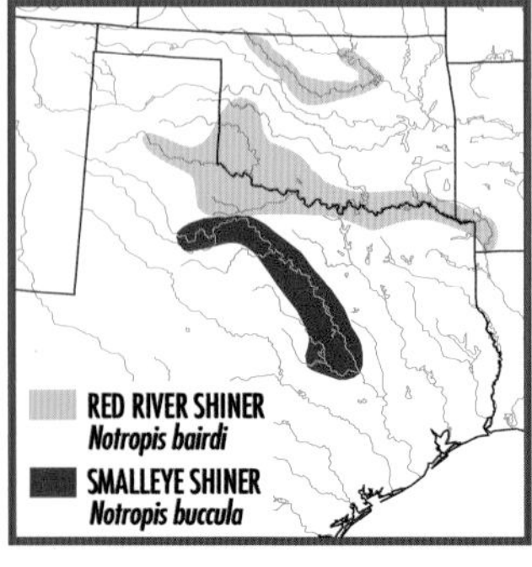

SMALLEYE SHINER *Notropis buccula* Not shown

IDENTIFICATION: Nearly identical to Red River Shiner, *N. bairdi*, but has longer snout, smaller eye (mouth ends in *front* of eye); shallower head. To 2¾ in. (7.1 cm). **RANGE:** Brazos R. drainage, TX. May be introduced into Colorado R. drainage, TX. Rare. **HABITAT:** Sandy, turbid channels of small to large rivers. **SIMILAR SPECIES:** See (1) Red River Shiner, *N. bairdi* (Pl. 8), and (2) Arkansas River Shiner, *N. girardi.*

ARKANSAS RIVER SHINER *Notropis girardi* Not shown

IDENTIFICATION: Similar to Red River Shiner, *N. bairdi*, and Smalleye Shiner, *N. buccula*, but has *fully scaled* breast and nape; usually 8 anal rays, 14 pectoral rays; larger and more falcate fins. To 3¼ in. (8 cm). **RANGE:** Formerly, Arkansas R. drainage from w. AR to w. KS, w. OK, TX panhandle, and ne. NM. Uncommon and apparently now restricted in native range to South Canadian R., OK; protected as *threatened species*. Introduced into Pecos R., NM. **HABITAT:** Flowing water over sand in creeks and small to large rivers. **SIMILAR SPECIES:** See (1) Red River Shiner, *N. bairdi* (Pl. 8), and (2) Smalleye Shiner, *N. buccula.*

ROUGH SHINER *Notropis baileyi* Pl. 17

IDENTIFICATION: *Red-brown above*; often dull red side and fin bases—some populations have yellow side and fin bases. Smooth, even-edged brown-black stripe along side *uniformly dark* from tip of lower jaw to caudal fin base with iridescence along upper edge; black caudal spot; silver green below. Breeding male has hump behind head, bright clear yellow fins (often reddish), is bright red above with gold dorsal and lateral stripes, red-orange below. Fairly deep, compressed body; deepest just in front of dorsal fin; dorsal fin origin over or in front of pelvic fin origin. Fairly long, rounded snout; terminal mouth. Has 33–38 lateral scales; usually 12 scales around caudal peduncle, 7 anal rays; pharyngeal teeth 2,4-4,2. To 3½ in. (9 cm). **RANGE:** Mobile Bay and Pascagoula R. drainages, AL and MS; Bear Creek system (Tennessee R. drainage),

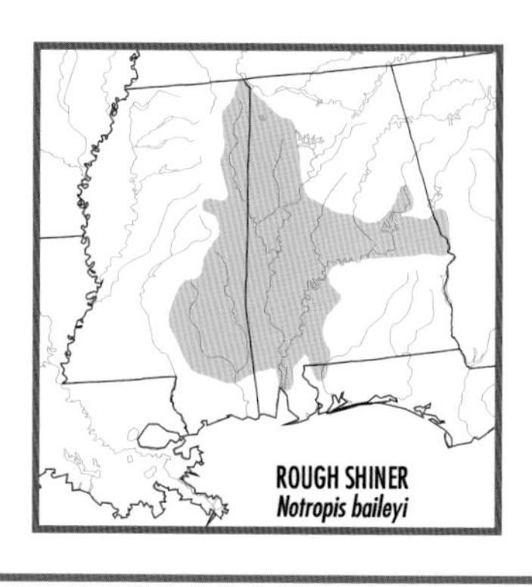

AL and MS. Also (possibly introduced) in Escambia R. drainage, AL and FL, and cen. Chattahoochee R. drainage, GA and AL. Common. **HABITAT:** Sand- and gravel-bottomed flowing pools of headwaters, creeks, and small rivers. **SIMILAR SPECIES:** See (1) Yellowfin Shiner, *N. lutipinnis*, and (2) Greenhead Shiner, *N. chlorocephalus* (both Pl. 17).

YELLOWFIN SHINER *Notropis lutipinnis* Pl. 17

IDENTIFICATION: Similar to Rough Shiner, *N. baileyi*, but lacks distinct caudal spot; has dorsal fin origin *behind* pelvic fin origin, longer snout, 8 anal rays, usually 14–16 scales around caudal peduncle. Breeding individuals have bright red body, yellow head, and opaque yellow or red fins. Gold dorsal and lateral stripes. Pharyngeal teeth 2,4-4,2. To 3 in. (7.5 cm). **RANGE:** Atlantic and Gulf drainages from Edisto R., SC, to Altamaha R., upper Chattahoochee R., and upper Etowah R. systems, GA. Little Tennessee R. system (Tennessee R. drainage), NC. Common to abundant in montane and upper Piedmont streams; less common on lower Piedmont. **HABITAT:** Clear rocky pools of headwaters, creeks, and small rivers. **SIMILAR SPECIES:** See (1) Rough Shiner, *N. baileyi*, and (2) Greenhead Shiner, *N. chlorocephalus* (both Pl. 17).

GREENHEAD SHINER *Notropis chlorocephalus* Pl. 17

IDENTIFICATION: Similar to Yellowfin Shiner, *N. lutipinnis*, but has *bright white fins* on male and female, scarlet red breeding male, pharyngeal teeth 1,4-4,1 (see Remarks). Has 36–39 lateral scales; 8 anal rays. To 2¾ in. (7.2 cm). **RANGE:** Upper Lynches R. (Peedee R. drainage) and Santee R. drainages, NC and SC. Abundant. **HABITAT:** Rocky flowing pools of clear headwaters, creeks, and small rivers. **REMARKS:** Individuals in Catawba, Broad, and Saluda river systems are geographically and morphologically intermediate (with yellow fins, both 1,4-4,1 and 2,4-4,2 pharyngeal tooth counts) between other Greenhead Shiners

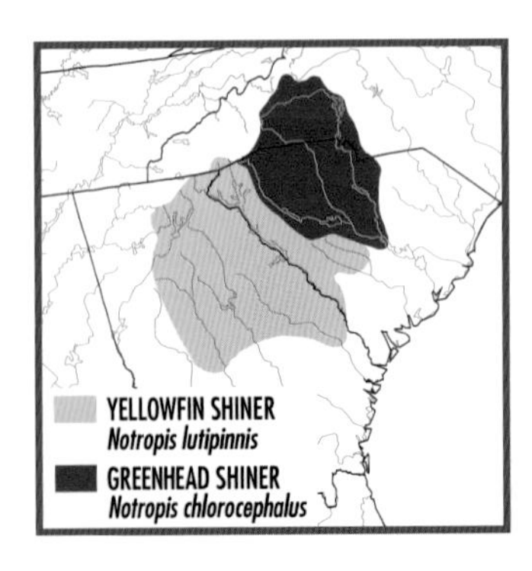

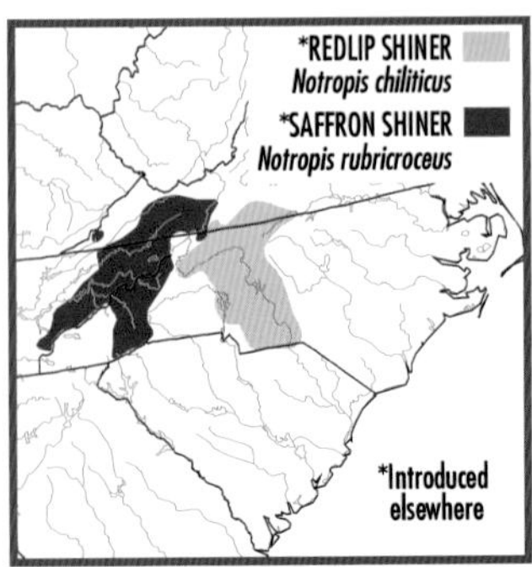

and Yellowfin Shiner, and may warrant recognition as a separate species. **SIMILAR SPECIES:** (1) See Yellowfin Shiner, *N. lutipinnis* (Pl. 17).

EDLIP SHINER *Notropis chiliticus* Pl. 17

IDENTIFICATION: *Bright red lips; scattered large black blotches* on side; black stripe on rear half of silver side leading to black caudal spot. Gold-green stripe along light green back and another along upper side; black specks on lower ⅓ of dorsal fin. Large individual has red body and eye, yellow fins; breeding male has scarlet red body with thick iridescent gold stripe, bright yellow-gold head and fins. Slender, compressed body; dorsal fin origin slightly behind pelvic fin origin. Large eye; rounded snout; slightly subterminal mouth. Complete lateral line; 34–37 lateral scales; 8 anal rays; pharyngeal teeth 2,4-4,2. To 2¾ in. (7.2 cm). **RANGE:** Dan R. (Roanoke R. drainage) and Peedee R. drainages, VA, NC, and SC. Introduced elsewhere, including New R. drainage, NC and VA, and Catawba R. system, NC. Abundant in montane streams; common on Piedmont. **HABITAT:** Rocky flowing pools of clear headwaters, creeks, and small rivers. **SIMILAR SPECIES:** See Pl. 17. (1) Greenhead Shiner, *N. chlorocephalus*, and (2) Yellowfin Shiner, *N. lutipinnis, lack* red lips, black blotches on side; large male Greenhead Shiner has red head, white fins. (3) Saffron Shiner, *N. rubricroceus, lacks* black blotches on side, has darkly outlined scales below lateral line.

AFFRON SHINER *Notropis rubricroceus* Pl. 17

IDENTIFICATION: *Darkly outlined scales* extend *below* lateral line. Dusky stripe on rear half of silver side leads to black caudal spot; no large black blotches on side; often red on snout, anterior body. Gold dorsal and lateral stripes. Breeding male is dark red-purple above, scarlet red below, has bright yellow fins and thin gold stripe along body. Slender, compressed body; dorsal fin origin slightly behind pelvic fin origin. Rounded snout; slightly subterminal mouth. Complete lateral line; 36–42 lateral scales; 8 anal rays; pharyngeal teeth 2,4-4,2. To 3¼ in. (8.4 cm). **RANGE:** Upper Tennessee R. drainage, VA, NC, and TN; headwaters of Santee and Savannah rivers, NC. Introduced into New R. drainage, VA and NC. Common. **HABITAT:** Bedrock- and rubble-bottomed pools of clear, fast headwaters and creeks. **SIMILAR SPECIES:** (1) Tennessee Shiner, *N. leuciodus* (Pl. 17), *lacks* darkly outlined scales below lateral line, has punctate lateral line. (2) Redlip Shiner, *N. chiliticus* (Pl. 17), *lacks* darkly outlined scales below lateral line, has large black blotches on side.

RAINBOW SHINER *Notropis chrosomus* Pl. 17

IDENTIFICATION: *Iridescent blue and pink head and body* (large individual). *Clear to red-purple stripe* above silver black stripe along side. Dusky purple above, scales darkly outlined; white below; faint to bright red-orange fins. Breeding male has bright purple head, back, and fins; red-purple stripe above, light blue below, silver stripe along side. Compressed body; dorsal fin origin over or slightly behind pelvic fin origin. Round snout, terminal mouth, moderate eye (shorter than snout length). Lateral line complete or missing on only last 4–5 scales; 34–38 lateral scales; usually 8 anal rays; pharyngeal teeth 2,4-4,2. To 3¼ in. (8.1 cm). **RANGE:** Alabama R. system, nw. GA, AL, and se. TN; upper Locust Fork of Black Warrior R. system, AL. Tennessee R. system in ne. AL, where probably introduced. Fairly common, especially in spring-fed streams. **HABITAT:** Gravelly riffles and pools (sometimes sandy pools) of creeks and small rivers. **SIMILAR SPECIES:** (1) Coosa Shiner, *N. xaenocephalus* (Pl. 11), *lacks* bright colors, has dark stripe along back, larger eye, 7 anal rays. (2) Burrhead Shiner, *N. asperifrons* (Pl. 11), *lacks* bright colors, has black stripe around longer snout, usually 7 anal rays; is more slender. (3) Tennessee Shiner, *N. leuciodus* (Pl. 17), is more slender; has pointed snout, larger eye, dark stripe along back midline, punctate lateral line, red on large male.

TENNESSEE SHINER *Notropis leuciodus* Pl. 17

IDENTIFICATION: *Dark wavy stripes* on olive back and upper side *meet* those of other side on caudal peduncle. Silver black stripe along side often faint at front, dark at rear, and followed by distinct *black rectangle* on caudal fin base. *Punctate lateral line* extends below dark stripe on front half of body. Black stripe along back darkest in front of dorsal fin, then narrower and lighter. Slender, fairly compressed body; dorsal fin origin behind pelvic fin origin. Pointed snout, terminal mouth; fairly large eye. Breeding male is *red overall*; light red on fin bases. Complete lateral line; 36–39 lateral scales; usually 8–9 anal

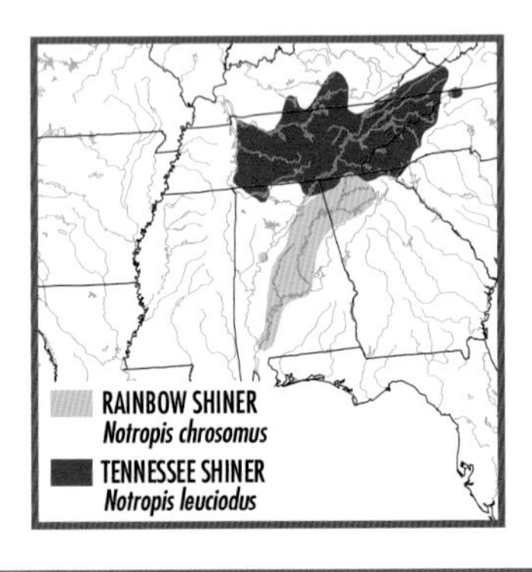

rays; pharyngeal teeth 1,4-4,1. To 3¼ in. (8.2 cm). **RANGE:** Green (upper Green and upper Barren systems), Cumberland (absent above Big South Fork), and Tennessee river drainages, VA, NC, KY, TN, GA, and AL. Also in extreme upper Savannah R. drainage (Atlantic Slope), NC, SC, and GA, Chattahoochee R. drainage (Gulf Slope), GA, and New R. drainage, VA and NC (possibly bait-bucket introductions). Generally common. **HABITAT:** Rocky pools and runs of creeks and small to medium rivers. **SIMILAR SPECIES:** (1) Saffron Shiner, *N. rubricroceus* (Pl. 17), *lacks* punctate lateral line, has darkly outlined scales below lateral line. (2) Rainbow Shiner, *N. chrosomus* (Pl. 17), *lacks* punctate lateral line, has more rounded snout; pink, purple, and blue on breeding male. (3) Telescope Shiner, *N. telescopus* (Pl. 9), *lacks* black rectangle at base of caudal fin and red on large male; has larger eye, deeper body, 10–11 anal rays.

ZARK MINNOW *Notropis nubilus* Pl. 17

IDENTIFICATION: *Strongly bicolored,* dark olive-brown above, white to yellow-orange below. Thin pale stripe above *black stripe* along silver side and around snout; punctate lateral line; often a small black spot on caudal fin base. *Black peritoneum* often visible through belly. Gold spots on dark stripe along back; clear fins. Large individual is yellow-orange below, has yellow fins (brighter on male). Slender body, barely compressed; slightly subterminal mouth; dorsal fin origin over pelvic fin origin; long, coiled gut (at least twice length of body). Has 33–38 lateral scales; 8 anal rays; pharyngeal teeth 0,4-4,0. To 3¾ in. (9.3 cm). **RANGE:** Disjunct. Upper Red Cedar R. system, n. WI; Mississippi R. tributaries, se. MN, e. IA, s. WI, and n. IL; Ozark Mt. drainages (Mississippi, Missouri, White, and Arkansas river drainages), s. MO, se. KS, n. AR, and ne. OK. Common (locally abundant) in southern part of range (except perhaps extinct in Boggy Creek system); uncommon in north. **HABITAT:** Gravel- and rubble-bottomed flowing pools and backwaters of creeks and small to medium rivers; rarely in large rivers.

LONGNOSE SHINER *Notropis longirostris* Pl. 17

IDENTIFICATION: *Small, upwardly directed eye; long, rounded snout;* large, *subterminal mouth.* Slender, compressed body, *flattened below;* dorsal fin origin in front of pelvic fin origin. Straw yellow to gray above, faintly outlined scales, dark streak along back; silver side (blue iridescence); *yellow fins.* Large male has bright yellow to orange fins and snout. Complete lateral line; 34–36 lateral scales. Usually 14–15 pectoral rays, 7 anal rays, pharyngeal teeth 1,4-4,1. To 2½ in. (6.5 cm). **RANGE:** Upper Altamaha R. drainage (Atlantic Slope), GA; Gulf Slope drainages from Apalachicola R., GA and FL, to Mississippi R., LA (including extreme lower Mobile Bay only); north in Mississippi R. basin to lower Yazoo R., MS; lower Ouachita R. drainage, LA. Common; locally abundant. **HABITAT:** Sandy, sometimes muddy, shallow runs and pools of creeks and small to medium rivers. **SIMILAR SPECIES:** (1) See Orangefin Shiner, *N. ammophilus* (Pl. 17), and (2) Yazoo Shiner, *N. rafinesquei.* (3) Sabine Shiner, *N. sabinae* (Pl. 17), has deep, arched body, no yellow or orange on fins or snout, pharyngeal teeth usually 0,4-4,0.

ORANGEFIN SHINER *Notropis ammophilus* Pl. 17

IDENTIFICATION: Similar to Longnose Shiner, *N. longirostris,* but has dorsal fin origin *over* pelvic fin origin; *orange snout and fins* on large male. Has 32–36 lateral scales; pharyngeal teeth 0,4-4,0. To 2¼ in. (6 cm). **RANGE:** Mobile Bay drainage (primarily below Fall Line), AL and MS; Yellow Creek (Tennessee R. drainage), MS; Hatchie R. system (Mississippi R.), TN and MS; Skuna R. system (Yazoo R.), MS. Common. **HABITAT:** Shallow sandy runs and pools of creeks and small to medium rivers. **SIMILAR SPECIES:** (1) See Longnose Shiner, *N. longirostris* (Pl. 17).

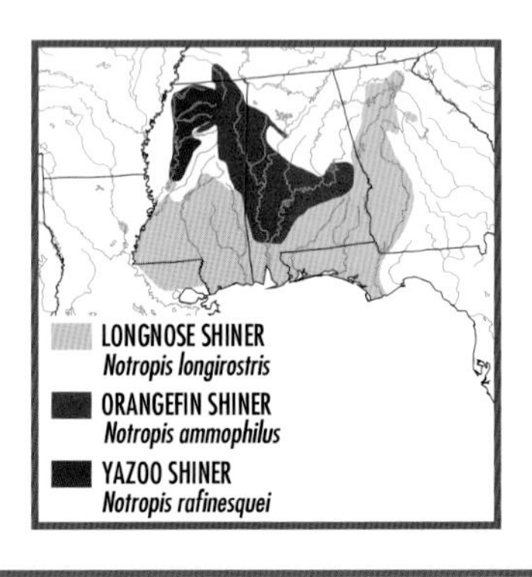

AZOO SHINER *Notropis rafinesquei* Not shown

IDENTIFICATION: Similar to Longnose Shiner, *N. longirostris,* but has *yellow or orange bar* behind head, yellow on fins *confined to fin bases,* usually *13 pectoral rays,* pharyngeal teeth 0,4-4,0. Has 33–39 lateral scales. To 2 in. (5.2 cm). **RANGE:** Yazoo R. system, MS. Common. **HABITAT:** Shallow sandy and gravelly runs and pools of creeks and small rivers. **SIMILAR SPECIES:** (1) See Longnose Shiner, *N. longirostris* (Pl. 17). (2) Orangefin Shiner, *N. ammophilus* (Pl. 17), has dorsal fin origin *over* pelvic fin origin, usually *14–15 pectoral rays, orange snout and fins* on large male.

ABINE SHINER *Notropis sabinae* Pl. 17

IDENTIFICATION: Deep, compressed, *strongly arched body;* deepest at dorsal fin origin, *flattened below.* Large, horizontal, *subterminal mouth; long, rounded snout.* Small eye directed upwardly. Dorsal fin origin in front of pelvic fin origin. Olive-yellow above, faintly outlined scales, dark streak along back; silver side; punctate (often faint) lateral line; head dark above and under eye; white cheek. Complete lateral line; 31–37 lateral scales; 7 anal rays; pharyngeal teeth 0,4-4,0 (often 1,4-4,1). To 2¼ in. (5.7 cm). **RANGE:** Disjunct. Big Black and Yazoo river systems, MS; White (including lower Black) R. and St. Francis drainages, se. MO and ne. AR; Little R. system (lower Red R. drainage), LA; Gulf drainages from Calcasieu R., LA, to San Jacinto R., TX. Locally common. **HABITAT:** Sandy runs and flowing pools of creeks and small to medium rivers. **SIMILAR SPECIES:** (1) Longnose Shiner, *N. longirostris,* (2) Orangefin Shiner, *N. ammophilus* (both Pl. 17), and (3) Yazoo Shiner, *N. rafinesquei*, have much more *slender body;* yellow or orange on snout and fins; lack dark area under eye. Longnose Shiner has pharyngeal teeth 1,4-4,1. (4) Chub Shiner, *N. potteri* (Pl. 8), has *pointed snout,* head flattened above and below, nearly terminal mouth, black specks on side.

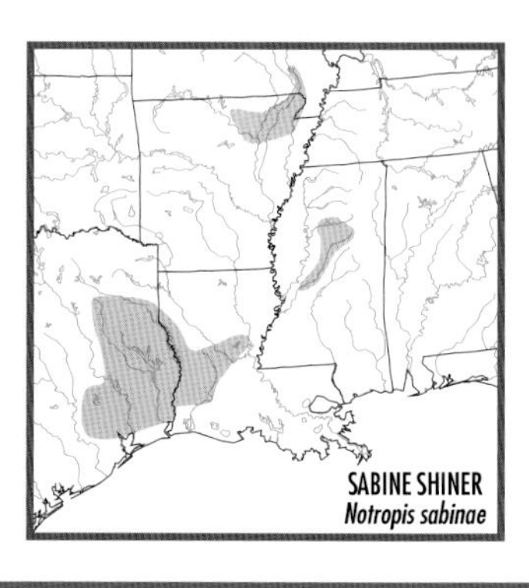

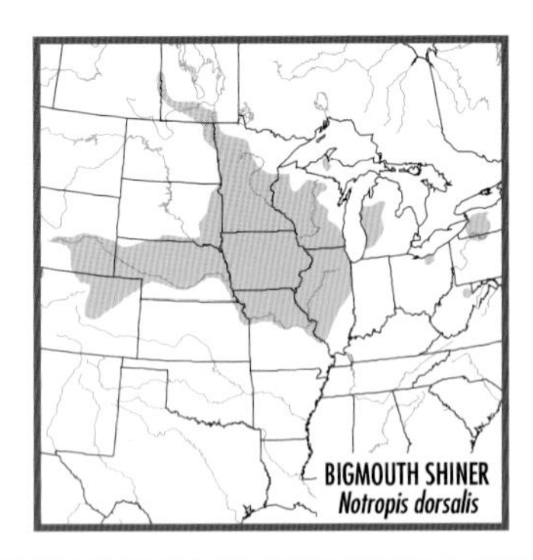

BIGMOUTH SHINER *Notropis dorsalis* Pl. 11

IDENTIFICATION: *Upwardly directed eye;* head *flattened* below; long snout; large, subterminal mouth. Slender, *strongly arched body;* dorsal fin origin over pelvic fin origin. Punctate lateral line (front half). Light tan to olive above, dark stripe along back; faintly outlined scales on back and upper side; silver side. Complete lateral line; 33–39 lateral scales; 8 anal rays (frequently 7 in west); pharyngeal teeth 1,4-4,1. To 3¼ in. (8 cm). **RANGE:** Great Lakes, Hudson Bay (Red R.), and Mississippi R. basins from MI to s. SK, and from e. IL to Platte R. system, WY and CO; highly disjunct populations in w. NY and PA, n. WV, and n. OH. Common over much of range. **HABITAT:** Shallow sandy and silty runs and pools of headwaters, creeks, and small to medium rivers. **REMARKS:** Three subspecies sometimes recognized but in need of study. *N. d. piptolepis,* in Platte R. system, WY and CO, has no, or embedded, scales on nape. *N. d. keimi,* in Lake Ontario and Allegheny R. drainages, NY and PA, has relatively short snout, small mouth. *N. d. dorsalis,* elsewhere in range, has exposed scales on nape, long snout, large mouth. **SIMILAR SPECIES:** (1) Silverjaw Minnow, *E. buccata* (Pl. 11), has large silver white chambers on cheek and underside of head. (2) Sand Shiner, *N. stramineus* (Pl. 11), *lacks* upwardly directed eye and flattened head; has dark wedge at dorsal fin origin, 7 anal rays.

BLACKCHIN SHINER *Notropis heterodon* Pl. 11

IDENTIFICATION: *Black stripe* along side and around short *pointed snout;* concentrations of blackest pigment at lateral-line pores give a *zigzag* appearance to black stripe, at least on front half of body. Dusky to black lips and chin (Fig. 30). Olive to pale yellow above, scales darkly outlined except pale stripe above dark stripe along side; dusky stripe along back darker and much wider in front of dorsal fin; silver side. Fairly compressed body; slender caudal peduncle; dorsal fin origin over to slightly in front of pelvic fin origin. Oblique mouth (rear edge to below nostril). Lateral line usually incomplete. Has 34–38 lateral scales; 8, often 7, anal rays; pharyngeal teeth 1,4-4,1. To 2¾ in. (7.1 cm). **RANGE:** S. QC and VT west to MB, MN, and IA. Mostly St. Lawrence-Great Lakes and upper Mississippi R. basins; localized in Susquehanna R. (Atlantic basin), upper Ohio R., and Hudson Bay basins. Generally common. **HABITAT:** Clear, vegetated lakes; pools and slow runs of creeks and small rivers. Usually over sand. **SIMILAR SPECIES:** (1) Bigeye Shiner, *N. boops* (Pl. 11), has larger mouth—rear edge almost to front of eye; larger eye, black peritoneum, complete lateral line. (2) Blacknose Shiner, *N. heterolepis* (Pl. 10), is more slender; has smaller subterminal mouth, no black on lower lip and chin, dorsal fin origin behind pelvic fin origin; pharyngeal teeth 0,4-4,0.

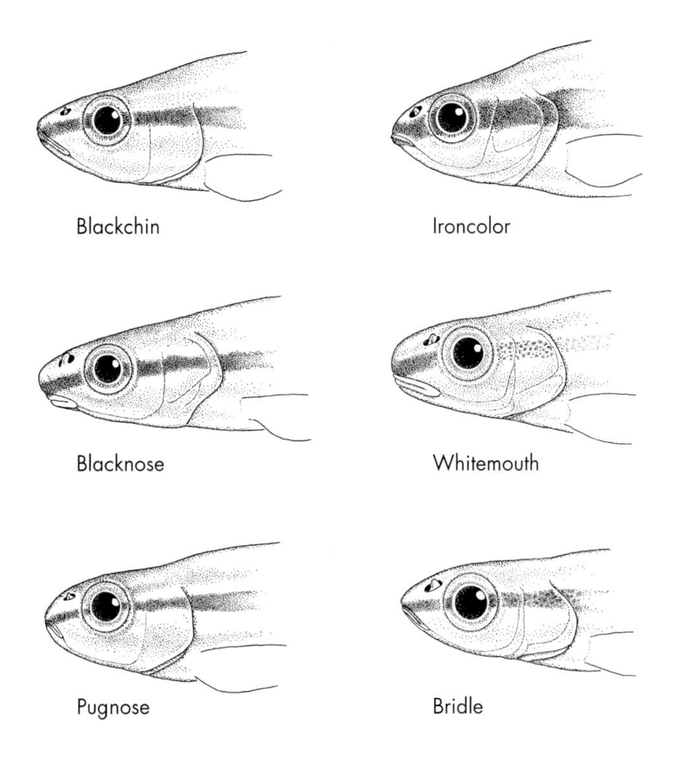

Fig. 30. Blackchin, Ironcolor, Blacknose, Whitemouth, Pugnose, and Bridle shiners.

BIGEYE SHINER *Notropis boops* Pl. 11

IDENTIFICATION: *Large eye,* much longer than snout. *Clear stripe* above *black stripe* along silver side and *around snout* (on both lips); lateral line punctate on front half of body. Slender, fairly compressed body; dorsal fin origin over pelvic fin origin. Fairly long, moderately pointed snout; large terminal mouth (rear edge almost to eye). Olive-yellow above, thin dark stripe along back; darkly outlined scales on back and upper side. Complete lateral line; 34–40 lateral scales; black peritoneum; 8 anal rays; pharyngeal teeth 1,4-4,1. To 3½ in. (9 cm). **RANGE:** Lake Erie drainage, nw. OH; Mississippi R. basin from cen. OH to e. KS and south to n. AL, n. LA, and s. OK. Mostly confined to uplands. Common; abundant in Ozark-Ouachita drainages; absent from most of Former Mississippi Embayment. Disappearing from large areas, including most of OH and IL. **HABITAT:** Flowing, usually clear and rocky, pools of creeks and small to medium rivers; often near emergent vegetation along stream margin. **SIMILAR SPECIES:** (1) Blackchin Shiner, *N. heterodon* (Pl. 11), has smaller mouth (to below nostril), smaller eye, incomplete lateral line, silver peritoneum. (2) Coosa Shiner, *N. xaenocephalus* (Pl. 11), has large black caudal spot, 7 anal rays, pharyngeal teeth 2,4-4,2.

COOSA SHINER *Notropis xaenocephalus* Pl. 11

IDENTIFICATION: *Clear to yellow stripe* above dark stripe along side; dark stripe extends onto side of head, expanded into *large black caudal spot. Large eye* (about as long as snout); fairly long, fairly blunt snout; black lips; terminal mouth. Fairly slender, compressed body; dorsal fin origin over pelvic fin origin. Dusky straw yellow above, darkly outlined scales; dark stripe along middle of back in front of dorsal fin, dark streaks behind dorsal fin; silver side. Complete lateral line; 33–38 lateral scales; 7 anal rays; pharyngeal teeth 2,4-4,2. To 3 in. (7.9 cm). **RANGE:** Above Fall Line in Coosa and Tallapoosa river systems (Mobile Bay drainage), TN, GA, and AL. Common. Possibly introduced in Chestatee R. (upper Chattahoochee R. drainage), GA. **HABITAT:** Clear, gravel-bottomed pools and runs; common in spring-fed streams. **SIMILAR SPECIES:** (1) Rainbow Shiner, *N. chrosomus* (Pl. 17), is bright purple and blue (adult); lacks dark stripe along back; has *smaller eye,* 8 anal rays. (2) Burrhead Shiner, *N. asperifrons* (Pl. 11), has black stripe around longer snout, is more slender, lacks (or has only faint) dark stripe along back.

BURRHEAD SHINER *Notropis asperifrons* Pl. 11

IDENTIFICATION: *Clear to yellow-orange stripe* above dark stripe along side; dark stripe extends around *long, rounded snout* (and lips and chin) and expands into black spot on caudal fin base. Slightly subterminal mouth; large eye. Slender, compressed body; dorsal fin origin over to slightly behind pelvic fin origin. Dusky yellow above, darkly outlined scales; black specks above and below lateral-line pores, silver white below. Complete lateral line; 33–39 lateral scales; usually 7 anal rays; pharyngeal teeth 2,4-4,2. To 3 in. (7.5 cm). **RANGE:** Alabama and Black Warrior river systems, se. TN, nw. GA, and AL. Mostly above Fall Line. Fairly common. **HABITAT:** Rocky and sandy pools and runs of creeks and small rivers. **SIMILAR SPECIES:** (1) Coosa Shiner, *N. xaenocephalus* (Pl. 11), *lacks* dark stripe around snout (but has black lips); has *shorter snout*, dark stripe along back; is deeper bodied. (2) Rainbow Shiner, *N. chrosomus* (Pl. 17), is bright purple and blue (when large); lacks dark stripe around *shorter snout,* has 8 anal rays.

HIGHSCALE SHINER *Notropis hypsilepis* Pl. 11

IDENTIFICATION: *Small black wedge on* caudal fin base. Large, round eye high on head. *Blunt snout;* small, subterminal, nearly horizontal mouth. Clear to white stripe above dusky to black stripe along side. Exposed area of scales near front of lateral line much deeper than wide. Slender, compressed body; dorsal fin origin over or slightly behind pelvic fin origin. Dusky above, darkly outlined scales, dark streaks along midline of back in front of dorsal fin; silver side. Complete lateral line; usually 35–36 lateral scales, 7 anal rays; pharyngeal teeth 2,4-4,2. To 2½ in. (6.4 cm). **RANGE:** Apalachicola R. drainage, GA and e. AL; one locality in upper Savannah R. drainage, ne. GA. Mostly on Piedmont. Uncommon. **HABITAT:** Sandy runs and pools of creeks and small rivers. **SIMILAR SPECIES:** (1) Clear Chub, *Hybopsis winchelli,* has dorsal fin origin in front of pelvic fin origin; longer, *more pointed snout;* smaller, more subterminal mouth; *lacks* black wedge on caudal fin base.

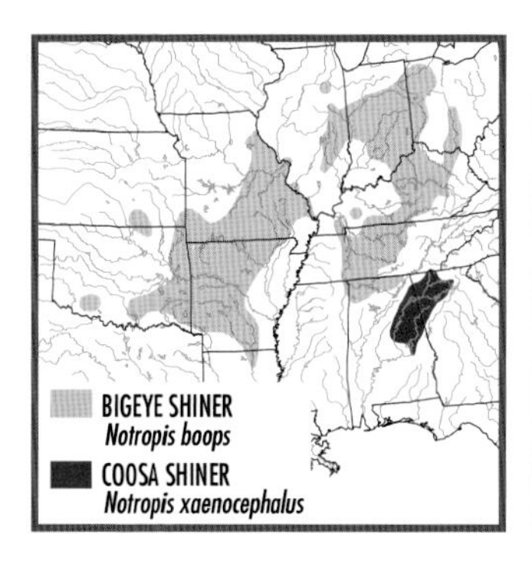

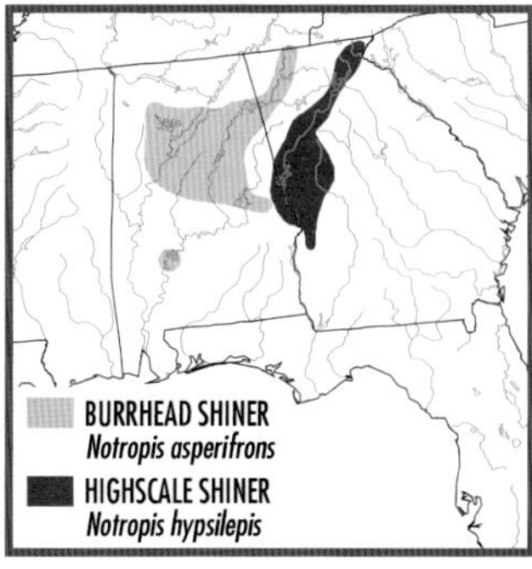

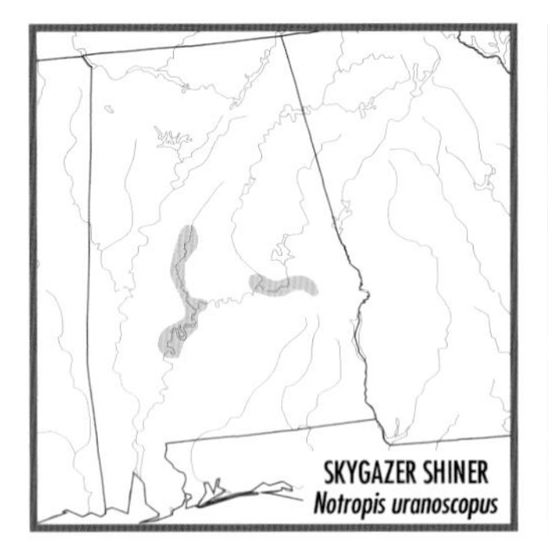

SKYGAZER SHINER *Notropis uranoscopus* Pl. 11

IDENTIFICATION: Large, *elliptical eye* high on head. Small *black wedge* on caudal fin base; punctate lateral line (obscured by dark stripe along rear half of side); 2 dark crescents between nostrils (see Fig. 28). Slender, compressed body; dorsal fin origin slightly behind pelvic fin origin. Somewhat flattened head; long, fairly pointed snout; slightly subterminal mouth. Straw yellow above, scales fairly darkly outlined; silver white below. Gold stripe above dark stripe along side in large individual. Complete lateral line; 34–37 lateral scales; 7 anal rays; pharyngeal teeth 0,4-4,0. To 2¾ in. (7.1 cm). **RANGE:** Cahaba, Tallapoosa, and Alabama river systems, AL. Locally common. **HABITAT:** Gravelly and sandy runs, often near riffles, of small to large rivers.

WEED SHINER *Notropis texanus* Pl. 11

IDENTIFICATION: *Black stripe* along side and around snout (on both lips); some *black-edged scales* below black stripe; often a clear stripe between black stripe and black-edged scales on upper side and back. Black spot on caudal fin base connected to, or barely separated from, black stripe; often streaking to end of caudal fin. Dark stripe along back much *wider in front of* than behind dorsal fin; often expanded into blotch in front of dorsal fin. Fairly compressed body; dorsal fin origin in front of pelvic fin origin. Fairly blunt snout; small (rear edge under nostril) terminal mouth. Olive-yellow above, silver side; clear to light red (in southern populations) fins; in Gulf Slope populations, *last 3–4 anal rays lined with black* (other rays clear). Complete or nearly complete lateral line; 32–39 lateral scales; usually 7 anal rays; pharyngeal teeth 2,4-4,2. To 3½ in. (8.6 cm). **RANGE:** Lowlands in Lake Michigan, Hudson Bay, and Mississippi R. basins from MI, WI, and MB south to Gulf; Gulf Slope drainages from Suwannee R., GA and FL, to Nueces R., TX. Common in south, uncommon and localized in north. **HABITAT:** Sandy runs and pools of creeks and small to medium rivers. Usually in clear water; in north often near vegetation. **REMARKS:** Individuals in north tend to be deeper bodied, more yellow; have black spot on caudal fin base continuous with black stripe along side; lack black along anal fin rays. **SIMILAR SPECIES:** See Pl. 11. (1) See Coastal Shiner, *N. petersoni*. (2) Coosa Shiner, *N. xaenocephalus*, is more slender, has large black spot on caudal fin base, no black along anal rays. (3) Blackchin Shiner, *N. heterodon*, and (4) Bigeye Shiner, *N. boops*, have pointed snout, dorsal fin origin over pelvic fin origin, usually 8 anal rays; pharyngeal teeth 1,4-4,1.

COASTAL SHINER *Notropis petersoni* Pl. 11

IDENTIFICATION: Similar to Weed Shiner, *N. texanus*, but *lacks* black-edged scales on side below black stripe; has *all anal rays* lined with black, wedge-shaped spot on caudal fin base, longer, more overhanging snout; is more compressed. Individuals from tannin-colored ("black") water are darkly pigmented, pink below; have light yellow to light pink stripe on upper side. To 3¼ in. (8.2 cm). **RANGE:** Atlantic and Gulf slope drainages from Onslow R., NC, to Jourdan R., MS. Restricted to lower Coastal Plain on Gulf Slope. Common. **HABITAT:** Over sand in pools and backwaters of creeks and small to large rivers, spring effluents, lakes. **SIMILAR SPECIES:** (1) See Weed Shiner, *N. texanus* (Pl. 11).

KIAMICHI SHINER *Notropis ortenburgeri* Pl. 9

IDENTIFICATION: Light olive above, *pale stripe* above silver black stripe along side; black stripe continues around snout (across chin); *dark-edged scales* on back. *Strongly upturned mouth.* Compressed body, deepest at dorsal fin origin; strongly tapering to narrow caudal peduncle. Dorsal fin origin slightly behind pelvic fin origin; large eye. Has 35–37 lateral scales; 9–10 anal rays; pharyngeal teeth 0,4-4,0. To 2¼ in. (5.5 cm). **RANGE:** Upper Ouachita, Arkansas, and Red river drainages; sw. AR and e. OK. Uncommon. **HABITAT:** Rocky, flowing pools of clear creeks and small rivers. **SIMILAR SPECIES:** (1) See Blackmouth

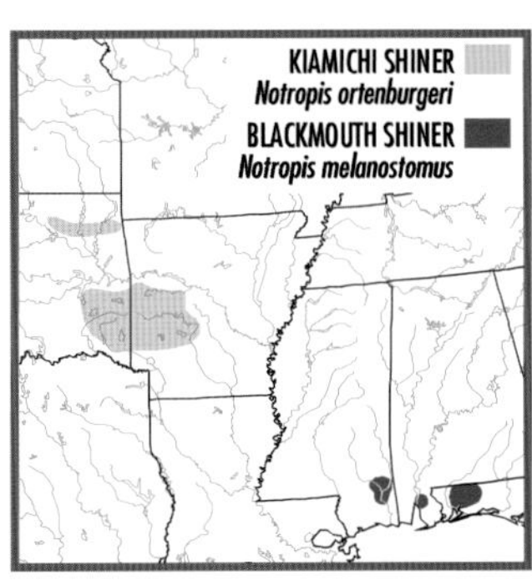

Shiner, *N. melanostomus*. (2) Bigeye Shiner, *N. boops* (Pl. 11), has less arched body, more *horizontal mouth,* bigger eye, 8 anal rays, pharyngeal teeth 1,4-4,1.

BLACKMOUTH SHINER *Notropis melanostomus* **Not shown**

IDENTIFICATION: Nearly identical to Kiamichi Shiner, *N. ortenburgeri*, but usually has *10–12 anal rays.* To 1½ in. (3.8 cm). **RANGE:** Blackwater-Yellow river system (Pensacola Bay drainage), FL; Bay Minette Creek (Mobile Bay), AL; lower Pascagoula R. system, MS. Rare and highly localized. **HABITAT:** Shallow vegetated muddy backwaters and quiet pools of creeks and small rivers. **SIMILAR SPECIES:** (1) See Kiamichi Shiner, *N. ortenburgeri* (Pl. 9).

PEPPERED SHINER *Notropis perpallidus* **Pl. 10**

IDENTIFICATION: Mostly translucent but with *many black spots* on back, head, and along side; 2 parallel *black dashes* just in front of dorsal fin; black on lips, base of dorsal (rear half) and anal fins, along rear half of side; some black along dorsal, caudal, and pectoral fin rays. Slender, compressed body; pointed snout; dorsal fin origin well behind pelvic fin origin. Has 9–10 anal rays, 32–35 lateral scales; pharyngeal teeth 2,4-4,2. To 2 in. (5 cm). **RANGE:** Ouachita and Red river drainages, s. AR and se. OK; in Red R. drainage restricted to Little and Kiamichi river systems. Uncommon. **HABITAT:** Pools and sluggish runs of small to medium rivers; often in quiet water near vegetation. **SIMILAR SPECIES:** Other small minnows with black spots are (1) Chihuahua Shiner, *N. chihuahua* (Pl. 10), with rounded snout, black wedge on caudal fin base, 7–8 anal rays, less compressed body; (2) Shoal Chub, *Macrhybopsis hyostoma,* and close relatives (Pl. 7), with long barbels, wide body, dorsal fin origin over or in front of pelvic fin origin.

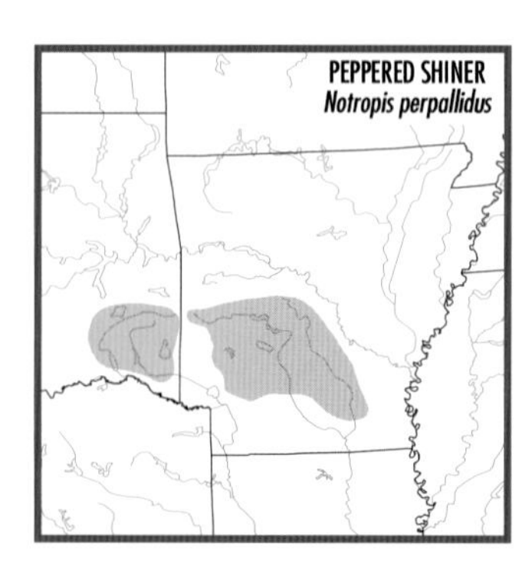

IRONCOLOR SHINER *Notropis chalybaeus* Pl. 18

IDENTIFICATION: *Black stripe* from spot on caudal fin base along side and *around snout,* covering both lips and chin (Fig. 30); *black inside mouth.* Large eye (longer than snout). Compressed body, deepest (and usually arched) at dorsal fin origin, which is over pelvic fin origin; slender caudal peduncle. Pointed snout; small, oblique terminal mouth. Straw yellow above, dusky stripe along back widest and darkest in front of dorsal fin. Scales above darkly outlined except just above black stripe where gold-orange streak may be present. Silver white below. Breeding male has orange-gold body and fins. Usually incomplete lateral line, 31–37 lateral scales; 8 anal rays; pharyngeal teeth 2,4-4,2. To 2½ in. (6.5 cm). **RANGE:** Lowlands of Atlantic, Gulf, and Mississippi R. basins, from Hudson R., NY, to s. FL, and across Gulf Slope to Pearl R., LA; Sabine R., LA and TX; north in Former Mississippi Embayment (on west side of Mississippi R. only) to se. MO; extends up Red R. drainage to se. OK. Isolated populations in San Marcos R., TX, Illinois R. drainage, IL and IN, Cedar R., IA (now extirpated), Wisconsin R., WI, Lake Winnebago system, WI, and Lake Michigan drainage of s. MI and n. IN. Generally common; highly localized in north. **HABITAT:** Clear, vegetated, sand-bottomed pools and slow runs of creeks and small rivers. **SIMILAR SPECIES:** (1) Often collected with Weed Shiner, *N. texanus* (Pl. 11), which also has black stripe on side but has black-edged scales above and *below* lateral line, less compressed and arched body, blunter snout, 7 anal rays.

DUSKY SHINER *Notropis cummingsae* Pl. 18

IDENTIFICATION: Wide *black stripe* along silver side from tip of snout and lips to caudal fin where expanded into black spot *streaking* backward on caudal fin; lower edge of stripe fuzzy (not sharply defined), extending below lateral line throughout. *Compressed* body; dorsal fin origin behind pelvic fin origin; slightly subterminal mouth. Dusky

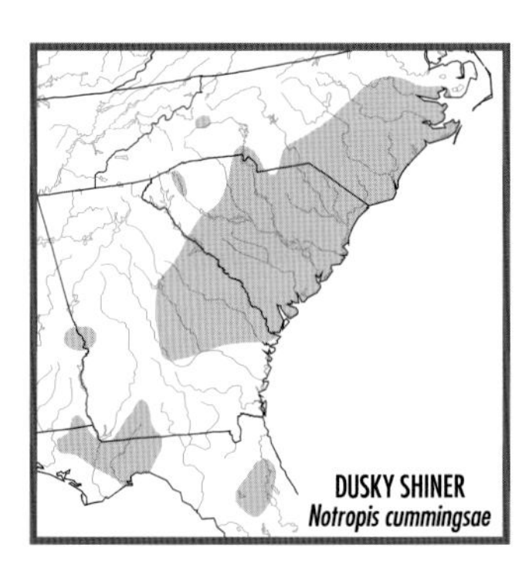

yellow-brown above, scales barely outlined; dark stripe along back, darkest in front of dorsal fin. Light yellow to orange (on large male) stripe above dark stripe along side. Fins clear to pale orange (on large male). Usually 37–39 lateral scales, 10–11 anal rays; pharyngeal teeth 1,4-4,1. To 2¾ in. (7.2 cm). **RANGE:** Atlantic and Gulf slopes in 4 areas: Tar R. drainage, NC, to Altamaha R. drainage, GA; St. Johns R. drainage, FL; Aucilla R. drainage to Choctawhatchee R. drainage, FL and AL; and middle Chattahoochee R. system, GA and AL. Generally common on Coastal Plain, uncommon on Piedmont. **HABITAT:** Pools and runs over sand and mud, usually in clear or tannin-stained creeks and small rivers. **SIMILAR SPECIES:** (1) Highfin Shiner, *N. altipinnis* (Pl. 18), has narrower, *more diffuse* dark stripe along side not extending below lowest part of lateral line; no dark stripe (thin streaks may be present) along back in front of dorsal fin; boldly outlined scales on back; pharyngeal teeth 2,4-4,2. (2) Ironcolor Shiner, *N. chalybaeus* (Pl. 18), has narrower blacker stripe along side with upper and lower edges sharply defined, is less compressed, has dorsal fin origin over pelvic fin origin; usually 8 anal rays.

REDEYE CHUB *Notropis harperi* Pl. 18

IDENTIFICATION: *Pink-tan above; red eye;* dark stripe along side and around snout and upper lip, followed by black spot on caudal fin base; light stripe above dark stripe extends onto snout as a *yellow arc;* dusky stripe along back; dark-edged scales on back and upper side; white to cream below. Small barbel (often absent) at corner of subterminal mouth. Slender, fairly compressed body; rounded snout; dorsal fin origin slightly behind pelvic fin origin. Has 30–38 lateral scales; 8 anal rays; pharyngeal teeth 0,4-4,0. To 2¼ in. (6 cm). **RANGE:** Below Fall Line in Atlantic and Gulf slope drainages from Altamaha R., GA, to Escambia R., AL; south in FL to St. Johns and Withlacoochee river drainages. Locally common; absent from some drainages within range; abundant in prime habitat. **HABITAT:** Springs, spring-fed headwaters and creeks. **SIMILAR SPECIES:** (1) Other minnows *lack* light-colored arc on snout.

HIGHFIN SHINER *Notropis altipinnis* Pl. 18

IDENTIFICATION: *Boldly outlined scales on* back, separated from *diffuse silver black stripe* along side of body by sharply edged, clear to yellow stripe. Dark stripe extends from lips to base of caudal fin, where slightly darkened into black spot streaking onto caudal fin. Stripe not extending below lowest part of lateral line. Compressed body; deepest at dorsal fin origin, which is slightly behind pelvic fin origin. *Large eye* (longer than snout); terminal mouth; somewhat rounded snout. Olive-brown above, thin dark streaks along midline of back; clear to amber fins. Yellow snout on large individual. Usually 34–38 lateral scales;

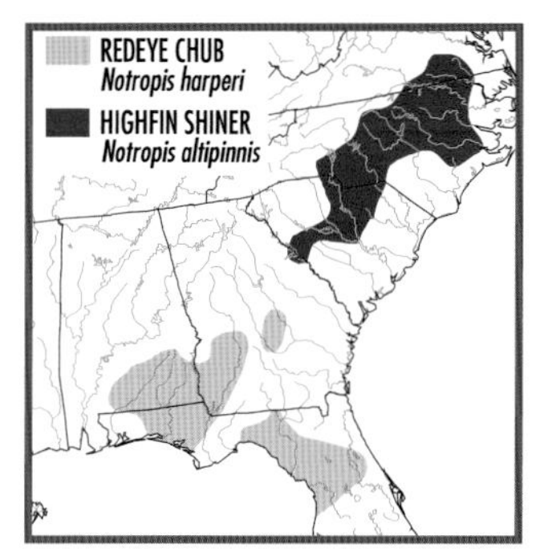

9–11 anal rays; pharyngeal teeth 2,4-4,2. To 2½ in. (6.1 cm). **RANGE:** Atlantic Piedmont and Coastal Plain drainages from Chowan R., VA, to Savannah R., GA. Common on Piedmont; uncommon on Coastal Plain. **HABITAT:** Pools, sometimes runs, of sandy and rocky creeks and small to medium rivers. **SIMILAR SPECIES:** (1) Dusky Shiner, *N. cummingsae* (Pl. 18), has *darker* stripe along side extending *below* lateral line throughout its length; dark stripe along back; scales on back barely outlined in black; pharyngeal teeth 1,4-4,1.

SAND SHINER *Notropis stramineus* Pl. 11

IDENTIFICATION: Complete, decurved, *punctate* (front half of body) *lateral line.* Dusky stripe along back expanded into dark *wedge* at dorsal fin origin. Slender, fairly compressed body; dorsal fin origin over to slightly behind pelvic fin origin. Rounded snout; small, slightly subterminal mouth; nipple at front of pupil (Fig. 29). Straw yellow above, faintly outlined scales on back and upper side; silver side (often a dusky stripe on rear half); small black caudal spot or wedge; clear fins tinged with white on large individual. Has 31–38 lateral scales; 7 anal rays; pharyngeal teeth 0,4-4,0. To 3¼ in. (8.1 cm). **RANGE:** St. Lawrence-Great Lakes, Hudson Bay (Red R.), and Mississippi R. basins from St. Lawrence R., s. QC, to e. SK, and south to TN and TX; west to e. MT, WY, CO, and NM; Gulf Slope drainages from Colorado R. to Rio Grande, TX, NM, and Mexico. Common; often abundant. **HABITAT:** Sand and gravel runs and pools of creeks and small to large rivers; sandy margins of lakes. Usually in clear creeks and small rivers; rarely in uplands. **SIMILAR SPECIES:** (1) See Swallowtail Shiner, *N. procne* (Pl. 11). (2) Palezone Shiner, *N. albizonatus*, has clear stripe above black stripe along side, less decurved lateral line; is more slender. (3) Mimic Shiner, *N. volucellus* (Pl. 11), *lacks* dark stripe along back, wedge at dorsal fin origin; has broader snout, wide scales along back, deep scales along side, 8 anal rays. (4) Bigmouth Shiner, *N. dorsalis* (Pl. 11), has upwardly directed eye, 8 anal rays; *lacks* dark wedge at dorsal fin origin.

SWALLOWTAIL SHINER *Notropis procne* **Pl. 11**

IDENTIFICATION: Similar to Sand Shiner, *N. stramineus*, but has *longer snout*, more subterminal mouth, nearly straight lateral line, blacker caudal wedge, often a darker stripe (black in some populations) along side and around snout, *yellow body and fins* on breeding male. To 2¾ in. (7.2 cm). **RANGE:** Atlantic drainages, above and below Fall Line, from Delaware and Susquehanna rivers, NY, to Santee R., SC; New R. drainage, VA (apparently introduced). Generally common. **HABITAT:** Sandy, sometimes rocky, pools and runs of creeks and small to large rivers. **SIMILAR SPECIES:** (1) See Palezone Shiner, *N. albizonatus*, and (2) Sand Shiner, *N. stramineus* (Pl. 11). (3) Cape Fear Shiner, *N. mekistocholas* (Pl. 11), has long coiled gut, black peritoneum, black lips, 8 anal rays.

PALEZONE SHINER *Notropis albizonatus* **Not shown**

IDENTIFICATION: Similar to Swallowtail Shiner, *N. procne*, but is more slender; has *clear stripe* above *black stripe* along side; black stripe extends around snout. To 2¾ in. (7.2 cm). **RANGE:** Little South Fork and Marrowbone Creek (middle Cumberland R. drainage), KY; upper Tennessee R. drainage, TN and AL. Rare and extremely localized; protected as an *endangered species*. **HABITAT:** Rocky and sandy, usually flowing, pools of creeks and small rivers. **SIMILAR SPECIES:** (1) See Swallowtail Shiner, *N. procne* (Pl. 11). (2) Sand Shiner, *N. stramineus* (Pl. 11), *lacks* well-defined clear and black stripes along side, is deeper bodied, has more decurved lateral line.

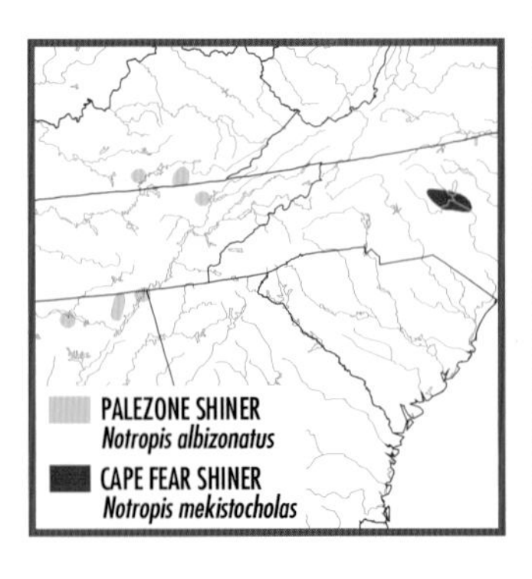

CAPE FEAR SHINER *Notropis mekistocholas* Pl. 11

IDENTIFICATION: *Long coiled dark gut* visible through belly wall; *black peritoneum.* Black stripe along side of body and side of snout, absent on front of snout; black lips. Black wedge on caudal fin base, usually detached from dark stripe along body. Olive above, scales outlined in black, thin dusky stripe along midline; light stripe above dark stripe along side. Clear to yellow fins (large individual). Compressed body; dorsal fin origin over or slightly in front of pelvic fin origin. Nearly horizontal, subterminal mouth. Complete lateral line; 34–37 lateral scales; 8 anal rays; pharyngeal teeth 0,4-4,0. To 3 in. (7.7 cm). **RANGE:** Cape Fear drainage near Fall Line, Chatham and Harnett counties, NC. Rare; recognized as an *endangered species.* **HABITAT:** Sandy and rocky pools and runs of small to medium rivers. **REMARKS:** A highly specialized detritus- and plant-eating species, Cape Fear Shiner has smallest range of any species of *Notropis.* **SIMILAR SPECIES:** (1) No other shiner in genus *Notropis* except Ozark Minnow, *N. nubilus* (Pl. 17), has long coiled gut; few have black peritoneum. (2) Swallowtail Shiner, *N. procne* (Pl. 11), and (3) Whitemouth Shiner, *N. alborus* (Pl. 10), have 7 anal rays, little or no black on lips.

CHIHUAHUA SHINER *Notropis chihuahua* Pl. 10

IDENTIFICATION: *Many black spots* on back and upper side of body and head; *black wedge* on caudal fin base. Lateral-line pores at front outlined in black. Yellow to pale orange lips and dorsal, caudal, and pectoral fins. Stout, barely compressed body, deepest under nape; dorsal fin origin over pelvic fin origin; rounded snout. Straw yellow above, often a dusky stripe along back; dusky (at front) to black (at rear) stripe on silver side; white below. Usually 7, often 8, anal rays. Has 33–37 lateral scales; pharyngeal teeth 0,4-4,0. To 3¼ in. (8 cm). **RANGE:** Rio Grande drainage in Big Bend region of sw. TX; Río Conchos system, n. Mexico. Uncommon in TX, common in Mexico. **HABITAT:** Sandy and rocky pools and runs of creeks and small rivers. **SIMILAR SPECIES:** (1) Speckled Chub, *Macrhybopsis aestivalis* (Pl. 7), has long barbel at corner of mouth, flattened body, no black wedge on caudal fin base.

TAMAULIPAS SHINER *Notropis braytoni* Pl. 10

IDENTIFICATION: *Dusky stripe* along side from opercle (where diffuse) to caudal peduncle, followed by *clear area,* then small *black wedge* on caudal fin base. Straw colored above; scales above dusky stripe along side darkly outlined, creating crosshatched appearance; often a dusky stripe along back. Silver side; white below; clear fins. Body compressed, deepest at origin or in front of dorsal fin; dorsal fin origin over pelvic fin origin. Bluntly rounded snout; subterminal mouth;

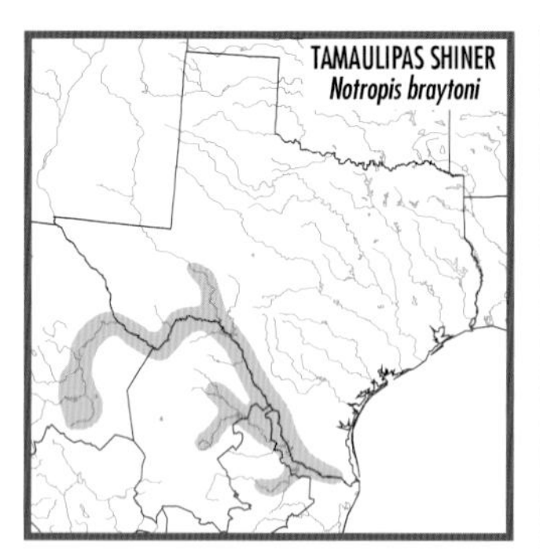

round eye. Complete lateral line; 32–39 lateral scales; usually 7 anal rays, 15–16 pectoral rays; pharyngeal teeth 1,4-4,1. To 2¾ in. (6.9 cm). **RANGE:** Rio Grande drainage from near mouth upstream to mouth of Río Conchos and lower Pecos R., TX. Also in Rio Grande drainage, n. Mexico. Common in Rio Grande mainstream. **HABITAT:** Rocky and sandy channels of large creeks and small to medium-sized rivers. **SIMILAR SPECIES:** (1) Within range, most similar to Phantom Shiner, *N. orca*, which has smaller eye, usually 8 anal rays, pharyngeal teeth 2,4-4,2, is more pallid with only faint stripe alongside.

GHOST SHINER *Notropis buchanani* Pl. 11

IDENTIFICATION: Aptly named, Ghost Shiner is *translucent milky white* overall. Body compressed, arched at front, deep at dorsal fin origin, *strongly tapering to thin caudal peduncle. Large, pointed fins;* depressed pelvic fins reach anal fin origin. In turbid water, lacks dark pigment; in clear water, scales on back may be faintly outlined; black specks may be present on snout, along lateral line, and along underside of caudal peduncle. Fairly large eye; rounded snout; small, subterminal mouth; dorsal fin origin over pelvic fin origin. Lateral-line scales on front half of body deeper than wide (Fig. 31). Complete lateral line; 30–35 lateral scales; 8 anal rays; no infraorbital canal (rarely a short segment); pharyngeal teeth 0,4-4,0. To 2½ in. (6.4 cm). **RANGE:** Mississippi R. basin from PA to se. NE and w. OK, and from MN and WI south to n. AL and LA; Lakes Erie and Huron drainages, ON and MI; Gulf Slope drainages from Calcasieu R., LA, to Rio Grande, TX and Mexico. Common in west, uncommon in east; absent in most of Ozarks and Ouachitas. **HABITAT:** Quiet pools and backwaters, usually over sand, of small to large rivers. **SIMILAR SPECIES:** (1) Mimic Shiner, *N. volucellus* (Pl. 11), *lacks* arched, deep body at dorsal fin origin; has infraorbital canal, *deeper caudal peduncle*; pelvic fins do not reach anal fin.

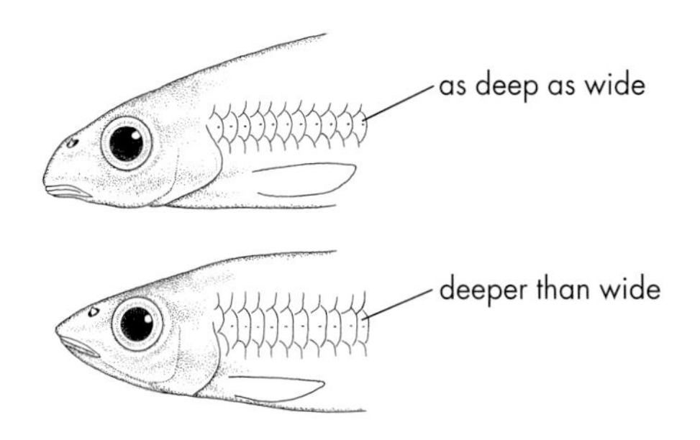

Fig. 31. Minnows—lateral-line scales.

MIMIC SHINER *Notropis volucellus* Pl. 11

IDENTIFICATION: Scales along back in front of dorsal fin *wider* than those on upper side. Scales along side (front half of body) *much deeper than wide* (Fig. 31). *Broad, rounded snout;* small, slightly subterminal mouth; large eye. Fairly slender, compressed body; dorsal fin origin over to slightly behind pelvic fin origin. Transparent gray to olive-yellow above, faintly to darkly outlined scales on back and upper side; dusky stripe (darkest at rear and on individuals from clear, vegetated habitats) along silver side; stripe expanded just in front of caudal fin (Fig. 32). Complete lateral line; 32–38 lateral scales; 8 anal rays; pharyngeal teeth 0,4-4,0. To 3 in. (7.6 cm). **RANGE:** St. Lawrence-Great Lakes, Hudson Bay, and Mississippi R. basins from QC and VT to MB, and south to Gulf; on Atlantic Slope in Susquehanna R., PA, and drainages from James R., VA, to Neuse R., NC (and introduced into Connecticut and Squannacook rivers, MA and CT); Gulf Slope drainages

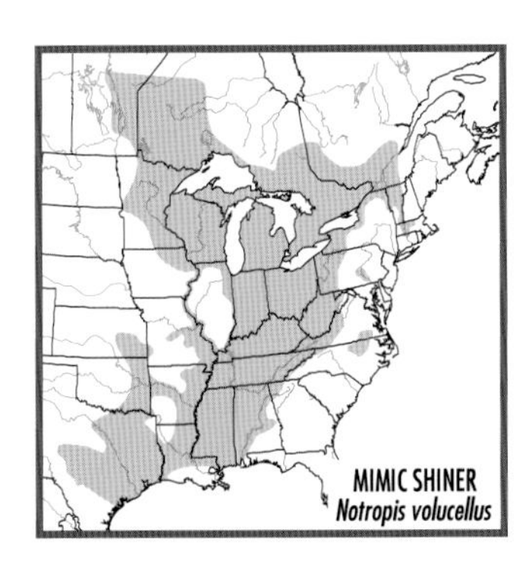

from Mobile Bay, GA and AL, to Nueces R., TX. Common. **HABITAT:** Sandy pools of headwaters, creeks, and small to large rivers, quiet areas of lakes. **REMARKS:** A highly variable form in need of study; almost certainly more than 1 species. Some populations recognized as Channel Shiner, *N. wickliffi*; however, descriptions vary regionally and do not seem to refer to same fish. **SIMILAR SPECIES:** (1) See Cahaba Shiner, *N. cahabae*. (2) Sand Shiner, *N. stramineus* (Pl. 11), has dusky stripe along back, black wedge at dorsal fin origin, narrower snout, 7 anal rays, *lacks* wide scales along back and deep scales along side.

CAHABA SHINER *Notropis cahabae* Not shown

IDENTIFICATION: Similar to Mimic Shiner, *N. volucellus,* but dark stripe along side *straight-edged* (rather than expanded) near caudal fin base (Fig. 32). To 3 in. (7.5 cm). **RANGE:** Cahaba R. and Locust Fork Black Warrior R., AL. Rare; protected as an *endangered species.* **HABITAT:** Flowing pools, usually over sand or gravel, in main channel of medium-sized rivers. **SIMILAR SPECIES:** (1) See Mimic Shiner, *N. volucellus* (Pl. 11).

OZARK SHINER *Notropis ozarcanus* Pl. 18

IDENTIFICATION: *Very slender,* usually arched body. *Small dusky to black spot* at dorsal fin origin. Rounded snout; small subterminal mouth; fairly large eye. Dorsal fin origin slightly behind pelvic fin origin. Light yellow above, dark-edged scales; thin black stripe along back; dusky

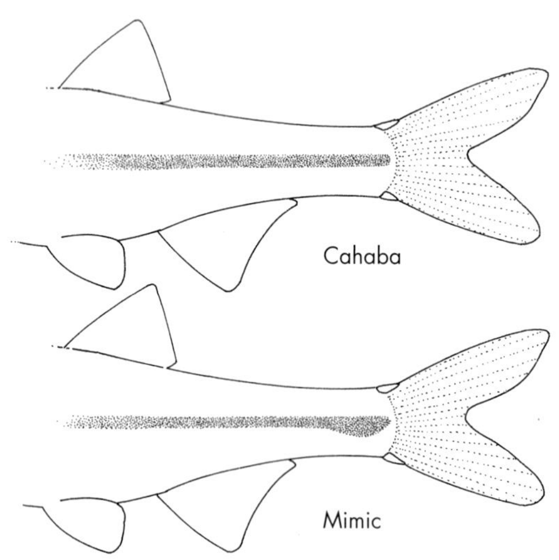

Fig. 32. Mimic and Cahaba shiners—dark stripe on caudal peduncle.

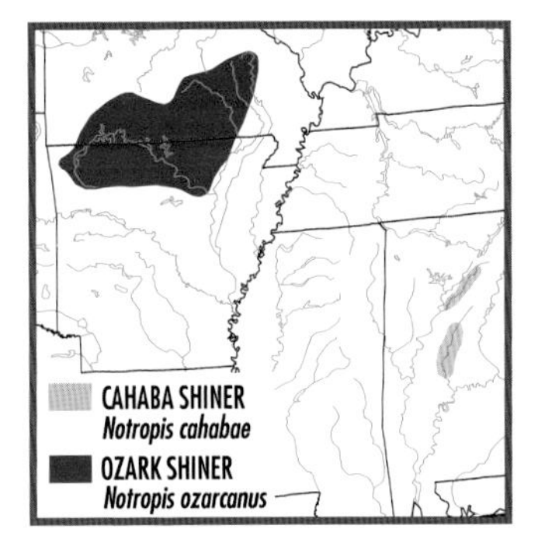

stripe (darkest at rear) along silver side; often a small black caudal spot. Fins clear to dusky in adult; black on breeding male. Complete lateral line; 34–38 lateral scales (those at front much deeper than adjacent scales); 8 anal rays; pharyngeal teeth 0,4-4,0. To 3 in. (7.5 cm). **RANGE:** Above Fall Line in White and Black river systems, MO and AR; Illinois R. system (Arkansas R. drainage), AR. Formerly in upper St. Francis R. drainage, MO, but now extirpated. Locally common. **HABITAT:** Rocky and sandy runs and flowing pools, often near riffles, of clear fast-flowing small to medium rivers. **SIMILAR SPECIES:** (1) Mirror Shiner, *N. spectrunculus* (Pl. 18), has no or crowded scales on nape, a broader head, black wedge at caudal fin base, no spot at dorsal fin origin.

MIRROR SHINER *Notropis spectrunculus* Pl. 18

IDENTIFICATION: Scales on front half of nape *absent* (skin "mirrorlike") or *small and crowded. Broad head;* rounded snout; somewhat upwardly directed eye; small subterminal mouth. Slender body, usually *arched* throughout; *black wedge* on caudal fin base. Dorsal fin origin behind pelvic fin origin. Olive above, thin black stripe along back; darkly outlined scales on back and upper side; dusky stripe (darkest at rear) on silver side. Large male has white edges on *red-orange fins.* Complete lateral line; 36–39 lateral scales; 8–9 anal rays; pharyngeal teeth 0,4-4,0. To 3 in. (7.5 cm). **RANGE:** Upper Tennessee R. drainage, VA, NC, TN, and GA; extreme upper Savannah and Santee river drainages, NC, SC, and GA. Common to abundant in NC; more localized elsewhere. **HABITAT:** Rocky, sandy, and muddy pools and backwaters of high-gradient creeks and small rivers. **SIMILAR SPECIES:** (1) See Sawfin Shiner, *N.* species (Pl. 18). (2) Ozark Shiner, *N. ozarcanus* (Pl. 18), has *normal scales* on nape, *black spot* at dorsal fin origin, *no* bold black wedge on caudal fin base, narrower head.

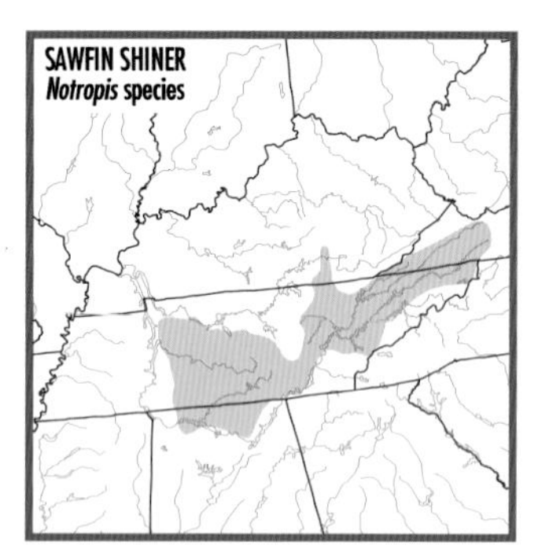

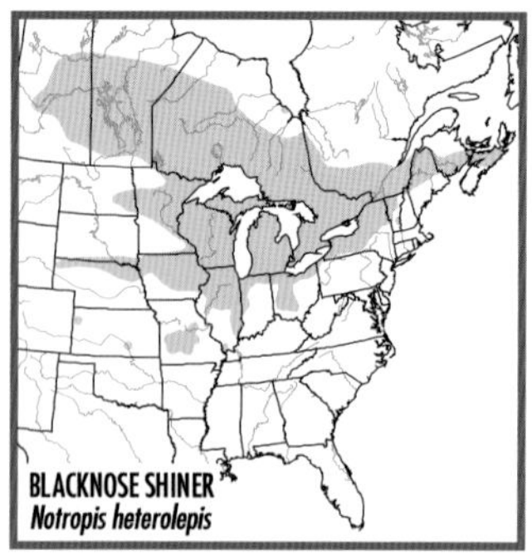

SAWFIN SHINER *Notropis* species **Pl. 18**

IDENTIFICATION: Similar to Mirror Shiner, *N. spectrunculus*, but has black specks along *only 1st 4,* rather than all, dorsal fin rays. Large male has red-orange and black specks on *front half* of dorsal fin, red-orange on anal and paired fins. To 2½ in. (6.6 cm). **RANGE:** Tennessee R. drainage, VA, TN, and AL; middle Cumberland R. drainage, KY and TN. Locally common. **HABITAT:** Rocky and sandy pools and backwaters of creeks and small to medium rivers. **SIMILAR SPECIES:** (1) See Mirror Shiner, *N. spectrunculus* (Pl. 18).

BLACKNOSE SHINER *Notropis heterolepis* **Pl. 10**

IDENTIFICATION: *Black stripe* along side and around snout, but barely onto upper lip and absent on chin (Fig. 30); *black crescents* within stripe. Rounded, somewhat elongated snout; small, nearly horizontal, subterminal mouth; round eye. Slender, slightly compressed body; dorsal fin origin slightly behind pelvic fin origin. Olive to straw colored above; often a faint streak in front of dorsal fin; scales darkly outlined except above dark stripe along silver side. Incomplete lateral line; 32–39 lateral scales; 13 or more predorsal scales; 8 anal rays; pharyngeal teeth 0,4-4,0. To 3¾ in. (9.8 cm). **RANGE:** Atlantic, Great Lakes, Hudson Bay, and Mississippi R. basins from NS to SK, south to OH, IL, s.-cen. MO, and KS (where extirpated). Common in north, but disappearing from southern part of range. **HABITAT:** Clear vegetated lakes and pools of creeks and small rivers; usually over sand. **SIMILAR SPECIES:** (1) See Bedrock Shiner, *N. rupestris* (Pl. 10). (2) Pallid Shiner, *Hybopsis amnis* (Pl. 7), has larger horizontally elliptical eye, dorsal fin origin over or in front of pelvic fin origin, body notably arched at dorsal fin origin, *no* black crescents in dark stripe along side, pharyngeal teeth 1,4-4,1.

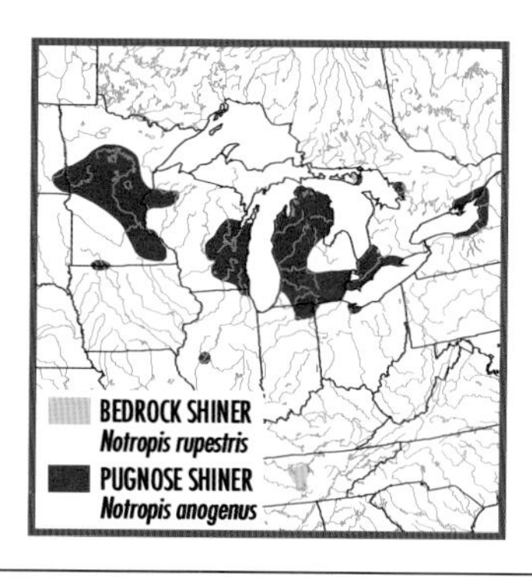

BEDROCK SHINER *Notropis rupestris* Pl. 10

IDENTIFICATION: Similar to Blacknose Shiner, *N. heterolepis*, but has *complete lateral line* (except small young), *more arched back,* dorsal fin origin over or only slightly behind pelvic fin origin; smaller, more terminal mouth, 10–14 (usually 11–12) predorsal scales. To 2½ in. (6.2 cm). **RANGE:** Lower Caney Fork system and nearby tributaries of cen. Cumberland R. drainage, TN; extreme upper Duck R., TN (where possibly introduced). Common. **HABITAT:** Bedrock pools of headwaters, creeks, and small rivers; usually near vegetation. **SIMILAR SPECIES:** (1) See Blacknose Shiner, *N. heterolepis* (Pl. 10).

PUGNOSE SHINER *Notropis anogenus* Pl. 10

IDENTIFICATION: Small shiner with very *small, sharply upturned mouth* (Fig. 30) and *black peritoneum.* Dark stripe along side and on chin, lower lip, and side of upper lip. *Black wedge* on caudal fin base. Fairly compressed body; dorsal fin origin over pelvic fin origin. Olive above, thin dark line along back; scales darkly outlined except above dark stripe along side; silver white below. Breeding male has yellow body and fins. Usually complete lateral line; 34–38 lateral scales; 8 (often 7) anal rays; pharyngeal teeth 0,4-4,0. To 2¼ in. (5.8 cm). **RANGE:** From Lake Ontario drainage of ON and NY to se. ND and cen. IL. Mostly restricted to Great Lakes and Mississippi R. basins but also in Red R. drainage (Hudson Bay basin) of MN and SD. Rare and disappearing over most of range; protected in Canada as an *endangered species.* **HABITAT:** Clear vegetated lakes; vegetated pools and runs of creeks and rivers. Usually over sand and mud. **SIMILAR SPECIES:** (1) Pugnose Minnow, *Opsopoeodus emiliae* (Pl. 8), also has strongly upturned mouth but has dark areas on dorsal fin (except in parts of FL), cross-hatched pattern on upper side, *silvery white peritoneum,* 9 dorsal rays (Pugnose Shiner has 8), pharyngeal teeth 0,5-5,0. (2) Bridle Shiner, *N. bifrenatus* (Pl. 10), has black stripe around snout and *less upturned mouth* (Fig. 30), incomplete lateral line, usually 7 anal rays.

WHITEMOUTH SHINER *Notropis alborus* Pl. 10

IDENTIFICATION: Body divided into straw yellow upper and white lower halves by *silver black jagged-edged stripe* along side ending at *black wedge* on caudal fin base. *Black bridle* around blunt snout; no black on lips (Fig. 30). Scales on back outlined in black; *no dark stripe* along back. Body compressed, deepest at dorsal fin origin (over or slightly in front of pelvic fin origin) and strongly tapering to thin caudal peduncle. Nearly horizontal mouth. Usually complete lateral line. Has 31–35 lateral scales, 7 anal rays; pharyngeal teeth 0,4-4,0. To 2¼ in. (6 cm). **RANGE:** Atlantic Slope drainages from Chowan R., VA, to Santee R., SC. Mostly on Piedmont, where fairly common. **HABITAT:** Sandy and rocky pools and runs of headwaters, creeks, and small rivers. **SIMILAR SPECIES:** (1) Swallowtail Shiner, *N. procne*, and (2) Cape Fear Shiner, *N. mekistocholas* (both Pl. 11), have black stripe on side but *not front* of snout, some black on lips, thin dark stripe along back, more slender body less highly arched at dorsal fin origin.

BRIDLE SHINER *Notropis bifrenatus* Pl. 10

IDENTIFICATION: *Black spot* at base of caudal fin usually joined to *brown-black stripe* along side and *around snout* (*narrower* on snout, mostly confined to upper lip—Fig. 30); light stripe above dark stripe. Scales on back darkly outlined; often a dusky stripe along midline of back; silver side. Body slightly compressed, deepest at dorsal fin origin, which is slightly behind pelvic fin origin. Moderately blunt snout; small oblique mouth. Incomplete lateral line except in large (2-in. [5-cm]) individual. Has 33–36 lateral scales; usually 7 anal rays; pharyngeal teeth 0,4-4,0. To 2½ in. (6.5 cm). **RANGE:** Atlantic Slope drainages from St. Lawrence-Lake Ontario drainage, s. QC, to Chowan R. system, VA; isolated populations in lower Neuse R., NC, and Santee R., SC, drainages. Uncommon and decreasing. **HABITAT:**

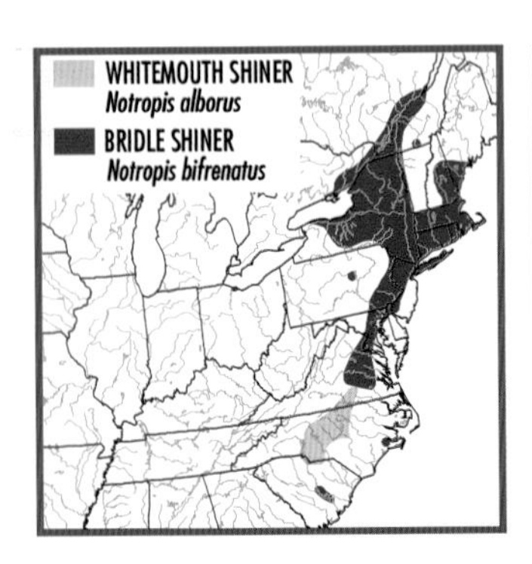

Vegetated ponds, lakes, and sluggish mud-bottomed pools of creeks and small to medium rivers. **SIMILAR SPECIES:** (1) Pugnose Shiner, *N. anogenus* (Pl. 10), has extremely small, upturned mouth; dark stripe *absent* from snout, confined to chin and lips; usually complete lateral line. (2) Whitemouth Shiner, *N. alborus* (Pl. 10), has more horizontal mouth; stripe around snout *as wide in front* as on side of snout, absent on exposed part of upper lip; no dark stripe along midline of back; usually complete lateral line.

'AILLIGHT SHINER *Notropis maculatus* Pl. 18

IDENTIFICATION: *Large black spot* at center, small black spots on upper and lower edge, of caudal fin base; *red* between spots. *Crosshatched pattern* on back and side. Large black *blotch* (darkest on male) along front of dorsal fin. Male has *black bands* near edges of dorsal, caudal, anal, and pelvic fins. Slender, compressed body; fairly long, rounded snout; subterminal mouth; large, pointed fins; dorsal fin origin behind pelvic fin origin. Light olive above, thin dusky stripe along back; dusky stripe along silver side and around snout; snout usually faint red. Breeding male has *bright red body and head, red-black edge on fins.* Incomplete lateral line, 8–10 pores; 34–39 lateral scales; 8 anal rays, 8 dorsal rays; pharyngeal teeth 0,4-4,0. To 3 in. (7.6 cm). **RANGE:** Below Fall Line in Atlantic, Gulf, and Mississippi R. basins from Cape Fear R., NC, to Red R. drainage, TX; north in Former Mississippi Embayment to s. IL. Locally common in se. U.S.; uncommon in Mississippi basin. **HABITAT:** Swamps, ponds, backwaters and pools of small to large rivers; usually near vegetation or debris. **SIMILAR SPECIES:** (1) Pugnose Minnow, *Opsopoeodus emiliae* (Pl. 8), *lacks* large black spot on caudal fin base; has nearly vertical mouth, 9 dorsal rays, pharyngeal teeth 0,5-5,0 or 0,5-4,0; large male lacks red on body, has bright white anal and pelvic fins.

POTTAIL SHINER *Notropis hudsonius* Pl. 10

IDENTIFICATION: *Large eye;* short, rounded snout; nearly horizontal, subterminal mouth. *Large black caudal spot* (inconspicuous in s. Atlantic drainages and often on large individuals elsewhere). Fairly slender, compressed body; dorsal fin origin over or slightly in front of pelvic fin origin. Olive-gray above, dusky stripe along back, dark-edged scales often form wavy lines on back and upper side; punctate (often faint) lateral line at front, dusky stripe along rear of silver side. Has 36–42 lateral scales; 8 anal rays; pharyngeal teeth usually 2,4-4,2 (1,4-4,1 or 0,4-4,0 on Atlantic Slope). To 5¾ in. (15 cm). **RANGE:** Atlantic and Gulf slope drainages from St. Lawrence R., QC, to Altamaha and upper Chattahoochee river, GA; Arctic, Great Lakes, and Mississippi R. basins from QC to Mackenzie R. drainage, NT and BC, and south to s. IL. Common; locally abundant. **HABITAT:** Sandy and rocky pools and runs of small to large rivers (and creeks on Atlantic Slope); sandy and rocky shores of lakes. **SIMILAR SPECIES:** (1) Silver Chub, *Macrhybopsis storeriana* (Pl. 6), has no black spot on caudal fin base, eye higher on head, barbel at corner of mouth.

BLACKSPOT SHINER *Notropis atrocaudalis* Pl. 10

IDENTIFICATION: *Narrow black stripe* along side and around snout (including on upper lip); *horizontally oblong black caudal spot, separated* from dark stripe along side and often streaking to end of caudal fin. *Stocky body,* slightly compressed; dorsal fin origin in front of pelvic fin origin. Small eye; rounded snout; nearly horizontal, subterminal mouth. Olive above, scales darkly outlined; dusky lines on back and upper side converging at rear; wide dusky stripe along back. Silver side, black specks above and below lateral-line pores. Usually complete lateral line, 35–40 lateral scales, 7 anal rays; pharyngeal teeth 0,4-4,0. To 3 in. (7.6 cm). **RANGE:** Red (Mississippi R. basin) and Calcasieu R. (Gulf Slope) drainages to Brazos R. drainage, sw. AR, se. OK, w. LA, and e. TX. Common in southern part of range; uncommon in north. **HABITAT:** Sandy and rocky runs and pools of creeks and small to medium rivers; usually in shallow (to 20 in. [50 cm]) water. **SIMILAR SPECIES:** (1) Topeka Shiner, *N. topeka* (Pl. 17), has black *wedge* on caudal base, *lacks* black stripe around snout, has red-orange fins on large male.

TOPEKA SHINER *Notropis topeka* Pl. 17

IDENTIFICATION: *Stocky,* compressed body; *small eye. Black wedge* on caudal fin base. Dusky to dark stripe along side, not extending around snout. Dorsal fin origin over pelvic fin origin, closer to tip of snout than to caudal fin base. Round snout; small, nearly terminal mouth. Olive above, large dark stripe along back in front of dorsal fin (often thin stripe behind dorsal fin); scales outlined in black. Silver white below. Breeding male is *orange* below and on side of head, has *red-orange*

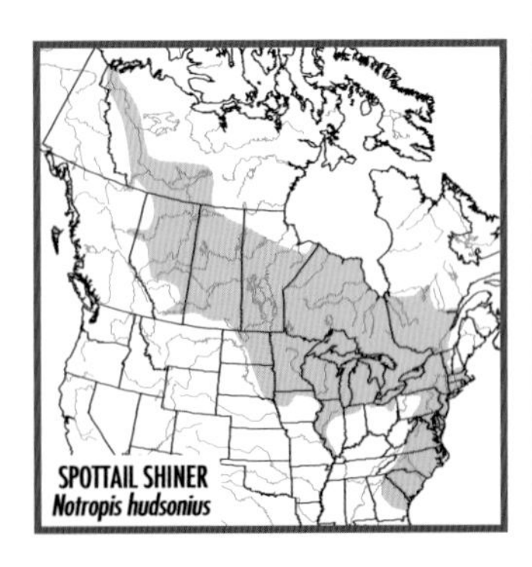

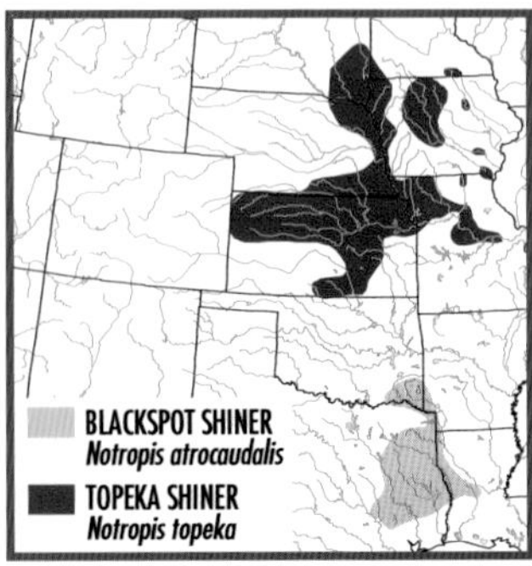

fins. Complete lateral line. Has 32–37 lateral scales; 7 anal rays; pharyngeal teeth 0,4-4,0. To 3 in. (7.6 cm). **RANGE:** Mississippi R. basin from s. MN and se. SD south to cen. MO and s. KS (Arkansas R. drainage). Generally uncommon; protected as an *endangered species.* **HABITAT:** Quiet gravel- and sand-bottomed pools of headwaters and creeks.

SILVERJAW MINNOW *Ericymba buccata* Pl. 11

IDENTIFICATION: Large silver white chambers (modified infraorbital and preoperculomandibular canals) on cheek and *flattened* underside of head; *upwardly* directed eye high on head. Slightly compressed body deepest at nape; long snout, subterminal mouth; dorsal fin origin over pelvic fin origin. Scales on breast. Light tan to olive-yellow above, dark streak along back, darkest in front of dorsal fin; scales may be darkly outlined; silver side, sometimes with dusky stripe. Complete lateral line; 31–37 lateral scales; 8 anal rays; pharyngeal teeth 1,4-4,1 or 0,4-4,0. To 3¾ in. (9.8 cm). **RANGE:** Atlantic Slope from Susquehanna R. drainage, PA, to Rappahannock R. drainage, VA; lower Great Lakes, and Mississippi R. drainages from w. NY, PA, and VA, to e. MO; south to upper Cumberland R. drainage, KY and TN. Generally common. **HABITAT:** Shallow sandy riffles and runs of creeks and small to medium rivers. **SIMILAR SPECIES:** (1) See Longjaw Minnow, *E. amplamala.* (2) Other minnows lack large silver white chambers on cheek and underside of head.

LONGJAW MINNOW *Ericymba amplamala* Not shown

IDENTIFICATION: Nearly identical to Silverjaw Minnow, *E. buccata,* but has no scales on breast, longer rakers on 1st gill arch (about twice as long as wide vs. about as long as wide), 5 (vs. 4) infraorbital ossicles (bones under eye). Has 34–38 lateral scales; pharyngeal teeth 1,4-4,1. To 3¾ in. (9.6 cm). **RANGE:** Gulf Slope drainages from Apalachicola R., GA and FL, to Mississippi R., s. MS; on Atlantic Slope in upper

Altamaha R. drainage. Common. **HABITAT:** Shallow sandy riffles and runs of creeks and small to medium rivers. **SIMILAR SPECIES:** (1) See Silverjaw Minnow, *E. buccata* (Pl. 11).

HYBOPSIS

GENUS CHARACTERISTICS: Shinerlike minnows living on or close to stream bottom. *Upwardly directed, horizontally elliptical eye;* long snout overhanging *subterminal mouth.* All but Pallid Shiner, *H. amnis*, usually have *barbel at corner of mouth* (Fig. 16).

BIGEYE CHUB *Hybopsis amblops* Pl. 7

IDENTIFICATION: *Black stripe* (faded in turbid water) along side and onto (often around) snout. *Large eye,* about equal to length of snout; small mouth; snout projecting well beyond upper lip. Slender, slightly compressed body; dorsal fin origin over or slightly behind pelvic fin origin. Light yellow above; dark streak along back in front of dorsal fin; scales dark-edged, producing wavy lines; silver side, often with a yellow streak above black stripe; sometimes a black caudal spot. Breeding male has many small tubercles scattered over head (Fig. 33). Complete lateral line; 33–38 lateral scales; 8 anal rays; pharyngeal teeth 1,4-4,1. To 3½ in. (9 cm). **RANGE:** Lakes Ontario and Erie drainages, NY, PA, OH, and MI; Ohio R. basin from NY to e. IL, and south to Tennessee R. drainage, GA and AL; Ozarks of s. MO, n. AR, and ne. OK (absent in Missouri R. drainage); 1 record for Cottonwood R., KS. **HABITAT:** Rocky pools with current, usually near riffles and vegetation. Common to abundant in south; declining in agricultural areas in north. **SIMILAR SPECIES:** (1) See Rosyface Chub, *H. rubrifrons.*

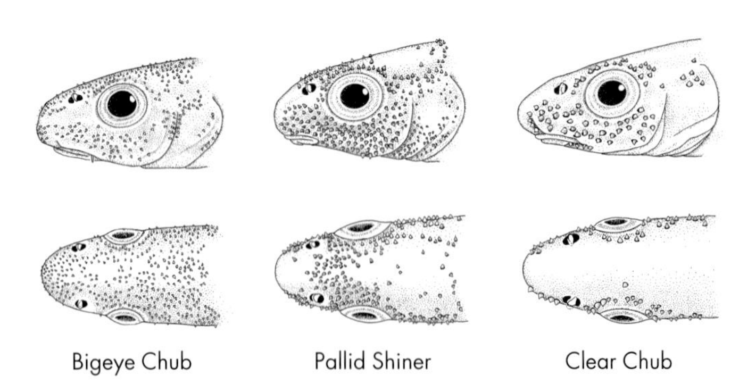

Fig. 33. Bigeye Chub, Pallid Shiner, and Clear Chub—head tubercles on breeding male.

(2) Pallid Shiner, *H. amnis* (Pl. 7), and (3) Clear Chub, *H. winchelli*, have arched, more strongly compressed body; fewer, larger tubercles on head of breeding male.

ROSYFACE CHUB *Hybopsis rubrifrons* Not shown

IDENTIFICATION: Similar to Bigeye Chub, *H. amblops*, but has smaller eye (less than snout length) and during (spring) breeding season develops red on front ⅓ of body (snout usually brightest). Has 35–39 lateral scales. To 3¼ in. (8.4 cm). **RANGE:** Saluda, Savannah, and Altamaha river drainages, SC and GA. Mostly above Fall Line, where common; locally abundant. Possibly introduced into Chattahoochee R. system, GA. **HABITAT:** Sand- and gravel-bottomed pools and runs of creeks and small rivers. **SIMILAR SPECIES:** (1) See Bigeye Chub, *H. amblops* (Pl. 7).

PALLID SHINER *Hybopsis amnis* Pl. 7

IDENTIFICATION: *Back arched* at dorsal fin origin. *Large eye,* about equal to length of snout; small mouth; snout projecting well beyond upper lip. Straw yellow above; scales usually dark-edged; black stripe along silver side and around snout (stripe darkest at rear, absent in turbid water); sometimes a black caudal spot. Fairly compressed body; dorsal fin origin over or in front of pelvic fin origin. Rarely a barbel at corner of mouth. Breeding male has tubercles concentrated on lower half of head (Fig. 33). Complete lateral line; 33–38 lateral scales; 8 anal rays; pharyngeal teeth 1,4-4,l. To 3¼ in. (8.4 cm). **RANGE:** Mississippi R. basin from WI and MN south to LA; mostly in lowlands but extends up Cumberland R. to s.-cen. KY, and in Arkansas and Red river drainages to e. OK; Gulf drainages to Guadalupe R., TX. Generally rare and declining. **HABITAT:** Sandy and silty pools of medium to large rivers; often in small rivers in e. TX. **SIMILAR SPECIES:** (1) See Clear Chub, *Hybopsis winchelli*. (2) Bigeye Chub, *H. amblops* (Pl. 7), has

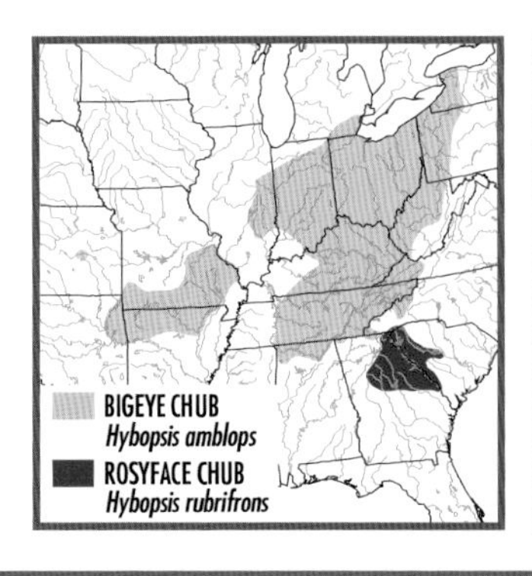

less arched, less compressed body; barbel at corner of mouth; more, smaller tubercles on head of breeding male. (3) Blacknose Shiner, *Notropis heterolepis* (Pl. 10), has *round* (not elliptical) eye, black crescents in dark stripe along side; pharyngeal teeth 0,4-4,0.

CLEAR CHUB *Hybopsis winchelli* **Not shown**

IDENTIFICATION: Similar to Pallid Shiner, *H. amnis*, but (almost always) has *barbel* at corner of mouth; *flatter, longer head,* black caudal spot; fewer, larger tubercles on head of breeding male (Fig. 33). Large adult has light orange caudal fin. Has 33–39 lateral scales. To 3¼ in. (8.4 cm). **RANGE:** Gulf drainages from Ocklockonee R., FL, and Flint R., GA, to Mississippi R., MS. Common. **HABITAT:** Sand- and silt-bottomed pools, often near riffles, in creeks and small to medium rivers. **SIMILAR SPECIES:** (1) See Pallid Shiner, *H. amnis* (Pl. 7).

LINED CHUB *Hybopsis lineapunctata* **Pl. 7**

IDENTIFICATION: Black stripe along side and around snout, *broad and diffuse at midbody, narrow and black on caudal peduncle; clear stripe* above black stripe; *large black spot* on caudal fin base. Slender, slightly compressed body; small mouth; snout projecting well beyond upper lip. Dorsal fin origin over or in front of pelvic fin origin. Gold yellow above, scales black-edged, producing wavy lines; dark streak along back in front of dorsal fin; silver side; thin dark bar behind head. Breeding male has many small tubercles scattered over head. Complete lateral line; 34–38 lateral scales; 8 anal rays (often 7 in Tallapoosa R. system); pharyngeal teeth 1,4-4,1. To 3 in. (7.9 cm). **RANGE:** Above Fall Line in Coosa and Tallapoosa river systems (Mobile Bay drainage), se. TN, GA, and AL. Fairly common. **HABITAT:** Rocky pools, often near riffles and vegetation. **SIMILAR SPECIES:** (1) Other *Hybopsis* species *lack* strongly demarcated clear stripe along side.

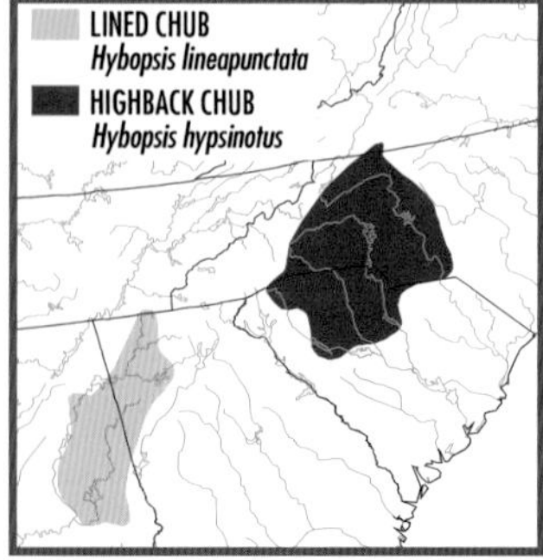

HIGHBACK CHUB *Hybopsis hypsinotus* Pl. 17

IDENTIFICATION: Large dorsal fin on *strongly arched back.* Dark olive above, scales darkly outlined; dull purple stripe along side and around snout; *red fins.* Barely compressed body, deepest at dorsal fin origin; head flattened above and below; dorsal fin origin in front of pelvic fin origin. Complete lateral line; 35–40 lateral scales; 8 anal rays; pharyngeal teeth 1,4-4,1. To 2¾ in. (7.2 cm). **RANGE:** Above Fall Line in Peedee and Santee river drainages, VA, NC, and SC. Common; localized. **HABITAT:** Sandy and rocky runs and pools of creeks and small to medium rivers. **SIMILAR SPECIES:** (1) Yellowfin Shiner, *Notropis lutipinnis* (Pl. 17), *lacks* barbel at corner of *terminal* mouth, has dorsal fin origin behind pelvic fin origin.

SUCKERS: Family Catostomidae (70)

Sixty-nine species of suckers live only in North America north of Mexico. The Longnose Sucker, *Catostomus catostomus*, lives in both North America and Siberia, and the Asiatic Sucker, *Myxocyprinus asiaticus,* lives in China. Suckers have *large thick lips,* protrusible premaxillae (except in Harelip Sucker, *Moxostoma lacerum*), soft rays in the fins, no teeth on the jaws, many *comblike* or *molarlike teeth in a single row* on the pharyngeal arches, *1 dorsal fin, 9 or more dorsal rays,* abdominal pelvic fins, the anal fin far back on the body, cycloid scales on the body, and no scales on the head. The sucker mouth with large lips is used in most species to "vacuum" and ingest invertebrates from stream and lake beds. Because of their abundance and large size, suckers often account for the largest biomass in streams and lakes.

ICTIOBUS

GENUS CHARACTERISTICS: *Long, falcate dorsal fin;* 24–31 rays. *Gray or dark olive* (not silver) *body; semicircular subopercle,* broadest at middle (Fig. 34); dusky gray pelvic fin. Complete lateral line; sum of pelvic and anal rays 18 or more; 2-chambered gas bladder.

BIGMOUTH BUFFALO *Ictiobus cyprinellus* Pl. 19

IDENTIFICATION: *Robust body; large head. Sharply oblique, terminal mouth;* front of upper lip *nearly level* with lower edge of eye. Upper jaw length *about equal to* snout length. *Faint grooves on thin upper lip.* Gray to olive-bronze above, green and copper sheen; black to olive-yellow side; white to pale yellow below; brown or black fins. Has 40 or more rakers on 1st gill arch. Usually 35–36 lateral scales, 10–11 pelvic rays, 8–9 anal rays. To 40 in. (100 cm). **RANGE:** Hudson Bay (Nelson and Red river drainages), Great Lakes, and Mississippi R. basins from ON

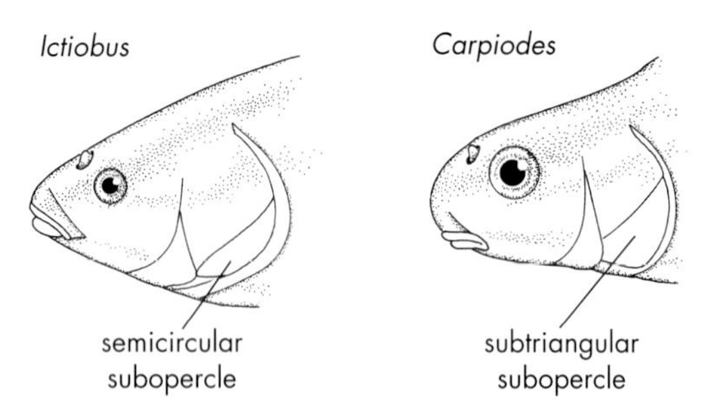

Fig. 34. *Ictiobus* and *Carpiodes*.

to SK and MT, and south to LA. Introduced in Leaf R., MS; possibly in impoundments in NC, AZ, and CA. Fairly common. **HABITAT:** Main channels, pools, and backwaters of small to large rivers; lakes and impoundments. **SIMILAR SPECIES:** (1) Smallmouth Buffalo, *I. bubalus*, and (2) Black Buffalo, *I. niger* (both Pl. 19), have more *conical head*, more *horizontal mouth; distinct grooves* on *thick upper lip*.

SMALLMOUTH BUFFALO *Ictiobus bubalus* Pl. 19

IDENTIFICATION: Deep body; *small head. Fairly small, horizontal, subterminal mouth;* front of upper lip well *below* lower edge of eye. Upper jaw length *much less* than snout length. Large eye. Distinct *grooves on thick upper lip.* Adult has moderately keeled nape. Gray, olive, or bronze above, dark blue to olive sheen; black to olive-yellow side; white to yellow below; olive to black fins. Has 35 or fewer rakers on 1st gill arch. Usually 36–37 lateral scales, 10 pelvic rays, 9 anal rays. To 31 in. (78 cm). **RANGE:** Lower Great Lakes, Red R. (Hudson Bay), and Mississippi R. basins from PA to MT and south to Gulf; Gulf Slope drainages from Mobile Bay, AL, to Rio Grande, TX and NM. Also in Mexico.

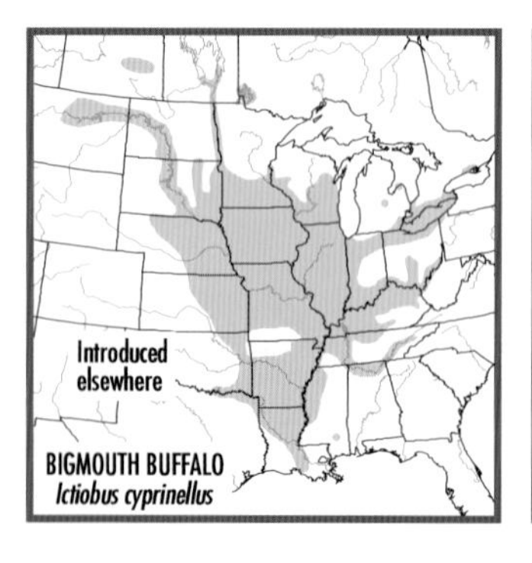

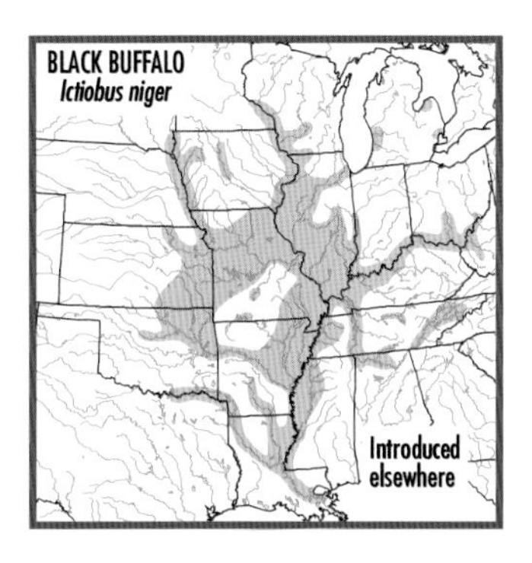

Introduced in impoundments in NC and AZ, possibly elsewhere. Common. **HABITAT:** Pools, backwaters, and main channels of small to large rivers; impoundments; lakes. **SIMILAR SPECIES:** (1) See Black Buffalo, *I. niger* (Pl. 19). (2) Bigmouth Buffalo, *I. cyprinellus* (Pl. 19), has *ovoid head; sharply oblique, terminal mouth; faint grooves on thin upper lip.*

BLACK BUFFALO *Ictiobus niger* Pl. 19

IDENTIFICATION: Similar to Smallmouth Buffalo, *I. bubalus*, but has *larger, more conical head;* nearly terminal, *slightly oblique mouth;* wider, somewhat shallower body, smaller eye; adult has *rounded or only weakly keeled nape.* Usually 37–39 lateral scales. To 37 in. (93 cm). **RANGE:** Lower Great Lakes and Mississippi R. basins from MI and OH to SD and south to LA. Related form occurs in Rio Grande drainage of Mexico and possibly TX. Uncommon. **HABITAT:** Pools and backwaters of small to large rivers; impoundments; lakes. Introduced in AZ impoundments. **SIMILAR SPECIES:** (1) See Smallmouth Buffalo, *I. bubalus* (Pl. 19).

CARPIODES

GENUS CHARACTERISTICS: *Long, falcate dorsal fin;* 23–30 rays. *Silver body; subtriangular subopercle,* broadest below middle (Fig. 34); white to orange pelvic fin (not densely covered with black specks). Complete lateral line; 2-chambered gas bladder.

QUILLBACK *Carpiodes cyprinus* Pl. 19

IDENTIFICATION: Deep body. *Long 1st dorsal ray*; usually *reaching beyond middle of dorsal fin. No nipple* on lower lip (Fig. 35). *Long, rounded snout,* about equal to distance from back of eye to upper end of gill opening. Upper jaw does *not* extend behind front of eye. Olive to gray above, silver or blue-green sheen; silver side; dusky gray median fins;

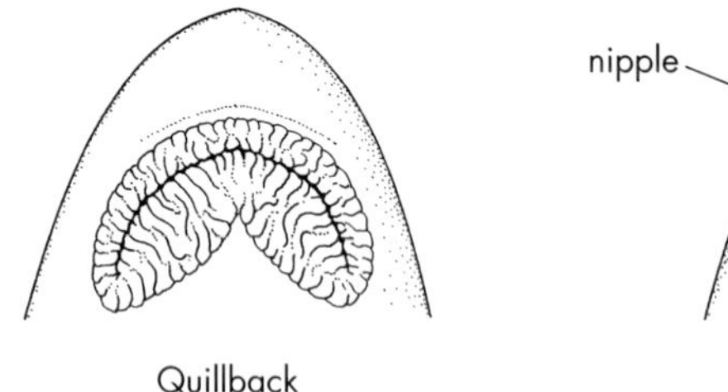

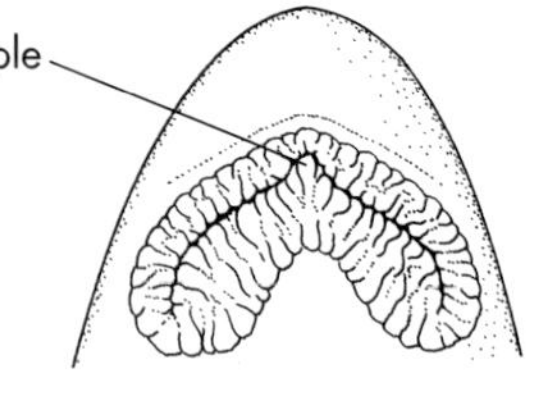

Fig. 35. Quillback and River Carpsucker—lips.

white to orange paired fins. Usually 36–37 lateral scales. Has 9–10 pelvic rays, 8–9 anal rays; sum of anal and pelvic rays 18 or more. To 26 in. (66 cm). **RANGE:** Great Lakes-St. Lawrence R., Hudson Bay, and Mississippi R. basins from QC to AB and south to LA; Atlantic Slope drainages from Delaware R., NY, to Savannah R., GA (absent from several drainages); Gulf Slope drainages from Apalachicola R., FL and GA, to Pearl R., LA. Common. **HABITAT:** Pools, backwaters, and main channels of creeks and small to large rivers; lakes. **SIMILAR SPECIES:** (1) River Carpsucker, *C. carpio* (Pl. 19), has *nipple* at middle of lower lip, 1st dorsal ray usually *not* reaching beyond middle of dorsal fin, usually 33–36 lateral scales. (2) Highfin Carpsucker, *C. velifer* (Pl. 19), has *nipple* on lower lip, *blunt snout,* long 1st dorsal ray *reaching to or beyond rear of dorsal fin*, usually 33–36 lateral scales.

RIVER CARPSUCKER *Carpiodes carpio* Pl. 19

IDENTIFICATION: Deep body. First dorsal ray usually *not reaching beyond middle of dorsal fin. Nipple* at middle of lower lip (Fig. 35). *Short, rounded snout;* length less than distance from back of eye to upper end

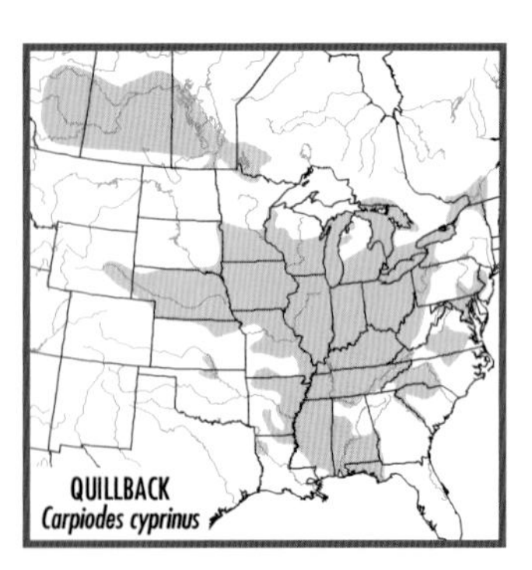

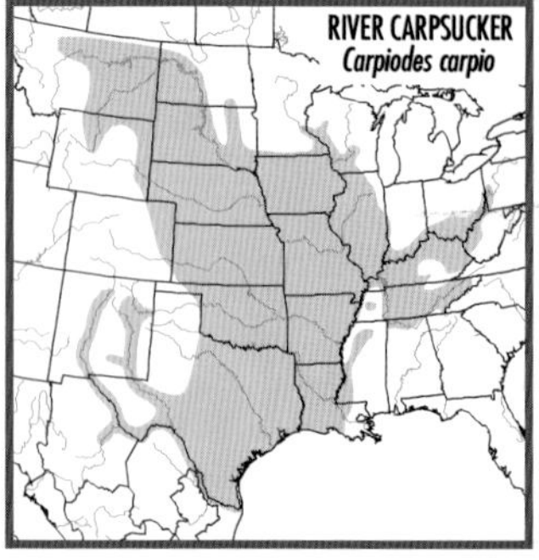

of gill opening. Upper jaw extends *behind* front of eye. Olive to bronze above; silver side; dusky gray median fins; white or pink-orange paired fins. Usually 33–36 lateral scales, 9 pelvic rays, 8 anal rays; sum of anal and pelvic rays 17 or fewer. To 25 in. (64 cm). **RANGE:** Mississippi R. basin from PA to MT and south to LA; Gulf Slope drainages from Mississippi R. to Rio Grande, TX and NM. Also in Mexico. Common. **HABITAT:** Pools and backwaters of creeks and small to large rivers; lakes. **SIMILAR SPECIES:** (1) Highfin Carpsucker, *C. velifer* (Pl. 19), has *long 1st dorsal ray reaching to or beyond rear of dorsal fin, blunt snout.* (2) Quillback, *C. cyprinus* (Pl. 19), *lacks* nipple at middle of lower lip; has 1st dorsal ray usually *reaching beyond middle of dorsal fin,* usually 36–37 lateral scales.

HIGHFIN CARPSUCKER *Carpiodes velifer* Pl. 19

IDENTIFICATION: Deep body. *Long 1st dorsal ray reaching to or beyond rear of dorsal fin. Blunt snout. Nipple* at middle of lower lip (Fig. 35). Olive-gray above, silver or blue-green sheen; silver side; dusky gray median fins; white to orange paired fins. Usually 33–36 lateral scales, 9–10 pelvic rays, 8–9 anal rays; sum of anal and pelvic rays usually 18 or more. To 19½ in. (50 cm). **RANGE:** Lake Michigan drainage and Mississippi R. basin from PA to MN, and south to LA; Atlantic Slope drainages from Cape Fear R., NC, to Altamaha R., GA; Gulf Slope drainages from Apalachicola R., GA and FL, to Pearl R., MS and LA. Generally common; rare on Atlantic Slope. **HABITAT:** Pools and backwaters of creeks and small to large rivers. **SIMILAR SPECIES:** (1) River Carpsucker, *C. carpio* (Pl. 19), has 1st dorsal ray usually *not reaching beyond middle of dorsal fin, rounded snout.* (2) Quillback, *C. cyprinus* (Pl. 19), *lacks* nipple at middle of lower lip; has *rounded snout*, 1st dorsal ray *not* reaching to or beyond rear of dorsal fin, usually 36–37 lateral scales.

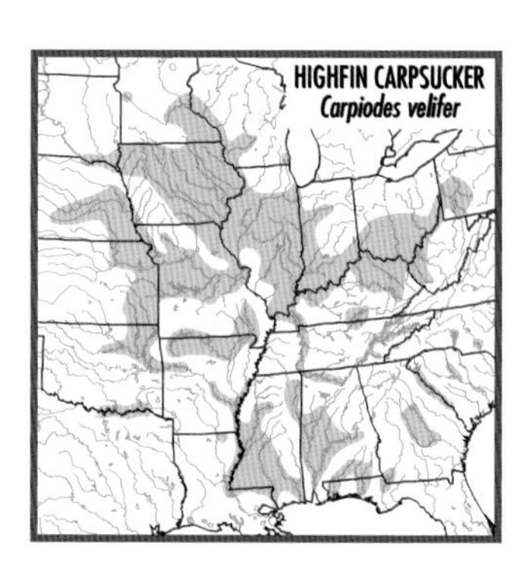

BLUE SUCKER *Cycleptus elongatus* Pl. 22

IDENTIFICATION: Long, somewhat compressed body; *small head* (length 5 or more times into standard length). *Long falcate dorsal fin;* 24–35 rays. Blunt snout overhangs small horizontal mouth; many *blunt papillae on lips.* Long caudal peduncle; forked caudal fin. Olive-blue or gray above; blue-white below; dark blue-gray fins. Large male is *blue-black;* small white tubercles cover head, body, and fins of breeding male. Large female is tan to light blue, has fewer tubercles. Usually 19–20 scales around caudal peduncle, 41–49 scales around body at dorsal fin origin, 53–58 lateral scales, 28–35 dorsal rays, 7 anal rays. To 39 in. (99 cm). **RANGE:** Mississippi R. basin from PA to cen. MT, and south to LA; Gulf Slope drainages from Sabine R., LA, to Rio Grande, TX, NM, and Mexico. Generally common but becoming less so at edges of range. **HABITAT:** Strong current in deep chutes and main channels of medium to large rivers over bedrock, sand, and gravel. **SIMILAR SPECIES:** (1) See Southeastern Blue Sucker, *C. meridionalis.*

SOUTHEASTERN BLUE SUCKER *Cycleptus meridionalis* Not shown

IDENTIFICATION: Similar to Blue Sucker, *C. elongatus*, but has *shorter snout*, shorter dorsal fin base, *larger scales*; usually 16 scales around caudal peduncle, 37–40 scales around body at dorsal fin origin, 49–53 lateral scales, and 25–29 dorsal rays. To 28 in. (71 cm). **RANGE:** Below Fall Line in Mobile Bay, Pascagoula, and Pearl river drainages, AL, MS, and LA (rare above Fall Line in Coosa R., AL). Locally common. **HABITAT:** Strong current in chutes and main channels of medium to large rivers over bedrock, sand, and gravel. **SIMILAR SPECIES:** (1) See Blue Sucker, *C. elongatus* (Pl. 22).

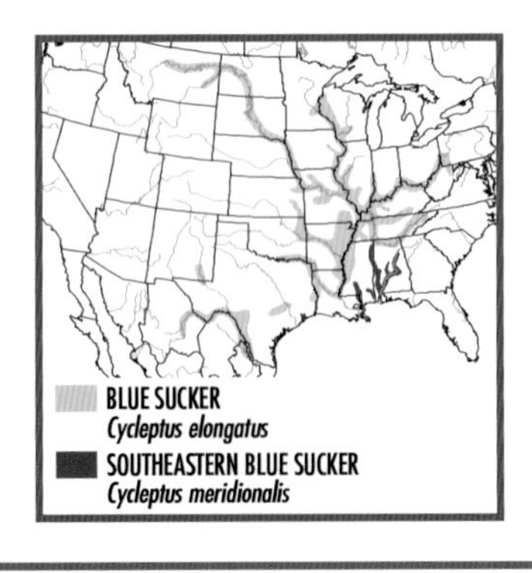

GENUS CHARACTERISTICS: *Large, stout body;* large head; often hump on snout. Large, *terminal or subterminal,* fairly to strongly *oblique mouth; thin, usually plicate* (sometimes sparsely papillose) *lips;* widely separated lower lip lobes (Fig. 36). Complete lateral line, 55–79 scales; 10–12 dorsal rays; 7 anal rays; 37–53 *thin, branched or fimbriate (i.e., broccoli-like) rakers* on 1st gill arch, flat and paddlelike on top; 2-chambered gas bladder.

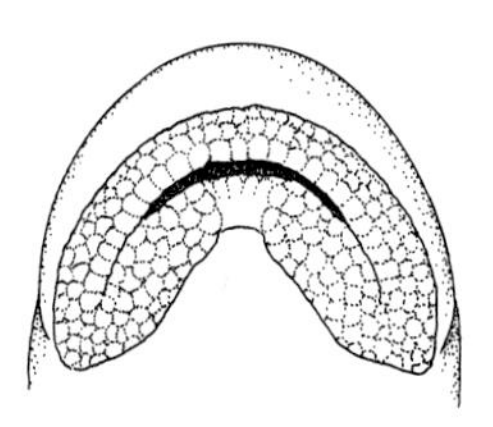

Fig. 36. *Chasmistes*—lips.

JUNE SUCKER *Chasmistes liorus* Pl. 20

IDENTIFICATION: *Large, terminal mouth.* Thin, smooth or plicate lips (rarely with sparse papillae). Dark gray, bronze, or copper above and on side; white to dark gray fins. Large male may have rosy stripe along side. Has 55–70 lateral scales, 29–35 predorsal scales, 52–61 scales around body at dorsal fin origin, 19–20 scales around caudal peduncle. To 20 in. (52 cm). **RANGE:** Utah Lake and its tributaries, UT. Original form of June Sucker, recognized in U.S. as an *endangered species,* probably is extinct (see Remarks). **HABITAT:** Formerly abundant in deep waters of Utah Lake; spawned in tributaries. **REMARKS:** Two subspecies. *C. l. liorus,* formerly confined to Utah Lake, had 55–64 lateral

scales, 45–53 rakers on 1st gill arch. *C. l. mictus,* with 60–70 lateral scales, 37–47 rakers on 1st gill arch, larger lip papillae, less oblique mouth, shorter, more slender head, smaller eye, may have arisen as a hybrid between *C. l. liorus* and Utah Sucker, *Catostomus ardens*, and replaced original form of June Sucker in Utah Lake. **SIMILAR SPECIES:** (1) See Snake River Sucker, *C. liorus*. See (2) Cui-ui, *C. cujus*, and (3) Shortnose Sucker, *C. brevirostris* (both Pl. 20).

SNAKE RIVER SUCKER *Chasmistes muriei* Not shown

IDENTIFICATION: Similar to June Sucker, *C. liorus*, but has *subterminal mouth, papillose lips, and smaller scales* (72 lateral, 40 predorsal). To 17¾ in. (45 cm). **RANGE:** Known from 1 specimen from Snake R. below Jackson Dam, WY. Extinct. **HABITAT:** Presumably deep pools in large rocky river. **SIMILAR SPECIES:** (1) See June Sucker, *C. liorus* (Pl. 20).

CUI-UI *Chasmistes cujus* Pl. 20

IDENTIFICATION: Similar to June Sucker, *C. liorus*, but has *larger, broader head;* less oblique mouth; 22–26 scales around caudal peduncle; more than 64 scales around body at dorsal fin origin. Blue-gray, brown, or black above and on side, copper or brassy sheen; white to dark gray fins. Breeding male has black and metallic red stripes on side, is silver or brassy below, has slate blue fins. Breeding female may have pink stripe along side. Has 59–68 lateral scales. To 26½ in. (67 cm). **RANGE:** Originally, Pyramid and Winnemucca lakes, NV. Winnemucca Lake is now dry. Cui-ui is declining in Pyramid Lake and protected as an *endangered species.* **HABITAT:** Deep water; formerly made spectacular spawning runs up Truckee R. but now spawns only around margin of Pyramid Lake. **SIMILAR SPECIES:** (1) See June Sucker, *C. liorus* (Pl. 20). (2) Shortnose Sucker, *C. brevirostris* (Pl. 20), has more oblique mouth; smaller (65–79 lateral) scales.

SHORTNOSE SUCKER *Chasmistes brevirostris* Pl. 20

IDENTIFICATION: Similar to June Sucker, *C. liorus*, but has *shorter head, smaller eye,* few or no papillae on lips, 21–25 scales around caudal peduncle. Slate gray, brown, or black above; white to dark gray fins. Large adult may have red cast to scales. Has 65–79 lateral scales. To 25 in. (64 cm). **RANGE:** Upper Klamath and Lost river basins, OR and CA. Rare, nearing extinction; recognized as an *endangered species.* **HABITAT:** Usually near vegetation around lake margin. Spawns in lake tributaries. **REMARKS:** Apparently no "pure" Shortnose Suckers persist. Recently captured individuals appear to be hybrids with either Klamath Largescale Sucker, *Catostomus snyderi*, or Klamath Smallscale Sucker, *Catostomus rimiculus.* **SIMILAR SPECIES:** (1) See June Sucker, *C. liorus* (Pl. 20). (2) Cui-ui, *C. cujus* (Pl. 20), has *broader, blunter head;* more oblique mouth; 59–68 lateral scales.

RAZORBACK SUCKER *Xyrauchen texanus* Pl. 20

IDENTIFICATION: *Sharp keel* ("humpback") on nape (absent on small young). Long head and body. Horizontal mouth; few papillae on lips; *lower lip widely separated into 2 lobes* by deep median groove. Olive to brown-black above; brown or pink side; white to yellow below; olive to yellow dorsal fin; white to yellow-orange anal and paired fins. Breeding male is black or brown above, *yellow to bright orange below;* sometimes has rosy fins. Has 68–87 lateral scales; 13–16 dorsal rays; 7 anal rays. To 36 in. (91 cm). **RANGE:** Formerly throughout medium to large rivers of Colorado R. basin from WY and CO to Baja California. Presently known only above Grand Canyon and in Lakes Mead, Mohave, and Havasu on lower Colorado R. Rare; protected as an *endangered species.* **HABITAT:** Silt- to rock-bottomed backwaters near strong current and deep pools in medium to large rivers; impoundments. **SIMILAR SPECIES:** No other sucker has large keel on nape. Until keel develops, young are difficult to distinguish from young *Catostomus* species.

LOST RIVER SUCKER *Deltistes luxatus* Pl. 20

IDENTIFICATION: *Distinct hump on snout.* Large, stout body; long head; subterminal mouth. *Thin, moderately papillose lips;* moderately deep lower lip notch; no deep indentations separate upper and lower lips. Eye on rear half of head. Dark olive to gray above; white or yellow below; fins similar in color to adjacent body. Has *82–88 lateral scales;* 11–12 dorsal rays; 7–8 anal rays. Usually 24–28 branched, triangular rakers on 1st gill arch. To 34 in. (86 cm). **RANGE:** Lost R. system (upper Klamath R. basin), OR and CA. Rare; protected as an *endangered species.* **HABITAT:** Lakes; impoundments; deep pools of small to medium rivers. **SIMILAR SPECIES:** (1) *Chasmistes* species (Pl. 20) have oblique mouth; thin, usually plicate, lips; 55–79 lateral scales; 37–53 rakers on 1st gill arch. (2) *Catostomus* species (Pls. 20 and 21) have thick, strongly papillose lips; long, slender body.

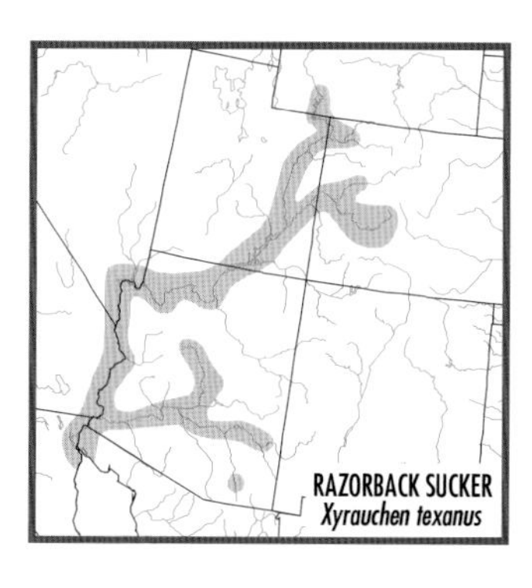

GENUS CHARACTERISTICS: Nearly *cylindrical body.* Large, *horizontal mouth; thick, strongly papillose lips* (Fig. 37); rounded snout. Shallow to deeply divided lower lip *joined at middle,* with or without notches (Fig. 37) at juncture of upper and lower lips. Complete lateral line, *54–124 scales;* 7–17 dorsal rays; usually 7 anal rays; 20–44 short to long, thin, unbranched rakers on 1st gill arch; 2-chambered gas bladder. Young of many species have *3 dark gray blotches* along side. *Catostomus* species readily hybridize with one another. Hybrids are especially common in streams where a non-native species has been introduced or where habitats have been altered.

NEXT 18 SPECIES: Outer edge of lips continuous, no deep indentations separating upper and lower lips (Fig. 37).

WHITE SUCKER *Catostomus commersonii* Pl. 20

IDENTIFICATION: Deep median lower lip notch; 0–3 rows of papillae at middle of lower lip; 2–6 rows of papillae on upper lip, lower lip about *twice as thick* as upper lip (Fig. 37). Caudal peduncle depth *more than half* dorsal fin base. No membrane connecting pelvic fin to body. Olive-brown to black above, often dusky-edged scales; clear to dusky fins. Breeding male is gold above, has scarlet stripe along side in most populations (cream to black stripe in some Canadian populations); tubercles on anal and caudal fins, caudal peduncle. Pale to slightly speckled peritoneum. Usually 10–12 dorsal rays, 53–74 lateral scales, 8–11 scale rows above lateral line. To 25 in. (64 cm). **RANGE:** Atlantic, Arctic, Great Lakes, and Mississippi R. basins from NL to Mackenzie R., NT, south to Tennessee R. drainage, n. AL, and Arkansas R. drainage, NM; south on Atlantic Slope to extreme upper Savannah R. drainage, GA. Upper Rio Grande drainage, NM; Skeena and Fraser river drainages (Pacific Slope), BC. Introduced into Colorado R. drainage,

WY, CO, NM, and UT. Common. **HABITAT:** Wide range of habitats from rocky pools and riffles of headwaters to large lakes. Usually in small, clear, cool creeks and small to medium rivers. **SIMILAR SPECIES:** (1) See Summer Sucker, *C. utawana,* and (2) Sonora Sucker, *C. insignis* (Pl. 20). (3) Utah Sucker, *C. ardens* (Pl. 20), has dorsal fin membranes densely speckled to edge; 9–14, usually 12, scale rows above lateral line.

SUMMER SUCKER *Catostomus utawana* Not shown

IDENTIFICATION: Similar to White Sucker, *C. commersonii,* but is smaller (to 9½ in. [24 cm]), has larger eye (diameter less than twice into snout length); breeding female has tubercles on anal and caudal fins, and on caudal peduncle (no tubercles on White Sucker female); breeding male has gold stripe along side. **RANGE:** St. Lawrence-Lake Ontario drainages, Adirondack Mts., NY. Common. **HABITAT:** Lakes; rocky pools and runs of tributary streams. **SIMILAR SPECIES:** (1) See White Sucker, *C. commersonii* (Pl. 20).

SONORA SUCKER *Catostomus insignis* Pl. 20

IDENTIFICATION: Similar to White Sucker, *C. commersonii,* but usually has dark-edged scales, sometimes forming *faint dashed lines* on upper side*;* lower lip about *3 times as thick* as upper lip. Sometimes *sharply bicolored,* olive-brown above, deep yellow below; white or yellow dusky fins. Usually 10–11 (rarely 12) dorsal rays; 54–67, usually fewer than 60, lateral scales. To 31½ in. (80 cm). **RANGE:** Gila and Bill Williams river systems (Colorado R. drainage), NM, AZ, and n. Sonora, Mexico. Common, but diminishing in southern half of range. **HABITAT:** Rocky pools of creeks and small to medium rivers; large individual in rocky riffles and runs at night. **REMARKS:** Some large individuals develop a massive, fleshy lower lip like that of Flannelmouth Sucker, *C. latipinnis.* **SIMILAR SPECIES:** (1) See White Sucker, *C. commersonii* (Pl. 20), and (2) Yaqui Sucker, *C. bernardini.*

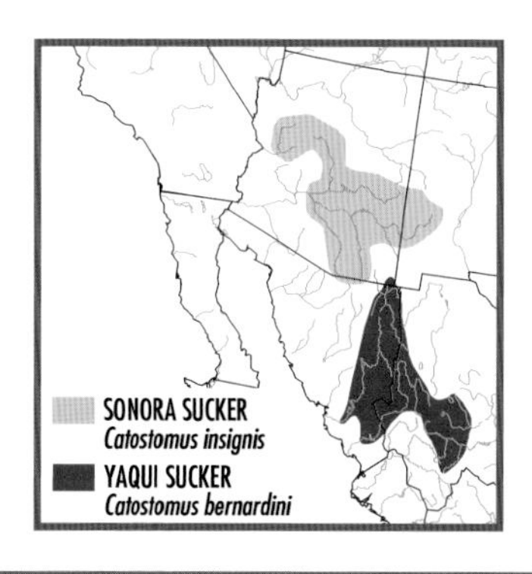

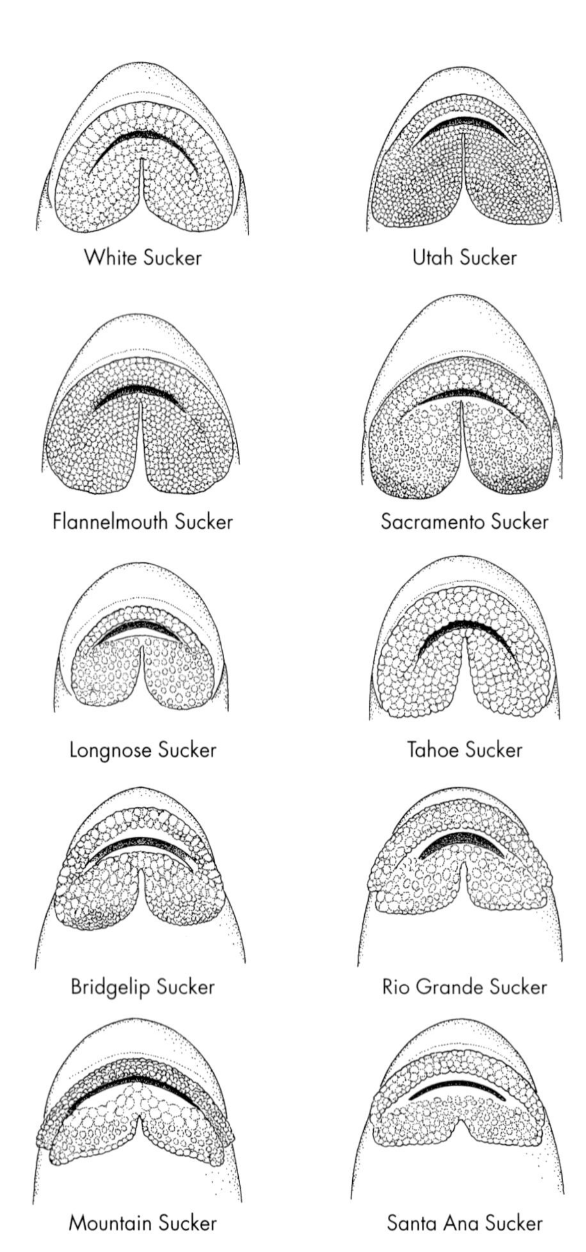

Fig. 37. *Catostomus* species—lips.

YAQUI SUCKER *Catostomus bernardini* Not shown

IDENTIFICATION: Nearly identical to Sonora Sucker, *C. insignis*, but usually has *12 dorsal rays;* 62–80, usually *more than 65,* lateral scales; usually *lacks* sharp bicoloration. Dark olive-brown above; white or yellow below; dusky dorsal and caudal fins, white or yellow anal and paired fins. Large male has enlarged lower fins. To 16 in. (40 cm). **RANGE:** Rio Yaqui basin, extreme se. AZ; also n. Mexico. Common in Mexico; extinct in U.S. **HABITAT:** Small montane and desert creeks; deep pools, runs, and rapids of medium rivers. **SIMILAR SPECIES:** (1) See Sonora Sucker, *C. insignis* (Pl. 20).

UTAH SUCKER *Catostomus ardens* Pl. 20

IDENTIFICATION: Very deep median lower lip notch; 0–1 row of papillae at middle of lower lip; 3–8 rows of papillae on upper lip (Fig. 37). Short, deep caudal peduncle (in adult). Dorsal fin membranes *densely speckled to edge.* No membrane connecting pelvic fin to body. Dark olive to copper above; dusky fins. Breeding male usually has red stripe along side. Has 60–79 lateral scales; usually 12 scale rows above lateral line, 12 dorsal rays; 28–36 predorsal scales. To 25½ in. (65 cm). **RANGE:** Snake R. system (Columbia R. drainage) above Shoshone Falls and adjacent endorheic drainages, WY, ID, UT, and south through Lake Bonneville basin, ID and UT. Introduced into upper Colorado R. system. Common. **HABITAT:** Lakes, impoundments, and streams over silt, sand, gravel, or rocks; often near vegetation. **SIMILAR SPECIES:** (1) See Largescale Sucker, *C. macrocheilus* (Pl. 20). (2) White Sucker, *C. commersonii* (Pl. 20), has dorsal fin membranes densely speckled on *bottom* ⅓ of fin only, 2–6 rows of papillae on upper lip, usually 8–11 scale rows above lateral line.

LARGESCALE SUCKER *Catostomus macrocheilus* Pl. 20

IDENTIFICATION: Similar to Utah Sucker, *C. ardens,* but has dorsal fin membranes *less densely speckled to edge*, usually *13–15 dorsal rays, mem-*

brane connecting pelvic fin to body, caudal peduncle depth less than ½ dorsal fin base, usually 36–52 predorsal scales. Olive to blue-gray above; white or yellow below; white to dusky fins. Breeding individual has iridescent olive green stripe above lateral line, dark bronze or *black stripe* along side, yellow stripe between black stripe and belly. To 24 in. (61 cm). **RANGE:** Arctic basin from Peace R. drainage, BC, to Smokey R. drainage, AB; Pacific Slope from Nass R., BC, to Snake R. drainage (below Shoshone Falls), ID and NV, and Coquille R., OR. Isolated record in Mackenzie R., NT. Common. **HABITAT:** Pools and runs of medium to large rivers; lakes. **SIMILAR SPECIES:** (1) See Utah Sucker, *C. ardens* (Pl. 20).

FLANNELMOUTH SUCKER *Catostomus latipinnis* Pl. 20

IDENTIFICATION: *Narrow caudal peduncle* (most populations). Very deep median lower lip notch; 0–1 row of papillae at middle of lower lip; 5–6 rows of papillae on upper lip (Fig. 37). Large adult has *large, fleshy lobes on lower lip. Large dorsal and caudal fins;* falcate dorsal fin. Green to blue-gray above, dusky scale edges; yellow to orange-red on lower side; white to dusky fins. Young is lighter with silver sheen. Has *90–116 lateral scales;* usually 12–13 dorsal rays. To 22 in. (56 cm). **RANGE:** Colorado R. drainage from sw. WY to s. AZ and CA. Locally common; extirpated from CA. **HABITAT:** Rocky pools, runs, and riffles of medium to large rivers; less often in creeks and small rivers. **SIMILAR SPECIES:** (1) See Little Colorado River Sucker, *C.* species. (2) Utah Sucker, *C. ardens* (Pl. 20), has *shorter, deeper caudal peduncle,* deeper body; *smaller lower lip; 60–79 lateral scales; smaller* dorsal and caudal fins.

LITTLE COLORADO RIVER SUCKER *Catostomus* species Not shown

IDENTIFICATION: Similar to Flannelmouth Sucker, *C. latipinnis,* but has *thicker, deeper caudal peduncle; smaller lower lip, slightly falcate to straight-edged dorsal fin,* usually 11–12 dorsal rays; 73–97, usually *fewer than 90,* lateral scales. Sharply bicolored, dark gray to blue-black above, white to yellow below; dusky fins. To 19¾ in. (50 cm). **RANGE:**

Little Colorado R. system, AZ. Introduced into Salt R., AZ. Common. **HABITAT:** Rocky pools and riffles of creeks and small to medium rivers; impoundments. **SIMILAR SPECIES:** (1) See Flannelmouth Sucker, *C. latipinnis* (Pl. 20).

SACRAMENTO SUCKER *Catostomus occidentalis* Pl. 21

IDENTIFICATION: Deep median lower lip notch; *1 row* of papillae at middle of lower lip; *4–6 rows* of papillae on upper lip (Fig. 37). Fairly blunt snout. Distance from pelvic fin origin to caudal fin base less than or equal to distance from eye to pelvic fin origin. Olive green, steel gray, or brown above; white to dirty *yellow gold below;* white to dusky fins. Young is gray, has 3–4 dark blotches along side. Breeding male has dark red stripe along side. Has 22–24 rakers in young, 25–30 in adult, on 1st gill arch; 56–75 lateral scales; 10–17 scale rows above, 8–10 below, lateral line; 18–22 scales around caudal peduncle; usually 12–15 dorsal rays. Dusky peritoneum. To 23½ in. (60 cm). **RANGE:** Pacific Slope from Mad R., n. CA, to Salinas R., cen. CA; throughout Sacramento-San Joaquin drainage from Goose Lake, OR, to Kern R., CA. Common. **HABITAT:** Usually in pools of clear, cool streams; lakes and impoundments. **REMARKS:** Four subspecies sometimes recognized, but in need of study. **SIMILAR SPECIES:** (1) See Klamath Largescale Sucker, *C. snyderi* (Pl. 21).

KLAMATH LARGESCALE SUCKER *Catostomus snyderi* Pl. 21

IDENTIFICATION: Similar to Sacramento Sucker, *C. occidentalis*, but distance from pelvic fin origin to caudal fin base equal to or *greater than* distance from eye to pelvic fin origin; *silver peritoneum;* thicker caudal peduncle; 10–12 scale rows below lateral line; 25–28 rakers in young, 30–35 in adult, on 1st gill arch; usually 11 (occasionally 12) dorsal rays. Has 67–77 lateral scales. To 21¾ in. (55 cm). **RANGE:** Klamath and Lost river drainages, OR and CA. Common. **HABITAT:** Rocky pools

and runs of creeks and small rivers; lakes; impoundments. **SIMILAR SPECIES:** (1) See Sacramento Sucker, *C. occidentalis* (Pl. 21).

LONGNOSE SUCKER *Catostomus catostomus* **Pl. 21**

IDENTIFICATION: *Long snout. Very deep median lower lip notch*; 0–1 row of papillae at middle of lower lip; *2* rows of papillae on upper lip (Fig. 37). Dark olive or gray with brassy sheen or dark gray irregular blotches above, white or cream below. Breeding male is nearly black; large female is green-gold to copper brown fading to white, yellow, orange, or pink below; both may have red stripe along side, dusky median fins (sometimes with pale red edges), amber-pink paired fins. Usually 9–11 dorsal rays; usually *95–120 lateral scales,* 26–34 scales around caudal peduncle; 16–18 pectoral rays. Silver to black peritoneum. To 25 in. (64 cm); some populations dwarfed. **RANGE:** Most widespread sucker of N. America. Atlantic, Arctic, and Pacific basins throughout most of Canada and AK; Atlantic Slope south to Delaware R. drainage, NY; Pacific Slope south to Columbia R., drainage, WA; Great Lakes basin; upper Monongahela R. drainage, PA, MD, and WV; upper Missouri R. drainage south to CO. One record in Mississippi R., n. IL. Introduced in upper Colorado R. drainage, WY and CO. Also in Arctic basin of e. Siberia. Common in northern cold waters; uncommon and sporadic in south. **HABITAT:** Usually in clear, cold, deep water of lakes and tributary streams; occasionally in brackish water in Arctic. To depth of 600 ft. (183 m) in Great Lakes. **REMARKS:** Subspecies often recognized, but in need of study. **SIMILAR SPECIES:** (1) See Salish Sucker, *C.* species. (2) Sacramento Sucker, *C. occidentalis* (Pl. 21), has 4–6 rows of papillae on upper lip; shorter snout; *56–75 lateral scales;* usually 12–15 dorsal rays. (3) Tahoe Sucker, *C. tahoensis*, usually has *82–95 lateral scales,* 14–16 pectoral rays.

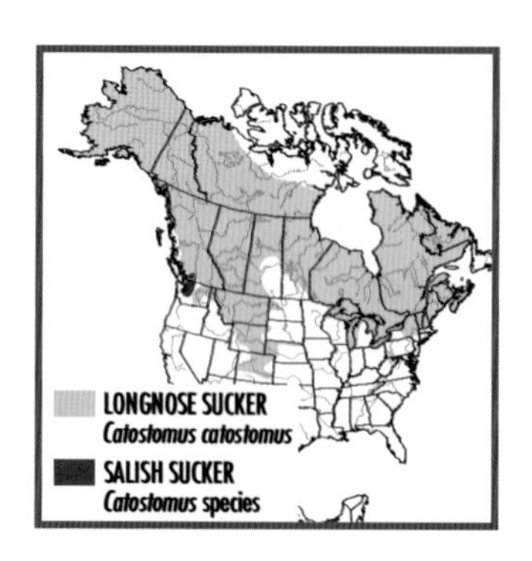

SALISH SUCKER *Catostomus* species Not shown

IDENTIFICATION: Nearly identical to Longnose Sucker, *C. catostomus*, but has *shorter snout;* deeper head; smaller mouth; *usually 81–88 lateral scales, 21–25 scales* around caudal peduncle; reaches about 8 in. (20 cm). **RANGE:** Puget Sound drainage, WA, and lower Fraser Valley, BC. Protected in Canada as an *endangered species.* Rare. **HABITAT:** Pools and runs of small, cool headwaters and creeks; lakes. **SIMILAR SPECIES:** (1) See Longnose Sucker, *C. catostomus* (Pl. 21).

TAHOE SUCKER *Catostomus tahoensis* Not shown

IDENTIFICATION: *Large head; long snout; usually thick caudal peduncle. Deep median lower lip notch;* usually 1 row of papillae at middle of lower lip; 2–4 rows of papillae on upper lip (Fig. 37). Dark olive above sharply contrasting with yellow or white below; dusky fins. Breeding male has bright red stripe along brassy side. Usually *82–95 lateral scales;* 16–19 scale rows above, *12–15 below,* lateral line; 40–50 predorsal scales; usually 9–11 dorsal rays; *14–16 pectoral rays.* Black peritoneum. *Open frontoparietal fontanelle.* To 24 in. (61 cm). **RANGE:** Native to Lahontan basin, se. OR, NV, and ne. CA. Introduced into upper Sacramento R. system, CA. Common. **HABITAT:** Variable but usually in large lakes such as Lake Tahoe and Pyramid Lake, NV; also in pools along lower reaches of streams. **SIMILAR SPECIES:** See Pl. 21. See (1) Owens Sucker, *C. fumeiventris,* (2) Warner Sucker, *C. warnerensis,* (3) Klamath Smallscale Sucker, *C. rimiculus,* and (4) Modoc Sucker, *C. microps.*

WARNER SUCKER *Catostomus warnerensis* Pl. 21

IDENTIFICATION: Similar to Tahoe Sucker, *C. tahoensis,* but has moderately deep median lower lip notch, *2–3 rows* of papillae at middle of lower lip, usually *73–79 lateral scales, 14–16 scale rows below* lateral line. Light to dusky peritoneum. Olive green or gray above; white or

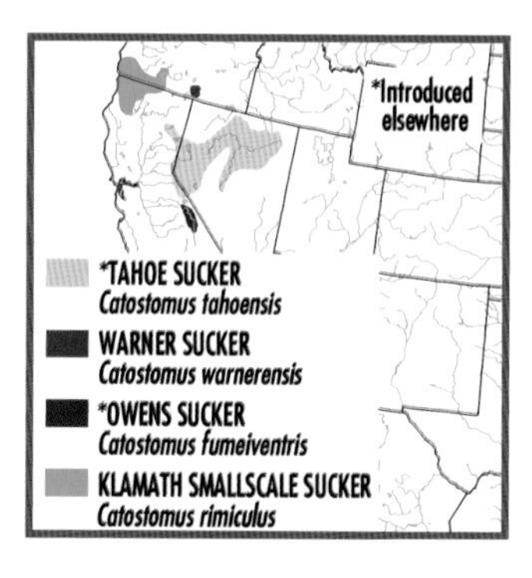

yellow below; white to dusky fins. Has 24–26 scales around caudal peduncle. To 13¾ in. (35 cm). **RANGE:** Endorheic Warner Lake basin, s. OR. Rare; protected as a *threatened species.* **HABITAT:** Lakes; pools and runs of streams and large irrigation canals. **SIMILAR SPECIES:** (1) See Tahoe Sucker, *C. tahoensis.* (2) Sacramento Sucker, *C occidentalis* (Pl. 21), has deep median lower lip notch, blunter snout, usually *1 row of papillae* at middle of lower lip, usually 12–15 dorsal rays, 18–22 scales around caudal peduncle.

OWENS SUCKER *Catostomus fumeiventris* Not shown

IDENTIFICATION: Similar to Tahoe Sucker, *C. tahoensis*, but usually has *75–78 lateral scales; 13–16 scale rows above, 9–11 below,* lateral line; 16–19 pectoral rays. Breeding male *lacks* red stripe along side. Slate above; pale blue sheen on side; *dusky below* (especially large male); dull olive paired fins; dull amber median fins. To 19½ in. (50 cm). **RANGE:** Owens R. drainage, CA. Introduced into June Lake (Mono Lake basin) and Santa Clara R. system, CA. Common, especially in reservoirs. **HABITAT:** Silty to rocky pools and runs of creeks. **SIMILAR SPECIES:** See Pl. 21. (1) See Tahoe Sucker, *C. tahoensis.* (2) Modoc Sucker, *C. microps,* usually has *80–89 lateral scales*; is *white to yellow below.* (3) Warner Sucker, *C. warnerensis,* usually has *14–16 scale rows below* lateral line, 2–3 rows of papillae at middle of lower lip; is *white or yellow below.*

KLAMATH SMALLSCALE SUCKER *Catostomus rimiculus* Not shown

IDENTIFICATION: Similar to Tahoe Sucker, *C. tahoensis*, but has moderately deep median lower lip notch; *2 or more rows* of papillae at middle of lower lip; *5–6 rows of papillae* on upper lip. Dusky olive-brown above and on side, white to yellow below; fins similar in color to surrounding body. Has 81–93 lateral scales; 16–18 pectoral rays. *Closed frontoparietal fontanelle.* To 19¾ in. (50 cm). **RANGE:** Rogue R. drainage, sw. OR, south to Trinity R. drainage, nw. CA. Common. **HABITAT:** Silt- to rock-bottomed pools and runs of small to medium rivers; occasionally impoundments. **SIMILAR SPECIES:** (1) See Tahoe Sucker, *C. tahoensis.* (2) Modoc Sucker, *C. microps* (Pl. 21), usually has *1 row of papillae* at middle of lower lip, *2 rows of papillae* on upper lip.

MODOC SUCKER *Catostomus microps* Pl. 21

IDENTIFICATION: Similar to Tahoe Sucker, *C. tahoensis*, but has shorter head, smaller eye, *small or closed frontoparietal fontanelle*; *9–13 scale rows* below lateral line. Has 1 row of papillae at middle of lower lip, 2 rows of papillae on upper lip. Gray to olive-brown above, fading to white or yellow below. Breeding male has red stripe along side, orange fins. Has 80–89 lateral scales. To 13¼ in. (34 cm). **RANGE:** Ash and Turner creeks (Pit R. system), n. CA; Goose Lake tributaries, s. OR.

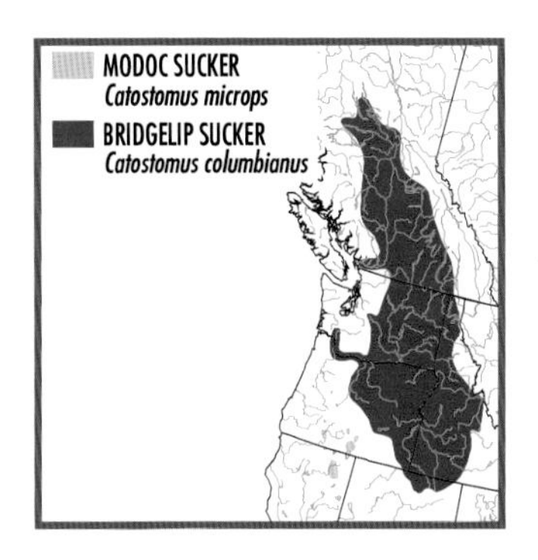

Uncommon in small area; protected as an *endangered species.* **HABITAT:** Shallow mud-bottomed pools of cool creeks. **SIMILAR SPECIES:** (1) See Tahoe Sucker, *C. tahoensis.* (2) Owens Sucker, *C. fumeiventris,* has dusky belly, usually 75–78 lateral scales, 16–19 pectoral rays. (3) Sacramento Sucker, *C. occidentalis* (Pl. 21), usually has 12–15 dorsal rays, 4–6 rows of papillae on upper lip, 56–75 lateral scales; reaches 23½ in. (60 cm).

BRIDGELIP SUCKER *Catostomus columbianus* Pl. 21

IDENTIFICATION: *Weak or no indentations* separate upper and lower lips at side of mouth; *front of upper lip often papillose;* fairly deep median lower lip notch; 2–3 rows of papillae at middle of lower lip (Fig. 37). Truncate lower jaw. Usually 11–12 dorsal rays. Dusky caudal fin membranes. Olive green, mottled brown, or blue-black above; white to yellow below; dusky olive or blue-black fins. Breeding male has orange stripe along side. Black peritoneum. Pelvic axillary process absent or a simple fold. Has 43–75, usually more than 50, predorsal scales. Usually more than 29 rakers on 1st gill arch (see Remarks). To 12 in. (30 cm). **RANGE:** Pacific Slope from Fraser R. drainage, BC, south through Columbia R. drainage to n. NV (only below Shoshone Falls in Snake R.); Harney R. basin, e. OR. Common. **HABITAT:** Lake margins; backwaters, rocky riffles, and sand/silt runs of creeks and small to medium rivers. **REMARKS:** Three subspecies. *C. c. hubbsi* in Wood R. system, ID, usually has 11 dorsal rays, relatively small lips with large papillae, 24–31 rakers on 1st gill arch. *C. c. palouseanus,* in Palouse R., w. ID and e. WA, and Crooked Creek, OR, usually has 11 dorsal rays, 30–39 rakers on 1st gill arch. *C. c. columbianus,* in rest of range, usually has 11–12 dorsal rays, relatively large lips with small papillae, 30–42 rakers on 1st gill arch. **SIMILAR SPECIES:** (1) Rio Grande Sucker, *C. plebeius* (Pl. 21), has rounded lower jaw, 27 or fewer rakers on 1st gill arch, silver peritoneum, usually fewer than 50 predorsal scales, usually 9 dorsal rays; clear caudal fin membranes. (2) Mountain Sucker, *C. platyrhynchus* (Pl. 21), usually *lacks* papillae on front of upper lip, has *deep indentations* separating upper and lower lips, clear caudal fin membranes, usually 10 dorsal rays.

NEXT 5 SPECIES: Upper lip separated from lower lip at corner of mouth by moderate to deep indentations (Fig. 37).

RIO GRANDE SUCKER *Catostomus plebeius* Pl. 21

IDENTIFICATION: *Small, papillose lips.* Moderate indentations separate upper and lower lips; deep median lower lip notch; 2–3 rows of papillae at middle of lower lip (Fig. 37). Rounded lower jaw. Often sharply bicolored, olive green to dusky brown above, white to yellow below; usually clear caudal fin. Breeding male has *red stripe* along side. Usually 9 dorsal rays; 32–55 (usually 40–50) predorsal scale rows; 20–27 rakers on 1st gill arch. No pelvic axillary process. Silver peritoneum. To 7¾ in. (20 cm). **RANGE:** Upper Rio Grande drainage, CO and NM; upper Gila R. system, NM and AZ. Introduced into Rio Hondo (Pecos R. drainage), NM. Also in Rio Grande and Pacific Slope drainages, Mexico. Common. **HABITAT:** Rocky pools, runs, and riffles of small to medium rivers. **SIMILAR SPECIES:** (1) Bridgelip Sucker, *C. columbianus* (Pl. 21), has *less papillose* front of upper lip, truncate lower jaw, usually more than 29 rakers on 1st gill arch (except *C. c. hubbsi*), black peritoneum, usually more than 50 predorsal scales, usually 11–12 dorsal rays, dusky caudal fin membranes. (2) Mountain Sucker, *C. platyrhynchus* (Pl. 21), usually *lacks* papillae on front of upper lip; has truncate lower jaw, pelvic axillary process, usually 10 dorsal rays, dusky or black peritoneum.

MOUNTAIN SUCKER *Catostomus platyrhynchus* Pl. 21

IDENTIFICATION: *Large papillae* on lower lip; *bare areas* on margins of median lip notch; 3–4 rows of papillae at middle of lower lip. *Deep*

indentations separate upper and lower lips; few or no papillae on front of upper lip (Fig. 37). Truncate lower jaw. Narrow caudal peduncle. Dusky gray to olive above, sometimes with dark stripe along side and blotches on back; white to yellow below; clear or pale red fins. Breeding male has moss green back, bright *red stripe* along side above green-black stripe. Usually 10 dorsal rays; 23–37 rakers on 1st gill arch. *Pelvic axillary process.* Dusky or black peritoneum. To 9¾ in. (25 cm). **RANGE:** W. Canada and U.S. from Saskatchewan R. system (Hudson Bay basin), SK and AB, and Fraser R. drainage, BC, south through upper Missouri and Colorado river drainages to CO and UT, and through Columbia R. drainage, OR; Lahontan basin, OR, NV, and CA; upper Sacramento R. system, ne. CA. Common in center of range. **HABITAT:** Variable; usually in rocky riffles and runs of clear montane creeks and small to medium rivers. **SIMILAR SPECIES:** (1) Rio Grande Sucker, *C. plebeius* (Pl. 21), has small, papillose lips (*no* blank areas), rounded lower jaw, *no* pelvic axillary process, usually 9 dorsal rays, silver peritoneum. (2) Bridgelip Sucker, *C. columbianus* (Pl. 21), has *weak or no indentations* separating upper and lower lips, papillae on front of upper lip, dusky caudal fin, usually 11–12 dorsal rays.

SANTA ANA SUCKER *Catostomus santaanae* **Not shown**

IDENTIFICATION: *Deep indentations* separate upper and lower lips; shallow median lower lip notch; 3–4 rows of papillae at middle of lower lip; front of upper lip often papillose (Fig. 37). Truncate lower jaw. Olive to dark gray above, dark blotches or faint dark stripes; silver below; dusky caudal fin membranes; other fins clear to dusky. Has 27–41 predorsal scales; 9–11 dorsal rays; 21–28 rakers on 1st gill arch; 67–86 lateral scales. Foldlike (poorly developed) pelvic axillary process. Black peritoneum. To 9¾ in. (25 cm). **RANGE:** Native to Los Angeles, San Gabriel, Santa Ana, and Santa Clara river drainages, s. CA. Uncommon; protected as a *threatened species.* **HABITAT:** Clear, cool rocky pools and runs of creeks and small to medium rivers. **SIMI-**

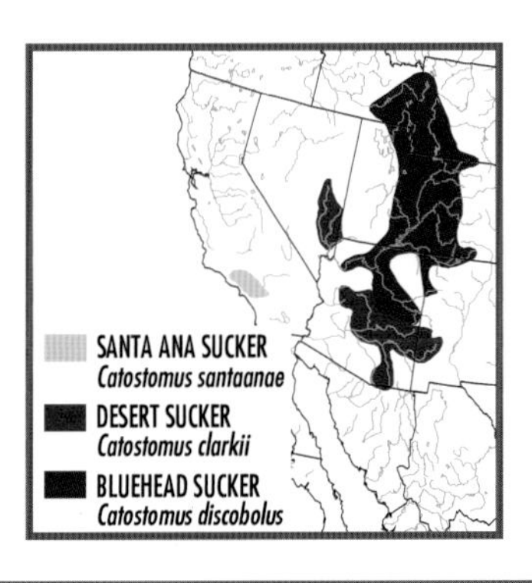

LAR SPECIES: (1) See Desert Sucker, *C. clarkii* (Pl. 21). (2) Bluehead Sucker, *C. discobolus* (Pl. 21), has blue head on adult, more than 43 predorsal scales, usually 86 or more lateral scales.

DESERT SUCKER *Catostomus clarkii* Pl. 21

IDENTIFICATION: Similar to Santa Ana Sucker, *C. santaanae*, but *lacks* papillae on front of upper lip; has *4–7 rows of papillae* at middle of lower lip, *27–43 rakers* on 1st gill arch, 13–52 predorsal scales, usually 10–11 dorsal rays. Silver tan to dark green above, silver to yellow below. To 13 in. (33 cm). **RANGE:** Lower Colorado R. drainage (downstream of Grand Canyon), including White R. and Meadow Valley Wash, NV; Virgin R., UT, AZ, and NV; Bill Williams R., AZ; and Gila R., NM, AZ, and n. Sonora, Mexico. Common. **HABITAT:** Small to medium rivers; small individual in riffles; adult in pools during day, riffles at night. **SIMILAR SPECIES:** See (1) Santa Ana Sucker, *C. santaanae*, and (2) Bluehead Sucker, *C. discobolus* (Pl. 21).

BLUEHEAD SUCKER *Catostomus discobolus* Pl. 21

IDENTIFICATION: Similar to Desert Sucker, *C. clarkii*, but has slender caudal peduncle, *usually 50 or more predorsal scales,* 78–122 lateral scales, *blue head*—darkest on adult (no blue in Little Colorado R., AZ, population). To 16 in. (41 cm). **RANGE:** Snake R. system (Columbia R. drainage), WY and ID; Lake Bonneville basin, ID, WY, and UT; south through upper Colorado R. drainage (Grand Canyon and above), WY, CO, UT, NM, and AZ. Common. **HABITAT:** Rocky riffles and runs of small to large rivers. **REMARKS:** Two subspecies. *C. d. jarrovii,* in Little Colorado R. system, NM and AZ, usually has 9 dorsal rays, papillae on anterior edge of upper lip, and faint red stripe along side of breeding male. *C. d. discobolus,* throughout rest of range, usually has 10 dorsal rays, no papillae on anterior edge of upper lip, no red stripe. **SIMILAR SPECIES:** (1) See Desert Sucker, *C. clarkii* (Pl. 21).

SPOTTED SUCKER *Minytrema melanops* Pl. 22

IDENTIFICATION: *Long, redhorselike body.* Has 8–12 parallel *rows of dark spots* (at scale bases) on back and side. Dusky to black, straight or concave, dorsal fin edge. *No lateral line* (rarely developed on a few scales). Small horizontal mouth; thin plicate lips, U-shaped lower lip edge. *Black edge* on lower caudal fin lobe. Dark green or olive-brown above; yellow to brown side; light yellow-orange to slate olive median fins, usually white to dusky paired fins. Young has pink median fins. Breeding male has dark lavender stripe over narrow pink stripe above brown to black stripe along side. Usually 42–47 lateral scales, 12 dorsal rays, 7 anal rays. To 19½ in. (50 cm). **RANGE:** Lower Great Lakes and Mississippi R. basins from PA to MN and south to Gulf; Atlantic and Gulf slope drainages from Cape Fear R., NC, to Brazos R., TX. Iso-

lated population in Llano R. (Colorado R. drainage), TX. Absent from most of peninsular FL. Common. **HABITAT:** Long deep pools of small to medium rivers over clay, sand, or gravel; occasionally creeks, large rivers, and impoundments. **SIMILAR SPECIES:** (1) Redhorses and jumprocks (*Moxostoma* and *Thoburnia*; Pls. 22 & 23) *lack* distinct rows of dark spots; have thicker lips, *lateral line.* (2) Chubsuckers (*Erimyzon*; Pl. 22) have *deeper body,* more oblique mouth, rounded or sharply pointed dorsal fin; snout extends only slightly beyond upper lip (vs. far beyond in Spotted Sucker).

'ASTERN CREEK CHUBSUCKER *Erimyzon oblongus* Not shown

IDENTIFICATION: Chubby body. *Small, slightly subterminal mouth* (upper lip well below level of eye); *plicate lips;* halves of lower lip meet at nearly right angle. *No lateral line. Rounded edge* on dorsal fin; 11–14, usually 12, rays. Narrow caudal peduncle. Olive to brown above, dark-edged scales; white to yellow below; yellow-orange to olive-gray fins. Young has broad yellow stripe above *5–8 confluent dark blotches* along side from snout tip to caudal fin base; amber or red caudal fin. Breed-

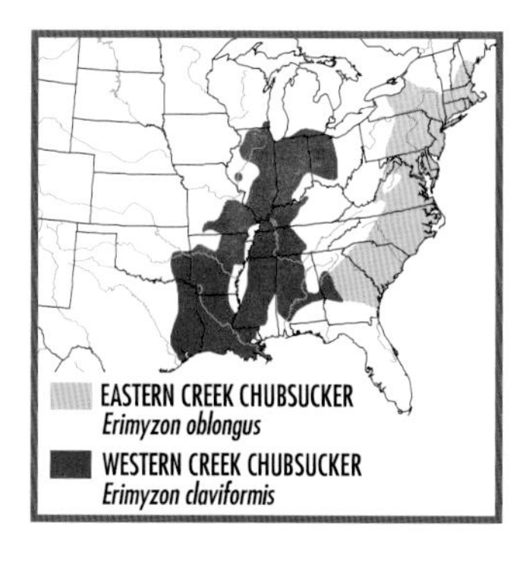

ing male is dark brown above, pink-yellow below; has orange paired fins, yellow median fins, 3 large tubercles on each side of snout, *bilobed anal fin.* Has 37–45 lateral scales, 14 or more predorsal scales, 7 anal rays. To 16½ in. (42 cm). **RANGE:** Atlantic Slope drainages from s. ME to Altamaha R., GA; Lake Ontario drainage, NY. Common on Seaboard Lowlands and Coastal Plain, less common in uplands. **HABITAT:** Sand- and gravel-bottomed pools of clear headwaters, creeks, and small rivers; often near vegetation. Occasionally in lakes. **SIMILAR SPECIES:** See Pl. 22. See (1) Western Creek Chubsucker, *E. claviformis*, (2) Lake Chubsucker, *E. sucetta*, and (3) Sharpfin Chubsucker, *E. tenuis*.

WESTERN CREEK CHUBSUCKER *Erimyzon claviformis* Pl. 22

IDENTIFICATION: Nearly identical to Eastern Creek Chubsucker, *E. oblongus*, but has *9–11, usually 10, dorsal rays.* To 9 in. (23 cm). **RANGE:** Lower Great Lakes and Mississippi R. basins from s. MI and se. WI (extirpated) south to Gulf; Gulf Slope drainages from Apalachicola R. drainage, GA, to San Jacinto R., TX. Common. **HABITAT:** Silt-, sand-, and gravel-bottomed pools of clear headwaters, creeks, and small rivers; often near submergent vegetation. Occasionally in lakes. **SIMILAR SPECIES:** (1) See Eastern Creek Chubsucker, *E. oblongus*.

LAKE CHUBSUCKER *Erimyzon sucetta* Pl. 22

IDENTIFICATION: Similar to Eastern Creek Chubsucker, *E. oblongus*, and Western Creek Chubsucker, *E. claviformis*, but adult has *deeper body*, usually *34–39 lateral scales, fewer than 14 predorsal scales*. Dark stripe along side extends *onto caudal fin* (darkest on young) but *not* around snout. Has 10–13, usually 11–12, dorsal rays. To 16 in. (41 cm). **RANGE:** Great Lakes and Mississippi R. basin lowlands from s. ON to WI and south to Gulf; Atlantic and Gulf slope drainages from s. VA to Brazos R., TX. Isolated population in Guadalupe R., TX. Sporadic in north; common on Coastal Plain. Protected in Canada as a *threatened species*. **HABITAT:** Lakes, swamps, ponds, sloughs, impoundments;

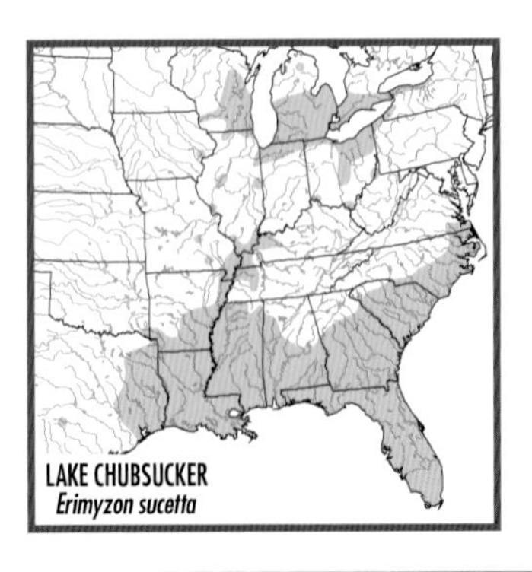

quiet pools of creeks and small rivers; usually over silt, sand, or debris. **SIMILAR SPECIES:** See (1) Eastern Creek Chubsucker, *E. oblongus,* and (2) Western Creek Chubsucker, *E. claviformis* (Pl. 22). (3) Sharpfin Chubsucker, *E. tenuis* (Pl. 22), has *pointed dorsal fin, more terminal mouth* (upper lip barely below level of eye), *40–45 lateral scales.* On young, dark stripe along side *extends around snout.*

SHARPFIN CHUBSUCKER *Erimyzon tenuis* Pl. 22

IDENTIFICATION: Similar to Eastern Creek Chubsucker, *E. oblongus,* and Western Creek Chubsucker, *E. claviformis,* but has *sharply pointed dorsal fin, nearly terminal mouth* (upper lip barely below level of eye); breeding male *lacks* bilobed anal fin. Dark stripe along side (darkest on young) extends onto caudal fin and around snout. Often black edge on light amber or olive dorsal and anal fins; yellow or olive caudal and paired fins. Has 40–45 lateral scales. To 13 in. (33 cm). **RANGE:** Gulf Slope below Fall Line from Pensacola Bay basin, AL and FL, to Amite R. system, MS and LA. Common. **HABITAT:** Pools and backwaters of creeks and small rivers over sand or silt; often near vegetation. **SIMILAR SPECIES:** See (1) Eastern Creek Chubsucker, *E. oblongus,* and (2) Western Creek Chubsucker, *E. claviformis* (Pl. 22). (3) Lake Chubsucker, *E. sucetta* (Pl. 22), has *rounded dorsal fin, more subterminal mouth* (upper lip well below level of eye), 34–39 lateral scales, dark stripe along side *not* extending around snout.

NORTHERN HOG SUCKER *Hypentelium nigricans* Pl. 22

IDENTIFICATION: *Large, rectangular head, broadly flat* (in young) or *concave* (in adult) *between eyes.* Body wide in front, abruptly tapering behind dorsal fin. Has 3–6 *dusky or brown saddles* (1 on nape) extending obliquely forward on upper side. Long, blunt snout; large fleshy lips on horizontal mouth; many large papillae on lips; halves of lower lip broadly joined at middle (Fig. 38). Dark olive or bronze to red-brown above; often light stripes along scale rows on side; pale yellow or white

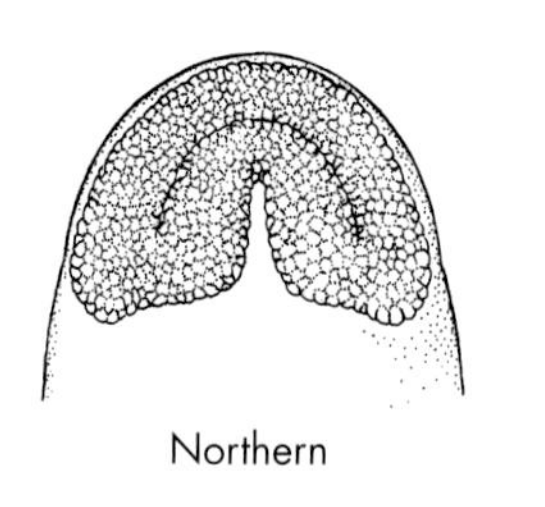

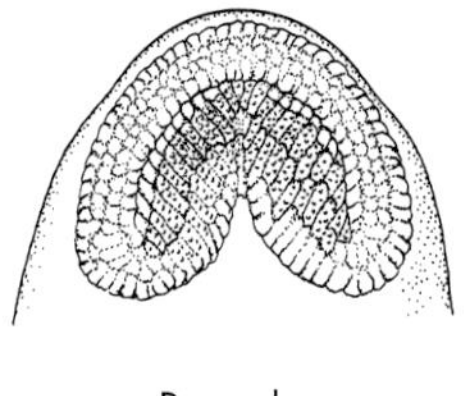

Fig. 38. Northern and Roanoke hog suckers—lips.

below; blue-black snout; olive to light orange fins; often black edge on dorsal and caudal fins. Large individual has black-tipped dorsal fin. Complete lateral line, 44–54 scales; usually 11 dorsal rays, 32–38 total (both sides) pectoral rays. To 24 in. (61 cm). **RANGE:** Great Lakes, Hudson Bay (Red R.), and Mississippi R. basins from NY and s. ON to MN, and south to n. AL, s. AR, and e. LA; Atlantic Slope drainages from Mohawk-Hudson river, NY, to Oconee R., n. GA; Gulf Slope drainages from Pascagoula R., MS, to Mississippi R., LA; Conasauga R. (Mobile Bay drainage), TN, and upper Chattahoochee R., GA. Common; generally avoids lowlands. **HABITAT:** Rocky riffles, runs, and pools of clear creeks and small rivers; occasionally large rivers, impoundments. **SIMILAR SPECIES:** See (1) Alabama Hog Sucker, *H. etowanum*, and (2) Roanoke Hog Sucker, *H. roanokense* (Pl. 22).

ALABAMA HOG SUCKER *Hypentelium etowanum* **Not shown**

IDENTIFICATION: Nearly identical to Northern Hog Sucker, *H. nigricans*, but usually has *10 dorsal rays*, head *only slightly concave* between eyes, light stripes on upper side usually more prominent, *red-orange anal and paired fins, orange snout and lips;* reaches 9 in. (23 cm). **RANGE:**

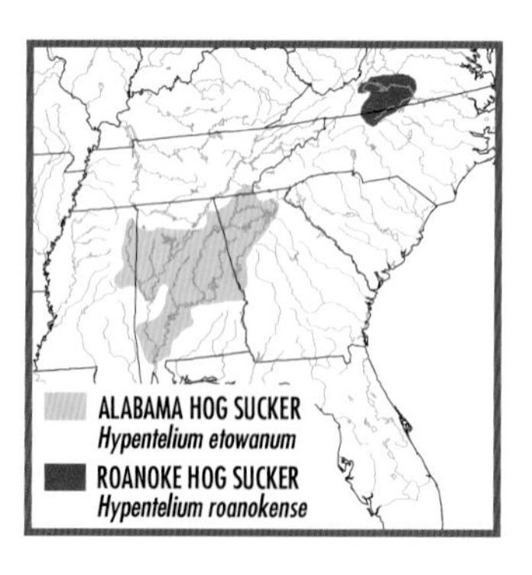

Chattahoochee R. and Mobile Bay drainages, GA, AL, MS, and se. TN; Baker Creek, TN, and Town Creek, AL (both Tennessee R. drainage). Common. **HABITAT:** Rocky riffles and runs of clear fast creeks and small rivers. **SIMILAR SPECIES:** (1) See Northern Hog Sucker, *H. nigricans* (Pl. 22).

ROANOKE HOG SUCKER *Hypentelium roanokense* Pl. 22

IDENTIFICATION: Similar to Northern Hog Sucker, *H. nigricans*, and Alabama Hog Sucker, *H. etowanum*, but has *plicate lips* (papillose on outer surfaces; Fig. 38); usually *lacks* or has *vague dark saddle* on nape; usually has *39–44 lateral scales, 28–32 total pectoral rays;* prominent light stripes on back and side. Copper to brown above, usually 4 black saddles; clear to light orange fins. Usually 11 dorsal rays. To 6½ in. (16 cm). **RANGE:** Upper Roanoke R. drainage, VA and NC. Common. **HABITAT:** Rocky riffles, runs, and pools of creeks and small rivers; often over sand in slow current. **SIMILAR SPECIES:** See (1) Northern Hog Sucker, *H. nigricans* (Pl. 22), and (2) Alabama Hog Sucker, *H. etowanum*.

MOXOSTOMA

GENUS CHARACTERISTICS: Robust to long slender body. *Large, horizontal mouth; thick papillose or plicate lips;* groove between upper lip and snout. Complete lateral line, 37–51 scales; 9–17 dorsal rays; 7 anal rays; 16–37 long, thin, flattened rakers on 1st gill arch; 3-chambered gas bladder. Young usually has 3 or 4 dark saddles and 4 dark blotches on side.

HARELIP SUCKER *Moxostoma lacerum* Not shown

IDENTIFICATION: Moderately stout body; slender caudal peduncle. *Upper lip bound to snout* (no groove between lip and snout); *lower lip halves separate* (Fig. 39). Concave dorsal fin edge. Dusky olive-brown above, lighter below; pale orange anal and paired fins. Has 42–46 lateral scales, 12 scales around caudal peduncle, 11–12 dorsal rays, 9 pelvic rays. To 14½ in. (37 cm). **RANGE:** Maumee (Lake Erie drainage) and Ohio river drainage from OH and IN south to GA and AL (including

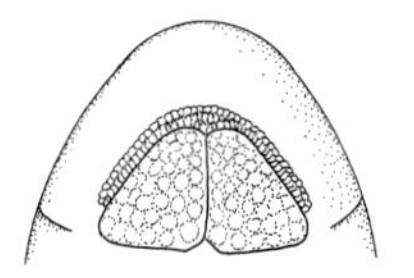

Fig. 39. Harelip Sucker—lips.

upper Tennessee R. system of VA and TN) and upper White R. drainage, AR. Extinct. **HABITAT:** Flowing pools of small to medium rivers. **SIMILAR SPECIES:** Other species of *Moxostoma* (Pl. 22 & 23) have groove between upper lip and snout, *undivided lower lip* (Fig. 40).

GREATER REDHORSE *Moxostoma valenciennesi* Pl. 23

IDENTIFICATION: Stout body; *large head* (about 25 percent of standard length in individual over 10 in. [25 cm]). *Red caudal fin. Thick plicate lips:* V-shaped rear edge on lower lip (Fig. 40). Usually convex dorsal fin edge. Pointed lobes on large, forked caudal fin. Bronze or coppery above, *dark spots on scales* on back and side; silver yellow or white below; yellow to red fins. Usually 42–45 lateral scales, 13–14 dorsal rays, 15–16 scales around caudal peduncle, 12–13 scales over back (in front of dorsal fin) from lateral line to lateral line (excluding lateral-line scales). Bladelike teeth on slender pharyngeal arch. To 31½ in. (80 cm). **RANGE:** Great Lakes-St. Lawrence R., Hudson Bay (Red R.), and Mississippi R. basins from QC and VT to ND, and south to Wabash R., IN. Uncommon. **HABITAT:** Sandy to rocky pools and runs of medium to large rivers; lakes. **SIMILAR SPECIES:** (1) Copper Redhorse, *M. hubbsi* (Pl. 23), has *shorter head*, highly arched back, 15 or 16 scales over back in front of dorsal fin, molarlike teeth on pharyngeal arch. (2) River Redhorse, *M. carinatum* (Pl. 23), usually has concave dorsal fin edge, 12–13 scales around caudal peduncle, molarlike teeth on pharyngeal arch

COPPER REDHORSE *Moxostoma hubbsi* Pl. 23

IDENTIFICATION: Stout, *deep body. Highly arched back* behind *short head* (about 20 percent of standard length in individual over 10 in. [25 cm]). *Red caudal fin.* Plicate lips; slightly V-shaped rear edge on lower lip. Usually convex dorsal fin edge. Pointed lobes on large, forked caudal fin. Gold to olive, often a coppery sheen above; dark spots on scales on back and side; yellow to red fins. Usually 44–47 lateral scales, 12–13

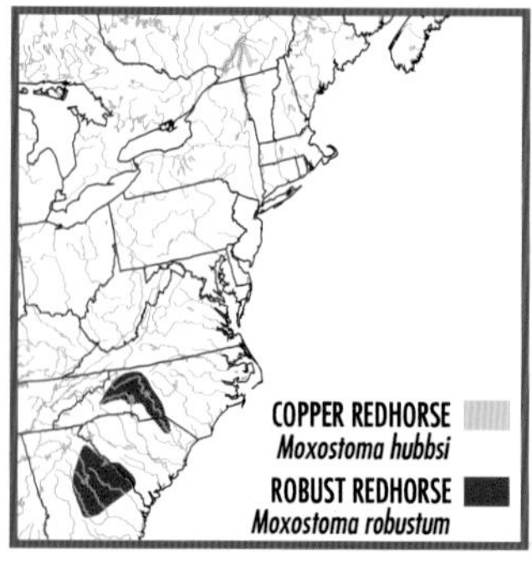

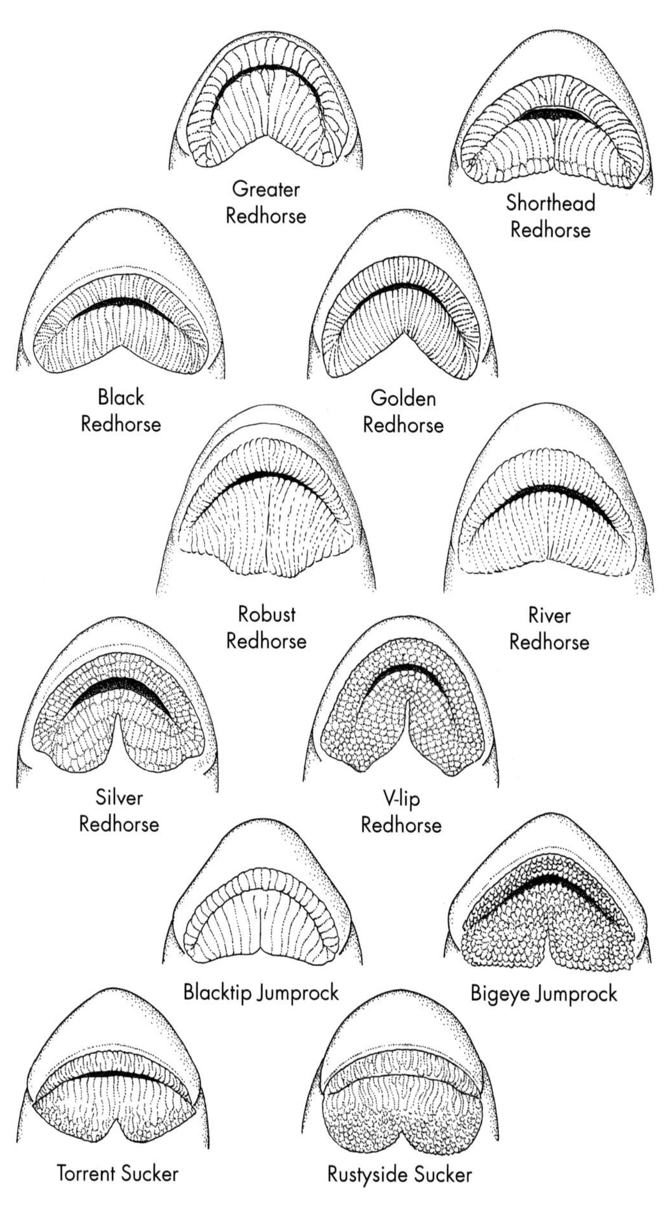

Fig. 40. Redhorses and related suckers—lips.

dorsal rays, 16 scales around caudal peduncle, 15 or 16 scales over back (in front of dorsal fin) from lateral line to lateral line. Has 4–6 molarlike teeth on stout pharyngeal arch. To 28 in. (72 cm). **RANGE:** St. Lawrence R. drainage, QC. Uncommon. Protected in Canada as a *threatened species.* **HABITAT:** Rocky pools, backwaters, and swift runs of medium to large rivers. **SIMILAR SPECIES:** (1) Greater Redhorse, *M. valenciennesi* (Pl. 23), has *more elongate body; longer head,* thicker lips, usually 12-13 scales across back in front of dorsal fin, bladelike teeth on pharyngeal arch. (2) River Redhorse, *M. carinatum* (Pl. 23), usually has concave dorsal fin edge, 12–13 scales around caudal peduncle, 42–44 lateral scales.

ROBUST REDHORSE *Moxostoma robustum* Pl. 23

IDENTIFICATION: *Stout, wide body; moderate-sized head.* Medium to large plicate lips; straight edge or medial flap on lower lip (Fig. 40). Slightly concave to straight dorsal fin. *Red-orange caudal fin* (often only along edge). Gold-brown above and on side, brass to copper sheen; *dusky to dark irregular stripes on side*; red-orange anal and paired fins; olive, sometimes partly orange, dorsal fin. Usually 41–44 lateral scales, 28–31 scales around body at dorsal fin origin, 11–12 scales around caudal peduncle, 12–14 dorsal rays. Molariform teeth on stout pharyngeal arch. To 28½ in. (72 cm). **RANGE:** Peedee R., NC and SC; Santee R. (extirpated but recently stocked), NC; Savannah R., SC, to Ocmulgee R., GA. Uncommon to rare. **HABITAT:** Deep flowing pools in medium to large rivers. **SIMILAR SPECIES:** (1) River Redhorse, *M. carinatum* (Pl. 23), *lacks* dusky to dark irregular stripes on lower side, has 30–37 scales around body at dorsal fin origin.

RIVER REDHORSE *Moxostoma carinatum* Pl. 23

IDENTIFICATION: *Stout body; large head* (about 25 percent of standard length in individual over 10 in. [25 cm]). *Large plicate lips;* slightly V-

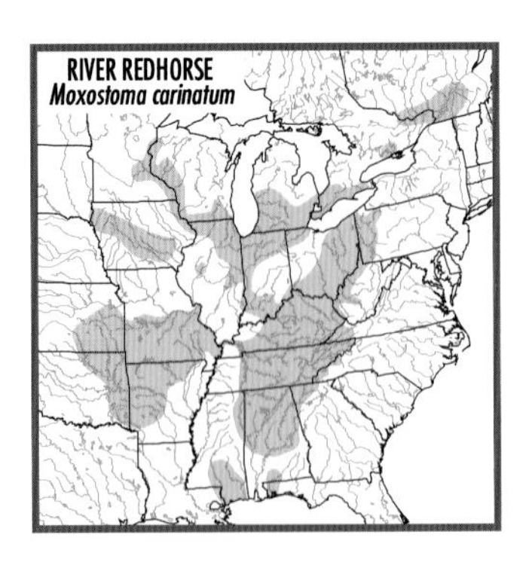

shaped rear edge on lower lip (Fig. 40). Usually concave dorsal fin. *Large, red, forked caudal fin*; pointed upper lobe usually longer than rounded lower lobe. Olive-bronze above, gold sheen; *crescent-shaped dark spots* on scales on back and side; deep yellow or orange anal and paired fins; red dorsal fin. Breeding male has dark stripe along side. Usually 42–44 lateral scales, 30–37 scales around body at dorsal fin origin, 12–13 scales around caudal peduncle, 12–13 dorsal rays. Has *6–8 molarlike teeth* on stout pharyngeal arch. To 30 in. (77 cm). **RANGE:** St. Lawrence R.-Great Lakes and Mississippi R. basins from QC to cen. MN and w. IA, and south to n. AL and e. OK; Gulf Slope from Escambia R., FL, to Pearl R., LA and MS. Locally common; disappearing from northern and western parts of range. **HABITAT:** Rocky pools and swift runs of small to large rivers; impoundments. **SIMILAR SPECIES:** See Pl. 23. (1) Robust Redhorse, *M. robustum*, has dusky to dark irregular stripes on lower side, less uniform red on dorsal and caudal fins. (2) Shorthead Redhorse, *M. macrolepidotum*, has *short head, papillae* on lower lip, straight rear edge on lower lip, *bladelike teeth* on pharyngeal arch. (3) Greater Redhorse, *M. valenciennesi*, has thicker lips, usually 15–16 scales around caudal peduncle, *bladelike teeth* on pharyngeal arch.

SHORTHEAD REDHORSE *Moxostoma macrolepidotum* Pl. 23

IDENTIFICATION: Fairly stout body; *short head* (about 20 percent of standard length in individual over 8 in. [20 cm]). *Red caudal fin.* Plicate lips; large papillae on lower lip; *straight rear edge* on lower lip (Fig. 40). Moderately *concave to falcate dorsal fin*. Large, moderately forked caudal fin; upper lobe usually distinctly longer than lower lobe (see Remarks). Olive or tan above; copper or silver cast on olive-yellow side; crescent-shaped dark spots on scales on back and side; white or yellow below; yellow to red fins. Usually 42–44 lateral scales, 12–13 scales around caudal peduncle, 12–13 dorsal rays. Bladelike teeth on

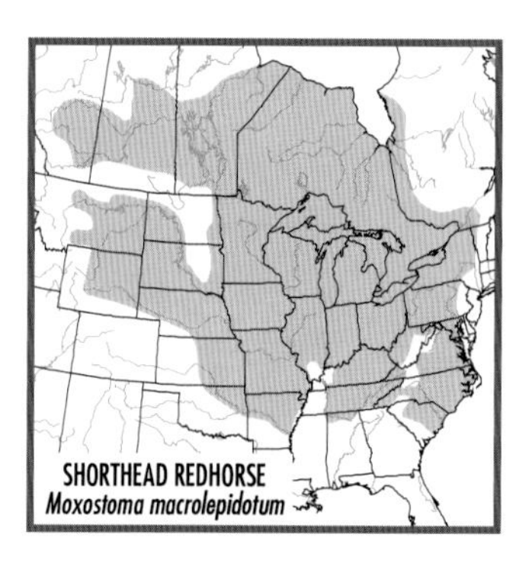
SHORTHEAD REDHORSE
Moxostoma macrolepidotum

slender pharyngeal arch. To 29½ in. (75 cm). **RANGE:** Great Lakes-St. Lawrence R., Hudson Bay, and Mississippi R. basins from QC to AB and south to n. AL and OK (1 record in Red R., OK); Atlantic Slope drainages from Hudson R., NY, to Santee R., SC. Locally common. **HABITAT:** Rocky pools, runs, and riffles in small to large rivers; lakes. **REMARKS:** Three subspecies. *M. m. pisolabrum,* in Ozark Uplands (Arkansas and Red river drainages), usually has a pea-shaped swelling at middle of upper lip; intergrades with *M. m. macrolepidotum* in lower Missouri R. drainage, MO, Mississippi R. tributaries, ne. MO, and Kaskaskia R., IL. *M. m. breviceps* (Pl. 23), in Ohio R. basin, usually has 12 dorsal rays (vs. 13 in other subspecies), 10 pelvic rays (vs. 9), smaller lips, longer upper caudal lobe, moderately concave dorsal fin, and is smaller (to 19 in. [48 cm]); intergrades with *M. m. macrolepidotum* in Wabash R. drainage, IN and IL. *M. m. macrolepidotum* (Pl. 23), throughout rest of range, usually has larger lips, equal caudal lobes, and straight to slightly concave dorsal fin. **SIMILAR SPECIES:** See Pl. 23. (1) See Sicklefin Redhorse, *M.* species. (2) River Redhorse, *M. carinatum*, has *larger head,* thicker lips, slightly *V-shaped rear edge* on lower lip, molarlike teeth on pharyngeal arch. (2) Greater Redhorse, *M. valenciennesi*, has *larger head,* thicker lips; *V-shaped rear edge* on lower lip; usually 15–16 scales around caudal peduncle.

SICKLEFIN REDHORSE *Moxostoma* species — Pl. 23

IDENTIFICATION: Similar to Shorthead Redhorse, *M. macrolepidotum*, but has *moderately to strongly falcate dorsal fin—rays 1–3 extend beyond tip of last ray* when depressed; usually *44–46 lateral scales*; slightly molariform teeth on moderate pharyngeal arch. To 21½ in. (55 cm). **RANGE:** Hiwassee and Little Tennessee river systems, Blue Ridge of NC and GA; locally common. **HABITAT:** Rocky riffles, runs, and pools of large creeks and small to medium rivers. **SIMILAR SPECIES:** (1) See Shorthead Redhorse, *M. macrolepidotum* (Pl. 23).

BLACKTAIL REDHORSE *Moxostoma poecilurum* Pl. 23

IDENTIFICATION: *Long, cylindrical body. Red caudal and lower fins; black stripe on* lower caudal fin lobe; dark and light stripes on side moderately developed. Plicate lips; slightly U-shaped rear edge on lower lip. Usually concave dorsal fin edge. Large, moderately forked caudal fin, lower lobe usually longer than upper lobe. Gold to bronze above, silver green iridescence; silver yellow or white below; dusky gray on lower half, red on upper half, of dorsal fin; white to red on other fins. Bladelike teeth on slender pharyngeal arch. Usually 41–44 lateral scales, 12 scales around caudal peduncle, 12–13 dorsal rays. To 20 in. (51 cm). **RANGE:** Mississippi R. tributaries on Former Mississippi Embayment from KY and AR south to LA; Gulf Slope drainages from Choctawhatchee R., AL and FL, to Galveston Bay, TX. Locally common. **HABITAT:** Sandy and rocky pools, runs, and riffles of small to medium rivers; impoundments. **SIMILAR SPECIES:** (1) See Apalachicola Redhorse, *M.* species.

APALACHICOLA REDHORSE *Moxostoma* species Not shown

IDENTIFICATION: Similar to Blacktail Redhorse, *M. poecilurum*, but has *dusky gray* caudal fin, *darker stripes* on side; *lacks* red on caudal and lower fins. To 20½ in. (52 cm). **RANGE:** Apalachicola R. drainage, GA, AL, and FL. Locally common. **HABITAT:** Mud- to rock-bottomed pools, sandy to rocky runs and riffles of small to large rivers; impoundments. **SIMILAR SPECIES:** (1) See Blacktail Redhorse, *M. poecilurum* (Pl. 23).

BLACK REDHORSE *Moxostoma duquesnii* Pl. 23

IDENTIFICATION: Long slender body; *long slender caudal peduncle. Gray caudal fin* (rarely pale red). Plicate lips; broadly V-shaped rear edge on lower lip (Fig. 40). Usually concave dorsal fin. Large, moderately forked caudal fin usually has equal, pointed lobes. Dusky olive above; gold to brassy side, green iridescence; white or yellow below; orange anal and paired fins; dusky or slate dorsal fin (sometimes red-tinged).

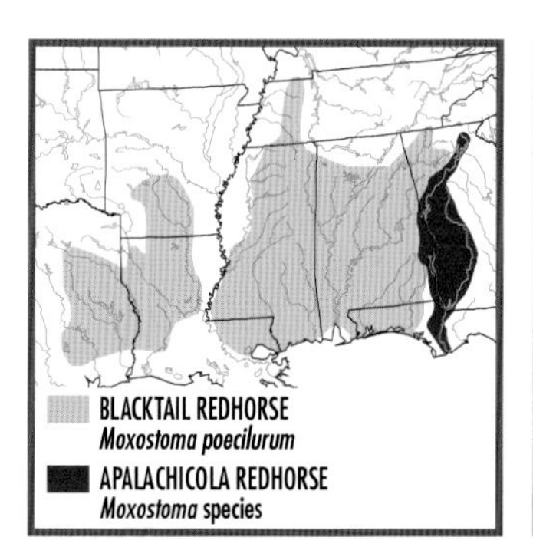

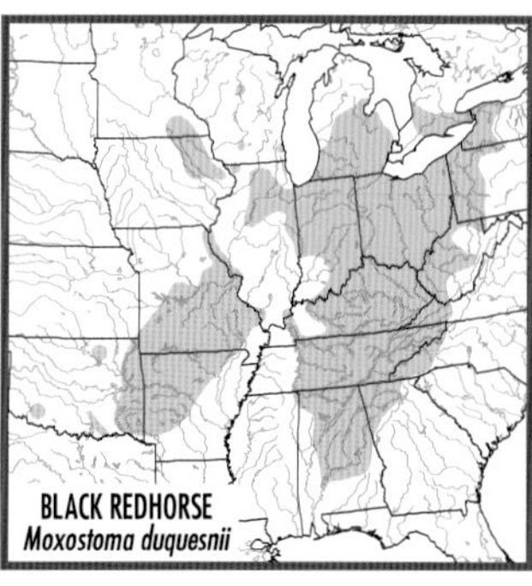

Breeding male has pink-orange stripe along side, pink-orange anal and paired fins, small tubercles on snout. Usually *44–47 lateral scales,* 12–14 scales around caudal peduncle, 12–14 dorsal rays, *10 pelvic rays* (vs. 9 in other redhorses). Many bladelike teeth on slender pharyngeal arch. To 20 in. (51 cm). **RANGE:** Lower Great Lakes and Mississippi R. basins from s. ON and NY to se. MN and south to n. AL and e. OK (absent on Former Mississippi Embayment); Mobile Bay drainage (absent in Tombigbee R. drainage except Black Warrior system), GA, AL, and se. TN. Common. Protected in Canada as a *threatened species.* **HABITAT:** Sand- to rock-bottomed pools and runs of creeks and small to medium rivers; impoundments. **SIMILAR SPECIES:** (1) Golden Redhorse, *M. erythrurum* (Pl. 23), has *shorter, deeper caudal peduncle; usually 40–42 lateral scales, 9 pelvic rays;* breeding male has large snout tubercles. (2) Silver Redhorse, *M. anisurum* (Pl. 23), has papillose lips, acutely V-shaped rear edge on lower lip, usually convex or straight dorsal fin edge, usually 14–16 dorsal rays.

GOLDEN REDHORSE *Moxostoma erythrurum* Pl. 23

IDENTIFICATION: Moderately stout body; fairly *stout caudal peduncle. Gray caudal fin* (pale orange in young). Plicate lips; V- or U-shaped rear edge on lower lip (Fig. 40). Usually concave dorsal fin. Large, moderately forked caudal fin has equal, pointed lobes. Olive to brass brown above; dark-edged scales on back and front half of side; yellow to brassy side, pale green sheen; yellow or white below; yellow to orange anal and paired fins; gray dorsal fin. Breeding male has dark stripe along side, bright salmon lower fins, large tubercles on snout. Usually *40–42 lateral scales,* 12 scales around caudal peduncle, 12–14 dorsal rays, 9 pelvic rays. Many *bladelike teeth* on slender pharyngeal arch. To 30½ in. (78 cm). **RANGE:** Great Lakes, Hudson Bay (Red R.), and Mississippi R. basins from NY and s. ON to MB, and south to n. AL and s. OK; isolated population in sw. MS; Atlantic Slope from Potomac R., PA, to Roanoke R., NC (absent in Rappahannock and York river drainages); Mobile Bay drainage, GA, AL, and se. TN. Common. **HABITAT:** Mud- to rock-bottomed pools, runs and riffles of creeks and small to large rivers; occasionally lakes. **SIMILAR SPECIES:** (1) See Carolina Redhorse, *M.* species. (2) Black Redhorse, *M. duquesnii* (Pl. 23), has *longer, more slender caudal peduncle; usually 44–47 lateral scales,* 10 pelvic rays; breeding male has small tubercles on snout. (3) Silver Redhorse, *M. anisurum* (Pl. 23), has papillose lips, acutely V-shaped rear edge on lower lip, usually convex or straight dorsal fin edge, usually 14–16 dorsal rays.

CAROLINA REDHORSE *Moxostoma* species Not shown

IDENTIFICATION: Similar to Golden Redhorse, *M. erythrurum,* but has *dusky black anal fin,* usually *44–45 lateral scales, 14–15 dorsal rays.*

Adult has 2–3 gold stripes on lower side; breeding male has small to medium tubercles on snout, *cheek, and opercle*. To 23 in. (59 cm). **RANGE:** Cape Fear and Peedee river drainages, NC and SC; uncommon and localized. **HABITAT:** Silty to rocky deep pools and slow runs of small to medium rivers. **SIMILAR SPECIES:** See Pl. 23. (1) See Golden Redhorse, *M. erythrurum*. (2) Silver Redhorse, *M. anisurum*, (3) Notchlip Redhorse, *M. collapsum*, and (4) V-Lip Redhorse, *M. pappillosum*, have acutely V-shaped rear edge on deeply divided lower lip.

SILVER REDHORSE *Moxostoma anisurum* Pl. 23

IDENTIFICATION: Stout body; *straight or slightly convex dorsal fin. Acutely V-shaped rear edge* on *deeply divided lower lip; many small papillae* on upper and lower lips (Fig. 40). *Slate gray dorsal and caudal fins.* Large, moderately forked caudal fin; upper lobe often longer than lower lobe. Iridescent blue-green to brown above; pale yellow-silver to brassy side, no dark spots on scales; pale yellow to red anal and paired fins. Usually 40–42 lateral scales, 12 scales around caudal peduncle, *14–16 dorsal rays.* To 28 in. (71 cm). **RANGE:** Great Lakes-St. Lawrence R., Hudson Bay, and Mississippi R. basins from QC to AB, and south to n. GA, AL, and AR. Uncommon. **HABITAT:** Mud- to rock-bottomed pools and runs of small to large rivers; occasionally lakes. **SIMILAR SPECIES:** See Pl. 23. (1) See Notchlip Redhorse, *M. collapsum*. (2) V-lip Redhorse, *M. pappillosum*, has more slender body, *concave to falcate dorsal fin,* usually *12–13 dorsal rays.* (3) Black Redhorse, *M. duquesnii*, and (4) Golden Redhorse, *M. erythrurum*, have *plicate lips,* usually *concave dorsal fin, 12–14 dorsal rays.*

NOTCHLIP REDHORSE *Moxostoma collapsum* Not shown

IDENTIFICATION: Nearly identical to Silver Redhorse, *M. anisurum*, but has *more elongate body*; smaller head; *straight rear margin* on dorsal fin; *equal-length caudal fin lobes*; more plicate, semipapillose

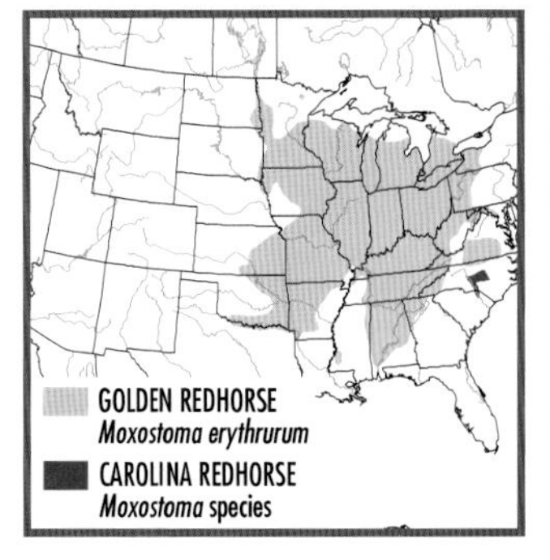

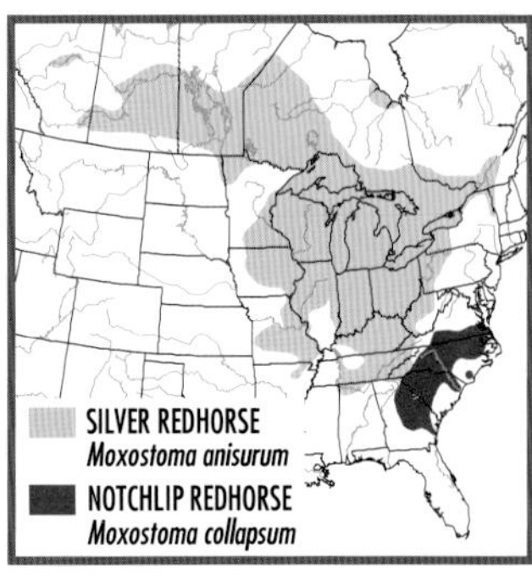

lips, clear to slightly dusky anal fin. To 23 in. (58 cm). **RANGE:** Atlantic drainages from Roanoke (including Chowan) R., VA, south to Altamaha R., GA. Locally common. **HABITAT:** Silty to rocky pools and slow runs of medium to large rivers; reservoirs. **SIMILAR SPECIES:** (1) See Silver Redhorse, *M. anisurum* (Pl. 23).

V-LIP REDHORSE *Moxostoma pappillosum* Pl. 23

IDENTIFICATION: *Long slender body. Papillose lips; acutely V-shaped rear edge* on *deeply divided lower lip* (Fig. 40). Moderately *concave to falcate dorsal fin.* Equal, pointed lobes on large, moderately forked caudal fin. Olive-tan above; brassy to yellow side, silver green iridescence; gray to salmon-orange fins. Usually 42–44 lateral scales, 12 scales around caudal peduncle, 12–13 dorsal rays. To 17¾ in. (45 cm). **RANGE:** Atlantic Slope from Chowan and Roanoke river drainages, VA, to Santee R. drainage, SC. Uncommon; rare in Peedee and Santee river drainages. **HABITAT:** Rocky runs and mud- to rock-bottomed pools of small rivers; impoundments. **SIMILAR SPECIES:** (1) Notchlip Redhorse, *M. collapsum,* has *straight rear margin* on dorsal fin, usually 14–16 dorsal rays. (2) Silver Redhorse, *M. anisurum* (Pl. 22), has *stouter body,* usually *straight or convex dorsal fin,* usually 14–16 dorsal rays. (3) Carolina Redhorse, *M.* species, has *plicate lips, broadly V- or U-shaped rear edge* on lower lip, 14–15 dorsal rays.

GRAY REDHORSE *Moxostoma congestum* Pl. 23

IDENTIFICATION: *Broad, U-shaped* (from above) *head.* Plicate lips; straight to rounded edge on lower lip. Long pectoral fin (length greater than head depth at rear). Straight to slightly concave dorsal fin. Moderately large, shallowly forked caudal fin; pointed upper lobe, rounded lower lobe. Olive to yellow-gray above, gray dorsal and caudal fins, yellow anal and paired fins. Breeding male is brassy gold, has *yellow to light orange fins.* Usually 42–46 lateral scales, 16 scales around cau-

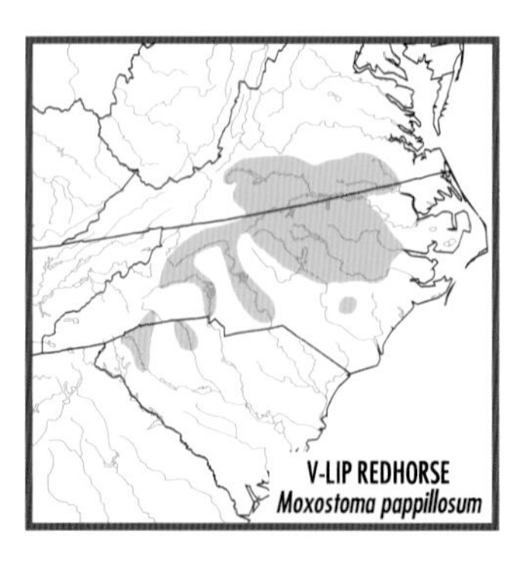

dal peduncle, 32–36 scales around body (at dorsal fin origin), 16–17 pectoral rays, 11–12 dorsal rays. To 25½ in. (65 cm). **RANGE:** Brazos R. drainage to Rio Grande drainage, TX and s. NM. Also in Mexico (south to Río Soto la Marina drainage). Locally common; extirpated from Rio Grande, NM. **HABITAT:** Deep runs and pools of small to medium rivers; lakes. **SIMILAR SPECIES:** (1) See Mexican Redhorse, *M. austrinum*. (2) Brassy Jumprock, *M.* species (Pl. 23), usually has 45–48 lateral scales; young has dark stripes on back.

MEXICAN REDHORSE *Moxostoma austrinum* Not shown

IDENTIFICATION: Similar to Gray Redhorse, *M. congestum,* but usually has *36–41 scales around body* (at dorsal fin origin), dark crescents on upper body scales. To 19 in. (49 cm). **RANGE:** Alamito Creek (Rio Grande drainage), Big Bend, TX; Rio Grande and Pacific drainages of Mexico. Rare in U.S.; uncommon in Mexico. **HABITAT:** Rocky runs and riffles of creeks and small to medium rivers; often near boulders in swift water. **SIMILAR SPECIES:** (1) See Gray Redhorse, *M. congestum* (Pl. 23).

BRASSY JUMPROCK *Moxostoma* species Pl. 23

IDENTIFICATION: *Moderate to deep head,* distinctly convex between eyes; *rounded snout* (viewed from above). Fairly deep caudal peduncle. Plicate lips; nearly straight lower lip edge. Small fins. Straight to slightly concave dorsal fin. Pointed upper lobe, rounded lower lobe, on moderately large, shallowly forked caudal fin. Gold-brown above and on side, dark-edged scales; dusky dorsal fin; olive caudal fin; orange to red on other fins. Young has dark stripes on back and side narrower than pale interspaces; 4–5 dark blotches along side may connect to 4 saddles. Usually 45–48 lateral scales, 34–37 scales around body at dorsal fin origin, 16 scales around caudal peduncle, 11–12 dorsal rays, 16–18 pectoral rays. To 16½ in. (42 cm). **RANGE:** Atlantic Slope from Peedee R. drainage, VA, and Cape Fear R. drainage, NC, to

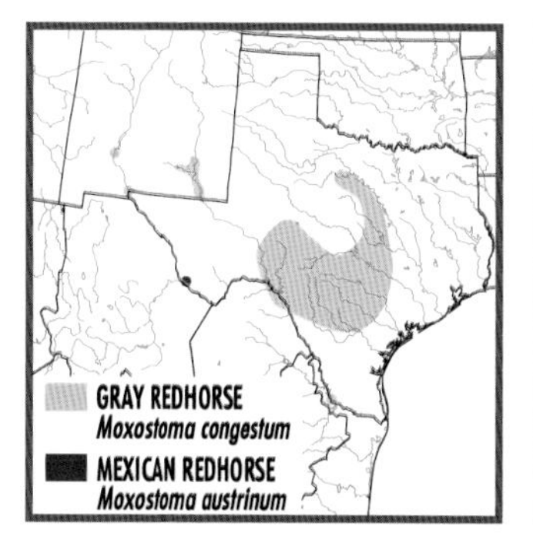

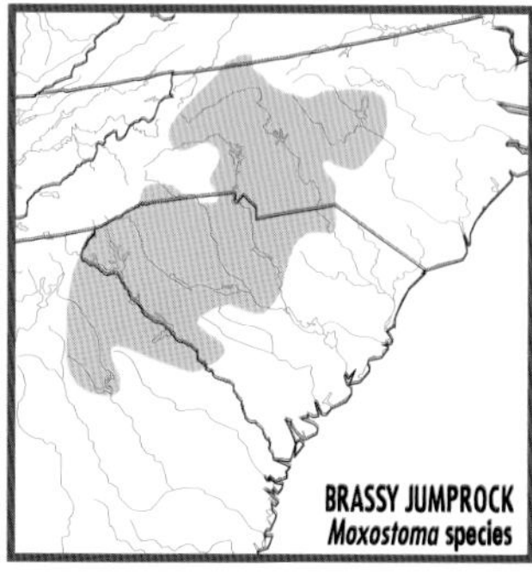

Oconee R. system (Altamaha R. drainage), GA. Uncommon in montane and Piedmont streams. **HABITAT:** Silty to rocky pools and slow runs of small to medium rivers; impoundments. **SIMILAR SPECIES:** See Pl. 22. (1) Striped Jumprock, *M. rupiscartes,* has more slender body, head *wider* than deep, usually 10–11 dorsal rays, plicate to semipapillose lower lip. (2) Blacktip Jumprock, *M. cervinum,* has black or dusky tips on dorsal and caudal fins; usually 40–44 lateral scales, 30–34 scales around body, 11 dorsal rays. (3) Greater Jumprock, *M. lachneri,* has more slender body, *longer head, blunt snout,* usually 30–34 scales around body.

STRIPED JUMPROCK *Moxostoma rupiscartes* Pl. 22

IDENTIFICATION: Cylindrical body; nearly straight upper and lower profiles; fairly deep caudal peduncle. Head flat or slightly convex between eyes; *wider than deep.* Plicate (plicae divided at rear) to semipapillose lower lip; *nearly straight lower lip edge.* Straight to slightly concave dorsal fin. *Dusky edge* on dorsal and caudal fins (in some populations). Pointed to rounded lobes on moderately large, slightly forked caudal fin. Yellow-olive to brown above; *prominent dark stripes* on back and side wider than or equal to pale interspaces; dusky olive to orange fins. Breeding male has faint yellow-brown stripe along side. Young has 4–5 blotches along side, yellow caudal fin. Usually 45–48 lateral scales, 32–37 scales around body at dorsal fin origin, 16 scales around caudal peduncle, 10–11 dorsal rays, 16–18 pectoral rays. To 11 in. (28 cm). **RANGE:** Atlantic Slope from Santee R. drainage, NC, to Altamaha R. drainage, GA; Gulf Slope in upper Chattahoochee R. drainage, GA. Possibly (introduced) in extreme upper Peedee R. drainage, NC. Common in montane and Piedmont streams. **HABITAT:** Sandy to rocky riffles and runs of small to medium rivers. **SIMILAR SPECIES:** (1) Brassy Jumprock, *M.* species (Pl. 23), has plicate lips, *moderate to deep head*; usually 11–12 dorsal rays. (2) Blacktip Jumprock, *M. cervinum* (Pl. 22), usually has 40–44 lateral scales, 30–34 scales around body. (3) Greater Jumprock, *M. lachneri* (Pl. 22), has *long narrow head, U-shaped lower lip edge,* usually 12 dorsal rays.

BLACKTIP JUMPROCK *Moxostoma cervinum* Pl. 22

IDENTIFICATION: Long, cylindrical body. *Black or dusky tips* on dorsal, caudal, and often anal fins. Light stripes on back and upper side. *Plicate lips;* straight lower lip edge (Fig. 40). Head slightly convex between eyes. Pointed to rounded lobes on small, slightly forked caudal fin. Olive to brown above, white on lower side; olive-yellow to orange fins. Young has about 6 dark blotches along side. Breeding male has brassy lower side, sometimes a light yellow stripe along side, red-orange caudal and paired fins. Usually 40–44 lateral scales, 30–34 scales around body at dorsal fin origin, 14–16 scales around caudal

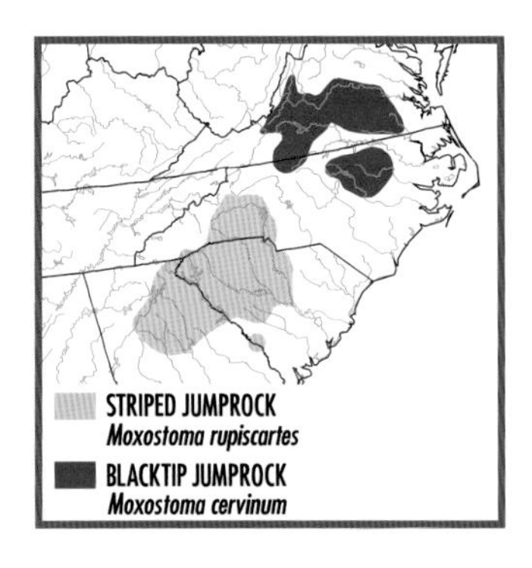

peduncle, 11 dorsal rays, 15 pectoral rays. To 7½ in. (19 cm). **RANGE:** Atlantic Slope drainages from James R., VA, to Neuse R., NC. Introduced into New R. drainage, VA. Common. **HABITAT:** Rocky riffles and runs of creeks and small rivers. **SIMILAR SPECIES:** (1) Greater Jumprock, *M. lachneri* (Pl. 22), *lacks* black tips on fins; has longer, more narrow head; broadly U-shaped lower lip edge; usually 45–46 lateral scales, 12 dorsal rays; large male has blue-gray anal and caudal fins. (2) Striped Jumprock, *M. rupiscartes* (Pl. 22), *lacks* black tips on fins; usually has 45–48 lateral scales, 16–18 pectoral rays. (3) Brassy Jumprock, *M.* species (Pl. 23), *lacks* black tips on fins; usually has 45–48 lateral scales, 11–12 dorsal rays, 34–37 scales around body.

GREATER JUMPROCK *Moxostoma lachneri* Pl. 22

IDENTIFICATION: Long, cylindrical body; *slender caudal peduncle. Long, narrow head* (deeper than wide); straight lower profile; blunt snout. Concave dorsal fin. *Plicate lips; broadly U-shaped lower lip edge. White lower ray* on gray caudal fin. Pointed upper lobe, rounded lower lobe on large, slightly forked caudal fin. Dark olive above (some individuals sharply bicolored; dark above, white below); about 8 dark stripes on back and upper side (obscure on large male); slate gray fins. Breeding male is iridescent silver black above; has blue-gray anal and caudal fins; other fins slate gray with some orange (especially pelvic fins). Light and dark stripes on back and side, about 5 dark blotches along side of young. Usually 45–46 lateral scales, 30–34 scales around body at dorsal fin origin, 16 scales around caudal peduncle, 12 dorsal rays, 17 pectoral rays. To 17½ in. (44 cm). **RANGE:** Apalachicola R. drainage, GA and AL. Uncommon. **HABITAT:** Rocky riffles and runs of clear small to medium rivers. **SIMILAR SPECIES:** (1) Blacktip Jumprock, *M. cervinum* (Pl. 22), has black or dusky dorsal and caudal fin tips; *straight lower lip edge;* usually 40–44 lateral scales, 11 dorsal rays; red-orange caudal and paired fins. (2) Striped Jumprock, *M. rupiscartes* (Pl. 22), has *nearly straight lower lip edge, head wider than deep,* usually 10–11 dorsal rays. (3) Brassy Jumprock, *M.* species (Pl. 23), has *rounded snout* (viewed from above), usually 34–37 scales around body.

BIGEYE JUMPROCK *Moxostoma ariommum* Pl. 22

IDENTIFICATION: Long cylindrical body flattened in front, compressed behind. *Very large eye.* Head flat or *slightly concave* between eyes. Flat, *flaring papillose lips* (Fig. 40). Straight to slightly concave dorsal fin. Slightly forked caudal fin. Olive to brown above (often with violet hue), vague light stripes (often absent) on back and upper side; sometimes an iridescent green streak on opercle and front of upper side; dusky to orange-red fins. Young has dark blotches along side. Usually 43–46 lateral scales, 15–16 scales around caudal peduncle, 11 dorsal rays. To 8½ in. (22 cm). **RANGE:** Upper Roanoke R. drainage, VA and NC. Uncommon. **HABITAT:** Deep rocky runs and pools of small to medium rivers; usually among large rubble and boulders, rarely in riffles. **SIMILAR SPECIES:** No other *Moxostoma* species has a large eye, slightly concave head, and flaring papillose lips.

TORRENT SUCKER *Thoburnia rhothoeca* Pl. 22

IDENTIFICATION: Cylindrical body; *2 large pale (sometimes dusky) areas* on caudal fin base highlighted by black streaks on adjacent caudal rays. Slightly concave or straight-edged dorsal fin. Small mouth; each half of lower lip edge *nearly triangular* (Fig. 40); *fewer papillae than plicae on lower lip.* Rounded lobes on small, slightly forked caudal fin. Dark olive-brown above; 5–6 vague to dark brown blotches along side, sometimes connected to 4 dark saddles; white to pale orange fins; dusky streaks on caudal and pectoral fins. Breeding male has rusty red *stripe* along side, bright yellow fins. Has 43–51 lateral scales; usually 16–18 scales around caudal peduncle, 10 dorsal rays. Black peritoneum. To 7¼ in. (18 cm). **RANGE:** Atlantic Slope from upper Po-

tomac R. drainage, VA and WV, to Roanoke R. drainage, VA (absent in York R.; probably introduced in Rappahannock R.); formerly in upper New R. drainage, VA and WV. Common. **HABITAT:** Rocky riffles and runs of moderate to swift creeks and small rivers; young in rocky runs and pools. **SIMILAR SPECIES:** (1) See Rustyside Sucker, *T. hamiltoni*. (2) Blackfin Sucker, *T. atripinnis* (Pl. 22), *lacks* pale areas on caudal fin base; has jet-black blotch on dorsal fin, 7–9 bold black stripes on back and upper side. (3) Blacktip Jumprock, *M. cervinum* (Pl. 22), has pronounced black or dusky dorsal and caudal fin tips, *plicate lips,* usually 11 dorsal rays.

RUSTYSIDE SUCKER *Thoburnia hamiltoni* Not shown

IDENTIFICATION: Nearly identical to Torrent Sucker, *T. rhothoeca*, but has *larger lower lip,* each half *square or broadly rounded; more papillae than plicae* on lower lip (Fig. 40). To 7 in. (18 cm). **RANGE:** Upper Dan R. system (Roanoke R. drainage), VA and NC. Locally common in small range. **HABITAT:** Rocky riffles and runs of swift montane creeks and small rivers; young in flowing pools. **SIMILAR SPECIES:** (1) See Torrent Sucker, *T. rhothoeca* (Pl. 22).

BLACKFIN SUCKER *Thoburnia atripinnis* Pl. 22

IDENTIFICATION: Cylindrical body. *Large jet-black blotch* on tip of white or light yellow dorsal fin; straight dorsal fin edge; *7–9 bold black stripes on back and upper side.* Small mouth; plicate lips; straight lower lip edge. Small, slightly forked caudal fin. Olive-gold stripes between black stripes on back and upper side; 2 dark saddles (often vague); shiny white below; pale yellow or olive anal and pelvic fins; pink-olive pectoral fin; pink-orange caudal fin. Young has 4–5 black blotches along side. Has 46–50 lateral scales; usually 16 scales around caudal peduncle, 10 dorsal rays. To 6¾ in. (17 cm). **RANGE:** Upper Barren R. system (Green R. drainage), KY and TN. Locally common in small

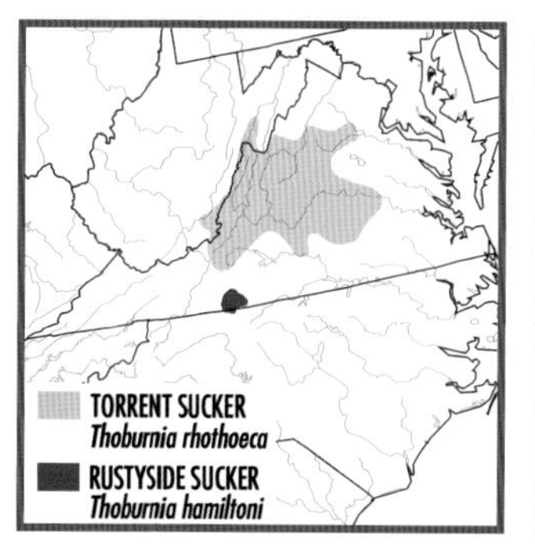

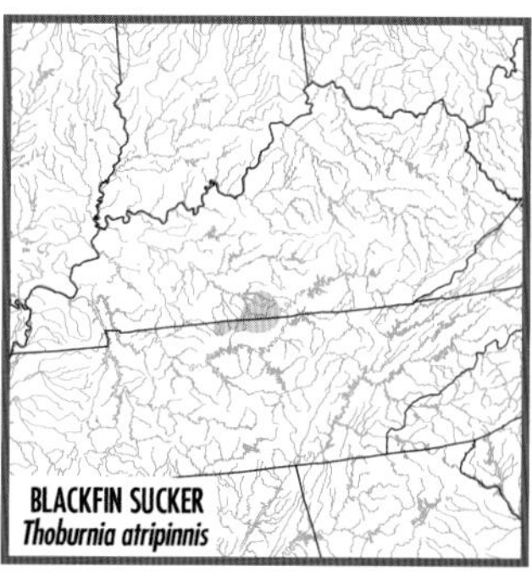

range. **HABITAT:** Rocky pools and adjacent riffles of creeks and small rivers; hides in bedrock crevices, near boulders, or under shoreline brush. **SIMILAR SPECIES:** (1) Torrent Sucker, *T. rhothoeca* (Pl. 22), and (2) Rustyside Sucker, *T. hamiltoni*, *lack* jet-black blotch on dorsal fin, bold black stripes on back and upper side; have 2 large pale areas on caudal fin base, more papillose lower lip.

LOACHES: Family Cobitidae (1 introduced)

Loaches, popular aquarium fishes, are native to fresh waters of Eurasia, Morocco, and Ethiopia. At least 170 species are known, and 1, Oriental Weatherfish, *Misgurnus anguillicdudatus*, was first introduced into the U.S. in the 1930s. Loaches have a *wormlike to fusiform body shape,* subterminal mouth, *3–6 pairs of barbels,* tiny or no scales, 1 row of pharyngeal teeth, a bifid suborbital spine, and a rounded to deeply forked caudal fin; some have an adipose fin.

ORIENTAL WEATHERFISH *Misgurnus anguillicaudatus* **Not shown**

IDENTIFICATION: *Long, cylindrical body; 10–12 barbels around mouth;* rounded caudal fin. *Stout spine* on pectoral fin. Dorsal fin origin above pelvic fin origin. Tiny scales. Suborbital spine subcutaneous. Faint dusky blotches on light olive or tan side; usually dark spots on dorsal and caudal fins; usually a small black spot at upper edge of caudal fin base. Has 9 dorsal rays, 6–7 pelvic rays, 7–8 anal rays. To 10 in. (25 cm). **RANGE:** Native to e. Asia. Established in streams near Tampa, FL; streams and canals in Cook Co., IL; Shiawassee R., MI; Snake-Columbia river drainage, ID, OR, and WA; Portage Bay drainage, WA. Possibly established elsewhere. Common. **HABITAT:** Mud-bottomed pools and backwaters; survives in poorly oxygenated water.

CHARACINS: Family Characidae (1)

Characidae includes 1122 species of mostly small fishes, similar in appearance to minnows and found mainly in Central America, South America, and Africa. Unlike minnows, characins have an *adipose fin* and *teeth* on the jaws. Many are brightly colored and are popular aquarium fishes. One species, Mexican Tetra, *Astyanax mexicanus,* is native to southern Texas and New Mexico.

MEXICAN TETRA *Astyanax mexicanus* **Pl. 36**

IDENTIFICATION: *Adipose fin. Black stripe* on caudal peduncle and fin; *2–3 dusky black spots* on silver side above pectoral fin. Deep, compressed body; blunt snout; terminal mouth; large, sharp teeth on

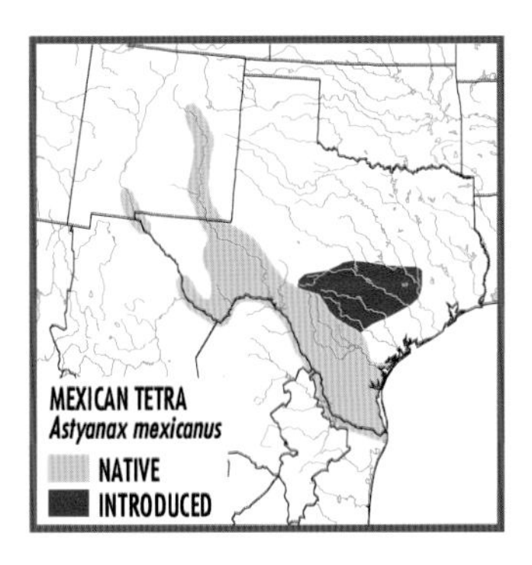

jaws. Large individual has yellow fins, red on caudal and at front of anal fin. Has 35–40 lateral scales; 10–11 dorsal rays; 21–23 anal rays. To 4¾ in. (12 cm). **RANGE:** Native to Mexico (mainly on Atlantic Slope), Guatemala, and Nueces, Rio Grande, and Pecos river drainages, TX and NM; established in streams on Edwards Plateau. Common. **HABITAT:** Rock- and sand-bottomed pools and backwaters of creeks and small to large rivers; springs, caves.

North American Catfishes: Family Ictaluridae (44)

Bullhead catfishes have *4 pairs of barbels* ("whiskers") around the mouth, *no scales, an adipose fin, stout spines* in the dorsal and pectoral fins, and abdominal pelvic fins. Members of this family are active mainly at night. Glandular cells in the skin surrounding the fin spines of madtoms, *Noturus*, contain venom; pain from the "sting" of a madtom is generally considered equivalent to a bee sting. Flathead Catfish, *Pylodictis olivaris*, and Blue Catfish, *Ictalurus furcatus*, the giants of the family at about 132 lb. (60 kg) and over 5 ft. (1.5 m) in length, and several other large species, especially the widely marketed Channel Catfish, *Ictalurus punctatus*, are of major commercial and angling value.

In addition to 44 species in the U.S. and Canada (7 of which range into Mexico, and 1 into Guatemala and Belize), 5 species are endemic to Mexico. About 34 other families of catfishes contain about 3000 species.

CTALURUS

GENUS CHARACTERISTICS: *Moderately to deeply forked caudal fin;* short base on *small adipose fin,* its rear edge free from back and far from caudal fin; anal fin long, with 23–35 rays; upper jaw projects beyond lower jaw; no backward projections on premaxillary tooth patch; eye fairly large, on side of head.

CHANNEL CATFISH *Ictalurus punctatus* Pl. 24

IDENTIFICATION: Usually scattered *dark spots* on silver back and side; white below; fins similar in color to adjacent body; white to dusky barbels. *Rounded anal fin, 24–32 rays.* Gently sloping, slightly rounded predorsal profile. Small young lacks spots, has black-tipped fins. Large individual is blue-black, lacks dark spots. Gas bladder without constrictions and chambers. To 50 in. (127 cm). **RANGE:** Native to St. Lawrence-Great Lakes, Hudson Bay (Red R. drainage), and Missouri-Mississippi river basins from s. QC to s. MB and MT south to Gulf. Possibly also native on Atlantic and Gulf slopes from Susquehanna R. to Neuse R., and from Savannah R. to Lake Okeechobee, FL, and west to n. Mexico and e. NM. Introduced throughout most of U.S. Common to abundant. **HABITAT:** Deep pools and runs over sand or rocks in small to large rivers; lakes. Avoids upland streams. **SIMILAR SPECIES:** (1) See Headwater Catfish, *I. lupus*. (2) Yaqui Catfish, *I. pricei*, has shorter pectoral spine, dorsal spine, and anal fin base and occurs only in Yaqui R. drainage. (3) Blue Catfish, *I. furcatus* (Pl. 24), *lacks* dark spots on body (except in Rio Grande), has *straight-edged anal fin* with *30–35 rays,* straight predorsal profile.

HEADWATER CATFISH *Ictalurus lupus* Not shown

IDENTIFICATION: Similar to Channel Catfish, *I. punctatus,* but has *20–27 (usually 22–26) anal rays*; *deeper caudal peduncle*; broader head, mouth, and snout. Has 13–17 rakers on 1st gill arch; 9–10 pectoral rays. To 19 in. (48 cm). **RANGE:** Restricted in U.S. to Rio Grande drainage, including Pecos R. system, NM, and Devils R., s. TX; formerly in Nueces, San Antonio Bay, and Colorado river drainages, TX. Also in

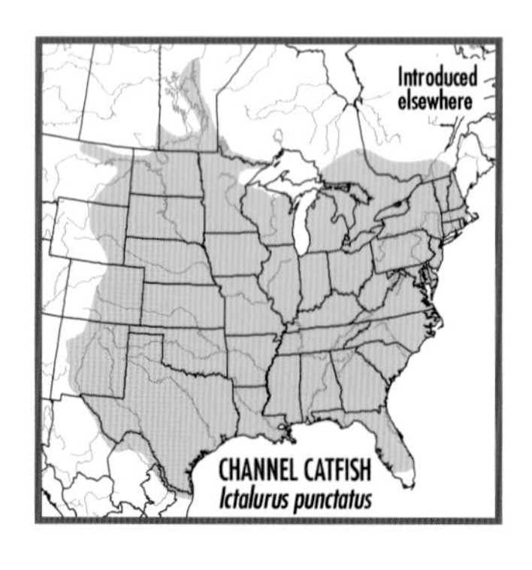

ne. Mexico. Locally common. **HABITAT:** Sandy and rocky riffles, runs, and pools of clear creeks and small rivers; springs. **SIMILAR SPECIES:** (1) See Channel Catfish, *I. punctatus* (Pl. 24). (2) Yaqui Catfish, *I. pricei*, has 16–24 rakers on 1st gill arch, 11 pectoral rays; is darker.

YAQUI CATFISH *Ictalurus pricei* Not shown

IDENTIFICATION: Similar to Channel Catfish, *I. punctatus*, but has *20–27 anal rays, shorter pectoral spine* (3–4 times into predorsal length; 2–3 times in Channel Catfish), and *shorter dorsal spine* (2.6–4 times into predorsal length; 2.1–2.6 times in Channel Catfish). Many round dark spots scattered on dark gray (juveniles, females) to black (large males) back and side; white to gray below. Has 16–24 rakers on 1st gill arch; 11 pectoral rays. To 22¼ in. (57 cm). **RANGE:** Río Yaqui and Río Casas Grandes drainages, nw. Mexico and (presumably) extreme se. AZ. Rare; protected in U.S. as a *threatened species.* **HABITAT:** Quiet water over sand-rock bottom in small to medium rivers. **SIMILAR SPECIES:** See (1) Channel Catfish, *I. punctatus* (Pl. 24). (2) Headwater Catfish, *I. lupus*, has 13–17 rakers on 1st gill arch, 9–10 pectoral rays.

BLUE CATFISH *Ictalurus furcatus* Pl. 24

IDENTIFICATION: *Long, straight-edged anal fin,* tapered like a barber's comb, *30–35 rays. No dark spots* on body (except in Rio Grande). Steeply sloping and straight predorsal profile. Pale blue to olive above and on side; white below; clear or white fins, except black or dusky borders on dorsal and caudal fins; white chin barbels. Large individual is blue-black above; silver blue below. Gas bladder with chambers. To 65 in. (165 cm). **RANGE:** Mississippi R. basin from w. PA to s. SD and Platte R., sw. NE, south to Gulf; Gulf Slope from Escambia R. drainage (where introduced), AL and FL, to Rio Grande drainage, TX and NM. Also in Mexico. Introduced in Atlantic Slope drainages, western states, and MN. Fairly common. **HABITAT:** Deep water of impoundments and main

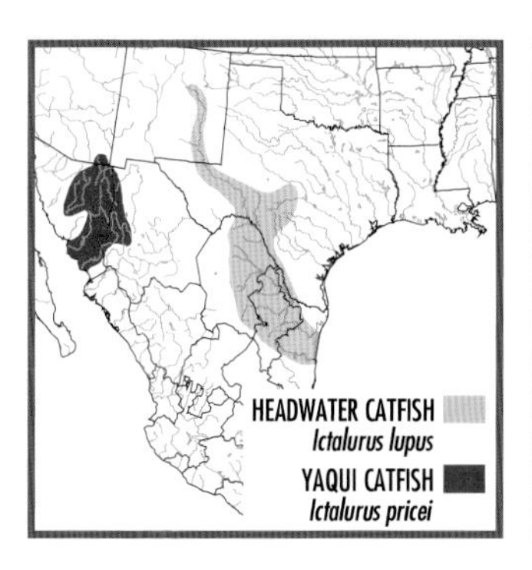

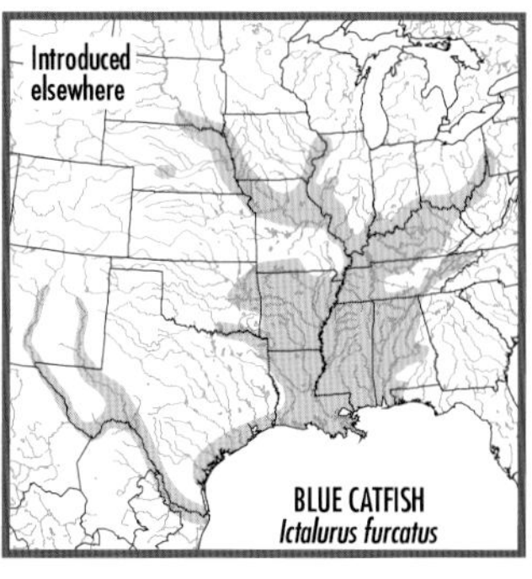

channels and backwaters of medium to large rivers, over mud, sand, and gravel. **SIMILAR SPECIES:** (1) Channel Catfish, *I. punctatus* (Pl. 24), has dark spots on body, *rounded anal fin* with 24–32 rays, no chambers in gas bladder.

AMEIURUS

GENUS CHARACTERISTICS: All but White Catfish, *A. catus*, have rear edge of caudal fin *rounded, straight,* or slightly *notched.* Short base on *small adipose fin,* its rear edge free from back and far from caudal fin; anal fin base usually shorter than in species of *Ictalurus,* with 17–27 rays. Upper jaw projecting beyond lower jaw or jaws nearly equal; eye relatively small, on side of head; no backward projections on premaxillary tooth patch.

WHITE CATFISH *Ameiurus catus* Pl. 24

IDENTIFICATION: *Moderately forked caudal fin.* Relatively short *anal fin base,* rounded in outline, *22–25 rays.* Has *11–15* moderately large *saw-like teeth* on rear of pectoral spine. Gray to blue-black above; white to light yellow below; *dusky to black adipose fin*; white or yellow chin barbels. No dark blotch at dorsal fin base. Large individual is blue-black on head and lips; dusky blue above; white or blue below. Has 18–21 rakers on 1st gill arch. To 24¼ in. (62 cm). **RANGE:** Atlantic and Gulf slope drainages from s. ME to Mobile Bay drainage, MS; south in peninsular FL to Peace R. drainage. Introduced widely outside native range. Common. **HABITAT:** Sluggish mud-bottomed pools, open channels, and backwaters of small to large rivers; lakes and impoundments. **SIMILAR SPECIES:** (1) Channel Catfish, *I. punctatus* (Pl. 24), and (2) Blue Catfish, *I. furcatus* (Pl. 24), *lack* dusky to black adipose fin, have *more deeply forked caudal fin* and either *straight-edged anal fin* (Blue Catfish) or scattered dark spots on a lighter body (Channel Catfish).

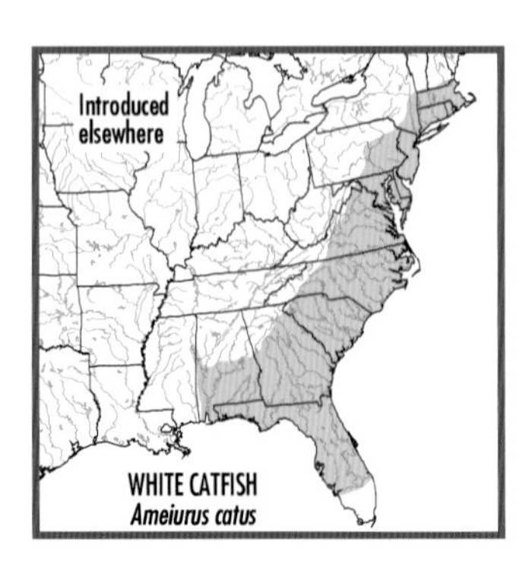

YELLOW BULLHEAD *Ameiurus natalis* Pl. 24

IDENTIFICATION: *White or yellow chin barbels.* Moderately long anal fin, nearly straight in outline, *24–27 rays;* rays at front only slightly longer than rear rays. Has *5–8 large sawlike teeth* on rear of pectoral spine. Rear edge of caudal fin rounded or nearly straight. Yellow-olive to slate black above; lighter, often yellow-olive, on side; bright yellow to white below; dusky fins; often a dark stripe in middle of anal fin. No dark blotch at dorsal fin base. Has 13–15 rakers on 1st gill arch. To 19 in. (47 cm). **RANGE:** Native to Atlantic and Gulf slope drainages from NY to n. Mexico; St. Lawrence-Great Lakes and Mississippi R. basins from s. QC west to cen. ND, and south to Gulf. Widely introduced elsewhere. Common in center of range. **HABITAT:** Pools, backwaters, and sluggish current over soft substrate in creeks and small to large rivers; oxbows, ponds, and impoundments. **SIMILAR SPECIES:** (1) Black Bullhead, *A. melas*, and (2) Brown Bullhead, *A. nebulosus* (both Pl. 24), have *dusky or black chin barbels*. Black Bullhead has short anal fin, rounded in outline, *19–23 rays; 15–21 rakers* on 1st gill arch. Brown Bullhead usually has brown or black mottling or spots on body.

BLACK BULLHEAD *Ameiurus melas* Pl. 24

IDENTIFICATION: *Dusky to black chin barbels.* Relatively short anal fin, rounded in outline, *19–23 rays;* rays at front distinctly longer than rear rays. Usually *no strong sawlike teeth* on rear of pectoral spine (Fig. 41). Rear edge of caudal fin slightly notched. Dark olive, yellow-brown, or slate-olive above; lighter, often shiny green-gold side; bright yellow to white below; dusky to black fins. Pale rays, black membranes on caudal and anal fins. No mottling on body; no dark blotch at dorsal fin base. Has *15–21 rakers* on 1st gill arch. To 24¼ in. (62 cm). **RANGE:** Native to Great Lakes, Hudson Bay, and Mississippi R. basins from NY to s. SK and MT, south to Gulf; Gulf Slope drainages from Mobile Bay, GA and AL, to n. Mexico. Apparently not native to Atlantic Slope. Widely introduced in U.S. and s. Canada. Common in center of range. **HABI-**

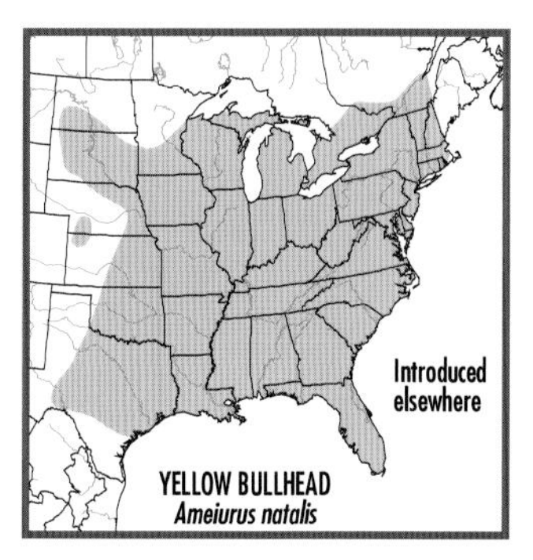

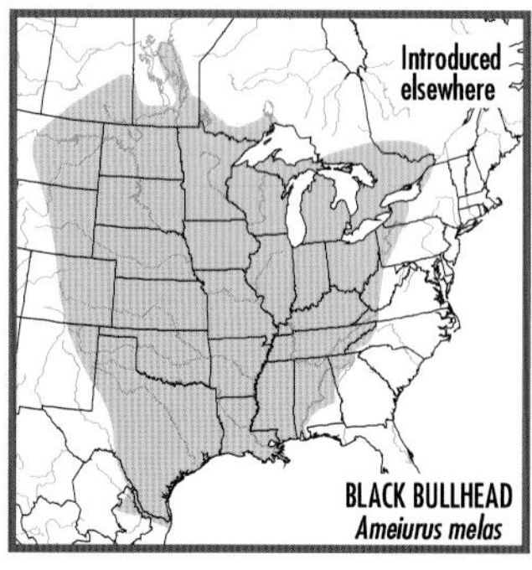

TAT: Pools, backwaters, and sluggish current over soft substrate in creeks and small to large rivers; impoundments, oxbows, and ponds. **SIMILAR SPECIES:** (1) See Brown Bullhead, *A. nebulosus* (Pl. 24). (2) Yellow Bullhead, *A. natalis* (Pl. 24), has *white or yellow chin barbels, 24–27 anal rays, 13–15 rakers* on 1st gill arch.

BROWN BULLHEAD *Ameiurus nebulosus* Pl. 24

IDENTIFICATION: Similar to Black Bullhead, *A. melas*, but has *5–8 large sawlike teeth on rear of pectoral spine* (sometimes eroded in large individual) (Fig. 41), *brown or black mottling or spots* on body (see Remarks); *lacks* black membranes contrasting with pale rays on caudal and anal fins. Has *11–15 rakers* on 1st gill arch. To 21 in. (50 cm). **RANGE:** Native to Atlantic and Gulf slope drainages from NS and NB to Mobile Bay, AL; St. Lawrence-Great Lakes, Hudson Bay, and Mississippi R. basins from QC west to se. SK, and south to LA. Widely introduced outside native range. Common in Northeast and on Atlantic and Gulf slopes; sporadic elsewhere. **HABITAT:** Pools and sluggish runs over soft substrate in creeks and small to large rivers; impoundments, lakes, and ponds. **REMARKS:** Many populations polymorphic; some individuals have white or black spots or blotches (Pl. 23). **SIMILAR SPECIES:** (1) See Black Bullhead, *A. melas* (Pl. 24).

SPOTTED BULLHEAD *Ameiurus serracanthus* Pl. 24

IDENTIFICATION: Many small *round gray-white spots* on *dark body*. Narrow *black edge on* fins. Has 15–20 *large sawlike teeth* on rear of pectoral spine. Large dark blotch at dorsal fin base. Gray or blue-black above; gray to white below; gold yellow cast to body and fins. Relatively short anal fin, rounded in outline, 20–23 rays. Has 12–14 rakers on 1st gill arch. To 13¼ in. (34 cm). **RANGE:** Gulf Coastal Plain drainages from Suwannee R. to Yellow R., n. FL, s. GA, and se. AL. Lo-

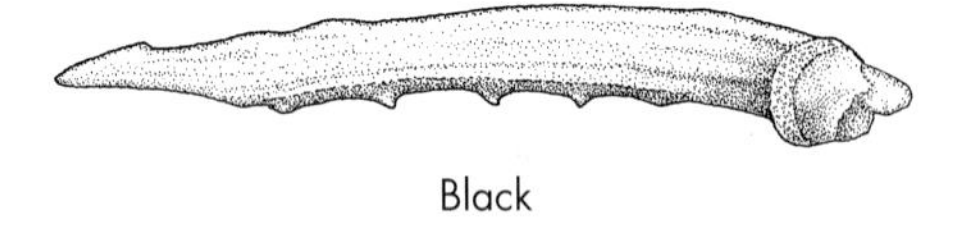

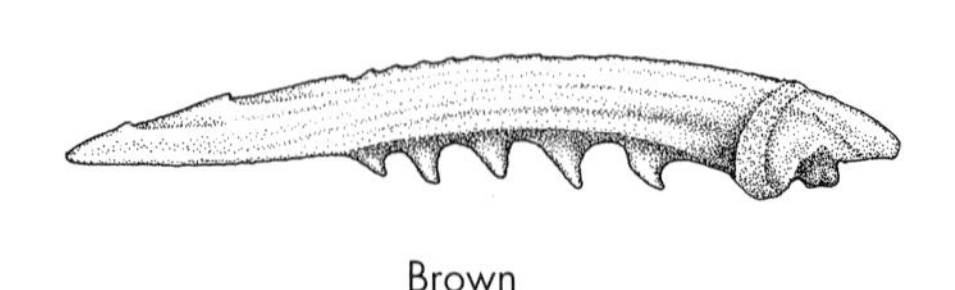

Fig. 41. Black and Brown bullheads—pectoral spine.

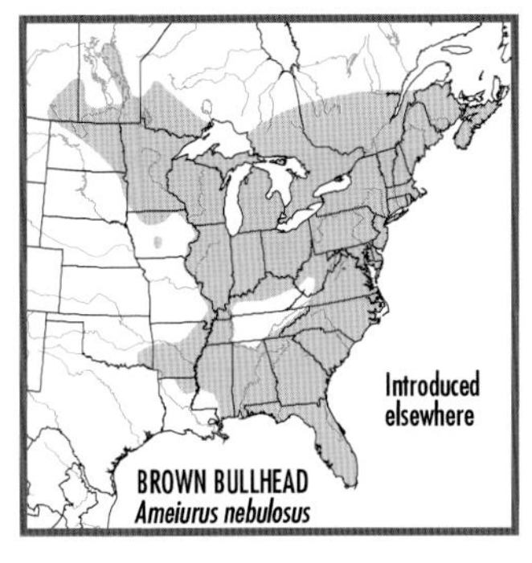

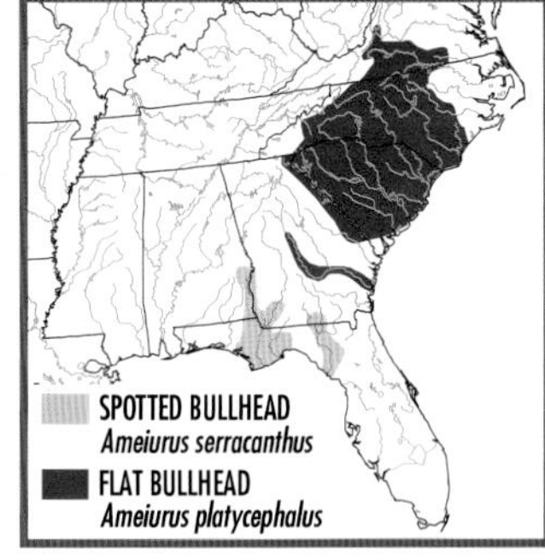

cally common. **HABITAT:** Deep rock- or sand-bottomed pools of small to medium swift rivers; impoundments. **SIMILAR SPECIES:** No other catfish has light round spots on otherwise dark body.

FLAT BULLHEAD *Ameiurus platycephalus* Pl. 24

IDENTIFICATION: Flat head, *relatively straight snout profile. Relatively short anal fin,* rounded in outline, *21–24 rays.* Large dark blotch at dorsal fin base. No large sawlike teeth on rear of pectoral spine. Gold yellow to dark brown above; *dark mottling* on side; dull cream below; dusky fins, narrow black edge on median fins; 11–13 rakers on 1st gill arch. To 11½ in. (29 cm). **RANGE:** Atlantic Slope drainages from Upper James R. and Roanoke R., VA, to Altamaha R., GA. Introduced into upper Tennessee R. system, NC. Uncommon. **HABITAT:** Mud-, sand-, or rock-bottomed pools of small to large rivers; lakes, impoundments, and ponds. **SIMILAR SPECIES:** (1) Snail Bullhead, *A. brunneus* (Pl. 24), has *rounded snout profile, short anal fin* with *17–20 rays, 14–17 rakers* on 1st gill arch. (2) Spotted Bullhead, *A. serracanthus* (Pl. 24), has light round spots on dark body, black edge on pectoral fin, large sawlike teeth on rear of pectoral spine.

SNAIL BULLHEAD *Ameiurus brunneus* Pl. 24

IDENTIFICATION: Flat head, *rounded snout profile. Short anal fin,* rounded in outline, *17–20 rays.* Large dark blotch at dorsal fin base. No large sawlike teeth on rear of pectoral spine. Yellow-green or olive above; gold to dusky yellow on side; white below; narrow black edge (except pectoral) on dusky olive-brown fins. Some populations (e.g., St. Johns R., FL) strongly mottled. Has *14–17 rakers* on 1st gill arch. To 11½ in. (29 cm). **RANGE:** Atlantic Slope from Dan R. system (where possibly introduced), s. VA, south to Altamaha R. drainage, GA, and St. Johns R. drainage, FL; Gulf Slope in Apalachicola R. drainage, GA, AL, and FL. Reported from upper Coosa R. system, n. GA, and French Broad and Nolichucky river systems, NC. Common. **HABITAT:** Rocky riffles, runs, and flowing pools of swift streams; rarely in backwaters. **SIMILAR SPECIES:** (1) Spotted Bullhead, *A. serracanthus* (Pl. 24), has light round spots on dark body, black edge on pectoral fin, large saw-like teeth on rear of pectoral spine. (2) Flat Bullhead, *A. platycephalus* (Pl. 24), has *longer anal fin* with *21–24 rays, 11–13* rakers on 1st gill arch, relatively *straight snout profile,* mottling on side.

FLATHEAD CATFISH *Pylodictis olivaris* Pl. 24

IDENTIFICATION: *White tip on upper lobe of caudal fin* (except in large adult). Wide, *flat head; lower jaw* projecting *beyond upper jaw* (except in small young). Slender, somewhat compressed body; small eye on top of head. Short, high adipose fin with rear end free from back and far from caudal fin. Yellow to dark purple-brown with *black or brown mottling above;* white to yellow below; white to yellow chin barbels; mottled fins. Short anal fin, rounded in outline, 14–17 rays. Rear edge of caudal fin rounded or slightly notched. Backward extension on each side of premaxillary tooth patch. To 61 in. (155 cm). **RANGE:** Lower Great Lakes and Mississippi R. basins from w. PA to White-Little Missouri river system, ND, and south to LA; Gulf Slope from Mobile Bay drainage, GA and AL, to Mexico. Introduced elsewhere in U.S. Fairly common. **HABITAT:** Pools with logs and other debris in low- to moderate-gradient, small to large rivers; lakes; impoundments. Young in rocky and sandy runs and riffles. **SIMILAR SPECIES:** (1) Other large catfishes (Pl. 24) *lack* white tip on upper lobe of caudal fin and projecting lower jaw.

WIDEMOUTH BLINDCAT *Satan eurystomus* Pl. 24

IDENTIFICATION: *No eyes;* pink body (from blood pigments). *Well-developed jaw teeth;* lips thick at corner of mouth; lower jaw normal in shape, slightly shorter than upper jaw. Separate branchiostegal membranes; obvious fold between them. Broad, flat head and snout. Long, high adipose fin. Relatively short anal fin, rounded in outline,

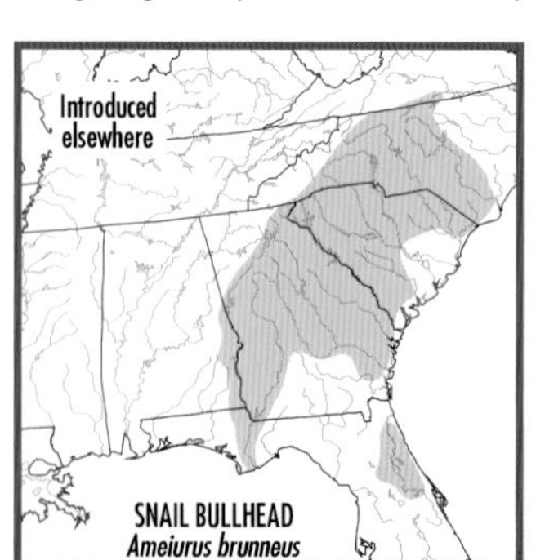

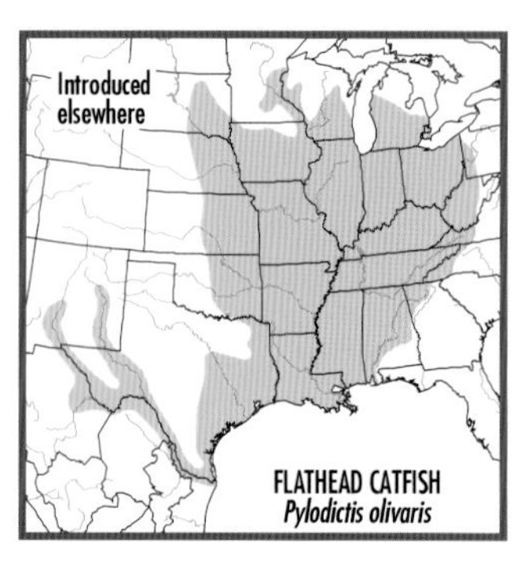

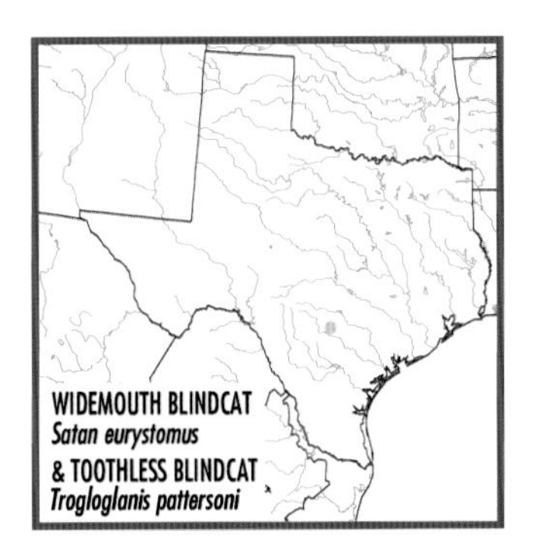

19–20 rays. Rear edge of caudal fin straight or slightly notched. Well-developed lateral-line canals and pores on head. To 5¼ in. (13.7 cm). **RANGE:** Known from artesian wells penetrating San Antonio Pool of Edwards Aquifer in and near San Antonio, Bexar Co., TX. Apparently common. **HABITAT:** Subterranean waters at depths of 975–1860 ft. (300–570 m). **SIMILAR SPECIES:** (1) Toothless Blindcat, *Trogloglanis pattersoni* (Pl. 24), *lacks* jaw teeth; has lower jaw curved into mouth, branchiostegal membranes connected.

TOOTHLESS BLINDCAT *Trogloglanis pattersoni* Pl. 24

IDENTIFICATION: *No eyes;* pink body; red mouth (from blood pigments). *No jaw teeth;* lips at corner of mouth thin; *short lower jaw curved upward and into mouth;* snout overhangs mouth. Connected branchiostegal membranes; fold between them barely visible. Rounded head and snout profile. Long, high adipose fin joined to caudal fin. Short anal fin, rounded in outline, 16–17 rays. Rear edge of caudal fin straight or slightly notched. Lateral-line canals and pores on head well developed. No gas bladder. To 4 in. (10.4 cm). **RANGE AND HABITAT:** Same as Widemouth Blindcat. **SIMILAR SPECIES:** (1) Widemouth Blindcat, *Satan eurystomus* (Pl. 24), has *jaw teeth,* lower jaw *not* curved into mouth, separate branchiostegal membranes, 19–20 anal rays.

NOTURUS

GENUS CHARACTERISTICS: Large genus (29 named, 1 unnamed species). *Small,* less than 12½ in. (32 cm); most species less than 4 in. (10 cm). *Long and low adipose fin joined to, or slightly separated from, caudal fin.* Madtoms can be divided into 2 groups for easier identification: (1) species with nearly uniform dark color pattern, no dark blotches or saddles on back, nearly straight to slightly curved pectoral spine without well-developed sawlike teeth along front edge; (2) species with dark blotches or saddles on lighter back and sides, a curved pectoral spine with sawlike teeth on front edge. Species accounts first cover species with uniform color (11 species; Stonecat, *N. flarus*, to Orangefin Madtom, *N. gilberti*), then banded species (19 species; Smoky Madtom, *N. bailey*, to Yellowfin Madtom, *N. flaripinnis*).

STONECAT *Noturus flavus* Pl. 25

IDENTIFICATION: Long, fairly slender body. *Light blotch on nape; cream white spot* at rear of dorsal fin base. *Cream white blotch* on upper edge of gray caudal fin. *Backward extension from each side of premaxillary tooth patch* (Fig. 42). No or few weak sawlike teeth on rear of pectoral spine. Rear edge of caudal fin straight or with slightly rounded corners. Yellow, slate, or olive above; gray pelvic and anal fins; pectoral, dorsal, and adipose fins dark at base, pale or white at edge. Has 15–18 anal rays. To 12 in. (31 cm). Larger in Lake Erie than in Mississippi R. basin. **RANGE:** St. Lawrence-Great Lakes, Hudson Bay (Red R.), and Mississippi R. basins from QC to AB, and south to n. AL, MS, and ne. OK; Hudson R. drainage, NY. Common. **HABITAT:** Rubble and boulder riffles and runs of creeks and small to large rivers; gravel shoals of lakes. **REMARKS:** Undescribed form in Cumberland and Tennessee river drainages has unique pigment pattern on nape (Fig. 43). Individuals in main channel of Missouri and lower Mississippi rivers are distinctly small-eyed. **SIMILAR SPECIES:** No other madtom has backward extension from each side of premaxillary tooth patch or exceeds 8 in. (20 cm).

TADPOLE MADTOM *Noturus gyrinus* Pl. 25

IDENTIFICATION: Chubby body. Terminal mouth with *equal jaws. No sawlike teeth* on rear of pectoral spine (Fig. 44). Uniformly light tan or gray above and on sides; lighter below; gray or brown fins; *dark gray veinlike line* along side. Rear edge of caudal fin rounded (more pointed on young). Has 13–18 anal rays. To 5 in. (13 cm). **RANGE:** Atlantic and

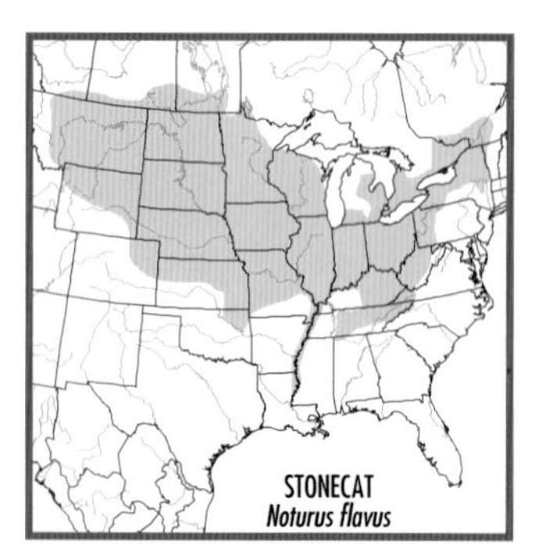

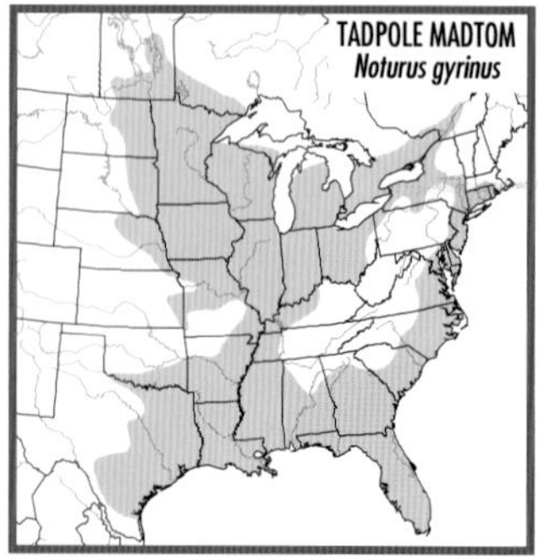

Gulf slope drainages from MA to Nueces R., TX; St. Lawrence-Great Lakes, Hudson Bay (Red R.), and Mississippi R. basins from s. QC to s. SK, and south to Gulf. Absent from Appalachian and Ozark highlands. Introduced into Snake R., ID and OR; and in Merrimack R. in NH. Usually common. **HABITAT:** Rock-, mud-, or detritus-bottomed pools and backwaters of lowland creeks and small to large rivers; lakes. **SIMILAR SPECIES:** See Pl. 25. (1) See Ouachita Madtom, *N. lachneri.* (2) Broadtail Madtom, *N.* species, has *upper jaw projecting beyond lower jaw.* (3) Speckled Madtom, *N. leptacanthus*, has dark specks scattered over upper body and fins; rear edge of caudal fin straight or only slightly rounded.

OUACHITA MADTOM *Noturus lachneri* Pl. 25

IDENTIFICATION: Similar to Tadpole Madtom, *N. gyrinus*, but has *shorter, flatter head; more slender body.* Uniform tan, dark gray, or brown above; light below; uniform light tan or gray fins sometimes with dark borders (especially median fins). Has 16–19 anal rays. To 4 in. (10 cm). **RANGE:** Upper Saline R. system and small unnamed tributary of Ouachita R., cen. AR. Rare to uncommon. **HABITAT:** Rocky pools, backwaters, and runs of clear swift creeks and small rivers. **SIMILAR SPECIES:** (1) See Tadpole Madtom, *N. gyrinus* (Pl. 25).

SPECKLED MADTOM *Noturus leptacanthus* Pl. 25

IDENTIFICATION: Slender body. *Black specks, some much larger than others, on upper body and fins. Short pectoral spine, no sawlike teeth* on rear edge (as in Tadpole Madtom, *N. gyrinus*; Fig. 44). Upper jaw projects beyond lower jaw. Rear edge of caudal fin straight or only slightly

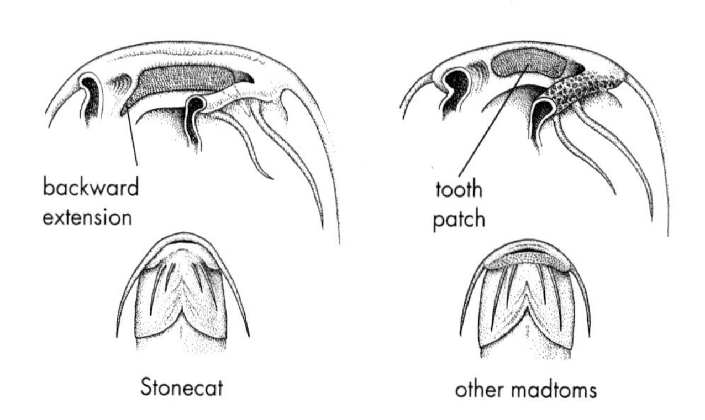

Fig. 42. *Noturus* species—premaxillary tooth patch.

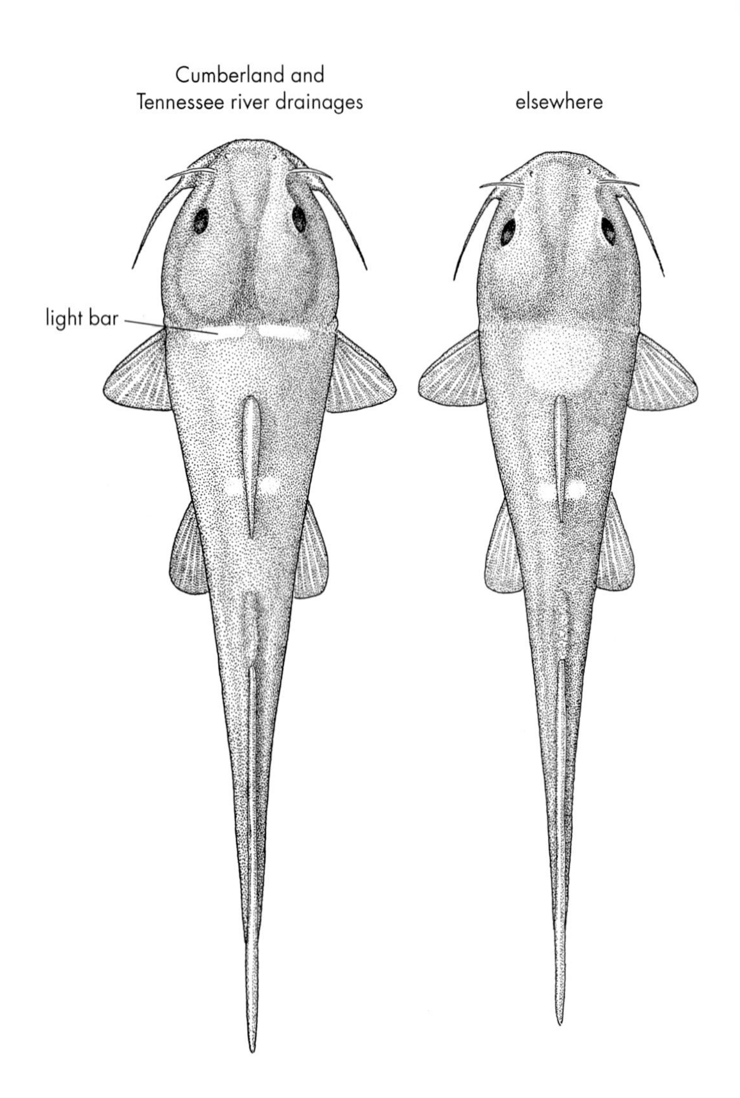

Fig. 43. Stonecat—nape pattern.

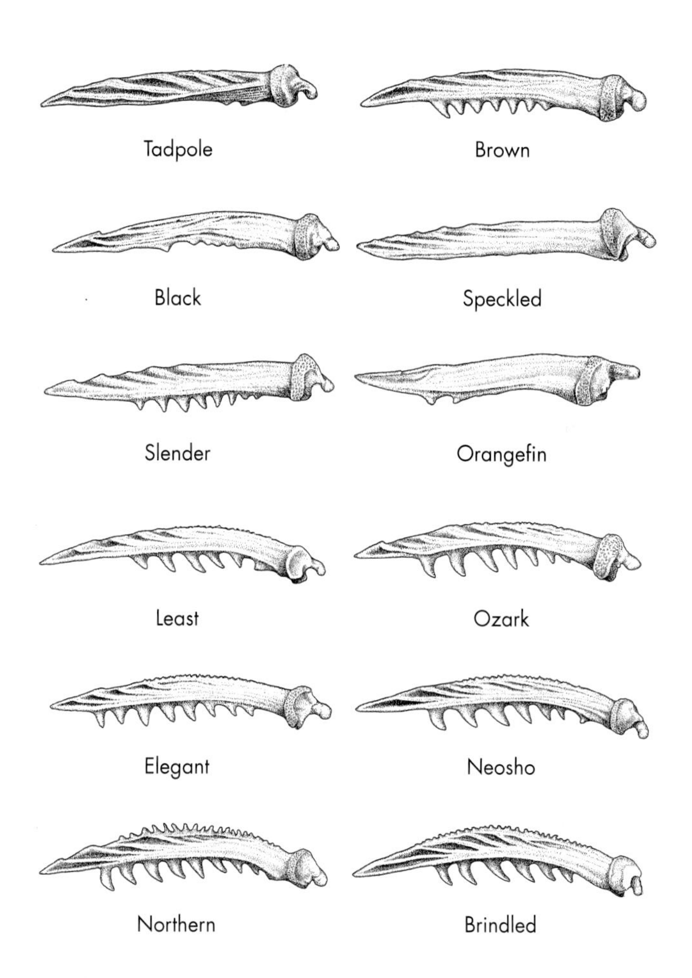

Fig. 44. Madtoms—left pectoral spine.

rounded. Red- or yellow-brown above; cream white below *without dark specks*; dark blotch at base of pectoral fin; dorsal, adipose, caudal, and anal fins dark or darkly blotched, with light edges. Has 14–19 anal rays. To 3½ in. (9.4 cm). **RANGE:** Atlantic and Gulf slope drainages from Santee R., SC, to Amite-Comite river, LA; south in peninsular FL to St. Johns R. drainage. Common. **HABITAT:** Gravel-sand runs and rocky riffles of creeks and small to medium rivers; near vegetation. **SIMILAR SPECIES:** (1) See Broadtail Madtom, *N.* species (Pl. 25). (2) No other madtom has scattered dark specks, some much larger than others, on upper body and fins.

BROADTAIL MADTOM *Noturus* species **Pl. 25**

IDENTIFICATION: Similar to Speckled Madtom, *N. leptacanthus*, but has *chubby body, rounded rear edge* on caudal fin, *fewer and smaller dark specks* on body, larger eye, *dark blotch* on caudal fin base. Tan to red-brown above; white or yellow below; all fins (except caudal) pale, dark at base; clear or pale border on light brown caudal fin. Has 12–16 anal rays. To 2¼ in. (6 cm). **RANGE:** Coastal Plain of NC and SC in South (Cape Fear drainage), Waccamaw, Lumber, and Lynches river systems (Peedee R. drainage). Locally common. **HABITAT:** Sand and gravel shoals, debris-laden pools, and main channels of medium rivers. In Lake Waccamaw at 3–6 ft. (1–2 m) over sand. **SIMILAR SPECIES:** (1) See Speckled Madtom, *N. leptacanthus* (Pl. 25). (2) Tadpole Madtom, *N. gyrinus* (Pl. 25), has *equal jaws*, reaches 5 in. (13 cm).

BROWN MADTOM *Noturus phaeus* **Pl. 25**

IDENTIFICATION: *Robust.* Uniform light to dark brown above; lighter below; *many brown dots*, most conspicuous on underside of head and belly. *Long anal fin, 20–22 rays;* narrow space between anal fin base and caudal fin. Upper jaw projects beyond lower jaw. About 6 distinct *sawlike teeth* on rear of pectoral spine (Fig. 44). Rear edge of caudal fin straight or slightly rounded. To 5¾ in. (15 cm). **RANGE:** Disjunct. Mississippi R. tributaries from Obion R., KY, to sw. MS, and from s. AR to cen. LA; Tennessee R. tributaries in sw. TN and ne. MS; Gulf Slope in Sabine R. and Bayou Teche drainages, LA, upper Yockanookany R., MS. Locally common. **HABITAT:** Sand-gravel riffles and runs among debris, rocks, and undercut banks of creeks and small rivers. **SIMILAR SPECIES:** See Pl. 25. (1) Black Madtom, *N. funebris*, is black or bluish above; has no or few small teeth on rear of pectoral spine, 21–27 anal

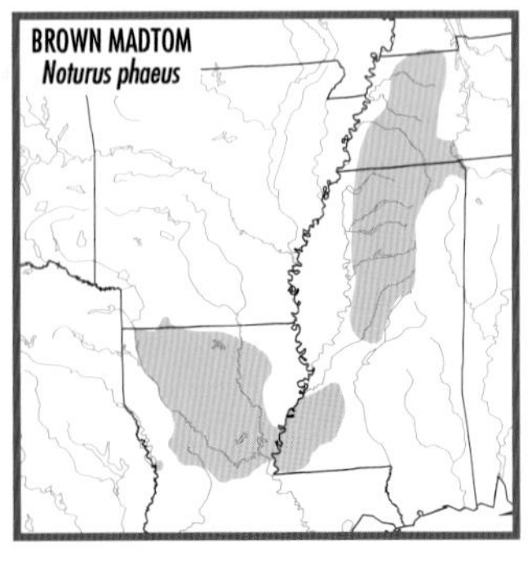

rays. (2) Freckled Madtom, *N. nocturnus,* lacks dark dots or specks on underside of head and belly; has 16–18 anal rays, 2–3 small teeth on rear edge of pectoral spine. (3) Speckled Madtom, *N. leptacanthus*, is more slender; has conspicuous black specks, some much larger than others, on back and sides; 14–19 anal rays; no teeth on pectoral spine.

BLACK MADTOM *Noturus funebris* Pl. 25

IDENTIFICATION: Similar to Brown Madtom, *N. phaeus*, but is *black* or steel blue above; has *no* or *few weak sawlike teeth* on rear of pectoral spine (Fig. 44), *longer anal fin* with *21–27 rays;* sometimes a diffuse black edge on median fins. To 5¾ in. (15 cm). **RANGE:** Gulf Slope drainages from Enconfina Creek, FL, to Pearl R., MS and LA. Yellow and Bear Creek systems (Tennessee R. drainage), nw. AL and ne. MS. Locally common. **HABITAT:** Near vegetation in moderate to fast clear water over gravel and sand in permanent springs, creeks, and small rivers. **SIMILAR SPECIES:** (1) See Brown Madtom, *N. phaeus* (Pl. 25).

FRECKLED MADTOM *Noturus nocturnus* Pl. 25

IDENTIFICATION: Light brown or gray to black above; *many tiny dark dots* on body and fins. Light yellow or white below; underside of head and belly mostly *without* dark dots or specks. Dusky black edge on anal fin; other fins dark at base, light at edge. Usually 16–18 rays, 2–3 weak sawlike teeth on rear of pectoral spine. Rear edge of caudal fin straight or slightly rounded. To 5¾ in. (15 cm). **RANGE:** Mississippi R. basin from n. IL to LA, and from e. KY to cen. KS and OK; Gulf Slope drainages from Mobile Bay, AL, to Brazos R., TX; isolated population in upper Guadalupe R., TX. Locally common. **HABITAT:** Sand-gravel riffles and runs near debris and among tree roots along undercut banks in creeks to large rivers. **SIMILAR SPECIES:** See Pl. 25. (1) Black Madtom, *N. funebris*, is blacker; has pattern of tiny black dots on underside of

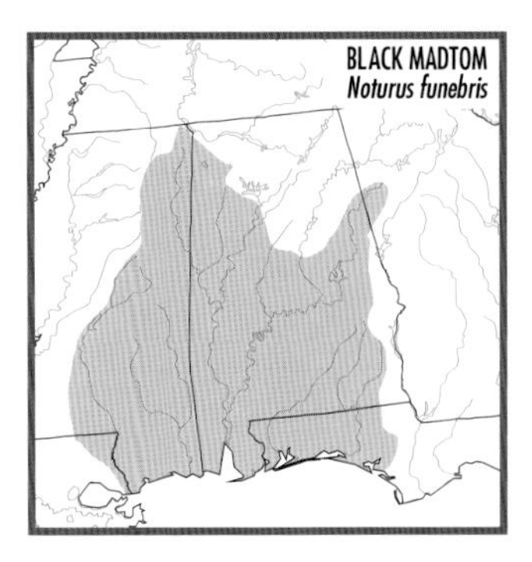

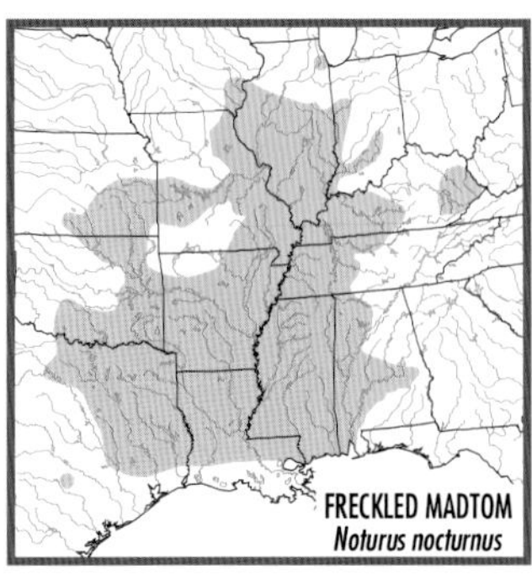

head and belly, 21–27 anal rays. (2) Brown Madtom, *N. phaeus,* has pattern of tiny brown dots on underside of head and belly, 20–22 anal rays, large sawlike teeth on rear edge of pectoral spine. (3) Speckled Madtom, *N. leptacanthus,* has conspicuous black specks, some much larger than others, on back and sides; no teeth on pectoral spine.

SLENDER MADTOM *Noturus exilis* Pl. 25

IDENTIFICATION: Long, slender body; flat head. *Black border* on median fins. Terminal mouth with *equal jaws.* About 6 strong *sawlike teeth* on rear edge of pectoral spine (Fig. 44). Large *light yellow spot* on nape; smaller spot at rear of dorsal fin base. Rear edge of caudal fin straight or with slightly rounded corners. Yellow-brown to gray-black above; light yellow below; pale or light yellow fins. Has 17–22 anal rays. To 5¾ in. (15 cm). **RANGE:** Disjunct. Green, Cumberland, and Tennessee river drainages, cen. KY to n. AL; upper Mississippi R. basin from s. WI and s. MN to Ouachita Highlands of AR, KS, and OK. Fairly common but localized in northern part of range. **HABITAT:** Rocky riffles, runs, and flowing pools of clear creeks and small rivers; rarely in springs. Rarely along wave-swept margins of large impoundments. **REMARKS:** Slender and Margined madtoms from cool clear streams are more slender and have more boldly edged fins than individuals from warm, turbid waters. Some individuals lack black fin borders. **SIMILAR SPECIES:** (1) See Margined Madtom, *N. insignis* (Pl. 25).

MARGINED MADTOM *Noturus insignis* Pl. 25

IDENTIFICATION: Similar to Slender Madtom, *N. exilis,* but *upper jaw projects beyond lower jaw,* longer caudal fin, *no light spot on nape or at rear of dorsal fin base.* Yellow to slate gray above; white or light below; yellow or light gray fins; black edge on median fins variable—darkest in clear water. Has 15–21 anal rays. To 6 in. (15 cm). **RANGE:** Atlantic Slope from St. Lawrence R.-Great Lakes, s. QC and ON, to upper Al-

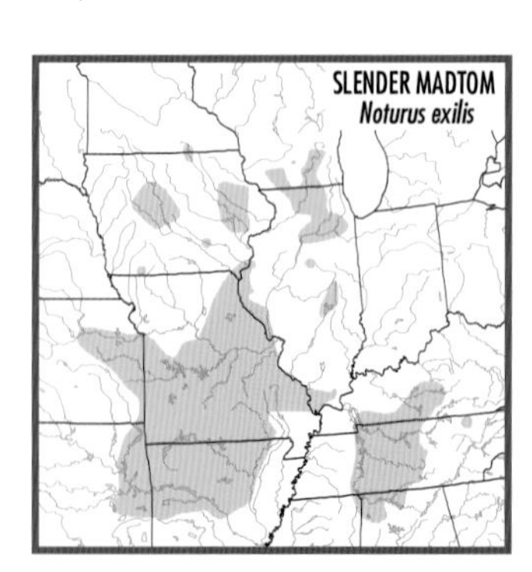

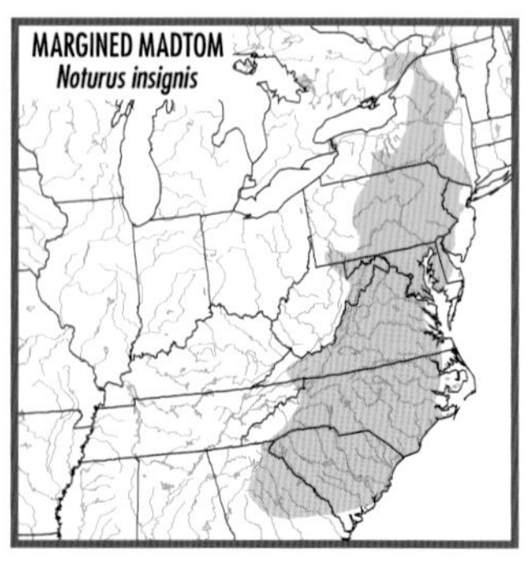

tamaha R. drainage, GA; upper Kanawha (New) R. system, VA and NC; upper Monongahela R. system, PA, WV, and MD. Introduced into Merrimack R., NH and MA, and upper Tennessee R. drainage, VA and TN. Common in U.S. **HABITAT:** Rocky riffles and runs of clear fast creeks and small to medium rivers. **REMARKS:** Often black spots on individuals from James, Chowan, and Roanoke river drainages, VA (Pl. 25). **SIMILAR SPECIES:** (1) See Slender Madtom, *N. exilis* (Pl. 25).

ORANGEFIN MADTOM *Noturus gilberti* Pl. 25

IDENTIFICATION: Long, slender body; flat head. *White to orange triangle* (widest at rear) on upper edge of dusky black caudal fin. Short dorsal and pectoral spines; irregular, sometimes large sawlike teeth on rear edge of pectoral spine (Fig. 44). *Short anal fin; 14–16 rays.* Upper jaw projects beyond lower jaw. Straight edge or slightly rounded corners on caudal fin. Olive or brown above; pale yellow or white below; light yellow or white edge on dark fins. To 3¾ in. (10 cm). **RANGE:** Upper Roanoke R. (including Dan R.) drainage, VA and NC; upper James R., VA, where possibly introduced. Rare to uncommon. **HABITAT:** Rocky riffles and runs of clear, swift, small rivers. **SIMILAR SPECIES:** (1) Stonecat, *N. flavus* (Pl. 25), has cream white spot at rear of dorsal fin base, backward projections on premaxillary tooth patch, no or few weak sawlike teeth on rear of pectoral spine.

SMOKY MADTOM *Noturus baileyi* Pl. 26

IDENTIFICATION: Slender body, deepest beneath dorsal fin. *Olive-brown above and on lower side;* white to yellow below. *Four pale yellow saddles*: on nape (often faint), at rear of dorsal fin base, and at front and rear of adipose fin. Dusky band on adipose fin *nearly to fin edge.* Other fins clear or yellow to dusky brown. Short pectoral spine with 4–5 large teeth on rear edge, small teeth on front edge (as in Least Madtom; Fig. 44). Rear edge of caudal fin straight. Adipose fin joined to caudal

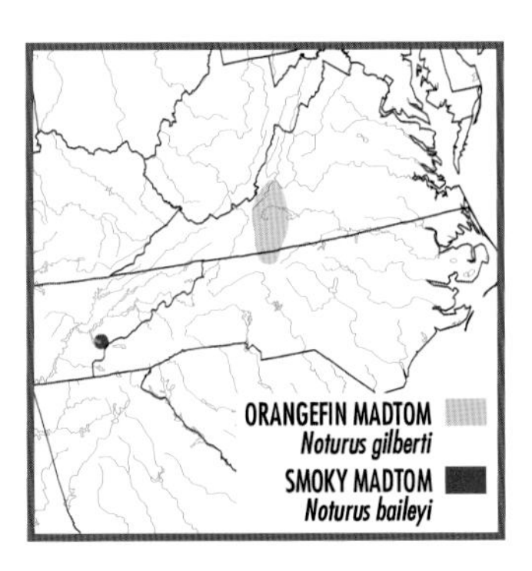

fin, but with notch between. Has 12–14 anal rays. To 2¾ in. (7.3 cm). **RANGE:** Citico Creek (Little Tennessee R. system), Monroe Co., TN. Formerly in but now extirpated from Abrams Creek (Little Tennessee R. system), Blount Co., TN. Rare, protected as an *endangered species.* **HABITAT:** Clear, cool, rocky riffles, runs, and flowing pools of creeks. **SIMILAR SPECIES:** (1) Least Madtom, *N. hildebrandi* (Pl. 26), has *mostly clear adipose fin* and white lower side (*N. h. lautus*), or is prominently *blotched above and on side* (*N. h. hildebrandi*). (2) Pygmy Madtom, *N. stanauli,* has white-tipped snout, *white lower side,* 14–17 anal rays.

LEAST MADTOM *Noturus hildebrandi* Pl. 26

IDENTIFICATION: Slender body; flat head. *White or clear adipose fin. Short pectoral spine* with 4–5 large sawlike teeth on rear edge; *no or small teeth on front edge* (Fig. 44). Rear edge of caudal fin straight or with slightly rounded corners. Adipose and caudal fins broadly joined with tiny notch between. Geographically variable in color (see Remarks); caudal fin dusky or has irregularly defined bars and clear edge. Has 12–17 anal rays. To 2½ in. (6.9 cm). **RANGE:** Tributaries of Mississippi R. from North Fork Obion R., sw. KY, to Homochitto R., s. MS. Common. **HABITAT:** Mixed rock and sand riffles and runs of clear lowland creeks and small rivers; often near debris. **REMARKS:** Two subspecies. *N. h. lautus,* in Obion, Forked Deer, and Hatchie rivers, w. KY and TN, has shorter head, 4 white or light yellow oval areas on uniform red-brown to black back; is strongly bicolored, with brown to black upper side, white or yellow below. *N. h. hildebrandi,* in Homochitto R. and Bayou Pierre, s. MS, has longer head, prominently blotched color pattern above and almost to belly. Intergrades occur in sw. TN and n. MS. **SIMILAR SPECIES:** (1) See Pygmy Madtom, *N. stanauli.* (2) Smoky Madtom, *N. baileyi* (Pl. 26), has more uniform olive-brown body, including well-pigmented lower side.

PYGMY MADTOM *Noturus stanauli* Not shown

IDENTIFICATION: Similar to Least Madtom, *N. hildebrandi,* but has *white snout* (in front of nostrils), *large teeth* on front of pectoral spine. Dark gray or brown-black above; olive-brown or pale yellow on upper side, white on lower side; light areas at back of head, back of dorsal fin base, front and rear of adipose fin; mostly white or pale yellow fins; 3 dark spots on caudal fin base, dark band or dusky blotches in middle of caudal fin. Has 14–17 anal rays. To 1½ in. (4.2 cm). **RANGE:** Tennessee R. drainage, TN; known only from Clinch R. at Frost Ford and Brooks I., Hancock Co., and Duck R. just above mouth of Hurricane Creek, Humphreys Co. Uncommon in Clinch R., rare in Duck R., protected as an *endangered species.* **HABITAT:** Moderate to swift gravel runs of clear medium-sized rivers. **SIMILAR SPECIES:** (1) See Least Madtom, *N. hildebrandi* (Pl. 26).

OZARK MADTOM *Noturus albater* Pl. 26

IDENTIFICATION: Stout body; *short head.* Large dark blotch beneath adipose fin usually *into lower half of fin* (sometimes to edge). *Dark bar* (sometimes diffuse) on caudal fin base. *White upper edge* on caudal fin. Rounded edge on high adipose fin. Distinct notch between adipose fin and caudal fin. Rear edge of caudal fin straight or with slightly rounded corners. Has 6–10 large teeth on rear, small teeth on front, of pectoral spine (Fig. 44). Dark mottling or blotches on yellow-brown back and side; 4 dusky saddles; yellow to cream white below; indistinct blotches on yellow or white fins; irregular dark bands on caudal fin. Has 9 pectoral rays, 9 pelvic rays, 13–17 anal rays. To 4¾ in. (12 cm). **RANGE:** Upper White and Little Red river systems, MO and AR. Locally common. **HABITAT:** Clear, cool, swift rocky riffles and pools of creeks and small to medium rivers. **SIMILAR SPECIES:** (1) See Black River Madtom, *N. maydeni.* (2) Least Madtom, *N. hildebrandi* (Pl. 26), is more slender; lacks white upper edge on caudal fin; has white lower side without dark pigment (*N. h. lautus*) or prominent dark blotches above and on side (*N. h. hildebrandi*), 8 pelvic rays. (3) Checkered Madtom, *N. flavater* (Pl. 26), also in White R. drainage, has 4 bold black saddles, saddle under adipose fin *extends to fin edge*; *broad black bar* on caudal fin base; black blotch on outer ⅓ of dorsal fin; 8 pectoral rays.

BLACK RIVER MADTOM *Noturus maydeni* Not shown

IDENTIFICATION: Genetically and karyotypically distinct but morphologically indistinguishable from Ozark Madtom, *N. albater* (Pl. 26). **RANGE:** Black and St. Francis river systems, MO and AR. Locally common. **HABITAT:** Clear, cool, swift rocky riffles and pools of creeks and small to medium rivers. **SIMILAR SPECIES:** (1) See Ozark Madtom, *N. albater* (Pl. 26).

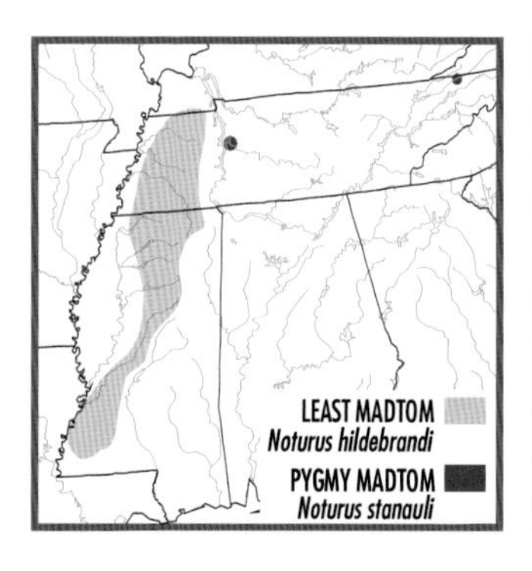

ELEGANT MADTOM *Noturus elegans* Pl. 26

IDENTIFICATION: Long, slender body. Yellow-gray to brown above; 3–4 dark saddles. *Brown blotch on dorsal fin base* extends up front edge of fin; *no black blotch* at top of dorsal fin. *Dark blotch in lower half* of adipose fin. Has 2 or 3 dark bands on caudal fin; 1 in middle, 1 (or 2) near edge. Origin of adipose fin over anal fin; adipose fin joined to caudal fin. Short pectoral spine with 5–9 sawlike teeth on rear edge, many small teeth on front edge (Fig. 44). Rear edge of caudal fin straight or with slightly rounded corners. Usually 15–17 anal rays. To 3 in. (7.4 cm). **RANGE:** Upper Green R. drainage, KY and TN; 1 record from Roaring R. (Cumberland R. drainage), TN. Locally common. **HABITAT:** Rocky riffles and runs of clear creeks and small rivers. **SIMILAR SPECIES:** (1) See Scioto Madtom, *N. trautmani.* (2) Saddled Madtom, *N. fasciatus* (Pl. 26), has large dark bar on caudal fin base, more prominent saddles, dark blotch *to edge* (or nearly to edge) of adipose fin. (3) Chucky Madtom, *N. crypticus* (Pl. 26), has long adipose fin beginning at dorsal fin, prominent black specks on cheek.

SCIOTO MADTOM *Noturus trautmani* Not shown

IDENTIFICATION: Similar to Elegant Madtom, *N. elegans*, but has *no dark blotch* in adipose fin; 13–16, usually 14, anal rays. To 2¼ in. (6.1 cm). **RANGE:** Big Darby Creek (Scioto R. system), s. OH. Rare; last seen in 1957. Protected as an *endangered species.* **HABITAT:** Known mainly from downstream end of 60-ft. (18-m) sand-gravel riffle. **SIMILAR SPECIES:** (1) See Elegant Madtom, *N. elegans* (Pl. 26).

SADDLED MADTOM *Noturus fasciatus* Pl. 26

IDENTIFICATION: Long, slender body. Dark brown above; *3–4 prominent ivory to yellow saddles* alternating with dark saddles. Brown blotch on dorsal fin base extends up front edge of fin; no black blotch at top of dorsal fin. Dark blotch to edge (or nearly to edge) of adipose fin.

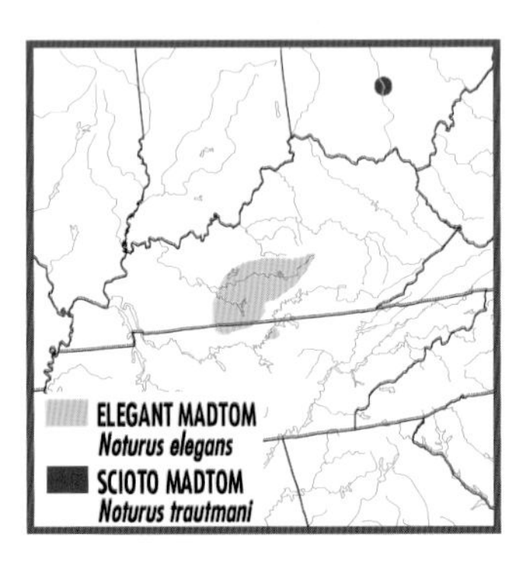

Large dark bar on caudal fin base; 2 or 3 dark bands on caudal fin: 1 in middle, 1 (or 2) near edge. Origin of adipose fin over anal fin; adipose fin joined to caudal fin. Short pectoral spine with 6–9 sawlike teeth on rear edge, many small teeth on front edge. Rear edge of caudal fin straight or with slightly rounded corners. Usually 17–18 anal rays. To 3¼ in. (8.5 cm). **RANGE:** Duck R. system and nearby Tennessee R. tributaries in Hardin and Wayne counties TN. Uncommon. **HABITAT:** Rocky riffles, runs, and flowing pools of clear creeks and small rivers. **SIMILAR SPECIES:** (1) Elegant Madtom, *N. elegans*, and (2) Chucky Madtom, *N. crypticus* (both Pl. 26), *lack* large dark bar on caudal fin base, have *less prominent saddles*, dark blotch confined to lower half of adipose fin. Chucky Madtom also has long adipose fin beginning at dorsal fin, prominent black specks on cheek, usually 16 anal rays.

CHUCKY MADTOM *Noturus crypticus* Pl. 26

IDENTIFICATION: Slender body. *Long, low adipose fin* begins at dorsal fin, reaches caudal fin. Yellow-gray to dark brown above; 3–4 light yellow saddles. Brown blotch on dorsal fin base extends up front edge of fin; no black blotch at top of dorsal fin. *Prominent black specks* on cheek. Dark blotch in lower half of adipose fin; 3 dark bands on caudal fin. Short pectoral spine with 6–8 sawlike teeth on rear edge, many small teeth on front edge. Rear edge of caudal fin rounded. Usually 16 anal rays. To 3 in. (7.4 cm). **RANGE:** Dunn Creek, Sevier Co., and Little Chucky Creek, Greene Co. (French Broad system), TN; Alabama Creek, Flint R., and Paint Rock R. (Tennessee R. drainage), AL. Rare. **HABITAT:** Slow rocky riffles and runs of clear creeks. **SIMILAR SPECIES:** See Pl. 26. (1) Saddled Madtom, *N. fasciatus*, has *shorter adipose fin* beginning over anal fin, large dark bar on caudal fin base, more prominent saddles, *no* prominent black specks on cheek, usually 17–18 anal rays. (2) Smoky Madtom, *N. baileyi*, has *shorter adipose fin* beginning over anal fin, *no* prominent black specks on cheek, dusky band on adipose fin

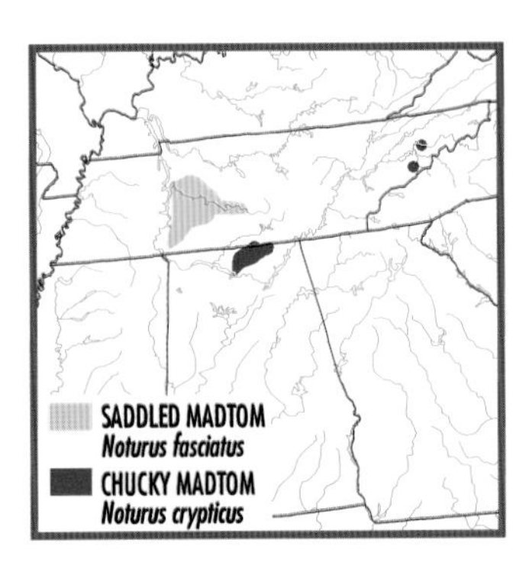

extending nearly to fin edge, 12–14 anal rays. (3) Elegant Madtom, *N. elegans*, has *shorter adipose fin* beginning over anal fin, *no* prominent black specks on cheek.

CADDO MADTOM *Noturus taylori* Pl. 26

IDENTIFICATION: Long, slender body. *Large black blotch* on front half of upper edge of dorsal fin. Brown band in *lower half* of adipose fin. Adipose fin broadly joined, with small notch, to caudal fin. Short pectoral spine with 5–9 sawlike teeth on rear edge, many small teeth on front edge (as in Elegant Madtom; Fig. 44). Rear edge of caudal fin straight or with rounded corners. Yellow to dark brown above; 4 dark brown saddles alternate with light yellow or cream ellipses; 2 (or 3) dusky brown crescent-shaped bands on caudal fin: 1 in middle of fin, 1 (or 2) near clear edge. Has 13–16 anal rays. To 3 in. (7.7 cm). **RANGE:** Caddo, Ouachita, and Little Missouri rivers, sw. AR. Locally common. **HABITAT:** Rocky riffles, pools, and shoals near shorelines of small to medium rivers. **SIMILAR SPECIES:** (1) Elegant Madtom (Pl. 26) *lacks* large black blotch at top of dorsal fin, has brown blotch on dorsal fin base extending up front edge of fin. (2) Brindled Madtom, *N. miurus* (Pl. 26), has dark band to *upper edge* of adipose fin.

NEOSHO MADTOM *Noturus placidus* Pl. 26

IDENTIFICATION: *White lower caudal rays.* Robust body, deepest beneath dorsal fin; relatively *deep caudal peduncle.* Faint to prominent blotches and saddles; no (or faint) pair of light spots in front of dorsal fin. Large pectoral spine with 6–10 large sawlike teeth on rear edge, *small teeth on front edge* (Fig. 44). Adipose fin joined to caudal fin. Dark crescent-shaped band in middle of caudal fin, another near white edge; dusky or gray band confined to lower half of adipose fin; usually no dark specks on belly. Rear edge of caudal fin straight or with slightly rounded corners. Light yellow-pink mottled with brown above;

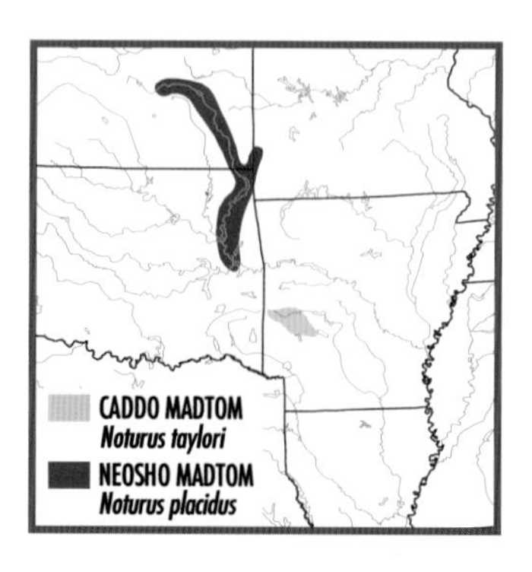

usually white to yellow below; fins blotched with white edges. Has 13–16 anal rays. To 3¼ in. (8.7 cm). **RANGE:** Arkansas R. drainage: Spring R., sw. MO and se. KS; Cottonwood and Neosho rivers, e. KS and ne. OK; lower few miles of Illinois R., e.-cen. OK. Uncommon; protected as a *threatened species*. **HABITAT:** Gravel riffles and runs of small to medium rivers. **SIMILAR SPECIES:** See Pl. 26. (1) Northern Madtom, *N. stigmosus*, has dark band extending into upper half of adipose fin, *lacks* lower white caudal rays. (2) Frecklebelly Madtom, *N. munitus*, has rear edge of adipose fin nearly free from caudal fin, *large teeth* on pectoral spine, dark band on adipose fin to fin edge, 4 dark saddles on back. (3) Mountain Madtom, *N. eleutherus*, has dark brown bar at base of caudal fin; no dark crescent in middle of caudal fin; rear edge of adipose fin nearly free from caudal fin; *large teeth* on pectoral spine.

NORTHERN MADTOM *Noturus stigmosus* Pl. 26

IDENTIFICATION: Robust body, deepest beneath dorsal fin; deep caudal peduncle. *Brown or black band into upper half of adipose fin* but not to edge. Front edge of 1st saddle irregular, usually enclosing *2 large light spots* in front of dorsal fin. Dark crescent-shaped band in middle of caudal fin extends forward across upper and lower caudal rays to *caudal peduncle;* another band near clear edge. Has 5–10 large saw-like teeth on rear edge of large pectoral spine; large teeth on front edge (Fig. 44). Rear edge of high adipose fin nearly free from caudal fin. Edge of caudal fin straight or with slightly rounded corners. Yellow or tan above, brown or black mottling; 4 saddles; white to yellow below; blotched or banded fins; dark bands at base and near edge of dorsal fin connected along front of fin. Has 13–16 anal rays. To 5 in. (13 cm). **RANGE:** Lake Erie and Ohio R. basins from w. PA, s. ON, and cen. WV to Ohio R., w. KY. Sporadic and uncommon; disappearing on edges of range; rare in main channels of Mississippi and Ohio rivers. Protected in Canada as an *endangered species*. **HABITAT:** Mixed sand and rock riffles and runs with debris in small to large, often swift rivers. **SIMILAR SPECIES:** See Pl. 26. (1) Piebald Madtom, *N. gladiator*, has bolder pattern; dark bar in adipose fin connected to bar in anal fin; 2 distinct yellow spots at base of caudal fin. (2) Frecklebelly Madtom, *N. munitus*, has black band *to adipose fin edge, lacks* 2 large light spots in front of dorsal fin. (3) Mountain Madtom, *N. eleutherus, lacks* 2 large light spots in front of dorsal fin, dark crescent in middle of caudal fin; has dark brown bar at caudal fin base.

PIEBALD MADTOM *Noturus gladiator* Pl. 26

IDENTIFICATION: *Bold dark brown or black mottling and saddles* on yellow to tan back and sides; *dark bar into upper half of adipose fin and across body into anal fin*. Blotched or banded fins; dark bands at base and near edge of dorsal fin connected along front of fin. Dark crescent-

shaped band in middle of caudal fin extends forward *at center to join dark pigment on caudal peduncle,* forming *2 yellow spots* at base of fin; another band near clear edge. Large pectoral spine with 6–12 large sawlike teeth on rear edge; large teeth on front edge (as in Northern Madtom; Fig. 44). Robust body, deepest beneath dorsal fin. Front edge of 1st saddle irregular, usually enclosing *2 large light spots* in front of dorsal fin. Rear edge of high adipose fin nearly free from caudal fin. Rear edge of caudal fin straight or with slightly rounded corners. Has 12–15 anal rays. To 5 in. (13 cm). **RANGE:** Tributaries of Mississippi R. from Obion R. system, w. TN, to Big Black R. system, cen. MS. Locally common. **HABITAT:** Sandy and clay-bottomed runs with debris in creeks and small rivers; usually near woody debris. **SIMILAR SPECIES:** (1) Northern Madtom, *N. stigmosus*, and (2) Frecklebelly Madtom, *N. munitus* (both Pl. 26), have less bold pattern; dark bar in adipose fin *not connected* to bar in anal fin; dark crescent in middle of caudal fin *weakly (or not) joined* to dark pigment on caudal peduncle; yellow spots at base of caudal fin indistinct.

FRECKLEBELLY MADTOM *Noturus munitus* Pl. 26

IDENTIFICATION: Robust body, deepest beneath or in front of dorsal fin. *Dark brown or black band to edge of adipose fin. Dark brown specks* on belly and base of pelvic fins. Large pectoral spine with 5–10 large sawlike teeth on rear edge, large teeth on front edge (as in Northern Madtom; Fig. 44). Rear edge of high adipose fin nearly free from caudal fin. Rear edge of caudal fin straight to slightly rounded. Yellow to dark brown above with dark mottling; 4 distinct saddles; large light spots in front of dorsal fin, often incompletely enclosed by dark pigment; blotched or mottled fins; broad dark band near clear edge of dorsal fin; 2 dark crescent-shaped bands on caudal fin, 1 in middle of fin, 1 near clear edge. Has 12–15 anal rays. To 3¾ in. (9.5 cm). **RANGE:** Gulf Slope drainages: disjunct populations in Conasauga R., se. TN;

Etowah R., n. GA; upper Alabama and Cahaba rivers, cen. AL; Tombigbee R., w. AL and e. MS; and Pearl R., s. MS and e. LA. Uncommon, declining in Mobile Bay drainage. **HABITAT:** Rocky riffles of small to large rivers. Often near vegetation. **SIMILAR SPECIES:** See Pl. 26. (1) Piebald Madtom, *N. gladiator*, has bolder pattern, dark bar in adipose fin connected to dark bar in anal fin, 2 distinct yellow spots at base of caudal fin. (2) Northern Madtom, *N. stigmosus, lacks* brown specks on belly; has 2 light spots in front of dorsal fin, dark band not to edge of adipose fin. (3) Neosho Madtom, *N. placidus, lacks* prominent saddles; has white lower caudal rays.

CAROLINA MADTOM *Noturus furiosus* Pl. 26

IDENTIFICATION: Robust body, deepest beneath dorsal fin; deep caudal peduncle. *No dark specks* on belly. Brown or black band nearly to *adipose fin edge*. Has 5–12 large sawlike teeth on rear edge of large pectoral spine; large teeth on front edge (as in Northern Madtom; Fig. 44). Rear edge of high adipose fin nearly free from caudal fin. Rear edge of caudal fin straight or slightly rounded. Yellow to dark brown above, dark mottling; 4 dark saddles; usually no pair of light spots in front of dorsal fin; white to yellow below; blotched fins; 2 crescent-shaped bands on caudal fin, 1 in middle of fin, 1 near clear edge. Has 14–17 anal rays. To 4¾ in. (12 cm). **RANGE:** On Piedmont and Coastal Plain in Neuse and Tar river drainages, NC. Locally common, but disappearing from upstream localities. **HABITAT:** Sand-, gravel-, and woody debris-bottomed riffles and runs of small to medium rivers. **SIMILAR SPECIES:** (1) Northern Madtom, *N. stigmosus* (Pl. 26), usually has 2 large light spots in front of dorsal fin; black band into upper half of adipose fin but *not to edge*; dark crescent in middle of caudal fin usually extends forward across upper and lower caudal rays to caudal peduncle. (2) Mountain Madtom, *N. eleutherus* (Pl. 26), lacks

dark crescent in middle of caudal fin; has dark brown band at base of caudal fin; usually is more mottled.

MOUNTAIN MADTOM *Noturus eleutherus* Pl. 26

IDENTIFICATION: Robust body; fairly deep caudal peduncle. *Dark brown bar* on caudal fin base. Usually 4 vague dorsal saddles; front edge of 1st saddle at dorsal spine. Long pectoral spine with 6–10 large saw-like teeth on rear edge, large teeth on front edge (as in Northern Madtom; Fig. 44). High adipose fin barely touches caudal fin. Dark *band on adipose fin irregular in outline,* usually *confined to lower half of fin.* Rear edge of caudal fin straight. Brown or gray above, usually with dark mottling; light below, usually no dark specks on belly; dark bands or mottling on other fins; dark band near clear edge of caudal fin. Has 12–16 anal rays. To 5 in. (13 cm). **RANGE:** Ohio R. basin from nw. PA to e. IL, and south to n. AL and n. GA; White and St. Francis river systems, MO and AR; Mississippi R., w. TN; Ouachita R. system, AR; and Red R. system, AR and OK. Locally common. Rare in main channels of Mississippi and Ohio rivers. **HABITAT:** Clean rocky riffles and runs of small to large rivers; often near vegetation or woody debris. **SIMILAR SPECIES:** See Pl. 26. (1) Most often confused with Northern Madtom, *N. stigmosus*, which has dark crescent in middle of caudal fin, 2 large light spots in front of dorsal fin; dark band on adipose fin into *upper half* of fin. (2) Neosho Madtom, *N. placidus*, has small teeth on front of pectoral spine; white lower caudal rays; dark crescent in middle of caudal fin. (3) Elegant Madtom, *N. elegans*, and (4) Saddled Madtom, *N. fasciatus*, are more slender; have bolder pattern, adipose fin joined to caudal fin.

CHECKERED MADTOM *Noturus flavater* Pl. 26

IDENTIFICATION: Robust body, deepest in front of dorsal fin. Dusky mottling on yellow back; 4 prominent black saddles; white to yellow

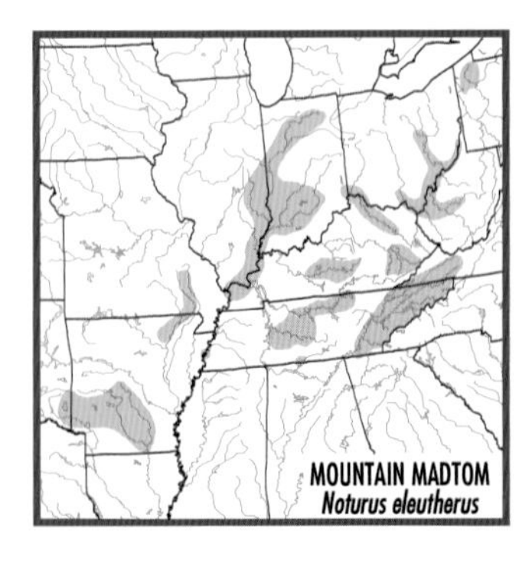

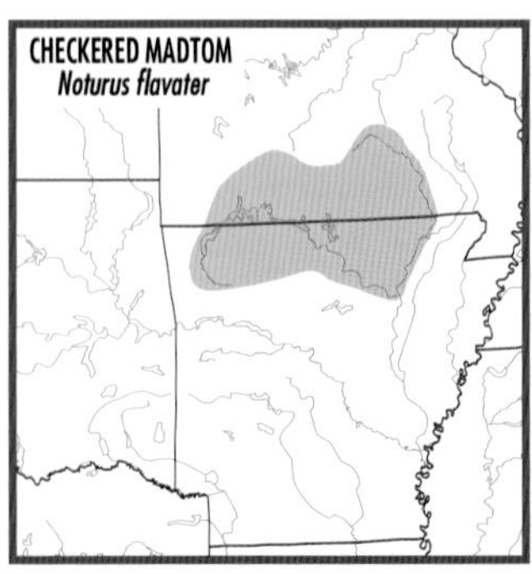

below. *Broad black bar* on caudal fin base from top to bottom of fin; prominent *black border on caudal fin; black blotch* on outer ⅓ of dorsal fin; black bands and blotches on other fins. Black saddle under adipose fin to fin edge. Rear edge of adipose fin free from caudal fin. Rear edge of caudal fin straight or with rounded corners. About 10 large sawlike teeth on rear of pectoral spine, large teeth on front edge (as in Brindled Madtom; Fig. 44). Has 14–17 anal rays. To 8 in. (20 cm). **RANGE:** Upper White R. drainage (excluding Black R. system), MO and AR. Uncommon. **HABITAT:** Pool margins and backwaters of clear small to medium rivers with moderate to high gradients. Often among leaves and woody debris. **SIMILAR SPECIES:** No other madtom has *both* a broad black bar on caudal fin base from top to bottom of fin and large black blotch on outer ⅓ of dorsal fin. (1) Brindled Madtom, *N. miurus* (Pl. 26), *lacks* broad black bar at caudal fin base, prominent saddles; has adipose fin broadly joined to caudal fin, rounded caudal fin. (2) Yellowfin Madtom, *N. flavipinnis* (Pl. 26), *lacks* black blotch on dorsal fin; has more mottled body

---

BRINDLED MADTOM *Noturus miurus* Pl. 26

IDENTIFICATION: Robust body, deepest in front of dorsal fin. *Black blotch* on outer ⅓ of dorsal fin *across 1st 3–5 rays;* dark saddle under adipose fin to fin edge. *Rounded rear edge on caudal fin.* Adipose and caudal fins broadly joined. Usually 5–9 large sawlike teeth on rear of pectoral spine; large teeth on front edge (Fig. 44). Light yellow or brown above, dark mottling; 4 vague saddles; white to yellow below; brown or black border on caudal fin; brown or black mottling on other fins. Has 13–17 anal rays. To 5 in. (13 cm). **RANGE:** Lower Great Lakes drainages, NY and ON, southwest through most of Ohio R. basin and lower Mississippi R. basin to Gulf (west to e. KS and OK); Mohawk R., NY; Pearl R. and Lake Pontchartrain drainages on Gulf Slope, MS and LA. Common. **HABITAT:** Riffles, runs, and flowing pools over gravel and

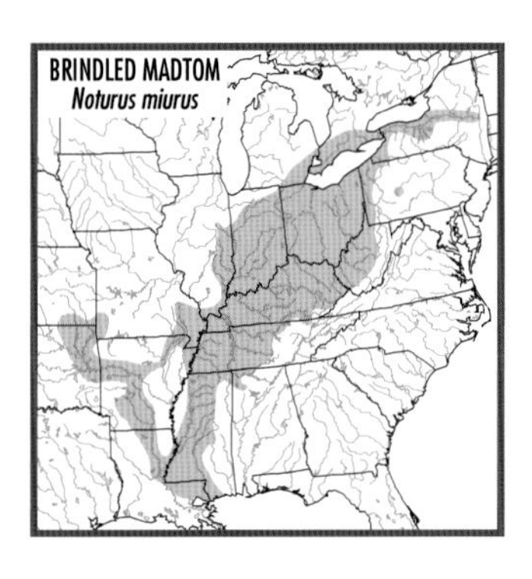

sand mixed with sticks and leaves in creeks and small rivers; rocky margins of lakes. **SIMILAR SPECIES:** (1) Checkered Madtom, *N. flavater* (Pl. 26), has broad black bar at caudal fin base from top to bottom of fin, *straight rear edge* on caudal fin, prominent saddles. (2) Yellowfin Madtom, *N. flavipinnis* (Pl. 26), *lacks* dark blotch on dorsal fin; has bold black bar on caudal fin base from top to bottom of fin, *straight rear edge* on caudal fin, 2 light spots in front of dorsal fin.

YELLOWFIN MADTOM *Noturus flavipinnis* Pl. 26

IDENTIFICATION: Robust body, deepest in front of dorsal fin; flat head. *Bold black bar* on caudal fin base from top to bottom of fin; *pale edge* on caudal fin. Brown saddle usually *encloses 2 light spots in front of dorsal fin;* dark saddle under adipose fin extends to fin edge (in adult). Rear edge of caudal fin straight or with slightly rounded corners. Rear edge of adipose fin nearly free from caudal fin. Has 4–10 large saw-like teeth on rear of pectoral spine; large teeth on front edge (as in Brindled Madtom; Fig. 44). Yellow above, dark mottling and specks; 4 saddles; yellow to white below; yellow fins with dark bands or mottling at middle and near edge. Has 14–16 anal rays. To 6 in. (15 cm). **RANGE:** Upper Tennessee R. drainage, VA, TN, and GA. Populations extant in Copper Creek (Clinch R. system), VA, Powell R., TN, and Citico Creek (Little Tennessee R. system), TN; apparently extirpated from North Fork Holston R., VA, Chickamauga Creek, n. GA, and Hines Creek, e. TN. Rare; protected as a *threatened species.* **HABITAT:** Pools and backwaters around slab rocks, bedrock ledges, and tree roots in clear creeks and small rivers. **SIMILAR SPECIES:** (1) Brindled Madtom, *N. miurus*, and (2) Checkered Madtom, *N. flavater* (both Pl. 26), have black blotch at top of dorsal fin, *dark border* on caudal fin.

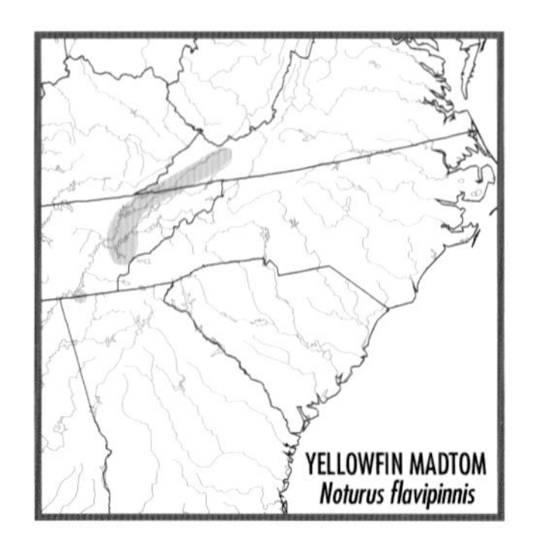

Labyrinth Catfishes: Family Clariidae (1 introduced)

This is a large family of catfishes with a *long dorsal fin,* usually with no spines and more than 30 rays; a rounded caudal fin, long anal fin, adipose fin frequently absent, no scales, 4 pairs of barbels, and wide gill openings. An unusual feature is an air-breathing ("labyrinth") organ made of modified gill filaments supported by a cartilaginous treelike structure. Labyrinth catfishes are native to Africa, Syria, and southwestern Asia (Philippines to Java). There are about 112 species; 1 was introduced to Florida in the late 1960s.

WALKING CATFISH *Clarias batrachus* **Not shown**

IDENTIFICATION: Long body, broad in front, narrow in back. Has *60–75 dorsal rays; 45–58 anal rays.* Dorsal and anal fins not joined to caudal fin. *No adipose fin.* Olive, dark brown, or purple-black above; blue-green on side; white below; white specks on rear half of side; gray-green fins, some yellow on dorsal fin; red borders on median fins. To 24 in. (61 cm); rarely over 14 in. (36 cm) in U.S. **RANGE:** Native to Sri Lanka through e. India to Malay Archipelago. Established in peninsular FL. Common. **HABITAT:** Slow creeks and small rivers, canals, lakes, swamps, and ponds with mud bottom and vegetation. **REMARKS:** Walking Catfish walks, including over land on rainy nights, using tips of its stout pectoral spines as pivots as it flexes its body.

Plated Catfishes: Family Callichthyidae (1 introduced)

These air-breathing catfishes, native to tropical fresh waters of Panama and South America, are easily distinguished by *2 rows of large overlapping bony plates* covering the side of the body. They have 2 or more pairs of barbels around a small mouth and stout spines on the dorsal, adipose, and pectoral fins. Air is swallowed, passed to the intestine (which serves as an accessory respiratory organ), and is expelled through the anus. Plated catfishes are popular aquarium fishes. There are about 196 species; 1 species is established in Florida.

BROWN HOPLO *Hoplosternum littorale* **Not shown**

IDENTIFICATION: *Two rows of overlapping plates on side*; large bony plates on wide head. Light brown or gray; dark spots or stripes on young. To 8¾ in. (22 cm). **RANGE:** Native to Trinidad and most of S. America. Established in peninsular FL. Abundant. **HABITAT:** Wetlands, canals, and other standing and sluggish water bodies. Often near vegetation.

SUCKERMOUTH ARMORED CATFISHES: Family Loricariidae (5 introduced)

Suckermouth catfishes are native to Costa Rica, Panama, and South America and occur in a variety of habitats ranging from marshes to montane streams to large rivers. Because of their popularity with aquarists, many species have been imported into North America; at least 5 species are established in North American streams. Taxonomy of the 780 or so species in this family is difficult, and accurate identification of introduced *Hypostomus* species is not possible. Suckermouth catfishes are characterized by usually having *bony plates* covering their body; *1 pair of barbels* on a large subterminal mouth; *papillose, sucking lips; and a spine* at the front of the adipose fin (when the fin is present). Body shape varies tremendously, but most species are *flat below.*

SUCKERMOUTH CATFISHES *Hypostomus* species **Not shown**

IDENTIFICATION: *Large bony plates above;* no or small plates below. Dorsal fin usually has 1 spine, *7 rays*. Large stout spine at front of dorsal, pectoral, and pelvic fins. Usually black or brown spots or stripes on olive-brown body and fins. To 18 in. (46 cm). **RANGE:** *Hypostomus* species are native to Middle and S. America from Costa Rica south to Río de la Plata drainage. One or more unidentified species established in s. and w.-cen. FL; springs and streams in cen. and s. TX; Indian Spring, Clark Co., NV. Locally common. **HABITAT:** Adult in rocky pools and runs; young near vegetation. **SIMILAR SPECIES:** (1) See sailfin catfishes, *Pterygoplichthys*.

PTERYGOPLICHTHYS

GENUS CHARACTERISTICS: *Pterygoplichthys* species, sailfin catfishes, are similar to *Hypostomus* species but have longer dorsal fin with 1 spine, *10–14 rays.* Four species are established in FL and TX. They are similar in appearance and readily hybridize when together. Specimens intermediate to descriptions below are common.

AMAZON SAILFIN CATFISH *Pterygoplichthys pardalis* **Not shown**

IDENTIFICATION: Coalesced black spots on gray back, side, and fins; discrete black spots on underside. To 30 in. (76 cm). **RANGE:** Native to Amazon R. basin, Brazil and Peru. Established in s. FL. Locally common. **HABITAT:** Canals, creeks, small rivers and lakes; usually near vegetation.

VERMICULATED SAILFIN CATFISH — Not shown

Pterygoplichthys disjunctivus

IDENTIFICATION: Similar to Amazon Sailfin Catfish, *P. pardalis*, but black spots on underside coalesced to form *vermiculations*. To 30 in. (76 cm). **RANGE:** Native to Madeira R. drainage, Brazil and Bolivia. Established in peninsular FL (St. Johns R. drainage and south) and upper San Antonio R. and San Marcos and Comal springs, TX. Locally common. **HABITAT:** Canals, creeks, small rivers and lakes; usually near vegetation.

ORINOCO SAILFIN CATFISH — Not shown

Pterygoplichthys multiradiatus

IDENTIFICATION: Similar to Amazon Sailfin Catfish, *P. pardalis*, but has discrete black spots on gray body (above and below) and fins. To 30 in. (76 cm). **RANGE:** Native to Orinoco R. basin, Venezuela. Established in s. FL. Locally common. **HABITAT:** Canals, creeks, small rivers and lakes; usually near vegetation.

SOUTHERN SAILFIN CATFISH *Pterygoplichthys anisitsi* — Not shown

IDENTIFICATION: Similar to Amazon Sailfin Catfish, *P. pardalis*, but has white spots and chevrons on black body and fins. To 30 in. (76 cm). **RANGE:** Native to Paraguay, Parana, Bermejo and Uruguay river basins, Argentina, Brazil, and Paraguay. Established in Buffalo Bayou drainage, se. TX. Uncommon. **HABITAT:** Creeks and small rivers; usually near vegetation.

PIKES AND MUDMINNOWS: Family Esocidae (8)

Pikes and mudminnows have 1 dorsal fin, no spines in the fins, cycloid scales, dorsal and anal fins *far back* on the body, and abdominal pelvic fins. Pikes (*Esox*) are large, predatory fishes with a long cylindrical green body, large *duckbill-like snout*, many small scales, and *forked caudal fin*. They live in North America, Europe, and northern Asia. One species, Northern Pike, *Esox lucius*, occurs in both North America and Eurasia. The Amur Pike, *E. reichertii*, is endemic to Asia.

Mudminnows—*Umbra, Novumbra*, and *Dallia*—are smaller fishes with a long, slender body, cylindrical in front and compressed at the rear. They have a disjunct distribution, with 4 species in North America, and the European mudminnow, *Umbra krameri,* in Europe, and the Russian Blackfish, *Dallia admirabilis,* in Russia. The Eastern Mudminnow, *U. pygmaea*, has been established in Europe. Mudminnows can breathe atmospheric oxygen and survive in poorly oxygenated water unsuitable for other fishes.

Pikes are voracious predators, feeding on a variety of fishes. Because of their large size and fighting behavior, Muskellunge, *Esox masquinongy*, and Northern Pike, *E. lucius*, are favorite sport fishes.

GRASS OR REDFIN PICKEREL *Esox americanus* **Pl. 31**

IDENTIFICATION: *Fully scaled cheek and opercle* (Fig. 45). Black (darker in female) teardrop (suborbital bar) *slanted* toward rear. Dark olive to brown above; 15–36 dark green to brown wavy bars along side of adult; amber to white below. Red, yellow, or dusky fins (see Remarks). Has 11–13 branchiostegal rays; 4 (rarely 3 or 5) submandibular pores; 92–118 lateral scales. To 15 in. (38 cm). **RANGE:** Atlantic Slope from St. Lawrence R. drainage, QC, to s. FL; Gulf drainages east to Brazos R., TX; Mississippi R. and Great Lakes basins north to NE, WI, MI, and s. ON. Common; absent in uplands. Introduced elsewhere, including into WA, CA, and CO. **HABITAT:** Lakes, swamps, and backwaters and sluggish pools of streams. Usually among vegetation in clear water. **REMARKS:** Two subspecies. *E. a. americanus,* Redfin Pickerel, has red fins, short broad snout with convex profile, occupies Atlantic Slope drainages to s. GA. Intergrades (with amber fins) occupy Gulf Slope drainages west to Pascagoula R., MS. *E. a. vermiculatus,* Grass Pickerel, has yellow-green to dusky fins, narrower snout with concave profile; occupies rest of range. **SIMILAR SPECIES:** See Pl. 31. (1) Chain Pickerel, *E. niger*, has chainlike pattern on side, *vertical teardrop*, longer snout, 14–17 branchiostegal rays. (2) Northern Pike, *E. lucius*, and (3) Muskellunge, *E. masquinongy, lack* scales on lower half of opercle; have 13 or more branchiostegal rays, 5 or more submandibular pores.

CHAIN PICKEREL *Esox niger* **Pl. 31**

IDENTIFICATION: *Fully scaled cheek and opercle* (Fig. 45). *Green chainlike pattern* on yellow side of adult, wavy bars on young. *Vertical* black teardrop. Very long snout, distance from tip of snout to middle of eye

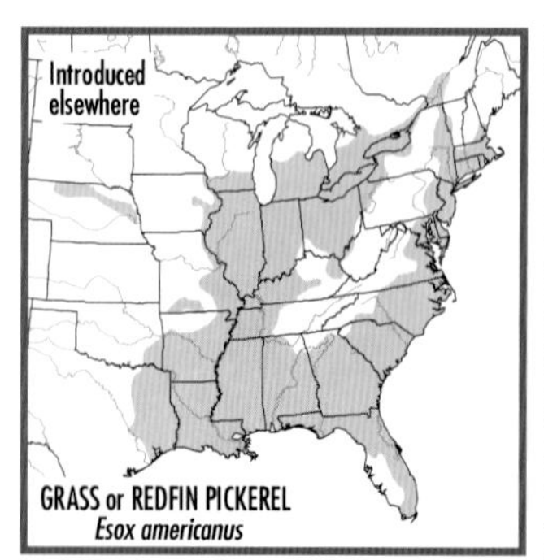

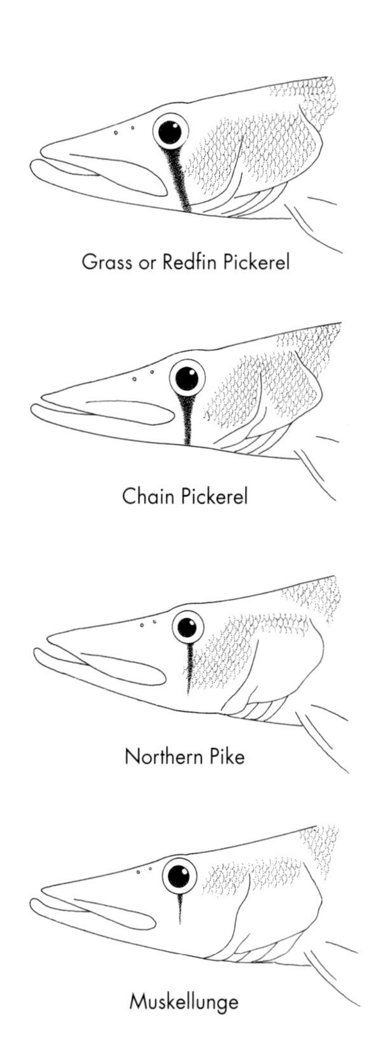

Fig. 45. ***Esox*** **species.**

greater than distance from middle of eye to rear edge of gill cover. Has 14–17 branchiostegal rays; 3–5 submandibular pores; 110–138 lateral scales. To 39 in. (99 cm). **RANGE:** Atlantic Slope from NS (where introduced) to s. FL; Gulf Coast west to Red R. drainage, OK, and Sabine R. drainage, TX; Mississippi R. basin north to KY and MO (mostly Former Mississippi Embayment but also upland streams in se. MO). Introduced into Lakes Ontario and Erie drainages, other scattered localities as far west as CO. Common. **HABITAT:** Vegetated lakes, swamps, and backwaters and quiet pools of creeks and small to medium rivers. **SIMILAR SPECIES:** (1) Grass or Redfin Pickerel, *E. americanus* (Pl. 31), *lacks* chainlike pattern on side, has *oblique teardrop*, shorter snout (distance from tip of snout to middle of eye about equal to distance from middle of eye to rear edge of gill cover), 11–13 branchiostegal rays.

NORTHERN PIKE *Esox lucius* Pl. 31

IDENTIFICATION: *Partly scaled opercle*; *fully scaled cheek* (Fig. 45). *Rows of yellow bean-shaped spots* (on adult), yellow to white wavy bars (on young) on green back and side; large individual dark gray above. No teardrop. Usually black spots on dusky green, yellow, or red fins. Has 13–16 branchiostegal rays; 5–6 submandibular pores; 105–148 lateral scales. To 56 in. (142 cm). **RANGE:** Atlantic, Arctic, Pacific, Great Lakes, and Mississippi R. basins from Labrador to AK and south to PA and NE. Widely introduced elsewhere. Also occurs in n. Eurasia as far south as n. Italy. Common. **HABITAT:** Clear vegetated lakes; quiet pools and backwaters of creeks and small to large rivers. **REMARKS:** "Tiger Musky" is a hybrid Northern Pike-Muskellunge (recognizable by strong barring pattern on side); considered a superior sport fish, often introduced in impoundments and lakes. **SIMILAR SPECIES:** (1) Muskellunge, *E. masquinongy* (Pl. 31), has pattern of *dark marks* on light background, *no scales* on lower half of cheek, 16–19 branchiostegal rays, 6–10 submandibular pores.

MUSKELLUNGE *Esox masquinongy* Pl. 31

IDENTIFICATION: *Partly scaled cheek and opercle* (Fig. 45; may be more scales on cheek than shown). *Dark spots, blotches, or bars* on light yellow-green back and side; cream to white below with small brown to gray blotches; large individuals dark gray above. No teardrop. Has 16–19 branchiostegal rays; 6–10 submandibular pores; 130–167 lateral

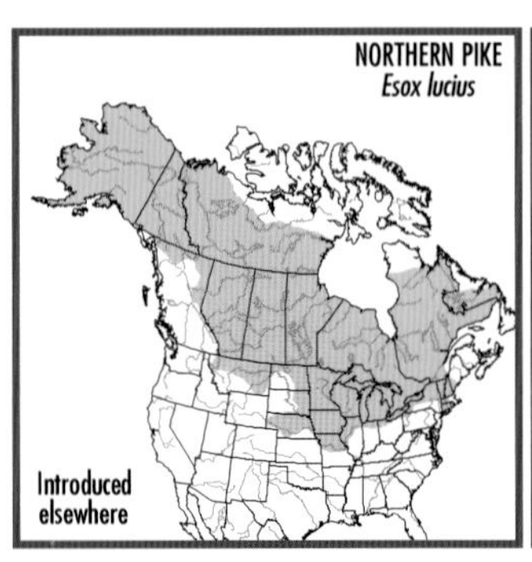

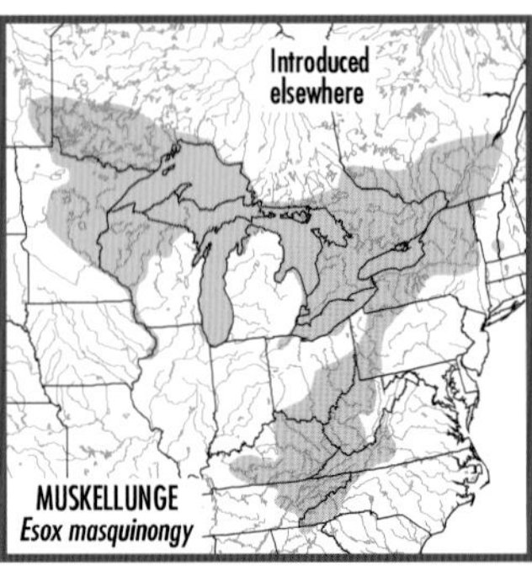

scales. To 6 ft. (2 m). **RANGE:** St. Lawrence R.-Great Lakes, Hudson Bay (Red R.), and Mississippi R. basins, from QC to se. MB and south in Appalachians to GA and west to IA. Introduced into Atlantic drainages as far south as s. VA; introduced elsewhere but seldom successfully. Locally common. **HABITAT:** Clear vegetated lakes; quiet pools and backwaters of creeks and small to large rivers. **REMARKS:** "Musky" is a solitary hunter, lurking about cover and lunging after suitable prey. It begins feeding on other fishes at about 4 days of age and may reach 12 in. (30 cm) in only 4 months. Large size and fighting spirit make it a favorite sport fish. **SIMILAR SPECIES:** (1) Northern Pike, *E. lucius* (Pl. 31), has *light marks* on *dark* background, *fully scaled cheek*, 13–16 branchiostegal rays, 5–6 submandibular pores.

OLYMPIC MUDMINNOW *Novumbra hubbsi* Pl. 31

IDENTIFICATION: Dorsal fin origin *above or slightly in front of* anal fin origin. Anal fin base about as long as dorsal fin base. *Small pelvic fins, 6–7 rays.* Straight-edged or slightly indented caudal fin. Fairly small pectoral fins, 18–25 rays. No lateral line. Green to dark brown above and on side; *10–15 cream to yellow interrupted narrow bars on side*; white to yellow below; pale olive to dark brown median fins, sometimes with light narrow bands. Breeding male is dark chocolate brown to black; has iridescent blue-green to white bars on side, blue edges on dorsal and anal fins. Has 52–58 lateral scales; 12–15 dorsal rays; 10–13 anal rays. To 3 in. (8 cm). **RANGE:** Coastal lowlands of Olympic Peninsula, WA, from Lake Ozette to Chehalis R. drainage; east side of Puget Sound in Cherry Creek and Peoples Creek drainages; lower Deschutes R. (Puget South drainage). Locally abundant. **HABITAT:** Quiet water with dense vegetation or other cover over mud and debris; cool, brown waters of bogs and swamps. **SIMILAR SPECIES:** (1) Central Mudminnow, *Umbra limi*, and (2) Eastern Mudminnow, *U. pygmaea* (both Pl. 31), have black bar on caudal fin base, *no* narrow yellow bars on side, dorsal fin origin *far in front* of anal fin origin, 30–37 lateral scales.

ALASKA BLACKFISH *Dallia pectoralis* Pl. 31

IDENTIFICATION: *Black mottling and blotches* on dark brown to olive back and side; yellow-white below with black specks; *black specks on red-brown fins*; white to clear edge on median fins of adult become pale red on large male. *Tiny pelvic fins, 2–3 rays.* Broadly rounded pectoral fins, 32–36 rays. Dorsal fin origin above anal fin origin. Anal fin base about as long as dorsal fin base. *Tiny (76–100 lateral) scales.* Short, flat snout. Rounded caudal fin. Lateral line present but inconspicuous. Has 10–14 dorsal rays; 11–16 anal rays. To 13 in. (33 cm). **RANGE:** AK; Colville R. delta south to cen. Alaska Peninsula; upstream in Yukon-Tanana drainage to near Fairbanks. Introduced into Hood and Spenard lakes near Anchorage. Also Bering Sea islands and ne. Siberia. Common to abundant. **HABITAT:** Usually in heavily vegetated swamps and ponds; occasionally in medium to large rivers and lakes with abundant vegetation.

CENTRAL MUDMINNOW *Umbra limi* Pl. 31

IDENTIFICATION: *Black bar* on caudal fin base. Dorsal fin origin *far in front* of anal fin origin. Anal fin base about *half as long* as dorsal fin base. No lateral line. Rounded caudal fin. Small pelvic fins, 6–7 rays. Fairly small pectoral fins, 11–16 rays. Green to brown-black above and on side; white to yellow below; up to *14 dark brown bars* on side; no spots on dusky brown fins. Breeding male has iridescent blue-green anal and pelvic fins. Fairly large (30–37 lateral) scales. Has 13–17 dorsal rays; 7–10 anal rays. To 6 in. (15 cm). **RANGE:** St. Lawrence-Great Lakes, Hudson Bay, and Mississippi R. basins from QC to MB and south to cen. OH, w. TN, and ne. AR; Hudson R. drainage (Atlantic Slope), NY. Isolated population in Missouri R. drainage of e.-cen. SD (possibly extinct). Introduced into Connecticut R., MA. Common. **HABITAT:** Quiet areas of streams, sloughs, swamps, and other wetlands over mud and debris. Often in dense vegetation. Tolerant of drought, low oxygen levels, and extremes in water temperature. **SIMILAR SPECIES:** (1) See Eastern Mudminnow, *U. pygmaea* (Pl. 31).

EASTERN MUDMINNOW *Umbra pygmaea* Pl. 31

IDENTIFICATION: Similar to Central Mudminnow, *U. limi*, but has *10–14 dark brown stripes* with pale interspaces (about as wide as stripes) on back and side. To 4½ in. (11 cm). **RANGE:** Atlantic and Gulf slopes from se. NY (including Long Island) to St. Johns R. drainage, FL, and west to Aucilla R. drainage, FL and GA. Common. **HABITAT:** Quiet streams, sloughs, swamps, and other wetlands over sand, mud, and debris; often in dense vegetation. **SIMILAR SPECIES:** (1) See Central Mudminnow, *U. limi* (Pl. 31).

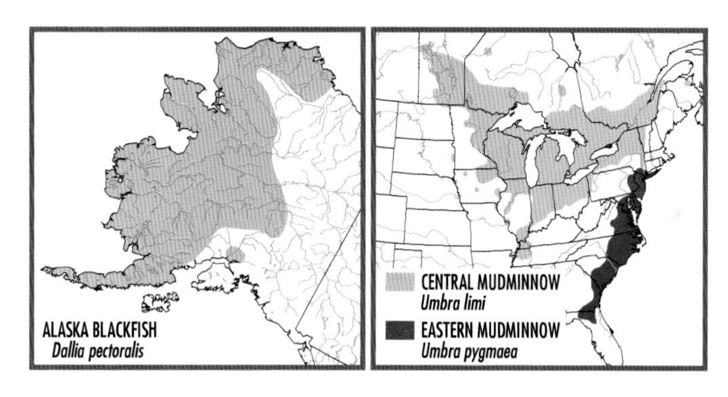

SMELTS: Family Osmeridae (5 native; 1 introduced)

Smelts (about 13 species) live in cold and temperate coastal waters, both marine and fresh, in the Northern Hemisphere. They are small, slender fishes with a large mouth, jaw teeth, cycloid scales, lateral line, abdominal pelvic fins, 1 rayed dorsal fin and an *adipose fin*, and no spines in the fins. Unlike similar-looking salmonids, they *lack* pelvic axillary process. They feed mainly on crustaceans and spawn in spring over gravel or sand in streams or gravel shoals of lakes.

POND SMELT *Hypomesus olidus* **Not shown**

IDENTIFICATION: *Small mouth*; upper jaw ends *in front of* middle of eye. Dorsal fin origin above or in front of pelvic fin origin. *Incomplete lateral line*; *51–62 lateral scales*. Small, pointed teeth on jaws; none enlarged. Olive to light brown above; purple iridescence on silver stripe along side; clear fins. Breeding male has gold cast. Has 8–10 dorsal rays, 12–16 anal rays, 10–13 pectoral rays, 31–36 rakers on 1st gill arch, 0–3 pyloric caeca. To 8 in. (20 cm). **RANGE:** Arctic and Pacific drainages from Rae R. (Coronation Gulf) and Great Bear Lake, NT, to Copper R., AK. Also n. Eurasia. Common to abundant. **HABITAT:** Middle and surface waters of ponds, lakes, and streams; over a variety of bottom types. Enters brackish water. **SIMILAR SPECIES:** (1) See Delta Smelt, *H. transpacificus*. (2) Wakasagi, *H. nipponensis,* has *4–7 pyloric caeca*.

DELTA SMELT *Hypomesus transpacificus* **Not shown**

IDENTIFICATION: Nearly identical to Pond Smelt, *H. olidus*, but has *9–10 dorsal rays*, *4–5 pyloric caeca.* Steel blue above and on side, almost translucent; silver white below; dusky fins. Faint black specks on silver stripe along side. Has 0–1 black speck between mandibles, 53–60 lateral scales, 15–17 anal rays, 10–12 pectoral rays; 27–33 rakers on 1st gill arch. To 4¾ in. (12 cm). **RANGE:** Sacramento-San Joaquin Delta

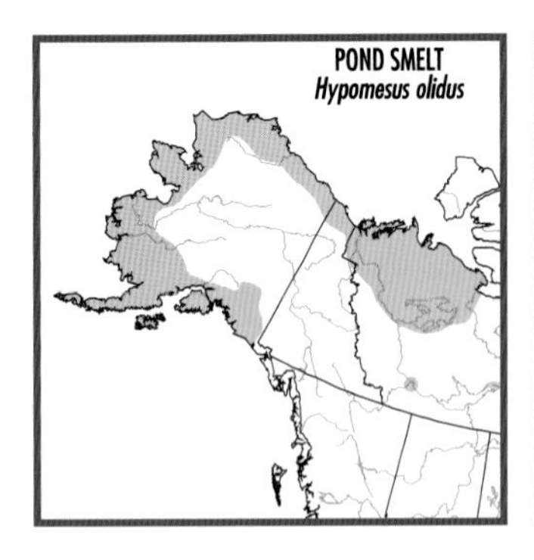

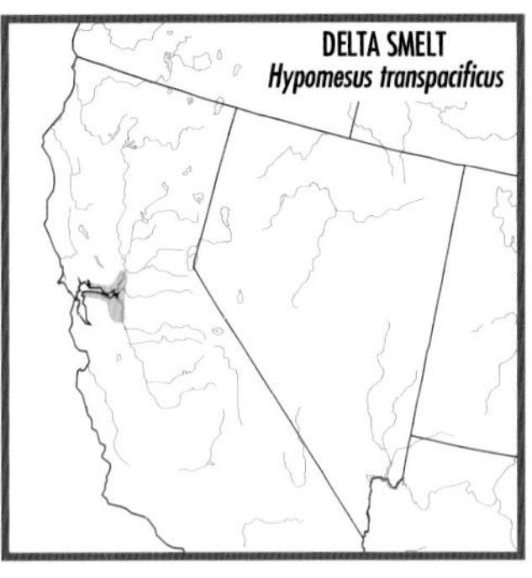

region, cen. CA. Uncommon; protected as a *threatened species.* **HABITAT:** Open brackish and fresh water of large channels. **SIMILAR SPECIES:** (1) See Pond Smelt, *H. olidus,* and (2) Wakasagi, *H. nipponensis.*

WAKASAGI *Hypomesus nipponensis* **Not shown**

IDENTIFICATION: Nearly identical to Delta Smelt, *H. transpacificus*, but has *10 or more black specks* between mandibles, 12–14 pectoral rays, *7–9 dorsal rays*, 13–15 anal rays. Yellow-brown above; black specks on back and side; 54–60 lateral scales, 4–7 pyloric caeca; 29–37 rakers on 1st gill arch. To 6¾ in. (17 cm). **RANGE:** Native to Japan. Introduced successfully into Sacramento and lower Klamath R. drainages, CA and OR, where common. **HABITAT:** Open brackish and fresh water of estuaries, small to large rivers, and impoundments. **SIMILAR SPECIES:** (1) See Delta Smelt, *H. transpacificus.*

RAINBOW SMELT *Osmerus mordax* **Pl. 4**

IDENTIFICATION: *Huge* mouth, upper jaw reaches *middle of eye* or beyond. *Two large canine teeth* (and sometimes smaller ones) in roof of mouth; large teeth on tongue. Dorsal fin origin *above or in front* of pelvic fin origin. Olive above; blue or pink iridescence on silver side; usually a silver stripe along side, dark specks on upper side; dusky fins. Landlocked individual may have black head, back, and fins. Incomplete lateral line, 62–72 lateral scales; 11–16 anal rays; usually 28–32 rakers on 1st gill arch. To 13 in. (33 cm). **RANGE:** Atlantic drainages from Lake Melville, NL, to Delaware R., PA, and west through Great Lakes; Arctic and Pacific drainages from Bathurst Inlet, NT, to Vancouver I., BC. Also Pacific drainages of Asia. Introduced into many impoundments and lakes in e. N. America including Great Lakes; seasonally present in main channels of Missouri, Mississippi, Ohio, and Illinois rivers from KY to MT and south to LA. Common; locally abundant. **HABITAT:** Cool clear lakes, medium to large rivers, and coastal

waters. Often in schools in midwater. Coastal populations are anadromous. **REMARKS:** Two subspecies: *O. m. mordax* of e. N. America and *O. m. dentex* of N. Pacific-Arctic drainages, sometimes recognized as genetically distinct species. Lake Utopia Dwarf Smelt, an undescribed form in NB, is protected as *threatened*. **SIMILAR SPECIES:** (1) Longfin Smelt, *Spirinchus thaleichthys*, has pectoral fin reaching to pelvic fin, *no* large teeth in roof of mouth, 15–21 anal rays.

LONGFIN SMELT *Spirinchus thaleichthys* Not shown

IDENTIFICATION: Pectoral fin almost *reaches or extends past* pelvic fin origin. *Huge mouth*; upper jaw *to middle of eye* or beyond. *Large canine teeth* on tongue; small jaw teeth. Dorsal fin origin above or behind pelvic fin base. Dusky or olive-brown to iridescent pink-green above; dusky fins. No silver stripe on side. Breeding male has large, rounded anal fin; moss green back; many black specks on head. Incomplete lateral line; 55–62 lateral scales; 15–21 anal rays; 38–47 rakers on 1st gill arch. To 6 in. (15 cm). **RANGE:** Pacific Coast from Prince Wil-

liam Sound, AK, to Monterey Bay, CA. Landlocked in Harrison Lake, BC, and Lakes Washington and Union, WA. Locally and seasonally abundant. **HABITAT:** Nearshore; bays and estuaries. Ascends coastal streams from October to December to spawn. **SIMILAR SPECIES:** (1) Other smelts have pectoral fin *not* reaching pelvic fin.

EULACHON *Thaleichthys pacificus* **Not shown**

IDENTIFICATION: Concentric bony *striations* on gill cover. *Huge mouth*; upper jaw to *rear of eye*. Usually no enlarged jaw teeth. Dorsal fin origin behind pelvic fin origin. Blue-brown above, small black specks on back; shading to silver white below; no silver stripe along side; dusky pectoral and caudal fins. Breeding male has thick ridge along side. *Complete lateral line*; 70–78 scales; 18–23 anal rays; 17–23 rakers on 1st gill arch. To 11¾ in. (30 cm). **RANGE:** Pacific Coast from Pribilof Is. (Bering Sea) and Nushagak R., AK, to Monterey Bay, CA. Seasonally abundant. **HABITAT:** Nearshore; coastal inlets. Young apparently in deeper water. Ascends rivers in spring to spawn, but seldom more than a few miles inland. **REMARKS:** Eulachons were a source of food and cooking oil for Native Americans, who also dried specimens, inserted wicks, and used them as torches; hence, alternate name "Candlefish." **SIMILAR SPECIES:** (1) Other smelts *lack* striations on gill cover.

Trouts, Salmons, and Whitefishes: Family Salmonidae (39 native; 1 introduced)

Salmonids live in cool to cold streams and lakes throughout Europe, northern Asia, North America as far south as northwestern Mexico, extreme northern Africa (north slope of Atlas Mountains), and Taiwan. Arctic Char, *Salvelinus alpinus*, occurs the farthest north of any freshwater fish. Of commercial importance, trout and salmon have been introduced into Africa south of their native range, South America, Australia, New Zealand, New Guinea, and India.

Salmonids are freshwater or migrate into fresh water to spawn. In migratory (anadromous) species, young migrate to sea, grow for several years, then migrate back to spawn in the stream where they hatched.

Salmonids are characterized by having many small cycloid scales, a lateral line, 1 dorsal fin plus *adipose fin*, abdominal pelvic fins, an axillary process at the base of the pelvic fin, and *no spines* in the fins. Many species are large; the Asiatic species Taimen, *Hucho taimen*, reaches 175 lb. (80 kg).

In North America, many populations of salmonids are not recognized as species even though they are reproductively isolated (often by geographic boundaries) and morphologically distinct. Whitefishes of the Great Lakes and trouts and chars of the western U.S. are among the most taxonomically difficult freshwater fishes in North America. Many introductions of salmonids throughout much of North America make identifications even more difficult; a species may appear suddenly where it previously did not occur, and introduced fishes often hybridize with species already present, producing intermediate forms not fitting any species description.

Young trouts, salmons, and chars have color patterns and characters differing from those of adults and from one another, but are difficult to identify. The best characters are shown in Fig. 46.

COREGONUS

GENUS CHARACTERISTICS: *Two small flaps* of skin between nostrils (Fig. 47); fairly *broad snout*; small subterminal to terminal mouth, body compressed to round in cross section; forked caudal fin; usually *long, slender rakers*, 18–64 on 1st gill arch; no or small teeth on jaws; 9–16 dorsal rays; *no parr marks*; usually no distinct black spots on body or fins; generally silver, silver green, or silver blue in color. Ciscoes and whitefishes are among the most difficult N. American freshwater fishes to identify.

BROAD WHITEFISH *Coregonus nasus* Not shown

IDENTIFICATION: *Compressed body*. Snout overhangs *subterminal mouth*. Head profile smoothly convex; no (or small) concavity between snout and nape (Fig. 48). Thick, white lower fins. Olive-brown to nearly black above; silver side, white to yellow below; white to gray fins. *Short, broad maxillary*, its length less than twice its width. Has 84–102 lateral scales; 18–25 *short rakers* (less than ⅕ as long as space between eyes) on 1st gill arch. To 28 in. (71 cm). **RANGE:** Arctic drainages from Perry R., NT, to Kuskokwim R. (Bering Sea tributary), AK. Also n.

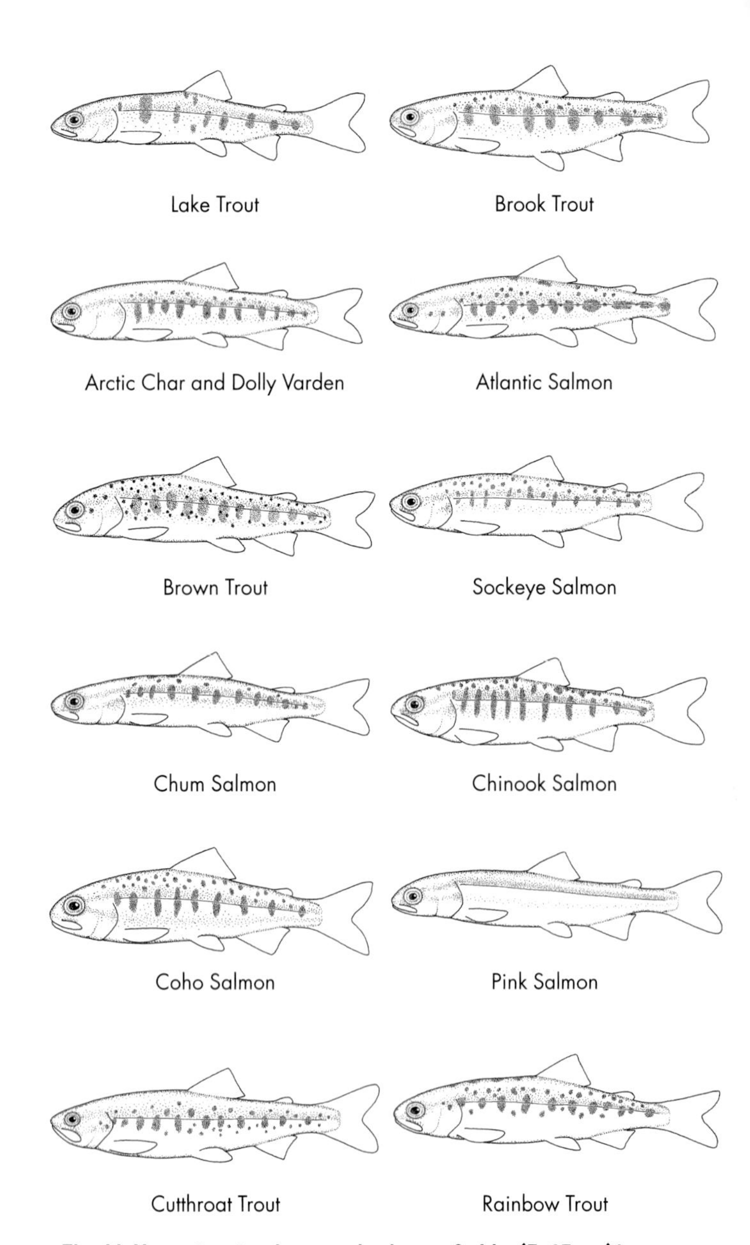

Fig. 46. Young trouts, chars, and salmons 2–6 in. (5–15 cm) long.

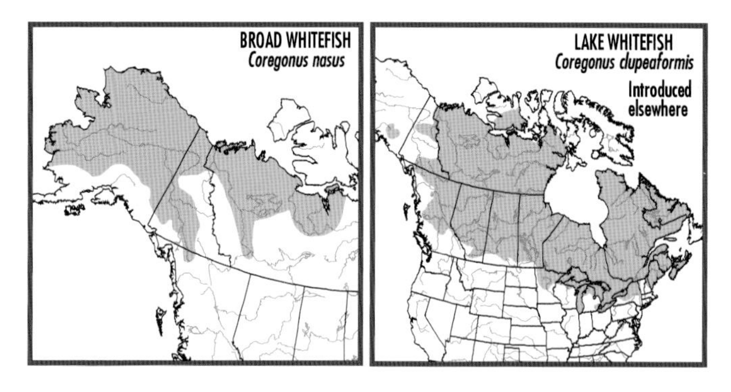

Eurasia. Common. **HABITAT:** Most frequently in streams; also lakes, brackish water. **SIMILAR SPECIES:** (1) See Humpback whitefishes, *Coregonus clupeaformis* complex (Pl. 27).

NEXT 3 SPECIES: Humpback whitefishes, *Coregonus clupeaformis* complex, are extremely difficult to distinguish from one another. They differ from Broad Whitefish, *C. nasus*, in having more distinct *concavity* between snout and nape, pronounced *hump* behind head in adult (Fig. 48), longer rakers on 1st gill arch (longest raker more than ⅕ of space between eyes), longer maxillary (length twice width or more), and more translucent lower fins. Humpback whitefishes differ from one another morphologically only in modal number of rakers on 1st gill arch and lateral scales.

AKE WHITEFISH *Coregonus clupeaformis* Pl. 27

IDENTIFICATION: Has 24–33, usually 26 or more, rakers on 1st gill arch. Dark brown to blue above, silver side; 70–97, usually fewer than 90, lateral scales. To 31 in. (80 cm). **RANGE:** Atlantic, Arctic, and Pa-

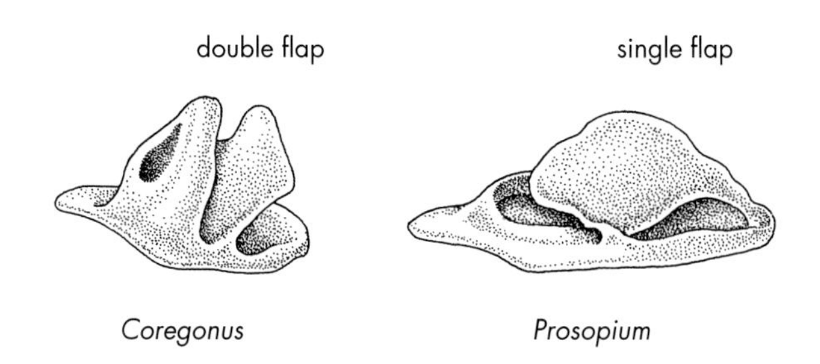

Fig. 47. *Coregonus* and *Prosopium*—nostril flaps.

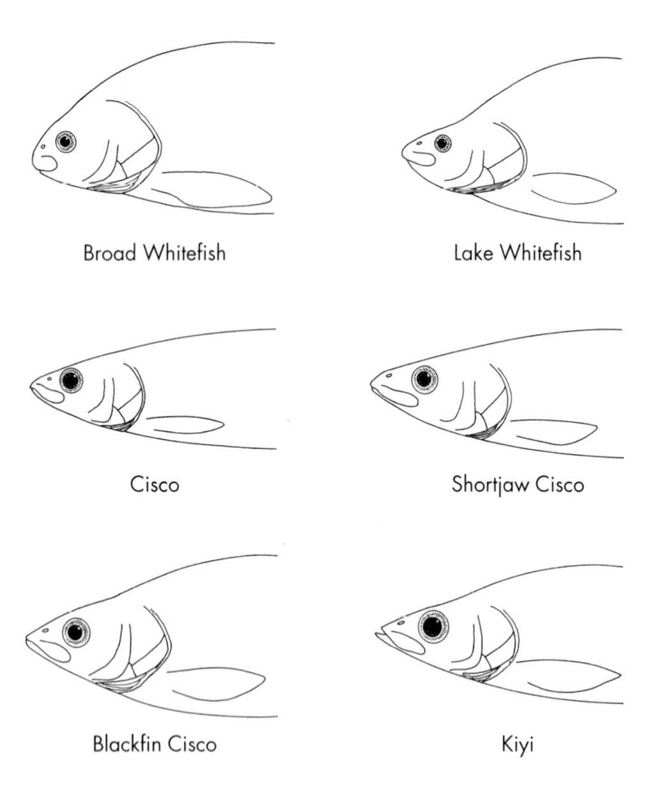

Fig. 48. *Coregonus* species.

cific basins through most of Canada south into New England, Great Lakes basin, and cen. MN; Copper and Susitna river drainages, AK. Introduced into nw. U.S. (MT, ID, and WA). Locally common. **HABITAT:** Lakes and large rivers; enters brackish water. **REMARKS:** Lake Simcoe Whitefish, *C. clupeaformis* subspecies, only in Lake Simcoe, ON, is unnamed genetically distinct form. **SIMILAR SPECIES:** See (1) Humpback Whitefish, *C. pidschian*, and (2) Atlantic Whitefish, *C. huntsmani*.

HUMPBACK WHITEFISH *Coregonus pidschian* **Not shown**

IDENTIFICATION: Nearly identical to Lake Whitefish, *C. clupeaformis*, but has *17–25*, usually 21–23, rakers on 1st gill arch. To 22 in. (56 cm). **RANGE:** Arctic basin in lower Anderson, Mackenzie, and nearby rivers, NT; coastal drainages, n. and w. AK; Yukon R. drainage, AK and YT. Also in n. Eurasia. Common. **HABITAT:** Lakes, large rivers, brackish

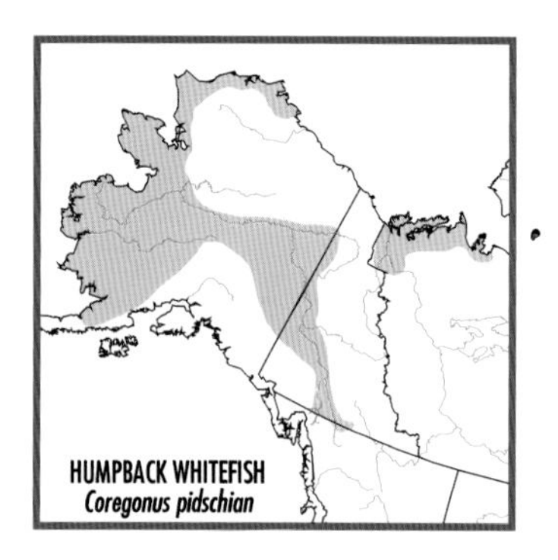

water; anadromous, migrates at least 800 mi. (1280 km) inland. **SIMILAR SPECIES:** (1) See Lake Whitefish, *C. clupeaformis* (Pl. 27).

ATLANTIC WHITEFISH *Coregonus huntsmani* Not shown

IDENTIFICATION: Similar to Lake Whitefish, *C. clupeaformis*, but has nearly *terminal mouth*, less compressed body, *91–100* lateral scales. To 16 in. (40 cm). **RANGE:** Known only from Tusket and Petite river drainages, s. NS. Protected as an *endangered species*. Rare. **HABITAT:** Nearshore coastal waters; open water of lakes and small to large rivers, often in current in rivers. Anadromous. **SIMILAR SPECIES:** (1) See Lake Whitefish, *C. clupeaformis* (Pl. 27).

NEXT 11 SPECIES: Cisco, *Coregonus artedi* complex, through Bloater, *C. hoyi*, are difficult to distinguish from one another using field characters. They differ from Humpback whitefishes (*C. clupeaformis* complex) in having *terminal mouth* or *projecting lower jaw* (except Shortjaw Cisco, *C. zenithicus*); *no* pronounced hump behind head, usually more than 35 rakers on 1st gill arch, dusky or black lower fins (except Arctic

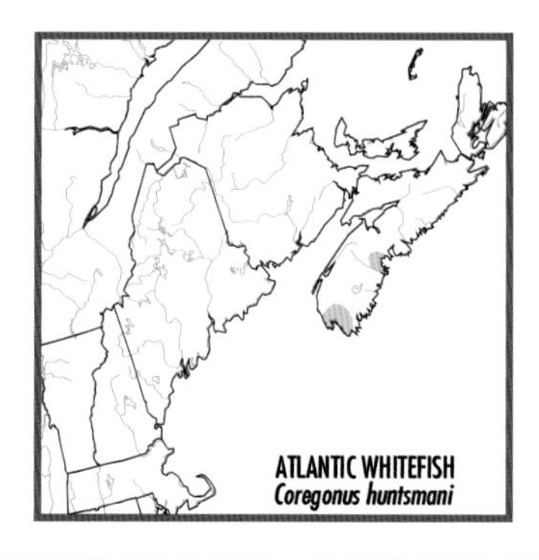

Cisco, *C. autumnalis*). Members of this complex are most easily distinguished from one by number of rakers on 1st gill arch.

CISCO *Coregonus artedi* **Pl. 27**

IDENTIFICATION: Long body, almost round in cross section, deepest at middle. Lower jaw equal to or projects slightly beyond upper jaw; upper jaw reaches front of pupil (Fig. 48). Symphyseal knob at tip of lower jaw. Often *dusky* or *black-tipped pelvic fins* on fish over 6 in. (15 cm). Pelvic fins *far back* on body; distance from snout to pelvic fin origin equal to distance from pelvic fin origin to caudal fin. Dark blue to green above; silver side. Has 36–64, usually *40–50 long slender rakers* on 1st gill arch. Has 62–94 lateral scales. To 22½ in. (57 cm). **RANGE:** Atlantic and Arctic basins from QC to NT and south to St. Lawrence-Great Lakes and upper Mississippi R. basins, n. OH, IL, and MN. Introduced elsewhere. Common. **HABITAT:** Open waters of lakes and large rivers; coastal waters of Hudson Bay. **SIMILAR SPECIES:** (1) See Deepwater Cisco, *C. johannae*, (2) Least Cisco, *C. sardinella*, (3) Arctic Cisco, *C. autumnalis,* and (4) Nipigon Cisco, *C. nipigon*. (5) Bloater, *C. hoyi*, and (6) Kiyi, *C. kiyi* (both Pl. 27), usually have *36–47 rakers* on 1st gill arch.

DEEPWATER CISCO *Coregonus johannae* **Not shown**

IDENTIFICATION: Nearly identical to Cisco, *C. artedi*, but has *25–36, usually 27–33, rakers* on 1st gill arch. To 16 in. (40 cm). **RANGE:** Lakes Huron and Michigan. Extinct. **HABITAT:** Deep water of lakes. **SIMILAR SPECIES:** Other species in Cisco, *C. artedi*, complex in Great Lakes *usually have 33 or more rakers* on 1st gill arch.

ARCTIC CISCO *Coregonus autumnalis* **Not shown**

IDENTIFICATION: Nearly identical to Cisco, *C. artedi*, but *lacks* dusky or black tips on pelvic fins. Light green to brown above; dusky dorsal and caudal fins, white lower fins. Has 41–48 long, slender rakers on 1st gill arch. To 25 in. (64 cm). **RANGE:** Arctic basin from Murchison R., NT,

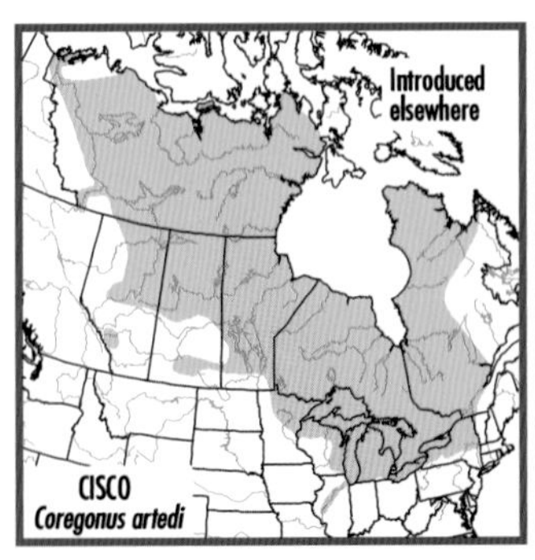

to Point Barrow, AK; up Mackenzie R. to BC. Also in n. Eurasia. Common. **HABITAT:** Large rivers, lakes, and brackish water; anadromous and landlocked forms. **SIMILAR SPECIES:** (1) See Cisco, *C. artedi* (Pl. 27), and (2) Bering Cisco, *C. laurettae*. (3) Least Cisco, *C. sardinella*, has higher, more falcate dorsal fin; dusky or black pelvic fins.

BERING CISCO *Coregonus laurettae* **Not shown**

IDENTIFICATION: Nearly identical to Arctic Cisco, *C. autumnalis*, but usually has *35–39 rakers* on 1st gill arch. To 19 in. (48 cm). **RANGE:** Arctic basin of AK from Beaufort Sea coast at Oliktok Point to Kenai Peninsula; up Yukon R. to Fort Yukon. Also in Asia. Common. **HABITAT:** Large rivers, lakes, and brackish water; anadromous and landlocked forms. **SIMILAR SPECIES:** (1) See Arctic Cisco, *C. autumnalis*.

LEAST CISCO *Coregonus sardinella* **Not shown**

IDENTIFICATION: Similar to Cisco, *C. artedi*, but distance from snout to pelvic fin origin equal to distance from pelvic fin origin to a point *on caudal peduncle*. Rather high, *falcate dorsal fin*. Brown to dark green above; black spots on back and dorsal fin (spots absent on landlocked populations). Has 41–53 long, slender rakers on 1st gill arch. Usually 78–98 lateral scales. To about 19 in. (47 cm). **RANGE:** Arctic basin from Bathurst Inlet and Cambridge Bay, NT, to Bristol Bay, AK. Also n. Eurasia. Common. **HABITAT:** Large rivers, lakes, and brackish water; anadromous and landlocked forms. **SIMILAR SPECIES:** (1) See Cisco, *C. artedi* (Pl. 27). (2) Arctic Cisco, *C. autumnalis,* has white pelvic fins; *lower*, less falcate dorsal fin.

NIPIGON CISCO *Coregonus nipigon* **Not shown**

IDENTIFICATION: Similar to Cisco, *C. artedi*, but has *45–70, usually more than 50, rakers* on 1st gill arch (Cisco has *usually fewer than 45* within range of Nipigon Cisco). To 19 in. (47 cm). **RANGE:** Uncertain (not mapped); in Lake Nipigon and adjacent lakes in ON, Lake Saganaga,

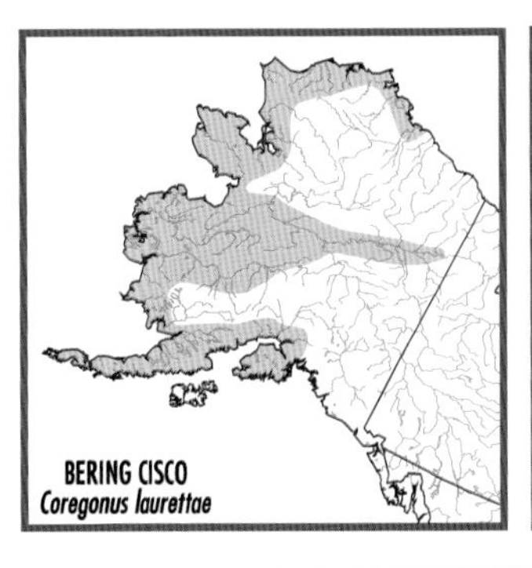

MN and ON, possibly other lakes in s.-cen. Canada. Locally common. **HABITAT:** Lakes; usually in relatively shallow water. **SIMILAR SPECIES:** See (1) Cisco, *C. artedi* (Pl. 27).

SHORTJAW CISCO *Coregonus zenithicus* **Not shown**

IDENTIFICATION: Stout lower jaw usually *shorter* than upper jaw, *lacks* symphyseal knob; upper jaw often reaches *middle of pupil* (Fig. 48). Body deepest near middle. Dark blue-green to pale pea green above; silver side; black on most fins (darkest on large individual). Usually 34–42 medium rakers (about length of gill filaments) on 1st gill arch. To 16 in. (40 cm). **RANGE:** Great Slave Lake, NT (Arctic Basin), southeast through Hudson Bay and Great Lakes basins (except Lakes Ontario and Erie). Declining in Great Lakes; uncommon elsewhere. Protected in Canada as a *threatened species.* **HABITAT:** Usually in deep (65–590 ft. [20–180 m]) water of large lakes; large rivers. **SIMILAR SPECIES:** (1) See Shortnose Cisco, *C. reighard*; and (2) Blackfin Cisco, *C. nigripinnis.* (3) Kiyi, *C. kiyi*, and (4) Bloater, *C. hoyi* (both Pl. 27), have slender lower jaw, *symphyseal knob* at tip of lower jaw, longer paired fins. (5) Cisco, *C. artedi* (Pl. 27), has slender lower jaw; lower jaw equal to or projecting beyond upper jaw.

SHORTNOSE CISCO *Coregonus reighardi* **Not shown**

IDENTIFICATION: Nearly identical to Shortjaw Cisco, *C. zenithicus*, but has *shorter snout and fins.* To 15 in. (38 cm). **RANGE:** Lakes Ontario, Huron, and Michigan. Possibly extinct; protected in Canada as a *threatened species.* **HABITAT:** Deep water of lakes. **SIMILAR SPECIES:** (1) See Shortjaw Cisco, *C. zenithicus.*

BLACKFIN CISCO *Coregonus nigripinnis* **Not shown**

IDENTIFICATION: Similar to Shortjaw Cisco, *C. zenithicus*, but has body *deepest under nape*, jaws *equal* in length, upper jaw reaches *front* of

pupil (Fig. 48). Dusky upper lip. Blue-green to blue-black above; silver side. Usually *blue-black fins*. Usually *46–50 long rakers* (longest usually longer than gill filaments) on 1st gill arch. To 15¼ in. (39 cm). **RANGE:** Endemic to Lakes Nipigon, Huron, and Michigan. Rare; extant only in Lake Nipigon. Protected in Canada as a *threatened species*. **HABITAT:** Formerly in deep waters (295–525 ft. [90–160 m]) of Lakes Michigan and Huron; in 6–325 ft. (2–100 m) in Lake Nipigon. **SIMILAR SPECIES:** (1) See Shortjaw Cisco, *C. zenithicus*. (2) Cisco, *C. artedi* (Pl. 27), has more elongate body *deepest at middle*, slender lower jaw, symphyseal knob at tip of lower jaw. (3) Kiyi, *C. kiyi*, and (4) Bloater, *C. hoyi* (both Pl. 27), have lower jaw projecting slightly beyond upper jaw, symphyseal knob at tip of lower jaw.

KIYI *Coregonus kiyi* Pl. 27

IDENTIFICATION: Pointed snout; lower jaw projecting slightly beyond upper jaw (Fig. 48); slender, darkly pigmented lower jaw; *black upper lip*; symphyseal knob at tip of lower jaw. *Long paired fins*; pelvic fin usually reaches to anus or beyond. *Large eye* nearly equal to snout length. Body deepest under nape. Silver with faint pink to purple iridescence; black edges on dorsal and caudal fins. Usually *36–41 medium rakers* (about length of gill filaments) on 1st gill arch. To 13¾ in. (35 cm). **RANGE:** Endemic to Great Lakes (except Lake Erie). Common in Lake Superior, rare in Lake Michigan, possibly extirpated from Lakes Huron and Ontario. **HABITAT:** Open water; generally at depths of 325–590 ft. (100–180 m) in Lake Superior. **SIMILAR SPECIES:** (1) See Bloater, *C. hoyi* (Pl. 27). (2) Cisco, *C. artedi* (Pl. 27), has more elongate body, rounder in cross section, deepest at middle; usually 40–50 long rakers on 1st gill arch.

BLOATER *Coregonus hoyi* Pl. 27

IDENTIFICATION: Similar to Kiyi, *C. kiyi*, but has *less dusky upper lip, smaller eye* (eye usually less than snout length), body deepest at middle, pelvic fin seldom reaching anus, usually 40–47 long rakers (longer than longest gill filament) on 1st gill arch. To 14½ in. (37 cm). **RANGE:** Endemic to Great Lakes (except Lake Erie) and Lake Nipigon. Probably extirpated from Lakes Ontario and Nipigon, rare in Lake Michigan, declining in Lakes Superior and Huron. **HABITAT:** Large lakes; generally at depths of 100–620 ft. (30–190 m). **REMARKS:** When brought up from depths, the gas bladder expands, giving this fish a bloated appearance; hence, its name. **SIMILAR SPECIES:** (1) See Kiyi, *C. kiyi* (Pl. 27).

INCONNU *Stenodus leucichthys* Fig. 50

IDENTIFICATION: *Large, wide mouth* extends to rear of pupil; *lower jaw projects* beyond upper jaw; *2 small flaps* of skin between nostrils.

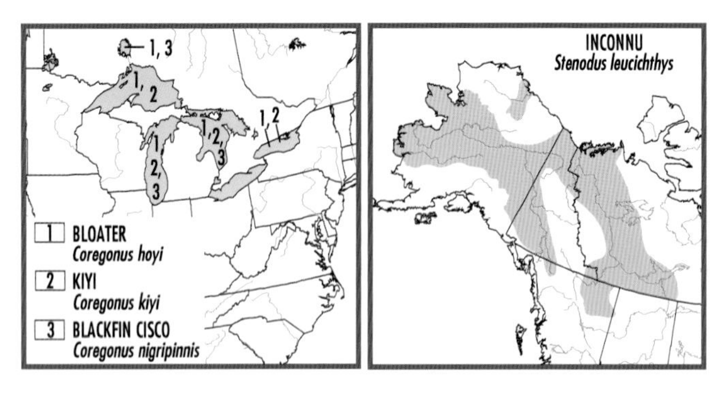

Forked caudal fin. High, pointed dorsal fin; 11–19 rays. Green, blue, or pale brown above; silver white side and below, dusky-edged dorsal and caudal fins. Has 90–115 lateral scales; 17–24 rakers on 1st gill arch, fewer than 18 on lower limb. Small, velvetlike bands of teeth on both jaws. To 55 in. (140 cm). **RANGE:** Arctic basin from Anderson R., NT, to Kuskokwim R. (Bering Sea basin), AK. Upstream in Mackenzie and Yukon river drainages to n. BC. Also in Eurasia. Abundant. **HABITAT:** Open water of small to large rivers. Anadromous near coasts. **SIMILAR SPECIES:** Whitefishes (Pl. 27) have *smaller mouth*, usually more than 24 rakers (and more than 20 on lower limb) on 1st gill arch.

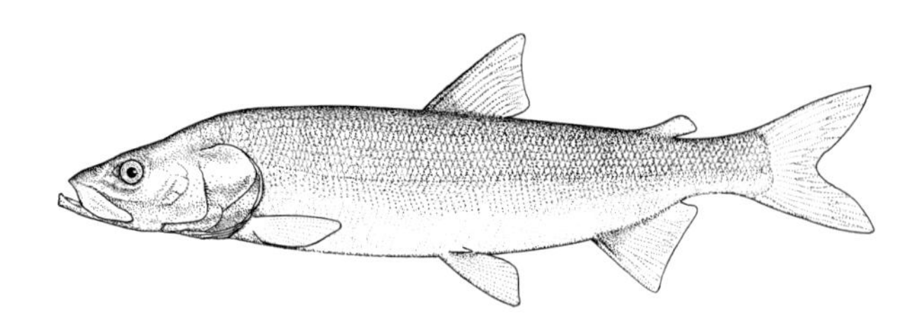

Fig. 49. Inconnu.

PROSOPIUM

GENUS CHARACTERISTICS: *One flap* of skin between nostrils (Fig. 47); *pinched snout* (when viewed from above); small, subterminal mouth, body round in cross section; forked caudal fin; *short, stout rakers*, 13–45 on 1st gill arch; no teeth on jaws; 10–14 dorsal rays; *parr marks* on young; silver to silver white.

BONNEVILLE CISCO *Prosopium gemmifer* Pl. 27

IDENTIFICATION: *Long, sharply pointed snout* (Fig. 50); long maxilla. *Long, slender body*; forked caudal fin. Pale moss green above; pearl iridescence on silver side. Brassy side on large male. Has 69–76 lateral scales; 10–11 dorsal rays; 9–12 anal rays; 37–45 rakers on 1st gill arch. To 8½ in. (22 cm). **RANGE:** Endemic to Bear Lake, se. ID and n. UT. Introductions elsewhere in western states apparently unsuccessful. Common. **HABITAT:** Open water. Widely scattered when lake is colder than 58°F (14°C); in deeper, colder water in warm season. **SIMILAR SPECIES:** (1) Other *Prosopium* species (Pl. 27; Fig. 50) have *short blunt snout*, 26 or fewer rakers on 1st gill arch.

NEXT 5 SPECIES: Short blunt snout (Fig. 50).

PYGMY WHITEFISH *Prosopium coulterii* Not shown

IDENTIFICATION: *Large eye*; diameter about same as snout length (Fig. 50). Transparent membrane surrounding eye has distinct notch below rear edge of pupil. Large scales; *54–70 lateral*, 31–40 around body, 16–20 around caudal peduncle. Dark brown above; silver or white below. Young has 7–14 oval parr marks on side. Usually 11–12 dorsal rays, 12–14 anal rays; 13–20 rakers on 1st gill arch; 13–33 pyloric caeca. To 11 in. (28 cm). **RANGE:** Disjunct. Lake Superior; Chignik, Naknek, and Wood river drainages, sw. AK; Copper R. drainage, se. AK; Great Bear Lake, NT; upper Athabasca R. drainage, AB; Yukon R. drainage, YT, to Columbia R. drainage, w. MT, and Puget Sound and Olympic Peninsula, WA. Abundant in west; uncommon in Lake Superior. Also in Russia. **HABITAT:** Swift, cold streams; below 20 ft. (6 m) in lakes; in Lake Superior at depths of 60–220 ft. (18–70 m). **SIMILAR SPECIES:** (1) Mountain Whitefish, *P. williamsoni* (Pl. 27), and (2) Round Whitefish, *P. cylindraceum* (Fig. 50), have *smaller eye* (diameter less than snout

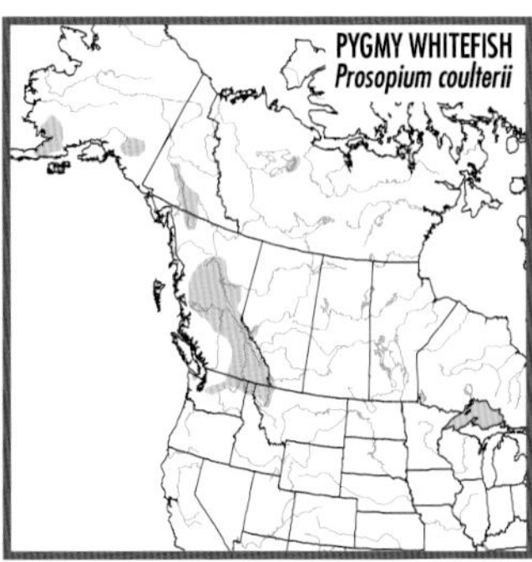

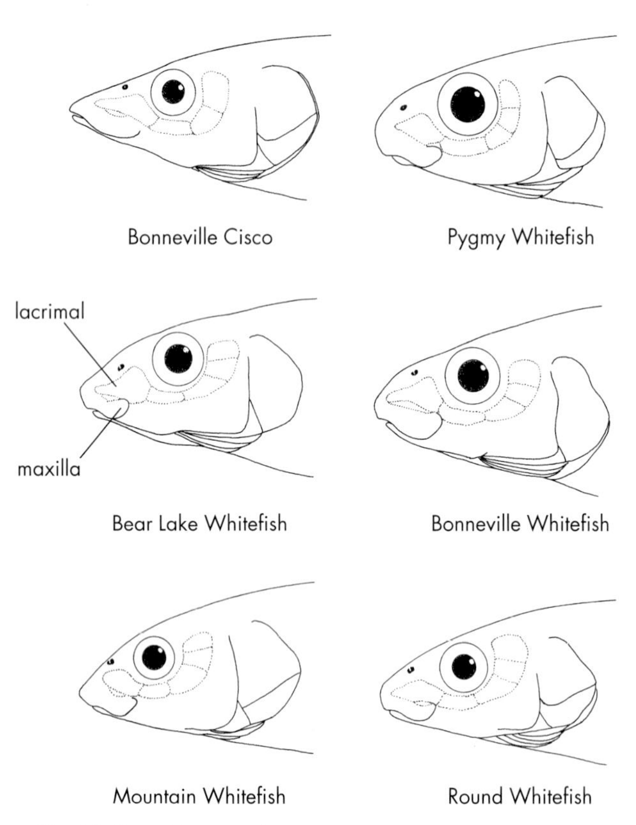

Fig. 50. *Prosopium* species.

length), *more than 70 lateral scales*, 20 or more scales around caudal peduncle, 50 or more pyloric caeca; reach 22 in. (56 cm).

BEAR LAKE WHITEFISH *Prosopium abyssicola* **Not shown**

IDENTIFICATION: *Short, deep maxilla* not reaching end of lacrimal (Fig. 50). Silver white; *no dark spots* on body. Has 67–78 (usually 69–74) lateral scales; *38–43 scales around body*; usually 11 dorsal rays, 10 anal rays; 19–23 rakers on 1st gill arch. To 11 in. (28 cm). **RANGE:** Endemic to Bear Lake, se. ID and n. UT. Common. **HABITAT:** Usually in deep water; rarely near shore. **SIMILAR SPECIES:** (1) See Bonneville Whitefish, *P. spilonotus* (Fig. 50). (2) Mountain Whitefish, *P. williamsoni* (Pl. 27), has maxilla reaching *end* of lacrimal, 76–89 lateral scales, usually 12–13 dorsal rays.

BONNEVILLE WHITEFISH *Prosopium spilonotus* Not shown

IDENTIFICATION: Similar to Bear Lake Whitefish, *P. abyssicola*, but has somewhat blunter snout; maxilla *reaches end of lacrimal* (Fig. 50); usually 76–86 lateral scales, *44 or more scales* around body. Upper jaw usually *more* than½ length of dorsal fin base. Young has *dark spots* on body (to 10 in. [25 cm]). Usually 10–12 dorsal rays, 9–11 anal rays, 18–22 rakers on 1st gill arch. To 22 in. (56 cm). **RANGE:** Endemic to Bear Lake, se. ID and n. UT. Common. **HABITAT:** Frequently at depths of 40–100 ft. (12–30 m); more often in shallow water than are Bonneville Cisco, *P. gemmiter*, and Bear Lake Whitefish, *P. abyssicola*. **SIMILAR SPECIES:** See (1) Bear Lake Whitefish, *P. abyssicola* (Fig. 50), and (2) Mountain Whitefish, *P. williamsoni* (Pl. 27).

MOUNTAIN WHITEFISH *Prosopium williamsoni* Pl. 27

IDENTIFICATION: Similar to Bonneville Whitefish, *P. spilonotus*, but has 12–13 dorsal rays, 11–13 anal rays; upper jaw *less* than½ length of dorsal fin base (Fig. 50). Olive green to dark brown above; silver or white below; dusky dorsal fin. Young has 7–11 large oval dark parr marks. Has 76–89 lateral scales; 20–23 scales around caudal peduncle; 19–26 rakers on 1st gill arch; 50–146 pyloric caeca. To 22½ in. (57 cm). **RANGE:** Mackenzie R. drainage (Arctic basin), NT, south in Pacific, Hudson Bay, and upper Missouri R. basins, to Truckee R. drainage, NV, and Sevier R. drainage, UT. Common. **HABITAT:** Lakes (to depth of at least 30 ft. [10 m]) and fast clear or silty streams. **SIMILAR SPECIES:** See Fig. 50. (1) See Bonneville Whitefish, *P. spilonotus*. (2) Bear Lake Whitefish, *P. abyssicola*, has short, deep maxilla not reaching end of lacrimal, usually 69–74 lateral scales, 11 dorsal rays, 10 anal rays.

ROUND WHITEFISH *Prosopium cylindraceum* Not shown

IDENTIFICATION: Brown to bronze above, *dark-edged scales*; silver white side; amber lower fins; dusky dorsal and caudal fins. Young (2–3 in. [5.0–7.5 cm]) are silver, have 2 or more rows of *black spots* on

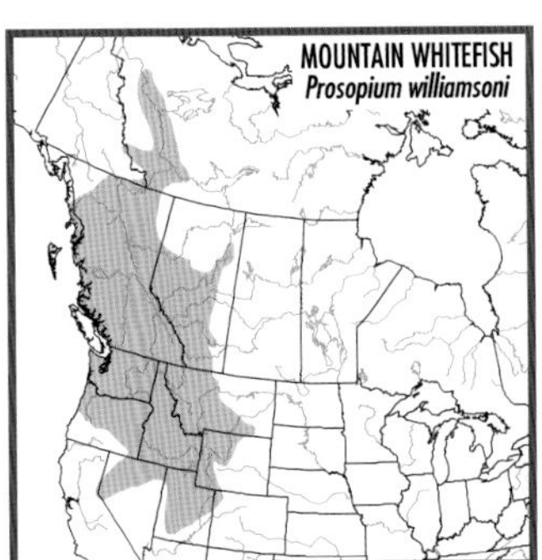

side that may coalesce with row of black spots on back. Fairly pointed snout (Fig. 50). Has 76–89 lateral scales; usually 22–24 scales around caudal peduncle; 14–21 rakers on 1st gill arch; 50–117 pyloric caeca. To 22 in. (56 cm). **RANGE:** Arctic and Pacific drainages from western shore of Hudson Bay to AK and n. BC; Arctic and Atlantic drainages from QC and ON to CT; St. Lawrence-Great Lakes basin except Lake Erie. Also in Arctic basin of Asia. Common. **HABITAT:** Shallow areas (usually less than 150 ft. [46 m]) of lakes and clear streams; rarely in brackish water. **SIMILAR SPECIES:** (1) Mountain Whitefish, *P. williamsoni* (Pl. 27), *lacks* dark-edged scales (may be dusky-edged); has larger, less distinct spots on side of young, 19–26 rakers on 1st gill arch. (2) Pygmy Whitefish, *P. coulterii* (Fig. 50), has larger eye, 54–70 lateral scales, 16–20 scales around caudal peduncle, 13–33 pyloric caeca; reaches 11 in. (28 cm).

ARCTIC GRAYLING *Thymallus arcticus* Pl. 27

IDENTIFICATION: Rows of red to green spots on *huge purple to black* (on adult) *dorsal fin*; 17–25 rays. *Small mouth*; small teeth on jaws. *Forked caudal fin*. Fairly large (77–103 lateral) scales. Dark blue-gray above; scattered black spots (darkest on young) on silver gray to blue, sometimes pink, side; sometimes a black stripe on lower side between paired fins; orange-yellow stripes on pelvic fin; other fins dusky to dark gray. Narrow, vertical parr marks on young. To 30 in. (76 cm). **RANGE:** Disjunct. Arctic and Pacific basins from Hudson Bay to AK and south to cen. AB; upper Missouri R. drainage, MT; formerly in Great Lakes basin, MI. Introduced elsewhere in w. U.S. Also in Asia. Locally common. **HABITAT:** Open water of clear, cold, medium to large rivers and lakes; enters rocky creeks to spawn (May–June).

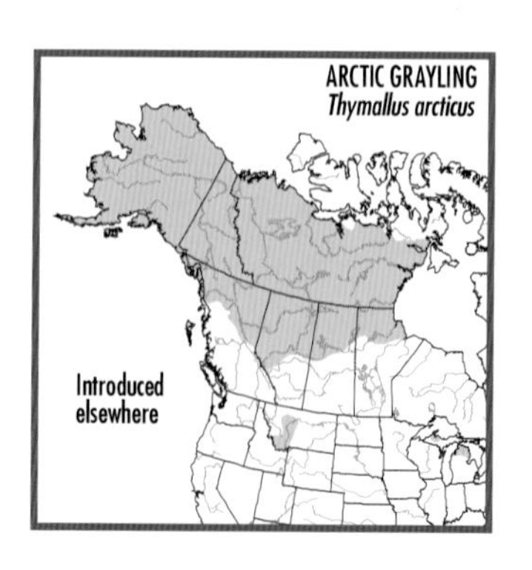

GENUS CHARACTERISTICS: Troutlike body. *Light* (pink, red, or cream) *spots* on body; *minute* (about 105–200 lateral) *scales*; *snow-white leading edge* on lower fins; scales along lateral line *smaller* than surrounding scales with little or no overlap with scales in front and behind; teeth on front (head) of vomer, not on shaft; 8–12 dorsal rays; 7–12 anal rays; dark parr marks on young.

LAKE TROUT *Salvelinus namaycush* Pl. 27

IDENTIFICATION: *Deeply forked caudal fin.* Many small, often bean-shaped, *cream or yellow spots* on dark green to gray head, body, and dorsal and caudal fins. Narrow white edge on orange-red lower fins. (Individuals in large lakes may be silver overall.) No bright orange or red on body. Breeding male has dark stripe along side. Has 7–12 narrow, often interrupted, parr marks on young (Fig. 46); 90–210 pyloric caeca. To 49½ in. (126 cm). **RANGE:** Widely distributed in Atlantic, Arctic, and Pacific basins from n. Canada and AK south to New England, Great Lakes, and n. MT. Introduced widely outside native range, including to w. U.S., New Zealand, S. America, and Sweden. Common in north; uncommon in Great Lakes except where maintained by artificial propagation. **HABITAT:** Restricted to relatively deep lakes in southern part of range; in shallow and deep waters of northern lakes and streams. **REMARKS:** Exceedingly fat form of Lake Trout from deep waters of Lake Superior is called a Siscowet. Hybrid between Lake Trout and Brook Trout, *S. fontinalis*, is called a Splake. Splakes have been successfully introduced into many parts of N. America. **SIMILAR SPECIES:** (1) Other *Salvelinus* species (Pl. 27) *lack* deeply forked caudal fin, have orange or red on body. (2) Pacific salmons and trouts, *Oncorhynchus* species (Pls. 28–30), have black spots on body or fins.

BROOK TROUT *Salvelinus fontinalis* Pl. 27

IDENTIFICATION: *Slightly forked* to nearly straight-edged caudal fin. *Light green or cream wavy lines or blotches* on back and dorsal fin, broken into spots on side; *blue halos* around *pink or red spots* on side. Olive to black above; black lines on caudal fin; black line behind white edge on red lower fins. Anadromous individual is dark green above; has pale pink spots on silver side. Breeding male is brilliant orange or red below, has black belly. Has 8–10 regularly arranged parr marks on side of young (Fig. 46). Has 14–22 (usually 16–21) rakers with marginal teeth on 1st gill arch; 23–55 pyloric caeca. To 28 in. (70 cm). **RANGE:** Native to most of e. Canada from NL to western side of Hudson Bay; south in Atlantic, Great Lakes, and Mississippi R. basins to MN and (in Appalachian Mts.) n. GA; headwaters of Chattahoochee R. (Gulf basin). Introduced widely in N. America and temperate regions of other continents. Common. **HABITAT:** Clear, cool, well-oxygenated creeks, small to medium rivers, and lakes. Some populations are anadromous. **REMARKS:** Aurora Trout, *S. f. timagamiensis,* a silvery form lacking characteristic green or cream wavy lines and maintained only as hatchery stocks in lakes in Temiskaming District of ON, is protected as an *endangered subspecies*. Extinct Silver Trout, *S. f. agassizii*, of Dublin and nearby Ponds, NH, also lacked wavy lines of Brook Trout. "Coaster" is a Brook Trout that moves into lakes to feed and back to streams to spawn. **SIMILAR SPECIES:** (1) Lake Trout, *S. namaycush* (Pl. 27), has *deeply forked caudal fin, cream spots* on dark green or gray body. (2) Other *Salvelinus* species (Pl. 27) *lack* wavy green or cream lines on dorsal fin.

ARCTIC CHAR *Salvelinus alpinus* Pl. 27

IDENTIFICATION: Slightly forked caudal fin. *Pink to red spots* (some larger than pupil of eye) *on back and side*. No dark, wavy lines on dorsal and caudal fins. Color highly variable; usually green to brown

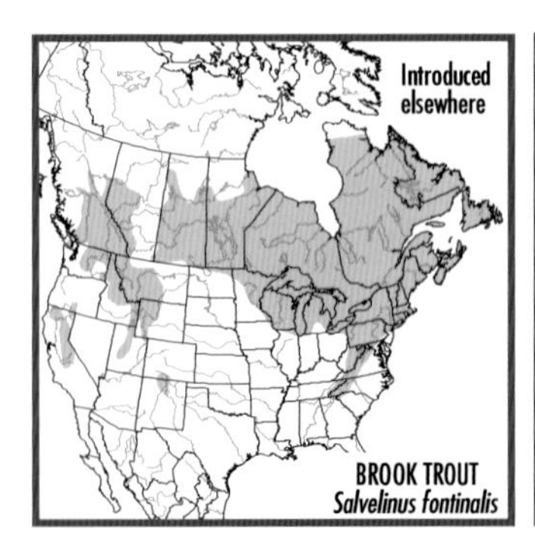

above, pink to red spots on side, lighter below. Silver overall in landlocked, nonspawning adult; some landlocked populations retain brilliant spawning colors year-round. Breeding male is dark green to blue-green above; has silver blue with scattered orange or red spots on side, white or brilliant orange-red below. Adult in estuaries is steel blue above; silver blue on side with many large red, pink, or cream spots. Fins similar in color to body; white front edge on lower fins. Young has 10–15 parr marks scattered on side (Fig. 46). Usually *20–30 rakers* on 1st gill arch; *35–50 pyloric caeca*; no marginal teeth on rakers. To 38 in. (96 cm). **RANGE:** Circumpolar; has most northern distribution of any N. American freshwater fish. Coastal areas in Atlantic, Arctic, and Pacific drainages from NL to AK; south along Atlantic Slope to ME. Seldom more than 186 mi. (300 km) from ocean. Also in Eurasia. Locally abundant. **HABITAT:** Deep runs and pools of medium to large rivers; lakes. Anadromous populations enter rivers to spawn in fall and winter. **REMARKS:** Three subspecies in N. America. *S. a. oquassa*, in n. New England and se. Canada, usually has 20–21 rakers on 1st gill arch, 35–40 pyloric caeca. *S. a. erythrinus*, in high Arctic region from n. QC and Labrador to n. AK, usually has 25–30 long rakers on 1st gill arch, 40–50 pyloric caeca. *S. a. taranetzi*, from nw. AK to Kodiak and Shumagin islands, usually has 23–25 short rakers on 1st gill arch, 40–50 pyloric caeca. **SIMILAR SPECIES:** (1) See Dolly Varden, *S. malma* (Pl. 27).

DOLLY VARDEN *Salvelinus malma* Pl. 27

IDENTIFICATION: Similar to Arctic Char, *S. alpinus*, but usually has *14–21 rakers* on 1st gill arch, usually *25–30 pyloric caeca* (rarely to 40). Color variable. In fresh water, olive green to gray above; yellow, orange, or red spots (largest spots usually *smaller* than pupil of eye) on side. Anadromous individual usually dark blue above, pale or pink spots on silver side. Breeding male is green-black above, has bright

red-orange lower side, white front edge followed by black or red line on lower fins; breeding female is similar but less brilliant. Young has 8–12 parr marks (Fig. 46). Usually 11–12 branchiostegal rays, 6 mandibular pores. To about 25 in. (63 cm); many dwarf populations under 12 in. (31 cm). **RANGE:** Arctic and Pacific drainages from lower Mackenzie R., NT, to Puget Sound and Quinault R., WA. Also in Asia. Common. **HABITAT:** Deep runs and pools of creeks and small to large rivers; lakes. Typically anadromous, but many populations landlocked. Anadromous individuals may spend 2–3 years at sea, evidently near shore; migrate upstream usually in fall, spawn in spring. **REMARKS:** Two subspecies in N. America. *S. m. malma*, from Mackenzie R. to Alaska Peninsula, usually has 20–24 rakers on 1st gill arch, 66–70 vertebrae. *S. m. lordi*, south of Alaska Peninsula, usually has 16–19 rakers on 1st gill arch, 62–65 vertebrae. **SIMILAR SPECIES:** See (1) Arctic Char, *S. alpinus*, and (2) Bull Trout, *S. confluentus* (both Pl. 27).

BULL TROUT *Salvelinus confluentus* Pl. 27

IDENTIFICATION: Similar to Dolly Varden, *S. malma*, but has flatter and *longer head*—length usually *less than* 4 times into standard length (more than 4 times in Dolly Varden); eye *higher* on head; usually 14–20 rakers on 1st gill arch, *large marginal teeth on rakers*; usually *13–15 branchiostegal rays, 7–9 mandibular pores*, 22–34 pyloric caeca. To about 3 ft. (1 m). **RANGE:** Coastal and montane streams of Arctic, Pacific, and Missouri R. drainages from upper Yukon, Mackenzie, and Athabasca rivers, YT, BC, and AB, to headwaters of Columbia R. drainage, n. NV, and (historically) McCloud R. drainage, n. CA. Locally common; rare in southern part of range and protected as a *threatened species* in U.S. **HABITAT:** Deep pools in large cold rivers and lakes; usually in montane areas with snow and glaciers. Rarely anadromous. **SIMILAR SPECIES:** (1) See Dolly Varden, *S. malma* (Pl. 27). (2) Arctic Char, *S. alpinus* (Pl. 27), *lacks* marginal teeth on gill rakers; usually has 20–30 rakers on 1st gill arch, 35–50 pyloric caeca.

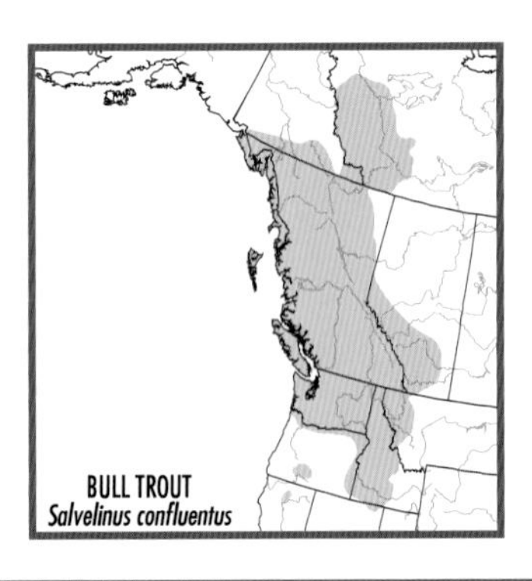

ATLANTIC SALMON *Salmo salar* Pl. 28

IDENTIFICATION: *Black spots* on head and body; *2–6 large spots* on gill cover. *X-shaped spots* on body of adult; usually no rows of large black spots on caudal fin. No white edge on pelvic and anal fins. Upper jaw reaches below *center of eye*; behind eye only in large male. At sea, brown, green, or blue above, silver on side. Darker in fresh water; breeding adult is bronze and dark brown (often with red spots). *Hooked lower jaw* on large male. Has 8–11 narrow parr marks along side of young (Fig. 46); *1 red spot* between each pair. Has 109–121 lateral scales; usually 12 branchiostegal rays, 11 dorsal rays; teeth on front (head) and shaft of vomer. To 55 in. (140 cm). **RANGE:** Atlantic drainages from n. QC to CT and NY; inland to Lake Ontario (where now extirpated). Widely introduced elsewhere but seldom successfully. Also e. Atlantic drainages from Arctic Circle to Portugal. Locally common; depleted or extirpated from western and southern parts of range. Protected as an *endangered species*. **HABITAT:** Rocky runs and pools of small to large rivers; lakes. Atlantic Salmon spawns in fall in rocky streams; most young remain in fresh water for 2–3 years, then migrate to ocean for 1 or more years before returning to fresh water to spawn. Unlike Pacific salmon, Atlantic Salmon does not die after spawning but returns to ocean. **REMARKS:** Few animals have attracted as much attention as Atlantic Salmon, a game fish par excellence and an overly exploited commercial species. It was one of the 1st N. American fishes to disappear from parts of its range because of our careless use of natural resources. **SIMILAR SPECIES:** (1) Brown Trout, *S. trutta* (Pl. 28), usually *lacks* X-shaped spots on body, usually has *red spots* on adult, red or orange adipose fin; upper jaw reaches *well beyond eye* in fish over 5 in. (13 cm). (2) Brook Trout, *Salvelinus fontinalis*, and (3) Lake Trout, *Salvelinus namaycush* (both Pl. 27), have *light pink, red, or cream spots* on dark body; white edge on pelvic and anal fins.

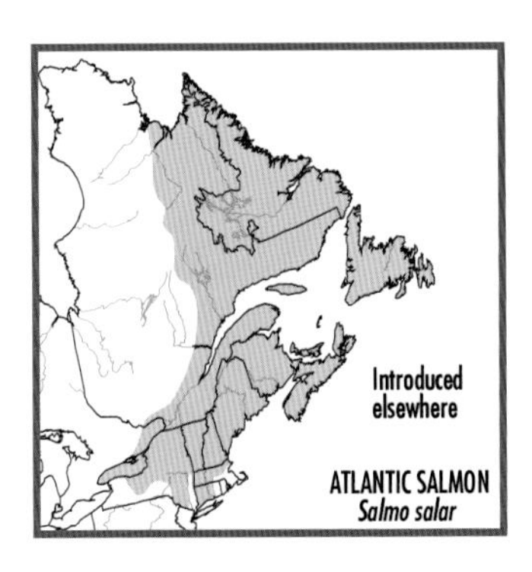

BROWN TROUT *Salmo trutta* Pl. 28

IDENTIFICATION: *Red and black spots on head and body, large black spots on gill cover*; usually no rows of black spots on caudal fin. No white edge on pelvic or anal fins. Upper jaw reaches below center of eye in 5-in. (13-cm) fish, *well beyond eye* in larger fish. In streams, olive to dark brown above; silver sheen on yellow-brown side, white to yellow below; bold black spots on head, back, dorsal, and adipose fins, and extending below lateral line on side (spots often surrounded by *pale halos*); rust red spots on side; usually orange or red adipose fin. Silver overall, sometimes with X-marks above in large lakes and ocean. Breeding male has hooked lower jaw, rounded anal fin, red lower side; female has falcate anal fin. Has 9–14 short, narrow parr marks along side of young (Fig. 46); few red spots along lateral line; 120–130 lateral scales; usually 10 branchiostegal rays, 9 dorsal rays. To 40½ in. (103 cm). **RANGE:** Native to Europe, n. Africa, and w. Asia. Introduced to N. America in 1883 (NY and MI) and now widely stocked throughout s. Canada and much of U.S. Locally common. **HABITAT:** Most stockings have been into cool, high-gradient streams and cold lakes. In streams, adults live in pools; young occupy pools and riffles. Some sea-run populations have become established. **SIMILAR SPECIES:** (1) Atlantic Salmon, *S. salar* (Pl. 28), *lacks* red spots on body, has no red or orange on adipose fin; upper jaw reaches only to below center of eye (except on large male).

ONCORHYNCHUS

GENUS CHARACTERISTICS: Salmon- or troutlike body. Tiny (about 100–200 lateral) scales; *no* snow-white leading edge on lower fins; scales along lateral line *as large* as or larger than scales in adjacent rows, overlap with scales in front and behind; teeth on front (head) and shaft of vomer; 8–16 dorsal rays; 8–12 anal rays in trouts, 13–19 in salmons; dark parr marks on young (except Pink Salmon, *O. gorbuscha*); to 58 in. (147 cm).

SOCKEYE SALMON *Oncorhynchus nerka* Pl. 28

IDENTIFICATION: *No large black spots* on back or caudal fin. At sea, metallic blue-green above; silver below. Breeding male has *green head, bright red body*, yellow-green caudal fin, white lower jaw, hooked upper jaw; female is similar but often has green or yellow blotches on side. Has 28–40 long, thin, serrated, closely spaced rakers and rudiments on 1st gill arch. Has 8–14 elliptical to oval parr marks along side of young (Fig. 46). To 33 in. (84 cm). **RANGE:** Arctic and Pacific drainages from Point Hope, AK, to Columbia R. drainage, OR and ID. Landlocked populations in AK, YT, BC, WA, OR, and CA. Also in ne. Asia. Widely stocked, but most transplants unsuccessful. Common in north; rare south of Columbia R. drainage; some populations protected as *endangered or threatened* in U.S. **HABITAT:** Open ocean; lakes; migrates up coastal streams to spawn. **REMARKS:** Landlocked lake populations known as Kokanee. **SIMILAR SPECIES:** (1) Chum Salmon, *O. keta* (Pl. 28), also lacks black spots, has 18–28 short, stout, smooth rakers and rudiments on 1st gill arch.

CHUM SALMON *Oncorhynchus keta* Pl. 28

IDENTIFICATION: *No large black spots* on back or caudal fin. At sea, metallic blue above; silver below. Breeding individual has *red, brown, and black bars and blotches* on *dull green side*, white-tipped (brightest on male) anal and pelvic fins; male has hooked upper jaw. Front teeth of males become more enlarged than in other salmons. Has 6–14 narrow, short parr marks along side of young (Fig. 46). Has 18–28 short, stout, smooth rakers and rudiments on 1st gill arch. To 40 in. (102 cm). **RANGE:** Arctic and Pacific drainages from Anderson and Mackenzie rivers, NT, through all of AK south to (historically) Sacramento R. drainage, CA. Also in ne. Asia. Once common, now protected as a *threatened species* in U.S. **HABITAT:** Anadromous; ocean and coastal streams. **SIMILAR SPECIES:** (1) Sockeye Salmon, *O. nerka* (Pl. 28), has no white tips on fins, usually more than 30 long, thin, serrated rakers and rudiments on 1st gill arch. (2) Other salmons (Pl. 28) have *large black spots* on back and caudal fin.

CHINOOK SALMON *Oncorhynchus tshawytscha* Pl. 28

IDENTIFICATION: *Largest salmon*; salmon over 30 lb. (14 kg) are almost always this species. *Irregular black spots* on back, *both lobes of caudal fin*, dorsal and adipose fins. *Gums black* at base of teeth. At sea, blue, green, or gray above; silver below. Small male often dull yellow; larger male often blotchy, *dull red* on side. Breeding individual dark olive-brown to purple. Has 6–12 large parr marks along side of young

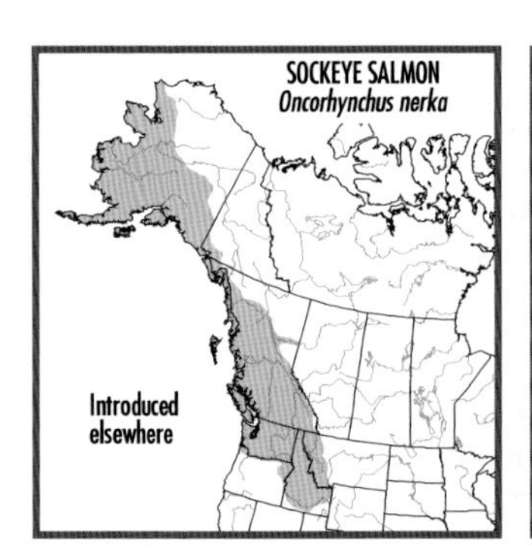

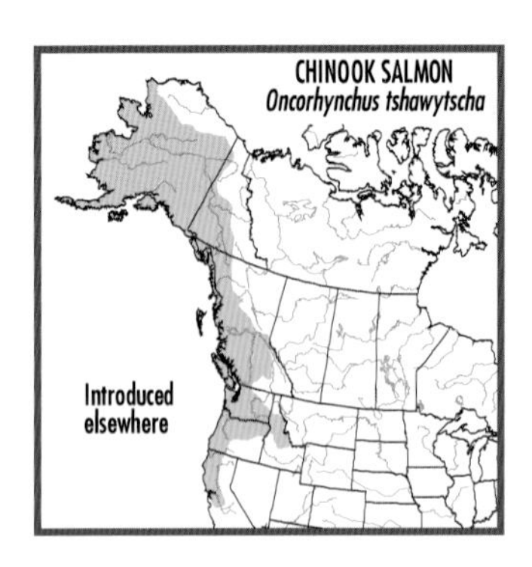

(Fig. 46). To 58 in. (147 cm); rarely over 50 lb. (23 kg); 1 specimen from AK reportedly weighed 126 lb. (57 kg). **RANGE:** Arctic and Pacific drainages from Point Hope, AK, to San Joaquin and King rivers, CA. Stocked outside native range, somewhat successfully in Great Lakes. Also in ne. Asia. Uncommon, least abundant of Pacific salmons; some populations protected as *endangered or threatened* in U.S. **HABITAT:** Anadromous; ocean and coastal streams. **SIMILAR SPECIES:** See Pl. 28. (1) Sockeye Salmon, *O. nerka*, and (2) Chum Salmon, *O. keta*, have no large black spots. (3) Coho Salmon, *O. kisutch*, has no black spots on *lower lobe* of caudal fin; *gums white* at base of teeth. (4) Pink Salmon, *O. gorbuscha*, has large *oval black spots* on back and caudal fin; does not exceed 30 in. (76 cm).

COHO SALMON *Oncorhynchus kisutch* Pl. 28

IDENTIFICATION: *Black spots* on *back* and *upper lobe* of caudal fin. *Gums white* at base of teeth. At sea, metallic blue above; silver below. Breeding male has dusky green-brown back and head, *red side*, hooked upper jaw; female has bronze to *pink-red side*. Has 8–12 narrow parr marks along side of young (Fig. 46). To 38½ in. (98 cm). **RANGE:** Arctic and Pacific drainages from Point Hope, AK, to San Lorenzo R., CA; infrequently as far south as Chamalu Bay, Baja California. Also in ne. Asia. Stocked outside native range, somewhat successfully in Great Lakes. Generally uncommon; rare south of cen. CA; some populations protected as *endangered or threatened* in U.S. **HABITAT:** Anadromous; ocean and coastal streams. **SIMILAR SPECIES:** (1) Chinook Salmon, *O. tshawytscha* (Pl. 28), has black spots on *both lobes* of caudal fin, *gums black* at base of teeth.

PINK SALMON *Oncorhynchus gorbuscha* Pl. 28

IDENTIFICATION: *Large*, mostly *oval*, *black spots* on back and *both lobes* of caudal fin. At sea, metallic blue or blue-green above; silver

below. Breeding male develops *humped back* and hooked upper jaw; *pink to brown stripe along side*. Female lacks hump, markedly hooked jaw. Young *lack* parr marks (Fig. 46). Usually more than 169 lateral scales (usually fewer than 155 in other salmons). To 30 in. (76 cm). **RANGE:** Arctic and Pacific drainages from Mackenzie R. delta, NT, to (historically) Sacramento R. drainage, CA. Also in ne. Asia. Introduced in upper Great Lakes and e. NL. Generally common but declining; no longer in Sacramento R. drainage. **HABITAT:** Anadromous; ocean and coastal streams. **SIMILAR SPECIES:** (1) Coho Salmon, *O. kisutch* (Pl. 28), *lacks* black spots on lower lobe of caudal fin. (2) Chinook Salmon, *O. tshawytscha* (Pl. 28), has small *irregular black spots* on back and caudal fin.

CUTTHROAT TROUT *Oncorhynchus clarkii* Pl. 29

IDENTIFICATION: *Red "cutthroat" mark* under lower jaw; *many black spots on body. No or faint red stripe* on side (some populations have narrow copper-orange stripe). Upper jaw reaches *well behind* eye in

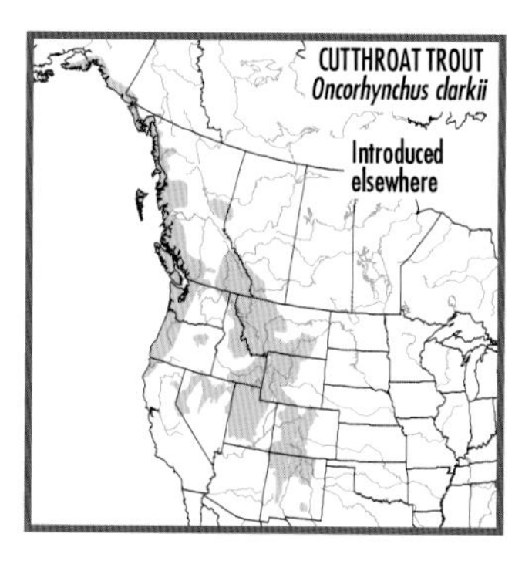

adult (except in some fast-growing females). Extremely variable in color. Inland populations have olive-yellow, blue-green, or red body and usually *no spots on top of head.* Coastal populations are olive to dark green above, silver blue to pale olive on side, silver olive to white below; have many black spots on body, top of head, and fins. Young has black spots on about 10 oval parr marks on side (Fig. 46). Has 130–215 lateral scales. Small teeth on floor of throat between gill arches. To 39 in. (99 cm). **RANGE:** Pacific drainages from Kenai Peninsula, AK, to Eel R., n. CA. Freshwater populations range through Rocky Mts. in Hudson Bay, Mississippi R., Great (including Lahontan, Bonneville, and Alvord basins), and Pacific basins from s. AB to Rio Grande drainage, NM. Established elsewhere in Canada and nw. U.S. Locally common. **HABITAT:** Gravel-bottomed creeks and small rivers; lakes. Anadromous in many coastal streams. **REMARKS:** Fourteen subspecies, some highly distinctive, described below. **SIMILAR SPECIES:** See Pl. 30. (1) Rainbow Trout, *O. mykiss*, usually *lacks* red "cutthroat" mark; has pink to red stripe on side (except in sea-run individual), usually no teeth on floor of throat between gill arches. (2) Gila Trout, *O. gilae*, and (3) Apache Trout, *O. apache*, have *yellow-gold side* and fins. (4) Golden Trout, *O. aguabonita*, has *10–12 dark purple parr marks* on adult, *bright yellow-gold side*, large black spots on caudal peduncle.

COASTAL CUTTHROAT TROUT *Oncorhynchus clarkii clarkii* Pl. 29

IDENTIFICATION: *Many black spots* over entire body, top of head, and lower side. Sea-run form has *silvery body*, dull orange "cutthroat" mark, pink to red-orange anal and pelvic fins. Stream-resident form has more distinct black spots, yellow-gold *lower side*, bright red "cutthroat" mark, *white tips* on anal and paired fins. Roundish violet or purple parr marks. To 22 in. (56 cm). **RANGE:** Pacific drainages from Kenai Peninsula, AK, to Eel R., n. CA. Locally common. **HABITAT:** Cool, gravel-bottomed headwaters, creeks, and small rivers; ponds and lakes. Anadromous in many coastal streams. Only sea-run subspe-

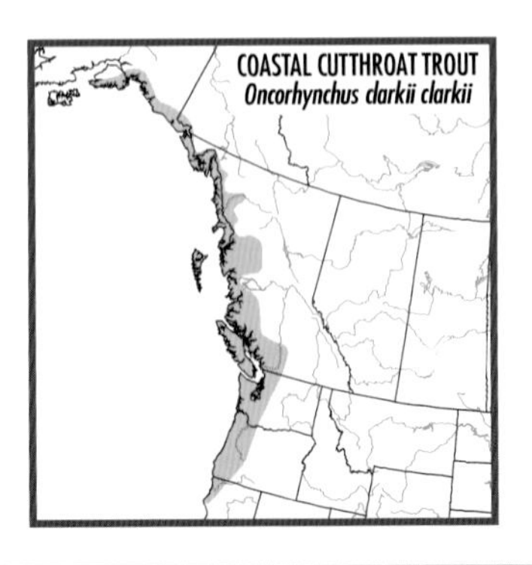

cies of Cutthroat Trout. **REMARKS:** Four life-history types are known: sea-run, stream-resident (nonmigrating), riverine (populations that migrate to small tributaries to spawn), and lake-adapted. Several Coastal Cutthroat Trout populations are under consideration for protection as endangered or threatened.

WESTSLOPE CUTTHROAT TROUT *Oncorhynchus clarkii lewisi* Pl. 29

IDENTIFICATION: Black spots on back and upper side, increasing in number and size toward caudal fin. *No black spots* on lower side between pectoral and anal fins. Deep red "cutthroat" mark; purple-blue gill cover. Breeding male has green, gold, and red-orange body. To 24 in. (61 cm). **RANGE:** Eastern and western slopes of Rocky Mts. in Hudson Bay and Columbia R. basins, s. BC and AB, south to Salmon and Clearwater rivers (Columbia R. basin), cen. ID, and headwaters of Madison and Jefferson river systems (Missouri R. drainage) of w. MT. Isolated populations in John Day R. system, cen. OR, and Columbia R. basin of n.-cen. WA and s. BC. Locally common, but lake populations declining. Protected in Canada as a *threatened subspecies*. **HABITAT:** Cool, rocky headwaters, creeks, and small rivers; lakes. **REMARKS:** Hybridizes with introduced Rainbow Trout, *O. mykiss*. Stream residents reach only 9 in. (23 cm). **SIMILAR SUBSPECIES:** (1) Coastal Cutthroat Trout, *O. c. clarkii* (Pl. 29), has many *black spots over entire body*. (2) Yellowstone Cutthroat Trout, *O. c. bouvieri* (Pl. 29), has larger spots, *black spots on lower side* in front of anal fin.

YELLOWSTONE CUTTHROAT TROUT Pl. 29
Oncorhynchus clarkii bouvieri

IDENTIFICATION: *Large black spots mainly on caudal peduncle and fin*; some spots on lower side. Red "cutthroat" mark and gill cover; red along lateral line. Yellow to bronze body, paler on lower side; red-orange pectoral, pelvic, and anal fins. To 24 in. (61 cm). **RANGE:** Both sides of Continental Divide; Snake R. system above Shoshone Falls,

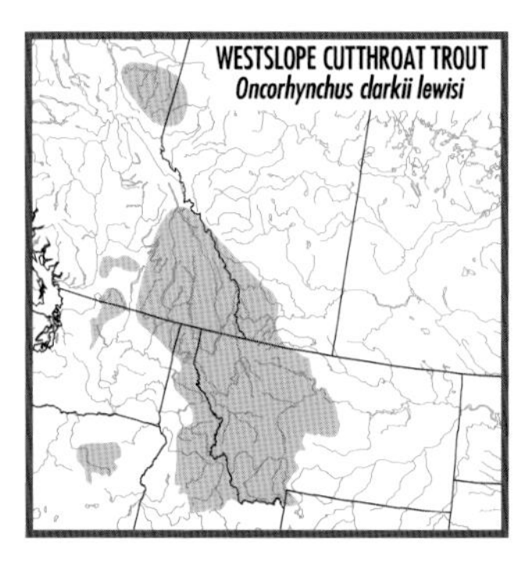

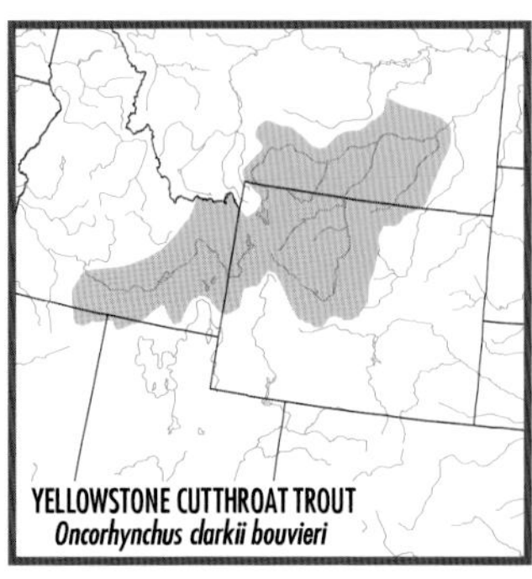

ID, ne. NV, and nw. UT; Yellowstone R. system, s. MT and nw. WY, downstream to Tongue R. Common in Yellowstone Lake, WY; stream populations greatly reduced by introductions of non-native trouts and salmons. **HABITAT:** Clear, cool, rocky creeks and small to medium rivers; ponds and lakes. **SIMILAR SUBSPECIES:** (1) Westslope Cutthroat Trout, *O. c. lewisi* (Pl. 29), has *smaller spots,* lacks black spots on lower side in front of anal fin. (2) Finespotted Cutthroat Trout, *O. c. behnkei* (Pl. 29), also in upper Snake R. system but in different tributaries, has *many more, much smaller spots.*

FINESPOTTED CUTTHROAT TROUT Pl. 29

Oncorhynchus clarkii behnkei

IDENTIFICATION: *Many small black spots*—essentially specks—on body and dorsal, adipose, and caudal fins. Red-orange "cutthroat" mark; *bright red or orange pectoral, pelvic, and anal fins.* Silver green or bronze sheen to yellow-brown body. To 28 in. (71 cm). **RANGE:** Upper Snake R. system, WY and ID. Most widely stocked Cutthroat Trout outside native range. Common. **HABITAT:** Small to large rocky rivers; lakes. **SIMILAR SUBSPECIES:** (1) Yellowstone Cutthroat Trout, *O. c. bouvieri* (Pl. 29), also in upper Snake R. system but in different tributaries, has *fewer, larger spots.*

BONNEVILLE CUTTHROAT TROUT *Oncorhynchus clarkii utah* Pl. 29

IDENTIFICATION: Two forms. Stream-resident form has *light blue elliptical parr marks* on yellow-green to silver gray body, *large, scattered black spots* on body and dorsal, adipose, and caudal fins, red-orange "cutthroat" mark; to 18 in. (46 cm). Bear Lake (ID and UT) form has *small, irregularly shaped spots* scattered on side and dorsal and caudal fins, *silvery blue-green above* fading to white below; orange fins; to 24 in. (61 cm). **RANGE:** Native to Lake Bonneville basin of ID, WY, UT, and NV. Possibly native to adjacent headwaters of Santa Clara R. (Colorado

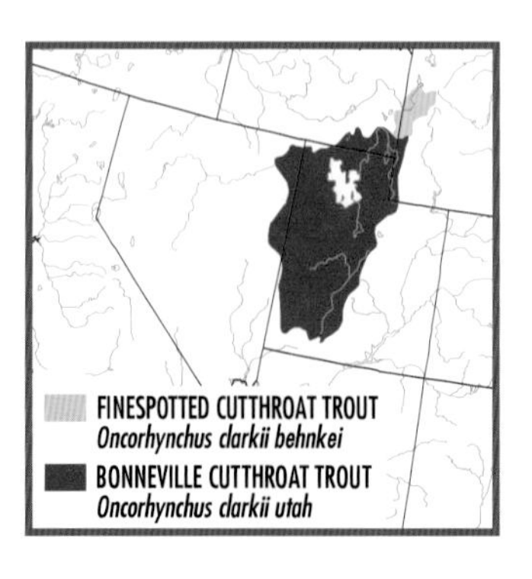

R. basin) in sw. UT. Common. **HABITAT:** Clear, cool, rocky creeks and small to medium rivers; lakes. **SIMILAR SUBSPECIES:** (1) Yellowstone Cutthroat Trout, *O. c. bouvieri* (Pl. 29), *lacks* parr marks on adult, has black spots more concentrated on rear of body, more red on gill cover and along lateral line.

COLORADO RIVER CUTTHROAT TROUT Pl. 29

Oncorhynchus clarkii pleuriticus

IDENTIFICATION: *Medium to large black spots* mainly on caudal peduncle and upper side (some populations have spots distributed more evenly on side). Red-orange "cutthroat" mark. *Bright yellow-gold body*; brassy green above, *orange lower side and belly*. Has *180–200 lateral scales*. To 20 in. (51 cm). **RANGE:** Upper Colorado R. basin, sw. WY, w. CO, and e. UT; formerly nw. NM and ne. AZ. Locally common. **HABITAT:** Clear rocky creeks and small to large rivers; lakes. **SIMILAR SUBSPECIES:** (1) Yellowstone, *O. c. bouvieri*, and (2) Bonneville *O. c. utah*, cutthroat trouts (both Pl. 29), have *160–175 lateral scales*, more subdued rose, yellow, and orange colors.

GREENBACK CUTTHROAT TROUT *Oncorhynchus clarkii stomias* Pl. 29

IDENTIFICATION: Nearly identical to Colorado River Cutthroat Trout, *O. c. pleuriticus*, but usually has 190–215 lateral scales, *white tip* on dorsal fin, more intense red-orange lower side, belly, and gill cover. Parr marks continue through maturity in small-stream populations. Breeding male has bright red underside. To 18 in. (46 cm). **RANGE:** Upper South Platte and Arkansas river systems, cen. CO and (historically) se. WY. Rare; protected as an *endangered subspecies*. **HABITAT:** Rocky montane creeks and small rivers; ponds and lakes. **SIMILAR SUBSPECIES:** (1) See Colorado River Cutthroat Trout, *O. c. pleuriticus* (Pl. 29).

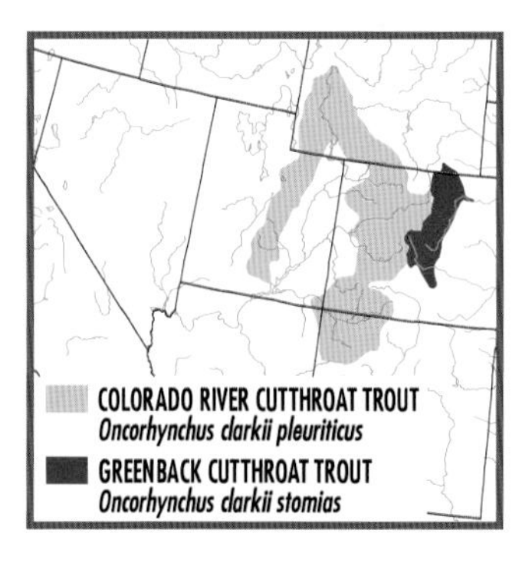

RIO GRANDE CUTTHROAT TROUT Pl. 29

Oncorhynchus clarkii virginalis

IDENTIFICATION: Similar to Colorado River Cutthroat Trout, *O. c. pleuriticus*, but has *150–180 lateral scales*, black spots mainly on upper side, caudal peduncle and fin; 4 or 5 black spots around eye in some populations. Pecos R. strain has *larger spots*, usually 170–180 lateral scales. To 15 in. (38 cm). **RANGE:** Rio Grande drainage, s. CO and NM, including upper Pecos R. system; upper Canadian R. system (Red R. drainage), n. NM. Locally common; greatly reduced by introductions of non-native trouts. **HABITAT:** Rocky montane headwaters, creeks, and small to large rivers. **SIMILAR SUBSPECIES:** (1) Colorado River, *O. c. pleuriticus*, and (2) Greenback Cutthroat Trout, *O. c. stomias* (both Pl. 29), *usually have 180–215 lateral scales*, some black spots on lower side.

YELLOWFIN CUTTHROAT TROUT Pl. 29

Oncorhynchus clarkii macdonaldi

IDENTIFICATION: *Silvery blue body*, yellow lower side; *bright yellow fins*. Bright red "cutthroat" mark. Small, irregularly shaped *black spots concentrated on rear half of body.* To 28 in. (71 cm). **RANGE:** Formerly Twin Lakes (upper Arkansas R. system), cen. CO. Extinct. **HABITAT:** Presumably deep waters of Twin Lakes.

LAHONTAN CUTTHROAT TROUT *Oncorhynchus clarkii henshawi* Pl. 29

IDENTIFICATION: Has *21–28 rakers* on 1st gill arch; *black spots on top of head*; 150–180 lateral scales. Two forms. Lake form has *white edge* on pelvic, anal, and caudal fins; small black spots scattered on side of body and top of head; greenish bronze back, pink tint on *pale copper side;* pale red "cutthroat" mark. Stream-resident form lacks white edge on caudal fin, has larger black spots evenly distributed on body and top of head; parr marks on adult; reaches only 9 in. (23 cm). To 39

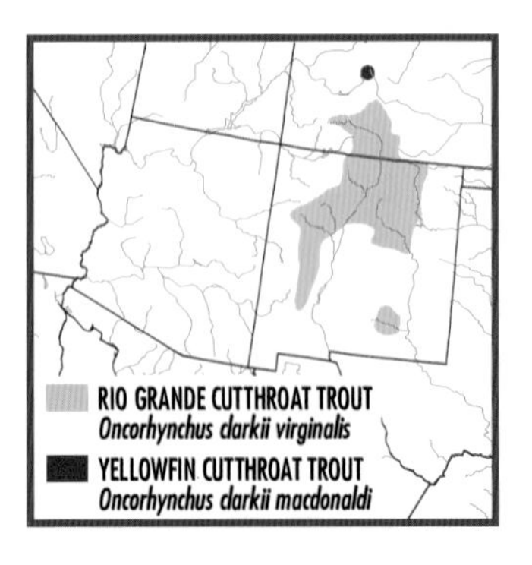

in. (99 cm). **RANGE:** Native to Lahontan basin, extreme se. OR (Quinn R.), nw. NV, and e. CA. Rare; protected in U.S. as a *threatened subspecies.* Maintained largely through propagation and stocking in CA, WA, and NV, often outside Lahontan basin. **HABITAT:** Small to large rocky rivers; ponds, lakes. **REMARKS:** Lake form is largest of cutthroat trouts. **SIMILAR SUBSPECIES:** See Pl. 29. (1) See Paiute Cutthroat Trout, *O. c. seleniris.* (2) Other Cutthroat Trout subspecies usually have *17–20 rakers* on 1st gill arch, usually *no black spots* on top of head (except Coastal Cutthroat Trout, *O. c. clarkii*).

PAIUTE CUTTHROAT TROUT *Oncorhynchus clarkii seleniris* Pl. 29

IDENTIFICATION: Similar to Lahontan Cutthroat Trout, *O. c. henshawi*, but has *no black spots* on body or caudal fin. Olive-bronze above and on side; light yellow below. Pale rose along lateral line and on gill cover; pale parr marks in some adults. To 12 in. (30 cm) in streams, 18 in. (46 cm) in lakes. **RANGE:** Silver King Creek (Carson R. system), Alpine Co., e. CA. Introduced into isolated creeks and lakes in s. CA. Rare; protected as a *threatened subspecies*. **HABITAT:** Rocky headwaters and creeks; lakes. **SIMILAR SUBSPECIES:** (1) See Lahontan Cutthroat Trout, *O. c. henshawi* (Pl. 29).

ALVORD CUTTHROAT TROUT *Oncorhynchus clarkii alvordensis* Pl. 29

IDENTIFICATION: Similar to Lahontan Cutthroat Trout, *O. c. henshawi*, but has only 25–50 medium-sized black spots, mostly on upper side; usually *125–150 lateral scales*. Bright red "cutthroat" mark; rose tint along side intense in adult male. To 20 in. (51 cm). **RANGE:** Formerly Trout Creek and Virgin-Thousand Creek basins (endorheic), s. OR and n. NV. Extinct; replaced by introduced Rainbow Trout, *O. mykiss*. **HABITAT:** Headwaters and creeks. **SIMILAR SUBSPECIES:** (1) See Lahontan Cutthroat Trout, *O. c. henshawi* (Pl. 29).

HUMBOLDT CUTTHROAT TROUT Pl. 29

Oncorhynchus clarkii subspecies

IDENTIFICATION: Nearly identical to Lahontan Cutthroat Trout, *O. c. henshawi*, but usually has *20–22 rakers* on 1st gill arch, 130–160 lateral scales, bright red "cutthroat" mark. Light blue parr marks in some adults. Has 50–60 pyloric caeca. To 18 in. (46 cm). **RANGE:** Humboldt R. drainage, NV. Uncommon; *protected as a form of Lahontan Cutthroat Trout*. **HABITAT:** Headwaters, creeks, and small rivers; adapts to reservoirs. **SIMILAR SUBSPECIES:** See (1) Lahontan Cutthroat Trout, *O. c. henshawi*, and (2) Whitehorse Cutthroat Trout, *O. c.* subspecies (both Pl. 29).

WHITEHORSE CUTTHROAT TROUT Pl. 29

Oncorhynchus clarkii subspecies

IDENTIFICATION: Nearly identical to Humboldt Cutthroat Trout, *O. c.* subspecies, but has *35–50 pyloric caeca*, pale red "cutthroat" mark. Has 147–150 lateral scales. To 14 in. (36 cm). **RANGE:** Whitehorse and Willow creek basins (endorheic), se. OR. Uncommon; *protected as a form of Lahontan Cutthroat Trout*. **HABITAT:** Small, rocky headwaters and creeks. **SIMILAR SUBSPECIES:** (1) See Humboldt Cutthroat Trout, *O. c.* subspecies (Pl. 29).

RAINBOW TROUT *Oncorhynchus mykiss* Pl. 30

IDENTIFICATION: Usually *small, irregular black spots* on back and most fins. *Pink to red stripe* on side (except in sea-run individual), radiating rows of black spots on caudal fin. Often black edge on adipose fin. Upper jaw reaches barely behind eye in young and female, well behind eye in large male. Highly variable: steel blue, yellow-green, or brown above; silver to pale yellow-green below. Stream and spawning fish have intense dark colors; lake fish are light and silvery. Young has 5–10 widely spaced, short, dark, oval parr marks (Fig. 46). Usually 120–170

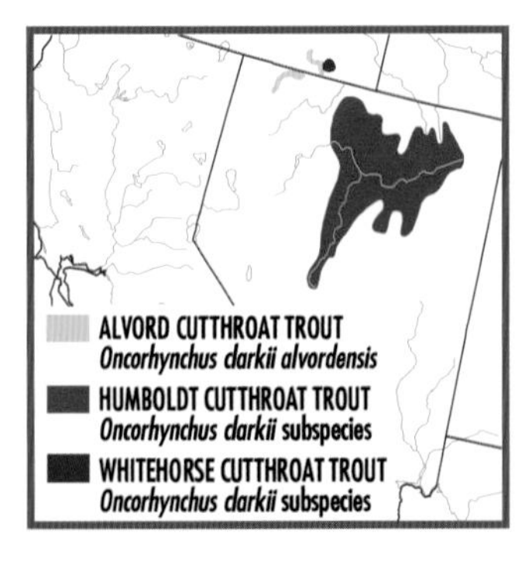

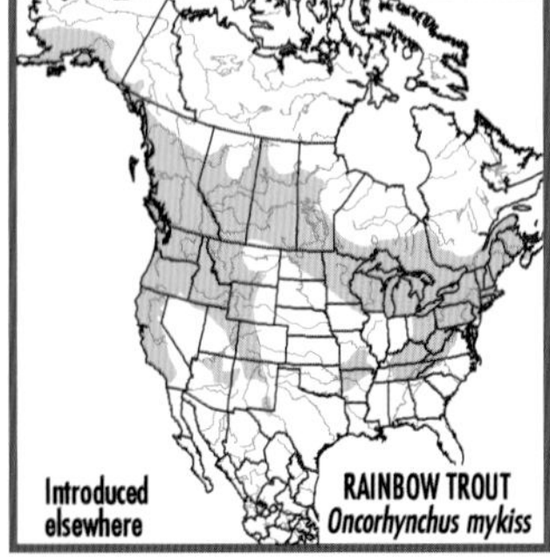

lateral scales; 8–12 anal rays. Usually no teeth on floor of throat between gill arches. To 45 in. (114 cm). **RANGE:** Native to Pacific Slope from Kuskokwim R. drainage, AK, to Otay R. drainage, CA. Widely established in Canada and U.S., including Arctic, Atlantic, Great Lakes, Mississippi R., and Rio Grande basins, and elsewhere in the world. Common. **HABITAT:** Clear, cold headwaters, creeks, and small to large rivers; lakes and intertidal areas. **REMARKS:** Six subspecies, some highly distinctive, described below. **SIMILAR SPECIES:** (1) Cutthroat Trout, *O. clarkii* (Pl. 29), usually has red "cutthroat" mark, *no* (or faint) pink to red stripe on side, usually teeth on floor of throat between gill arches. (2) Gila Trout, *O. gilae*, and (3) Apache Trout, *O. apache* (both Pl. 30), have *yellow-gold side* and fins; white tips on fins. (4) Golden Trout, *O. aguabonita* (Pl. 30), has *10–12 dark purple parr marks* on adult, *bright yellow-gold side*, large black spots on caudal peduncle.

COASTAL RAINBOW TROUT *Oncorhynchus mykiss irideus* Pl. 30

IDENTIFICATION: Two forms. Stream-resident form has *bright red-pink stripe* along side, most vivid on spawning male. Many black spots on top of head, body, dorsal and caudal fins. Snout blunt, rounded. Usually 120–140 lateral scales; 50–60 pyloric caeca. No parr marks on adult. To 16 in. (41 cm). Sea-run (anadromous) form ("Steelhead") has smaller black spots, is silver or pink on side. To 43 in. (110 cm). **RANGE:** Native to Pacific Slope from Kuskokwim R. drainage, AK, to Otay R. drainage, CA. Generally common; several populations in decline and protected in U.S. as evolutionarily significant populations. **HABITAT:** Clear, cold headwaters, creeks, and small to large rivers; lakes and intertidal areas. Anadromous in coastal streams. **SIMILAR SUBSPECIES:** (1) Columbia Rainbow Trout, *O. m. gairdnerii* (Pl. 30), has larger black spots on body and fins, *including anal and pelvic fins*; usually 140–170 lateral scales, 35–45 pyloric caeca.

COLUMBIA RAINBOW TROUT *Oncorhynchus mykiss gairdnerii* Pl. 30

IDENTIFICATION: Two forms. Stream-resident form has *many fairly large black spots on body and all (except pectoral) fins. Pink to red stripe* along side; often light yellow or orange on lower side. Yellow-orange "cutthroat" mark. Straight-edged caudal fin. Usually 140–170 lateral scales; 18–20 rakers on 1st gill arch; 35–45 pyloric caeca. To 18 in. (46 cm). Sea-run form ("Redband Steelhead") has silvery sheen masking spots on side and fins; adult male is bronze, develops bright red side of head and lower side of body, hooked lower jaw. To 40 in. (100 cm). **RANGE:** Finlay R. basin, BC, south to Columbia R. basin, ID and n. NV; also in upper Mackenzie R. drainage (Arctic basin), AB. **HABITAT:** Clear, cool creeks, rivers, and lakes. Common (often as result of stocking). "Redband Steelhead" is protected as an evolutionarily significant population in Columbia R. basin. **REMARKS:** "Kamloops Trout" (lake form of *O. m. gairdnerii*) originally in lakes in BC reached 52 lbs. 8 oz. (24 kg); introduced into lakes elsewhere but rarely reaches 15 lbs. (7 kg). Stocking of Coastal Rainbow Trout, *O. m. irideus*, into range of Columbia Rainbow Trout has resulted in extensive hybridization. **SIMILAR SUBSPECIES:** (1) See Great Basin Rainbow Trout, *O. m. newberrii* (Pl. 30). (2) Coastal Rainbow Trout, *O. m. irideus* (Pl. 30), *lacks* black spots on anal and pelvic fins, usually has 120–140 lateral scales, 50–60 pyloric caeca.

GREAT BASIN RAINBOW TROUT Pl. 30

Oncorhynchus mykiss newberrii

IDENTIFICATION: Similar to Columbia Rainbow Trout, *O. m. gairdnerii*, but usually has *white tip* on dorsal, pelvic, and anal fins; *elliptical purple parr marks on adult*; forked caudal fin. Usually 21–23 rakers on 1st gill arch (upper Klamath Lake form has 20–22 rakers). Lake form (e.g., Goose Lake, OR and CA) has fewer spots, subdued color on side; dark green or blue above, silver to white below. Form in upper

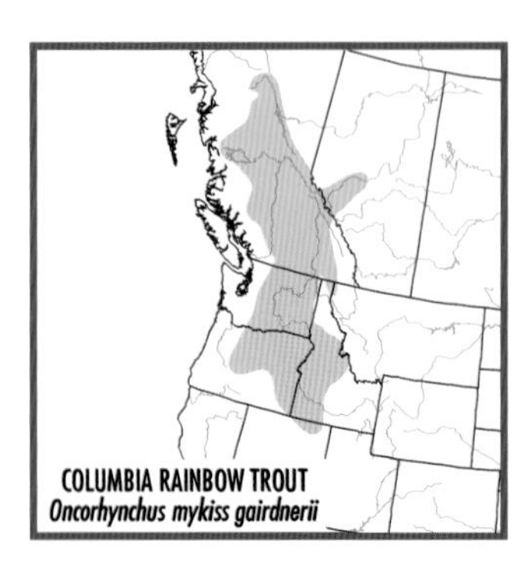

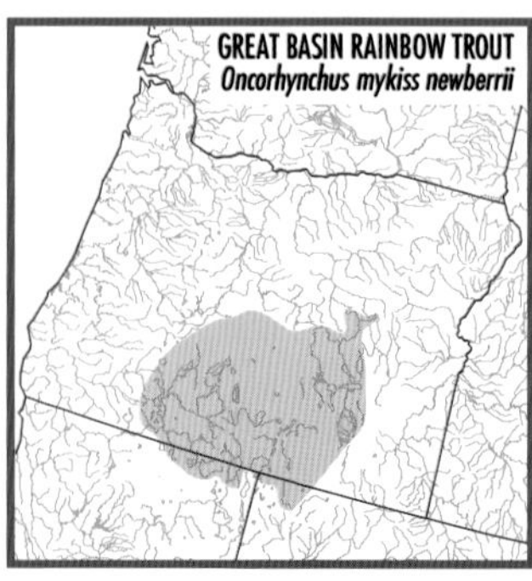

Klamath Lake and Williamson R., OR, nearly identical to "Steelhead," *O. m. irideus*, but has rounded snout, bullet-shaped head. To 20 in. (51 cm) in streams; to 36 in. (91 cm) in upper Klamath Lake and Williamson R., OR. **RANGE:** Upper Klamath R., upper Pit R., and endorheic basins to the east, including Fort Rock, Harney-Malheur, Catlow, Warner Lakes, Goose Lake, and Chewaucan basins, s. OR, ne. CA, and nw. NV. Common. **HABITAT:** Rocky headwaters, creeks, and small rivers; lakes. **REMARKS:** Originally extremely variable; at least 7 forms in isolated drainages. Most populations now genetically contaminated with hatchery raised Coastal Rainbow Trout, *O. m. irideus*. **SIMILAR SUBSPECIES:** (1) See Columbia Rainbow Trout, *O. m. gairdnerii* (Pl. 30). (2) Sacramento Rainbow Trout, *O. m. stonei* (Pl. 30), has few black spots on lower side, *less persistent parr marks*, orange "cutthroat" mark.

SACRAMENTO RAINBOW TROUT *Oncorhynchus mykiss stonei* Pl. 30

IDENTIFICATION: Many black spots on back and upper side, *few below. Bright red stripe* along lateral line; rose on gill cover. *Orange "cutthroat" mark. White tip* on anal and pelvic fins; orange and white tip on dorsal fin. Forked caudal fin. To 20 in. (51 cm). **RANGE:** Upper Sacramento R. drainage, CA, including McCloud, Pit, and Feather river systems. Common. **HABITAT:** Clear, cool creeks and small to medium rivers. **REMARKS:** "Sheepheaven Redband Trout," in Sheapheaven Creek, CA, has yellow on body and fins, dark parr marks on adult, fewer spots, only 16 rakers on 1st gill arch; reaches only 8 in. (20 cm). **SIMILAR SUBSPECIES:** (1) Great Basin Rainbow Trout, *O. m. newberrii* (Pl. 30), has *more black spots on lower side*, more persistent parr marks; *lacks* orange "cutthroat" mark.

KERN RAINBOW TROUT *Oncorhynchus mykiss gilberti* Pl. 30

IDENTIFICATION: Similar to Sacramento Rainbow Trout, *O. m. stonei*, but has *purple parr marks* on adult (as well as young), *duller colors*, less intense red stripe along side. To 28 in. (71 cm). **RANGE:** Kern R. system (San Joaquin R. basin), CA (replaced in upper river system by Golden Trout, *O. aguabonita*). Uncommon. **HABITAT:** Small, cool, high-elevation headwaters and creeks; lakes. **SIMILAR SPECIES:** (1) See Sacramento Rainbow Trout, *O. m. stonei* (Pl. 30). (2) Golden Trout, *O. aguabonita* (Pl. 30), has *bright yellow-gold side*; bright red belly, cheek, and branchiostegal membranes; large black spots on caudal peduncle.

EAGLE LAKE RAINBOW TROUT Pl. 30

Oncorhynchus mykiss aquilarum

IDENTIFICATION: Many black, *irregularly shaped spots on pink side*, dorsal and caudal fins. *Straight-edged caudal fin*. To 30 in. (76 cm). **RANGE:** Eagle Lake (Lahontan basin), ne. CA. Rare; propagated and stocked in

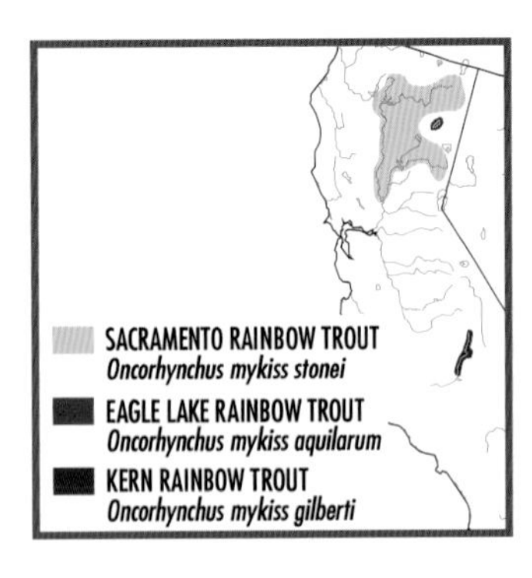

lakes and reservoirs of several western states. **HABITAT:** Open water of Eagle Lake. **SIMILAR SUBSPECIES:** (1) Sacramento Rainbow Trout, *O. m. stonei* (Pl. 30), in adjacent upper Sacramento R. drainage, has bright red stripe on side, *rounder black spots, forked caudal fin.*

GOLDEN TROUT *Oncorhynchus aguabonita* Pl. 30

IDENTIFICATION: Has *10–12 dark purple parr marks* on adult (as well as young). Copper above; *red stripe along bright yellow-gold side*; bright red belly, cheek, and branchiostegal membranes; large black spots on caudal peduncle, dorsal and caudal fins; white tips preceded by dusky black stripe on orange anal and pelvic fins; white to orange tip on dorsal fin. To 12 in. (30 cm). **RANGE:** Upper Kern R. system, Tulare and Kern counties, CA. Common. Introduced into lakes and streams in Sierra Nevada, CA, and other Rocky Mt. states and provinces. Most transplanted populations hybridize with Rainbow Trout, *O. mykiss*, or Cutthroat Trout, *O. clarkii*. **HABITAT:** Clear, cool headwaters, creeks, and lakes at elevations above 6900 ft. (2100 m). **REMARKS:** Two subspecies. Trout Creek Golden Trout, *O. a. aguabonita,* in South Fork Kern R. and Golden Trout Creek, has bright gold, orange, and red colors, few spots on upper side, usually 170–200 lateral scales, 59–60 pyloric caeca. Little Kern River Golden Trout, *O. a. whitei*, in Little Kern R., has more subdued colors, more black spots on upper side, usually 155–160 lateral scales, and 35–40 pyloric caeca. **SIMILAR SPECIES:** See Pl. 30. Other trout usually *lack* obvious parr marks on adult; most *lack* bright yellow-gold side.

GILA TROUT *Oncorhynchus gilae* Pl. 30

IDENTIFICATION: Many *small black spots* (mostly above lateral line) on *yellow-gold side*, head, dorsal and caudal fins; no spots on anal and paired fins; *large spots on adipose fin*, rose stripe along side of adult; white or yellow tips on dorsal, anal, and pelvic fins. No or faint parr marks on juvenile (as well as adult). To 9 in. (23 cm). **RANGE:** Upper

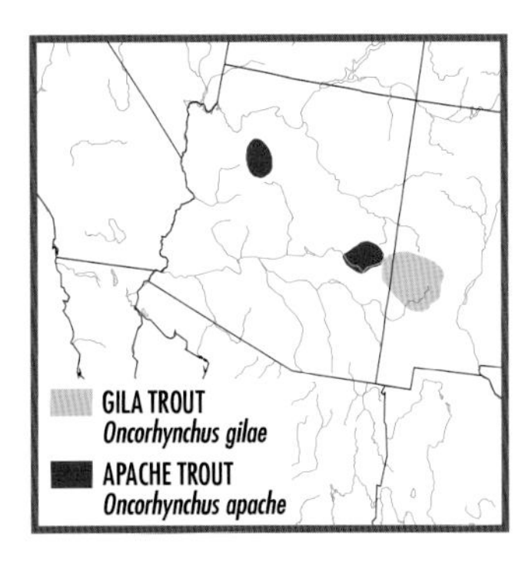

Gila R. system, NM and AZ; introduced to other streams in w. NM and cen. AZ. Rare; protected as an *endangered species*. **HABITAT:** Clear, cool headwaters and creeks (above 6500 ft. elevation [2000 m]). **SIMILAR SPECIES:** (1) See Apache Trout, *O. apache* (Pl. 30). (2) Golden Trout, *O. aguabonita* (Pl. 30), has 10–12 dark purple parr marks on juvenile and adult, large black spots on caudal peduncle. (3) Cutthroat Trout, *O. clarkii* (Pl. 29), usually has red "cutthroat" mark, usually *lacks* yellow-gold side and fins. (4) Rainbow Trout, *O. mykiss* (Pl. 30), *lacks* yellow-gold side and fins, white tips on fins.

APACHE TROUT *Oncorhynchus apache* Pl. 30

IDENTIFICATION: Similar to Gila Trout, *O. gilae*, but has *brighter yellow body and fins*; *larger black spots* on body, head, and dorsal, adipose, and caudal fins; often *2 small black spots* on either side of pupil, creating black stripe through eye; no rose stripe on side. To 9 in. (23 cm). **RANGE:** Upper Salt R. system, AZ. An extinct population in Verde R. (Salt R. system), AZ, was closely related to Gila and Apache trouts. Introduced into several surrounding streams and lakes but rare in pure (nonhybrid) form. Protected as a *threatened species*. **HABITAT:** Clear, cool montane headwaters and creeks (generally above 8175 ft. elevation [2500 m]); stocked in montane lakes. **REMARKS:** Hybridization with Rainbow Trout, *O. mykiss*, and competition from Brook Trout, *Salvelinus fontinalis*, and Brown Trout, *Salmo trutta*, have resulted in 95 percent reduction in range. **SIMILAR SPECIES:** (1) See Gila Trout, *O. gilae* (Pl. 30).

TROUT-PERCHES: Family Percopsidae (2)

Trout-perches, endemic to North America, have a *large unscaled head,* cycloid and ctenoid scales, subthoracic pelvic fins, 1 large dorsal fin, an *adipose fin,* and *spines* in the dorsal, anal, and pelvic fins.

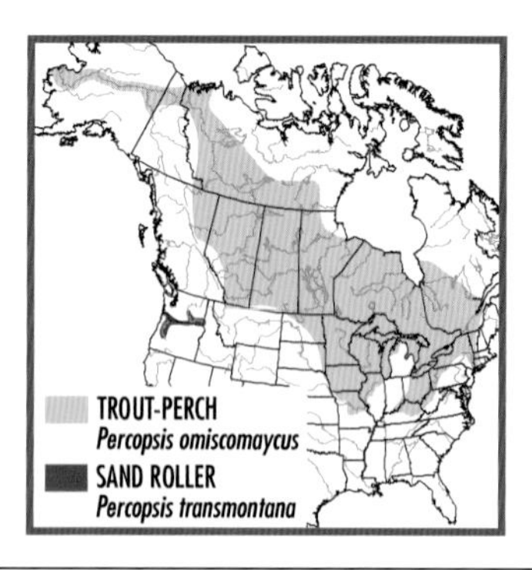

TROUT-PERCH *Percopsis omiscomaycus* Pl. 31

IDENTIFICATION: Transparent yellow-olive with silver flecks above; rows of *7–12 dusky spots* along back, upper side, and side. Large head, flattened below; *large silver white chambers* on cheek and underside of head on lower jaw and edge of cheek. Fairly deep body, slender caudal peduncle; large eye high on head; large forked caudal fin. Complete lateral line; no or small spines on preopercle. To 7¾ in. (20 cm). **RANGE:** Atlantic and Arctic basins throughout most of Canada from QC to YT and BC, and south to Potomac R. drainage, VA; Yukon R. drainage, YT and AK; Great Lakes and Mississippi R. basins south to WV and cen. MO. Locally common in lakes; uncommon throughout most of range. **HABITAT:** Lakes; deep flowing pools of creeks and small to large rivers; usually over sand. **SIMILAR SPECIES:** (1) See Sand Roller, *P. transmontana* (Pl. 31).

SAND ROLLER *Percopsis transmontana* Pl. 31

IDENTIFICATION: Similar to Trout-Perch, *P. omiscomaycus*, but is *darker blue-green* above; has *more arched back,* incomplete lateral line, few large spines on preopercle. To 3¾ in. (9.6 cm). **RANGE:** Columbia R. drainage, w. ID, s. WA, and nw. OR. Uncommon. **HABITAT:** Quiet backwaters and pool margins of small to large rivers. Usually near vegetation over sand. **SIMILAR SPECIES:** (1) See Trout-Perch, *P. omiscomaycus* (Pl. 31).

Pirate Perch: Family Aphredoderidae (1)

The Pirate Perch, *Aphredoderus sayanus*, is the only living member of its family. It has a *large mouth, ctenoid scales* on the head and body, *1 dorsal fin* with both spines and rays, thoracic pelvic fins, and *anus and urogenital openings between the branchiostegal membranes* on

the adult. The anus and urogenital opening are positioned just in front of the anal fin in the juvenile but migrate to the throat during development.

PIRATE PERCH *Aphredoderus sayanus* Pl. 31

IDENTIFICATION: Short, deep body; *large head; large mouth,* lower jaw protruding; truncate or barely notched caudal fin. *Gray to black above,* often speckled with black; yellow-white below; black teardrop; black bar on caudal fin base; dusky to black fins. Large individual has *purple sheen.* No or incomplete lateral line. To 5½ in. (14 cm). **RANGE:** Atlantic and Gulf slopes from Long Island, NY, to Brazos R. drainage, TX; Great Lakes and Mississippi R. basins from MI, WI, and s. MN to Gulf. Isolated population in Niagara R.-Lake Ontario drainage, NY. Common; primarily restricted to Coastal Plain and other lowland regions. **HABITAT:** Swamps, sloughs, ponds, lakes, backwaters, and quiet pools of creeks and small to large rivers. Usually near vegetation over mud. **REMARKS:** Two subspecies. *A. s. sayanus,* on Atlantic Slope south to n. FL, has dark stripe along side; usually 3 anal spines, 4 dorsal spines, 11 pectoral rays, and fewer than 42 lateral scales. *A. s. gibbosus,* in Great Lakes, Mississippi R., and Gulf Slope drainages from Pearl R., MS, to Brazos R., TX, lacks dark stripe along side; usually has 2 anal spines, 3 dorsal spines, 12 pectoral rays, and more than 45 lateral scales. Intergrades are intermediate in characters and occur from Altamaha R., GA, to Pascagoula R., MS.

CAVEFISHES: Family Amblyopsidae (6)

These small cave-, spring-, and swamp-inhabiting fishes have *no or very tiny pelvic fins, small or rudimentary eyes,* a strongly protruding lower jaw, flattened head, tubular anterior nostrils, anus and urogenital openings located between the branchiostegal membranes, small and embedded cycloid scales, 1 dorsal fin, 0–2 spines in the dorsal and anal fins, and sensory papillae in rows on the head, body, and caudal fin. Although cave-adapted fishes in other families occur elsewhere in the world, the 6 members of this family occur only in the unglaciated eastern U.S. The evolutionary transition from surface to spring to cave living is apparent among living amblyopsids, making them an especially intriguing group. Many large sensory papillae scattered over the head, body, and caudal fin of cavefishes are sensitive to touch and compensate for lack of sight.

SWAMPFISH *Chologaster cornuta* Pl. 36

IDENTIFICATION: *Strongly bicolored;* dark brown above, white to yellow below. *Small eye. No pelvic fins.* Has 3 narrow black stripes on side; one on lower side *wide at front, narrow at rear. Two* black streaks along back in front of dorsal fin separate to encircle dorsal fin; pink gills visible. Caudal fin clear near base, often black in center; dusky black band in dorsal fin. Has 0–2 rows of papillae on caudal fin; 9–12 (usually 11) dorsal rays; 9–10 anal rays; 9–11 branched caudal rays. To 2¾ in. (6.8 cm). **RANGE:** Atlantic Coastal Plain from Chickahominy Creek (James R. drainage), VA, to Altamaha R. drainage, GA. Common. **HABITAT:** Near vegetation and debris in swamps, sloughs, and quiet pools and backwaters of streams. **SIMILAR SPECIES:** (1) Spring Cavefish, *Forbesichthys agassizii* (Pl. 36), *lacks* strong bicoloration, wide black stripe on lower side; has 11–16 branched caudal rays.

SPRING CAVEFISH *Forbesichthys agassizii* Pl. 36

IDENTIFICATION: Small eye. Extremely long, slender (salamanderlike) body. No pelvic fins. Dark brown above; 2 black streaks along back in front of dorsal fin separate to encircle dorsal fin; 3 thin axial stripes along side; pink gills visible; white to yellow-brown below; black bar

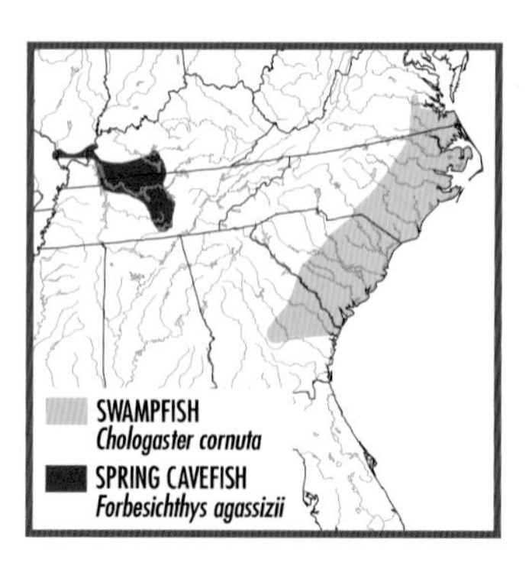

at base of dusky black caudal fin; other fins clear. Has 0–2 rows of papillae on caudal fin; 9–11 dorsal rays; 9–11 anal rays; 11–16 branched caudal rays. To 3¼ in. (8.4 cm). **RANGE:** Green R. system, s.-cen. KY, west across Shawnee Hills to se. MO; south to Elk R. system, s. TN. Common in a few localities; uncommon to rare elsewhere. **HABITAT:** Springs and caves (but almost always near surface). **SIMILAR SPECIES:** (1) Swampfish, *Chologaster cornuta* (Pl. 36), is bicolored with bold black stripes on side, less slender; has 9–11 branched caudal rays.

SOUTHERN CAVEFISH *Typhlichthys subterraneus* Pl. 36

IDENTIFICATION: *Pink-white. No eyes* (vestigial eye tissues under skin). *No pelvic fins.* Large, broad head; 0–2 rows of papillae on caudal fin. Has 7–10 (usually 8–9) dorsal rays; 7–10 anal rays; 10–15 branched caudal rays. To 3½ in. (9 cm). **RANGE:** Disjunct. East of Mississippi R. in Cumberland and Interior Low plateaus from extreme s. IN to nw. GA and n. AL (to upper Coosa R. system); west of Mississippi R. in Ozark Plateau of s. MO and ne. AR. Uncommon. **HABITAT:** Subterranean water. **SIMILAR SPECIES:** (1) Northern Cavefish, *Amblyopsis spelaea* (Pl. 36), (2) Ozark Cavefish, *Troglichthys rosae,* and (3) Alabama Cavefish, *Speoplatyrhinus poulsoni* (Pl. 36), have 4–6 rows of papillae on caudal fin.

NORTHERN CAVEFISH *Amblyopsis spelaea* Pl. 36

IDENTIFICATION: *Pink-white. No eyes* (vestigial eye tissues under skin). *Very small pelvic fins* (rarely absent). Large, broad head; 4–6 rows of papillae on caudal fin. Has 9–11 dorsal rays; 8–11 (usually 9–10) anal rays; 11–13 branched caudal rays. To 4¼ in. (11 cm). **RANGE:** S.-cen. IN to Mammoth Cave area, cen. KY. Common in a few localities; uncommon to rare elsewhere. **HABITAT:** Subterranean water. **SIMILAR SPECIES:** (1) Ozark Cavefish, *Troglichthys rosae, lacks* pelvic fins, has 7–8 dorsal rays. (2) Southern Cavefish, *Typhlichthys subterraneus* (Pl. 36), *lacks* pelvic fins, has 0–2 rows of papillae on caudal fin.

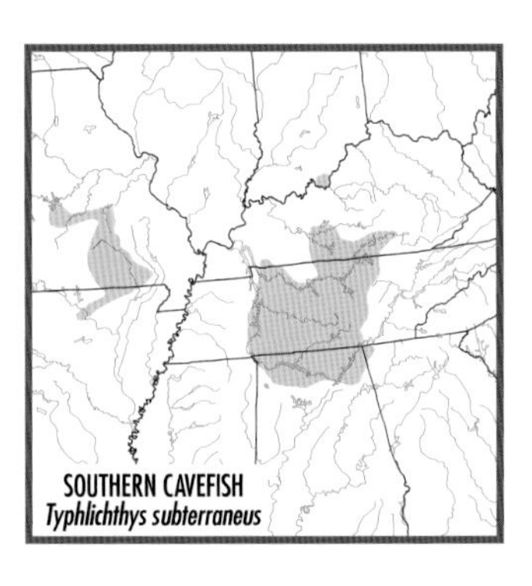

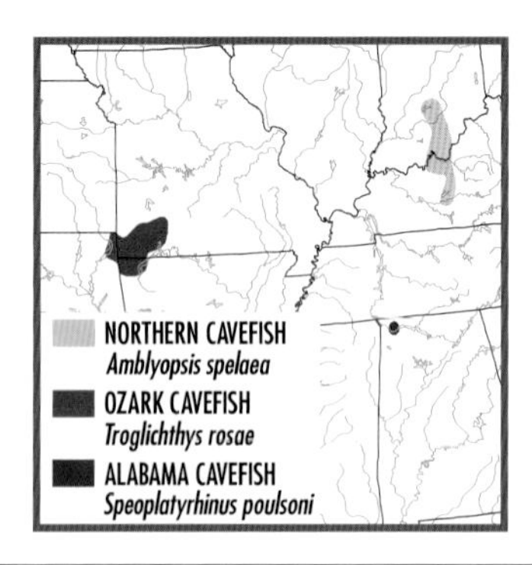

OZARK CAVEFISH *Troglichthys rosae* **Not shown**

IDENTIFICATION: *Pink-white. No eyes. Lacks pelvic fins.* Large, broad head; 4–6 rows of papillae on caudal fin. Has 7–8 dorsal rays, usually 8 anal rays, 9–11 branched caudal rays. To 2½ in. (6.2 cm). **RANGE:** Springfield Plateau, sw. MO, ne. OK, and nw. AR (Missouri, Arkansas, and upper White river drainages). Rare; protected as a *threatened species.* **HABITAT:** Subterranean water. **SIMILAR SPECIES:** (1) Northern Cavefish, *A. spelaea* (Pl. 36), *has pelvic fins*; 9–11 dorsal rays.

ALABAMA CAVEFISH *Speoplatyrhinus poulsoni* **Pl. 36**

IDENTIFICATION: *Long flat head constricted* behind snout. *White. No eyes. No pelvic fins.* No branched fin rays; incised fin membranes. Has 4 rows of papillae on caudal fin; 9–10 dorsal rays; 8–9 anal rays; 21–22 caudal rays (all unbranched). To 3 in. (7.4 cm). **RANGE:** Known only from Key Cave, Lauderdale Co., AL (Tennessee R. drainage). Rare; protected as an *endangered species.* **HABITAT:** Subterranean water. **SIMILAR SPECIES:** (1) Other cavefishes (Pl. 36) have *shorter, less flattened head lacking constriction* behind snout; branched fin rays.

Cuskfishes: Family Lotidae (1)

All 22 members of this family are marine except for the Burbot, *Lota lota*, which occurs in fresh waters of North America and Eurasia. Cuskfishes have a *long slender body*, large head; *1 or 2 dorsal fins, 1 anal fin*; *long barbel at the tip of the chin*, no barbels on the snout; no spines in the fins; thoracic or jugular pelvic fins; and a round caudal fin.

BURBOT *Lota lota* **Pl. 36**

IDENTIFICATION: Two dorsal fins, *1st short* with 8–16 rays, *2d very long* with 60–80 rays. Long, thin *barbel at tip of chin.* Large, wide head; small eye. Long, slender body, strongly compressed posteriorly. Long

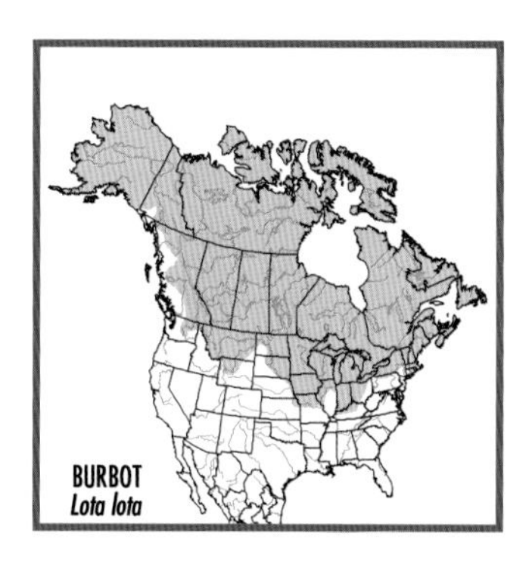

anal fin; small pelvic fin located in front of pectoral fin. Scales small, embedded. Light brown to yellow with dark brown to black mottling on back and side. Second dorsal and anal fins may have dark edge. To 33 in. (84 cm). **RANGE:** Throughout Canada, AK, and n. U.S. (south to PA, KY, MO, WY, and WA). Also n. Eurasia. Common. **HABITAT:** Deep water, to 300 ft. (90 m), of large rivers and lakes.

Cods: Family Gadidae (1)

Important food fishes, cods have *3 dorsal fins, 2 anal fins*; usually a *long barbel at the tip of the chin*; no spines in the fins; thoracic or jugular pelvic fins; and a truncate or slightly forked caudal fin. Twenty-four species are recognized, 1 of which is established in a few freshwater lakes in our area.

ATLANTIC TOMCOD *Microgadus tomcod* **Not shown**

IDENTIFICATION: *Three dorsal fins* of nearly equal length, 2 anal fins, 2d ray of pelvic fin almost *twice* length of other rays. First dorsal fin has 11–15 rays, 2d has 15–19, 3d has 16–21; 1st anal fin has 12–21 rays, 2d has 16–20. Brown with yellow or green sheen above, black blotches on back, side, and fins, white lateral line; white to yellow below. To 15 in. (38 cm). **RANGE:** Marine species that frequently enters Atlantic drainages from s. Labrador to VA. More common in northern part of range where some populations have become permanent residents of freshwater lakes.

New World Silversides: Family Atherinopsidae (3)

These are small, silvery, translucent, strongly compressed fishes with scales on the head, a large eye, terminal mouth, *long snout, no lateral line,* long sickle-shaped anal fin, abdominal pelvic fins, and *2 widely separated dorsal fins,* the first one *small* and with spines.

Atherinids swim, often in large schools, near the surface of the water and correspondingly have the top of the head flattened, an upturned mouth, and pectoral fins located high on the body. They may leave the water and glide through the air for short distances when spawning or disturbed. They are worldwide in distribution, and most are marine. Of the 110 species, 3 occur in fresh waters of North America.

BROOK SILVERSIDE *Labidesthes sicculus* Pl. 36

IDENTIFICATION: *Long beaklike snout,* length about 1½ times eye diameter. *Two widely separated dorsal fins, 1st small* and with spines. First dorsal fin origin *above* anal fin origin. Long sickle-shaped anal fin. Pale green above, scales faintly outlined; bright silver stripe along side. In se. U.S., breeding male has red snout, bright yellow-green body. Usually 74–87 lateral scales, 22–25 anal rays. To 5 in. (13 cm). **RANGE:** St. Lawrence-Great Lakes (except Lake Superior) and Mississippi R. basins from s. QC to e. MN and south to LA; Atlantic and Gulf slopes from Peedee R. drainage, SC, to Galveston Bay drainage, TX. Introduced elsewhere, usually into impoundments as forage for sport fishes. Common; locally abundant. **HABITAT:** Near surface of lakes, ponds, and quiet pools of creeks and small to large rivers. Usually in open water. **REMARKS:** Two subspecies: *L. s. vanhyningi,* on Atlantic and Gulf slopes east of Mississippi R., and *L. s. sicculus,* elsewhere, are probably recognizable but need study. **SIMILAR SPECIES:** (1) Mississippi Silverside, *Menidia audens* (Pl. 36), and (2) Waccamaw Silverside, *M. extensa, lack* beaklike snout (length equal to or less than diameter of eye); have 1st dorsal fin origin *in front of* anal fin origin, 36–50 lateral scales, 14–21 anal rays.

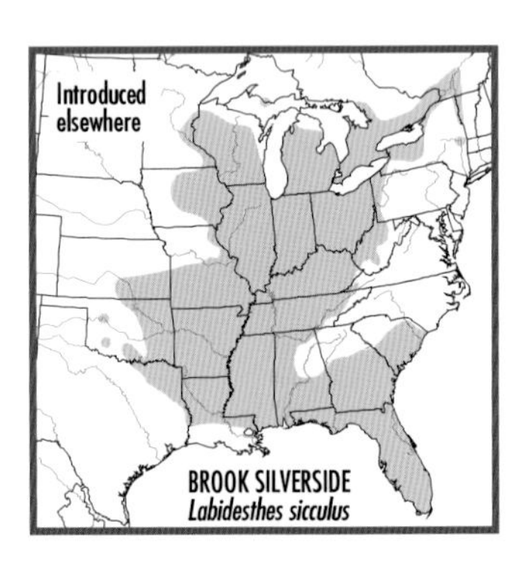

MISSISSIPPI SILVERSIDE *Menidia audens* Pl. 36

IDENTIFICATION: *Two widely separated dorsal fins, 1st small* and with spines. First dorsal fin origin *in front of* anal fin origin. Long snout, length about equal to eye diameter. Long sickle-shaped anal fin. Pale yellow-green above, scales faintly outlined; bright silver stripe along side. Has 36–49 (usually 40–45) lateral scales, 17–28 (usually 19–24) predorsal scales, usually 16–18 anal rays. To 6 in. (15 cm). **RANGE:** See Remarks. Atlantic and Gulf slopes (mostly near coast) from MA to Rio Grande drainage, TX and se. NM; north in Mississippi R. and major tributaries (mainly Arkansas and Red rivers) to s. IN and e. OK. Also in Mexico. Common; expanding range as stocked in impoundments as forage for game fish. **HABITAT:** Usually at surface of clear, quiet water over sand or gravel. **REMARKS:** Often confused with Inland Silverside, *M. beryllina*, which has 34–40 (usually 36–38) lateral scales, 14–21 (usually 15–18) predorsal scales. Inland Silverside is a brackish/marine species that ascends rivers. Range given here is probably combined range of both species. **SIMILAR SPECIES:** (1) See Waccamaw Silverside, *M. extensa*. (2) Brook Silverside, *Labidesthes sicculus* (Pl. 36), has long beaklike snout, 1st dorsal fin origin *over* anal fin origin, usually 74–87 lateral scales, 22–25 anal rays.

WACCAMAW SILVERSIDE *Menidia extensa* Not shown

IDENTIFICATION: Similar to Mississippi Silverside, *M. audens*, but has *much more slender body* (depth about 1½ times width), *darkly outlined scales* on back, usually 44–50 lateral scales, 19–20 (often 18) anal rays. To 3 in. (8 cm). **RANGE:** Lake Waccamaw, NC. Common; usually in large schools. Protected as a *threatened species* because of extremely small range. **HABITAT:** Near surface in open water. **SIMILAR SPECIES:** (1) See Mississippi Silverside, *M. audens* (Pl. 36).

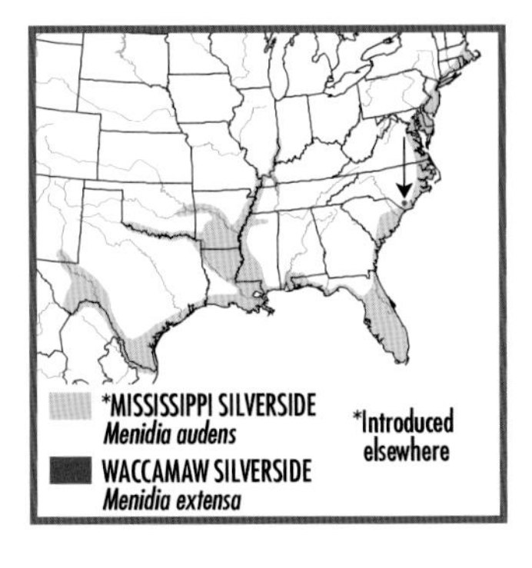

New World Rivulines: Family Rivulidae (1)

New World rivulines occur in Florida, the West Indies, Middle America, and South America south to Uruguay. Most species occupy fresh or brackish water; some are in coastal marine waters. There are 324 recognized species; many are popular aquarium fishes. Rivulines are characterized by having a *killifish-body shape, tubular nostrils,* and opercular and branchiostegal membranes *united* and often covered with scales. Some have *rounded fins.* Males of many species are brilliantly colored. Rivulines are usually less than 4 in. (10 cm).

MANGROVE RIVULUS *Kryptolebias marmoratus* **Not shown**

IDENTIFICATION: Brown to maroon above and on side; many small *black spots* on side; *cream-colored ring* around *large black spot* on upper half of caudal fin base; dusky to marbled median fins. Large male (males are exceedingly rare in wild FL populations) has orange body and fins, obscure caudal spot. To 2¼ in. (6 cm). **RANGE:** Atlantic and Gulf coasts, cen. and s. FL; also in West Indies, Cen. and S. America. Common. **HABITAT:** Mangrals ("mangrove swamps") and other brackish-water environments; usually over mud; often in crab burrows. Aestivates during long low-water periods in leaf litter and burrows. Over 99 percent of FL individuals are hermaphrodites; rest are males. **SIMILAR SPECIES:** (1) Topminnows and killifishes *lack* tubular nostrils.

Topminnows: Family Fundulidae (28)

As the name suggests, many topminnows swim at or near the surface of the water and are easy to observe from shore. Several species have bright gold or silver marks on the top of the head and body. Topminnows and killifishes are small, often brightly colored fishes with a *flattened head and back, upturned mouth*, large eyes, spineless fins, *1 dorsal fin located far back* on the body, *no lateral line*, and abdominal pelvic fins. About 46 species are known. They occur in fresh, brackish, and salt water of North America south to Yucatán, and in Bermuda and Cuba.

SEMINOLE KILLIFISH *Fundulus seminolis* **Pl. 32**

IDENTIFICATION: *Metallic green side;* interrupted rows of *many small black spots,* dusky crosshatching on large male; fewer spots, *15–20 dark green bars* (often faint) on female. Dark green above; white below; clear to dusky orange fins. Breeding male has bright pink to red anal fin, dark brown wavy lines in dorsal and caudal fins, black border on anal and pelvic fins. Long, fairly slender body; dorsal fin origin in front of anal fin origin. Has 50–55 lateral scales; 16–17 dorsal rays; 13

anal rays. To 6½ in. (16 cm). **RANGE:** Peninsular FL from St. Johns (Atlantic) and New (Gulf) river drainages south to Everglades. Common. **HABITAT:** Open areas of lakes and quiet stream pools. Young usually in schools near vegetation. **SIMILAR SPECIES:** (1) Banded Killifish, *F. diaphanus* (Pl. 32), and (2) Waccamaw Killifish, *F. waccamensis, lack* rows of small black spots, have more distinct bars on side, more slender caudal peduncle. (3) Banded Topminnow, *F. cingulatus* (Pl. 32), and (4) Redface Topminnow, *F. rubrifrons, lack* metallic green side, dark brown wavy lines in fins; are smaller; have red fins on adult, 6–8 dorsal rays, 9–10 anal rays, 28–32 lateral scales.

BANDED KILLIFISH *Fundulus diaphanus* Pl. 32

IDENTIFICATION: Has *10–20 green-brown bars* along silver side; bars more numerous and wider in male. *Long, slender body;* dorsal fin origin in front of anal fin origin. Color variable (see Remarks). Dark olive to tan above, brown stripe along back; white to yellow below; clear to dusky olive-yellow fins. Breeding male has wide green bars along side, yellow throat and fins. Has 35–50, usually 40–49, lateral scales; 13–15 dorsal rays; 10–12 anal rays. To 5 in. (13 cm). **RANGE:** Atlantic Slope drainages from NL to Peedee R., SC; St. Lawrence-Great Lakes and Mississippi R. basins from QC to MB, and south to s. PA, n. IL, and ne. NE. Common; locally abundant. Enters brackish water. **HABITAT:** Shallow, quiet margins of lakes, ponds, and sluggish streams; usually over sand or mud; often near vegetation. Swims in schools a few inches below surface of water. **REMARKS:** Two subspecies, with intergrades in St. Lawrence and Lake Erie drainages. *F. d. diaphanus,* on Atlantic Slope, lacks brown spots on back and upper side, dark stripe along caudal peduncle; has dusky crosshatching on back and side, usually 45–49 lateral scales and 16–17 pectoral rays. *F. d. menona,* elsewhere, has many brown spots on back and upper side, bars on caudal peduncle partially fused at middle into dark stripe, no crosshatching,

usually 40–44 lateral scales, 14–15 pectoral rays. **SIMILAR SPECIES:** (1) See Waccamaw Killifish, *F. waccamensis.*

WACCAMAW KILLIFISH *Fundulus waccamensis* Not shown

IDENTIFICATION: Nearly identical to Banded Killifish, *F. diaphanus*, but is more slender, usually has 54–64 lateral scales. To 4 in. (10 cm). **RANGE:** Lake Waccamaw, Columbus Co., NC. Common. Also Phelps Lake, Washington Co., NC, where probably introduced. **HABITAT:** Over sand in lakes; often near vegetation. Swims a few inches below surface. **SIMILAR SPECIES:** (1) See Banded Killifish, *F. diaphanus* (Pl. 32).

NORTHERN STUDFISH *Fundulus catenatus* Pl. 32

IDENTIFICATION: Light yellow-brown above, short *gold stripe* in front of dorsal fin; silver blue side, *rows of small brown* (female and young) or *red-brown spots* (male) on side. Adult has rows of small brown spots on dorsal and caudal fins. Breeding male (color variable geographically) has *bright blue side,* red spots on head and fins, usually yellow paired fins; orange edge, often submarginal black band on caudal fin. Dorsal fin origin over or slightly in front of anal fin origin. Breeding male has tubercles on side of head, body, and caudal peduncle and on dorsal, anal, and paired fins. Usually 19 scales around caudal peduncle; 12–17 dorsal rays, 13–18 anal rays, 30–52 (usually 41–50) lateral scales; peglike pharyngeal teeth in adult. To 7 in. (18 cm). **RANGE:** Disjunct. (1) Upper East Fork White R. system, IN; upper Salt and Kentucky river drainages, KY; upper Green, middle and lower Cumberland, and Tennessee river drainages, VA, KY, TN, and AL. (2) West of Mississippi R., primarily in Ozark and Ouachita uplands, in cen. and s. MO, e. OK, and AR. (3) Sw. MS in Mississippi R. (Coles Creek, Homochitto R., and Buffalo Bayou), Lake Pontchartrain, and Pearl R. drainages. Introduced into Licking R. system, KY. Common. **HABITAT:** Margins, pools, and backwaters of creeks and small to medium rivers; usually in shallow sandy backwaters of clean rocky creeks. **SIMILAR SPECIES:** See Pl. 32. See (1) Stippled Studfish, *F. bifax*, and (2) Southern Studfish, *F. stellifer*. (3) Barrens Topminnow, *F. julisia*, *lacks* interrupted rows of spots on side, black band on caudal fin; has yellow-orange fins, 9–12 dorsal rays, 10–12 anal rays, usually 36–41 lateral scales.

STIPPLED STUDFISH *Fundulus bifax* Pl. 32

IDENTIFICATION: Similar to Northern Studfish, *F. catenatus*, but has short *interrupted* rows of red or brown spots on side. No black band on dorsal or caudal fins. Has 13–14 dorsal rays, 13–15 anal rays, 42–52 lateral scales. To 4¾ in. (12 cm). **RANGE:** Tallapoosa R. system, GA and AL; Sofkahatchee Creek (lower Coosa R. system), AL. Locally common. **HABITAT:** Same as Northern Studfish. **SIMILAR SPECIES:** (1) See Northern Studfish, *F. catenatus* (Pl. 32). (2) Southern Studfish, *F. stellifer* (Pl. 32), *lacks* rows of brown or red spots on side, often has black edge on caudal fin.

SOUTHERN STUDFISH *Fundulus stellifer* Pl. 32

IDENTIFICATION: Similar to Stippled Studfish, *F. bifax*, and Northern Studfish, *F. catenatus*, but has few to many brown or red spots *scattered* over side (rarely in irregular rows on upper side), *black edge* on dorsal and caudal fins of some large males (about 60 percent), usually 21–22 scales around caudal peduncle, molariform pharyngeal teeth in adult. Has 12–16 dorsal rays, 13–17 anal rays, 38–53 (usually 41–50) lateral scales. To 4¾ in. (12 cm). **RANGE:** Alabama (except Tallapoosa R. system) and upper Chattahoochee river drainages, GA, AL, and se. TN; 1 record for Chickamauga Creek, Tennessee R. drainage, nw. GA. Fairly common. **HABITAT:** Same as Northern Studfish. **SIMILAR SPECIES:** See (1) Stippled Studfish, *F. bifax*, and (2) Northern Studfish, *F. catenatus* (both Pl. 32).

BARRENS TOPMINNOW *Fundulus julisia* Pl. 32

IDENTIFICATION: Many *scattered brown* (female and young) or *red-orange spots* (male) on side of head and body, *yellow-orange fins*, yellow eye on large male. Olive above, *iridescent white-gold stripe* along back to dorsal fin; gray-brown side; white below. Breeding male has red-orange spots on iridescent blue-yellow side; small orange spots on yellow anal and paired fins; yellow edge, orange spots on dusky dorsal and caudal fins. Breeding male has tubercles on cheek and opercle. Dorsal fin origin over anal fin origin. Has 9–12 dorsal rays, 10–12 anal rays, 35–43 lateral scales. To 3¾ in. (9.4 cm). **RANGE:** Upper Caney Fork (Cumberland R. drainage) and upper Duck R. and Elk R. (Tennessee R. drainage) systems, cen. TN. Rare. **HABITAT:** Vegetated pools and margins of springs and spring-fed headwaters and creeks. **SIMILAR SPECIES:** (1) See Whiteline Topminnow, *Fundulus albolineatus.* (2) Northern Studfish, *F. catenatus* (Pl. 32), has rows of spots on side, black band on caudal fin of large male, usually 12–14 dorsal rays, 13–16 anal rays, 41–50 lateral scales; *lacks* orange spots on yellow fins of large male.

WHITELINE TOPMINNOW *Fundulus albolineatus* Pl. 32

IDENTIFICATION: Nearly identical to Barrens Topminnow, *F. julisia*, but has *interrupted white streaks* on rear half of side of large male. To 3¾ in. (8.4 cm). **RANGE:** Big Spring, Madison Co., AL. Extinct. **HABITAT:** Big Spring was impounded, lined with concrete, and stocked with Common Carp, *Cyprinus carpio*, and Goldfish, *Carassius auratus*. **SIMILAR SPECIES:** (1) See Barrens Topminnow, *F. julisia* (Pl. 32).

SPECKLED KILLIFISH *Fundulus rathbuni* Pl. 32

IDENTIFICATION: *Black line* (absent on juvenile) from mouth to bottom of eye. Yellow-brown above and on side, *many dark brown spots on* back and side of juvenile and female; silver cheek, opercle, and throat;

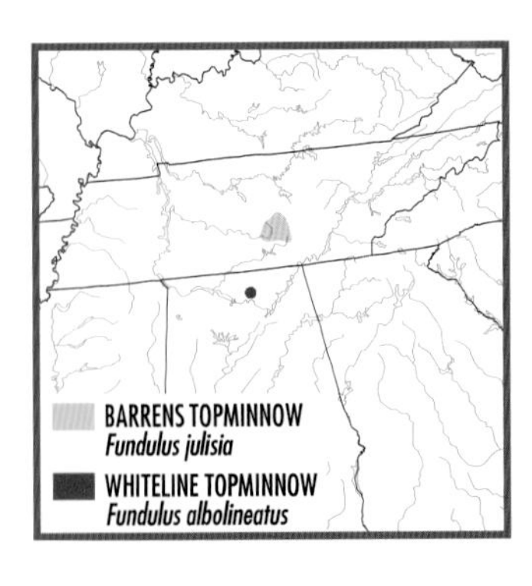

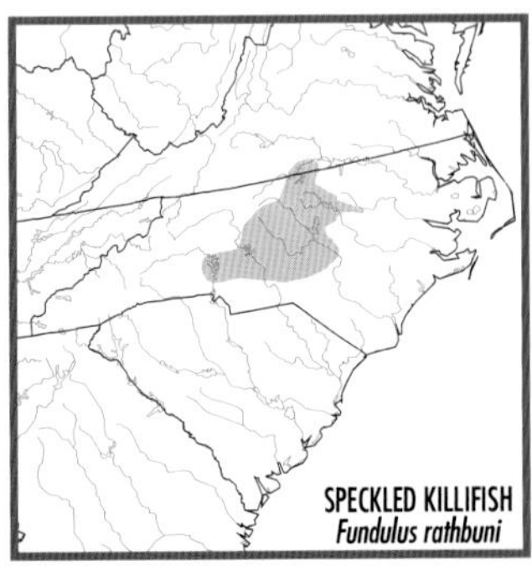

white to yellow below; clear to yellow fins. Adult male is gold-brown with black spots on side of iridescent gold head; has dusky fins, yellow edge on caudal fin. Dorsal fin origin in front of anal fin origin. Has 10–12 dorsal rays; 10–12 anal rays; 34–40 (usually 35–37) lateral scales. To 3¾ in. (9.6 cm). **RANGE:** Dan (Roanoke), Neuse, Cape Fear, Peedee, and Catawba (Santee) river drainages, VA and NC. Primarily on Piedmont and upper Coastal Plain. Common but localized. **HABITAT:** Backwaters and pools of creeks and small to medium rivers. Usually over sand or mud. **SIMILAR SPECIES:** (1) Other topminnows (Pls. 32 & 33) *lack* black line on side of head, yellow-brown to gold head and body.

PLAINS TOPMINNOW *Fundulus sciadicus* Pl. 32

IDENTIFICATION: Bronze flecks and dark crosshatching on *blue-green back and upper side;* narrow *gold stripe* in front of dorsal fin; silver blue dashes on side of head; white below; clear to yellow-orange fins. Breeding male has red-orange band on dusky caudal fin, red-orange edge on dorsal and anal fins. Dorsal fin origin behind anal fin origin; deep caudal peduncle. Has 9–11 dorsal rays, 12–15 anal rays, 33–37 lateral scales. To 2¾ in. (7 cm). **RANGE:** Disjunct. Missouri R. basin from w. IA and MN (where extirpated) to e. WY; Missouri R. drainage, cen. MO; Neosho R. system, sw. MO, se. KS, and ne. OK. Common but localized. **HABITAT:** Springs and their effluents; quiet to flowing pools and backwaters of creeks and small to medium rivers; usually near vegetation. **SIMILAR SPECIES:** (1) Golden Topminnow, *F. chrysotus* (Pl. 32), has green bars, red spots on body and fins of large male; 7–9 dorsal rays, 9–11 anal rays, 30–34 lateral scales; *lacks* dark crosshatching on blue-green back and upper side. (2) Northern Studfish, *F. catenatus* (Pl. 32), has rows of small brown spots on side, dorsal fin origin above or in front of anal fin origin.

GOLDEN TOPMINNOW *Fundulus chrysotus* PL. 32

IDENTIFICATION: *Gold flecks* on side; usually *8–11 green bars* (often faint) on side of large male. Yellow-green above; white below; clear to yellow fins. Breeding male has *bright red to red-brown spots* on rear half of body and on dorsal, caudal, and anal fins; red caudal fin. Dorsal fin origin behind anal fin origin; long snout, large eye. Usually 7–8 long rakers on 1st gill arch, *7 preopercular pores*; 7–9 dorsal rays; 9–11 anal rays; 30–34 lateral scales. To 3 in. (7.5 cm). **RANGE:** Atlantic and Gulf coastal plains from Waccamaw R. drainage, SC, to Trinity R. drainage, TX; Former Mississippi Embayment north to KY and MO. East of Mississippi R. mostly restricted to lower Coastal Plain. Common in FL; uncommon elsewhere. **HABITAT:** Swamps, sloughs, vegetated pools and backwaters of sluggish creeks and small to medium rivers. **SIMILAR SPECIES:** (1) Banded Topminnow, *F. cingulatus* (Pl. 32), and (2) Redface Topminnow, *F. rubrifrons*, *lack* gold flecks on side; usually have 9–10 short rakers on 1st gill arch, 5–6 preopercular pores, 12–15 dark bars along side of male, smaller eye, shorter snout, dorsal fin origin in front of anal fin origin. (3) Plains Topminnow, *F. sciadicus* (Pl. 32), *lacks* dark bars on side, red spots on body and fins; has dark crosshatching on blue-green back and upper side, 9–11 dorsal rays, 12–15 anal rays, 33–37 lateral scales.

BANDED TOPMINNOW *Fundulus cingulatus* PL. 32

IDENTIFICATION: Olive above, *rows of small brown to red spots* on side; white to orange below; clear to *light red-orange fins.* Large male has red-orange fins, *12–15 green bars* along side. Dorsal fin origin in front of anal fin origin; moderate eye. Usually 9–10 short rakers on 1st gill arch, *6 preopercular pores*. Has 6–8 dorsal rays; 9–10 anal rays; 28–32 lateral scales. To 3 in. (7.8 cm). **RANGE:** Suwannee and Waccasassa river systems, peninsular FL; lower Coastal Plain from Ochlockonee R. system, FL, to lower Mobile Bay drainage, AL. Fairly common. **HABI-**

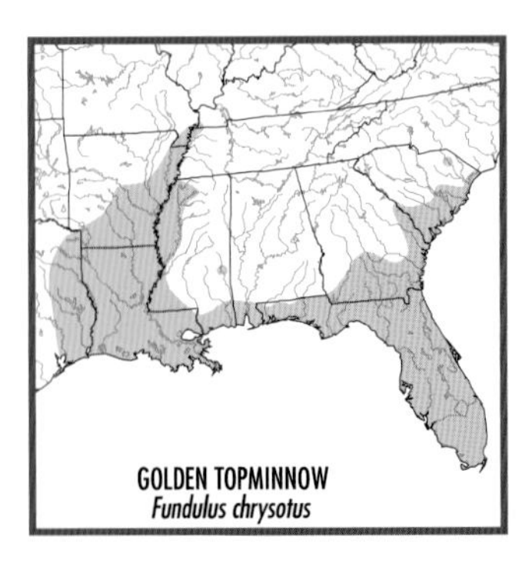

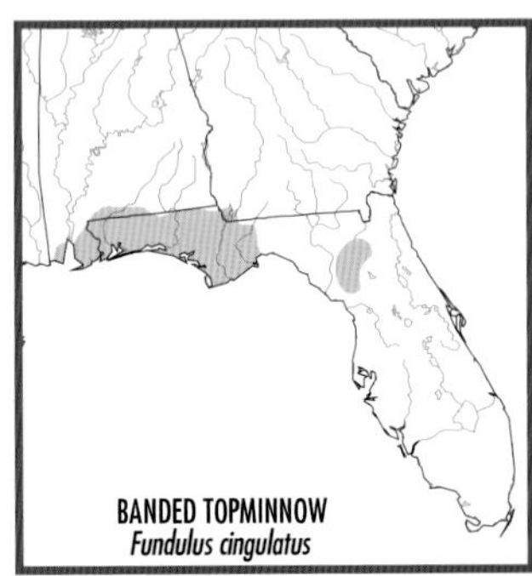

TAT: Shallow, often ephemeral, water bodies, including ditches, ponds, and backwaters of sluggish creeks and small to medium rivers; usually near vegetation. **SIMILAR SPECIES:** (1) See Redface Topminnow, *F. rubrifrons*. (2) Golden Topminnow, *F. chrysotus* (Pl. 32), has *gold flecks* on side, dorsal fin origin behind anal fin origin, usually *8–11 dark bars* along side of male, longer snout, larger eye, usually 7–8 long rakers on 1st gill arch, *7 preopercular pores*. (3) Lined Topminnow, *F. lineolatus*, and (4) Russetfin Topminnow, *F. escambiae* (both Pl. 33), have large blue-black bar under eye, large gold spot on top of head, and smaller spot at dorsal fin origin; *lack* red fins.

REDFACE TOPMINNOW *Fundulus rubrifrons* Not shown

IDENTIFICATION: Nearly identical to Banded Topminnow, *F. cingulatus*, but has *5 preopercular pores, red side of head and jaws* on large male. To 3 in. (7.8 cm). **RANGE:** Lower Coastal Plain from Altamaha R. drainage (1 record) and Okefenokee Swamp, se. GA, to s. FL; isolated population in Fenholloway R. system, w. FL. Fairly common but localized. **HABITAT:** Shallow, often ephemeral, water bodies, including ditches, marshes, and backwaters of sluggish creeks and small to medium rivers; usually near vegetation. **SIMILAR SPECIES:** (1) See Banded Topminnow, *F. cingulatus* (Pl. 32).

PLAINS KILLIFISH *Fundulus zebrinus* Pl. 33

IDENTIFICATION: Has *12–26 gray-green bars* (fewer, wider bars on male) on silver white side. Breeding male has *bright orange to red dorsal, anal, and paired fins*. Tan-olive above, white to yellow below; clear to dusky fins, yellow pectoral fins. Deep caudal peduncle; dorsal fin origin in front of anal fin origin. Has 38–68 lateral scales; 11–17 (usually 13–15) dorsal rays; 9–16 (usually 13–14) anal rays. To 4 in. (10 cm). **RANGE:** Native to Mississippi R. and Gulf Slope basins from n.-cen. MO to WY, and south to Colorado and Pecos river drainages,

REDFACE TOPMINNOW
Fundulus rubrifrons

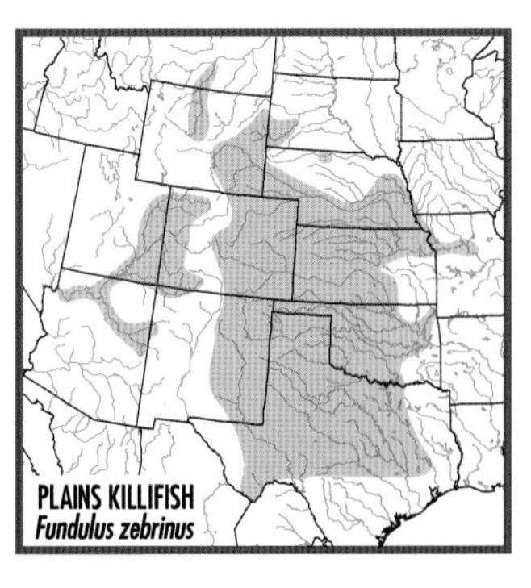
PLAINS KILLIFISH
Fundulus zebrinus

TX and NM. Mostly on Great Plains. Introduced into upper Missouri R. basin, SD, MT, and WY; Colorado R. drainage, CO, UT, NM, and AZ; and Rio Grande, NM and TX. Common; locally abundant. **HABITAT:** Shallow sandy runs, pools, and backwaters of headwaters, creeks, and small to medium rivers. Tolerant of extremely alkaline and saline streams, often where few other fishes can survive. **REMARKS:** Sometimes placed in monotypic genus, *Plancterus.* Plains Killifishes bury in sand with only mouth and eyes visible. **SIMILAR SPECIES:** (1) Plains Topminnow, *F. sciadicus* (Pl. 32), *lacks* bars on side; is smaller; has dorsal fin origin behind anal fin origin, 33–37 lateral scales, 9–11 dorsal rays.

NEXT 5 SPECIES: Starhead topminnows have *large blue-black bar* under eye, *6–8 thin brown to red-brown stripes (on female and juvenile) or rows of dots (on male)* along side of body, *large iridescent gold spot* on top of head, and *small iridescent gold spot* at dorsal fin origin. Olive above; green, red, and blue flecks on silver yellow side; white below. *Dark green bars* (usually absent on Western Starhead Topminnow, *F. blairae*) along side of male (faint bars, usually only on rear half of side, may be present on female); small red-brown spots in dorsal, caudal, and anal fins of male. Dorsal fin origin behind anal fin origin; 6–8 dorsal rays; 8–11 anal rays; 30–36 lateral scales.

Starhead topminnows differ from one another primarily in color pattern, arrangement of scales on top of head, and configuration of supraorbital canal pores (Fig. 51). To escape predators, these fishes may jump on land, wait a short time, then jump back into water.

LINED TOPMINNOW *Fundulus lineolatus* Pl. 33

IDENTIFICATION: Has 11–15 dark green bars on side of male, *thickest at middle;* no or few dark dashes and specks between *6–8 black stripes* on side of female. Head red from eye to opercle. Black blotch on pectoral fin base. Has 16–19, usually 16, scales around caudal peduncle. To

3¼ in. (8.4 cm). **RANGE:** Atlantic and Gulf coastal plains from extreme lower James R. drainage, s. VA, to Ocklockonee R. drainage, GA and FL; south in FL to Lake Okeechobee. Common. **HABITAT:** Swamps and other vegetated standing water bodies; quiet pools and backwaters of streams. **SIMILAR SPECIES:** (1) Other starhead topminnows (Pl. 33) have dark bars along side of male *uniform in width* or *lack* them, usually have 17–20 scales around caudal peduncle.

BAYOU TOPMINNOW *Fundulus nottii* Pl. 33

IDENTIFICATION: Has 9–15 dark bars of *nearly uniform width* on side of male extending forward to (or near) pectoral fin base and ventrally below lowest dark stripe; many dark dashes and specks between 6–8 dark stripes on side of female. Has 18–20 scales around caudal peduncle. To 3 in. (7.8 cm). **RANGE:** Gulf drainages from Choctawhatchee R. drainage, FL, to Lake Pontchartrain drainage, LA and MS; Yazoo, Big Black, and Homochitto river systems, MS. Common. **HABITAT:** Swamps, sloughs, and vegetated backwaters and pools of streams. **SIMILAR SPECIES:** (1) See Russet Topminnow, *F. escambiae* (Pl. 33). (2) Lined Topminnow, *F. lineolatus* (Pl. 33), has bars along side of male *thickest at middle,* usually 16 scales around caudal peduncle.

RUSSETFIN TOPMINNOW *Fundulus escambiae* Pl. 33

IDENTIFICATION: Nearly identical to Bayou Topminnow, *F. nottii,* but *lacks* many dark dashes and specks between dark stripes on side of female, has 0–14 dark bars on side of male extending *forward only to between paired fins* and below only as far as lowest dark stripe. To 3 in. (7.8 cm). **RANGE:** Gulf Coastal Plain in Suwannee R. drainage, FL, and from Aucilla R., FL, to Perdido R. drainage, AL. Common. **HABITAT:** Vegetated sloughs, swamps, and quiet pools and backwaters of streams. **SIMILAR SPECIES:** (1) See Bayou Topminnow, *F. nottii* (Pl. 33).

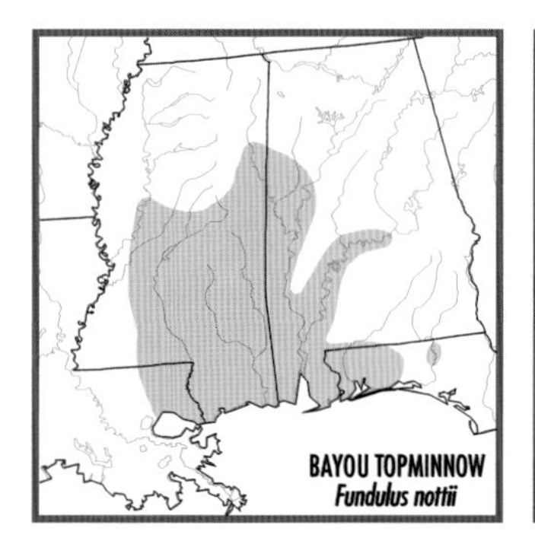

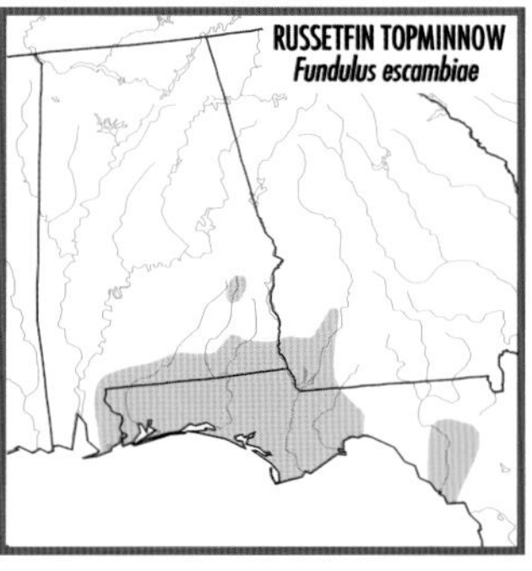

STARHEAD TOPMINNOW *Fundulus dispar* Pl. 33

IDENTIFICATION: *No or few dark dashes and specks* between 6–8 thin dark stripes on side of female; *3–13 dark bars* on side of male, those at front thin and restricted to midside. Has 16–20, usually 18–20, scales around caudal peduncle. To 3 in. (7.8 cm). **RANGE:** Lake Michigan and Mississippi R. basins from s. MI and WI south to Ouachita R. drainage, LA, and Big Black R., MS; Gulf Slope drainages from Mobile Bay, AL, to Pearl R., MS. Locally common but less so as wetlands are drained. **HABITAT:** Vegetated standing water bodies; quiet pools and backwa-

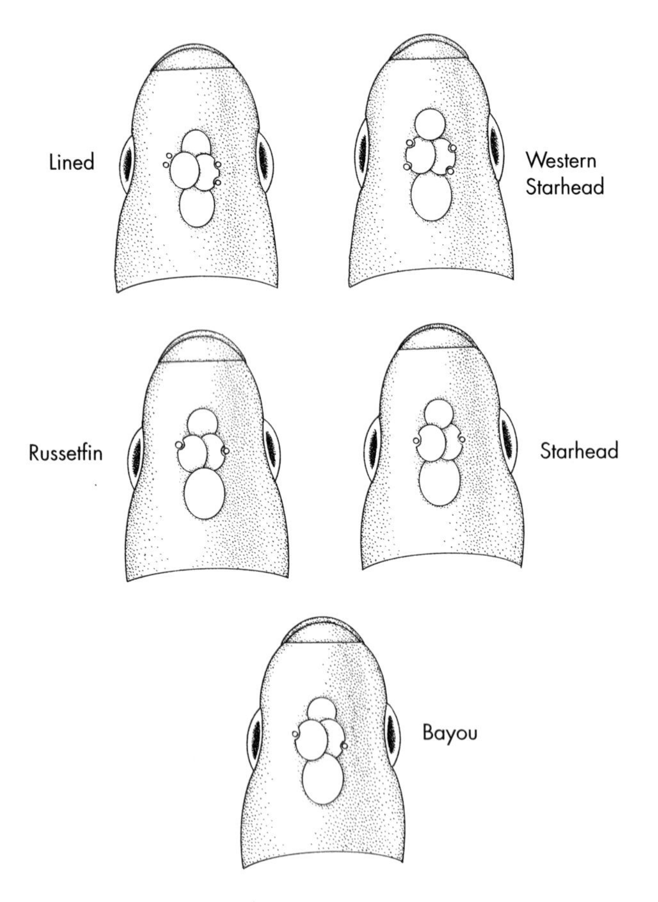

Fig. 51. Starhead topminnows—scale and supraorbital pattern.

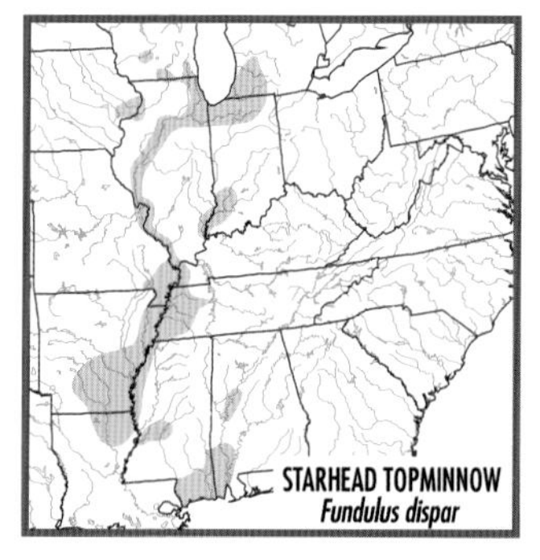

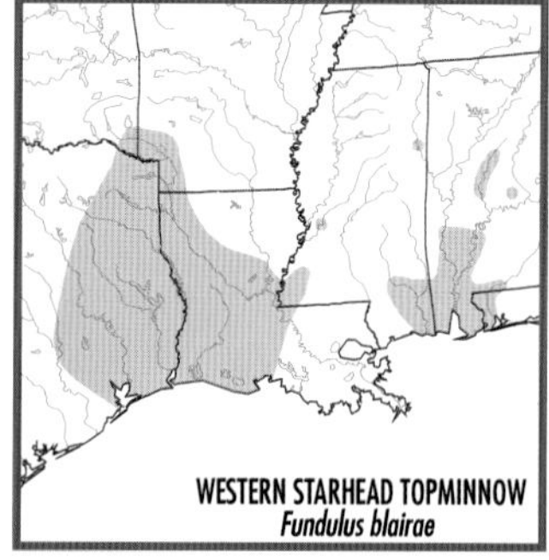

ters of streams. **SIMILAR SPECIES:** See Pl. 33. (1) Western Starhead Topminnow, *F. blairae*, *lacks* dark bars along side of body of male, has *many dark dashes and specks* between dark stripes along side of female. (2) Bayou Topminnow, *F. nottii*, and (3) Russetfin Topminnow, *F. escambiae*, have *many dark dashes and specks* between dark stripes along side of female.

VESTERN STARHEAD TOPMINNOW *Fundulus blairae* Pl. 33

IDENTIFICATION: *No dark bars* (rarely present on large male) on side of body; *many dark dashes and specks* between 7–9 dark stripes on side of female. Has 16–20, usually 17–18, scales around caudal peduncle. To 3 in. (7.8 cm). **RANGE:** Gulf Slope drainages from Escambia R., AL and FL, to Brazos R., TX; north in Red R. drainage (Mississippi R. basin) to sw. AR and se. OK. Common west of Mississippi R.; rare and sporadic east of Mississippi R. **HABITAT:** Vegetated sloughs, swamps, and quiet pools and backwaters of streams. **SIMILAR SPECIES:** See Pl. 33. (1) Starhead Topminnow, *F. dispar*, has *dark bars* along side of body of male, has *no or few* dark dashes and specks between dark stripes along side of female. (2) Bayou Topminnow, *F. nottii*, (3) Russetfin Topminnow, *F. escambiae*, and (4) Lined Topminnow, *F. lineolatus,* have *dark bars* along side of body of male.

LACKSTRIPE TOPMINNOW *Fundulus notatus* Pl. 33

IDENTIFICATION: *Wide blue-black stripe* along side, around snout, and onto caudal fin. Olive-tan above, *silver white spot* on top of head; usually few dusky to dark (rarely black) spots (often absent) on upper side; light blue along upper edge of stripe; yellow fins; dorsal, caudal, and anal fins amber at base and heavily spotted with black; white to light yellow below. Male has crossbars on stripe, larger dorsal and anal fins; is deeper bodied. To 3 in. (7.4 cm). **RANGE:** Great Lakes (Erie, Huron, and Michigan) and Mississippi R. basins from s. ON, MI, WI, and

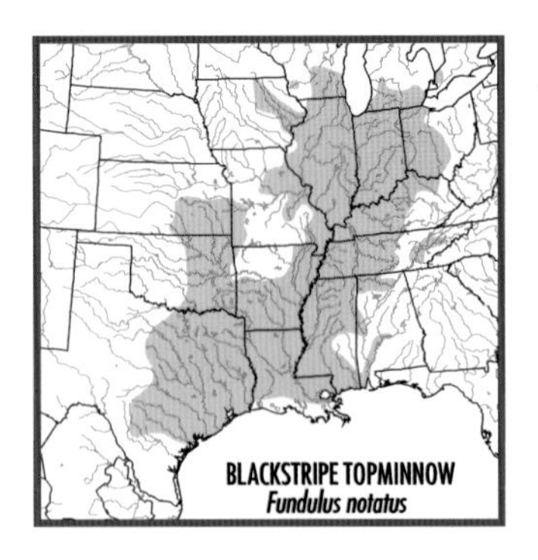

IA south to Gulf (west to cen. KS and OK); Gulf Slope drainages from Mobile Bay, AL, to San Antonio Bay, TX. Common in lowlands, rare to absent in uplands. **HABITAT:** Quiet surface water, usually near margins of creeks and small rivers, ponds, and lakes. **SIMILAR SPECIES:** See (1) Blackspotted Topminnow, *F. olivaceus*, and (2) Broadstripe Topminnow, *F. euryzonus* (both Pl. 33).

BLACKSPOTTED TOPMINNOW *Fundulus olivaceus* Pl. 33

IDENTIFICATION: Similar to Blackstripe Topminnow, *F. notatus*, but has few to many (male has more) *discrete, intensely black spots* on light tan upper side, slightly longer snout. To 3¾ in. (9.7 cm). **RANGE:** Mississippi R. basin from e. TN, w. KY, s. IL, and cen. MO south to Gulf (west to e. OK); Gulf Slope drainages from Chattahoochee R., GA and FL, to Galveston Bay, TX. Common. **HABITAT:** Near surface of quiet to flowing water, usually near margins of clear, sandy to gravelly headwaters, creeks, and small rivers. **REMARKS:** Hybrids between Blackspotted and Blackstripe and Broadstripe topminnows are common and intermediate in morphology. **SIMILAR SPECIES:** See (1) Blackstripe Topminnow, *F. notatus*, and (2) Broadstripe Topminnow, *F. euryzonus* (both Pl. 33).

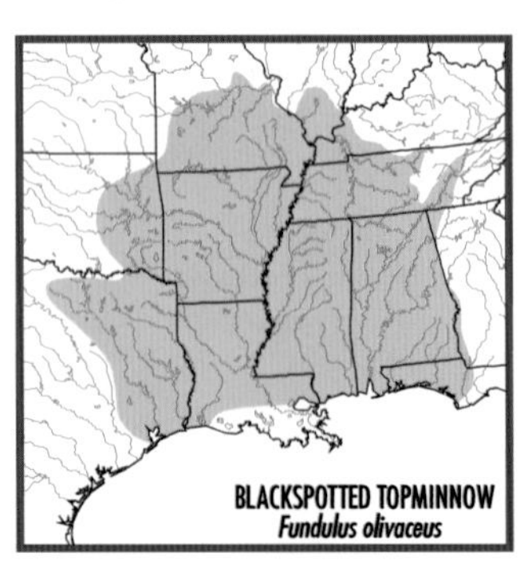

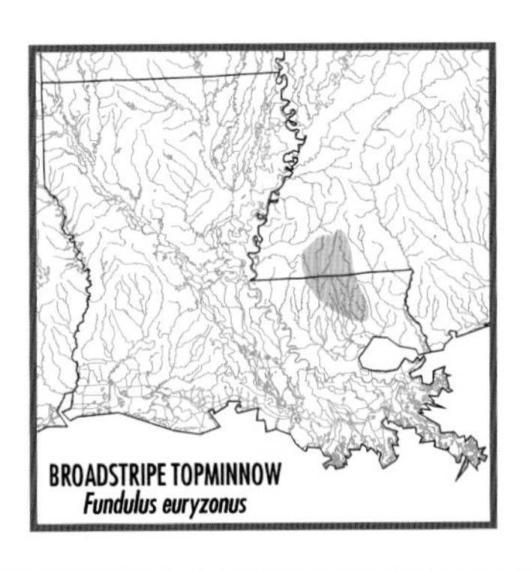

BROADSTRIPE TOPMINNOW *Fundulus euryzonus* Pl. 33

IDENTIFICATION: Similar to Blackstripe Topminnow, *F. notatus*, and Blackspotted Topminnow, *F. olivaceus*, but has *extremely wide, purple-brown stripe* along side, lower edge reaching to middle of pectoral fin base and to within 1 scale row of lower edge of caudal peduncle; *no crossbars* on stripe along side of male; usually 8 dorsal rays (Blackstripe and Blackspotted topminnows usually have 9 and 10, respectively, within range of Broadstripe Topminnow). Spots on upper side like those of Blackstripe Topminnow. To 3¼ in. (8.3 cm). **RANGE:** Tangipahoa and upper Amite river systems (Lake Pontchartrain drainage), MS and LA. Uncommon. **HABITAT:** Usually at surface near shoreline vegetation in quiet pools and backwaters of creeks and small rivers. **SIMILAR SPECIES:** See (1) Blackstripe Topminnow, *F. notatus*, and (2) Blackspotted Topminnow, *F. olivaceus* (both Pl. 33).

NEXT 3 SPECIES: Brackish water species that enter freshwater bodies, especially in FL.

MARSH KILLIFISH *Fundulus confluentus* Not shown

IDENTIFICATION: Dorsal fin origin slightly in front of anal fin origin. *Ocellus* at rear of dorsal fin of female. Has 12–24 dark bars wider than interspaces on side of male; narrow dark bars separated by wide interspaces and dark spots on female. Large male is olive-blue, has *light blue to yellow spots* on side and median fins; orange or yellow paried fins. Has 33–35 lateral scales; usually *4 pores* on each mandible, 11 dorsal rays, 10 anal rays. To 3¾ in. (9.4 cm). **RANGE:** Atlantic and Gulf coasts from St. Johns R., FL, to FL Keys and west to s. AL. Common. **HABITAT:** Brackish and freshwater marshes, bayous, tidal streams; near vegetation. **SIMILAR SPECIES:** (1) Golden Topminnow, *F. chrysotus*, and (2) Banded Topminnow, *F. cingulatus* (both Pl. 32), *lack* ocellus on dorsal fin, have red-brown spots on side, red fins, 6–9 dorsal rays; Golden Topminnow has dorsal fin origin behind anal fin origin. (3) Gulf Killifish, *F. grandis*, *lacks* ocellus on dorsal fin, has *5 pores* on mandible, chubbier body.

GULF KILLIFISH *Fundulus grandis* **Pl. 57**

IDENTIFICATION: Blunt head, short snout; robust body, deep caudal peduncle. Dorsal fin origin in front of anal fin origin. Silver gray to blue-green above, white to orange-yellow below; *light blue to yellow spots* on side and median fins, brighter on male. Has 32–37 lateral scales; 10–14 dorsal rays; usually 11 anal rays; *5 pores* on each mandible. To 7 in. (18 cm). **RANGE:** Atlantic and Gulf coasts from St. Johns R., FL, to Veracruz, Mexico; Cuba. Uncommon in fresh water except far inland in Brazos R. and Rio Grande, TX. **HABITAT:** Grassy bays, canals, and nearby fresh water. Usually over mud, near vegetation. **SIMILAR SPECIES:** (1) See Mummichog, *F. heteroclitus*.

MUMMICHOG *Fundulus heteroclitus* **Not shown**

IDENTIFICATION: Nearly identical to Gulf Killifish, *F. grandis*, but has more *convex upper profile, dark bars alternating with silvery interspaces* on side, small ocellus at rear of dorsal fin of male, *4 pores* on each mandible. To 5 in. (12.5 cm). **RANGE:** Atlantic Coast from Gulf of St. Lawrence to ne. FL. Sporadic in fresh water. Abundant. **HABITAT:** Saltwater marshes, tidal creeks, and nearby fresh water. **REMARKS:** Two subspecies: *F. h. heteroclitus* in south appears to intergrade with *F. h. macrolepidotus*, northern form, in NJ. **SIMILAR SPECIES:** (1) See Gulf Killifish, *F. grandis* (Pl. 57).

RAINWATER KILLIFISH *Lucania parva* **Pl. 33**

IDENTIFICATION: *Large, dark-edged scales* on back and side. Large male has *black spot* at front of dusky orange dorsal fin, thin *black edges* on all but pectoral fins. Deep compressed body; small upturned mouth. Dorsal fin origin in front of anal fin origin. Light brown to olive above, dusky stripe along silver side; white below; orange-yellow

anal and paired fins. Has 23–29 lateral scales; 9–13 dorsal rays; 5–6 branchiostegal rays. To 2¾ in. (7 cm). **RANGE:** Marine; enters fresh water from MA to Mexico. Ascends Rio Grande and Pecos R., TX and NM. Introduced in OR, UT, NV, and CA. Common along coast and in St. Johns R., FL. **HABITAT:** Vegetated quiet water; usually swims several inches below surface of water. **SIMILAR SPECIES:** (1) Bluefin Killifish, *L. goodei* (Pl. 33), has dark stripe along side; blue, yellow, orange, or red dorsal and anal fins on male.

BLUEFIN KILLIFISH *Lucania goodei* Pl. 33

IDENTIFICATION: *Wide zigzag black stripe* from tip of snout to *black spot* on caudal fin base. Dorsal fin of large male *bright iridescent blue at front; yellow, orange, red, or blue at rear; black band on base.* Anal fin of large male yellow, orange, red, or blue with black base, thin black edge; caudal fin red-orange. Fairly slender, compressed body; small upturned mouth. Dorsal fin origin in front of anal fin origin. Dusky brown to olive back; light stripe along upper side; dark-edged scales on side; silver white below. Has 29–32 lateral scales; 9–12 dorsal rays; 5–6 branchiostegal rays. To 2 in. (5 cm). **RANGE:** Peninsular FL and FL panhandle as far west as Choctawhatchee R. drainage; Chipola R. drainage, se. AL; sporadic along Atlantic Coast as far north as Ogeechee R., GA. Isolated populations near Wilmington, NC, and Charleston, SC, probably introduced. Common. **HABITAT:** Vegetated sloughs, ponds, lakes; pools and backwaters of streams; common near springs; usually swims well below surface of water. **SIMILAR SPECIES:** (1) Rainwater Killifish. *L. parva* (Pl. 33), *lacks* dark stripe along side, brightly colored dorsal and anal fins; is deeper bodied.

PYGMY KILLIFISH *Leptolucania ommata* Pl. 33

IDENTIFICATION: *Cream-yellow halo* around *large black spot* on caudal peduncle. Male has 5–7 *faint bars* on rear half of side; female has dusky brown stripe along side, *black spot* on midside. Slender body;

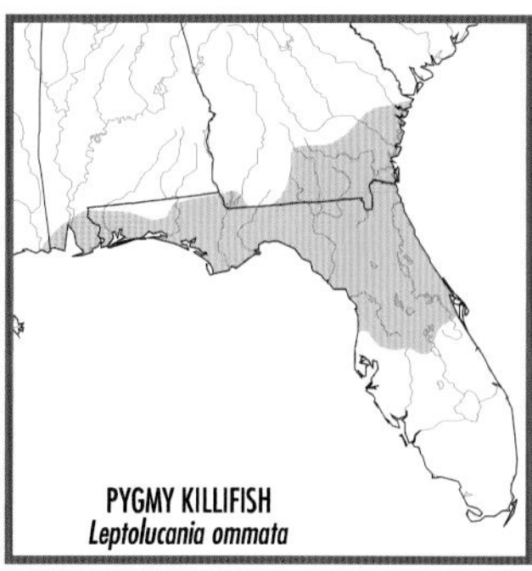

small upturned mouth; large eye. Dorsal fin origin behind anal fin origin. Dusky green to straw yellow above; dusky stripe through eye and across lower jaw; yellow below. Large male has yellow-orange median and pelvic fins. Has 26–32 lateral scales; 6–8 dorsal rays; 9–10 anal rays; 3 branchiostegal rays. To 1¼ in. (2.9 cm). **RANGE:** Atlantic and Gulf slope drainages from Ogeechee R., GA, to Escatawpa R., MS; south to cen. FL. Common; locally abundant. **HABITAT:** Surface waters of swamps, vegetated sloughs, and quiet-water areas of creeks and small rivers. **SIMILAR SPECIES:** (1) Least Killifish, *Heterandria formosa* (Pl. 35), has red around black spot at front of dorsal fin, black spot at front of anal fin of female, gonopodium on male, 6–9 anal rays on female.

LIVEBEARERS: Family Poeciliidae (16 native; 8 introduced)

Males of this North and South American family have the front rays of their anal fin *elongated and modified into an intromittent organ* (gonopodium) for internal fertilization. Poeciliids occur in both fresh and brackish water. Most of the 340 species are in the tropics. Western and Eastern mosquitofishes, *Gambusia affinis* and *G. holbrooki,* respectively, have been introduced into many parts of the world to control mosquitoes. The Guppy, *Poecilia reticulata*, and several other species have been introduced outside their native ranges through the aquarium trade.

Livebearers are similar in appearance to topminnows (Fundulidae), with the top of the *head flat,* the mouth strongly upturned, *no lateral line,* 1 dorsal fin, and abdominal pelvic fins. They differ in that the third anal fin ray is unbranched (branched in topminnows), the male has the anal fin modified into an intromittent organ, and females give birth to live young (except the Loanguma, *Tomeurus gracilis*, in South America).

Some populations consist only of females (see Amazon Molly, *Poecilia formosa*). In some but not all of these populations, the female mates with a male of another species; sperm stimulates development but does not contribute genetic material to offspring.

SAILFIN MOLLY *Poecilia latipinna* Pl. 35

IDENTIFICATION: Large male has *huge sail-like dorsal fin* with orange edge, black spots on outer half, black wavy lines on lower half. Small head; deep, compressed body; extremely deep caudal peduncle; dorsal fin origin *in front of* pelvic fin origin. About *5 rows of dark brown spots,* iridescent yellow flecks on olive side. Olive above; white to yellow below; brown spots on dorsal and caudal fins. Large male has iridescent blue back; orange on lower head and breast; black edge (often absent), orange spots on purple-blue caudal fin. Has 23–28 lateral scales; 13–16 dorsal rays. To 6 in. (15 cm). **RANGE:** Atlantic and Gulf drainages from Cape Fear drainage, NC, to Veracruz, Mexico. Restricted to coastal areas in most of range; found farther inland in FL, LA, and TX. Introduced into Colorado R. drainage, AZ, and elsewhere in w. U.S. and Canada. Abundant in peninsular FL, less common elsewhere. **HABITAT:** Ponds, lakes, sloughs, and quiet, often vegetated, backwaters and pools of streams; fresh and brackish water. **SIMILAR SPECIES:** (1) See Amazon Molly, *P. formosa* (Pl. 35). (2) Shortfin Molly, *P. mexicana*, and (3) Mexican Molly, *P. sphenops*, have dorsal fin origin *behind* pelvic fin origin, 8–11 dorsal rays; large male *lacks* sail-like dorsal fin.

AMAZON MOLLY *Poecilia formosa* Pl. 35

IDENTIFICATION: Similar to Sailfin Molly, *P. latipinna*, but *lacks* males (see Remarks); lacks rows of brown spots on side (may have rows of dusky black spots); has smaller dorsal fin, *10–12 rays.* To 3¾ in. (9.6 cm). **RANGE:** Gulf Coastal Plain of TX from Brazos R. to Rio Grande; also in Mexico south to Veracruz. Common. **HABITAT:** Backwaters and quiet pools of streams, sloughs, and ditches, usually over mud; fresh and brackish water. **REMARKS:** Amazon Molly is an all-female species thought to have originated from hybridization between Sailfin and Shortfin mollies, and is intermediate between these 2 species in morphology. Eggs develop in Amazon Mollies following stimulation by sperm from either parental species. **SIMILAR SPECIES:** (1) See Sailfin Molly, *P. latipinna* (Pl. 35). (2) Shortfin Molly, *P. mexicana*, *has males,* dorsal fin origin *behind* pelvic fin origin, rows of orange spots on side, *8–11 dorsal rays.*

SHORTFIN MOLLY *Poecilia mexicana* **Not shown**

IDENTIFICATION: Dorsal fin origin *behind* pelvic fin origin. Dark olive above; usually 5–6 rows of orange spots on side; white to pale orange below; orange anal and pelvic fins. Large male is dark green (rarely black); has yellow-orange spots on dorsal fin, broad orange band, blue spots on black caudal fin. Usually 25–29 lateral scales, 8–11 dorsal rays. Tricuspid teeth. To about 4 in. (10 cm). **RANGE:** Native to Atlantic Slope from Mexico to Costa Rica. Locally established in CA, CO, ID, and NV. Common at some localities. **HABITAT:** Warm springs and their effluents, canals, weedy ditches, and stream pools. **SIMILAR SPECIES:** (1) See Mexican Molly, *P. sphenops*. (2) Sailfin Molly, *P. latipinna* (Pl. 35), has dorsal fin origin *in front of* pelvic fin origin, 13–16 dorsal rays; large male has sail-like dorsal fin. (3) Amazon Molly, *P. formosa* (Pl. 35), *lacks males,* lacks rows of orange spots on side; has 10–12 dorsal rays.

MEXICAN MOLLY *Poecilia sphenops* **Not shown**

IDENTIFICATION: Similar to Shortfin Molly, *P. Mexicana*, but has *unicuspid teeth*, large male has black edge on dorsal and caudal fins. To 4 in. (10 cm). **RANGE:** Native to Atlantic and Pacific slope streams of s. Mexico and Guatemala. Locally established in MT and NV, possibly FL. **HABITAT:** Creeks with little current; warm springs, ponds, ditches, and lagoons. **SIMILAR SPECIES:** (1) See Shortfin Molly, *P. mexicana*.

GUPPY *Poecilia reticulata* **Not shown**

IDENTIFICATION: Similar to Western and Eastern mosquitofishes, *Gambusia attinis* and *G. holbrooki*, respectively, but *lacks* black teardrop; large male usually has *black spot(s)* on lower side, red or blue blotches in fins, usually on body. Gray body; slightly outlined scales; 6 or 7 dorsal rays. To 2 in. (5.1 cm); males usually less than 1 in. (2.5 cm). **RANGE:** Native to Trinidad and Tobago, and n. S. America from w. Venezuela to Guyana. Established in AB, AZ, ID, NV, TX, and WY. Generally uncommon. **HABITAT:** Warm springs and their effluents, weedy ditches, and canals. **SIMILAR SPECIES:** (1) Western Mosquitofish, *Gambusia affinis* (Pl. 35), lacks red or blue blotches, has *black teardrop*.

GREEN SWORDTAIL *Xiphophorus hellerii* **Not shown**

IDENTIFICATION: Lower caudal fin rays elongated into a *black-edged "sword"* on large male. Green above; dark-edged scales on back and side; yellow below. Large male is usually red below; has *black stripe along side,* yellow-orange to red "sword"; *scythe-shaped claw* at tip of ray 5a (as large as ray 3 hook) on gonopodium; usually small, blunt serrae at tip of ray 4p. Usually 26–27 lateral scales, *12–14 dorsal rays*; 1 row of jaw teeth. To 5 in. (13 cm). **RANGE:** Native from Río Nantla, Veracruz, Mexico, to nw. Honduras. Established in canals near Tampa, Hillsborough Co., FL; Warm Springs Creek, Clark Co., ID; Trudeau and Beaverhead Rock ponds, Madison Co., MT; Kelly Warm Spring, Teton Co., WY; and (as hybrid with Southern Platyfish, *X. maculatus*) Indian Spring, Clark Co., NV. Abundant at some localities. **HABITAT:** Warm springs and their effluents, weedy canals, and ponds. **SIMILAR SPECIES:** (1) Variable Platyfish, *X. variatus*, and (2) Southern Platyfish, *X. maculatus, lack* "swordtail" on large male, usually have *9–12 dorsal rays*.

VARIABLE PLATYFISH *Xiphophorus variatus* **Not shown**

IDENTIFICATION: Olive; *black spots or marbling* on side of caudal peduncle; often a green edge on caudal fin. Large male has dusky or black blotches on dorsal fin. No claw at tip of ray 5a; well-developed serrae at tip of ray 4p. Usually *24–25 lateral scales, 10–12 dorsal rays*; 2 rows of jaw teeth. To 2¾ in. (7 cm). **RANGE:** Native to Atlantic Slope of Mexico from s. Tamaulipas to n. Veracruz. Established in Alachua Co., FL, and in springs in Beaverhead, Granite, and Madison counties, MT. Uncommon. **HABITAT:** Warm springs; weedy canals and ditches. **SIMILAR SPECIES:** (1) See Southern Platyfish, *X. maculatus.*

SOUTHERN PLATYFISH *Xiphophorus maculatus* **Not shown**

IDENTIFICATION: Similar to Variable Platyfish, *X. variatus,* but usually has *23–24 lateral scales, 9–10 dorsal rays.* Extremely variable in color: olive, yellow, orange, red, black, or combinations of those colors on body and fins. To 2¼ in. (6 cm). **RANGE:** Native to Atlantic Slope from Río Nautla basin, Mexico, to n. Belize. Established in Hillsborough and Brevard counties, FL; Beaverhead Rock Pond, Madison Co., WY. Uncommon. **HABITAT:** Warm springs and their effluents; weedy canals. **SIMILAR SPECIES:** (1) See Variable Platyfish, *X. variatus.*

GAMBUSIA

GENUS CHARACTERISTICS: Upper 4–6 pectoral fin rays of male distinctly *thickened* and usually *curved upward* to form a bow or notch. Dark scale outlines (usually darkest on upper side) produce a crosshatched appearance (variable among species); white to light yellow below. Deep caudal peduncle; small dorsal fin, origin behind anal fin origin. Females much larger than males; large females usually pregnant and potbellied. Well-developed spines on 3d ray of gonopodium (Fig. 52). A dark spot (anal spot), thought to provide a target at which males aim their gonopodia during copulation, is present near urogenital opening on female of some *Gambusia.* Intensity of spot varies with female's reproductive condition. Eleven freshwater species of *Gambusia* in N. America north of Mexico are similar in appearance; however, only Western Mosquitofish, *G. affinis*, and Eastern Mosquitofish, *G. holbrooki*, occur outside TX and NM.

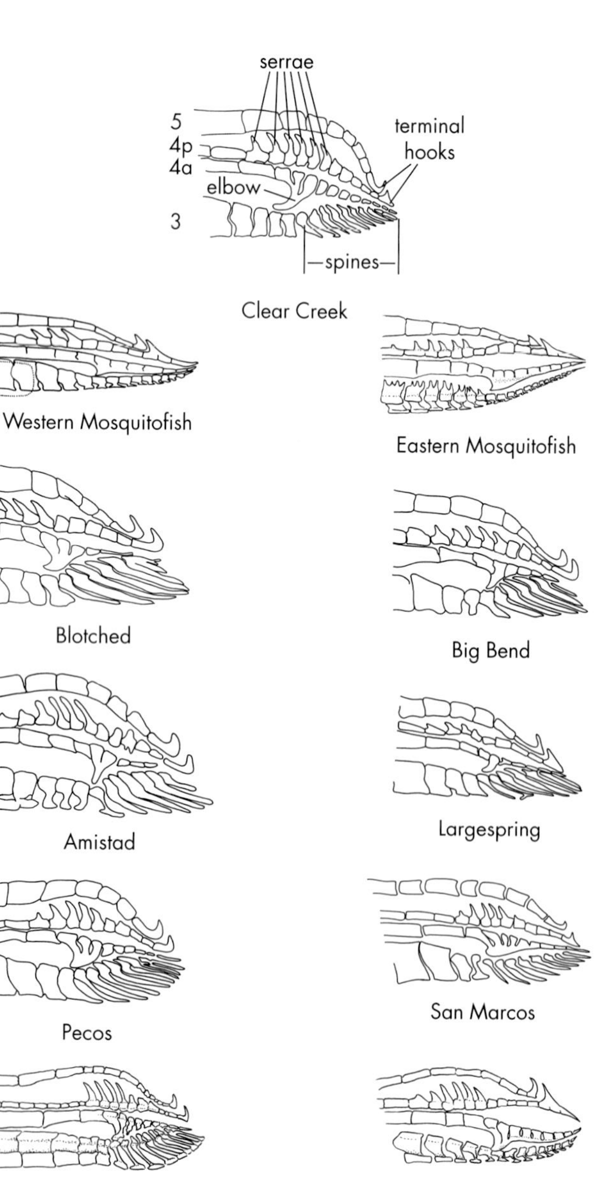

Fig. 52. ***Gambusia*** **species—tips of gonopodia.**

WESTERN MOSQUITOFISH *Gambusia affinis* Pl. 35

IDENTIFICATION: *Large dusky to black teardrop* (often reduced). Has 1–3 rows of *black spots* on dorsal and caudal fins. No discrete dark spots or stripes on side. Light olive-gray to yellow-brown above, dark stripe along back to dorsal fin; yellow and blue iridescence on transparent silver gray side. Black anal spot on pregnant female. Usually 6 dorsal, 9 anal rays. Gonopodium has elbow on ray 4a composed of 2 or more segments, 8–11 short spines on ray 3, no large teeth on gonopodial ray 3 (Fig. 52). To 2½ in. (6.5 cm). **RANGE:** Gulf Slope drainages from Mobile Bay to Mexico; Mississippi R. basin from KY, IL, and MO south to Gulf. Possibly native in upper Savannah and middle Chattahoochee rivers. Widely transplanted elsewhere, including many w. U.S. drainages (see Remarks). Common; often abundant. **HABITAT:** Standing to slow-flowing water; usually in vegetated ponds and lakes, backwaters and quiet pools of streams. Frequents brackish water. **REMARKS:** Introductions of this species and Eastern Mosquitofish, *G. holbrooki*, often for mosquito control, have caused or contributed to elimination of many populations of fishes with similar ecological requirements. Introductions into western drainages have been especially detrimental to survival of rare fishes. **SIMILAR SPECIES:** See (1) Eastern Mosquitofish, *G. holbrooki*, and (2) Tex-Mex Gambusia, *G. speciosa*. See Pl. 35. (3) Pecos Gambusia, *G. nobilis,* is deeper bodied; has dark edges on median fins, usually 8 dorsal rays; *lacks* row of black spots on middle of caudal fin. (4) Blotched Gambusia, *G. senilis*, (5) Big Bend Gambusia, *G. gaigei*, and (6) Amistad Gambusia, *G. amistadensis,* have black spots and crescents on side, dark stripe along side. (7) Largespring Gambusia, *G. geiseri,* has many black spots on side; *lacks* black to dusky teardrop, black anal spot on female. (8) Clear Creek Gambusia, *G. heterochir,* has distinct notch in pectoral fin of male, no dark stripe along back, *no row* of black spots on caudal fin, is deeper bodied. (9) San Marcos Gambusia, *G. georgei,* has lemon yellow fins, *no row* of black spots on middle of caudal fin.

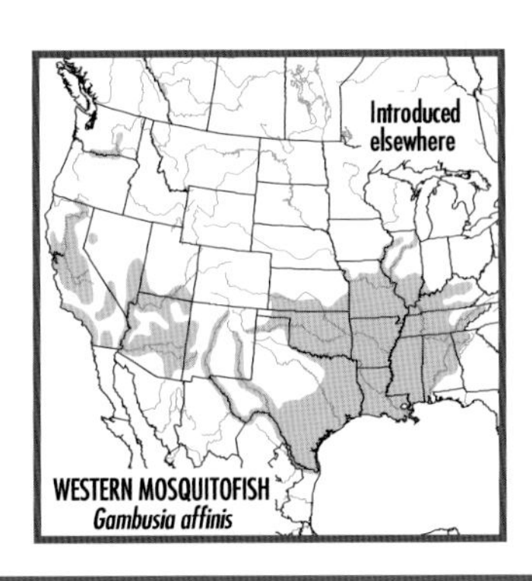

EASTERN MOSQUITOFISH *Gambusia holbrooki* **Not shown**

IDENTIFICATION: Nearly identical to Western Mosquitofish, *G. affinis*, but has *large teeth* on ray 3 of gonopodium (Fig. 52), *usually 7 dorsal, 10 anal rays*. Black specks may be present on side; black-blotched individuals common in FL. To 2½ in. (6.5 cm). **RANGE:** Atlantic and Gulf slope drainages from s. NJ to Pascagoula R. and nearby streams, se. MS. Absent from Mobile Bay drainage except extreme southern part. Transplanted elsewhere in U.S. Common; locally abundant. **HABITAT:** Same as Western Mosquitofish. **SIMILAR SPECIES:** (1) See Western Mosquitofish, *G. affinis* (Pl. 35).

TEX-MEX GAMBUSIA *Gambusia speciosa* **Not shown**

IDENTIFICATION: Nearly identical to Western Mosquitofish, *G. affinis*, but *gonopodial ray 4a arching posteriorly at tip toward ray 4p arching anteriorly* (Fig. 52), *bright orange dorsal, caudal fins* on male. To 2¼ in. (5.8 cm). **RANGE:** Devils R., TX. Also in Mexico. Common. **HABITAT:** Small to medium streams and springs over silt, sand, or rock; often near vegetation. **SIMILAR SPECIES:** (1) Western Mosquitofish, *G. affinis* (Pl. 35), has relatively straight gonopodial rays 4a and 4p, ray 4p hook ending in acute point (Fig. 52).

SAN FELIPE GAMBUSIA *Gambusia clarkhubbsi* **Not shown**

IDENTIFICATION: *No teardrop*. Body with *black spots; dark crosshatching* on back and upper ⅔ of side. Dusky to clear edges on dorsal, caudal, and anal fins. Deep bodied. Tan above, dark stripe along back to dorsal fin; dusky silver side. Orange on lower caudal fin of large male. No or small black anal spot on female. No or 1 row of dark spots on dorsal fin. Usually 9 dorsal rays. Gonopodium has distal segment of 4a, spines of ray 3 projecting past hook of ray 4p (Fig. 52). To 1½ in. (4.2 cm). **RANGE:** San Felipe Creek (Rio Grande drainage), Val Verde Co, TX. Uncommon. **HABITAT:** Vegetated margins and quiet pools of spring-fed creek. **SIMILAR SPECIES:** See Pl. 35. (1) Other Gambusia usually have

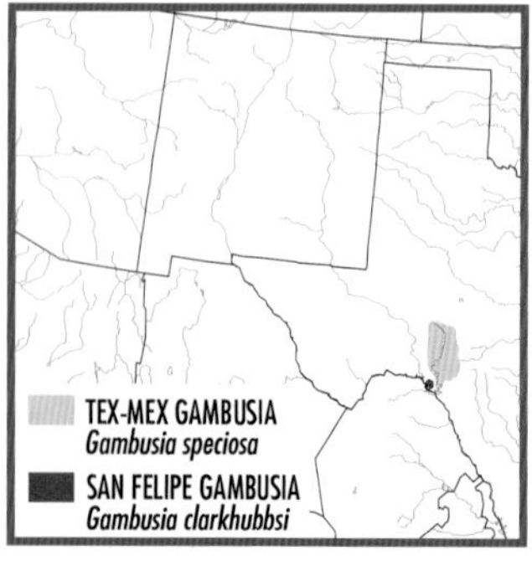

6–8 dorsal rays. (2) Most similar to Pecos Gambusia, *G. nobilis*, which has *large black teardrop*, black edges on dorsal and caudal fins, black streak behind anal fin.

PECOS GAMBUSIA *Gambusia nobilis* Pl. 35

IDENTIFICATION: *Black edges* on dorsal and caudal fins, anal fin of female. Black edges on scales produce *crosshatching* on back and upper ⅔ of side. Deep bodied. Olive above, dark stripe along back to dorsal fin; iridescent blue and yellow on silver side; large black teardrop. Black streak behind anal fin; black anal spot on female. One row of dark spots on dorsal fin, often a few dusky spots on middle of caudal fin. Usually 8 dorsal rays. Gonopodium has large elbow of 2–4 (usually 3) segments on ray 4a (Fig. 52). To 2 in. (4.8 cm). **RANGE:** Pecos R. system, NM and TX. Uncommon and localized; protected as an *endangered species.* **HABITAT:** Vegetated springs and effluents. **SIMILAR SPECIES:** See Pl. 35. (1) San Felipe Gambusia, *G. clarkhubbsi, lacks* teardrop, black streak behind anal fin, black edges on dorsal, caudal, and anal fins. (2) Blotched Gambusia, *G. senilis*, (3) Big Bend Gambusia, *G. gaigei*, and (4) Amistad Gambusia, *G. amistadensis,* have black spots and crescents on side, dark stripe along side. (5) Largespring Gambusia, *G. geiseri,* has row of discrete black spots on caudal fin. (6) Clear Creek Gambusia, *G. heterochir,* has distinct notch in pectoral fin, no dark stripe along back. (7) San Marcos Gambusia, *G. georgei,* has lemon yellow fins. (8) Western Mosquitofish, *G. affinis*, is more slender, *lacks* black edges on median fins, has less distinct scale outlines, usually 6 dorsal rays.

BLOTCHED GAMBUSIA *Gambusia senilis* Pl. 35

IDENTIFICATION: *Dusky stripe* (about 1 scale deep) along side, dark scale outlines (often appearing as black crescents) and *black spots* (often poorly developed on male) on lower side. Usually a large black teardrop. Dusky olive above, dark stripe along back to dorsal fin;

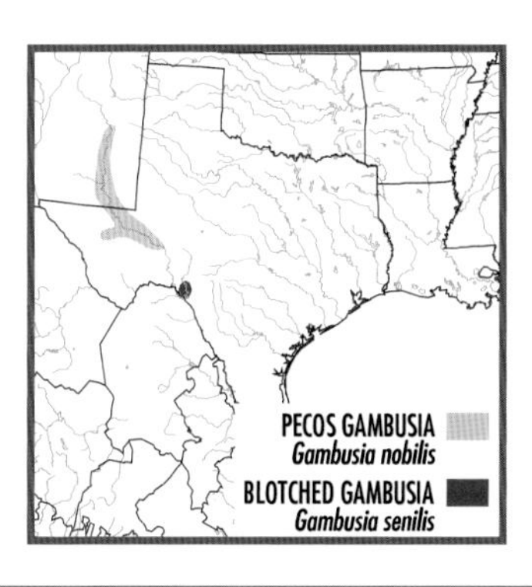

silver side; light yellow median fins, small dusky spots near base of dorsal fin. No dark anal spot; no spots on caudal fin. Some individuals have scattered, large black spots on body. Usually 8 dorsal rays. Gonopodium has ray 4a much shorter than 4p; elbow on ray 4a of 2–4 segments (Fig. 52). To 2¼ in. (5.5 cm). **RANGE:** Devils R. (Rio Grande drainage), TX; Río Conchos drainage, Mexico. Extinct in U.S.; common in Mexico. **HABITAT:** Springs; vegetated quiet pools and backwaters. **SIMILAR SPECIES:** (1) Big Bend Gambusia, *G. gaigei*, and (2) Amistad Gambusia, *G. amistadensis* (both Pl. 35), *lack* prominent black marks on lower side; have dark anal spot on female.

BIG BEND GAMBUSIA *Gambusia gaigei* Pl. 35

IDENTIFICATION: *Prominent black spots, crescents* on upper side (absent on belly), *dark stripe* along side. Dusky teardrop; dusky edge on dorsal fin. Golden olive above, dark stripe along back to dorsal fin; faint orange stripes on side. Dark anal spot on female; row of dusky spots on lower ⅓ of dorsal fin; dark specks on outer ⅓ of anal fin of female. Large male has orange snout; black edge, basal row of black spots on red-orange dorsal fin; dusky edge, orange base on caudal fin. Usually 8 dorsal rays; 6–15 teeth in anterior row on each jaw. Gonopodium has elbow of 1–2 segments on ray 4a; ray 4a much shorter than 4p (Fig. 52). To 2¼ in. (5.4 cm). **RANGE:** Springs in Big Bend National Park, TX. Protected as an *endangered species.* **HABITAT:** Vegetated, spring-fed sloughs and ponds. **SIMILAR SPECIES:** See Pl. 35. (1) See San Marcos Gambusia, *G. georgei*, and (2) Amistad Gambusia, *G. amistodensis*. (3) Blotched Gambusia, *G. senilis*, has black spots and crescents *on lower side;* lacks dark anal spot. (4) Other *Gambusia* species *lack* black spots and crescents on upper side, dark stripe on side.

SAN MARCOS GAMBUSIA *Gambusia georgei* Pl. 35

IDENTIFICATION: Similar to Big Bend Gambusia, *G. gaigei*; has *dark edges* on dorsal and caudal fins, distinctly crosshatched side, *5–6*

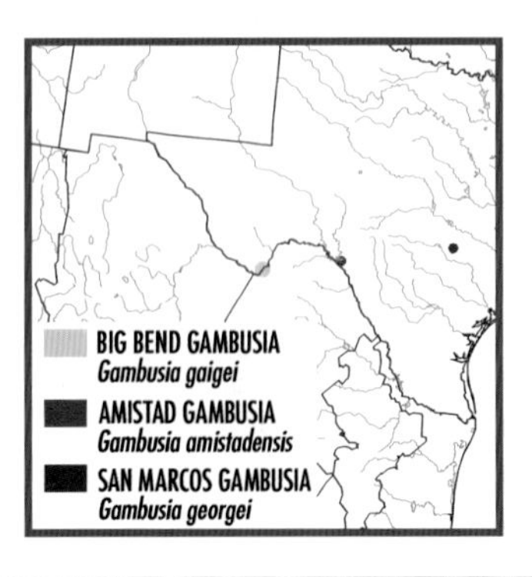

segments fused in elbow on ray 4a of gonopodium (Fig. 52). Only *Gambusia* species with *lemon yellow median fins.* To 2 in. (4.8 cm). **RANGE:** San Marcos Spring and River, TX. Protected as an *endangered species* but probably extinct. **HABITAT:** Large vegetated spring and its effluent. **SIMILAR SPECIES:** (1) See Big Bend Gambusia, *G. gaigei* (Pl. 35).

AMISTAD GAMBUSIA *Gambusia amistadensis* Pl. 35

IDENTIFICATION: Similar to Big Bend Gambusia, *G. gaigei*, but is *more slender,* has *more terminal mouth* and longer serrae on ray 4p of gonopodium (Fig. 52). To 2¼ in. (5.8 cm). **RANGE:** Goodenough Spring (Rio Grande drainage), TX. Extinct. Amistad Reservoir inundated only known locality. **HABITAT:** Large vegetated spring and its effluent. **SIMILAR SPECIES:** (1) See Big Bend Gambusia, *G. gaigei* (Pl. 35).

LARGESPRING GAMBUSIA *Gambusia geiseri* Pl. 35

IDENTIFICATION: *Distinct row of black spots* on middle of dorsal and caudal fins (indistinct on juvenile); often additional faint rows. Dark scale outlines (some appearing as crescents) produce crosshatching; scattered *black spots* on side. Olive above, dark stripe along back to dorsal fin; iridescent blue and yellow on silver side. No teardrop, no black anal spot. Gray edge, row of dusky spots on middle of anal fin of female. Usually 7 dorsal rays. Gonopodium has hook on ray 3, angular terminal hooks on rays 4p and 5 (Fig. 52). To 1¾ in. (4.4 cm). **RANGE:** San Marcos and Comal river systems, TX. Introduced into Colorado and Rio Grande (including Pecos R.) drainages, TX. Highly localized and uncommon. **HABITAT:** Large springs. **SIMILAR SPECIES:** See Pl. 35. (1) Western Mosquitofish, *G. affinis*, and (2) Pecos Gambusia, *G. nobilis, lack* black spots on side; have black teardrop, black anal spot on female. (3) Other *Gambusia* species *lack* row of discrete black spots on caudal fin.

CLEAR CREEK GAMBUSIA *Gambusia heterochir* Pl. 35

IDENTIFICATION: *Deep notch* at top of pectoral fin of male. *No dark stripe* along back. Deep bodied. Dark anal spot on female. Dusky teardrop. Olive above; iridescent blue and yellow on silver side. Dusky to clear fins; row of faint spots on middle of dorsal fin. Usually 7–8 dorsal rays. Gonopodium has long elbow on ray 4a (Fig. 52). To 2¼ in. (5.4 cm). **RANGE:** Headwater springs of Clear Creek (San Saba R. system), Menard Co., TX. Headwater springs of Clear Creek are impounded, and Clear Creek Gambusia is extremely rare or extinct; recognized as an *endangered species.* **HABITAT:** Vegetated springs. **SIMILAR SPECIES:** (1) Other species of *Gambusia* (Pl. 35) *lack* deep notch in pectoral fin; have *dark* or *dusky stripe* along back, shorter elbow on gonopodium (Fig. 52).

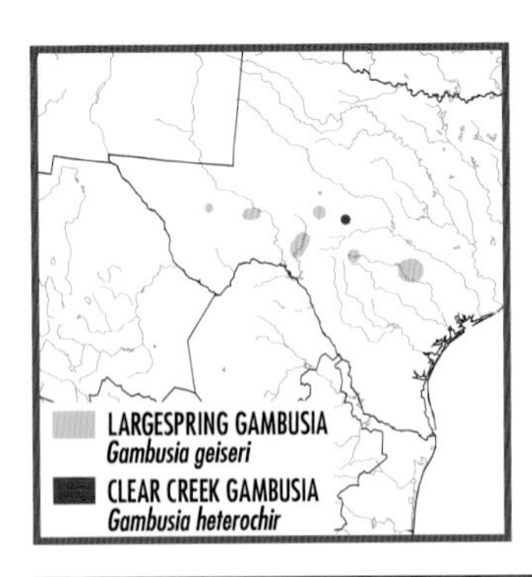

PIKE KILLIFISH *Belonesox belizanus* **Not shown**

IDENTIFICATION: Long jaws form *pointed beak; large teeth.* Dorsal fin origin well *behind* anal fin origin. Dark olive above fading to white below; *several rows of black spots on side; black spot* at base of caudal fin. Small (usually 52–63 lateral) scales; usually 8–9 dorsal rays. *Giant* of livebearer family; females reach 6¾ in. (17 cm), males about 4½ in. (11 cm). **RANGE:** Native from Laguna San Julian, ne. of Ciudad Veracruz, Mexico, to Costa Rica. Established in s. FL. Common. **HABITAT:** Weedy canals and streams. **REMARKS:** Two subspecies. *B. b. maxillosus,* from Yucatán Peninsula, has stout body and jaws. *B. b. belizanus,* occupying rest of range, is more slender, has smaller jaws. FL population is *B. b. maxillosus.*

LEAST KILLIFISH *Heterandria formosa* **Pl. 35**

IDENTIFICATION: *Red around black spot* on front of dorsal fin. *Black spot* at front of anal fin of female. Olive above; black to dusky stripe, series of *black bars* along side; *black spot* on caudal fin base; silver white to light yellow below. Deep, chubby body; large eye. Long gonopodium,

more than 1/3 body length. Has 24–30 lateral scales; 6–9 anal rays (in female). To 1½ in. (3.6 cm). **RANGE:** Coastal Plain from Cape Fear R. drainage, NC, to se. TX. Common; abundant throughout peninsular FL. **HABITAT:** Heavily vegetated standing to slow-flowing fresh and brackish water. **SIMILAR SPECIES:** (1) Pygmy Killifish, *Leptolucania ommata* (Pl. 33), *lacks* black spots at front of dorsal and anal fins, gonopodium; has large black spot on side of female, 9–10 anal rays.

GILA TOPMINNOW *Poeciliopsis occidentalis* Pl. 35

IDENTIFICATION: *Extremely long gonopodium,* more than 1/3 body length. *Dark to dusky stripe* along side forward to opercle; small black spot at rear of dorsal fin. Light olive-tan above; large, darkly outlined scales on back and upper side, black specks (often absent) on lower side; white to yellow below. Short snout; nearly terminal mouth. Large male is *black;* has orange at base of gonopodium, sometimes at base of dorsal and caudal fins. Slender body; small eye. To 2¼ in. (6 cm). **RANGE:** Gila R. system, NM and AZ; south on Pacific Slope of Mexico. Once considered most abundant "low desert" fish in U.S. (still common in Mexico); now extinct in NM, rare in AZ. Protected as an *endangered species.* **HABITAT:** Vegetated springs; shallow pools and backwaters of creeks and small to medium rivers. **SIMILAR SPECIES:** (1) See Sonora Topminnow, *P. sonoriensis*, and (2) Porthole Livebearer, *P. gracilis.* (3) Species of *Gambusia* (Pl. 35) are deeper bodied; have larger eye, gonopodium usually much *less* than 1/3 body length, often a black anal spot on female, upper pectoral fin rays of male thickened and curved upward.

SONORA TOPMINNOW *Poeciliopsis sonoriensis* Not shown

IDENTIFICATION: Nearly identical to Gila Topminnow, *P. occidentalis*, but has longer snout (longer than diameter of eye), more upturned mouth, dark stripe usually forward only to above pelvic fin. To 2¼ in.

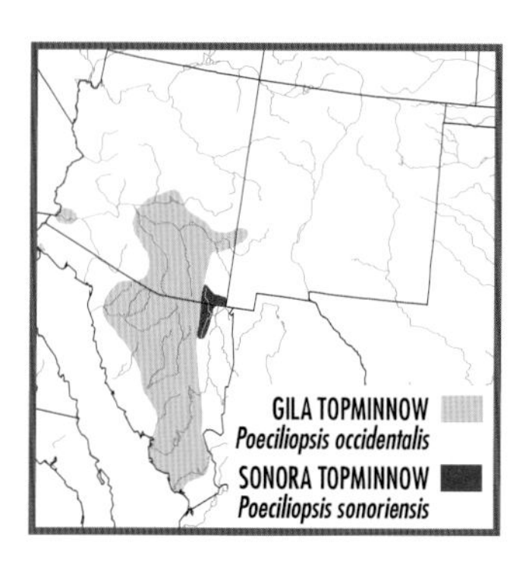

(6 cm). **RANGE:** Rio Yaqui system, AZ. Also in Mexico. Locally common. **HABITAT:** Vegetated springs; shallow pools and backwaters of creeks and small rivers. **SIMILAR SPECIES:** (1) See Gila Topminnow, *P. occidentalis* (Pl. 35).

PORTHOLE LIVEBEARER *Poeciliopsis gracilis* **Not shown**

IDENTIFICATION: Similar to Gila Topminnow, *P. occidentalis*, but has 3–6 black spots along side, *no* dark to dusky stripe. To 2 in. (5.1 cm). **RANGE:** Native from s. Mexico to Honduras on Atlantic and Pacific slopes. Established south of Mecca, Riverside Co., CA. Common in ditches on north side of Salton Sea. **HABITAT:** Irrigation canals. **SIMILAR SPECIES:** See Gila Topminnow, *P. occidentalis* (Pl. 35).

SPLITFINS: Family Goodeidae (4)

Most of the 48 species of splitfins are on the Mesa Central of Mexico. Mexican species are live-bearing fishes with anterior rays of the anal fin of the male shortened and slightly separated from the rest of the fin, possibly functioning as a gonopodium. Species in the U.S., unlike their Mexican relatives, lay eggs and are easily recognized by their *thick body, no pelvic fins,* no lateral line, and dorsal and anal fins *far back* on the body. *Empetrichthys* species feed on both plant and animal material and have a relatively short gut. *Crenichthys* species feed more heavily on plant matter and have a long coiled intestine.

WHITE RIVER SPRINGFISH *Crenichthys baileyi* **Pl. 34**

IDENTIFICATION: *Row of black spots* (or black stripe) along side; *2d row of black spots* along lower side from midbody to caudal fin. Dark olive above; silver white below; bright silver cheek and opercle; clear to light olive fins, sometimes with dusky edges. Jaws about equal in length; bicuspid jaw teeth. To 3½ in. (9 cm). **RANGE:** White R. system, NV. Common but threatened by human encroachment and introductions of non-native fishes. Protected as an *endangered species.* **HABITAT:** Warm springs and their effluents. **REMARKS:** Five subspecies differ in body size and shape, ray counts, and color: *C. b. albivallis*, in Preston Big and nearby springs; *C. b. thermophilus* in Mormon and nearby springs; *C. b. grandis* in Hiko and Crystal springs; *C. b. baileyi*, in Ash Spring; *C. b. moapae*, in headwater springs of Moapa R. **SIMILAR SPECIES:** (1) See Railroad Valley Springfish, *C. nevadae* (Pl. 34). (2) Pahrump Poolfish, *Empetrichthys latos* (Pl. 34), and (3) Ash Meadows Poolfish, *E. merriami*, have distinctly *mottled side,* lower jaw projecting beyond upper jaw, conical jaw teeth.

AILROAD VALLEY SPRINGFISH *Crenichthys nevadae* Pl. 34

IDENTIFICATION: Similar to White River Springfish, *C. baileyi*, but has *1 row of dark spots* along side. To 2¼ in. (6 cm). **RANGE:** Native to springs in Railroad Valley, Nye Co., NV; introduced into springs in se. Mineral Co., NV. Common in extremely small areas; protected as a *threatened species.* **HABITAT:** Warm springs (about 95°F [35°C]). **SIMILAR SPECIES:** (1) See White River Springfish, *C. baileyi* (Pl. 34).

AHRUMP POOLFISH *Empetrichthys latos* Pl. 34

IDENTIFICATION: *Black mottling,* usually a black streak, on silver side. *Wide mouth,* sides of snout barely converging at front. Green-brown above; white to light yellow below. Breeding male has silver blue side, orange eye, yellow-orange dorsal, caudal, and anal fins. Lower jaw projects beyond upper jaw. Conical jaw teeth. Has 29–33, usually 31–32, lateral scales. To 2¼ in. (6 cm). **RANGE:** Native to 3 springs in Pahrump Valley, Nye Co., NV, where it was only native fish. Once common, but removal of water for irrigation eliminated it from native habitats; now only outside Pahrump Valley where transplanted. Protected as an *endangered species.* **HABITAT:** Warm springs (about 77°F [25°C]); usually in deep holes. **REMARKS:** Three subspecies, differing slightly in fin lengths and body proportions, sometimes recognized: *E. l. latos* (only surviving subspecies), *E. l. pahrump,* and *E. l. concavus.* **SIMILAR SPECIES:** (1) See Ash Meadows Poolfish, *E. merriami.*

SH MEADOWS POOLFISH *Empetrichthys merriami* Not shown

IDENTIFICATION: Similar to Pahrump Poolfish, *E. latos*, but has *interrupted black stripe* along side, sides of snout *converging* to narrow mouth. To 2¾ in. (7.2 cm). **RANGE:** Ash Meadows, NV. Extinct. **HABITAT:** Warm springs. **SIMILAR SPECIES:** (1) See Pahrump Poolfish, *E. latos* (Pl. 34).

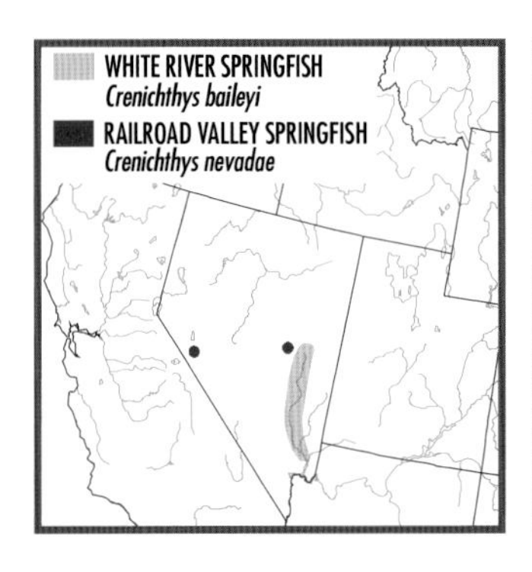

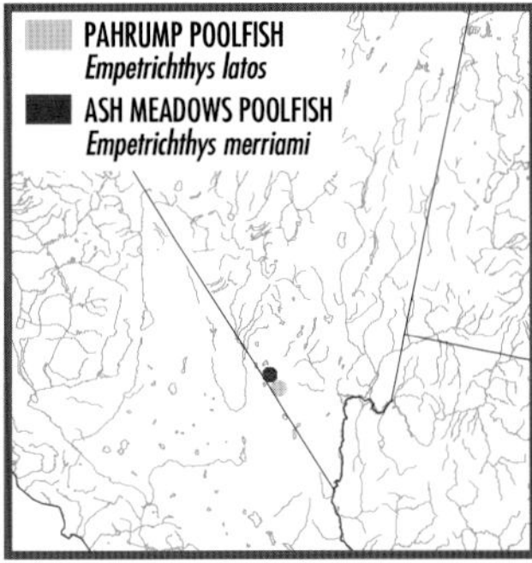

Pupfishes: Family Cyprinodontidae (15)

Like topminnows and killifishes, pupfishes are small and have an *upturned mouth,* 1 dorsal fin, no lateral line, abdominal pelvic fins (if present), and cycloid scales. Unlike topminnows and killifishes, pupfishes are *deep-bodied,* and most species have a *deep, strongly compressed caudal peduncle.* The top of the head is flat, but the back is usually *highly arched* (especially on the male). There are about 120 species in North America, South America, southern Eurasia, and Africa. Most of our pupfishes inhabit southwestern deserts, and several are endangered. Pupfishes survive extreme environmental conditions and tolerate temperatures from 32 to 113°F (0–45°C), salinity as high as 142 ppt (ocean water is typically 35 ppt), and oxygen concentrations as low as 0.13 mg/liter (the lowest known for any fish restricted to gill breathing).

CYPRINODON

GENUS CHARACTERISTICS: Robust body, deep in male, more slender in female; arched back; 1 row of *tricuspid teeth* in each jaw; large (22–34 lateral scales); long, coiled intestine. Breeding male is brightly colored, develops contact organs on scales and a narrow to broad black edge on caudal fin. Only other freshwater fish in our area with tricuspid teeth is Flagfish, *Jordanella floridae.*

SHEEPSHEAD MINNOW *Cyprinodon variegatus* Pl. 34

IDENTIFICATION: Extremely deep-bodied. Has *5–8 triangular-shaped, dark gray-brown bars* along silver olive side, wide at top, narrow on lower side. Green to blue-gray above; large dark brown blotches on rear half of upper side; white below; clear to light orange fins. Breeding male is blue above, has wide dark gray bars along side (best developed at rear); brass-salmon cheek, breast, and belly; dusky orange fins; black edge on caudal fin. Fully scaled belly. Has 22–28 lateral scales; 10–12 dorsal rays; 9–11 anal rays; 6–8 pelvic rays; 18–22, usually 19–21, rakers on 1st gill arch. To 3 in. (7.5 cm). **RANGE:** Coastal waters from MA to ne. Mexico. Also in West Indies. Common; locally abundant. Rarely far inland except in peninsular FL. Introduced in Pecos R., TX, where displacing native Pecos Pupfish, *C. pecosensis.* **HABITAT:** Salt, brackish, and fresh water; usually near vegetation. **REMARKS:** Lake-inhabiting fishes are often more elongate than their counterparts in streams. Although Sheepshead Minnows in Lake Eustis and other headwater lakes of Oklawaha R. in cen. FL have been recognized taxonomically (Lake Eustis Pupfish, *Cyprinodon hubbsi* or *C. variegatus hubbsi*), the differences used to separate them—a more slender body and caudal peduncle—may be developmental responses to the lake habitat. **SIMILAR SPECIES:** See Pl. 34. (1) Red River Pupfish, *C. rubrofluviatilis*, and (2) Pecos Pupfish, *C. pecosensis*, have unscaled or partly scaled belly and yellow paired fins, head, and belly on large male. (3) Leon Springs Pupfish, *C. bovinus*, and (4) White Sands Pupfish, *C. tularosa,* have many small dark blotches on lower side of female, yellow on dorsal and caudal fins of male.

ED RIVER PUPFISH *Cyprinodon rubrofluviatilis* Pl. 34

IDENTIFICATION: *Unscaled belly; 5–8 large triangular brown blotches* along silver side, *no dark blotches on lower side.* Green-brown above, dusky dorsal and caudal fins; white below. Breeding male has iridescent blue nape; wide black bars on side; yellow paired fins, head, and belly; narrow black edge on caudal fin; clear to white edges on black dorsal and anal fins. Has 25–29 lateral scales; 9–12 dorsal rays; 8–11 anal rays; usually 6–7 pelvic rays; usually 21–23 rakers on 1st gill arch; no (rarely 1 or 2) mandibular pores. To 2¼ in. (5.8 cm). **RANGE:** Upper Red and Brazos river drainages, OK and TX. Common. Introduced into Canadian R. and headwaters of Colorado R., TX. **HABITAT:** Shallow, sandy pools and runs of headwaters, creeks, and small to medium rivers. Often in extremely shallow and hot water. **SIMILAR SPECIES:** See Pl. 34. (1) See Pecos Pupfish, *C. pecosensis*. (2) Leon Springs Pupfish, *C. bovinus*, and (3) White Sands Pupfish, *C. tularosa,* have *fully scaled belly, many small dark brown blotches* on lower side of female; yellow on dorsal and caudal fins of male. (4) Sheepshead Minnow, *C. variegatus*, has *fully scaled belly,* orange fins on male; is deeper bodied.

ECOS PUPFISH *Cyprinodon pecosensis* Pl. 34

IDENTIFICATION: Similar to Red River Pupfish, *C. rubrofluviatilis*, but has *partly unscaled belly,* bars on side of female usually broken into blotches on lower side; usually has 2–3 mandibular pores; has less iridescent blue on nape, clear to pale yellow belly and pectoral fins on large male. To 2¼ in. (6 cm). **RANGE:** Formerly widespread in Pecos R. system, TX and NM; now restricted in TX to upper Salt Creek. Replaced by introduced Sheepshead Minnow, *C. variegatus*. **HABITAT:** Springs, sinkholes, and pools of streams. **SIMILAR SPECIES:** (1) See Red River Pupfish, *C. rubrofluviatilis* (Pl. 34). (2) Leon Springs Pupfish, *C. bovinus* (Pl. 34), has *fully scaled belly; many small brown blotches on lower side* of female; yellow in dorsal and caudal fins, wide black edge on caudal fin of male.

ON SPRINGS PUPFISH *Cyprinodon bovinus* Pl. 34

IDENTIFICATION: *Dark brown* (rectangular or triangular) *blotches* along silver side, *many small brown blotches* on lower side of female (rarely

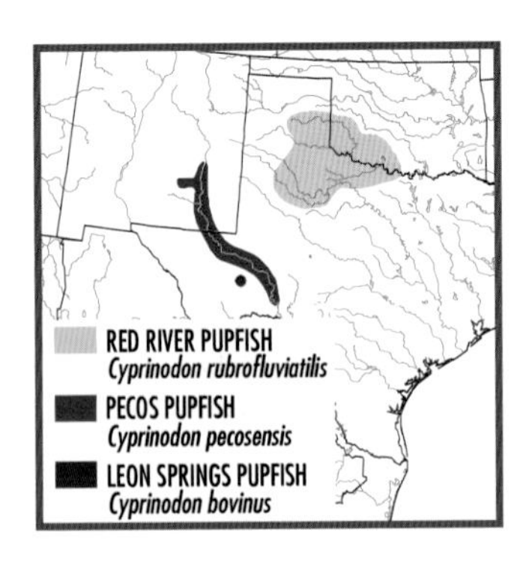

on male). *Usually 25 or fewer lateral scales.* Fully scaled belly. Gray-brown above; white below; dusky dorsal and caudal fins; pale yellow bar on caudal fin base. Large male has yellow dorsal fin edge, wide jet-black edge on yellow caudal fin. Has 23–26 lateral scales; 9–11 dorsal rays; 9–12 anal rays; usually 7 pelvic rays; 18–23, usually 19–21, rakers on 1st gill arch; 2 mandibular pores. To 2¼ in. (5.6 cm). **RANGE:** Leon Springs (where extinct) and Diamond Y Draw (Pecos R. system), Pecos Co., TX. Common in extremely small area; protected as an *endangered species.* **HABITAT:** Margins of spring-fed pools. **SIMILAR SPECIES:** See Pl. 34. Other pupfishes in Rio Grande drainage usually have *26 or more lateral scales.* (1) See White Sands Pupfish, *C. tularosa.* (2) Pecos Pupfish, *C. pecosensis,* has mostly unscaled belly; *few dark blotches* on lower side of female; no yellow in dorsal and caudal fins, narrow black bar on caudal fin edge of male. (3) Red River Pupfish, *C. rubrofluviatilis,* has unscaled belly; *no dark blotches on lower side;* yellow pectoral fins, head, and belly, narrow black bar on caudal fin edge, no yellow in dorsal and caudal fins of male

WHITE SANDS PUPFISH *Cyprinodon tularosa* Pl. 34

IDENTIFICATION: Similar to Leon Springs Pupfish, *C. bovinus,* but has dark bars on side of female *joined at bottom,* yellow to orange *outer half* of dorsal fin on large male, *26–28 lateral scales;* usually 21–24 rakers on 1st gill arch, 6 pelvic rays; 0–5 mandibular pores. To 2 in. (5 cm). **RANGE:** Tularosa Valley, NM. Abundant in small area. **HABITAT:** Clear, shallow spring-fed marsh pools and saline creeks. **SIMILAR SPECIES:** (1) See Leon Springs Pupfish, *C. bovinus* (Pl. 34).

COMANCHE SPRINGS PUPFISH *Cyprinodon elegans* Pl. 34

IDENTIFICATION: *Slender body* (relative to other pupfishes); *long slender caudal peduncle.* Brown-black blotches form *"stripe"* (often faint on male) along silver side; additional blotches on upper and lower sides of female. Gray-green above; pale yellow to white below; clear to light orange fins. Large male has *black specks on silver side,* black edge on caudal fin. Usually fully scaled belly. Has 25–28 lateral scales; 10–12 dorsal rays; 9–11 anal rays; usually 6–7 pelvic rays; 19–22 rakers on 1st gill arch. To 2½ in. (6.2 cm). **RANGE:** Restricted to Toyah Creek and effluents (including irrigation canals) of San Solomon, Phantom Cave, and Griffin springs, Reeves Co., TX; formerly in Comanche Springs, Pecos Co., TX. Common in small area but threatened by removal of water for agriculture. Protected as an *endangered species.* **HABITAT:** Springs, spring-fed canals, and ditches; usually over mud in current. **SIMILAR SPECIES:** (1) Other pupfishes (Pl. 34) have less distinct black stripe along side, *lack* pattern of black specks on silver side.

DESERT PUPFISH *Cyprinodon macularius* **Pl. 34**

IDENTIFICATION: Breeding male has *light blue body, lemon yellow to orange caudal peduncle and fin,* black edge on median fins. Dark olive above; dark brown blotches or bars along silver side; smaller dark blotches on upper and lower sides; white below. Deep body; dorsal fin origin equidistant between tip of snout and caudal fin base. Usually 25–26 lateral scales, 7 pelvic rays. To 2¾ in. (7.2 cm). **RANGE:** Lower Colorado R. drainage, including Gila R. system, s. AZ and s. CA; Salton Sea, CA. Also in n. Mexico. Uncommon; protected as an *endangered species.* **HABITAT:** Springs, marshes, lakes, and pools of creeks; usually over mud or sand. **SIMILAR SPECIES:** See Pl. 34. Other pupfishes *lack* yellow to orange caudal peduncle and fin on large male. See (1) Santa Cruz Pupfish, *C. arcuatus*, (2) Sonoyta Pupfish, *C. eremus*, and (3) Owens Pupfish, *C. radiosus*. (4) Amargosa Pupfish, *C. nevadensis*, (5) Salt Creek Pupfish, *C. salinus*, and (6) Devils Hole Pupfish, *C. diabolis*, have dorsal fin origin *nearer to caudal fin base than to tip of snout,* small or no pelvic fins.

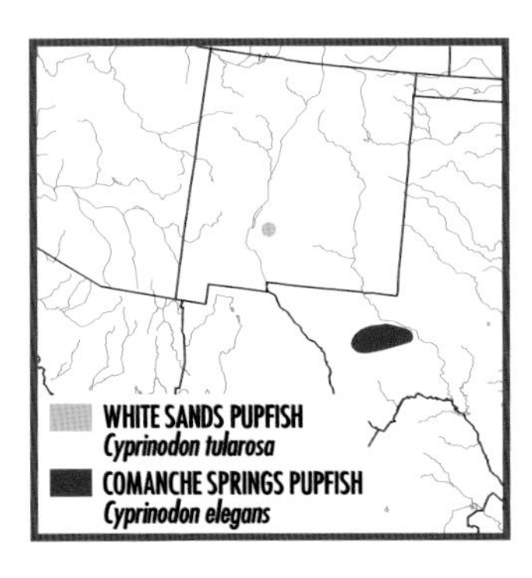

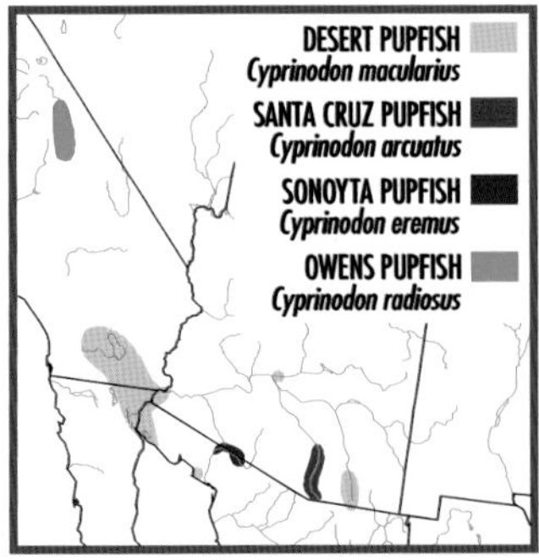

SANTA CRUZ PUPFISH *Cyprinodon arcuatus* Not shown

IDENTIFICATION: Similar to Desert Pupfish, *C. macularius*, but has highly *convex predorsal profile, no yellow or orange* on caudal fin or peduncle of breeding male. To 2¼ in. (5.5 cm). **RANGE:** Upper Santa Cruz R. system (Gila R. drainage), AZ. Extinct. **HABITAT:** Springs, and their effluents. **SIMILAR SPECIES:** (1) See Desert Pupfish, *C. macularius* (Pl. 34).

SONOYTA PUPFISH *Cyprinodon eremus* Not shown

IDENTIFICATION: Nearly identical to Desert Pupfish, *C. macularius*, but has larger head; female has longer dorsal fin base. To 2¾ in. (7.2 cm). **RANGE:** Quitobaquito Springs (Gila R. system) and adjacent Rio Sonoyta (Gulf of California basin), s. AZ and n. Mexico. Uncommon. **HABITAT:** Springs, marshes, lakes, and pools of creeks; usually over mud or sand. **SIMILAR SPECIES:** (1) See Desert Pupfish, *C. macularius* (Pl. 34).

OWENS PUPFISH *Cyprinodon radiosus* Pl. 34

IDENTIFICATION: Similar to Desert Pupfish, *C. macularius*, but breeding male has deep blue body, purple-gray bars along side, *orange edge on blue dorsal and anal fins.* Usually 26–27 lateral scales. To 2¾ in. (7.2 cm). **RANGE:** Owens Valley, s. CA. Formerly common within small range, now restricted to refuges. Protected as an *endangered species.* **HABITAT:** Marshes, vegetated sloughs and backwaters of Owens R. **SIMILAR SPECIES:** (1) See Desert Pupfish, *C. macularius* (Pl. 34).

AMARGOSA PUPFISH *Cyprinodon nevadensis* Pl. 34

IDENTIFICATION: Deep body; *dorsal fin origin nearer to caudal fin base than to tip of snout.* Dark olive above; teardrop-shaped blotches or bars (often faint) on yellow-brown to silver blue side; smaller dark blotches on upper and lower sides; white below. Breeding male has gray bars on deep blue side, black edges on blue-gray median fins. Small pelvic fins (occasionally absent), usually 6 pelvic rays. Usually 25–26 lateral scales. To 3 in. (7.8 cm). **RANGE:** Amargosa R. basin, NV and CA. **HABITAT:** Springs, their effluents, and spring-fed ponds and lakes. **REMARKS:** Six subspecies, often distinguished from one another by only average differences. *C. n. nevadensis*, restricted to Saratoga Springs and adjoining lakes, Death Valley National Monument, CA, has deep, broad body, intensely blue breeding male. *C. n. amargosae,* restricted to Amargosa R., CA, has small scales (usually 25 or more scales around body at dorsal fin origin; other subspecies usually have 25 or fewer). *C. n. calidae,* now extinct, occurred in outlets of North and South Tecopa Hot Springs, CA, had large scales, wide body, short caudal peduncle. *C. n. shoshone,* in outlet of Shoshone Springs, CA, has slender body, large scales. *C. n. mionectes* (protected as an *en-*

dangered subspecies), in large springs in lower Ash Meadows, NV, has low scale and fin ray counts; short, deep body; silver blue side, yellow nape on breeding male. *C. n. pectoralis* (protected as an *endangered subspecies*), in small springs in hills around Devils Hole, upper Ash Meadows, NV, usually has 17 pectoral rays (other subspecies usually 16), bright yellow nape on breeding male. **SIMILAR SPECIES:** See Pl. 34. (1) Salt Creek Pupfish, *C. salinus*, has more slender body, usually 28–29 lateral scales, 24–27 predorsal scales (Amargosa Pupfish usually has 17–19 predorsal scales). (2) Devils Hole Pupfish, *C. diabolis*, lacks pelvic fins; has more slender body; yellow-gold dorsal, caudal, and anal fins on large male. (3) Desert Pupfish, *C. macularius*, and (4) Owens Pupfish, *C. radiosus*, have dorsal fin *equidistant* between tip of snout and caudal fin base, usually 7 pelvic rays, yellow or orange on median fins of large male.

SALT CREEK PUPFISH *Cyprinodon salinus* Pl. 34

IDENTIFICATION: *Slender body; dorsal fin far back on body*, origin closer to caudal fin base than to tip of snout. *Small, crowded scales on nape;* 22–30, usually 24–27, predorsal scales. Olive-brown above; dark brown bars or blotches (largest at top) along silver brown (female) to blue (male) side; smaller dark blotches on upper and lower sides; white below. Breeding male has gray bars along silver turquoise side; gray dorsal fin; black edges on caudal, anal, and paired fins. Small pelvic fins (occasionally absent), usually 6 pelvic rays. Usually 28–29 lateral scales; 4–8, usually 6, preorbital pores. To 3 in. (7.8 cm). **RANGE:** Salt Creek, Death Valley, CA. Extreme population fluctuations in harsh environment, but typically abundant in small range. **HABITAT:** Vegetated spring-fed pools and marshes. **REMARKS:** Cottonball Marsh Pupfish, sometimes considered a species (*Cyprinodon milleri*) or a subspecies (*C. salinus milleri*) of Salt Creek Pupfish, occupies Cottonball Marsh, adjacent to Salt Creek in Death Valley, CA. It differs from Salt Creek

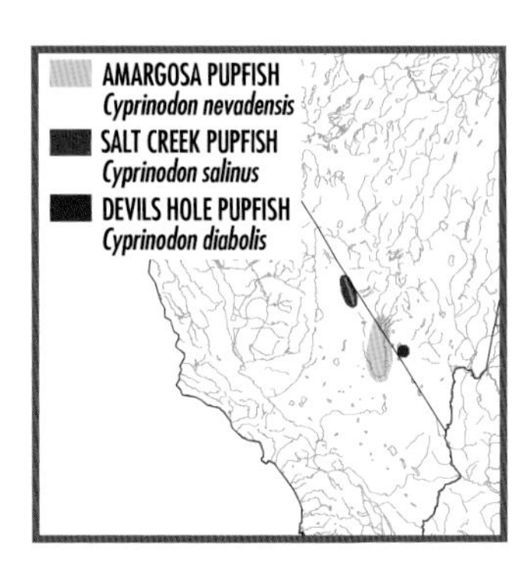

Pupfish by having shorter, more slender caudal peduncle; usually 3–5 pelvic rays; 0–7, usually 0, preorbital pores. **SIMILAR SPECIES:** See Pl. 34. (1) Amargosa Pupfish, *C. nevadensis*, and (2) Devils Hole Pupfish, *C. diabolis,* usually have 15–19 predorsal scales, 24–26 lateral scales. (3) Desert Pupfish, *C. macularius*, and (4) Owens Pupfish, *C. radiosus*, are *deeper bodied,* have dorsal fin origin equidistant between tip of snout and caudal fin base, usually 17–19 predorsal scales, 27 or fewer lateral scales.

DEVILS HOLE PUPFISH *Cyprinodon diabolis* Pl. 34

IDENTIFICATION: A *dwarf species;* rarely more than 1 in. (2.5 cm) long. *No pelvic fins.* Fairly slender body; large head and eye; dorsal fin *far back* on body, origin closer to caudal fin base than to tip of snout. Dark brown above; many black specks, no dark bars on silver side; white below. Breeding male has blue side, black edges on yellow-gold dorsal, caudal, and anal fins. Usually 24–25 lateral scales. To 1¼ in. (3.4 cm). **RANGE:** Restricted to Devils Hole, Ash Meadows, Nye Co., NV. Rare; population endangered by lowering water table. Protected as an *endangered species.* **HABITAT:** Deep limestone pool; entire population feeds and breeds in an area 215 ft. (20 m) square, a limestone ledge along 1 side of Devils Hole. This is smallest known range of any vertebrate animal. **SIMILAR SPECIES:** (1) Other pupfishes (Pl. 34) are *larger; have pelvic fins* (rarely absent in Amargosa Pupfish, *C. nevadensis,* and Salt Creek Pupfish, *C. salinus)*, dark bars on side of female, smaller head and eye, deeper body, dorsal fin farther forward.

CONCHOS PUPFISH *Cyprinodon eximius* Pl. 34

IDENTIFICATION: *Faint brown blotches* on silver side, *rows of small brown spots* on upper side. Gray-brown above; white below; dusky median fins. Breeding male has *yellow-orange dorsal fin;* dark brown bars along side; *black spots, dashes* on front half, wide black edge on caudal fin. Usually 26–27 lateral scales; 12–18 rakers on 1st gill arch; 6–7 pelvic rays; 0 mandibular pores. To 2 in. (5 cm). **RANGE:** Devils R., Terlingua Creek, and Alamito Creek (Rio Grande drainage), TX. Also in upper Río Conchos system and endorheic Río Sáuz basin, Mexico. Localized and uncommon in U.S.; common in Mexico. **HABITAT:** Sloughs, backwaters, and margins of small to medium rivers (avoids hot springs). **SIMILAR SPECIES:** (1) Other pupfishes (Pl. 34) *lack* rows of small brown spots on upper side, black spots and dashes on caudal fin.

FLAGFISH *Jordanella floridae* Pl. 34

IDENTIFICATION: *Large black spot* on midside; often several smaller spots along side. *Alternating thin black and red-orange lines, gold*

flecks, on side. Long dorsal fin, 14–18 rays. Gray-green above; silver white below; clear to dusky red fins. Large male is bright red-orange, has red wavy lines, white spots in dusky black dorsal fin; juvenile and female have white halo around small black spot at rear of dorsal fin. Has 25–27 lateral scales; 11–13 anal rays. To 2½ in. (6.5 cm). **RANGE:** Peninsular FL north to St. Johns and Ochlockonee river drainages. Common. **HABITAT:** Vegetated sloughs, ponds, lakes, and sluggish streams; enters brackish water.

NEEDLEFISHES: Family Belonidae (1)

The aptly named needlefishes have long slender bodies, small scales, dorsal and anal fins far back on the body, and *very long jaws with many sharp teeth.* Needlefishes often swim near the surface of the water. Thirty-four species occur worldwide in marine, brackish, and freshwater environments. About 12 species are restricted to fresh water, and a few marine species, including 1 in our area, enter fresh water.

ATLANTIC NEEDLEFISH *Strongylura marina* Pl. 57

IDENTIFICATION: *Extremely slender body,* round in cross section; *long jaws* with needlelike teeth. *Tiny scales.* Falcate dorsal (12–17 rays) and anal (16–20 rays) fins far back on body. Light green above, white below, silver on cheek and gill cover, dusky or clear fins. To 24 in. (61 cm). **RANGE:** Atlantic Coast from ME to Brazil. Ascends rivers far inland on Atlantic and Gulf slopes, especially in Mobile Bay drainage, AL, lower Mississippi R., LA, and rivers in FL and TX; north to lower Ohio R., KY. Common in FL. **HABITAT:** Shallow coastal waters, especially grassy areas; medium to large rivers. Known to reproduce in fresh water. **SIMILAR SPECIES:** (1) Gars (Pl. 3), Lepisosteidae, are more robust, have large ganoid (armorlike) scales; more rounded fins.

STICKLEBACKS: Family Gasterosteidae (4)

These highly distinctive, small, *scaleless* fishes have *3–16 isolated dorsal spines* followed by a normal dorsal fin with 14–16 rays and an *extremely narrow caudal peduncle.* They have a small thoracic or subthoracic pelvic fin (sometimes absent) with 1 spine and 1–2 rays. Some populations have *large bony plates* along the side. Only 12 species are recognized in this family, but numerous taxonomic problems remain (see Threespine Stickleback, *Gasterosteus aculeatus*). Sticklebacks inhabit both marine and fresh waters of North America and Eurasia. The male builds an oblong nest of plant material held together with a sticky kidney secretion. Through complex courting maneuvers of the male, the female is enticed to enter the nest and deposit eggs. She leaves the nest, and the male enters and fertilizes the eggs. The male guards the eggs and young.

NINESPINE STICKLEBACK *Pungitius pungitius* Pl. 36

IDENTIFICATION: Has 7–12, *usually 9, short dorsal spines* angled alternately to left and right. *Slender, compressed body;* usually well-developed *keel* on caudal peduncle. No large bony plates on side (except in some Atlantic Coast populations); 0–8 small plates on lateral-line pores on front half of body. Branchiostegal membranes joined to one another but free from isthmus. Pelvic fin with 1 spine, 1 ray. Gray to olive above, dark mottling on back and side; silver below. Breeding male has black belly, white pelvic fins, sometimes red on head. To 3½ in. (9 cm). **RANGE:** Arctic and Atlantic drainages across Canada and AK, and as far south as NJ; Pacific Coast of AK; Great Lakes basin except Lake Erie. Also in Eurasia. Common. **HABITAT:** Shallow vegetated areas of lakes, ponds, and pools of sluggish streams; open water over sand. Marine populations live near shore, move into fresh water to spawn. **REMARKS:** Five subspecies; only *P. p. occidentalis* occurs in N.

America. **SIMILAR SPECIES:** (1) Other N. American sticklebacks (Pl. 36) have *fewer* than 6 dorsal spines, are *deeper bodied.*

BROOK STICKLEBACK *Culaea inconstans* Pl. 36

IDENTIFICATION: Has *4–6 short dorsal spines. Deep, compressed body;* no large bony plates on side (small plates on lateral-line pores). No keel on *short caudal peduncle.* Branchiostegal membranes joined to one another but free of isthmus. Pelvic fin (absent in AB and SK) with 1 spine and 1 ray. Olive above with pale green flecks, dark green mottling; often a pale stripe along side; silver white to light green below. Breeding male is dark green to black, sometimes with red on pelvic fins. To 3½ in. (8.7 cm). **RANGE:** Atlantic and Arctic drainages from NS to NT; Great Lakes-Mississippi R. basins south to s. OH and NE, and west to MT and e. BC. Isolated population in Canadian R. system, ne. NM. Introduced elsewhere. Common; abundant in central part of range. **HABITAT:** Vegetated lakes, ponds, quiet to flowing pools and backwaters of headwaters, creeks, and small rivers; usually over sand or mud. Rarely in brackish water. **SIMILAR SPECIES:** (1) Fourspine Stickleback, *Apeltes quadracus*, usually has 1st and 2d dorsal spines *longer* than 3d and 4th; *long, slender caudal peduncle.* (2) Threespine Stickleback, *Gasterosteus aculeatus* (Pl. 36), usually has *3 dorsal spines,* branchiostegal membranes broadly united to isthmus, keel on caudal peduncle.

FOURSPINE STICKLEBACK *Apeltes quadracus* Not shown

IDENTIFICATION: Has 4 (range 3–5) *dorsal spines of various lengths*, angled alternately right and left; 1st and *2d longer* than 3d and 4th; *wide gap* before last spine. No bony plates on side. No keel on *long, slender caudal peduncle.* Branchiostegal membranes broadly united to isthmus. Pelvic fin with 1 spine and 2 rays. Olive-brown above, dark brown mottling on side; silver white below. Large male is black; breeding male has red pelvic fins. To 2½ in. (6.4 cm). **RANGE:** Atlantic Slope from

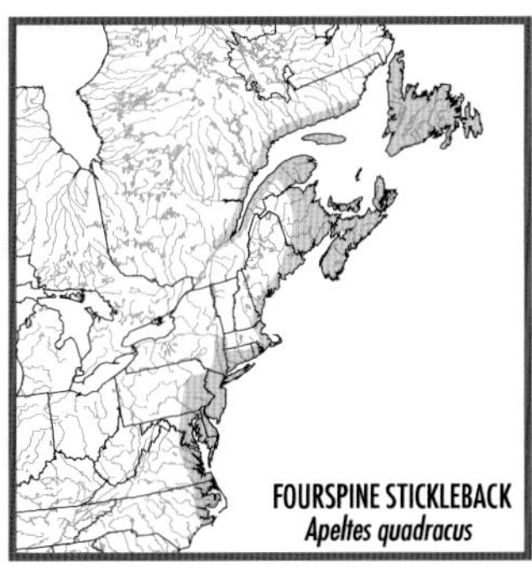

Gulf of St. Lawrence to Trent R. system, NC. Mostly nearshore marine, but inland populations in lakes in NS, and in Hudson, Delaware, and Susquehanna river drainages. Introduced to Lake Superior drainage, ON. Common. **HABITAT:** Vegetated, quiet water areas. **SIMILAR SPECIES:** (1) Threespine Stickleback, *Gasterosteus aculeatus* (Pl. 36), has *3 dorsal spines, short caudal peduncle.*

THREESPINE STICKLEBACK *Gasterosteus aculeatus* Pl. 36

IDENTIFICATION: Has *3* (rarely 2 or 4) *dorsal spines, last very short;* 0–30 bony plates on side (fewer in freshwater populations). *Bony keel* along side of caudal peduncle. Branchiostegal membranes broadly united to isthmus. Pelvic fin with 1 spine, 1 ray. Silver green to brown above; silver side, often with dark mottling. Large male is black, often with red on fins; breeding male has blue side, bright red belly and lower side, bright blue eye. Large female has pink throat and belly. To 4 in. (10 cm). **RANGE:** Marine and fresh water. Arctic and Atlantic drainages from Baffin I. and west side of Hudson Bay to Chesapeake Bay, VA; Pacific drainages from AK to Baja California. Found far inland, including Lake Ontario. Introduced to other Great Lakes. Also in Europe, Iceland, Greenland, and Pacific Coast of Asia. Common; locally abundant. **HABITAT:** Shallow vegetated areas, usually over mud or sand. **REMARKS:** "Threespine Stickleback" may refer to more than 1 species. Although hybridization occurs where fully plated marine (anadromous) form and partially plated freshwater form come together, the zone of hybridization is usually very narrow and suggests reproductive isolation. Subspecies have been proposed, but ranges are poorly demarcated. *G. a. williamsoni* is a s. CA plateless (or nearly) form protected as an *endangered subspecies.* Several scientifically undescribed forms are protected as *endangered* in Canada (e.g., Enos Lake Stickleback, Paxton Lake Stickleback). **SIMILAR SPECIES:** (1) Other sticklebacks (Pl. 36) have *4 or more dorsal spines.*

Pipefishes: Family Syngnathidae (2)

Pipefishes have a long, slender body *encased in bony rings*, a long tubular snout, 1 dorsal fin, a tiny anal fin, and no pelvic fins. Eggs are carried by the male on his trunk or tail, either on the surface or in a brood pouch in which developing embryos receive nutrition. Nearly 300 species of pipefishes occur worldwide in marine, brackish, and fresh water; a few marine and brackish species enter fresh water, including 3 in our area.

GULF PIPEFISH *Syngnathus scovelli* Pl. 57

IDENTIFICATION: Ridge along side *not continuous* (at anal fin) with ridge along underside; *fewer than 16 pectoral rays.* Has 15–18 silver bars on brown side. To 7 in. (18 cm). **RANGE:** Atlantic and Gulf coasts from St. Johns R., FL, to s. S. America. Locally common; freshwater populations in St. Johns and Suwannee river drainages, FL, and in Lake St. John (Mississippi R. basin), LA. **HABITAT:** In fresh water, near vegetation in current. **REMARKS:** Chain Pipefish, *Syngnathus louisianae*, rarely in fresh water along coast of MS, differs from Gulf Pipefish in having *longer snout* (less than twice into head length; more than twice in Gulf Pipefish), 19–21 rings around body from head to anal fin (16–17 in Gulf Pipefish). **SIMILAR SPECIES:** (1) See Opossum Pipefish, *Microphis brachyurus.*

OPOSSUM PIPEFISH *Microphis brachyurus* Not shown

IDENTIFICATION: Similar to Gulf Pipefish, *Syngnathus scovelli*, but ridge along side *continuous* (at anal fin) with ridge along underside, *17–23 pectoral rays*. Only W. Atlantic pipefish with brood pouch entirely in front of anal fin. Dark bars on nearly transparent or light brown body (in young) to dark red blotches, silver stripe on side (adult); large individual has black bars on bright red snout, red caudal fin. To 8 in. (20 cm). **RANGE:** Atlantic and Gulf coasts from NJ to Brazil; also Indo-West Pacific Ocean. Uncommon in fresh water except in s. FL; apparently not on TX or LA coast. **HABITAT:** In fresh water, near dense vegetation in quiet water. **SIMILAR SPECIES:** (1) See Gulf Pipefish, *Syngnathus scovelli* (Pl. 57).

Swamp Eels: Family Synbranchidae (1 introduced)

Swamp eels are not true eels, but are *eel-like fishes lacking paired fins*; their dorsal and anal fins are reduced to ridges. Gill openings are reduced to a single pore or slit under the head or throat. Scales are usually absent. Swamp eels occur largely in tropical and subtropical fresh waters (occasionally in brackish water) in western Africa, Asia, the Indo-Australian Archipelago, Mexico, and Central and South America. The 22 known species are adapted for air-breathing and live in burrows or dense vegetation. Most species begin life as females and change to males years later. They are popular food fishes; 1 species has been introduced into the U.S. from Asia.

ASIAN SWAMP EEL *Monopterus albus* **Not shown**

IDENTIFICATION: Eel-like body; no fins, tiny eyes; triangular-shaped gill slit on throat. Red-brown above, often dark vermiculations; light tan to orange below. To 40 in. (102 cm). **RANGE:** Native to Asia from n. India and Burma to China, probably Russia, Japan, and Indo-Malay Archipelago. Established in n. GA and s. FL. Uncommon, but increasing. **HABITAT:** Pools of tannic, mud-bottomed streams; ponds, impoundments. **REMARKS:** More than 1 species possibly established in U.S. **SIMILAR SPECIES:** (1) American Eel, *Anguilla rostrata* (Pl. 1), has pectoral fins; gill slits on side of head; dorsal, caudal, and anal fins (all continuous).

FRESHWATER SPINY EELS: Family Mastacembelidae (1 introduced)

Spiny eels are not true eels, but are long slab-sided fishes with *9–42 isolated spines* in front of the dorsal fin with 52–131 rays, no pelvic fins, and anterior nostrils at the end of a fleshy appendage on the snout. Scales are very small or absent. Spiny eels live in fresh water in tropical and subtropical Africa and Asia. Several of 80 known species are popular aquarium fishes, and 1 has been introduced into the U.S. from Asia.

SPOTFIN SPINY EEL *Macrognathus siamensis* **Not shown**

IDENTIFICATION: Brown with *large black and white ocelli* along 2d dorsal fin base, white-yellow stripe along upper side. Has 13–19 dorsal spines, 50–55 dorsal rays. To 16 in. (40 cm). **RANGE:** Native to Asia. Established in extreme s. FL. Uncommon, but increasing. **HABITAT:** Vegetated canals in FL; buries in substrate.

SCULPINS: Family Cottidae (33)

Most of the more than 250 species of sculpins are marine, but several inhabit fresh waters of North America and northern Eurasia. Sculpins have a *suborbital stay* (bony connection under the cheek uniting the bones under the eye with the front of the gill cover); *large mouth; large fanlike pectoral fins, 1–4 preopercular spines* (at the front of the gill

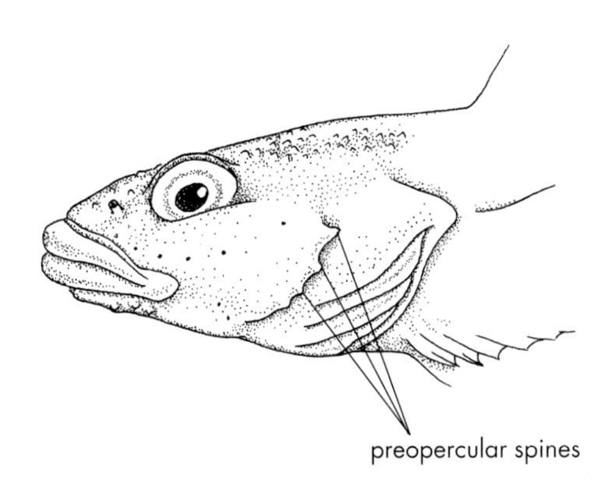

Fig. 53. Sculpin.

cover; Fig. 53), thoracic pelvic fins with 1 hidden spine (i.e., a fleshy sheath covers the small spine and the first much larger ray), 3–4 rays; and *no or few scales*. They have 2 dorsal fins, the first with spines, the second longer and with rays. The long anal fin lacks spines. The body is wide at the front and tapers to a slender, compressed caudal peduncle; it often has small spines or prickles.

Sculpins are among the most difficult freshwater fishes to identify. Most are drab and variably mottled, and the color pattern is of less use than in other groups. Development of prickles and preopercular spines, useful characters in identifying some species, varies with habitat. Individuals in small cold streams tend to have poorly developed prickles and spines; those from lakes and other quiet waters have larger prickles and spines. Other useful (often necessary) traits, such as the presence of palatine teeth, require close, often microscopic, examination. Some forms recognized as species appear to grade into one another; other "species" may contain several reproductively isolated populations that warrant recognition as species. Monikers such as *"confusus"* and *"perplexus"* are appropriate for sculpins.

DEEPWATER SCULPIN *Myoxocephalus thompsonii* Pl. 37

IDENTIFICATION: *Large gap* between dorsal fins. *Bony plates* along lateral line. *Extremely wide, flat head;* huge mouth extends to beneath eye. Large disklike scales (often absent on juvenile) on back and side above lateral line. Long body, wide at front, tapered to extremely slender caudal peduncle. Large male has huge 2d dorsal, pectoral fins. Complete or nearly complete (in eastern populations) lateral line; 4 preopercular spines, upper 2 large, directed upward, appearing as 1

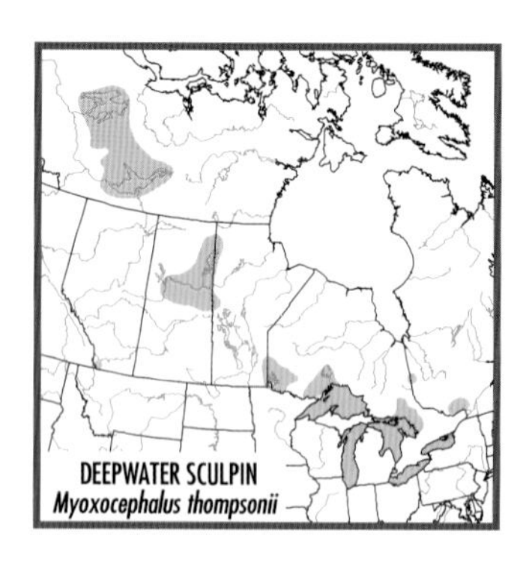

divided spine; lower 2 small, directed downward. Dark brown to green mottling, often 4–7 green saddles, on gray-brown back and side; white below; dark bars on fins except pelvics; 3 (rarely 4) pelvic rays. To 9 in. (23 cm). **RANGE:** St. Lawrence-Great Lakes and Arctic basins from w. QC and NY west to AB and north to NT. Localized; common in some lakes (e.g., Lake Michigan). Protected in Canada as a *threatened species.* **HABITAT:** Bottoms of deep (to 1200 ft. [366 m]) cold lakes. **SIMILAR SPECIES:** (1) See Fourhorn Sculpin, *M. quadricornis.*

FOURHORN SCULPIN *Myoxocephalus quadricornis* **Not shown**

IDENTIFICATION: Nearly identical to Deepwater Sculpin, *M. thompsonii,* but has *4 large rough and warty projections* on top of head (2 just behind eye, 2 at rear of head); large disklike scales *below,* as well as above, lateral line. To 14¾ in. (37 cm). **RANGE:** Arctic drainages of N. America and Eurasia; only in Arctic Archipelago in N. America (not mapped). Common. **HABITAT:** Near shore (to 60 ft. [20 m]) in brackish water. Enters coastal rivers and may occur 120 miles (190 km) inland. **SIMILAR SPECIES:** (1) See Deepwater Sculpin, *M. thompsonii* (Pl. 37).

COTTUS

GENUS CHARACTERISTICS: *Joined or narrowly separated dorsal fins.* No bony plates along lateral line; upper preopercular spine not branched. Usually many thin dark brown bands on 2d dorsal, caudal, and pectoral fins.

SPOONHEAD SCULPIN *Cottus ricei* **Pl. 37**

IDENTIFICATION: *Very wide flat head; extremely slender caudal peduncle. Prickles* on most of body. Dorsal fins separate to base. Complete lateral line; 33–36 pores. Light to dark brown above, dark brown specks and mottling on back and side, often 4 dark saddles on back. No large black spots on dorsal fins. Has 3 preopercular spines, uppermost spine long, curved inward; 4 pelvic rays, 7–10 dorsal spines, 16–18 dorsal rays, 12–15 anal rays, 14–16 pectoral rays, 1 pore at tip of chin. No palatine teeth. To 5 in. (13 cm). **RANGE:** St. Lawrence-Great Lakes and Arctic basins from s. QC to Mackenzie R. drainage, NT, YT, and BC; south to Great Lakes and n. MT. Milk R. system (Missouri R. drainage), s. AB. Common but disappearing from lower Great Lakes. **HABITAT:** Variable. Rocky areas of swift creeks and small to medium rivers; shores and deep water of lakes. To depth of 450 ft. (137 m). **SIMILAR SPECIES:** (1) Other *Cottus* species (Pls. 37 & 38) have *narrower, deeper head; deeper caudal peduncle.*

TORRENT SCULPIN *Cottus rhotheus* Pl. 37

IDENTIFICATION: *"Pinched" caudal peduncle* (extremely slender just behind dorsal and anal fins). Robust body; large head (about ⅓ body length); *prickles on entire dorsal surface*; prickles with fimbriate bases. Dorsal fins separate to base. Complete lateral line; 23–27 pores. Brown to black above, dark mottling on back and side of some individuals; 2 broad dark bars under 2d dorsal fin (slanted forward); white below. No large black spots on dorsal fins. Large male has orange edge on 1st dorsal fin. Has 3–4 preopercular spines (upper 1 large, others small), 4 pelvic rays, 7–9 dorsal spines, 15–17 dorsal rays, 11–13 anal rays, 15–17 pectoral rays, 2 pores at tip of chin. Palatine tooth patch broadly connected to vomerine tooth patch. To 6 in. (15 cm). **RANGE:** Pacific Slope drainages from upper Fraser R. drainage, BC, to Nehalem R., OR (including Columbia R. drainage of BC, WA, OR, ID, and MT). Common. **HABITAT:** Rubble and gravel riffles of small to large rivers; rocky lake shores. **SIMILAR SPECIES:** (1) Other *Cottus* species (Pls. 37 & 38) *lack* pinched caudal peduncle; peduncle is slender in Coastrange Sculpin, *C. aleuticus* (Pl. 37), which has pelvic fin reaching anus, tubular posterior nostril, 1 preopercular spine.

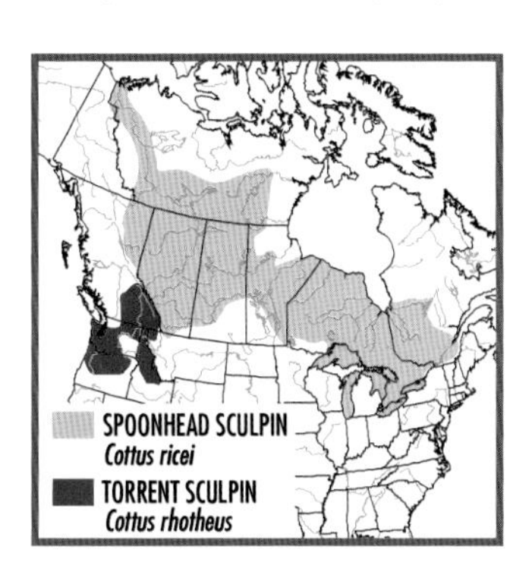

BANDED SCULPIN *Cottus carolinae* Pl. 37

IDENTIFICATION: Olive, tan, or red-brown above; 4–5 brown-black saddles, last 3 extending onto side as *sharply defined bars* (darker near edges; often vague in upper Tennessee R. drainage); mottled chin (see Remarks). No large black spots on dorsal fins. No prickles (or only in patch behind pectoral fin base). Dorsal fins separate to base. Complete lateral line (except *C. c. zopherus*; see Remarks); usually 29–34 pores. Has 3 preopercular spines (see Remarks), 4 pelvic rays, 6–9 dorsal spines, 15–18 dorsal rays, 11–14 anal rays, 15–17 pectoral rays, 2 pores at tip of chin. Palatine teeth. To 7¼ in. (18 cm). **RANGE:** Upland streams in Mississippi R. basin from upper Tennessee R. drainage, VA, across KY, TN, s. IN, and s. IL to Ozark Mt. drainages, s. MO, se. KS, n. AR, and ne. OK; south to n. AL; upland and lowland streams in Alabama R. drainage, GA, TN, and AL. Common. **HABITAT:** Gravel and rubble riffles of headwaters, creeks, and small rivers; springs and their effluents. Enters caves; populations in Perry Co., MO, have reduced eyes, light color, enlarged cephalic pores. **REMARKS:** Three subspecies recognized, but populations in Mobile Bay drainage are highly variable and in need of study. *C. c. infernatis,* in Mobile Bay drainage below Fall Line, and above Fall Line in Cahaba R. system, has complete and uninterrupted lateral line (29–34 pores), uniformly speckled chin, 2 preopercular spines (sometimes 3d knob), usually 15 pectoral rays, broad dark bars on body. *C. c. zopherus,* above Fall Line in Mobile Bay drainage (Coosa, Cahaba, and Black Warrior river systems), usually has incomplete and interrupted lateral line (22–34 pores), mottled chin, 3 preopercular spines, usually 15 pectoral rays, and narrow dark bars on body. *C. c. carolinae,* in rest of species' range, usually has complete lateral line (28–34 pores), strongly mottled chin, 3 (rarely 4) preopercular spines, usually 16–17 pectoral rays, and narrow dark bars on body. **SIMILAR SPECIES:** See (1) Kanawha Sculpin, *C. kanawhae,* and (2) Potomac Sculpin, *C. girardi* (Pl. 37). (3) Often in same streams as Mottled Sculpin, *C. bairdii* (Pl. 38), which has black spots on 1st dorsal fin; black base, orange edge on 1st dorsal fin of large male; dorsal fins joined at base; *less distinct bars* on side, not darker at edges.

KANAWHA SCULPIN *Cottus kanawhae* Not shown

IDENTIFICATION: Similar to Banded Sculpin, *C. carolinae*, but usually has *26–29 lateral-line pores* (usually 30 or more in Banded Sculpin); less defined bars on body, lighter mottling on chin. Usually 16–17 pectoral rays. To 6 in. (15 cm). **RANGE:** New R. system, WV and VA. Common. **HABITAT:** Rubble and gravel riffles of headwaters, creeks, and small rivers; springs and their effluents. **SIMILAR SPECIES:** (1) See Banded Sculpin, *C. carolinae* (Pl. 37). (2) Potomac Sculpin, *C. girardi* (Pl. 37), has 17–25 lateral-line pores, usually 15 pectoral rays, 1 pore at tip of chin.

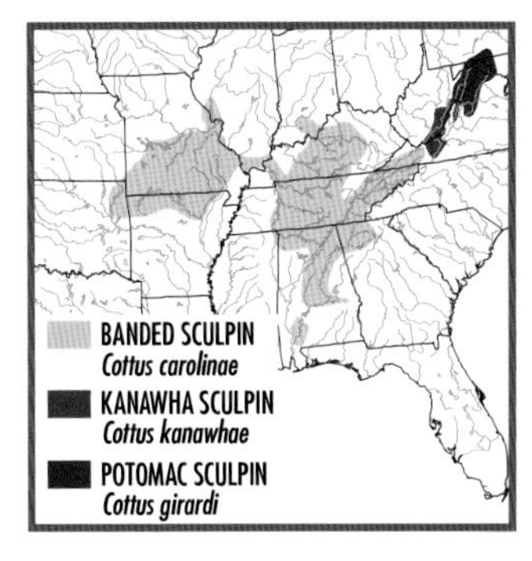

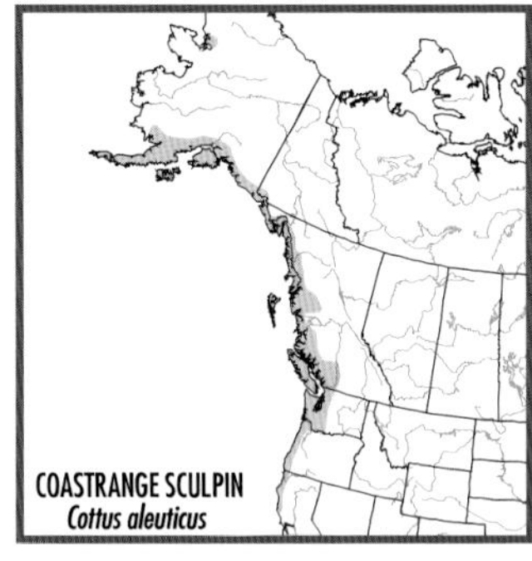

POTOMAC SCULPIN *Cottus girardi* Pl. 37

IDENTIFICATION: Similar to Banded Sculpin, *C. carolinae*, but has *incomplete lateral line* (17–25 pores), *1 pore* at tip of chin, *irregular borders* on bars on side; large male is dark overall, has narrow orange edge on 1st dorsal fin. To 5¼ in. (14 cm). **RANGE:** Mountain and Piedmont streams in upper Potomac R. drainage, PA, MD, VA, and WV; Cowpasture R. (James R. drainage), VA. Common. **HABITAT:** Rocky runs and pools of creeks and small to medium rivers, often near vegetation. **SIMILAR SPECIES:** (1) See Banded Sculpin, *C. carolinae* (Pl. 37). (2) Kanawha Sculpin, *C. kanawhae*, usually has 26–29 lateral-line pores, 16–17 pectoral rays; *2 pores* at tip of chin.

COASTRANGE SCULPIN *Cottus aleuticus* Pl. 37

IDENTIFICATION: *Long pelvic fin* reaches anus. *Long, tubular posterior nostril.* Fairly slender body; slender caudal peduncle. Prickles restricted to patch behind pectoral fin. Dorsal fins joined at base. Complete lateral line, 32–44 pores. Brown to gray above, dark brown mottling on back; 2–3 dark bars below 2d dorsal fin. No large black spots on dorsal fins. Large male has black base, orange edge, on 1st dorsal fin. Has 1 large preopercular spine (often 2d small spine), 4 pelvic rays, 8–10 dorsal spines, 17–20 dorsal rays, 12–15 anal rays, 13–15 pectoral rays, 1 pore at tip of chin. No palatine teeth. To 6¾ in. (17 cm). **RANGE:** Pacific Slope drainages from Aleutian Is. and Bristol Bay, AK, to Oso Flaco Creek, n. CA; isolated population in lower Kobuk R., AK. Most populations near coast. Locally common. Protected in Canada as a *threatened species*. **HABITAT:** Gravel and rubble riffles of medium to large rivers; rocky shores of lakes; occasionally estuaries. **SIMILAR SPECIES:** (1) Klamath Lake Sculpin, *C. princeps* (Pl. 37), has broadly joined dorsal fins; large pores on flat head; many prickles on body; *short posterior nostril;* incomplete lateral line, 15–25 lateral-line pores; 5–8 dorsal spines; 20–25 dorsal rays; 15–19 anal rays. (2) Other *Cottus* species (Pls. 37 & 38) have pelvic fins too *short* to reach anus, *short posterior nostril.*

PRICKLY SCULPIN *Cottus asper* Pl. 37

IDENTIFICATION: *Long anal* (usually *16–19 rays*) and *dorsal* (*19–23 rays*) *fins.* Often *many prickles* on body (greatly reduced in some populations). Dorsal fins joined at base. Complete lateral line, 28–43 pores. Red-brown to brown above, large *dark brown blotches,* mottling on back and upper side, 5 small black saddles on back, often 3 black bars under 2d dorsal fin; large black spot at rear (only) of 1st dorsal fin; white to yellow below. Orange edge on 1st dorsal fin of large individual. Large male is dark brown overall. Has 2–3 preopercular spines (1 large, 1–2 small), 4 pelvic rays, 7–10 dorsal spines, 15–18 pectoral rays, 1 pore at tip of chin. Large palatine teeth. To 12 in. (30 cm). **RANGE:** Pacific Slope drainages from Kenai Peninsula, AK, to Ventura R., CA; east of Continental Divide in upper Peace R. (Arctic basin), BC. Common; locally abundant. **HABITAT:** Usually over sand in pools and quiet runs of small to medium rivers; sandy and rocky shores of lakes; brackish tidepools and estuaries. **SIMILAR SPECIES:** (1) Klamath Lake Sculpin, *C. princeps* (Pl. 37), has broadly joined dorsal fins; long, slender body; flatter, wider head; no black spot on 1st dorsal fin. (2) Other *Cottus* species (Pls. 37 & 38) usually have *fewer* than 16 anal rays, fewer than 19 dorsal rays.

KLAMATH LAKE SCULPIN *Cottus princeps* Pl. 37

IDENTIFICATION: *Long, slender body; long fins* (pelvic fins reach anus on adult). *Broadly joined dorsal fins. Large pores* on flat head. *Many prickles* on body. Incomplete lateral line ends under 2d dorsal fin; 15–25 lateral-line pores. Olive to purple above, 6–7 dark brown saddles (4 under 2d dorsal fin); dark blotches on side; white to gray below. Has 0–1 preopercular spines, 4 pelvic rays, 5–8 dorsal spines, 20–25 dorsal rays, 15–19 anal rays, 14–16 pectoral rays, 1 pore at tip of chin. No palatine teeth. To 2¾ in. (7 cm). **RANGE:** Upper Klamath and Agency lakes, OR. Abundant. Formerly in Lost R., OR; now extirpated. **HABITAT:** Rocky and sandy shores of lakes. **SIMILAR SPECIES:** (1) Spoonhead

Sculpin, *C. ricei* (Pl. 37), has *separate dorsal fins*; wider, flatter head; complete lateral line. (2) Other *Cottus* species (Pls. 37 & 38) are *deeper bodied,* have deeper heads, *separate to narrowly joined dorsal fins, smaller pores* on head.

SHOSHONE SCULPIN *Cottus greenei* Pl. 37

IDENTIFICATION: *Deep body; deep caudal peduncle.* Prickles absent or only behind pectoral fin base. Dorsal fins joined at base. Incomplete lateral line, 20–28 pores. Brown to gray above; 4–5 black saddles (darkest on juvenile), 3 under 2d dorsal fin; large black spot at rear (only) of 1st dorsal fin. Has 2 preopercular spines (upper large, lower small), 3 pelvic rays; usually 6 dorsal spines, 18–19 dorsal rays, 12–13 anal rays, 14–15 pectoral rays, 2 pores at tip of chin. Palatine teeth. To 3½ in. (9 cm). **RANGE:** Hagerman Valley (Snake R. system), ID. Uncommon in extremely small area. **HABITAT:** Rocky springs and their effluents. **SIMILAR SPECIES:** (1) Other *Cottus* species (Pls. 37 & 38) have *more slender caudal peduncle.*

SLENDER SCULPIN *Cottus tenuis* Pl. 37

IDENTIFICATION: *Strongly bicolored;* chestnut brown above, silver white to brassy below. Has 1 or more *branched pelvic rays* (except in young); *3 large preopercular spines.* Prickles on upper side of body (best developed in lakes; often absent on stream-living individuals). Dorsal fins separate to base. Incomplete lateral line extends to end of 2d dorsal fin; 23–32 pores. Fairly long, slender body. No large black spots on dorsal fins; 4–5 dark brown blotches on side. Has 3 pelvic rays, 5–7 dorsal spines, 17–19 dorsal rays, 13–17 anal rays, 12–16 pectoral rays, 2 pores at tip of chin. Usually no palatine teeth. To 3½ in. (9 cm). **RANGE:** Upper Klamath R. drainage (Upper Klamath Lake and upstream), OR. Uncommon. **HABITAT:** Over mud, sand, and gravel on lake shores; riffles, runs, and pools of creeks and small to medium rivers. **SIMILAR SPECIES:** (1) See Rough Sculpin, *C. asperrimus.* (2)

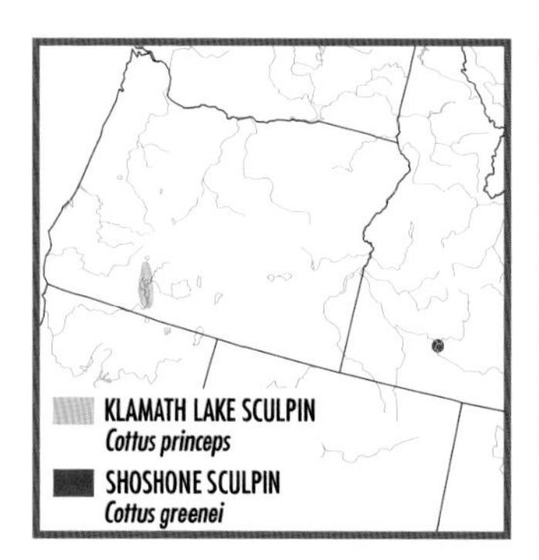

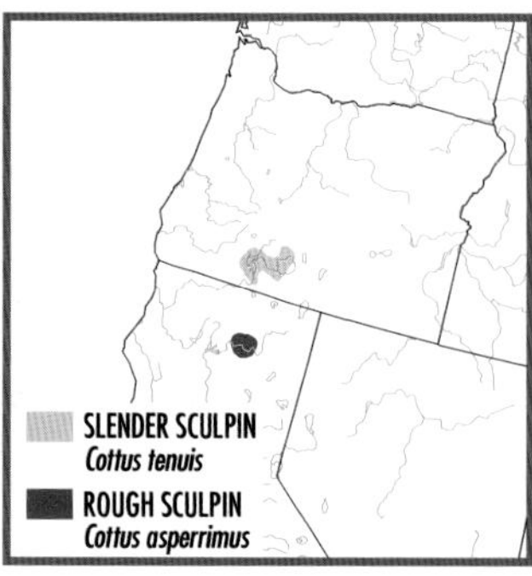

Other *Cottus* species (Pls. 37 & 38) are less bicolored, *lack* branched pelvic rays.

ROUGH SCULPIN *Cottus asperrimus* **Not shown**

IDENTIFICATION: Similar to Slender Sculpin, *C. tenuis*, but has only *1–2 large preopercular spines* (often 2d or 3d blunt knob), lateral line usually extending only to *middle* of 2d dorsal fin, *many black specks* on underside. To 3¾ in. (9.6 cm). **RANGE:** Pit R. system, Shasta and Lassen counties, CA. Abundant in small range. **HABITAT:** Vegetated runs and riffles of creeks and small to medium rivers; usually over mud in clear, fairly deep water (3–6 ft. [1–2 m]). **SIMILAR SPECIES:** (1) See Slender Sculpin, *C. tenuis* (Pl. 37).

BEAR LAKE SCULPIN *Cottus extensus* **Pl. 37**

IDENTIFICATION: *Many prickles* on *long, slender body* (none on breast and belly); usually *3 preopercular spines.* Light brown above; dark mottling on side but *no bold saddles or bars;* black band on 1st dorsal fin of large male. Dorsal fins separate to base. Incomplete lateral line, 26–29 pores. Has 4 pelvic rays, 7–8 dorsal spines, 16–19 dorsal rays, 13–15 anal rays, 15–18 pectoral rays, 2 pores at tip of chin. Palatine teeth. To 5¼ in. (13 cm). **RANGE:** Bear Lake, ID and UT. Common. **HABITAT:** Rocky bottom from near shore to deep (at least 175 ft. [53 m]) water. **SIMILAR SPECIES:** (1) See Utah Lake Sculpin, *C. echinatus* (Pl. 37).

UTAH LAKE SCULPIN *Cottus echinatus* **Pl. 37**

IDENTIFICATION: Similar to Bear Lake Sculpin, *C. extensus*, but has *prickles on breast and belly*, larger head, slightly less slender body, usually 4 preopercular spines. To 4¼ in. (11 cm). **RANGE:** Utah Lake, UT. Extinct. **HABITAT:** Nearshore rocky areas. **SIMILAR SPECIES:** (1) See Bear Lake Sculpin, *C. extensus* (Pl. 37).

SLIMY SCULPIN *Cottus cognatus* Pl. 38

IDENTIFICATION: Usually *3* (if 4, 4th is greatly reduced) *pelvic rays.* Long, fairly slender body. Prickles often on head and behind pectoral fin base. Dorsal fins separate to base. Usually incomplete lateral line (to 2d dorsal fin); 12–26 pores. Dark brown, green, or gray above; dark gray mottling on back and upper side, often 2 dark saddles under 2d dorsal fin; large black spots at front and rear of 1st dorsal fin (often joined into black band). Breeding male is dark gray to black overall, has orange edge on 1st dorsal fin. Has 2–3 preopercular spines (uppermost large), 7–9 (rarely 10) dorsal spines, 14–19 dorsal rays, 10–13 (usually 10–11) anal rays, 12–15 pectoral rays, 2 pores at tip of chin. No palatine teeth. To 4½ in. (12 cm). **RANGE:** Atlantic, Arctic, and Pacific basins throughout most of mainland Canada (except NS, much of s. SK, s. AB, w. BC) and AK; Atlantic Slope drainages south to Potomac R., VA (see Remarks); St. Lawrence-Great Lakes basin; upper Mississippi R. basin in w. WI, e. MN, and ne. IA; upper Columbia R. drainage, BC, MT, ID, and WA. Also in e. Siberia. Common. **HABITAT:** Rocky riffles of cold streams; rocky areas of lakes (usually at depth of 300–350 ft. [90–105 m]); springs and their effluents. **REMARKS:** Potomac R. population usually has 1 pore at tip of chin and palatine teeth. **SIMILAR SPECIES:** (1) Mottled Sculpin, *C. bairdii* (Pl. 38), has *4 pelvic rays*, deeper body, darker saddles and bars, dorsal fins joined at base, usually palatine teeth, 11–14 anal rays. (2) Shorthead Sculpin, *C. confusus* (Pl. 38), has 3 saddles under 2d dorsal fin, usually palatine teeth, *4 pelvic rays,* 12–14 anal rays.

MOTTLED SCULPIN *Cottus bairdii* Pl. 38

IDENTIFICATION: *Robust body;* large head; *incomplete lateral line*, 14–27 (usually 21–23) pores. Dorsal fins *joined* at base. Light to dark brown above, dark brown to black mottling on back and side; 2–4 dark brown to black saddles extending onto side as bars; uniformly speckled chin.

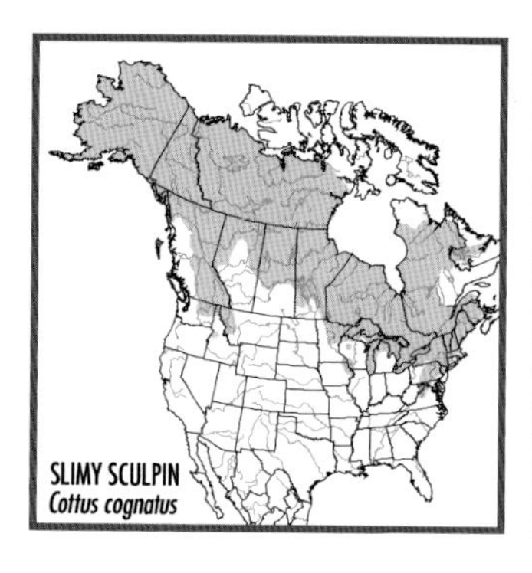

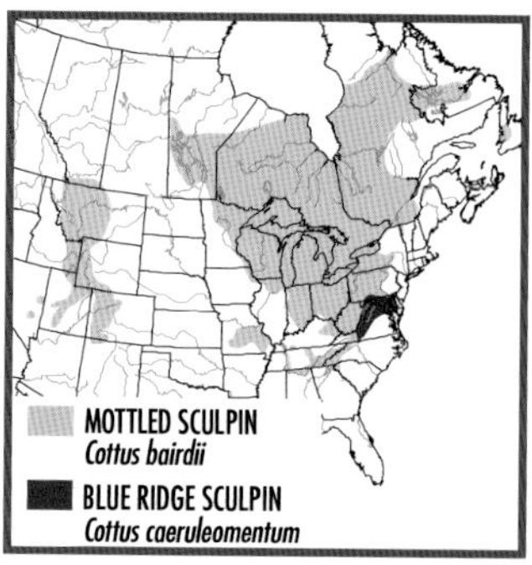

Large black spots at front and rear of 1st dorsal fin. Large male has black band, orange edge on 1st dorsal fin. Has *3 preopercular spines,* 4 pelvic rays, 6–9 dorsal spines, 15–18 dorsal rays, 11–14 anal rays, 14–16 (usually 15) pectoral rays, usually 2 pores at tip of chin, 9 infraorbital canal pores, 6 suborbital bones. *Palatine teeth* usually present. To 6 in. (15 cm). **RANGE:** Widespread with highly disjunct eastern and western distributions. In e. N. America in Arctic, Atlantic, and Mississippi R. basins from Labrador and n. QC west to w. MB, and south to Susquehanna R. drainage, PA, and Tennessee R. drainage, n. GA and AL; Missouri R. and streams in e. Ozarks, MO; isolated populations on Atlantic and Gulf slopes in extreme upper Santee (NC), Savannah (SC and GA), and Coosa (GA) river systems. In w. N. America in upper Missouri, Colorado, and Columbia river basins, AB to NM; endorheic basins in UT and NV. Common. **HABITAT:** Rubble and gravel riffles, less often sand-gravel runs, of headwaters, creeks, and small rivers; springs and their effluents; rocky shores of lakes. **SIMILAR SPECIES:** See Pl. 38. See (1) Blue Ridge Sculpin, *C. caeruleomentum*, (2) Black Sculpin, *C. baileyi*, (3) Tallapoosa Sculpin, *C. tallapoosae*, and (4) Ozark Sculpin, *C. hypselurus*. (5) Shorthead Sculpin, C. *confusus,* is more *slender,* has dorsal fins *separate to* base, 11–15 (usually 13–14) pectoral rays, *2 preopercular spines.* (6) Paiute Sculpin, *C. beldingii,* has dorsal fins *separate* to base, *no or small* palatine teeth, *1–2 preopercular spines.* (7) Margined Sculpin, *C. marginatus,* has 3 pelvic rays, 1 pore at tip of chin, *no or small* palatine teeth, 14–16 anal rays.

BLUE RIDGE SCULPIN *Cottus caeruleomentum* **Not shown**

IDENTIFICATION: Nearly identical to Mottled Sculpin, *C. bairdii*, but usually has 14 pectoral rays, caudal base bar unnotched on at least 1 side (both sides notched in Mottled Sculpin). To 3¾ in. (9.6 cm). **RANGE:** Atlantic Slope drainages from Elk R., PA, to Roanoke R., VA and NC. Common in uplands; uncommon in lowlands. **HABITAT:** Rocky riffles of headwaters and creeks; springs. **SIMILAR SPECIES:** (1) See Mottled Sculpin, *C. bairdii* (Pl. 38).

BLACK SCULPIN *Cottus baileyi* **Pl. 38**

IDENTIFICATION: Nearly identical to Mottled Sculpin, *C. bairdii*, but is *smaller,* usually *lacks* palatine teeth. To 3¼ in. (8.4 cm). **RANGE:** Extreme upper Clinch and Holston river systems (Tennessee R. drainage), VA and TN. Common in small range. **HABITAT:** Rocky riffles of headwaters and creeks; springs. **SIMILAR SPECIES:** (1) See Mottled Sculpin, *C. bairdii* (Pl. 38).

PYGMY SCULPIN *Cottus paulus* **Pl. 38**

IDENTIFICATION: *Small; to 1¾ in. (4.5 cm). Boldly patterned* (especially juvenile) with black head, white nape, black saddle under 1st dorsal

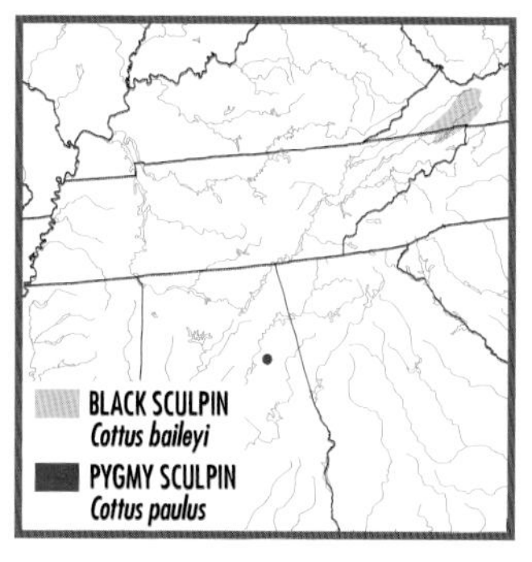

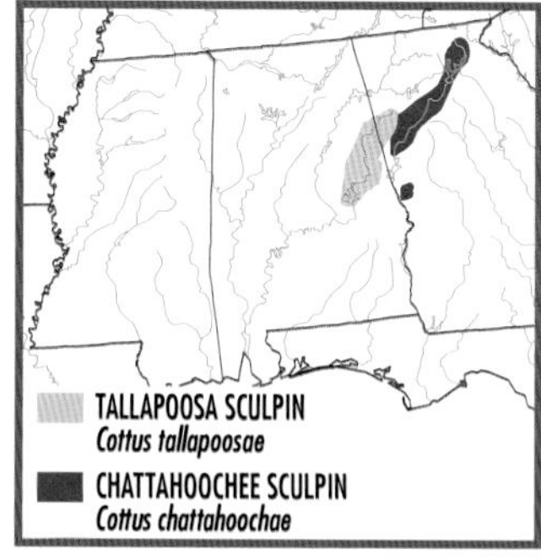

fin, 2 black saddles under 2d dorsal fin extending onto side as wide bars, wide black bar at caudal fin base. With or without prickles. Dorsal fins joined at base. Incomplete lateral line, 20–24 pores. First dorsal fin of large male has large black spots at front and rear, orange edge. Has 1–2 preopercular spines, *3 pelvic rays,* 7–8 dorsal spines, 14–16 dorsal rays, 11–13 anal rays, 13–15 pectoral rays, 2 pores at tip of chin. No palatine teeth. **RANGE:** Known only from Coldwater Spring (Coosa R. system), Calhoun Co., AL. Common within extremely small range; protected as a *threatened species.* **HABITAT:** Rocky spring runs. **SIMILAR SPECIES:** (1) Mottled Sculpin, *C. bairdii* (Pl. 38), is *larger* (to 6 in. [15 cm]), *less boldly patterned;* has *4 pelvic rays, 3* preopercular spines, usually palatine teeth.

TALLAPOOSA SCULPIN *Cottus tallapoosae* Not shown

IDENTIFICATION: Similar to Mottled Sculpin, *C. bairdii,* but usually has *8 infraorbital canal pores, 5 suborbital bones.* Usually 1–11 prickles behind pectoral fin base. Dorsal fins separate to base. Incomplete lateral line; 21–33 pores. Palatine teeth in 1–2 rows in elongate patches. To 3½ in. (9.2 cm). **RANGE:** Tallapoosa R. system above Fall Line, GA and AL. Common. **HABITAT:** Gravel and rubble riffles of headwaters, creeks, and small rivers; springs and their effluents. **SIMILAR SPECIES:** (1) See Mottled Sculpin, *C. bairdii* (Pl. 38). Only other sculpins with 8 (vs. 9) infraorbital canal pores, 5 (vs. 6) suborbital bones are (2) Chattahoochee Sculpin, *C. chattahoochee,* which usually has complete lateral line, large patch of prickles (5–130) behind pectoral fin base, and (3) Pygmy Sculpin, *C. paulus* (Pl. 38), which has 3 pelvic rays, broadly joined dorsal fins, no palatine teeth.

CHATTAHOOCHEE SCULPIN *Cottus chattahoochee* Not shown

IDENTIFICATION: Nearly identical to Tallapoosa Sculpin, *C. tallapoosae,* but usually has *complete lateral line* with 28–33 pores, *more prickles—*

usually 45–50 in patch behind pectoral fin base. To 4 in. (10 cm). **RANGE:** Chattahoochee R. system above Fall Line, GA. Common. **HABITAT:** Gravel and rubble riffles of headwaters, creeks, and small rivers; springs and their effluents. **SIMILAR SPECIES:** (1) See Tallapoosa Sculpin, *C. tallapoosae*.

OZARK SCULPIN *Cottus hypselurus* Pl. 38

IDENTIFICATION: Similar to Mottled Sculpin, *C. bairdii*. but has *wide, wavy black bands* on dorsal and caudal fins (within range of Ozark Sculpin, Mottled Sculpin lacks mottling on median fins); sharper snout; deeper body (width under 2d dorsal fin enters depth 2 or more times). Has 17–26 lateral-line pores. To 5½ in. (14 cm). **RANGE:** Meramec, Missouri, and White (including Black) river drainages in Ozark Uplands of MO and AR. Abundant. **HABITAT:** Gravel riffles of creeks and small to medium rivers; often near vegetation in springs and their effluents. **SIMILAR SPECIES:** (1) See Mottled Sculpin, *C. bairdii* (Pl. 38); Ozark and Mottled sculpins occur together only in Osage R. system, MO. (2) Banded Sculpin, *C. carolinae* (Pl. 37), has complete lateral line, *no* large black spots or wide bands on dorsal fins, 29 or more lateral-line pores, darkly mottled chin, broad dark band on caudal fin base.

COLUMBIA SCULPIN *Cottus hubbsi* Not shown

IDENTIFICATION: *Robust body;* large head; narrow caudal peduncle. *Complete lateral line* not deflected down below origin of 2d dorsal fin; 23–34 (usually 28 or more) pores. *Many prickles on side of body.* Light to dark brown above, dark brown to black mottling on back and side; 2–3 dark brown to black bars on body under 2d dorsal fin; darkly spotted and mottled chin. Mottling on 1st dorsal fin; often large black spots at front and rear. Large male has black band, orange edge on 1st dorsal fin. Has 3 preopercular spines, 4 pelvic rays, 6–9 dorsal spines, 15–18 (usually 17) dorsal rays, 11–14 anal rays, 14–17 (usually 15–16) pectoral rays, usually 2 pores at tip of chin. Palatine teeth narrowly or

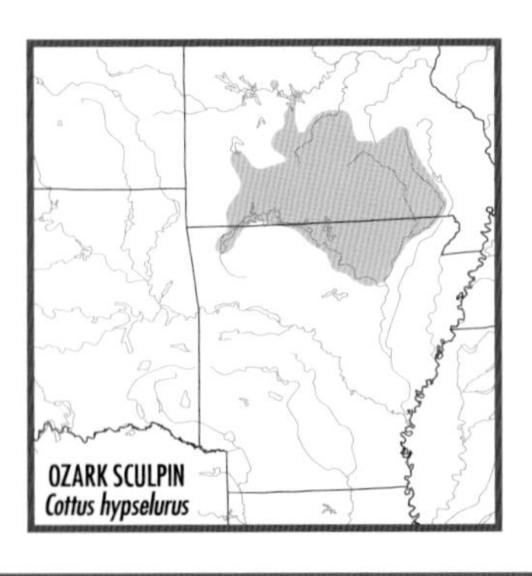

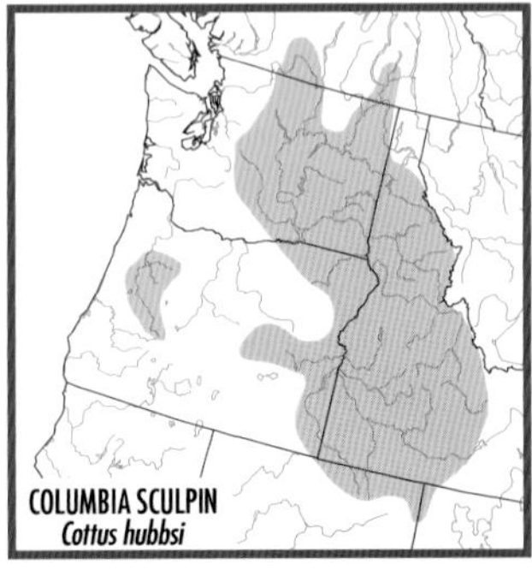

not connected to vomerine teeth. To 3¼ in. (8 cm). **RANGE:** Columbia R. system, BC, ID, WA, OR, and NV; Harney basin, OR. Common. **HABITAT:** Rocky riffles of creeks and small rivers. **SIMILAR SPECIES:** (1) See Malheur Sculpin, *C. bendirei*. (2) Torrent Sculpin, *C. rhotheus* (Pl. 37), has palatine tooth patch broadly connected to vomerine tooth patch; *prickles on back*; prickles with fimbriate bases (smooth in Columbia Sculpin), usually 16 (range 15–17) dorsal rays.

MALHEUR SCULPIN *Cottus bendirei* Not shown

IDENTIFICATION: Similar to Columbia Sculpin, *C. hubbsi*, but has *no or few prickles* on side of body; *incomplete lateral line* usually ending on caudal peduncle, deflected down below dorsal fin; 23–34 (usually 25 or more) pores. To 3¼ in. (8 cm). **RANGE:** Harney basin, OR. Common within small range. **HABITAT:** Rocky riffles of headwaters and creeks. **SIMILAR SPECIES:** (1) See Columbia Sculpin, *C. hubbsi*.

PAIUTE SCULPIN *Cottus beldingii* Pl. 38

IDENTIFICATION: No prickles. Dorsal fins *separate* to base. Complete or incomplete lateral line; 23–35 pores. Brown to gray above, dark mottling on back and side; 5–7 black saddles often extending onto side as dark bars. *Black spots* at front and rear of 1st dorsal fin, often joined into black bar. Orange edge on 1st dorsal fin. Has 1–2 (upper large) preopercular spines, 3–4 pelvic rays, 7–8 dorsal spines, 15–16 dorsal rays, 11–13 anal rays, 14–15 pectoral rays, usually 2 pores at tip of chin. No or small palatine teeth. To 5¼ in. (13 cm). **RANGE:** Columbia R. drainage from ID, w. WY, and ne. NV to w. WA and OR; western endorheic basins, including Lake Tahoe, NV and CA, Humboldt R., NV, and Bear R., UT. Common; abundant in Lake Tahoe. **HABITAT:** Rubble and gravel riffles of cold creeks and small to medium rivers; rocky shores of lakes. **SIMILAR SPECIES:** See Pl. 38. (1) See Margined Sculpin, *C. marginatus*. (2) Mottled Sculpin, *C. bairdii*, has dorsal fins *joined* at base, 3 preopercular spines, usually palatine teeth. (3) Shorthead

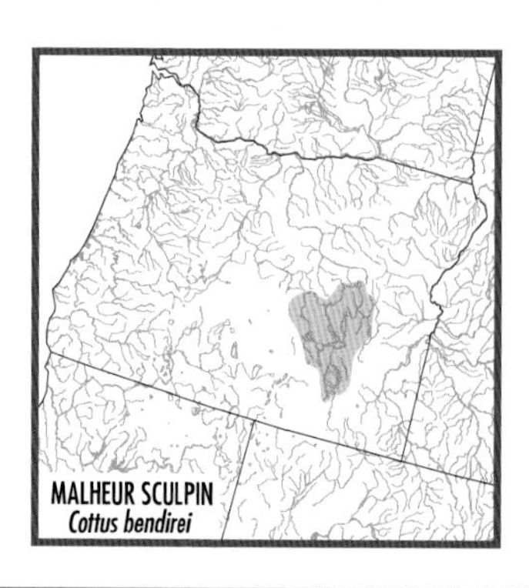

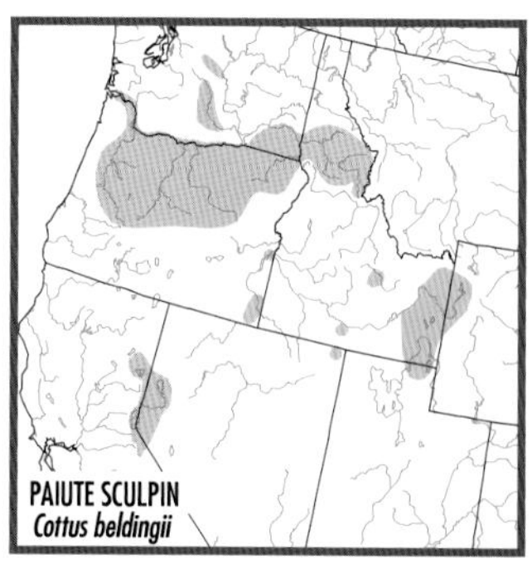

Sculpin, *C. confusus,* is more slender; has patch of prickles behind pectoral fin, usually palatine teeth.

MARGINED SCULPIN *Cottus marginatus* Pl. 38

IDENTIFICATION: Similar to Paiute Sculpin, *C. beldingii*, but has dorsal fins *joined* at base; usually *3 pelvic rays,* 1 pore at tip of chin; 14–16 anal rays. To 5 in. (13 cm). **RANGE:** Columbia R. drainage from Walla Walla R. system, WA, to Umatilla R. system, OR. Fairly common in small range. **HABITAT:** Rubble and gravel riffles. **SIMILAR SPECIES:** See Pl. 38. (1) See Paiute Sculpin, *C. beldingii*. (2) Mottled Sculpin, *C. bairdii*, has 2 pores at tip of chin, palatine teeth, *4 pelvic rays*, 12–14 anal rays. (3) Shorthead Sculpin, *C. confusus,* has dorsal fins *separate* to base, 2 pores at tip of chin, usually palatine teeth, 11–14 anal rays, *4 pelvic rays,* usually no prickles behind pectoral fin.

SHORTHEAD SCULPIN *Cottus confusus* Pl. 38

IDENTIFICATION: *Long, slender body;* short head. Prickles restricted to area behind pectoral fin. Dorsal fins *separate* to base. Usually incomplete lateral line, usually 22–37 pores. Light yellow-brown above, dark mottling on back and side; often 5–6 black saddles, 3 under 2d dorsal fin. Black band (or spots at front and rear), orange edge, on 1st dorsal fin of large male. Has *2 preopercular spines,* 4 pelvic rays, 7–9 dorsal spines, 15–19 dorsal rays, 12–14 anal rays, 11–15 (usually 13–14) pectoral rays, 2 pores at tip of chin. *Palatine teeth* usually present. To 5¾ in. (15 cm). **RANGE:** Columbia R. drainage, BC, MT, ID, WA, and OR; Puget Sound drainage, WA; upper Oldman (Hudson Bay basin) and Milk R. (Missouri R. basin), sw. AB. Common in parts of range; protected in Canada as a *threatened species.* **HABITAT:** Fast, rocky riffles of cold headwaters, creeks, and small to large rivers. **SIMILAR SPECIES:** See Pl. 38. (1) Mottled Sculpin, *C. bairdii*, is *deeper bodied,* has dorsal fins *joined* at base, 14–16 (usually 15) pectoral rays, *3 preopercular spines*. (2) Paiute Sculpin, *C. beldingii,* is *deeper bodied;*

usually *lacks* palatine teeth, patch of prickles behind pectoral fin. (3) Margined Sculpin, *C. marginatus, lacks* palatine teeth, patch of prickles behind pectoral fin; has 14–16 anal rays, 3 pelvic rays.

RIFFLE SCULPIN *Cottus gulosus* Pl. 38

IDENTIFICATION: *Deep, compressed caudal peduncle.* Large mouth (width more than body width just behind pectoral fins). Large black spot at rear (only) of 1st dorsal fin; none at front (often some black but no distinct spot). Prickles confined to area behind pectoral fin base. Dorsal fins joined at base. Complete or incomplete lateral line, 21–38 (usually 22–36) pores. Brown above, dark brown mottling on back and upper side, 5–6 small black saddles on back, no black bars under 2d dorsal fin; white to yellow below. Black band, orange edge, on 1st dorsal fin of large individual. Breeding male is dark brown overall. Has 2–3 preopercular spines, 4 pelvic rays, 8-9 dorsal spines, 16–19 dorsal rays, 12–17 anal rays, 15–16 pectoral rays, usually 1 pore at tip of chin. Palatine teeth. To 4¼ in. (11 cm). **RANGE:** Pacific Slope drainages from Puget Sound, WA, to Coquille R., OR; Sacramento-San Joaquin drainage, CA; streams in San Francisco Bay region. Absent in Rogue, Klamath, and Trinity river drainages. Common. **HABITAT:** Sand and gravel riffles of headwaters and creeks; sand-gravel runs and backwaters of small to large rivers. **SIMILAR SPECIES:** See (1) Reticulate Sculpin, *C. perplexus,* and (2) Pit Sculpin, *C. pitensis* (Pl. 38). (3) Marbled Sculpin, *C. klamathensis* (Pl. 38), has dark marbled pattern on fin rays, 14–22 lateral-line pores, 5–8 dorsal spines.

RETICULATE SCULPIN *Cottus perplexus* Not shown

IDENTIFICATION: Nearly identical to Riffle Sculpin, *C. gulosus*, but has mouth *narrower* than body just behind pectoral fins, no palatine teeth. Has 1–2 pores at tip of chin. To 4 in. (10 cm). **RANGE:** Pacific Slope drainages from Snohomish R. and Puget Sound, WA, to Rogue R. system, OR and CA. Abundant. **HABITAT:** Rubble- and gravel-bot-

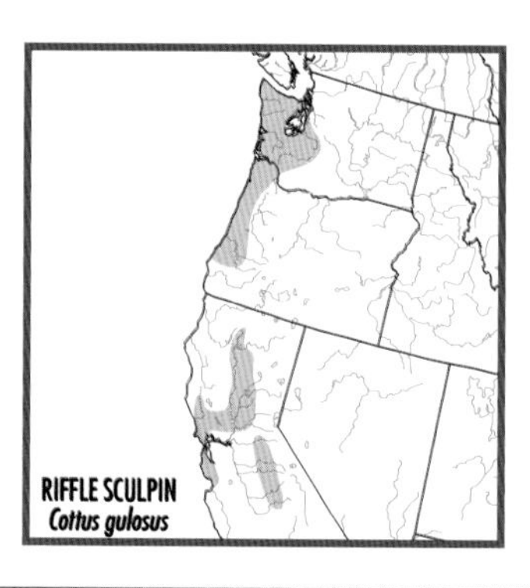

tomed pools and riffles of headwaters, creeks, and small rivers. See Remarks. **REMARKS:** Although fishes usually show a distinct preference for a particular habitat, many are able to shift to other habitats in response to certain environmental conditions, such as presence of related species. Where Torrent and Coastrange sculpins do not occur, Reticulate Sculpin occupies riffles; where all occur together, Reticulate Sculpin shifts to pools. **SIMILAR SPECIES:** (1) See Riffle Sculpin, *C. gulosus* (Pl. 38).

PIT SCULPIN *Cottus pitensis* Pl. 38

IDENTIFICATION: Similar to Riffle Sculpin, *C. gulosus*, but has dark vermiculations and *small blotches* on back and side, dorsal fins separate to base, *no* palatine teeth, usually *complete lateral line* with 31–39 (usually 33–37) pores, usually 2 pores at tip of chin. To 5 in. (13 cm). **RANGE:** Pit and upper Sacramento rivers (above mouth of Pit R.) systems, OR and CA. Common. **HABITAT:** Rocky riffles of headwaters, creeks, and small rivers. **SIMILAR SPECIES:** (1) See Riffle Sculpin, *C. gulosus* (Pl. 38). (2) Reticulate Sculpin, *C. perplexus*, has 22–32 lateral-line pores, dorsal fins joined at base.

MARBLED SCULPIN *Cottus klamathensis* Pl. 38

IDENTIFICATION: *Deep body. Black spot* at rear (only) of 1st dorsal fin. Prickles on small individuals (to 2½ in. [6.5 cm]) only. Dorsal fins joined at base. Usually incomplete lateral line, 14–22 pores. Brown above, black-brown mottling on back and upper side, 5–6 small black saddles on back; white to yellow below; *marbled pattern* of alternating dark and light spots on fin rays. Breeding male is dark brown overall. Has 1 large preopercular spine (1–2 blunt spines also may be present), 4 pelvic rays, 5–8 dorsal spines, 18–20 dorsal rays, 13–15 anal rays, 14–16 pectoral rays, 1–2 pores at tip of chin. No palatine teeth. To 3½ in. (9 cm). **RANGE:** Klamath R. drainage, OR and CA; Pit R. system from

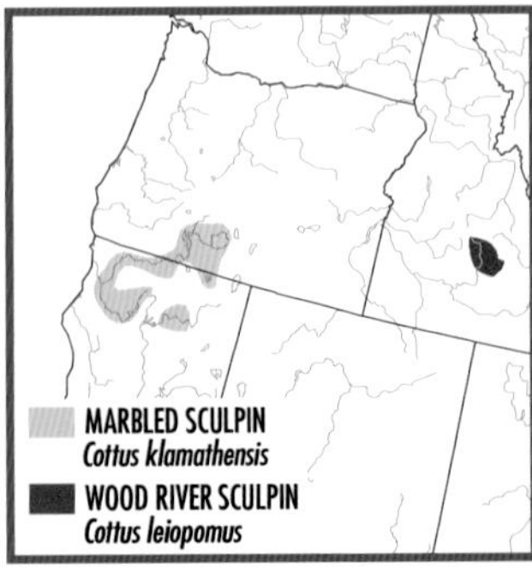

Fall R. to Hat Creek, CA. Abundant in Klamath R. drainage; common in Pit R. system. **HABITAT:** Soft-bottomed runs of clear, cold creeks and small to medium rivers. **REMARKS:** Three subspecies: *C. k. klamathensis*, in Klamath R. above Klamath Falls; *C. k. polyporus*, in lower Klamath R.; and *C. k. macrops*, in Pit R. system. Marbled Sculpins in Klamath R. have strongly contrasting, marbled appearance and black-banded fins; those in Pit R. are much darker overall, lack marbled appearance. **SIMILAR SPECIES:** (1) Riffle Sculpin, *C. gulosus*, and (2) Pit Sculpin, *C. pitensis* (both Pl. 38), *lack* prominent marbling on fins, have 22 or more lateral-line pores, 8–9 dorsal spines.

WOOD RIVER SCULPIN *Cottus leiopomus* Pl. 38

IDENTIFICATION: *Short head,* about 3 times into *long, slender body.* No prickles. Dorsal fins separate at base. Incomplete lateral line; 30–36 pores. Gray-olive above, 4 black saddles (3 under 2d dorsal fin); mottled with dark brown or black. Dark spots at front and rear of 1st dorsal fin. Has 1 preopercular spine, 4 pelvic rays, 6–8 dorsal spines, 17–19 dorsal rays, 12–14 anal rays, 12–14 pectoral rays, 2 pores at tip of chin. No or small palatine teeth. To 4¼ in. (11 cm). **RANGE:** Little and Big Wood river systems (Snake R. system), ID. Fairly common in small area. **HABITAT:** Rubble and gravel riffles of creeks and small rivers. **SIMILAR SPECIES:** (1) Slimy Sculpin, *C cognatus* (Pl. 38), has 12–26 lateral-line pores, usually 3 pelvic rays, 2 saddles under 2d dorsal fin. (2) Other slender sculpins (Pl. 38) lack black spots on 1st dorsal fin.

TEMPERATE BASSES: Family Moronidae (4)

Temperate basses (6 species) are in North America, Europe, and northern Africa. They are compressed, deep-bodied fishes with 2 dorsal fins, the first with usually 9 spines, the second with 1 spine and 11–14 rays; 3 anal spines; a large mouth; ctenoid scales; thoracic pelvic fins; a complete lateral line; a large spine on the gill cover; a *small gill* (pseudobranch) on the underside of the gill cover; and a strongly saw-toothed preopercle.

STRIPED BASS *Morone saxatilis* Pl. 39

IDENTIFICATION: Has *6–9 dark gray stripes* (on adult) on silver white side. Smoothly arched dorsal profile; body deepest between dorsal fins. Dark olive to blue-gray above; silver white side, often with brassy specks; clear to gray-green fins. Large adult has white pelvic fin, white edge on anal fin. Young lacks dark stripes, has dusky bars, on side. Second anal spine distinctly shorter than 3d; 9–13 (usually 11) anal rays. Has 1–2 patches of teeth on rear of tongue. To 6½ ft. (2 m). **RANGE:** Atlantic and Gulf slope drainages from St. Lawrence R., QC, to Lake Pontchartrain, LA; south in FL to St. Johns and Suwannee river drainages. Widely introduced in U.S., including Pacific drainages and freshwater impoundments far inland. Fairly common, but less so than formerly because of pollution of major spawning areas. **HABITAT:** Marine; ascends large rivers far upstream to spawn. Channels of medium to large rivers during spring spawning runs; lakes, impoundments, and connecting rivers. **REMARKS:** Hybrids with White Bass, *M. chrysops*, are called "Wipers" and are common where "Stripers" (Striped Bass) have been introduced. **SIMILAR SPECIES:** (1) White Bass, *M. chrysops* (Pl. 39), has deeper body, strongly arched behind head; reaches only 17¾ in. (45 cm). (2) White Perch, *M. americana* (Pl. 39), *lacks* dark stripes on side, has body deepest under 1st dorsal fin, 2d anal spine about as long as 3d, usually 9–10 anal rays, no teeth on tongue.

WHITE BASS *Morone chrysops* Pl. 39

IDENTIFICATION: Has *4–7 dark gray-brown stripes* on silver white side. Deep body, strongly arched behind head; deepest between dorsal fins. Blue-gray above; silver white side; yellow eye; clear to gray dorsal and caudal fins, clear to white paired fins. Second anal spine distinctly shorter than 3d; 11–13 anal rays. Has 1–2 patches of teeth on rear of tongue. To 17¾ in. (45 cm). **RANGE:** St. Lawrence-Great Lakes, Hudson Bay (Red R.), and Mississippi R. basins from QC to SD and south to LA; Gulf Slope drainages from Mississippi R., LA, to Rio Grande, TX and NM. Widely introduced elsewhere in U.S. and in MB. Common. **HABITAT:** Lakes, ponds, and pools of small to large rivers. **SIMILAR SPECIES:** See Pl. 39. (1) Striped Bass, *M. saxatilis*, is more slender, not strongly arched behind head; reaches 6½ ft. (2 m). (2) White Perch, *M. americana*, *lacks* dark stripes on side; has body deepest under 1st dorsal fin, 2d anal spine about as long as 3d, usually 9–10 anal rays, no teeth on tongue. (3) Yellow Bass, *M. mississippiensis*, has broken and offset stripes on yellow side, 2d anal spine about as long as 3d, usually 9 anal rays, no teeth on tongue.

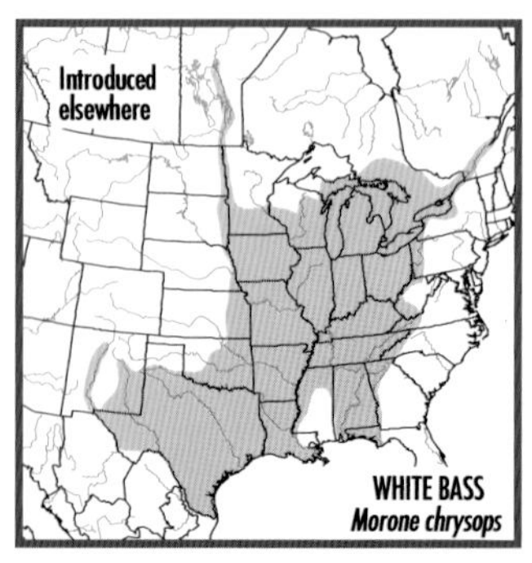

YELLOW BASS *Morone mississippiensis* **Pl. 39**

IDENTIFICATION: Has *5–7 black stripes* on side, *broken and offset on lower side.* Olive-gray above; *silver yellow side,* often with green cast; clear to blue-gray fins. Body deepest between dorsal fins; 2d anal spine about as long as 3d; usually 9 anal rays. No teeth on tongue. To 18 in. (46 cm). **RANGE:** Lake Michigan drainage and Mississippi R. basin from WI and MN south to Gulf; east to w. IN and e. TN, and west to w. IA and e. OK. On Gulf Slope in lower Mobile Bay drainage, AL, to Galveston Bay drainage, TX. Introduced elsewhere in U.S. Fairly common; mostly restricted to lowlands. **HABITAT:** Pools and backwaters of small to large rivers; ponds and lakes. **SIMILAR SPECIES:** (1) Other temperate basses (Pl. 39) *lack* broken stripes on side; have *silver white side.*

WHITE PERCH *Morone americana* **Pl. 39**

IDENTIFICATION: *No dark stripes* along side (of adult). Body deepest under *1st dorsal fin.* Olive to dark green-brown above; silver green, often brassy, side; white below; dusky fins. Young has interrupted dark lines, bars on sides. Large adult has blue cast on head. Second anal spine about as long as 3d; usually 9–10 anal rays. *No teeth* on tongue. To 22¾ in. (58 cm). **RANGE:** Atlantic Slope drainages from St. Lawrence drainage, QC, south to Savannah R., GA. Introduced to Great Lakes (except Superior) and Ohio-Missouri-Mississippi river systems of midwestern U.S.; elsewhere as far west as CO. Common; rapidly expanding in Mississippi and Missouri rivers. **HABITAT:** Brackish and fresh water; pools and other quiet-water areas of medium to large rivers; usually over mud. **SIMILAR SPECIES:** (1) White Bass, *M. chrysops,* and (2) Striped Bass, *M. saxatilis* (both Pl. 39), have *dark stripes* on side, body deepest *between* dorsal fins, 2d anal spine distinctly shorter than 3d, usually 11–12 anal rays, *1–2 patches of teeth* on rear of tongue.

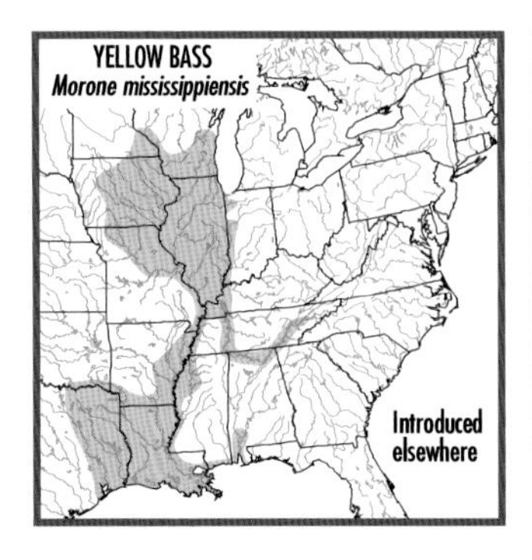

SUNFISHES: Family Centrarchidae (32)

Centrarchids occur naturally only in fresh waters of North America. They have been introduced throughout the world, including parts of North America where they are not native. Included in this family are some of the most popular sport fishes.

Sunfishes and basses are laterally compressed with 2 dorsal fins, the first with spines, the second with rays, so *broadly joined* that they appear as 1 fin. They have 3–8 anal spines, thoracic pelvic fins, no sharp spine near the back of the gill cover, and ctenoid scales (except Mud Sunfish, *Acantharchus pomotis*). Sunfishes and basses build nests and guard young. The male constructs a circular pit of gravel or vegetation by vigorously fanning his fins. After spawning, the male remains to guard the eggs and young.

SACRAMENTO PERCH *Archoplites interruptus* Pl. 40

IDENTIFICATION: Has *12–13 dorsal spines, 6–7 anal spines.* Dorsal fin base about *twice* as long as anal fin base. Deep, fairly compressed body; large mouth, upper jaw extending under eye pupil. Olive-brown above; 6–7 irregular dark bars on upper side down to lateral line; silver green to purple sheen on mottled black and white side; white below; black spot on ear flap. Has 38–48 lateral scales; 10–11 anal rays. To 24 in. (61 cm). **RANGE:** Native to Sacramento-San Joaquin, Pajaro, and Salinas river drainages, and Clear Lake, CA; introduced elsewhere in w. U.S. Uncommon, declining in native range where unable to compete with introduced sunfishes. **HABITAT:** Originally vegetated sloughs, pools of sluggish rivers, and lakes; now most common in ponds and impoundments. **SIMILAR SPECIES:** (1) White Crappie, *Pomoxis annularis*, and (2) Black Crappie, *P. nigromaculatus* (both Pl. 40), have dorsal fin base *about as long* as anal fin base, 6–8 dorsal spines.

FLIER *Centrarchus macropterus* Pl. 40

IDENTIFICATION: *Large black teardrop; interrupted rows of black spots* along side; *7–8 anal spines.* Deep, extremely compressed body; small mouth. *Red-orange* around black spot near rear of 2d dorsal fin on young. Dorsal fin base about as long as anal fin base. Dusky gray back; silver side, many green and bronze flecks; brown-black spots, often in wavy bands, on dorsal, caudal, and anal fins; black edge on front half of anal fin. Has 4 broad dark brown bars, widest at top, on side of young. Has 36–44 lateral scales; 11–13 dorsal spines, 12–15 rays; 13–17 anal rays. To 7½ in. (19 cm). **RANGE:** Coastal Plain from Potomac R. drainage, MD, to cen. FL, and west to Trinity R., TX; north in Former Mississippi Embayment to s. IL and s. IN. Common; uncommon above Fall Line. **HABITAT:** Swamps; vegetated lakes, ponds, sloughs, and backwaters and pools of creeks and small rivers; usually over mud. **SIMILAR SPECIES:** (1) White Crappie, *Pomoxis annularis*, and (2) Black Crappie, *P. nigromaculatus* (both Pl. 40), *lack* bold black teardrop, rows of black spots on side; have 6–8 dorsal spines.

BLACK CRAPPIE *Pomoxis nigromaculatus* Pl. 40

IDENTIFICATION: *Long predorsal region* arched with *sharp dip* over eye; dorsal fin base about as long as distance from eye to dorsal fin origin, about as long as anal fin base. Has *7–8 dorsal spines,* 1st much shorter than last. Large mouth; upper jaw extends under eye. Deep, extremely compressed body. Gray-green above; wavy black lines, blotches, green flecks on silver blue side; white below; many wavy black bands, spots on dorsal, caudal, and anal fins. Has 15–16 dorsal rays, 6 anal spines, 17–19 rays. To 19¼ in. (49 cm). **RANGE:** So widely introduced throughout U.S. that native range is difficult to determine; presumably Atlantic Slope from VA to FL, Gulf Slope west to TX, St. Lawrence-Great Lakes, Hudson Bay (Red R.), and Mississippi R. basins from QC to MB south to Gulf. Common in lowlands; rare in uplands. **HABITAT:**

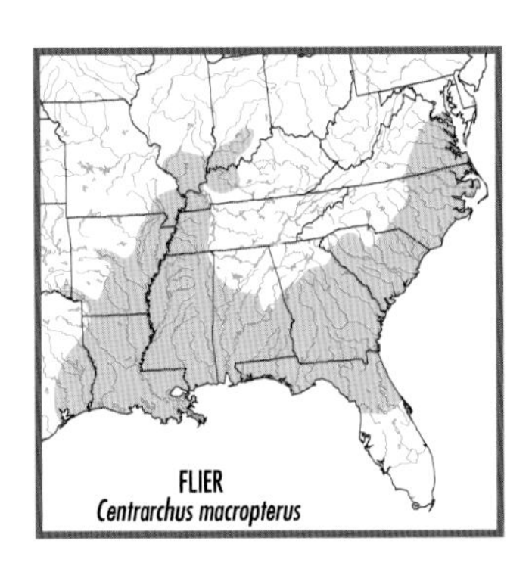

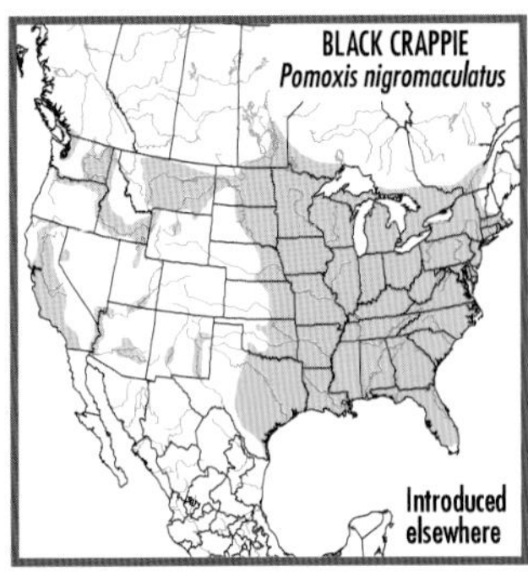

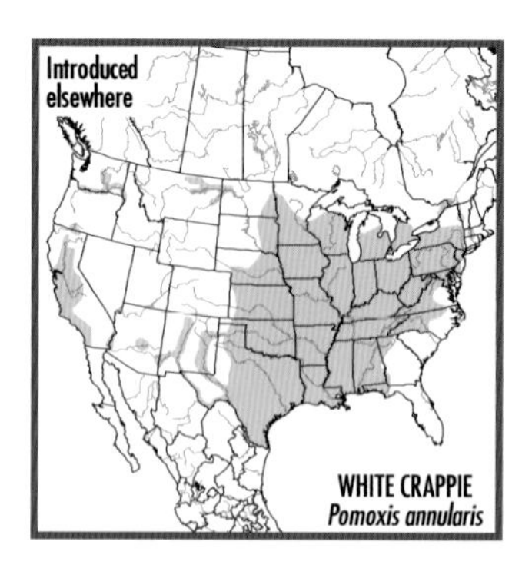

Lakes, ponds, sloughs, and backwaters and pools of streams. Usually among vegetation over mud or sand; usually in clear water. **SIMILAR SPECIES:** (1) White Crappie, *P. annularis* (Pl. 40), has *6 dorsal spines*, dark bars on side, dorsal fin base shorter than distance from eye to dorsal fin origin.

WHITE CRAPPIE *Pomoxis annularis* Pl. 40

IDENTIFICATION: *Very long predorsal region* arched with *sharp dip* over eye; dorsal fin base shorter than distance from eye to dorsal fin origin, about as long as anal fin base; *6 dorsal spines,* 1st much shorter than last. Large mouth; upper jaw extends under eye. Deep, extremely compressed body. Gray-green above; silver side with 6–9 dusky chain-like bars on side (widest at top), black blotches, green flecks; white below; wavy black bands, spots on dorsal, caudal, and anal fins. Has 14–15 dorsal rays, 6 anal spines, 17–19 rays. To 21 in. (53 cm). **RANGE:** Native to Great Lakes, Hudson Bay (Red R.), and Mississippi R. basins from NY and s. ON west to MN and ND, and south to Gulf; Gulf drainages from Mobile Bay basin, GA and FL, to Nueces R., TX. Widely introduced elsewhere in U.S. Common. **HABITAT:** Sand- and mud-bottomed pools and backwaters of creeks and small to large rivers; lakes, ponds. Often in turbid water. **SIMILAR SPECIES:** (1) Black Crappie, *P. nigromaculatus* (Pl. 40), has *7–8 dorsal spines*, dorsal fin base about same length as distance from eye to dorsal fin origin.

AMBLOPLITES

GENUS CHARACTERISTICS: Usually *6* (range 5–7) *anal spines. Red eye.* Dusky spots, brown wavy lines on dorsal, caudal, and anal fins; white edge on ear flap; dusky to black teardrop. Compressed as young; thicker bodied as adult. Large mouth, upper jaw extending under eye pupil. Short, rounded pectoral fin. Fairly long, slender rakers on 1st gill arch. Complete lateral line.

ROCK BASS *Ambloplites rupestris* Pl. 40

IDENTIFICATION: Adult has *rows of brown-black spots* along side, largest and darkest below lateral line. Young has *brown marbling* on gray side. Light green above, brassy yellow flecks on side; about 5 wide dark saddles over back and down to midside; white to bronze breast and belly; black edges on dorsal, caudal, and anal fins of adult. Has 36–47 lateral scales; usually 7–8 scales above lateral line, 21–25 scale rows across breast from pectoral fin to pectoral fin; usually 11–13 dorsal spines, 10–11 anal rays. To 17 in. (43 cm). **RANGE:** Native to St. Lawrence R.-Great Lakes, Hudson Bay (Red R.), and Mississippi R. basins, from QC to SK, and south to n. GA, n. AL, and MO (native in MO only to Meramec R.). Introduced in Atlantic drainages as far south as Roanoke R., VA; in Missouri and Arkansas river drainages, MO, AR, se. KS, and ne. OK; and in some western states. Common. **HABITAT:** Vegetated, brushy stream margins and pools of creeks and small to medium rivers; rocky, vegetated margins of lakes. Usually in clear, silt-free rocky streams. **SIMILAR SPECIES:** See Pl. 40. See (1) Ozark Bass, *A. constellatus*, and (2) Shadow Bass, *A. ariommus*. (3) Roanoke Bass, *A. cavifrons*, has unscaled or partly scaled cheek, iridescent gold to white spots on upper side and head. (4) Warmouth, *Lepomis gulosus*, has only 3 (vs. 6) anal spines, dark lines radiating from eye.

OZARK BASS *Ambloplites constellatus* Pl. 40

IDENTIFICATION: Similar to Rock Bass, *A. rupestris*, but has *freckled pattern* (scattered dark brown spots) on side of body and head, more slender body, no black edge on anal fin of large male, usually 41–46 lateral scales (range 38–48), 8–9 scale rows above lateral line. To 7½ in. (19 cm). **RANGE:** Native to White R. drainage, MO and AR; introduced in Osage R., MO. Common. **HABITAT:** Clear rocky pools of creeks and small to medium rivers; usually near stream bank, large boulders, or brush. **SIMILAR SPECIES:** See Pl. 40. (1) See Rock Bass, *A. rupestris*. (2)

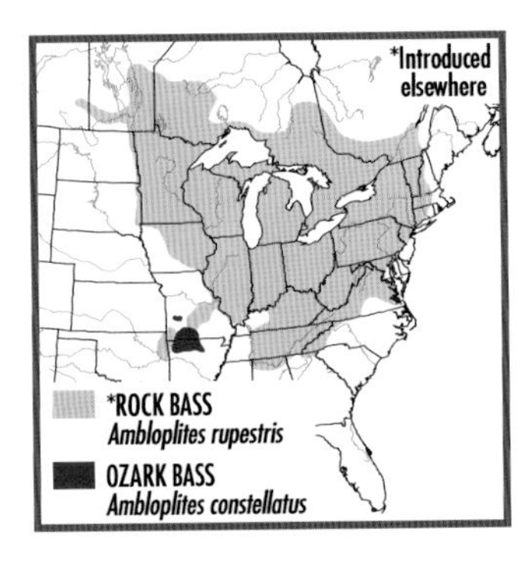

Shadow Bass, *A. ariommus*, and (3) Roanoke Bass, *A. cavifrons*, *lack* freckled pattern.

SHADOW BASS *Ambloplites ariommus* Pl. 40

IDENTIFICATION: Young and adult similar to young Rock Bass, *A. rupestris,* in color pattern with *irregular marbling* of brown or gray on light green or brown side, *large eye,* and compressed body, but usually have 15–18 scale rows across breast (pectoral fin to pectoral fin). To 8¾ in. (22 cm). **RANGE:** Gulf Slope from Apalachicola R. drainage, GA and FL, to lower Mississippi R. basin, LA; St. Francis, Black, Red, and upper Ouachita river drainages, MO and AR; also in upper Guadalupe R. drainage, TX (where probably introduced). Common. **HABITAT:** Brushy, gravel-, sand-, and mud-bottomed pools of creeks and small to medium rivers. **SIMILAR SPECIES:** (1) Rock Bass, *A. rupestris* (Pl. 40), has rows of dark spots along side of adult, smaller eye, usually 21–25 scale rows across breast. (2) Ozark Bass, *A. constellatus* (Pl. 40), has freckled pattern on side, smaller eye, 20–26 scale rows across breast.

ROANOKE BASS *Ambloplites cavifrons* Pl. 40

IDENTIFICATION: *Unscaled or partly scaled cheek.* Many *iridescent gold to white spots* on upper side and head. Olive to tan above; dark and light marbling on side, often with rows of black spots; white to bronze breast and belly. Has 39–49 lateral scales, 11 anal rays, 27–35 scale rows across breast (pectoral fin to pectoral fin). To 14½ in. (36 cm). **RANGE:** Chowan, Roanoke, Tar, Neuse, and Cape Fear river drainages, VA and NC. Uncommon; reduced populations in part due to competition with introduced Rock Bass, *A. rupestris*. **HABITAT:** Rocky and sandy pools of creeks and small to medium rivers; usually in clear streams above Fall Line. **SIMILAR SPECIES:** (1) Other *Ambloplites* species (Pl. 40) have *fully scaled cheek, lack* gold to white spots; Rock Bass, *A. rupestris*, has black edges on median fins of adult.

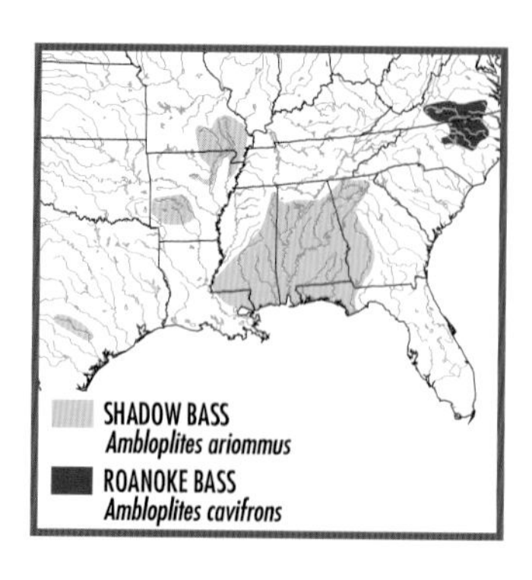

MUD SUNFISH *Acantharchus pomotis* Pl. 43

IDENTIFICATION: *Rounded caudal fin. Oblong, compressed body; 3–4 parallel brown to black stripes* across face (above eye, through eye, along upper jaw) and along side of body. Large eye; short snout; large mouth, upper jaw extending under eye. Light to dark green body; 4–5 dark brown stripes (often faint) along side; black spot, orange edge (on large individual) on ear flap; clear to dusky olive fins, black edge on anal fin. *Cycloid scales.* Has 32–45 lateral scales; 20–30 (usually 24–28) scales around caudal peduncle; 10–12 dorsal spines, 9–13 rays; usually 5 (4–6) anal spines, 9–11 rays. To 8¼ in. (21 cm). **RANGE:** Atlantic Coastal Plain and lower Piedmont drainages from Hudson R., NY, to St. Johns R., FL; Gulf Coastal Plain drainages of n. FL and s. GA from Suwannee R. to St. Marks R. Isolated population, apparently native, in lower Tombigbee R. drainage, AL. Uncommon. **HABITAT:** Vegetated sloughs, lakes, pools, and backwaters of creeks and small to medium rivers; usually over mud and detritus. **SIMILAR SPECIES:** (1) All other sunfishes have *ctenoid scales.* (2) Only Green Sunfish, *Lepomis cyanellus* (Pl. 41), has 23 or more scales around caudal peduncle.

BANDED SUNFISH *Enneacanthus obesus* Pl. 43

IDENTIFICATION: *Rounded caudal fin; 5 or more dark bars* on side (darkest on large individual). Dusky olive above; rows of indistinct purple-gold spots along side; light olive below; black teardrop; median fins dark with rows of pale spots. Dark spot on ear flap larger than eye pupil. Deep, compressed body. Usually *19–22 scales around caudal peduncle.* Usually *incomplete or interrupted lateral line.* Has 29–35 lateral scales; 8–9 (rarely 10) dorsal spines, 10–13 rays; 3 anal spines; 9–12 rays. To 3¾ in. (9.5 cm). **RANGE:** Below Fall Line in Atlantic and Gulf slope drainages from s. ME to Perdido R., AL; south to cen. FL. Common. **HABITAT:** Heavily vegetated lakes, ponds, sluggish sand- or mud-bottomed pools and backwaters of creeks and small to large

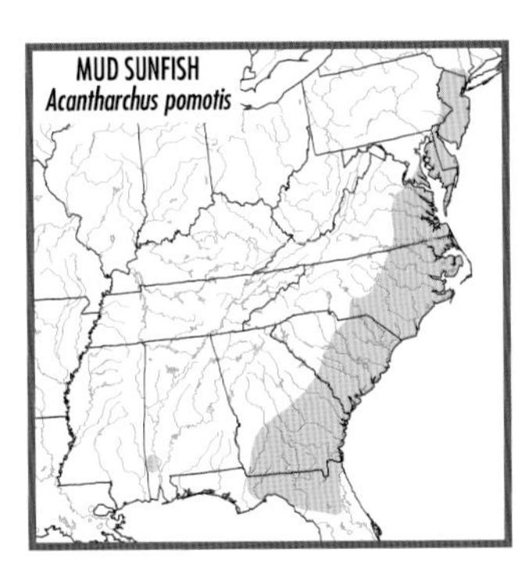

rivers. **SIMILAR SPECIES:** (1) See Bluespotted Sunfish, *E. gloriosus* (Pl. 43). (2) Blackbanded Sunfish, *E. chaetodon* (Pl. 43), has bold black bars on side, black blotch at front of dorsal fin, red/pink and black on pelvic fins; usually 10 dorsal spines, 12–13 anal rays.

BLUESPOTTED SUNFISH *Enneacanthus gloriosus* Pl. 43

IDENTIFICATION: Similar to Banded Sunfish, *E. obesus*, but usually has *16–18 scales* around more slender caudal peduncle; usually *complete* lateral line; rows of *bright blue* or *silver spots* along side of large young and adult; dark spot on ear flap about ⅔ as large as eye pupil; fewer than *5 indistinct bars* on side of adult. Has 28–33 lateral scales. To 3¾ in. (9.5 cm). **RANGE:** Atlantic and Gulf slope drainages from s. NY to s. MS; mostly below Fall Line. Lake Ontario drainage, NY. Common. **HABITAT:** Vegetated lakes, ponds, sluggish sand- and mud-bottomed pools and backwaters of creeks and small to large rivers. **SIMILAR SPECIES:** (1) See Banded Sunfish, *E. obesus* (Pl. 43).

BLACKBANDED SUNFISH *Enneacanthus chaetodon* Pl. 43

IDENTIFICATION: *Six bold black bars* on side, 1st through eye, 6th (often faint) on caudal peduncle. *First 2–3 membranes* of dorsal fin black; middle spines longest. *Pink to red* (spine and 1st membrane), then black, on pelvic fin. Deep, extremely compressed body; small mouth; rounded caudal fin. Dusky yellow-gray above, light below; yellow flecks on side; black spot on ear flap; dorsal, anal, and caudal fins with black mottling. Usually 19–21 scales around caudal peduncle; 26–30 lateral scales; usually 10 dorsal spines, 11–12 rays; 3 anal spines, 11–14 (usually 12–13) rays. To 3¼ in. (8 cm). **RANGE:** Below Fall Line in Atlantic and Gulf slope drainages from NJ to cen. FL, west to Flint R., GA. Locally common but absent from several drainages within range. **HABITAT:** Vegetated lakes, ponds, quiet sand- and mud-bottomed pools and backwaters of creeks and small to medium rivers. **SIMILAR**

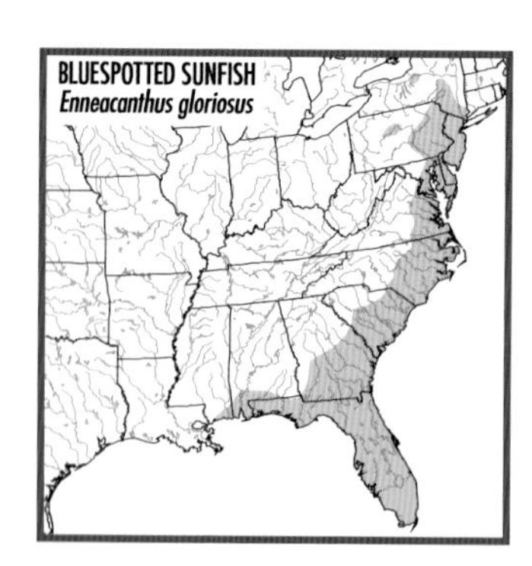

SPECIES: (1) Banded Sunfish, *E. obesus*, and (2) Bluespotted Sunfish, *E. gloriosus* (both Pl. 43), *lack* bold black bars on side, black at front of dorsal fin, red/pink and black on pelvic fin; usually have 8–9 dorsal spines, 9–11 anal rays.

MICROPTERUS

GENUS CHARACTERISTICS: *Large* (all species reach at least 14 in. (36 cm)), feisty fishes that are among the most popular sport fishes in the world. Moderately compressed, *elongate body,* becoming deeper with age. *Large mouth* extends under or past eye. Anal fin base less than half dorsal fin base; shallowly forked caudal fin; smooth (not serrated) opercle. Black spot at rear of gill cover (no long flap); dark brown lines radiating from snout and back of eye to edge of opercle; clear to yellow-olive fins, dusky spots on median fins; *3 anal spines, 55 or more lateral scales,* 9–11 dorsal spines.

LARGEMOUTH BASS *Micropterus salmoides* Pls. 39 & 42

IDENTIFICATION: First dorsal fin *highest at middle,* low at rear; 1st (spinous) and 2d (soft) dorsal fins *nearly separate. Very large mouth,* upper jaw extending well past eye (except in young). Silver to brassy green (brown in dark water) above, dark olive mottling; *broad black stripe* (often broken into series of blotches) along side and onto snout; brown eye; scattered black specks on lower side; white below. Usually no patch of teeth on tongue. Has 14–53 *branches on pyloric caeca.* Usually 58–73 lateral scales, 8 rakers on 1st gill arch. To 38 in. (97 cm). **RANGE:** Native range thought to be St. Lawrence-Great Lakes and Mississippi R. basins from s. QC to MN and south to Gulf; Atlantic and Gulf drainages from NC (probably Tar R.) into n. Mexico. Now introduced over much of U.S. and s. Canada. Also introduced to Eurasia and Africa. Common. **HABITAT:** Lakes, ponds, swamps, backwaters

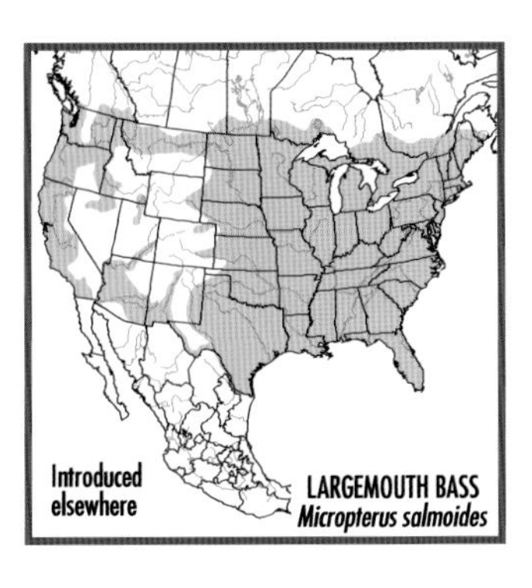

and pools of creeks and small to large rivers; usually over mud or sand. Common. **REMARKS:** Two subspecies. Florida Bass, *M. s. floridanus,* endemic to peninsular FL north to the mouth of St. Johns R. in the east and to, but not including, Suwannee R. in the west, attains a larger size, has 26–53, usually 31 or more, branches on pyloric caeca; 65–77, usually 69–73, lateral scales; 10–14, usually 11–13, scale rows on cheek; 16–18, usually 17–18, scale rows below lateral line; 27–34, usually 29–31, scales around caudal peduncle; and darker midlateral stripe often broken into blotches at front. Northern Largemouth Bass, *M. s. salmoides,* throughout rest of range except for broad area of intergradation from Choctawhatchee R. drainage on Gulf Slope to Savannah R. drainage on Atlantic Slope, has 14–35, usually fewer than 28, branches on pyloric caeca; 58–69, usually 59–67, lateral scales; 9–13, usually 10–11, scale rows on cheek; 16–18, usually 14–17, scale rows below lateral line; 24–32, usually 27–28, scales around caudal peduncle. Intergrades have intermediate counts. **SIMILAR SPECIES:** (1) Other species of *Micropterus* (Pl. 39) have more *confluent* dorsal fins, upper jaw to or barely past eye, *simple (unbranched) pyloric caeca.*

SUWANNEE BASS *Micropterus notius* **Pls. 39 & 42**

IDENTIFICATION: Large mouth, upper jaw extending to rear of eye. Color similar to Largemouth Bass, *M. salmoides*, except usually brown overall, *black vermiculations* or *rows of small black spots* in dorsal, anal, and caudal fins, often rows of black spots on lower side. Large male has bright *turquoise* cheek, breast, and belly. Usually patch of teeth on tongue. First dorsal fin of nearly uniform height throughout. Unbranched pyloric caeca. Usually 59–64 lateral scales, *usually 5 rakers on 1st gill arch.* To 14¼ in. (36 cm). **RANGE:** Suwannee and Ochlockonee river drainages, FL and GA. Introduced into St. Marks and Wacissa rivers, FL. Fairly common. **HABITAT:** Rocky riffles, runs, and pools of small to medium rivers; large springs and spring runs. **SIMILAR SPECIES:** (1) Largemouth Bass, *M. salmoides* (Pl. 39), *lacks* vermiculations or spots in fins; *usually has 8 rakers on 1st gill arch,* branched pyloric caeca.

SPOTTED BASS *Micropterus punctulatus* **Pls. 39 & 42**

IDENTIFICATION: *Rows of small black spots* on lower side; black stripe (or series of partly joined blotches) along side; black caudal spot (darkest on young). Light gold-green above, dark olive mottling; yellow-white below. Young has *3-colored* (yellow, black, white edge) caudal fin. Large mouth, upper jaw extending under rear half of eye. Patch of teeth on tongue. Unbranched pyloric caeca. Usually 61–73 lateral scales, 54–71 pored lateral-line scales, 21–28 scales around caudal peduncle, 12–13 dorsal rays, 9–11 anal rays, 15–16 pectoral rays. To 24 in. (61 cm). **RANGE:** Mississippi R. basin from s. PA and

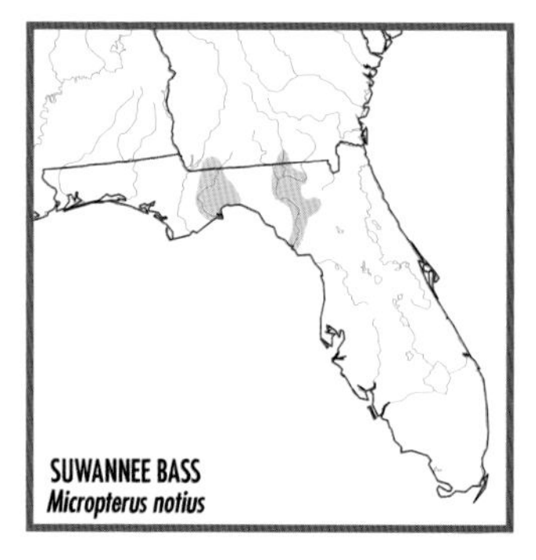

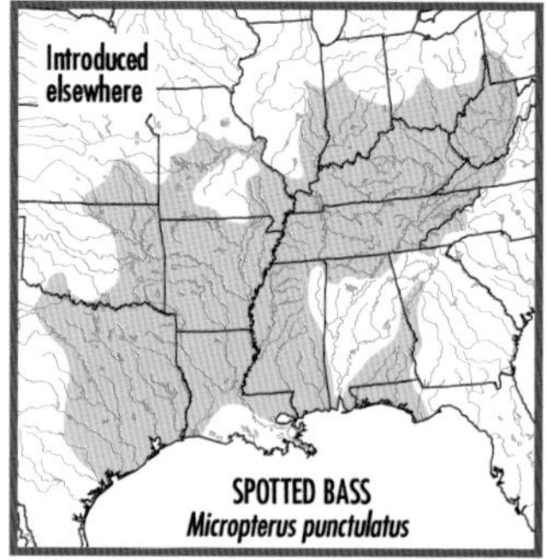

WV to se. KS, and south to Gulf; Gulf drainages from Chattahoochee (where possibly introduced), GA, to Guadalupe R., TX; absent in Mobile Bay basin. Introduced elsewhere, including w. U.S. Common. **HABITAT:** Clear, gravel-bottomed runs and flowing pools of creeks and small to medium rivers; impoundments in southern part of range. **SIMILAR SPECIES:** See Pl. 39. See (1) Alabama Bass, *M. henshalli*, and (2) Guadalupe Bass, *M. treculii*. (3) Redeye Bass, *M. coosae*, and (4) Shoal Bass, *M. cataractae*, have *dusky to dark bars* along side, usually 27 or more scales around caudal peduncle; Redeye Bass has white tips on orange caudal fin, red fins on young; Shoal Bass has no patch of teeth on tongue, 27–35 scales around caudal peduncle. (4) Largemouth Bass, *M. salmoides, lacks* rows of black spots; has deep notch between dorsal fins, *2-colored* (white, black edge) caudal fin on young.

ALABAMA BASS *Micropterus henshalli* Not shown

IDENTIFICATION: Similar to Spotted Bass, *M. punctulatus*, but has *68–84 (usually 71 or more) pored lateral-line scales, 27 or more scales around caudal peduncle;* black blotches along upper back *do not reach dorsal*

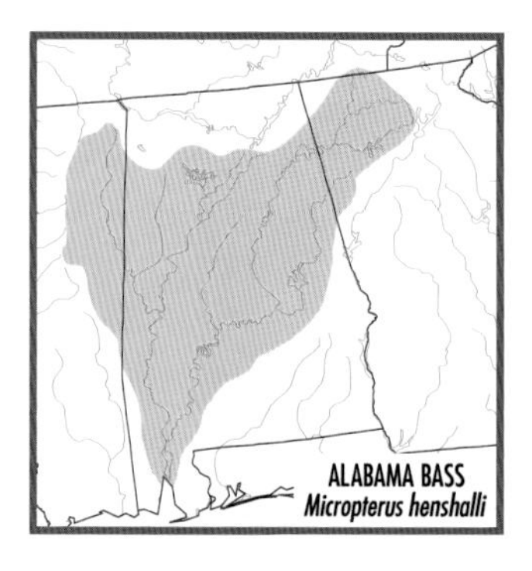

fin base, blotches along midside *do not coalesce into black stripe* on caudal peduncle. To 24 in. (61 cm). **RANGE:** Mobile Bay drainage, GA, AL, and MS. Common. **HABITAT:** Flowing pools and runs of small to medium rivers over silt, sand, or gravel; impoundments. **SIMILAR SPECIES:** (1) See Spotted Bass, *M. punctulatus* (Pl. 39).

GUADALUPE BASS *Micropterus treculii* Pls. 39 & 42

IDENTIFICATION: Similar to Spotted Bass, *M. punctulatus*, but has *10–12 dark bars* along side (darkest in young), usually 16 pectoral rays, 26–27 scales around caudal peduncle. To 16 in. (40 cm). **RANGE:** Edwards Plateau in Brazos, Colorado, Guadalupe, San Antonio, and upper Nueces (where introduced) river drainages, TX. Common. **HABITAT:** Gravel riffles, runs, and flowing pools of creeks and small to medium rivers. **SIMILAR SPECIES:** (1) See Spotted Bass, *M. punctulatus* (Pl. 39).

REDEYE BASS *Micropterus coosae* Pls. 39 & 42

IDENTIFICATION: *White tips* on orange caudal fin. *Rows of brown spots* on lower side; *dusky bars* or blotches (often absent) along side, diamond-shaped with *light centers* on caudal peduncle in large adult. Second dorsal, caudal, and front of anal fin *brick red* on young. Bronze-olive above, dark olive mottling; yellow-white to blue below; dusky spot on caudal fin base (darkest on young). Large mouth, upper jaw extending under rear half of eye. Usually patch of teeth on tongue. Unbranched pyloric caeca. Has 62–73 (usually 64–69) lateral scales, 25–31 (usually 27–29) scales around caudal peduncle, 13–16 scale rows below lateral line. To 19 in. (47 cm). **RANGE:** Saluda, Savannah, Chattahoochee, and Mobile Bay drainages, NC, SC, GA, TN, and AL; mostly above Fall Line. Introduced elsewhere, including Altamaha R. drainage, GA, e. TN, and upper Cumberland R. drainage, KY. Common; apparently extinct in Saluda R. drainage. **HABITAT:** Rocky runs and pools of creeks and small to medium rivers. **SIMILAR SPECIES:** See Pl. 39. (1) Shoal Bass,

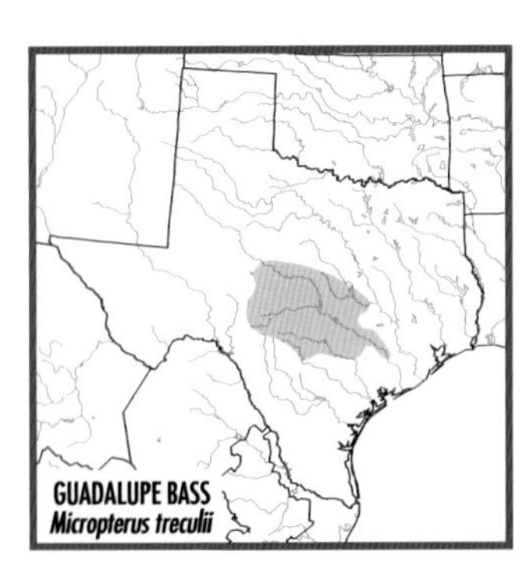

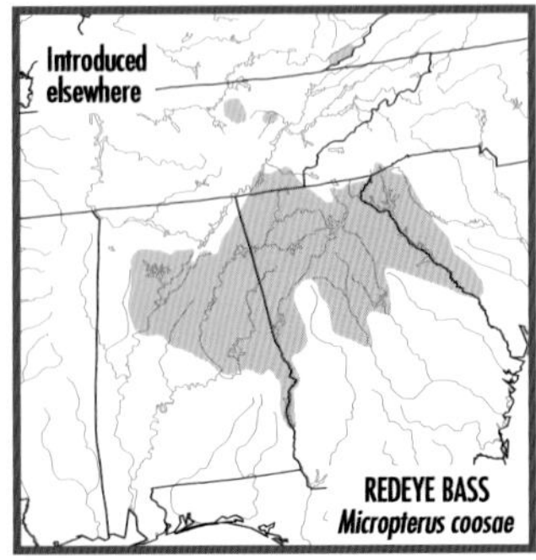

M. cataractae, lacks white tips on caudal fin, teeth on tongue; has *10–15 dark bars* along side, 15–21 (usually 18–20) scale rows below lateral line. (2) Spotted Bass, *M. punctulatus*, and (3) Alabama Bass, *M. henshalli, lack* white tips on caudal fin, red fins; have *rows of black spots* on lower side, *black stripe or blotches* along side.

SHOAL BASS *Micropterus cataractae* Pls. 39 & 42

IDENTIFICATION: Has *10–15 dark bars along side* (best developed on front half of body) and 6–8 bars along upper side create *tiger-stripe-like pattern* (less obvious on young). Olive above, dark olive to black mottling; yellow-white below; dusky spot on caudal fin base (darkest on young). *Rows of dark spots* along lower side. Yellow-green base, *dark edge* on caudal fin. Red eye; large mouth, upper jaw extending under rear half of eye. No teeth on tongue (rarely present). Unbranched pyloric caeca. Has 27–35 (usually 30-33) scales around caudal peduncle, 67–81 (usually 72–77) lateral scales, 15–21 (usually 18–20) scale rows below lateral line. To 25¼ in. (64 cm). **RANGE:** Native to Apalachicola R. system, GA, FL, and AL; introduced into Ocmulgee R. system, GA. Locally common; eliminated from much of native range by channelization and impoundments. **HABITAT:** Rocky riffles and runs of creeks and small to large rivers. **SIMILAR SPECIES:** See Pl. 39. (1) Redeye Bass, *M. coosae*, has white tips on orange caudal fin, teeth on tongue, 13–16 scale rows below lateral line. (2) Spotted Bass, *M. punctulatus*, and (3) Alabama Bass, *M. henshalli,* have teeth on tongue, *black stripe or blotches* along side; Spotted Bass has 21–28 scales around caudal peduncle.

SMALLMOUTH BASS *Micropterus dolomieu* Pls. 39 & 42

IDENTIFICATION: Has *8–16 dark brown bars, bronze specks,* on yellow-brown to olive green side; red eye; yellow-white below. Young has *3-colored* (yellow, black, white edge) caudal fin. Large male is green-brown to bronze with black mottling on back, bars on side. Large

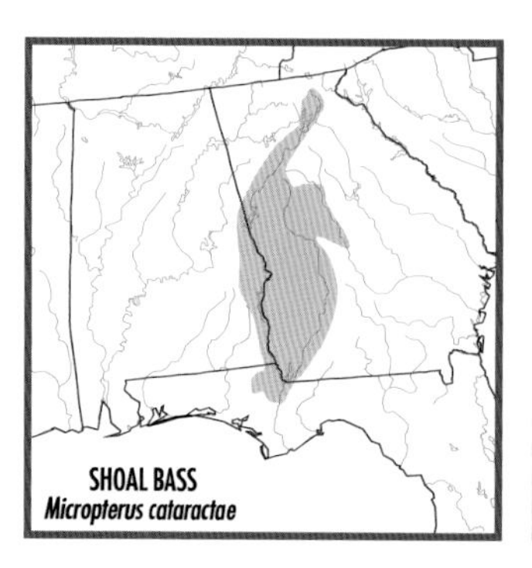

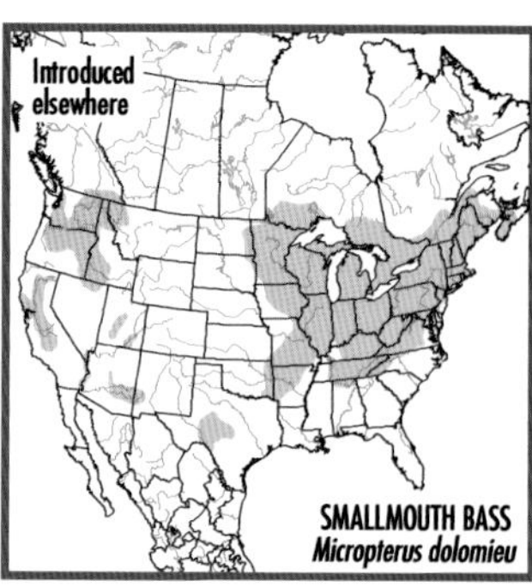

mouth; upper jaw extends under eye. Usually no patch of teeth on tongue. Unbranched pyloric caeca. Usually 69–77 lateral scales, 29–32 scales around caudal peduncle. To 27¼ in. (69 cm). **RANGE:** Native to St. Lawrence-Great Lakes, Hudson Bay (Red R.), and Mississippi R. basins from s. QC to ND and south to n. AL and e. OK. Widely introduced on northern Atlantic Slope and in w. U.S. Fairly common. **HABITAT:** Clear, gravel-bottom runs and flowing pools of small to large rivers; shallow rocky areas of lakes. **REMARKS:** Two subspecies often recognized. *M. d. velox,* in Arkansas R. drainage, se. MO, se. KS, nw. AR, and ne. OK, is slender, has protruding lower jaw (teeth visible from above), usually 13 dorsal rays. Intergrades (usually 14 dorsal rays) occupy rest of s. Ozark and Ouachita uplands. *M. d. dolomieu,* elsewhere, is stouter, has less protruding lower jaw, usually 14 dorsal rays. **SIMILAR SPECIES:** See Pl. 39. (1) Spotted Bass, *M. punctulatus,* has rows of black spots along lower side, black stripe (*no bars*) along side. (2) Redeye Bass, *M. coosae,* has rows of brown spots along lower side, white tips on caudal fin. (3) Largemouth Bass, *M. salmoides,* has black stripe (*no bars*) along side, larger mouth, 1st and 2d dorsal fin nearly separate, usually 58–73.

LEPOMIS

GENUS CHARACTERISTICS: Deep, *strongly compressed body* ("pan fish"); *3 anal spines;* shallowly forked caudal fin; smooth (not serrated) opercle edge; fewer than 55 lateral scales. Adult males are among the most brightly colored fishes in N. America. Colors of adults are among best characters to distinguish species. Other characters include size and shape of rakers on 1st gill arch (Fig. 54), viewed with gill cover lifted; "ear flap," a fleshy extension at rear of gill cover; and gill cover itself, which may be stiff to its edge or have thin and flexible edge. *Lepomis* species readily hybridize with one another. Hybrids are especially characteristic of turbid or polluted waters, where conditions hinder accurate species recognition. Spawning behavior in some species (e.g., Bluegill, *Lepomis macrochirus*) involves individuals that have been referred to as "sneakers" and "satellites." A small male, unable to command a good territory, may successfully spawn in territory of a larger male by "sneaking" in and fertilizing eggs when larger male is attending a female, or by mimicking a female (a "satellite") and being allowed in by the territorial male.

WARMOUTH *Lepomis gulosus* Pls. 40 & 42

IDENTIFICATION: *Dark red-brown lines* radiating from back of *red eye. Large mouth,* upper jaw extending under or beyond eye pupil. Fairly slender, *thick body. Patch of teeth* on tongue. Short, rounded pectoral fin, usually not reaching past eye when bent forward. Short ear flap;

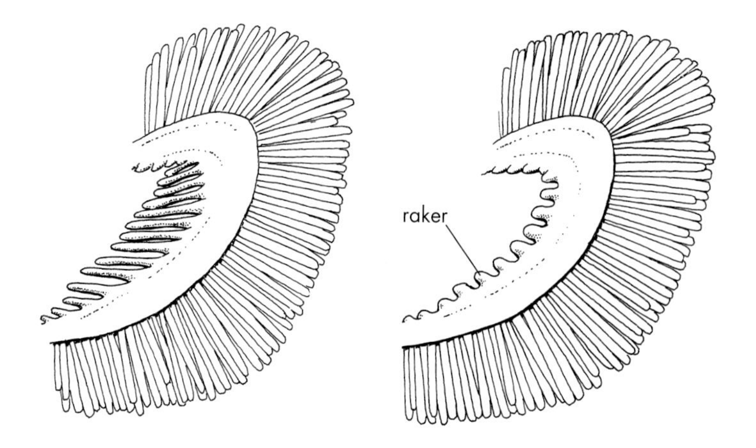

Fig. 54. Long and short gill rakers of sunfishes.

stiff rear edge on gill cover (excluding ear flap). Olive-brown above, often with purple sheen overall; dark brown mottling on back and upper side; often 6–11 chainlike dark brown bars on side; red spot (on adult) on yellow edge of ear flap; cream to bright yellow below; dark brown spots (absent on young) and wavy bands on fins. Breeding male has bold pattern on body and fins, bright red-orange spot at base of 2d dorsal fin, black pelvic fins. Complete lateral line; 36–44 lateral scales; usually 14 pectoral rays, 9–10 anal rays. Long, thin rakers on 1st gill arch. To 12 in. (31 cm). **RANGE:** Native to Great Lakes and Mississippi R. basins from ON and PA to MN, and south to Gulf; Atlantic and Gulf drainages from James R., VA, to Rio Grande, TX. Introduced

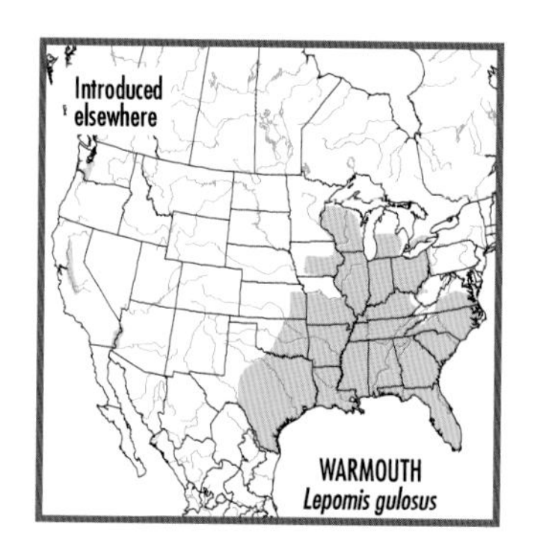

to Pacific drainages in w. U.S. Common in lowlands; uncommon in uplands. **HABITAT:** Vegetated lakes, ponds, swamps, and quiet-water areas of streams; usually over mud. **SIMILAR SPECIES:** (1) Green sunfish, *L. cyanellus* (Pl. 41), *lacks* dark lines behind eye, patch of teeth on tongue; has large black spot at rear of 2d dorsal and anal fins; yellow or orange edges on median fins. (2) Rock Bass, *Ambloplites rupestris* (Pl. 40), has 6 (vs. 3) anal spines, *no* dark lines behind eye.

GREEN SUNFISH *Lepomis cyanellus* Pls. 41 & 42

IDENTIFICATION: *Large mouth,* upper jaw extending beneath eye pupil. Fairly *slender, thick body.* Adult has *large black spot* at rear of 2d dorsal and anal fin bases; *yellow or orange edges* on 2d dorsal, caudal, and anal fins. Blue-green back and side; often yellow-metallic green flecks, sometimes dusky bars on side; green wavy lines on cheek and opercle; white to yellow edge (some red on young) on black ear flap; white to yellow belly. No teeth on tongue. Short, rounded pectoral fin, usually not reaching past front of eye when bent forward. Short ear flap; stiff rear edge of gill cover (excluding ear flap). Complete lateral line; 41–53 lateral scales; 23 or more scales around caudal peduncle; usually 13–14 pectoral rays, 9 anal rays. Long, slender rakers on 1st gill arch. To 12 in. (31 cm). **RANGE:** Native to Great Lakes, Hudson Bay, and Mississippi R. basins from NY and ON to MN, and south to Gulf; Gulf Slope drainages from Escambia R., FL, and Mobile Bay, GA and AL, to Pecos R., NM. Also n. Mexico. Introduced over much of U.S., including Pacific drainages. Common to abundant; one of most frequently encountered N. American fishes. **HABITAT:** Quiet pools and backwaters of sluggish streams; lakes and ponds. Often near vegetation. **SIMILAR SPECIES:** (1) Other sunfishes (Pl. 41) *lack* yellow-orange edges on fins and black spot at rear of anal fin (except Bluegill, *L. macrochirus*); have *smaller mouth* (except Warmouth). (2) Warmouth,

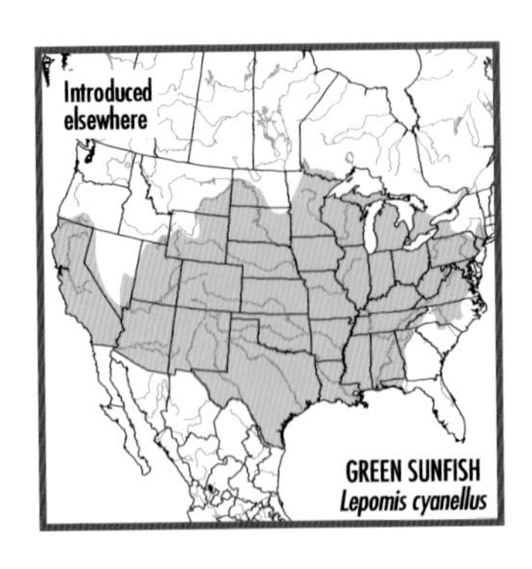

L. gulosus (Pl. 40), has dark brown lines radiating from eye, mottling on side, teeth on tongue.

BANTAM SUNFISH *Lepomis symmetricus* Pls. 41 & 42

IDENTIFICATION: Red-brown reticulations around *bold black spot* at rear of dorsal fin on young; spot diminishes as fish grows (absent in large adult). *Lacks* bright coloration of other sunfishes. *Chubby body* (less compressed). *Usually interrupted, incomplete lateral line.* Short, rounded pectoral fin, usually not reaching past eye when bent forward. Fairly large mouth; upper jaw extending under eye pupil. Short ear flap; stiff rear edge on gill cover (excluding ear flap). Dusky green back and side; yellow flecks and scattered small dark brown spots on side of adult, chainlike bars on side of young; white edge on black ear flap; yellow-brown below; red dorsal and anal fins on young, clear to dusky fins on adult. Has 30–40 lateral scales; usually 12–13 pectoral rays, 10 anal rays. Long, thin rakers on 1st gill arch. To 3½ in. (9 cm). **RANGE:** Former Mississippi Embayment from s. IL to Gulf; Gulf Coastal Plain drainages from Biloxi R., MS, to Colorado R., TX; historically above Fall Line in Illinois and Wabash river drainages, IL. Common in s.-cen. part of range (especially LA). **HABITAT:** Swamps; mud-bottomed, heavily vegetated ponds, lakes, and sloughs. **SIMILAR SPECIES:** Other sunfishes except (1) Green Sunfish, *L. cyanellus,* and (2) Bluegill, *L. macrochirus* (both Pl. 41), *lack* black spot at rear of dorsal fin. Green Sunfish is more slender; has larger mouth, yellow-orange edges on fins; Bluegill has long pectoral fin, dark edge on ear flap; is more compressed.

SPOTTED SUNFISH *Lepomis punctatus* Pls. 41 & 42

IDENTIFICATION: *Small black spots* on side of head; *rows of red-orange* (in male) or *yellow-brown spots* (female) along side (see Remarks). Short, rounded pectoral fin, usually not reaching past eye when bent

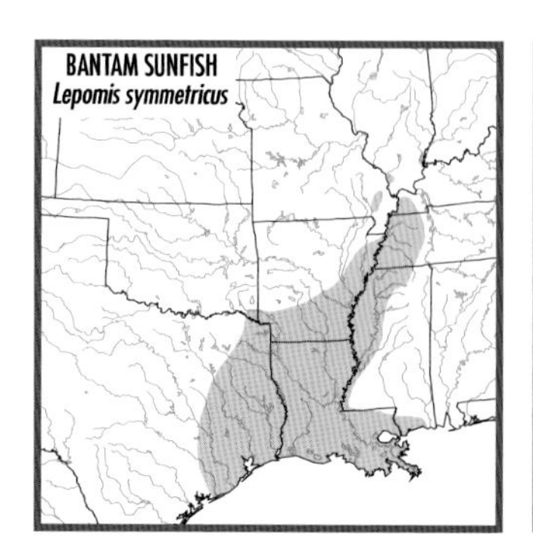

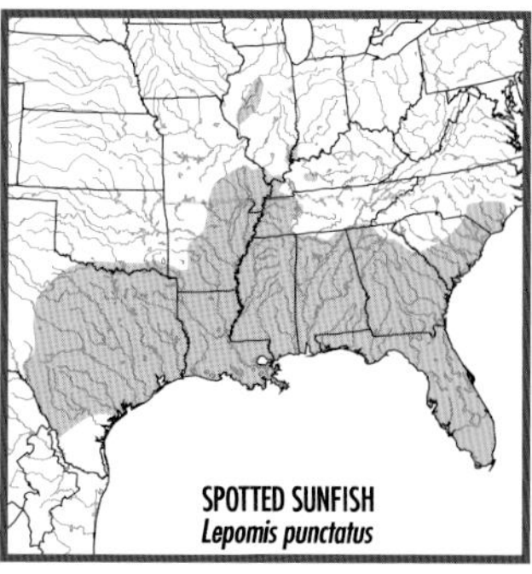

forward. Short ear flap; stiff rear edge of gill cover (excluding ear flap). Dark olive above; white to yellow edge on black ear flap; white, yellow, or red-orange below; clear to dusky fins. Complete lateral line; 34–45 lateral scales; usually 13–14 pectoral rays, 10 anal rays. Moderately long, slender rakers on 1st gill arch. To 8 in. (20 cm). **RANGE:** Atlantic and Gulf slope drainages from Cape Fear R., NC, to Rio Grande, TX; north in Mississippi R. basin to cen. IL. Common; locally abundant in southern part of range. **HABITAT:** Heavily vegetated ponds, lakes, pools of creeks and small to medium rivers, swamps. Usually over mud or sand. **REMARKS:** Two subspecies. *L. p. punctatus,* in Atlantic drainages and peninsular FL, has many black specks on body and head, no rows of red/yellow spots on side, usually 38–44 lateral scales. *L. p. miniatus,* throughout rest of range, has pale to bright red-orange patch on side just above ear flap, rows of red/yellow spots on side, no black specks on body, usually 35–41 lateral scales. Intergrades on Gulf Slope from Apalachicola R., GA and FL, to Perdido R., AL and FL; upper Coosa R. drainage, GA and TN. **SIMILAR SPECIES:** See Pl. 41. (1) Bantam Sunfish, *L. symmetricus, lacks* black specks on head, rows of red or yellow spots, red patch near ear flap; has black spot at rear of 2d dorsal fin, interrupted lateral line. (2) Longear Sunfish, *L. megalotis*, (3) Dollar Sunfish, *L. marginatus*, and (4) Redbreast Sunfish, *L. auritus,* have wavy blue lines on cheek and opercle, short rakers on 1st gill arch; *lack* black spots on head.

BLUEGILL *Lepomis macrochirus* Pls. 41 & 42

IDENTIFICATION: *Large black spot* at rear of dorsal fin (faint on young); often a dusky spot at rear of anal fin. *Dark bars* (absent in turbid water; thin and chainlike on young) on deep, *extremely compressed body. Long pointed pectoral fin;* usually extends far past eye when bent forward. Ear flap black to edge, fairly long in adult; thin, flexible rear edge on gill cover. Small mouth, upper jaw not extending under eye pupil. Olive back and side with yellow and green flecks; adult with blue sheen

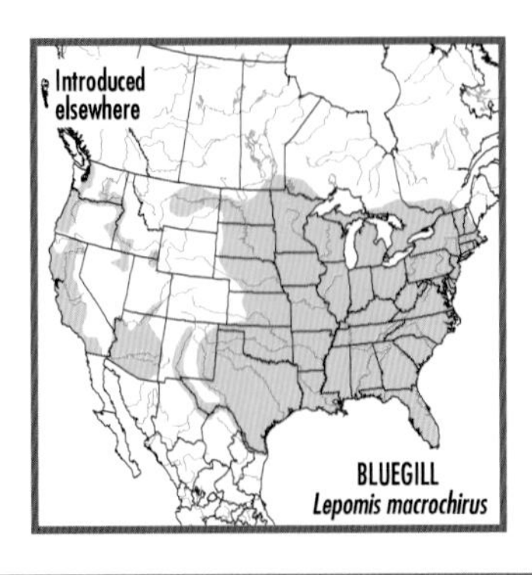

overall, 2 blue streaks from chin to edge of gill cover; white to yellow below; clear to dusky red fins. Breeding male has blue head, back; bright red-orange breast and belly; black pelvic fins. Complete lateral line; 38–48 lateral scales; usually 13 pectoral rays, 11 anal rays. Long, thin rakers on 1st gill arch. To 16¼ in. (41 cm). **RANGE:** Native to St. Lawrence-Great Lakes and Mississippi R. basins from QC and NY to MN and south to Gulf; Atlantic and Gulf slope drainages from Cape Fear R., NC, to Rio Grande, TX and NM. Also in n. Mexico. Now introduced over most of U.S. Also introduced to Eurasia and Africa. Common. Often abundant in lakes and impoundments. **HABITAT:** Vegetated lakes, ponds, swamps, and pools of creeks and small to large rivers. **REMARKS:** Three subspecies generally recognized, but ranges uncertain. *L. m. purpurescens,* on Atlantic Slope and peninsular FL, usually has 12 anal rays, red fins and broader bars in young, cream-colored bar on nape of adult, larger size than other subspecies. *L. m. speciosus*, in w. TX and Mexico, with usually 10 anal rays, intergrades with *L. m. macrochirus* in Arkansas and Red river drainages, AR, OK, and TX. *L. m. macrochirus,* throughout rest of range, usually has 11 anal rays, many narrow dark bars on side, no red on fins. Stocking programs have mixed populations. **SIMILAR SPECIES:** (1) Redear Sunfish, *L. microlophus* (Pl. 41), *lacks* large spot in 2d dorsal fin, has red spot, white edge on ear flap, short gill rakers.

REDEAR SUNFISH *Lepomis microlophus* Pls. 41 & 42

IDENTIFICATION: *Bright red or orange spot, white edge* on black ear flap (best developed on large adult). *Long, pointed pectoral fin* usually extends *far* past eye when bent forward. Fairly pointed snout; small mouth, upper jaw not extending under eye pupil. Light gold-green above; dusky gray spots (on adult) or bars (on young) on side; white to yellow below; mostly clear fins, some dark mottling on 2d dorsal fin of adult. Breeding male is brassy gold, has dusky pelvic fins. Short ear flap; thin, flexible rear edge on gill cover. Complete lateral line; 34–47

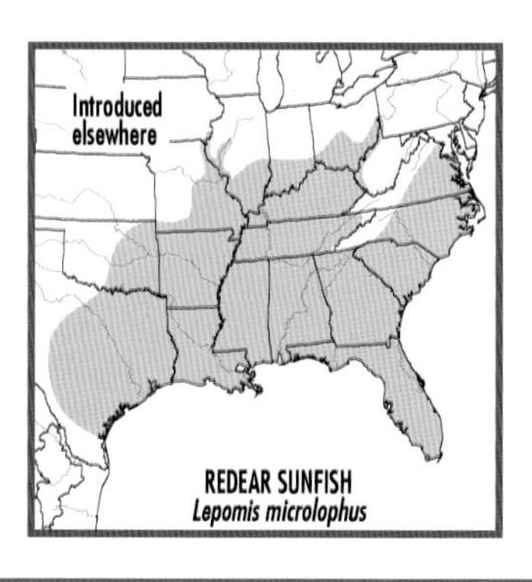

lateral scales; usually 13–14 pectoral rays, 10 anal rays. Short, thick rakers on 1st gill arch. To 10 in. (25 cm). **RANGE:** Native to Atlantic and Gulf slope drainages from about Savannah R., SC, to Nueces R., TX; north in Mississippi R. basin to s. IN and se. IA. Introduced elsewhere in U.S. Common. **HABITAT:** Ponds, swamps, lakes; vegetated pools, usually with mud or sand bottoms, of small to medium rivers. **REMARKS:** Two subspecies: *L. m. microlophus*, in FL, GA, and s. AL, has 40–47 lateral scales and 5–6 scale rows on cheek; *L. m.* subspecies, throughout rest of range, has 34–39 lateral scales, 3–4 scale rows on cheek. Stocking programs have mixed populations. **SIMILAR SPECIES:** See Pl. 41. (1) Pumpkinseed, *L. gibbosus,* has bold pattern on 2d dorsal fin, wavy blue lines on cheek and opercle, stiff rear edge on gill cover. (2) Longear Sunfish, *L. megalotis,* and (3) Dollar Sunfish, *L. marginatus,* have short, rounded pectoral fins, wavy blue lines on cheek and opercle, long ear flap.

PUMPKINSEED *Lepomis gibbosus* Pls. 41 & 42

IDENTIFICATION: *Bright red or orange spot,* light-colored edge on black ear flap. Many *bold dark brown wavy lines* or *orange spots* on 2d dorsal, caudal, and anal fins. *Wavy blue lines* on cheek and opercle of adult. *Long, pointed pectoral fin* usually extends far past eye when bent forward. Small mouth, upper jaw not extending under eye pupil. Olive back and side, many gold and yellow flecks; adult blue-green, spotted with orange; dusky chainlike bars on side of young and adult female; white to red-orange below. Short ear flap; stiff rear edge on gill cover (excluding ear flap). Complete lateral line; 35–47 lateral scales; usually 12–13 pectoral rays, 10 anal rays. Short, thick rakers on 1st gill arch. To 16 in. (40 cm). **RANGE:** Native to Atlantic drainages from NB to Savannah R., GA; Great Lakes, Hudson Bay (Red R.), and upper Mississippi R. basins from QC west to se. MB and ND, and south to n. KY and MO. Introduced to Pacific drainages from BC to CA. Common. **HABITAT:** Vegetated lakes, ponds and quiet pools of creeks and small rivers. **SIMILAR SPECIES:** (1) Redear Sunfish, *L. microlophus* (Pl. 41), *lacks* wavy lines or orange spots on 2d dorsal fin, wavy blue lines on cheek, stiff rear edge on gill cover. (2) Longear Sunfish, *L. megalotis* (Pl. 41), *lacks* wavy lines or orange spots on 2d dorsal fin; has short, rounded pectoral fin, flexible rear edge on gill cover.

LONGEAR SUNFISH *Lepomis megalotis* Pls. 41 & 42

IDENTIFICATION: *Long ear flap* (especially in adult male) usually bordered above and below by *blue line,* horizontal to slanted downward in adult, upward in young. *Wavy blue lines* on cheek and opercle. Short, rounded pectoral fin usually not reaching past eye when bent forward. Fairly large mouth, upper jaw extending under eye pupil. Thin, flexible rear edge on gill cover. Young has olive back and side speckled with

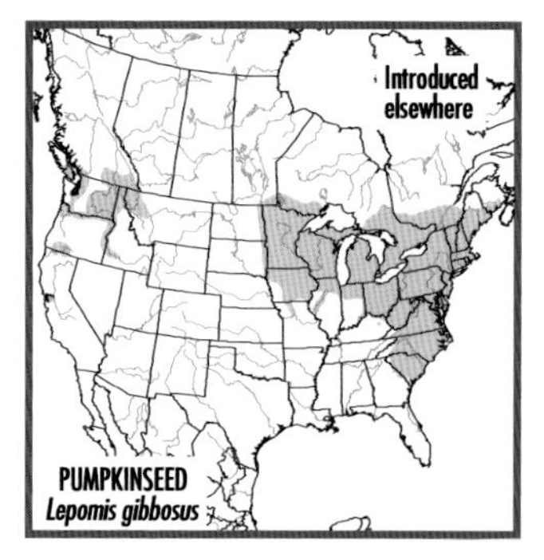

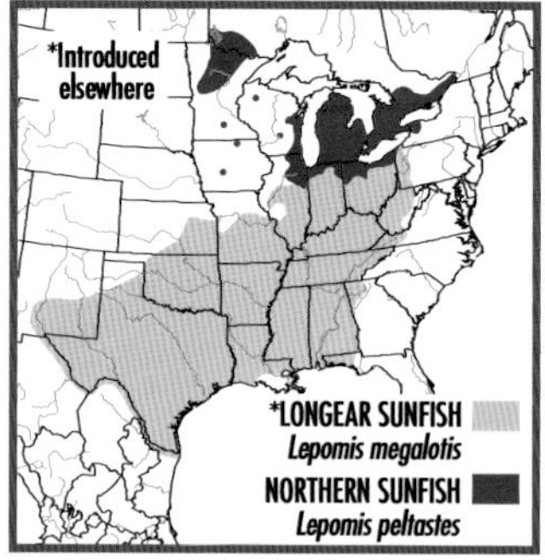

yellow flecks, often with chainlike bars on side, white below. Adult is dark red above, *bright orange* below, marbled and spotted with *blue,* sometimes with dusky bars on side; white (rarely orange) edge on black ear flap; clear to orange and blue, unspotted fins. Breeding male is a brilliant contrast in orange and blue, has red eye, orange to red median fins, blue-black pelvic fins. Complete lateral line; 33–46 (usually 39 or more) lateral scales; usually 13–14 pectoral rays, 9–10 anal rays, 5–7 scale rows on cheek from eye to lower angle of preopercle. Very short, thick rakers on 1st gill arch. To 9½ in. (24 cm). **RANGE:** Mississippi R. basin from PA to IL, MO, and KS, and south to Gulf; Gulf Slope drainages from Apalachicola R., GA, to Rio Grande, TX and NM. Also in ne. Mexico. Introduced sporadically elsewhere in U.S. Common; locally abundant in uplands and clear streams throughout range. **HABITAT:** Rocky and sandy pools of headwaters, creeks, and small to medium rivers; usually near vegetation or debris. **REMARKS:** Variable with several subspecies, but ranges and variation poorly understood. **SIMILAR SPECIES:** See Pl. 41. (1) See Northern Sunfish, *L. peltastes*, and (2) Dollar Sunfish, *L. marginatus*. (3) Redbreast Sunfish, *L. auritus*, *lacks* blue spots on side; has rows of red-brown to orange spots on upper side; longer, narrower ear flap, black to edge. (4) Pumpkinseed, *L. gibbosus*, has bold pattern on 2d dorsal, caudal, and anal fins; long, pointed pectoral fin; stiff rear edge on gill cover.

NORTHERN SUNFISH *Lepomis peltastes* Pls. 41 & 42

IDENTIFICATION: Similar to Longear Sunfish, *L. megalotis*, but reaches only 5 in. (13 cm); has large *red spot* on *upwardly slanted ear flap*, usually *12 pectoral rays, 40 or fewer lateral scales*. **RANGE:** St. Lawrence-Great Lakes, Hudson Bay (Red R.), and upper Mississippi R. basins from ON to MN, and south to OH, IN, IL, and IA (where extirpated). Common. **HABITAT:** Rocky and sandy pools of headwaters, creeks, and small to medium rivers; usually near vegetation. **SIMILAR SPECIES:** (1) See Longear Sunfish, *L. megalotis* (Pl. 41).

DOLLAR SUNFISH *Lepomis marginatus* Pl. 41

IDENTIFICATION: Similar to Longear Sunfish, *L. megalotis*, but reaches only 4¾ in. (12 cm); has shorter, *upwardly slanted ear flap, red streak* along lateral line, usually *12 pectoral rays,* 4 scale rows on cheek. Breeding male is bright red, marbled and spotted with blue-green, and has *large silver green specks* on ear flap. Has 34–44 lateral scales. **RANGE:** Atlantic and Gulf slope drainages (mostly below Fall Line) from Tar R., NC, to Brazos R., TX; Former Mississippi Embayment from w. KY and se. MO to Gulf. Common in se. U.S., especially FL; generally uncommon in western part of range. **HABITAT:** Sand- and mud-bottomed, usually brushy, pools of creeks and small to medium rivers; swamps. **REMARKS:** Can be difficult to distinguish from Longear Sunfish, *L. megalotis*, except when habitat is considered. Longear-like sunfish in a swamp or swampy stream habitat is probably a Dollar Sunfish. **SIMILAR SPECIES:** (1) See Longear Sunfish, *L. megalotis* (Pl. 41). (2) Redbreast Sunfish, *L. auritus* (Pl. 41), *lacks* blue spots on side; has rows of red-brown to orange spots on upper side; longer, narrower ear flap black to its edge; usually 14 pectoral rays.

REDBREAST SUNFISH *Lepomis auritus* Pls. 41 & 42

IDENTIFICATION: *Very long, narrow* (no wider than eye) *ear flap,* black to edge, usually bordered above and below by *blue line; wavy blue lines* on cheek and opercle. Dark olive back and side, yellow flecks, rows of *red-brown to orange spots* on upper side, orange spots scattered on lower side; white to orange below; clear to dusky orange fins. Breeding male has bright orange breast and belly, orange fins. Short, rounded pectoral fin, usually not reaching past eye when bent forward. Large mouth, upper jaw extending under eye pupil. Thin, flexible rear edge on gill cover. Complete lateral line; 39–54 lateral scales; usually 14 pectoral rays, 9–10 anal rays, 6–9 scale rows on cheek. Short, thick rakers on 1st gill arch. To 9½ in. (24 cm). **RANGE:** Native to Atlantic and Gulf slopes, from NB to cen. FL and west to Apalachicola and

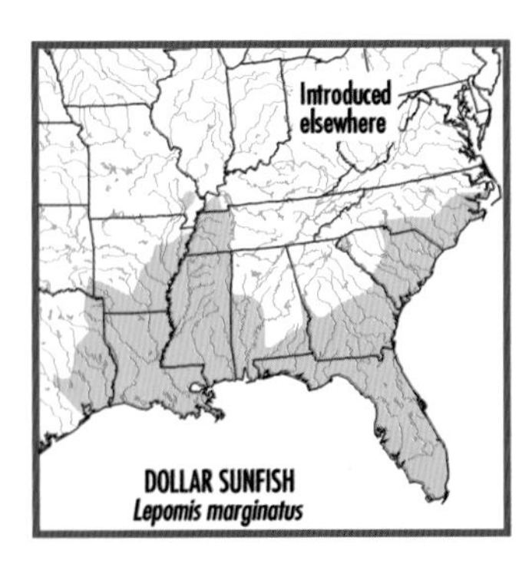

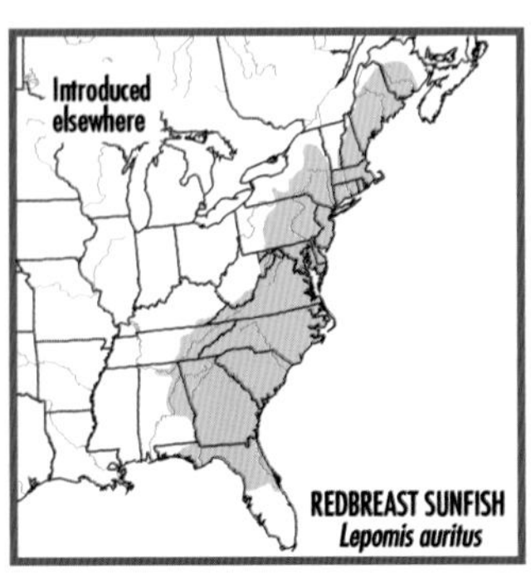

Choctawhatchee drainages, GA and FL. Introduced elsewhere but rarely successfully, except in upper Tennessee R., upper Cumberland R., upper Alabama R., and several drainages in TX. Locally common. **HABITAT:** Rocky and sandy pools of creeks and small to medium rivers; rocky and vegetated lake margins. **SIMILAR SPECIES:** (1) Longear Sunfish, *L. megalotis*, and (2) Dollar Sunfish, *L. marginatus* (both Pl. 41), have *shorter* ear flap with white or orange edge, blue marbling and spots on side of adult; *lack* distinct rows of red-brown to orange spots.

ORANGESPOTTED SUNFISH *Lepomis humilis* Pls. 41 & 42

IDENTIFICATION: *Bright orange* (on large male) or *red-brown* (female) spots on silver green side; young has vertical dusky bars, not spots, on side. Long black ear flap in adult has *wide white edge. Greatly elongated pores* along preopercle edge; large sensory pits between eyes. Short, rounded pectoral fin usually not reaching past eye when bent forward. Fairly long snout; large mouth, upper jaw extending under eye pupil. Thin, flexible rear edge on gill cover. Olive above, silver blue flecks on side; white to orange below; orange (on male) or red-brown (female) wavy lines on cheek and opercle; unspotted fins. Young has chainlike dark bars on side. Breeding male is brilliantly colored, with red eye, bright red-orange spots on side, red belly and edges on anal and dorsal fins; black edge on white pelvic fins. Complete lateral line; 32–41 lateral scales; usually 14 pectoral rays, 9 anal rays. Fairly long, thin rakers on 1st gill arch. To 6 in. (15 cm). **RANGE:** Lower Great Lakes (southern ends of Lakes Erie and Michigan), Hudson Bay (Red R.), and Mississippi R. basins from OH to ND, and south to LA; Gulf Slope drainages from Pearl R. basin, AL and MS, to Colorado R., TX; isolated populations in s. TX. Sporadically introduced elsewhere in Canada and U.S.; now widespread in Mobile Bay basin. Common. **HABITAT:** Quiet pools of creeks and small to large, often turbid, rivers; usually near brush. **SIMILAR SPECIES:** (1) Other sunfishes (Pl. 41) with orange spots have dark (blue or olive-brown) side; *lack* wide white edge on ear flap, elongated pores on preopercle.

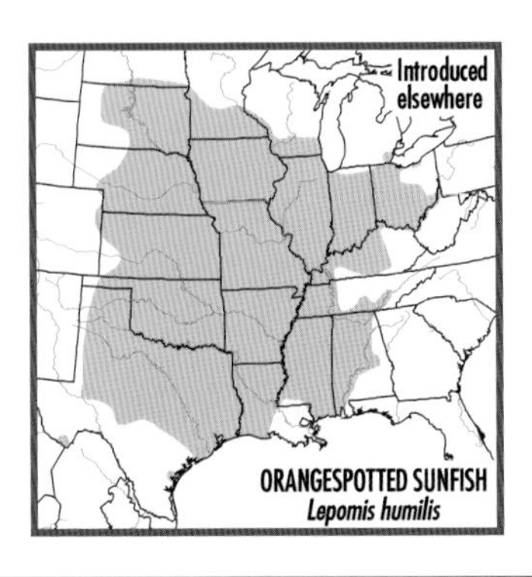

DARTERS AND PERCHES: Family Percidae (201 native; 1 introduced)

Percidae is the second-most diverse family (after Cyprinidae) of North American freshwater fishes. All but 3 species of North American percids (Walleye, *Sander vitreus*; Sauger, *S. canadensis*; and Yellow Perch, *Perca flavescens*) are darters. Darters are small—most are less than 4 in. (10 cm) long—and found throughout the U.S. and Canada except in Pacific and Arctic drainages (1 species has been introduced to California). Most have lost the gas bladder and, as their name implies, dart about on the bottoms of streams and lakes and eat small invertebrates. Some darters are drab, but many are extremely colorful, especially as breeding males. In addition to species in the U.S. and Canada, 17 percids are endemic to Eurasia and 4 darters are endemic to Mexico.

Percids have 2 dorsal fins, separate or slightly joined (broadly joined in some Eurasian species), the first with spines, the second with rays; thoracic pelvic fins with *1 spine and 5 rays;* and ctenoid scales. Characters useful in identification of some groups of darters include shape and completeness of lateral line and head canals (Fig. 55), connection of branchiostegal membranes (Fig. 56), presence or absence of a premaxillary frenum (Fig. 57), and number of anal spines. Unless stated otherwise, the *lateral line is straight and complete, branchiostegal membranes are separate or narrowly joined, anal spines number 2.*

WALLEYE *Sander vitreus* Pl. 43

IDENTIFICATION: *Huge mouth* extends beyond middle of eye; *large canine teeth. Opaque, silver eye. Large black spot* (on adult) on rear of 1st dorsal fin. Long, slender body; long pointed snout; forked caudal fin. Yellow-olive to brown above, dark green vermiculations; brassy yellow-blue side; 5–12 dusky saddles extend onto side as short bars

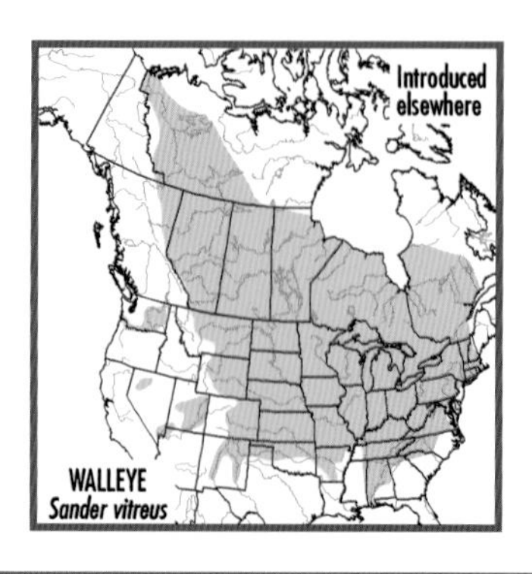

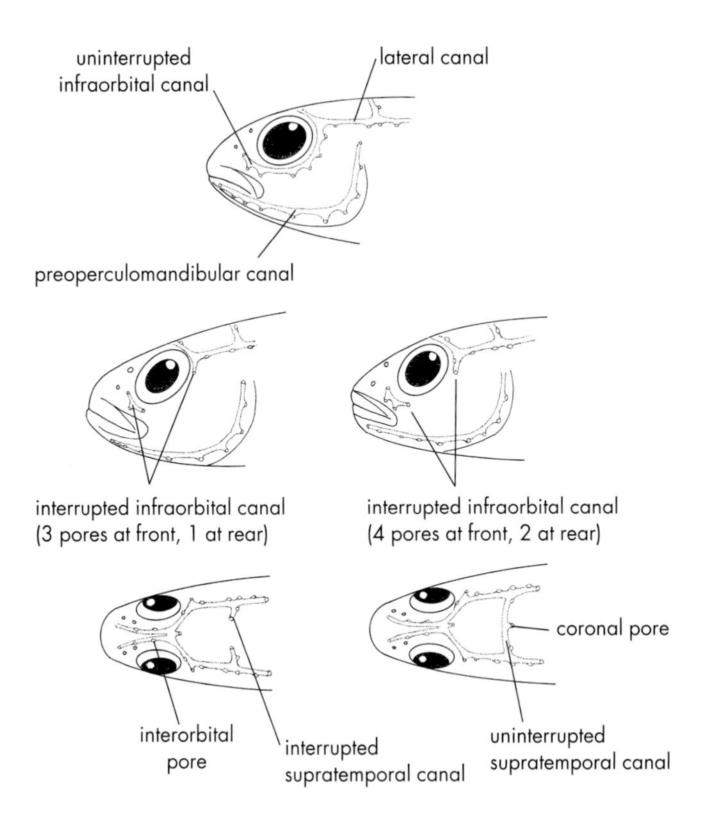

Fig. 55. Darters—head canals and pores.

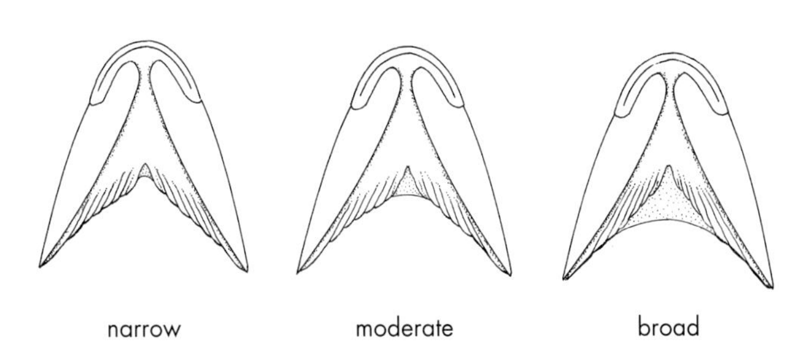

Fig. 56. Darters—branchiostegal membrane connection.

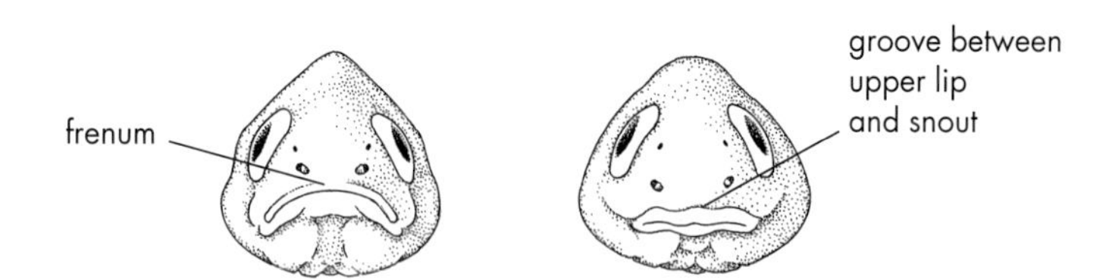

Fig. 57. Darters—presence and absence of premaxillary frenum.

(faint on adult); wavy dark brown bands on yellow fins; white tips on anal fin, lower lobe of caudal fin. Has 77–104 lateral scales; usually 19–22 dorsal rays, 12–14 anal rays; 3 pyloric caeca. To 36 in. (91 cm). **RANGE:** Native to St. Lawrence-Great Lakes, Arctic, and Mississippi R. basins from QC to NT, and south to AL and AR; possibly native to Mobile Bay basin. Widely introduced elsewhere in U.S., including Atlantic and Pacific drainages. Fairly common. **HABITAT:** Lakes; pools, backwaters, and runs of medium to large rivers. Usually in clear water, often near brush. **REMARKS:** Because of its fighting behavior and large size, Walleye is one of our most popular sport fishes. The color morph known as Blue Pike, sometimes considered a subspecies (*S. v. glaucus*), formerly occurred in Lakes Erie and Ontario and lower Niagara R. but is probably extinct. Blue Pike lacked brassy yellow color of other Walleyes, was gray-blue with blue-white lower fins, had larger eye. **SIMILAR SPECIES:** (1) See Sauger, *S. canadensis* (Pl. 43). (2) Yellow Perch, *Perca flavescens* (Pl. 43), is deeper bodied; *lacks* large canine teeth, opaque silver eye; has triangular bars on side, 52–61 lateral scales, 12–14 dorsal rays, 6–8 anal rays.

SAUGER *Sander canadensis* Pl. 43

IDENTIFICATION: Similar to Walleye, *S. vitreus*, but has *many black half-moons* on 1st dorsal fin of adult, *3–4 dusky brown saddles extending onto side as broad bars,* usually 17–19 dorsal rays, 11–12 anal rays, 5–8 pyloric caeca; *lacks* large black spot at rear of 1st dorsal fin, white tip on caudal fin. To 30 in. (76 cm). **RANGE:** Native to St. Lawrence-Great Lakes, Hudson Bay, and Mississippi R. basins from QC to AB, and south to n. AL and LA. Introduced into Atlantic, Gulf, and southern Mississippi R. drainages but rarely successfully. Fairly common. **HABITAT:** Sand and gravel runs, sandy and muddy pools and backwaters of small to large rivers; less often in lakes and impoundments. **REMARKS:** Opaque silver color of eyes of Walleye and Sauger is due to *tapetum lucidum,* a special layer of light-gathering tissue that enables these fishes to be active in low light. **SIMILAR SPECIES:** (1) See Walleye, *S. vitreus* (Pl. 43).

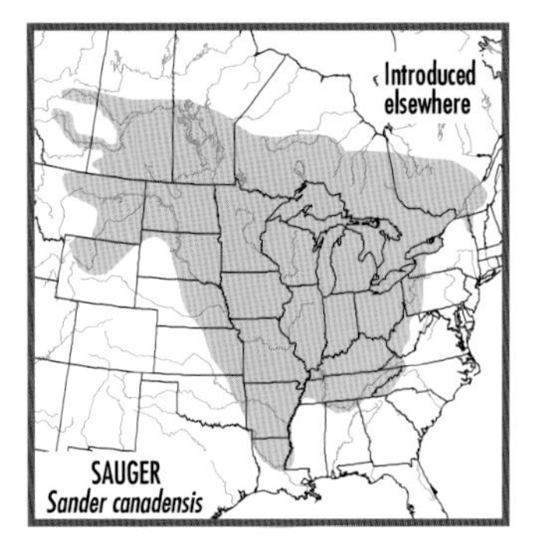

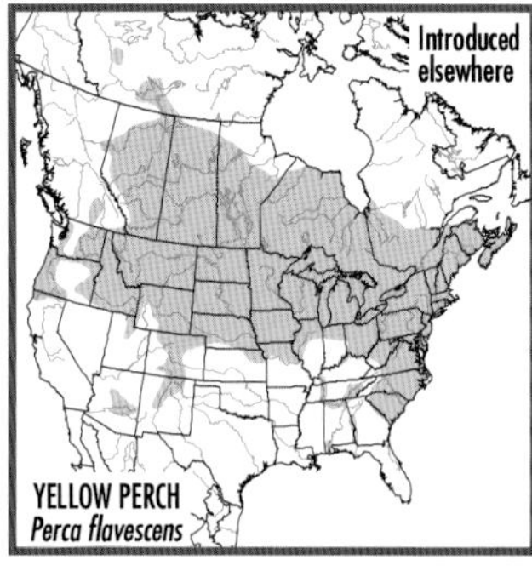

YELLOW PERCH *Perca flavescens* Pl. 43

IDENTIFICATION: *Fairly deep, compressed body;* forked caudal fin. Green above, *6–9 green-brown saddles extend down yellow side* (often as triangular bars); *black blotch* at rear (and often another at front) of dusky 1st dorsal fin; yellow to red paired fins. Large mouth extends to middle of eye; no canine teeth. Has 52–61 lateral scales; 12–14 dorsal rays; 6–8 anal rays. To 16 in. (40 cm). **RANGE:** Native to Atlantic, Arctic, Great Lakes, and Mississippi R. basins from NS to Mackenzie R. drainage, NT, and south to OH, IL, and NE; south in Atlantic drainages to Savannah R., GA. Widely introduced elsewhere in U.S. Common. **HABITAT:** Lakes, ponds, and pools of creeks and small to large rivers. Usually in clear water near vegetation. **SIMILAR SPECIES:** (1) Walleye, *Sander vitreus*, and (2) Sauger, *S. canadensis* (both Pl. 43), are *more slender; lack* triangular bars on side; have large canine teeth, opaque silver eye, more than 77 lateral scales, 11–14 anal rays.

RUFFE *Gymnocephalus cernua* Not shown

IDENTIFICATION: *Broadly joined dorsal fins; many small black spots* on dorsal and caudal fins. Green-brown above; many small dark blotches on light brown side; yellow below; clear to pink pectoral fins; 35–40 lateral scales. To 10 in. (25 cm). **RANGE:** Native to Eurasia. Established in Lakes Superior, Huron, and Michigan, where uncommon and localized. **HABITAT:** Lakes; quiet pools and margins of streams.

PERCINA

GENUS CHARACTERISTICS: *Percina* species are generally larger than other darters, retain a small gas bladder, and spend more time swimming off the stream bottom than do other darters. Two anal spines; *complete lateral line; no interruptions in head canals* (Fig. 55); *scutes* on breast and, in all species except Bluestripe Darter, *P. cymatotaenia*,

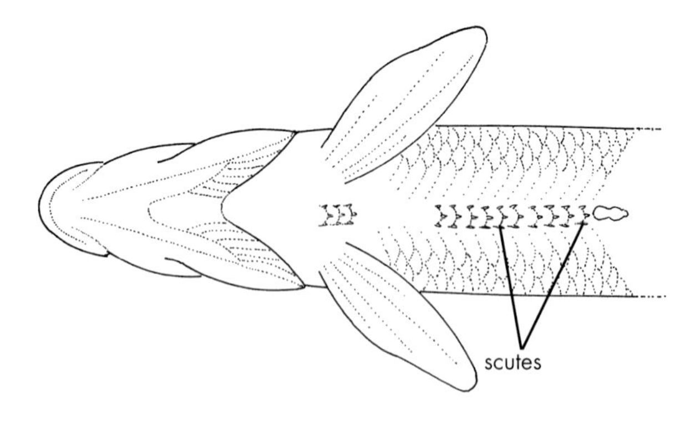

Fig. 58. *Percina* species—breast and belly.

and Frecklebelly Darter, *P. stictogaster*, in row along midline of belly of male (Fig. 58). Darters other than *Percina* species *lack* scutes.

BLUESTRIPE DARTER *Percina cymatotaenia* Pl. 44

IDENTIFICATION: Highly distinctive, with *broad, scallop-edged black stripe* along side, uninterrupted light stripe on upper side, 3 dark brown stripes along back to 2d dorsal fin. Yellow below with *many black blotches. Black wedge* on caudal fin base. No scutes on belly; no teardrop; no large black bands on fins. Fully scaled breast. Has 62–78 lateral scales. "Bluestripe" refers to blue-green squares along side of large male. To 3½ in. (9 cm). **RANGE:** Gasconade and Osage river drainages, s.-cen. MO. Rare. **HABITAT:** Quiet pools and backwaters with submerged or emergent vegetation, usually over mud or sand, in small to medium rivers. Swims in midwater. **SIMILAR SPECIES:** (1) See Frecklebelly Darter, *P. stictogaster*.

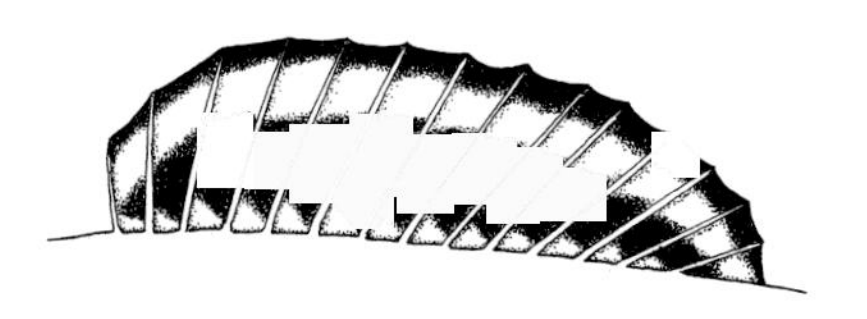

Fig. 59. Frecklebelly Darter—first dorsal fin.

FRECKLEBELLY DARTER *Percina stictogaster* Not shown

IDENTIFICATION: Similar to Bluestripe Darter, *P. cymatotaenia*, but has *black teardrop*. Large male has *black bands* on 1st dorsal fin (Fig. 59), black bar on chin. To 3¼ in. (8.5 cm). **RANGE:** Upper Kentucky and upper Green river (including Barren R.) drainages, e. and cen. KY, and n.-cen. TN. Fairly common in Kentucky R. drainage; uncommon and localized in Green R. drainage. **HABITAT:** Quiet water areas, especially vegetated marginal pools, of creeks and small to medium rivers. **SIMILAR SPECIES:** (1) See Bluestripe Darter, *P. cymatotaenia* (Pl. 44). (2) Longhead Darter, *P. macrocephala* (Pl. 44), lacks black chin bar and fully scaled breast, has midbelly row of scutes.

LONGHEAD DARTER *Percina macrocephala* Pl. 44

IDENTIFICATION: *Long snout (longer than eye). Sickle-shaped teardrop* curved back and down onto underside of head. *Black bar* below medial black caudal spot. Olive above, wavy black lines, 10–15 dark saddles; *9–15 fused black blotches* along side; light yellow below; black edge and base on 1st dorsal fin. Unscaled or partly scaled cheek and opercle; unscaled breast (except large scutes near pelvic fins). Has 72–90 lateral scales. To 4¾ in. (12 cm). **RANGE:** Ohio R. basin from Allegheny R. system, NY, to Duck R. system, w.-cen. TN. Rare and highly localized. **HABITAT:** Rocky flowing pools, usually above and below rubble riffles, of clear, small to medium rivers. **SIMILAR SPECIES:** (1) See Sickle Darter, *P. williamsi*. (2) Bridled Darter, *P. kusha* (Pl. 44), *lacks* teardrop, medial black caudal spot, black bar on caudal fin base; has scaled breast. (3) Blackside Darter, *P. maculata* (Pl. 44), has *shorter snout, straight* (not curved) *teardrop, no bar* on caudal fin base, fully scaled opercle. (4) Frecklebelly Darter, *P. stictogaster*, has *straight teardrop, shorter snout,* black chin bar, fully scaled breast.

SICKLE DARTER *Percina williamsi* Not shown

IDENTIFICATION: Similar to Longhead Darter, *P. macrocephala*, but has *shorter snout (length = eye), smaller scales* (usually 24–26 scales around caudal peduncle, 21–23 transverse scales, 70–77 lateral scales, vs. usually 27–31, 23–26, and 76–86 in Longhead Darter). To 4 in. (10 cm). **RANGE:** Upper Tennessee R. drainage, VA, NC, and TN.

Rare and highly localized. **HABITAT:** Flowing pools of clear creeks and small rivers; over gravel or sand. **SIMILAR SPECIES:** (1) See Longhead Darter, *P. macrocephala* (Pl. 44).

BRIDLED DARTER *Percina kusha* Pl. 44

IDENTIFICATION: Has 7–11 black blotches along side, fused into *black stripe* on adult, followed by rectangular *brown to black blotch* on caudal fin base centered *below* black stripe; *small dark brown spot* at top, often another at bottom, of caudal fin base. Olive to light brown above, usually without wavy dark brown lines or saddles; white below. Black edge and base on 1st dorsal fin of large adult. No teardrop. Snout about as long as eye. Scaled nape and breast. Has 58–73 lateral scales; 0–1 pored scales on caudal fin. To 3 in. (7.8 cm). **RANGE:** Conasauga and Etowah river systems (upper Coosa R. drainage), GA and TN. Rare. **HABITAT:** Rocky flowing pools and runs of creeks and small rivers. **SIMILAR SPECIES:** (1) Longhead Darter, *P. macrocephala* (Pl. 44), and (2) Sickle Darter, *P. williamsi*, have large teardrop, medial black caudal spot, black bar on caudal fin base, unscaled breast. (3) Muscadine Darter, *P. smithvanizi*, and (4) Bankhead Darter, *P. sipsi*, have less confluent (more discrete) blotches along side, dark dorsal saddles, large diffuse brown caudal spot *extending onto caudal rays;* Bankhead Darter has few or no scales on nape and breast.

MUSCADINE DARTER *Percina smithvanizi* Not shown

IDENTIFICATION: Has 7–11 black blotches along side connected in middle by *black stripe*; stripe ends at *large diffuse brown caudal spot extending onto caudal rays*. Olive above, brown wavy lines and dorsal saddles; white below. Dusky to black base on 1st dorsal fin of large adult more intense posteriorly. No (or faint) teardrop. Nape and breast usually fully scaled; no serrae on preopercle. Has 57–71 lateral scales;

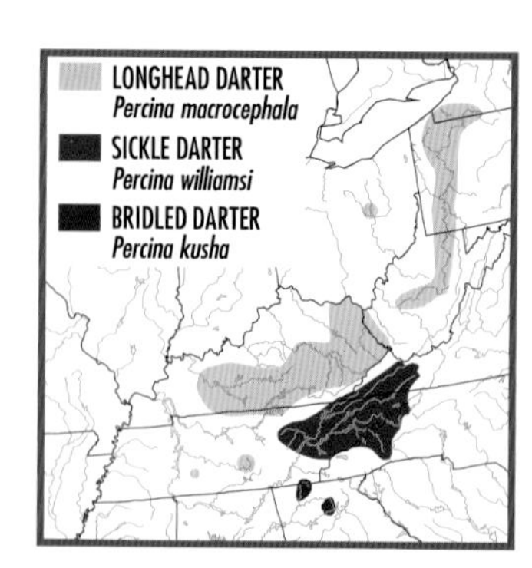

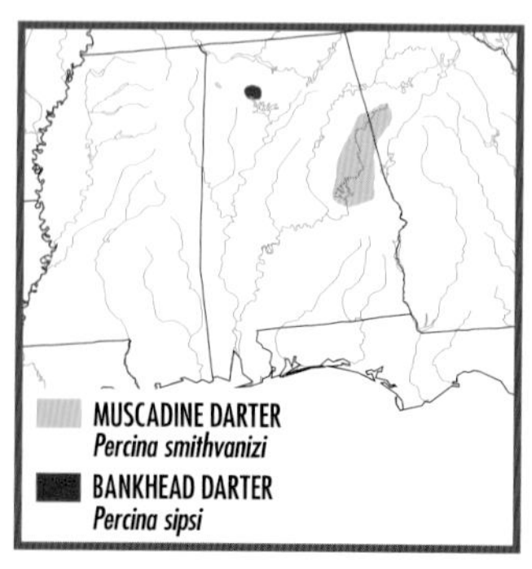

0–1 pored scales on caudal fin. To 3 in. (7.6 cm). **RANGE:** Tallapoosa R. drainage above Fall Line, GA and AL. Common. **HABITAT:** Rocky riffles and flowing pools and runs of creeks and small rivers. **SIMILAR SPECIES:** See Pl. 44. (1) See Bankhead Darter, *P. sipsi*. (2) Bridled Darter, *P. kusha,* has more confluent (less discrete) blotches along side, rectangular *black blotch* on caudal fin base centered *below* black stripe. (3) Blackside Darter, *P. maculata*, has black teardrop, *discrete medial black caudal spot.* (4) Dusky Darter, *P. sciera*, and (5) Goldline Darter, *P. aurolineata*, have vertical row of *3 dark brown spots* on caudal fin base, lower 2 often *fused;* serrated preopercle.

BANKHEAD DARTER *Percina sipsi* — Not shown

IDENTIFICATION: Nearly identical to Muscadine Darter, *P. smithvanizi*, but has *no or few scales on nape and breast*. Has 56–72 lateral scales. To 2¼ in. (6 cm). **RANGE:** Sipsey Fork (Black Warrior R. drainage), AL. Uncommon. **HABITAT:** Rocky flowing pools and runs of creeks and small rivers. **SIMILAR SPECIES:** (1) See Muscadine Darter, *P. smithvanizi*.

LEOPARD DARTER *Percina pantherina* — Pl. 44

IDENTIFICATION: Named for *10–14 round black spots* along side, *black or dusky spots* on upper side and back. Black teardrop. Medial round or vertically oval black caudal spot. First dorsal fin dusky, black along base and at front; 81–96 lateral scales. To 3½ in. (9.2 cm). **RANGE:** Little R. system (Red R. drainage) of sw. AR and se. OK. Rare; protected as a *threatened species.* **HABITAT:** Gravel and rubble runs of clear, small to medium rivers. **SIMILAR SPECIES:** See Pl. 44. (1) Blackside Darter, *P. maculata*, has *6–9 large blotches* along side, no round spots on upper side, 81 or fewer lateral scales. (2) Dusky Darter, *P. sciera*, and (3) Channel Darter, *P. copelandi,* common within range of Leopard Darter, *lack* spotted pattern.

BLACKSIDE DARTER *Percina maculata* — Pl. 44

IDENTIFICATION: *Discrete medial black caudal spot.* Prominent teardrop. Olive above, wavy black lines, 8–9 dark saddles. Has *6–9 large oval black blotches* along side. First dorsal fin dusky, black at front and along base. Fully scaled opercle. Has 56–81 lateral scales. To 4¼ in. (11 cm). **RANGE:** Great Lakes, Hudson Bay, and Mississippi R. basins from s. ON and NY to se. SK and south to LA; Gulf drainages from Mobile Bay, AL, to Neches R. (Sabine R. drainage), TX. One of most common and widespread darters. **HABITAT:** Pools of creeks and small to medium rivers, usually with moderate current and gravel or sand bottoms. **SIMILAR SPECIES:** See Pl. 44. Often confused with (1) Dusky Darter, *P. sciera,* which has vertical row of *3 diffuse dark brown spots* on caudal fin base. (2) Longhead Darter, *P. macrocephala*, has longer snout, curved teardrop, black bar below caudal spot. (3) Leopard

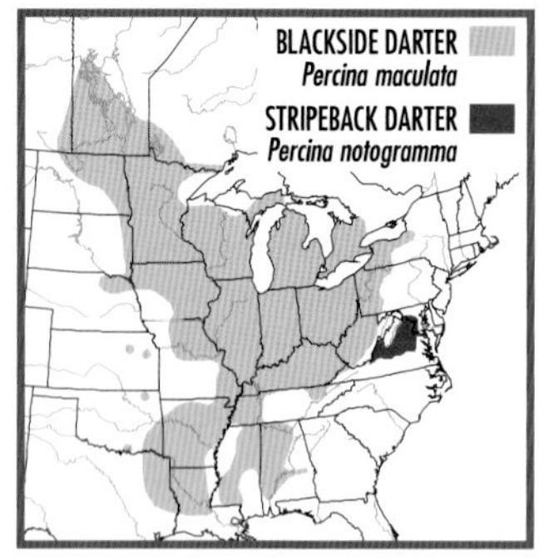

Darter, *P. pantherina*, has *10–14 round spots* along side, spots on upper side and back, 81–96 lateral scales.

STRIPEBACK DARTER *Percina notogramma* Pl. 44

IDENTIFICATION: *Pale yellow stripe* along upper side (on large individual). *First dorsal fin dusky throughout,* darkest at front and along base; fin lacks yellow or orange band, black crescents or ovals. No black bar on chin. Light brown back, dark saddles joined by wavy lines to 6–8 *horizontally oval black blotches* along side. Usually a discrete black caudal spot. Black teardrop. Has 49–67 lateral scales. To 3½ in. (8.4 cm). **RANGE:** Patuxent, Potomac, Rappahannock, York, and James river drainages (all tributaries of Chesapeake Bay) in MD, VA, and WV. Fairly common. **HABITAT:** Rocky pools, usually near riffles, of creeks and small to medium rivers. **REMARKS:** Two subspecies. *P. n. montuosa,* endemic to upper James R. system, VA and WV, has higher scale counts (usually 59 or more lateral scales) than *P. n. notogramma* (usually 58 or fewer lateral scales), occupying rest of range. **SIMILAR SPECIES:** (1) Shield Darter, *P. peltata* (Pl. 44), only similar species within range, has black bar on chin, *black crescents* on 1st dorsal fin, no yellow stripe on upper side. (2) Blackside Darter, *P. maculata* (Pl. 44), *lacks* yellow stripe, has more discrete black caudal spot.

SHIELD DARTER *Percina peltata* Pl. 44

IDENTIFICATION: *Black or dusky bar on chin;* often a large black spot on breast; *row of black crescents* on 1st dorsal fin (Fig. 60). Olive to tan above; *6–7 horizontally rectangular black blotches* along side, usually joined by narrow black stripe. Wavy brown lines join 8–11 dark saddles on back to dark blotches along side. Large black teardrop. Large black blotch *below* center of caudal fin base. Unscaled or partly scaled opercle; usually unscaled cheek. Has 16–21, usually 18, scales around caudal peduncle; 48–66 lateral scales. To 3½ in. (9 cm). **RANGE:** Atlantic Slope from Hudson R. and Susquehanna R. drainages, NY, to

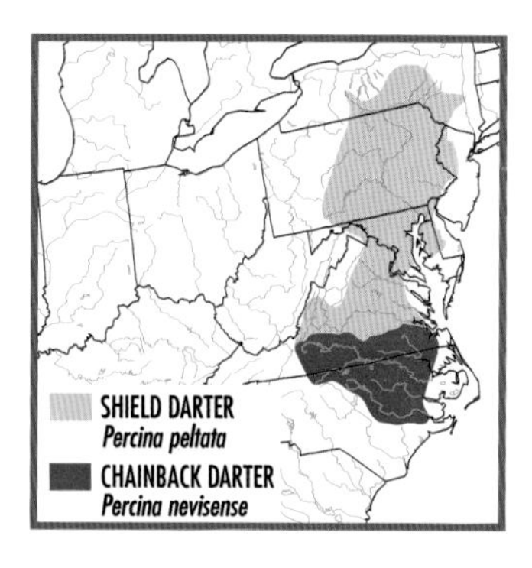

James R., VA. Common. **HABITAT:** Gravel and sand riffles and runs of creeks and small to medium rivers. **SIMILAR SPECIES:** (1) See Chainback Darter, *P. nevisense*, and (2) Appalachia Darter, *P. gymnocephala*. (3) Piedmont Darter, *P. crassa* (Pl. 46), only other darter with crescents on dorsal fin and black chin bar, is deeper bodied, usually has *7–9 vertically oval blotches* along side, bright yellow band on 1st dorsal fin of large male.

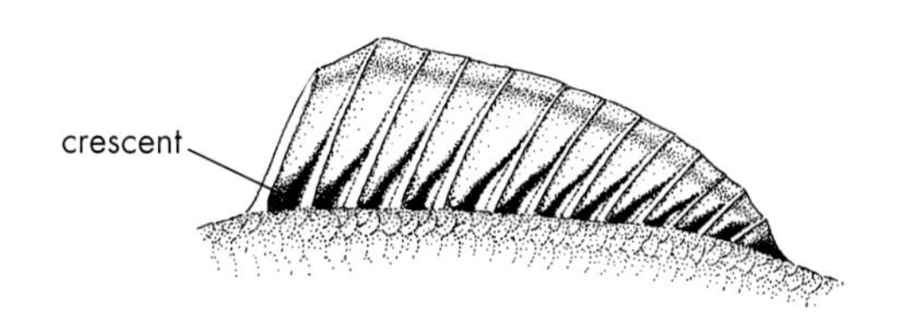

Fig. 60. Shield Darter—first dorsal fin.

CHAINBACK DARTER *Percina nevisense* Not shown

IDENTIFICATION: Nearly identical to Shield Darter, *P. peltata*, but usually has *scales on cheek*, 16–22 (usually 20) scales around caudal peduncle, narrower caudal peduncle. Has 45–60 lateral scales. To 3¾ in. (9.4 cm). **RANGE:** Atlantic Slope from Roanoke R. drainage, VA, to Neuse R., NC. Common. **HABITAT:** Gravel and sand riffles and runs of creeks and small to medium rivers. **SIMILAR SPECIES:** (1) See Shield Darter, *P. peltata* (Pl. 44).

APPALACHIA DARTER *Percina gymnocephala* Not shown

IDENTIFICATION: Similar to Shield Darter, *P. peltata*, but *lacks* black bar on chin, has *black ovals (not crescents)* on 1st dorsal fin. Has 56–72 lateral scales. To 3¾ in. (9.6 cm). **RANGE:** New R. drainage above

Kanawha Falls, WV, VA, and NC. Fairly common. **HABITAT:** Gravel and rubble runs and riffles of small to medium rivers. **SIMILAR SPECIES:** (1) See Shield Darter, *P. peltata* (Pl. 44). (2) Chainback Darter, *P. nevisense*, has black bar on chin, scales on cheek. (3) Blackside Darter, *P. maculata* (Pl. 44), has dusky 1st dorsal fin (*no* black ovals), fully scaled opercle.

PIEDMONT DARTER *Percina crassa* Pl. 46

IDENTIFICATION: *Black bar* on chin. *Large black spot* on breast. *Row of black crescents* (as on Shield Darter; Fig. 60), yellow band (best developed on large male) on 1st dorsal fin. Olive to tan above; 7–9 vertically oval black blotches along side joined by wavy brown lines to 8–10 dark saddles. Large black teardrop. Black bar below medial black caudal spot. Has 44–58 lateral scales. To 3½ in. (9 cm). **RANGE:** Cape Fear, Peedee, and Santee river drainages, VA, NC, and SC. Common in Cape Fear R. and on Piedmont; uncommon on Coastal Plain. **HABITAT:** Rocky riffles and runs of small to medium rivers. **SIMILAR SPECIES:** (1) Shield Darter, *P. peltata* (Pl. 44), and (2) Chainback Darter, *P. nevisense*, *lack* yellow band on 1st dorsal fin, are more slender, have 6–7 horizontally rectangular black blotches along side. (3) Roanoke Darter, *P. roanoka* (Pl. 46), *lacks* black bar on chin, has *orange band* on 1st dorsal fin, 8–14 black bars along side; large male is bright blue and orange.

ROANOKE DARTER *Percina roanoka* Pl. 46

IDENTIFICATION: *Bright blue side, orange breast and belly, orange band on 1st dorsal fin* of large male; *8–14 black bars* (on adult) or oval blotches (on young) along side. Olive-brown above, wavy lines join 7–9 dark saddles on back to bars or blotches along side; white to orange below. Black bar on lower half of caudal fin base. Has 38–54 lateral scales. To 3 in. (7.8 cm). **RANGE:** Roanoke, Neuse, and Tar river drainages, VA and NC; James R. drainage, VA; New R. drainage, WV and VA. James and New river populations probably are recent introduc-

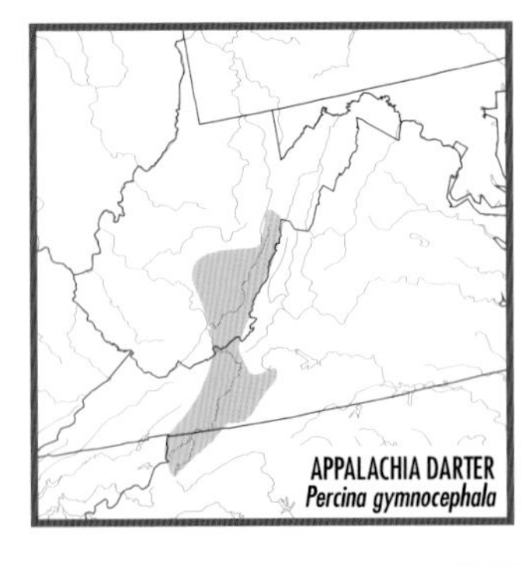

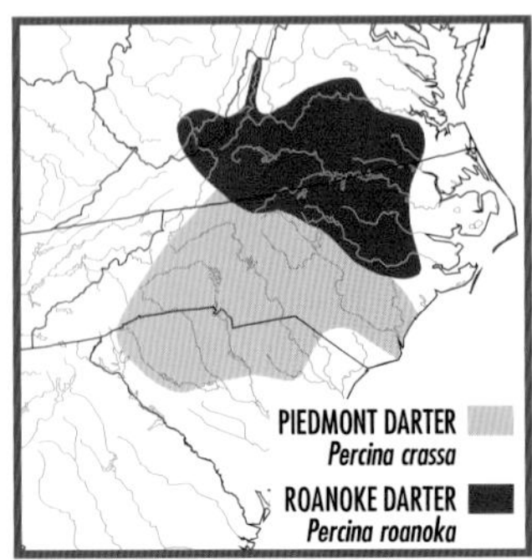

tions. Common; locally abundant. **HABITAT:** Fast gravel and rubble riffles (adult), gravel-bottomed pools and runs near riffles (young), of small to medium rivers. **SIMILAR SPECIES:** (1) Young of Roanoke Darter and Piedmont Darter, *P. crassa*, are difficult to separate, but Roanoke Darter usually has 10–12 dorsal spines, 10–11 dorsal rays; Piedmont Darter usually has 11–12 dorsal spines and 12 dorsal rays. Adults easy to distinguish; see Piedmont Darter (Pl. 46).

DUSKY DARTER *Percina sciera* Pl. 44

IDENTIFICATION: Olive to dusky black above, 8–9 dark brown saddles on back; wavy dark brown lines on upper side; *8–12 oval black blotches* along side; vertical row of *3 dark brown spots* on caudal fin base, lower 2 often *fused*. Usually no teardrop. Fins mostly clear; often a dusky spot at rear of 1st dorsal fin. Usually 5–16 serrae on preopercle. Has 56–76 lateral scales. To 5 in. (13 cm). **RANGE:** Mississippi R. basin from OH and WV to e. IL and south to LA; Gulf drainages from Alabama R., AL, to Colorado R., TX. Common; locally abundant. **HABITAT:** Fast gravel runs, sometimes riffles, of creeks and small to medium rivers; often near brush. **SIMILAR SPECIES:** See Pl. 44. See (1) Guadalupe Darter, *P. apristis*, and (2) Goldline Darter, *P. aurolineata*. (3) Blackside Darter, *P. maculata*, has *1 black caudal spot*. (4) Blackbanded Darter, *P. nigrofasciata*, has *12–15 black bars* along side, usually no serrae on preopercle. (5) Freckled Darter, *P. lenticula*, has large dark spot at front of 2d dorsal fin, 77–93 lateral scales.

GUADALUPE DARTER *Percina apristis* Not shown

IDENTIFICATION: Similar to Dusky Darter, *P. sciera*, but usually has 0–1 (rarely 2–6) serrae on preopercle, 65–81 (usually more than 67) lateral scales, usually 24 or more scutes on belly of male (Dusky Darter has 10–23 scutes). To 4½ in. (11 cm). **RANGE:** Guadalupe R. system, TX. Common. **HABITAT:** Fast rocky runs of small to medium rivers; often near brush. **SIMILAR SPECIES:** (1) See Dusky Darter, *P. sciera* (Pl. 44).

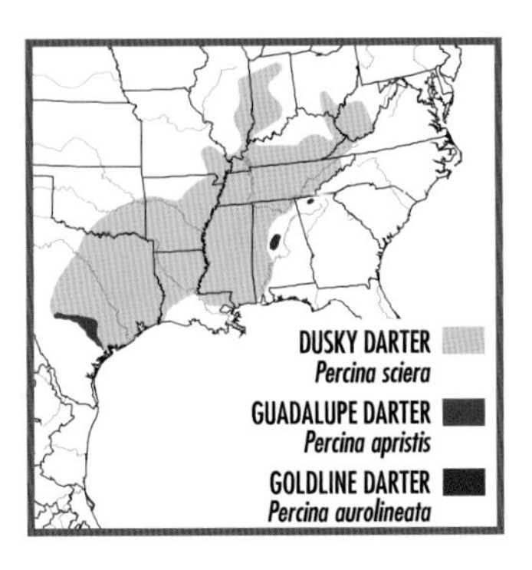

GOLDLINE DARTER *Percina aurolineata* Pl. 44

IDENTIFICATION: Similar to Dusky Darter, *P. sciera*, but has *thin amber or brown stripe* (often interrupted) on upper side. Has 59–74 lateral scales. To 3½ in. (9 cm). **RANGE:** Coosawattee R. (Coosa R. system), GA, and Cahaba R. system (Mobile Bay drainage), AL. Rare and localized; protected as a *threatened species.* **HABITAT:** Fast rocky runs of small to medium rivers. **SIMILAR SPECIES:** (1) See Dusky Darter, *P. sciera* (Pl. 44).

BLACKBANDED DARTER *Percina nigrofasciata* Pl. 44

IDENTIFICATION: Olive to dusky black above, 6–8 dark saddles on back; dark wavy lines on upper side; *12–15 dusky to black bars* along side; vertical row of *3 dark brown spots* on caudal fin base, lower 2 often *fused.* Usually no serrae on preopercle. No or faint teardrop. Fins clear or with thin dark bands. Has 46–71 lateral scales. To 4½ in. (11 cm). **RANGE:** Atlantic and Gulf slope drainages from Santee R., SC, to Mississippi R. tributaries, LA; south in peninsular FL to Lake Okeechobee. Most common darter throughout most of range, but absent in Satilla and St. Marys rivers in se. GA and n. FL; rare in Altamaha R., GA. **HABITAT:** Headwaters to medium-sized rivers; usually over gravel or sand, often over mud on Coastal Plain. **SIMILAR SPECIES:** See Pl. 44. (1) See Halloween Darter, *P. crypta.* (2) Dusky Darter, *P. sciera*, (3) Goldline Darter, *P. aurolineata*, and (4) Freckled Darter, *P. lenticula*, have *fewer* (usually 8–10), *more oval blotches* along side, serrated edge on preopercle.

HALLOWEEN DARTER *Percina crypta* Not shown

IDENTIFICATION: Similar to Blackbanded Darter, *P. nigrofasciata*, but has dark dorsal saddles wider (front to back) than light interspaces (most obvious on caudal peduncle), dark bands on pectoral fin, 1 modified scale between pelvic fins. Has 50–68 lateral scales. To 4½

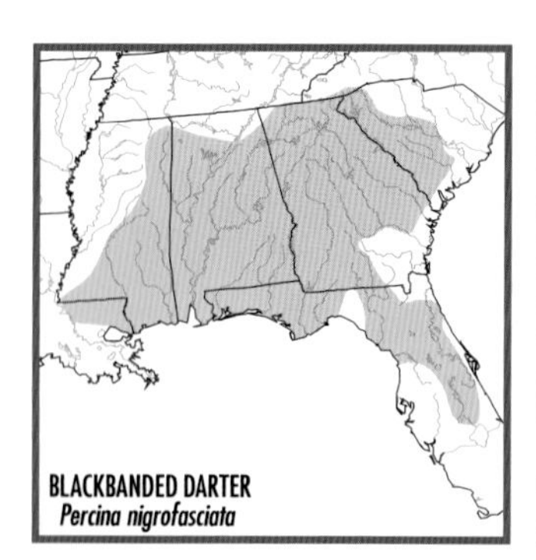

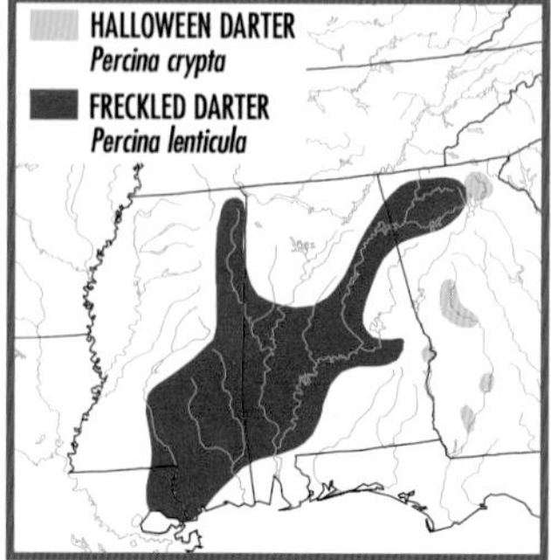

in. (12 cm). **RANGE:** Apalachicola R. drainage, GA and AL. Locally common. **HABITAT:** Fast rocky shoals of creeks and small rivers; often near vegetation. **SIMILAR SPECIES:** (1) Blackbanded Darter, *P. nigrofasciata* (Pl. 44), has dark dorsal saddles narrower than light interspaces, no or dusky bands on pectoral fin, 2 or more modified scales between pelvic fins.

FRECKLED DARTER *Percina lenticula* Pl. 44

IDENTIFICATION: *Giant* of darters—to 8 in. (20 cm). *Large black spot* at front of 2d dorsal fin; smaller spot at front of 1st dorsal fin. Olive to dusky overall, 8 dark brown saddles on back, brown wavy lines on upper side, 8 black vertical blotches along side; vertical row of *3 dark brown spots* on caudal fin base, lower 2 often *fused*. Has 77–93 lateral scales. **RANGE:** Mobile Bay, Pascagoula R., Pearl R., and Lake Pontchartrain drainages, GA, AL, MS, and LA. Rare. **HABITAT:** Fast, deep rocky riffles of small to medium rivers. **SIMILAR SPECIES:** See Pl. 44. (1) Dusky Darter, *P. sciera,* (2) Blackbanded Darter, *P. nigrofasciata*, and (3) Goldline Darter, *P. aurolineata*, *lack* large black spot at front of 2d dorsal fin, have 76 or fewer lateral scales; none exceeds 5 in. (13 cm).

CHANNEL DARTER *Percina copelandi* Pl. 44

IDENTIFICATION: *Blunt snout. No premaxillary frenum* (if present, extremely narrow; Fig. 57). Olive above; *9–10 horizontally oblong to square black blotches* along side; black Xs and Ws on back and upper side. *Medial black caudal spot.* Dusky teardrop often reduced to spot. Black along edge and base of 1st dorsal fin of male. Breeding male has black lower head and body; may have tubercles, tubercular ridges on anal and pelvic fin rays. Cheek usually unscaled to partly scaled east of Mississippi R., partly to fully scaled west of Mississippi R. Has 42–67 lateral scales. To 2½ in. (6.2 cm). **RANGE:** Wide-ranging but highly localized in St. Lawrence-Great Lakes and Mississippi R. basins from s. QC and VT south to n. TN; west of Mississippi R. in Arkansas, Ouachita, and Red river drainages, MO, KS, AR, OK, and LA. Common in west; uncommon and declining in east. Protected in Canada as a *threatened species*. **HABITAT:** Flowing pools and margins of riffles of small to medium rivers over sand or rocky bottoms; shores of lakes. **SIMILAR SPECIES:** (1) See Pearl Darter, *P. aurora*, and Coal Darter, *P. brevicauda*. (2) Johnny Darter, *Etheostoma nigrum*, and (3) Tessellated Darter, *E. olmstedi* (both Pl. 45), have *brown* (not black) marks, no black edge on 1st dorsal fin, no scutes (Fig. 58). (4) Blackside Darter, *P. maculata*, and (5) Dusky Darter, *P. sciera* (both Pl. 44), have *larger, rounder black blotches* along side, *wide* premaxillary frenum.

PEARL DARTER *Percina aurora* Not shown

IDENTIFICATION: Nearly identical to Channel Darter, *P. copelandi*, but has mostly to fully scaled cheek; breeding male lacks tubercles and tubercular ridges on anal and pelvic fin rays. Has 50–61 lateral scales. To 3 in. (7.7 cm). **RANGE:** Pascagoula and Pearl river drainages, MS and LA. Rare. **HABITAT:** Gravelly runs and margins of riffles of small to medium rivers. **SIMILAR SPECIES:** (1) See Channel Darter, *P. copelandi* (Pl. 44). (2) Coal Darter, *P. brevicauda*, lacks scales on cheek; breeding male may have tubercles and usually has tubercular ridges on anal and pelvic fin rays.

COAL DARTER *Percina brevicauda* Not shown

IDENTIFICATION: Nearly identical to Channel Darter, *P. copelandi*, but has *unscaled cheek*. Has 43–61 lateral scales. To 2¼ in. (5.5 cm). **RANGE:** Mobile Bay drainage in lower Coosa, upper Cahaba, and upper Black Warrior river systems, AL. Locally common. **HABITAT:** Gravelly runs and riffles of small to medium rivers. **SIMILAR SPECIES:** (1) See Channel Darter, *P. copelandi* (Pl. 44). (2) Pearl Darter, *P. aurora*, *has scales* on cheek; breeding male lacks tubercles and tubercular ridges.

NEXT 5 SPECIES: *Anal fin extends to caudal fin* on large male (see Pl. 44).

RIVER DARTER *Percina shumardi* Pl. 44

IDENTIFICATION: *Small black spot at front, large black spot near rear,* of 1st dorsal fin. Dusky olive above; *8–15 black bars* along side; small black caudal spot; black teardrop. Moderately blunt snout; premaxillary frenum *absent or narrow* (Fig. 57). Has 46–62 lateral scales. To 3 in. (7.8 cm). **RANGE:** Hudson Bay basin, ON and MB, and south in Great Lakes and Mississippi R. basins to LA; Gulf drainages from Mobile Bay, AL, to Neches R., TX; isolated population in San Antonio Bay drainage,

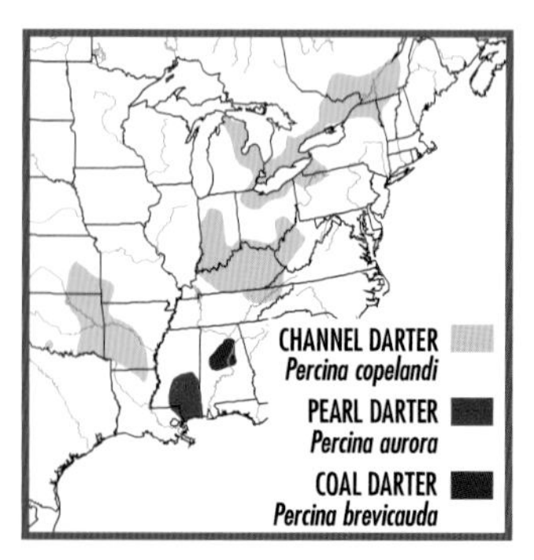

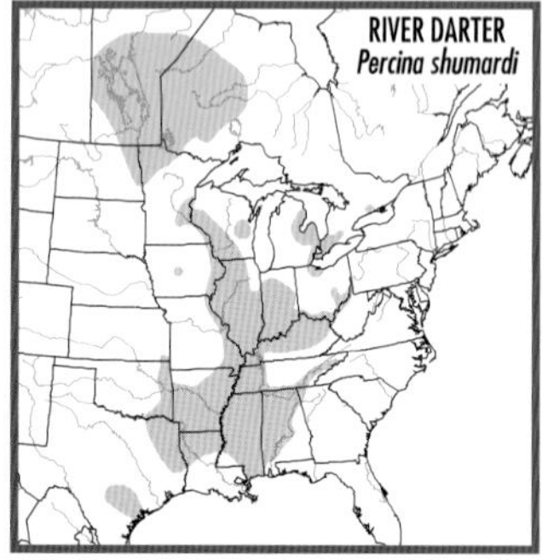

TX. Locally common; most common darter in very large rivers, including Mississippi. **HABITAT:** Rocky riffles of small to large rivers. Adult usually in deep swift riffles; young in shallow riffles and runs. **SIMILAR SPECIES:** See Pl. 44. Similar darters *lack* black spots at front and rear of 1st dorsal fin. (1) Channel Darter, *P. copelandi*, has *horizontally* oriented blotches along side, no long anal fin on male. (2) Saddleback Darter, *P. vigil*, (3) Stargazing Darter, *P. uranidea*, and (4) Snail Darter, *P. tanasi*, have 4–5 dark saddles, larger blotches (*not bars*) on side.

STARGAZING DARTER *Percina uranidea* Pl. 44

IDENTIFICATION: *Red-brown above; 4 dark brown saddles (1st under 1st dorsal fin)* extend down side to lateral line; yellow below. Small black spot at front of 1st dorsal fin; 9–12 dark brown blotches along side. Large black teardrop. Stout body; *blunt snout;* premaxillary frenum *absent or narrow* (Fig. 57). Has 46–58 lateral scales. To 3 in. (7.8 cm). **RANGE:** St. Francis, White, and Ouachita river drainages, MO, AR, and LA. Formerly in lower Wabash R., IN and IL. Localized and uncommon. **HABITAT:** Fast gravel runs of clear, medium-sized rivers. **SIMILAR SPECIES:** See Pl. 44. (1) See Snail Darter, *P. tanasi*. (2) River Darter, *P. shumardi*, is dusky olive above; *lacks* bold dark saddles; has large black spot at rear of 1st dorsal fin, 8–15 black bars along side. (3) Saddleback Darter, *P. vigil*, has 5 dark saddles.

SNAIL DARTER *Percina tanasi* Pl. 44

IDENTIFICATION: Nearly identical to Stargazing Darter, *P. uranidea*, but has *gray edge and base on 1st dorsal fin.* Has 48–57 lateral scales. To 3½ in. (9 cm). **RANGE:** Upper Tennessee R. drainage of e. TN and nw. GA. Localized and uncommon; protected as a *threatened species.* **HABITAT:** Gravel and sand runs of medium-sized rivers. **REMARKS:** Made famous by 1970s conflict over choice between completing Tellico Dam and saving Little Tennessee R., then believed to support only popu-

lation of Snail Darter. **SIMILAR SPECIES:** (1) See Stargazing Darter, *P. uranidea* (Pl. 44).

SADDLEBACK DARTER *Percina vigil* Pl. 44

IDENTIFICATION: *Five dark brown saddles; 1st under 1st dorsal fin;* 5th near front of caudal fin, small and often indistinct. *Small black spot* on caudal fin base. Over sandy bottom, saddles fade and dark hatch-marks and wavy lines cover back. Has 8–10 rectangular dark blotches along side, often fused together but not joined to dorsal saddles. Dusky edge and base on 1st dorsal fin. Teardrop variable. Fairly slender body; moderately pointed snout; premaxillary frenum *absent or narrow* (Fig. 57). Has 46–62 lateral scales. To 3 in. (7.8 cm). **RANGE:** Mississippi R. basin from sw. IN and se. MO to LA; Gulf Slope from Escambia R. drainage, AL and FL, to Mississippi R., LA. Mostly confined to Coastal Plain. Common but localized. **HABITAT:** Sand and gravel runs of creeks and small to medium rivers; sometimes in very shallow water. **SIMILAR SPECIES:** (1) River Darter, *P. shumardi*, and (2) Channel Darter, *P. copelandi* (both Pl. 44), *lack* large dark saddles (but may have about 8 small dark blotches on back).

AMBER DARTER *Percina antesella* Pl. 46

IDENTIFICATION: Yellow-brown above, *4 dark brown saddles (1st in front of 1st dorsal fin)* extending down side to lateral line; 8–12 small dark brown blotches along side; yellow to white below. Black along edge and base of 1st dorsal fin. Black teardrop. Slender body; *pointed snout;* premaxillary frenum *absent or narrow* (Fig. 57). Breeding male has semicircular *keel* below front part of caudal fin. Has 51–66 lateral scales. To 2¾ in. (7.2 cm). **RANGE:** Conasauga, Coosawattee, and Etowah rivers (Coosa R. system) in se. TN and n. GA. Rare; may be extirpated from Etowah R., extremely localized in Conasauga R. Protected as an *endangered species.* **HABITAT:** Swift gravel and sand riffles and runs of medium-sized rivers. **SIMILAR SPECIES:** (1) Saddleback Darter, *P. vigil* (Pl. 44), has 5 dark saddles on back, *1st under* 1st dorsal fin.

BRONZE DARTER *Percina palmaris* Pl. 46

IDENTIFICATION: Yellow-brown above, *8–10 brown saddles; 8–11 brown blotches* along side. On large individual, *wide dusky brown* (*bronze* on large male) *bars join* blotches along side to saddles. Two large white to yellow areas on caudal fin base. Teardrop weak or absent. First dorsal fin of male has amber base and white edge, of female is mottled with brown. Has 57–73 lateral scales. To 3½ in. (9 cm). **RANGE:** Above Fall Line in Coosa and Tallapoosa river systems (Mobile Bay drainage) in GA, AL, and se. TN. Locally common. **HABITAT:** Swift riffles over gravel

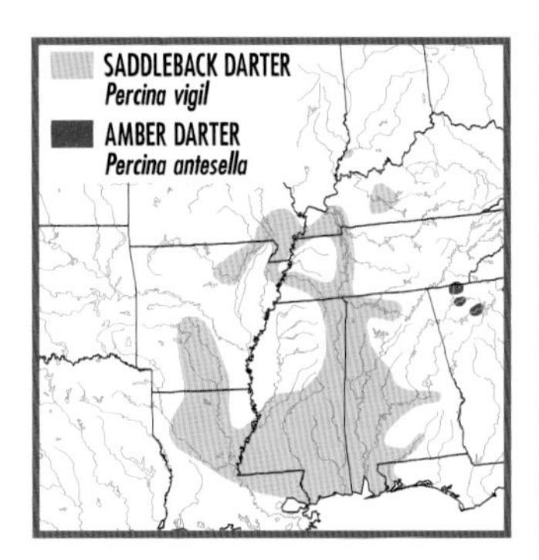

and rubble in small to medium rivers. **SIMILAR SPECIES:** (1) Gilt Darter, *P. evides* (Pl. 46), has dusky *green bars* over back, *yellow to orange* breast and belly. (2) Blackbanded Darter, *P. nigrofasciata* (Pl. 44), has 12–15 *narrow dark bars* on side, clear or mostly clear 1st dorsal fin.

ILT DARTER *Percina evides* Pl. 46

IDENTIFICATION: *Wide dusky green bars* (darkest on adult) *join* 6–8 dark saddles to dark blotches along side. *Yellow to bright orange belly, breast, and underside of head.* First dorsal fin often orange or amber at base; has orange band below edge. Black teardrop. Two large white or yellow areas on caudal fin base. Breeding male has blue-green bars over back, bright orange breast and belly. Has 51–77 lateral scales. To 3¾ in. (9.6 cm). **RANGE:** Mississippi R. basin from NY to MN and south to n. AL and n. AR; Maumee R. system (Lake Erie drainage), OH and IN. Locally common but extirpated from much of former range, including all of OH, IA, and IL. **HABITAT:** Rocky riffles of small to medium rivers. **REMARKS:** Unnamed subspecies: Population west of Mississippi R. has partly scaled cheek, orange 1st dorsal fin and yellow belly on large male. Population in upper Tennessee R. drainage in w. NC, n. GA, and e. TN usually has unscaled cheek, orange edge on 1st dorsal fin and orange belly on large male. Elsewhere, species has partly scaled cheek, orange 1st dorsal fin and belly on large male. **SIMILAR SPECIES:** (1) Bronze Darter, *P. palmaris* (Pl. 46), has *brown* marks on side and back; *lacks* orange breast. Often confused with (2) Blackside Darter, *P. maculata*, and (3) Dusky Darter, *P. sciera* (both Pl. 44), which *lack* bright colors, broad bands over back, large white or yellow spots on caudal fin base.

ANGERINE DARTER *Percina aurantiaca* Pl. 46

IDENTIFICATION: Large and colorful. *Thin black stripe* along back breaks into spots at rear. Broad black stripe of 8–12 fused blotches along side, longitudinal *row of small dark brown spots* on upper side. No teardrop. Underside of young white, of large female yellow, of large male orange. Breeding male has bright red-orange belly, blue breast, orange edge on 1st dorsal fin, red branchiostegal membranes. Has 84–99 lateral scales. To 7¼ in. (18 cm). **RANGE:** Upper Tennessee R. drainage, VA, NC, TN, and GA. Fairly common locally. **HABITAT:** Clear, fairly deep, rocky pools (usually below riffles) of creeks and small rivers. Large male often in rocky riffles.

OLIVE DARTER *Percina squamata* Pl. 46

IDENTIFICATION: *Long, pointed snout.* Olive-brown above, 13–15 small dark brown saddles; dark wavy lines on upper side; *10–12 dark brown rectangles* along side; *orange band* on 1st dorsal fin; small distinct *black spot* on caudal fin base; white or yellow below. Teardrop usually present. *Fully scaled breast.* Has 71–88 lateral scales. To 5¼ in. (13 cm). **RANGE:** Middle Cumberland R. drainage (Big South Fork and Rockcastle rivers), KY and TN; upper Tennessee R. drainage, NC, TN, and GA. Localized but relatively common in a few streams. **HABITAT:** Moderately deep boulder riffles and runs of small to medium rivers. **REMARKS:** Snouts of this and related species (Sharpnose, Slenderhead, and Longnose, *P. nasuta*, darters) become proportionally longer as fish grows. **SIMILAR SPECIES:** (1) See Sharpnose Darter, *P. oxyrhynchus*. (2) Slenderhead Darter, *P. phoxocephala* (Pl. 46), has *shorter snout, 10–16 round blotches* along side, *unscaled or partly scaled breast.*

SHARPNOSE DARTER *Percina oxyrhynchus* Not shown

IDENTIFICATION: Nearly identical to Olive Darter, *P. squamata*, but has *unscaled or only partly scaled breast.* Has 68–80 lateral scales. To 4½ in. (12 cm). **RANGE:** Southern tributaries of Ohio R. from Monongahela R., PA, to Kentucky R., KY; south in New R. drainage to NC. Locally common. **HABITAT:** Fast boulder riffles and runs of small to medium rivers. **SIMILAR SPECIES:** (1) See Olive Darter, *P. squamata* (Pl. 46).

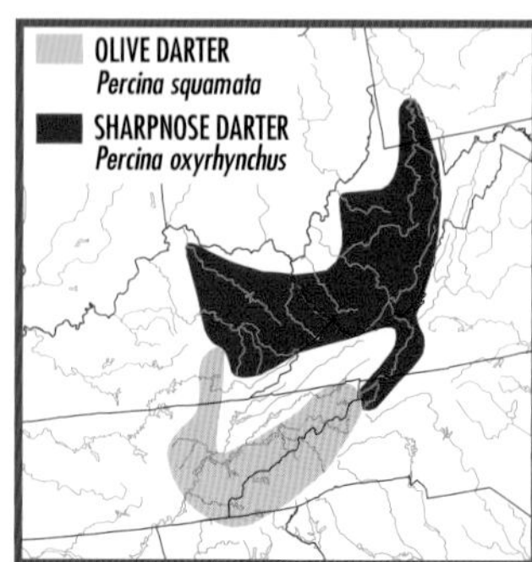

SLENDERHEAD DARTER *Percina phoxocephala* Pl. 46

IDENTIFICATION: *Moderately long, pointed snout.* Yellow-brown above with dark brown wavy lines; *10–16 round brown-black blotches* along side; *orange band* on 1st dorsal fin; small distinct *black spot* on caudal fin base; white to yellow below. Black teardrop. *Unscaled* (except large scutes near pelvic fins) or partly scaled breast. Has 58–80 lateral scales. To 3¾ in. (9.6 cm). **RANGE:** Mississippi R. basin from WV to ne. SD and south to n. AL and ne. TX; Lake Winnebago system (Lake Michigan drainage), WI. Generally common. **HABITAT:** Gravel runs and riffles of creeks and small to medium rivers. **SIMILAR SPECIES:** (1) Olive Darter, *P. squamata* (Pl. 46), and (2) Sharpnose Darter, *P. oxyrhynchus,* have *longer snout, 10–12 rectangular blotches* along side; Olive Darter has *fully scaled* breast. (3) Longnose Darter, *P. nasuta* (Pl. 46), has *bars* along side, *longer snout.*

LONGNOSE DARTER *Percina nasuta* Pl. 46

IDENTIFICATION: *Long, pointed snout* (extremely so in some populations). Yellow-brown above; dark brown wavy lines on upper side; *12–15 brown-black bars* along side; *orange band* on 1st dorsal fin; small distinct *black spot* on caudal fin base; white to yellow below. No teardrop. *Unscaled* (except large scutes near pelvic fins) or partly scaled breast. Has 64–83 lateral scales. To 4½ in. (11 cm). **RANGE:** Ozark and Ouachita uplands (St. Francis, White, Arkansas, and Ouachita river drainages), MO, AR, and OK. Uncommon. **HABITAT:** Gravel runs and riffles of small to medium clear rivers. **SIMILAR SPECIES:** (1) Slenderhead Darter, *P. phoxocephala* (Pl. 46), has *10–16 round blotches along side, shorter snout.*

NEXT 10 SPECIES (through Bigscale Logperch, *Percina macrolepida*): A distinctive group of darters (logperches), yellow-brown above, yellow to white below, with *bulbous snout* extending well beyond upper jaw (especially on large individual), *wide flat area* between eyes, *medial*

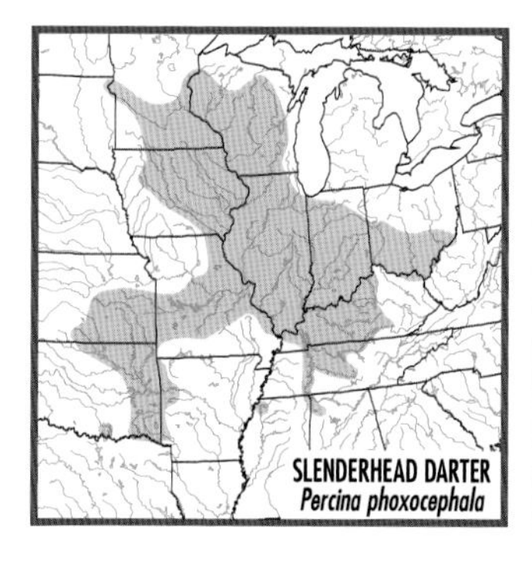
SLENDERHEAD DARTER
Percina phoxocephala

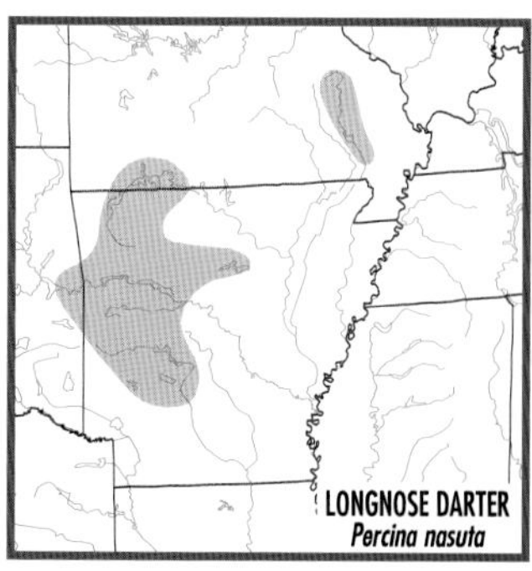
LONGNOSE DARTER
Percina nasuta

black spot on caudal fin base. All but Blotchside Logperch, *P. burtoni*, have *alternating long and short bars* along side that usually extend over back and join those of other side. Logperches use long snout to flip over stones and root in gravel to expose food organisms.

BLOTCHSIDE LOGPERCH *Percina burtoni* Pl. 47

IDENTIFICATION: *No (or few) scales* on nape. Has *8–10 dark green to black round or oval blotches* along side; dark blotches and bars on back extend down upper side. Large male has *orange edge, large black blotch* at rear of 1st dorsal fin. Black teardrop. Has 79–92 lateral scales, 32–38 transverse scales, 34–37 scales around caudal peduncle. To 6½ in. (16 cm). **RANGE:** Tennessee and Cumberland river drainages, VA, NC, KY, TN, GA, and AL. Rare and sporadic in Tennessee R. drainage; probably extirpated from Cumberland R. drainage. **HABITAT:** Gravel runs and riffles of clear, small to medium rivers. **SIMILAR SPECIES:** (1) Other logperches (Pl. 47) have *many alternating long and short bars* along side that usually extend over back and join those of other side; all but northern populations of Logperch (*P. caprodes semifasciata* and *P. c. manitou*) have *scaled nape.*

LOGPERCH *Percina caprodes* Pl. 47

IDENTIFICATION: *Many alternating long and short bars* along side extend over back and join those of other side; bars relatively uniform, not constricted at middle. Dusky teardrop. Nape usually scaled (unscaled in northern populations). *No orange band* on 1st dorsal fin (except in Ozark Logperch, *P. c. fulvitaenia*). *No scales* on top of head, usually none on area in front of pectoral fin. Has 67–100 lateral scales. To 7¼ in. (18 cm). **RANGE:** St. Lawrence-Great Lakes, Hudson Bay, and Mississippi R. basins from e. QC to AB and south to LA; Hudson R. drainage, VT and NY. Generally common. **HABITAT:** Usually over gravel and sand in medium-sized rivers but can be found almost anywhere

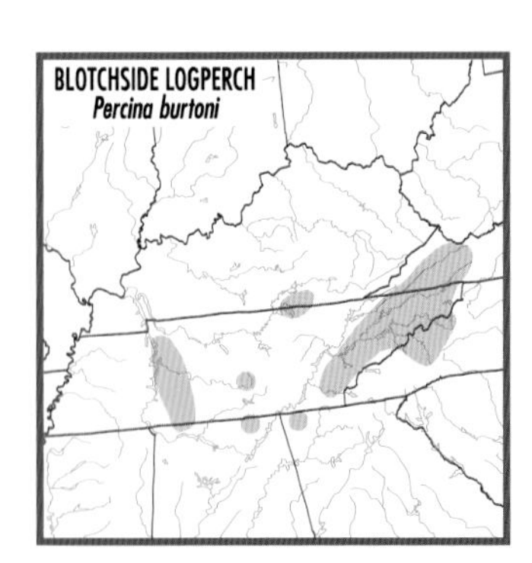

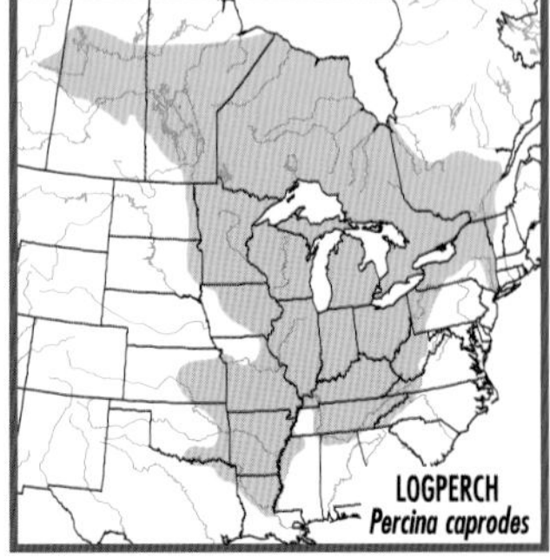

from small, fast-flowing rock-bottomed streams to vegetated lakes. **REMARKS:** Highly variable; 4 described subspecies with uncertain distributions. Northern Logperch, *P. c. semifasciata* (Pl. 47), throughout Canada and northern states, has no orange band on 1st dorsal fin, unscaled nape. Ozark Logperch, *P. c. fulvitaenia* (Pl. 47), in tributaries of Missouri and Arkansas rivers in MO, AR, KS, and OK, and in Blue R., OK, has orange band on 1st dorsal fin, fully scaled nape. Indiana Logperch, *P. c. manitou* (Pl. 47), in glacial lakes region (Wabash R. and Lake Michigan drainages) of n. IN and possibly s. MI, has interrupted bars on rear half of side forming light stripe along upper side, bars along side joining on back to form dark circles, unscaled nape, no orange band on 1st dorsal fin; intergrades with *P. c. caprodes* in upper Tippecanoe R. *P. c. fulvitaenia* intergrades with *P. c. semifasciata* in Illinois, Kaskaskia, and Mississippi rivers south to s. IL and se. MO, and with *P. c. caprodes* in St. Francis, Black, and White rivers, s. MO and n. AR. Central Logperch, *P. c. caprodes* (Pl. 47), throughout rest of range, has no orange band, fully scaled nape. **SIMILAR SPECIES:** See Pl. 47. Only other logperches within range are (1) Blotchside Logperch, *P. burtoni,* which *lacks* long bars extending over back to join those of other side, and (2) Bigscale Logperch, *P. macrolepida*, which is lighter, has smaller head, scales on top of head.

ROANOKE LOGPERCH *Percina rex* Pl. 47

IDENTIFICATION: Has *10–12 short black bars* along side, *not* joined over back with those of other side. Large male has *red-orange band* below black edge on 1st dorsal fin. Olive to yellow-brown above, wavy dark blotches on back. Black teardrop. Fully scaled nape. Has 83–89 lateral scales. To 6 in. (15 cm). **RANGE:** Upper Roanoke, upper Dan, and upper Chowan river systems (Roanoke R. drainage), VA. Uncommon; protected as an *endangered species.* **HABITAT:** Gravel and boulder runs of small to medium rivers. **SIMILAR SPECIES:** (1) Blotchside Logperch, *P.*

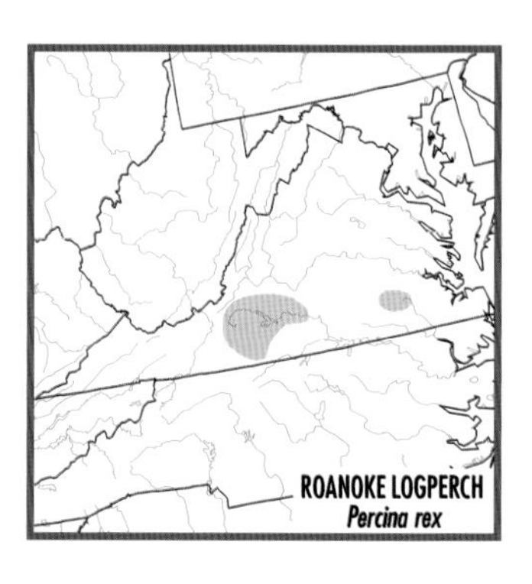

burtoni (Pl. 47), has *oval blotches* along side. (2) Other logperches (Pl. 47) have long bars extending over back to *join* those of other side.

CHESAPEAKE LOGPERCH *Percina bimaculata* Pl. 47

IDENTIFICATION: *Long and short bars along side of body wavy, broken into blotches.* Long bars extend over back and join those of other side. Dusky teardrop. Nape unscaled in adult; partly scaled in young. *No orange band* on 1st dorsal fin. *No scales* on top of head, usually none on area in front of pectoral fin. Has 68–80 lateral scales. To 6 in. (15 cm). **RANGE:** Susquehanna and Potomac river drainages, PA, MD, and VA. Rare in Susquehanna; extirpated from Potomac R. **HABITAT:** Gravel runs and riffles of clear, small to medium rivers. **SIMILAR SPECIES:** (1) Logperch, *P. caprodes* (Pl. 47), has bars along side relatively uniform, *not* broken into blotches. (2) Roanoke Logperch, *P. rex* (Pl. 47), has *10–12 short bars* along side, not joined over back with those of other side.

CONASAUGA LOGPERCH *Percina jenkinsi* Pl. 47

IDENTIFICATION: Short bars on side *broken into spots and short wavy lines. Black spot* at pectoral fin origin; *no red-orange band* on 1st dorsal fin. Has 32–38 (usually 35–37) transverse scales, 32–37 (usually 34–36) scales around caudal peduncle, 87–97 lateral scales. To 5½ in. (14 cm). **RANGE:** Conasauga R. (Alabama R. system), TN and GA. Uncommon in extremely small range; protected as an *endangered species.* **HABITAT:** Rocky runs and flowing pools of Conasauga R., a small, fast-flowing river. **SIMILAR SPECIES:** See Pl. 47. (1) Other logperches in Gulf drainages *lack* black spot at pectoral fin origin; have *red-orange band* on 1st dorsal fin, long and short bars on side. (2) Gulf Logperch, *P. suttkusi*, and (3) Southern Logperch, *P. austroperca,* also have 32 or fewer transverse scales, usually fewer than 34 scales around caudal peduncle.

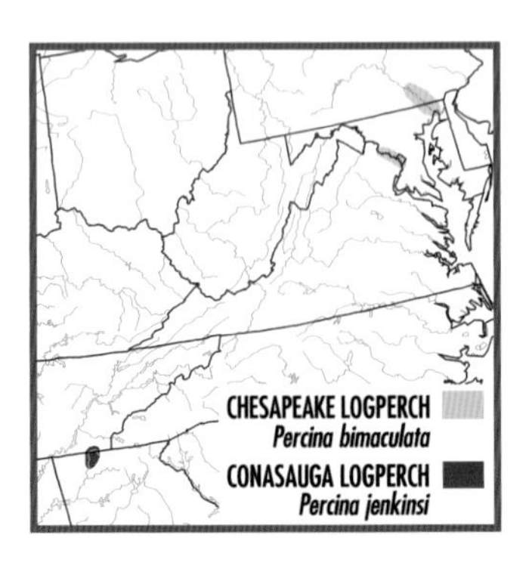

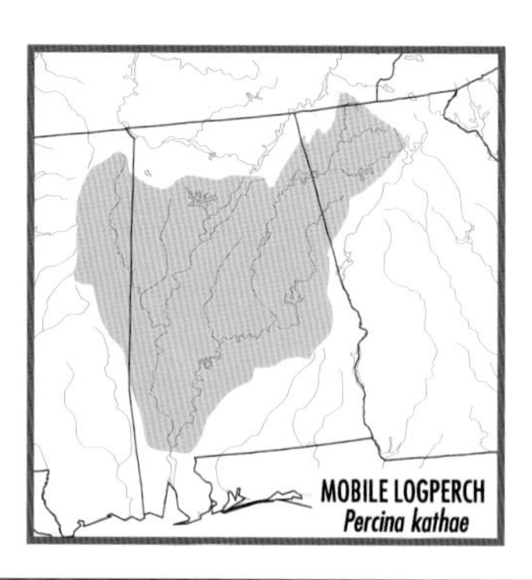

MOBILE LOGPERCH *Percina kathae* Pl. 47

IDENTIFICATION: *Wide red-orange band* below black edge on 1st dorsal fin. Long and half bars on side; long bars *expanded into blotches*. Fully scaled nape; few or no scales on breast. Has *31–37 (usually 33–37) scales around caudal peduncle*, 87–98 (usually 88–94) lateral scales, 29–37 (usually 30–36) transverse scales, 10–12 anal rays. To 6¾ in. (17 cm). **RANGE:** Mobile Bay drainage, GA, AL, MS, and se. TN. Generally common above Fall Line; rare below. **HABITAT:** Gravel runs of creeks and small rivers; rarely in headwaters, large rivers, and impoundments. **SIMILAR SPECIES:** (1) Southern Logperch, *P. austroperca* (Pl. 47), has narrow red-orange band on 1st dorsal fin, whole, half, and quarter bars on side, 25–32 (usually 27–31) transverse scales. (2) Gulf Logperch, *P. suttkusi* (Pl. 47), has 8–10 (rarely 11) anal rays, narrow red-orange band on 1st dorsal fin, long bars along side *not* expanded into blotches, *24–33 (usually 28–31) scales around caudal peduncle*, 19–28 (usually 21–26) transverse scales.

SOUTHERN LOGPERCH *Percina austroperca* Pl. 47

IDENTIFICATION: *Narrow red-orange band* below black edge on 1st dorsal fin. Narrow bars on side; long, half, and *quarter bars*; long bars *expanded into blotches*. Fully scaled nape; few or no scales on breast. Large male has *dusky gray breast, branchiostegal membranes, anal and pelvic fins*. Has 86–96 (usually 88–93) lateral scales, 25–32 (usually 27–31) transverse scales, 30–35 (usually 31–34) scales around caudal peduncle, 10–12 anal rays. To 6¾ in. (17 cm). **RANGE:** Choctawhatchee and Escambia river drainages, AL and FL. Common in Escambia drainage; uncommon in Choctawhatchee drainage. **HABITAT:** Gravel and rubble riffles and runs of creeks and small to medium rivers. **SIMILAR SPECIES:** See Pl. 47. (1) Other logperches except Texas Logperch, *P. carbonaria*, and Ozark Logperch, *P. c. fulvitaenia*, *lack* dark breast, branchiostegal membranes, anal and pelvic fins. (2) Mobile Logperch,

P. kathae, has *wide red-orange band* on 1st dorsal fin; long, half, usually *no quarter bars* on side, 29–37 (usually 30–36) transverse scales, 31–37 (usually 33–37) scales around caudal peduncle. (3) Gulf Logperch, *P. suttkusi*, has 8–10 (rarely 11) anal rays, long bars along side *not expanded into blotches*, 19–28 (usually 21–26) transverse scales.

TEXAS LOGPERCH *Percina carbonaria* Pl. 47

IDENTIFICATION: Large male has *black breast, branchiostegal membranes, anal and pelvic fins*; orange band below black edge on 1st dorsal fin. Has 15–21 alternating long and short bars along side; long bars *constricted* near middle, producing series of round blotches along lower side; bars extend over back and join those of other side. Black teardrop. No scales on top of head. Has 80–93 lateral scales. To 5¼ in. (13 cm). **RANGE:** Brazos, Colorado, Guadalupe, and San Antonio river drainages, TX. Nearly confined to Edwards Plateau. Common. **HABITAT:** Rocky riffles and runs of small to medium rivers. **SIMILAR SPECIES:** (1) Only other logperch within range is Bigscale Logperch, *P. macrolepida* (Pl. 47), which *lacks* orange band on 1st dorsal fin, black breast, branchiostegal membranes, and pelvic fins; is lighter in color; has scales on top of head and in front of pectoral fin.

GULF LOGPERCH *Percina suttkusi* Pl. 47

IDENTIFICATION: *Slender. Narrow red-orange band* below black edge on 1st dorsal fin. Long bars along side *not expanded into blotches*. Fully scaled nape; few or no scales on breast. Has *8–10 (rarely 11) anal rays*; 79–90 (usually 82–87) lateral scales; 19–28 (usually 21–26) transverse scales; 24–33 (usually 28–31) scales around caudal peduncle. To 6¼ in. (16 cm). **RANGE:** Below Fall Line in Gulf drainages from Mobile Bay to Lake Pontchartrain. Common, but declining in areas subject to siltation. **HABITAT:** Gravel runs of creeks and small to large rivers; often in fairly deep (2–6.5 ft. [1–2 m]) water. **SIMILAR SPECIES:** See Pl. 47. (1)

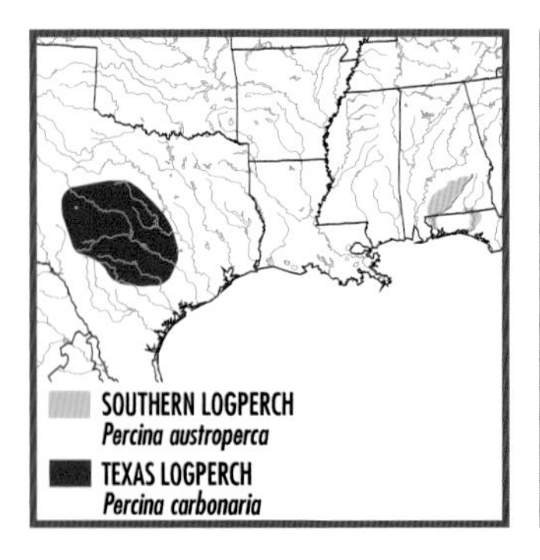

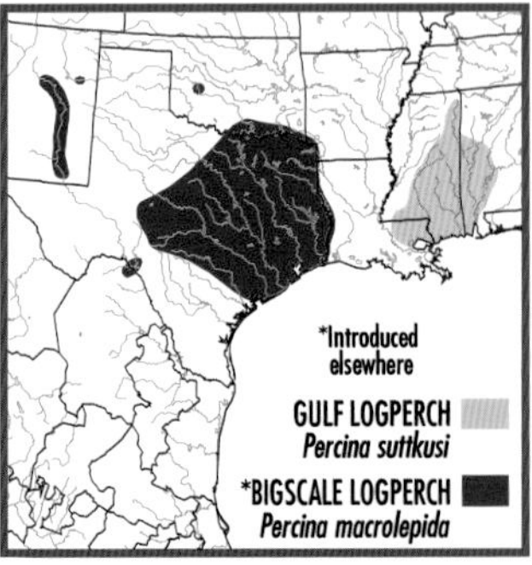

Other logperches (except Bigscale Logperch, *P. macrolepida*) usually have *10–12 anal rays.* (2) Mobile Logperch, *P. kathae,* has wide red-orange band on 1st dorsal fin, long bars on side *expanded into blotches,* 31–37 (usually 33–37) scales around caudal peduncle, 29–37 (usually 30–36) transverse scales. (3) Southern Logperch, *P. austroperca*, has long bars on side *expanded into blotches,* 25–32 (usually 27–31) transverse scales, 30–35 (usually 31–34) scales around caudal peduncle.

BIGSCALE LOGPERCH *Percina macrolepida* Pl. 47

IDENTIFICATION: *Light-colored. Scales on top of small head, breast, and nape and in front of pectoral fin.* Has 15–20 long dark bars along side extending over back and joining those of other side. No orange band on 1st dorsal fin. Teardrop reduced to dusky spot. Has 77–90 lateral scales; 18–27 (usually 20–25) transverse scales; 7–10 (usually 9) anal rays. To 4½ in. (11 cm). **RANGE:** From Sabine R., LA, and Red R., OK, to Rio Grande drainage, TX, NM, and Mexico. Introduced into Sacramento-San Joaquin river drainage, CA, and Canadian R. drainage, NM. Locally common; localized and uncommon in Rio Grande drainage. **HABITAT:** Gravel and sand runs and pools of small to medium rivers; impoundments. **SIMILAR SPECIES:** (1) Other logperches (Pl. 47) are darker, have larger head, *lack* scales on top of head, usually lack scales on breast.

CRYSTAL DARTER *Crystallaria asprella* Pl. 45

IDENTIFICATION: *Very slender body; wide flat head,* closely set eyes. *Forked caudal fin;* large anal and dorsal fins; pelvic fin not falcate. Brown mottling, *4 dark brown saddles* (1st 3 large) on back and upper side; dark brown oblong blotches along side. Black stripe around snout continuous from eye to eye. Has 4–14 (usually 11) rows of scales on cheek; 12–16 anal fin rays; 77–96 lateral scales. To 6¼ in. (16 cm). **RANGE:** Mississippi R. basin from Wabash R., IN, to se. MN and south

to s. MS, n. LA, and se. OK; Gulf Slope in Escambia, Mobile Bay, Pascagoula, and Pearl river drainages, FL, AL, and MS. Localized and rare. **HABITAT:** Clean sand and gravel runs of small to medium rivers. Usually in fairly fast water 2 ft. (0.6 m) or more deep. **SIMILAR SPECIES:** (1) See Diamond Darter, *C. cincotta.* (2) Sand darters, *Ammocrypta* species (Pl. 45), *lack* dark dorsal saddles, have straight-edged or shallowly forked caudal fin.

DIAMOND DARTER *Crystallaria cincotta* **Not shown**

IDENTIFICATION: Similar to Crystal Darter, *C. asprella*, but has larger gape (width about equal to pelvic fin base; less in Crystal Darter), dark blotch in front of eye not touching eye (usually touches eye in Crystal Darter), falcate pelvic fin, usually 2 rows of scales on cheek. Has 11–13 anal fin rays; 83–93 lateral scales. To 3½ in. (9.2 cm). **RANGE:** Ohio R. drainage from OH and WV to Cumberland R. system, TN and KY (absent in Wabash R. system). Extremely rare; extant only in Elk R., WV. **HABITAT:** Clean sand, gravel, and cobble runs of small to medium rivers. **SIMILAR SPECIES:** (1) See Crystal Darter, *C. asprella* (Pl. 45).

AMMOCRYPTA

GENUS CHARACTERISTICS: Sand darters are *long, slender, glass-clear* darters that bury in sandy streams with only eyes and snout protruding. Adaptations for living in sand include transparent bodies (usually yellow tint above, iridescent green on back and side, silver white below) to blend in with sand; loss of scales; long slender shape to facilitate burying; eyes near top of head for viewing above while buried. Protruding snout, *no premaxillary frenum*, complete lateral line, *1 anal spine.*

WESTERN SAND DARTER *Ammocrypta clara* **Pl. 45**

IDENTIFICATION: *Spine* on opercle. No black on dorsal or pelvic fins, weakly developed dark green blotches along back and side. Has 63–84 lateral scales. Lateral line slants down at rear. To 2¾ in. (7.1 cm). **RANGE:** Mississippi R. basin from WI and MN south to VA, MS, and TX; Lake Michigan drainage, MI and WI; Sabine and Neches river drainages, TX. Generally sporadic and uncommon; extremely rare in east. **HABITAT:** Sandy runs of medium to large rivers. **SIMILAR SPECIES:** (1) Other sand darters (Pl. 45) *lack* spine on opercle, have well-developed dark blotches along side or black bands on dorsal fins.

NAKED SAND DARTER *Ammocrypta beanii* **Pl. 45**

IDENTIFICATION: *Middle black band* (most prominent near front) on each dorsal fin (Fig. 61). No spine on opercle. No black pigment on pelvic fin. "Naked" refers to lack of scales except along lateral line and on caudal peduncle. Has 57–77 lateral scales. Lateral line slants down at rear. To 2¾ in. (7.2 cm). **RANGE:** Mississippi R. and w. Gulf drainages in w. TN, AL, MS, and LA; on Gulf Slope from Mobile Bay drainage, AL, to Mississippi R., LA. Common. **HABITAT:** Sandy runs of creeks and small to medium rivers. **SIMILAR SPECIES:** (1) See Florida Sand Darter, *A. bifascia*. (2) Scaly Sand Darter, *A. vivax* (Pl. 45), has *dark blotches* along side, black pigment on pelvic fin of male.

FLORIDA SAND DARTER *Ammocrypta bifascia* Not shown

IDENTIFICATION: Nearly identical to Naked Sand Darter, *A. beanii*, but has *2 black bands on* each dorsal fin (Fig. 61). Has 63–78 lateral scales. To 3 in. (7.7 cm). **RANGE:** Gulf Slope drainages from Apalachicola R. to Perdido R., s. AL and FL. Common. **HABITAT:** Sandy runs of small to medium rivers. **SIMILAR SPECIES:** (1) See Naked Sand Darter, *A. beanii* (Pl. 45).

EASTERN SAND DARTER *Ammocrypta pellucida* Pl. 45

IDENTIFICATION: Has 12–17 small dark green blotches along back; *10–19 horizontal dark green blotches* along side. No spine on opercle. No black bands on dorsal fins. Black pigment on pelvic fin of male. Has 62–84 lateral scales. To 3¼ in. (8.4 cm). **RANGE:** St. Lawrence R. drainage, s. QC, VT, and NY; Great Lakes (Lakes Ontario, Erie, and Huron) and Ohio R. basins from w. NY to e. IL, and south to KY. Locally common but decreasing in abundance because of siltation and pollution. Protected in Canada as a *threatened species*. **HABITAT:** Sandy runs of small to medium rivers; usually in water 2 ft. (0.6 m) or more deep. **SIMILAR SPECIES:** (1) See Southern Sand Darter, *A. meridiana*. (2) Other sand darters (Pl. 45), except Scaly Sand Darter, *A. vivax*, *lack* dark blotches on side; Scaly Sand Darter has *vertical blotches*.

WESTERN SAND DARTER
Ammocrypta clara

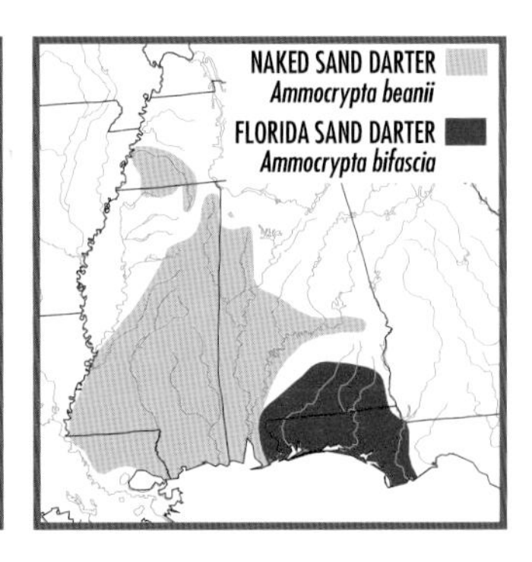

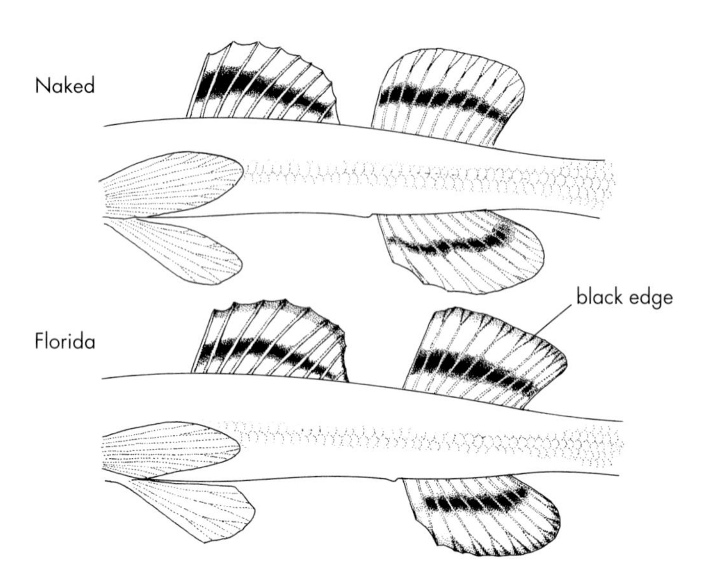

Fig. 61. Naked and Florida sand darters.

SOUTHERN SAND DARTER *Ammocrypta meridiana* Not shown

IDENTIFICATION: Similar to Eastern Sand Darter, *A. pellucida*, but usually has 14–16 transverse scales (vs. 11–13 in Eastern Sand Darter). Has 63–79 lateral scales. To 2¾ in. (7 cm). **RANGE:** Mobile Bay drainage, AL and MS. Common. **HABITAT:** Sandy runs of small to medium rivers. **SIMILAR SPECIES:** (1) See Eastern Sand Darter, *A. pellucida* (Pl. 45).

SCALY SAND DARTER *Ammocrypta vivax* Pl. 45

IDENTIFICATION: *Vertical dark green blotches* along side. *Dusky edge* and *middle band* on dorsal fins. No spine on opercle. Black pigment on pelvic fin of male. Has 10–15 dark green blotches along back; 58–79 lateral scales. To 2¾ in. (7.3 cm). **RANGE:** Mississippi R. basin from w. KY and se. MO, south to s. MS, and west to e. OK and TX; Gulf drainages from Pascagoula R., AL, to San Jacinto R., TX. Primarily on Coastal Plain. Locally common. **HABITAT:** Sandy runs of creeks and small to medium rivers. **SIMILAR SPECIES:** (1) Other sand darters (Pl. 45) have *horizontal* (not vertical) or *no* dark blotches along side.

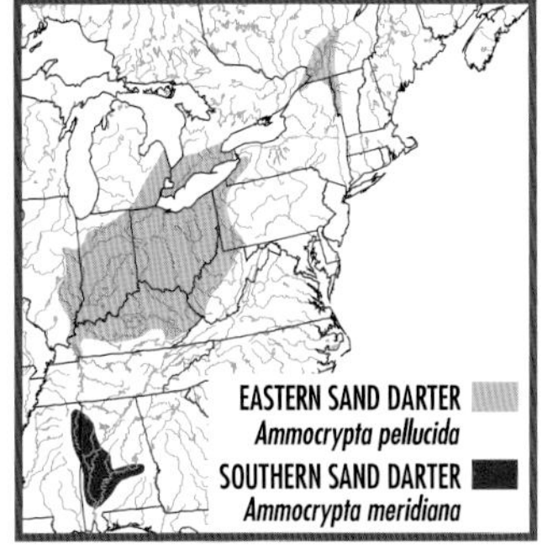

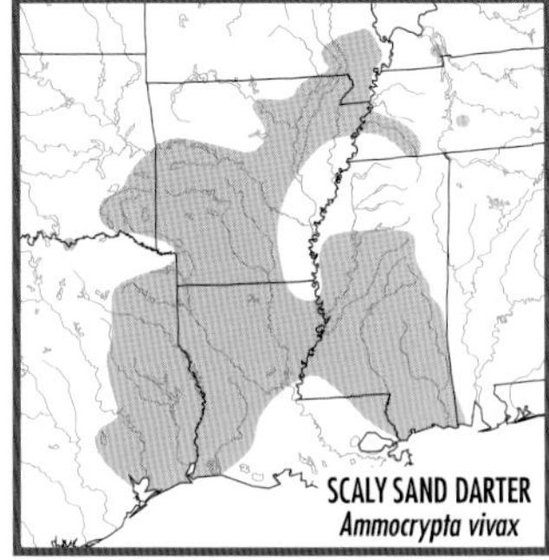

ETHEOSTOMA

GENUS CHARACTERISTICS: *Etheostoma* is largest genus of N. American fishes (144 species in U.S. and Canada plus 4 endemic to Mexico). Included in *Etheostoma* are some of our rarest fishes (e.g., Maryland Darter, *E. sellare*), some of our most common fishes (e.g., Johnny Darter, *E. nigrum,* and Tessellated Darter, *E. olmstedi*), and some of our most spectacularly colorful fishes. Species of *Etheostoma lack* scutes of *Percina* (Fig. 58); all but Glassy Darter, *E. vitreum*, lack transparency of *Ammocrypta*.

GLASSY DARTER *Etheostoma vitreum* Pl. 45

IDENTIFICATION: *Translucent.* Has 7–8 dark brown saddles, many dark brown specks and *small black spots* on back and upper side; *dark brown dashes* along lateral line. No teardrop. Protruding snout, no premaxillary frenum, closely set eyes, 1–2 anal spines, narrowly joined branchiostegal membranes (Fig. 56). Fleshy villi around anus. Complete lateral line; 50–65 lateral scales. To 2½ in. (6.6 cm). **RANGE:** Atlantic Slope drainages from Bush R., MD, to Neuse R., NC. Common. **HABITAT:** Sandy runs of creeks and small to medium rivers. **SIMILAR SPECIES:** (1) Johnny Darter, *E. nigrum*, and (2) Tessellated Darter, *E. olmstedi* (both Pl. 45), are much more *opaque;* have blunter snout, Xs and Ws along side, teardrop, black spot on dorsal fin of male, no villi around anus.

JOHNNY DARTER *Etheostoma nigrum* Pl. 45

IDENTIFICATION: *Dark brown Xs and Ws along side*, wavy brown lines on upper side. Light brown above, 6 dark brown saddles; black preorbital bar extends onto upper lip (Fig. 62). Breeding male has black head and anal and pelvic fins; black spot at front of 1st dorsal fin; often has white knobs on tips of pelvic spine and rays. *One anal spine, no premax-*

illary frenum, interrupted infraorbital and supratemporal canals (Fig. 55), narrowly joined branchiostegal membranes (Fig. 56), moderately blunt snout, slender caudal peduncle. Complete lateral line; 35–56 lateral scales. To 2¾ in. (7.2 cm). **RANGE:** St. Lawrence-Great Lakes, Hudson Bay, and Mississippi R. basins from Hudson Bay to s. MS, and from QC and VA to SK and CO; on Atlantic Slope in James, Roanoke, Tar, and Neuse river drainages, VA and NC; on Gulf Slope in Mobile Bay drainage, AL and MS. Absent in White R. drainage (AR and MO), most of Arkansas R. drainage, upper Tennessee R. drainage, and middle Cumberland R. drainage. Introduced into Colorado R. drainage, CO. Common to abundant. **HABITAT:** Sandy and muddy, sometimes rocky, pools of headwaters, creeks, and small to medium rivers; sandy shores of lakes. **REMARKS:** Two subspecies. *E. n. susanae,* endemic to Cumberland R. drainage above and just below Cumberland Falls, KY and TN, is extremely rare and, unlike *E. n. nigrum,* has break in eye stripe (Fig. 62), no scales on top of head, opercle, or along midbelly. *E. n. susanae* sometimes is recognized as a species (Cumberland Darter) but intergrades with *E. n. nigrum* in headwaters of Kentucky R., KY. Where they co-occur, Johnny Darter and Tessellated Darter, *E. olm-*

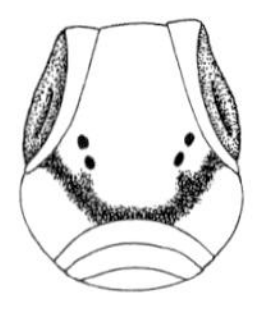

Bluntnose

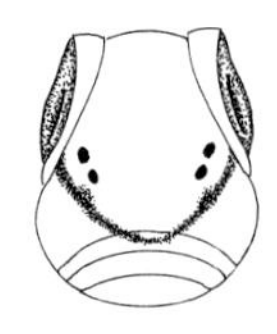

Choctawhatchee

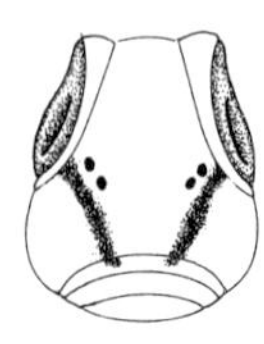

Johnny (*E. n. nigrum*)

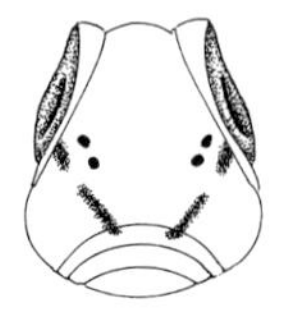

Johnny (*E. n. susanae*)

Fig. 62. Bluntnose, Choctawhatchee, and Johnny darters.

stedi, interbreed and lose their distinctiveness, especially in e. Canada, and James and Roanoke river drainages, VA. **SIMILAR SPECIES:** (1) See Tessellated Darter, *E. olmstedi* (Pl. 45). (2) Speckled Darter, *E. stigmaeum*, and (3) Blueside Darter, *E. jessiae* (both Pl. 55), have *2 anal spines,* row of dark squares just below lateral line (*no* Xs and Ws along lateral line), orange and blue on body and fins of large male. (4) Longfin Darter, *E. longimanum* (Pl. 48), and (5) Riverweed Darter, *E. podostemone*, have deep caudal peduncle, broadly joined branchiostegal membranes, *2 anal spines.* (6) Bluntnose Darter, *E. chlorosoma* (Pl. 45), has bridle (Fig. 62) around extremely blunt snout, horizontal black blotches along side.

TESSELLATED DARTER *Etheostoma olmstedi* Pl. 45

IDENTIFICATION: Similar to Johnny Darter, *E. nigrum*, but has *uninterrupted infraorbital and supratemporal canals* (Fig. 55), enlarged 2d dorsal fin on breeding male. One anal spine, except *E. o. maculaticeps* has 2. Has 34–64 (usually 36–56) lateral scales. To 4½ in. (11 cm). **RANGE:** Atlantic drainages from St. Lawrence R., QC and ON (absent in ME) to Altamaha R., GA; St. Johns R. drainage, FL; Lake Ontario drainage, NY. Introduced into upper New R. drainage, NC. Abundant in central part of range, generally common elsewhere; mostly restricted to Coastal Plain in VA. **HABITAT:** Sandy and muddy pools of headwaters, creeks, and small to medium rivers; shores of lakes. **REMARKS:** Three subspecies. *E. o. maculaticeps,* from Cape Fear R., NC, to FL, has 2 anal spines. *E. o. vexillare,* in upper Rappahannock R. drainage, VA, has 1 anal spine, 42 or fewer lateral scales. *E. o. olmstedi,* in rest of range, has 1 anal spine, usually 43 or more lateral scales. See Remarks for Johnny Darter, *E. nigrum*. **SIMILAR SPECIES:** See Pl. 45. See (1) Johnny Darter, *E. nigrum,* and (2) Waccamaw Darter, *E. perlongum*. (3) Glassy Darter, *E. vitreum*, is transparent and has protruding snout, many dark specks on back and side.

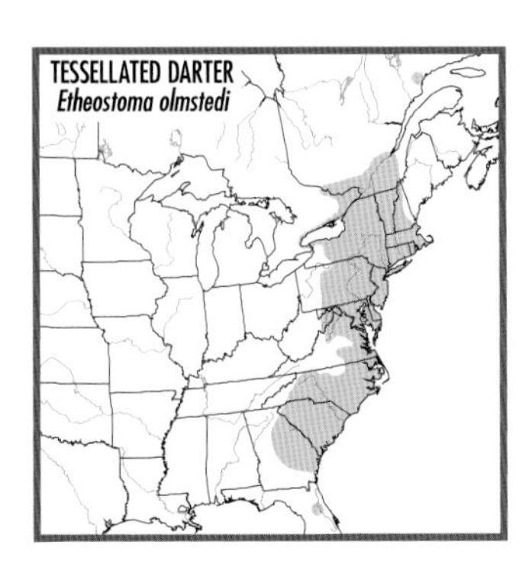

WACCAMAW DARTER *Etheostoma perlongum* Pl. 45

IDENTIFICATION: Nearly identical to Tessellated Darter, *E. olmstedi*, but has *more slender body*, 58–66 (usually 60–62) lateral scales. To 3½ in. (9 cm). **RANGE:** Lake Waccamaw, Columbus Co., NC (not mapped). Moderately common. **HABITAT:** Shallow, sandy areas of lake. **REMARKS:** May be only a lake form of Tessellated Darter, *E. olmstedi*. **SIMILAR SPECIES:** (1) See Tessellated Darter, *E. olmstedi* (Pl. 45).

BLUNTNOSE DARTER *Etheostoma chlorosoma* Pl. 45

IDENTIFICATION: *Black bridle* (Fig. 62) around extremely *blunt snout*, no premaxillary frenum, *1 anal spine*, slender body, scaled breast. Olive above, 6 dusky saddles; brown Xs and Ws on side and back; *horizontal dusky to black blotches* along side; black teardrop. Middle black band (darkest at front), dusky edge on 1st dorsal fin of breeding male. Incomplete lateral line; 49-60 lateral scales. To 2¼ in. (6 cm). **RANGE:** Mississippi R. basin from s. WI and MN to LA; Gulf Slope from Mobile Bay drainage, AL, to San Antonio R. drainage, TX. Formerly Lake Michigan drainage, IL. Mostly restricted to lowlands. Common; locally abundant. **HABITAT:** Muddy (sometimes sandy) pools and backwaters of creeks and small to medium rivers; weedy lakes and ponds; swamps. **SIMILAR SPECIES:** (1) See Choctawhatchee Darter, *E. davisoni*. (2) Johnny Darter, *E. nigrum* (Pl. 45), and (3) Speckled Darter, *E. stigmaeum* (Pl. 55), *lack* snout bridle and horizontal blotches along side, have more *pointed snout*.

CHOCTAWHATCHEE DARTER *Etheostoma davisoni* Not shown

IDENTIFICATION: Similar to Bluntnose Darter, *E. chlorosoma*, but *snout bridle faint at front* and mostly on upper lip (Fig. 62), *2 anal spines, unscaled breast*. Has 44–57 lateral scales. To 2½ in. (6.1 cm). **RANGE:** Choctawhatchee and Pensacola Bay drainages, s. AL and FL. Fairly

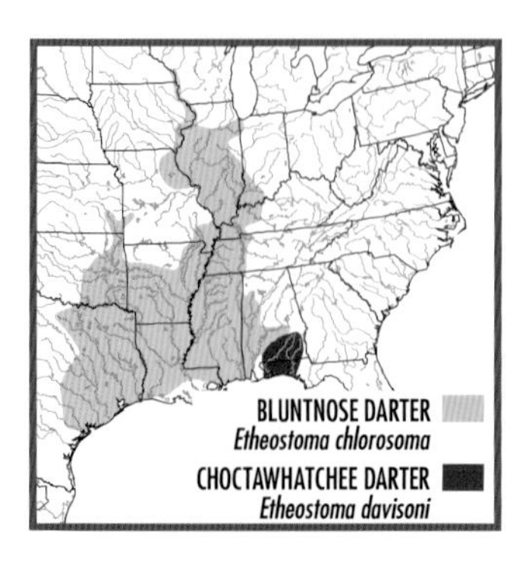

common. **HABITAT:** Sandy and muddy pools of creeks and small rivers. **SIMILAR SPECIES:** (1) See Bluntnose Darter, *E. chlorosoma* (Pl. 45).

LONGFIN DARTER *Etheostoma longimanum* Pl. 48

IDENTIFICATION: *Deep caudal peduncle, broadly joined branchiostegal membranes* (Fig. 56), 2 anal spines. Large pectoral and pelvic fins. No premaxillary frenum. Light brown above, 6 dark brown saddles; wavy lines on upper side; 9–14 dark squares (Ws when examined closely) along side. Small black teardrop often reduced to spot. Breeding male has enlarged 2d dorsal fin; orange belly, caudal and dorsal fins; black spot at front of 1st dorsal fin; blue-green head and lower fins. Breeding male may be almost black. Complete lateral line; 39–51 (usually 42–46) lateral scales. To 3½ in. (8.9 cm). **RANGE:** Upper James R. drainage, VA and WV. Common. **HABITAT:** Rocky riffles of creeks and small rivers. **SIMILAR SPECIES:** (1) See Riverweed Darter, *E. podostemone*. (2) Johnny Darter, *E. nigrum,* and (3) Tessellated Darter, *E. olmstedi* (both Pl. 45), have *slender* caudal peduncle, *narrowly* joined branchiostegal membranes, 1 anal spine (except *E. o. maculaticeps*), no orange or blue-green on large male.

RIVERWEED DARTER *Etheostoma podostemone* Not shown

IDENTIFICATION: Nearly identical to Longfin Darter, *E. longimanum*, but has *rows of dark spots* on side (best developed on upper side) and caudal peduncle; 35–41 (usually 37–40) lateral scales. To 3 in. (7.6 cm). **RANGE:** Upper Roanoke R. drainage, VA and NC. Common. **HABITAT:** Rocky riffles of creeks and small rivers. **SIMILAR SPECIES:** (1) See Longfin Darter, *E. longimanum* (Pl. 48).

SPECKLED DARTER *Etheostoma stigmaeum* Pl. 55

IDENTIFICATION: Has *7–11 dark brown squares* (Ws if examined closely) along side *just below lateral line* on female and young; *7–11 turquoise*

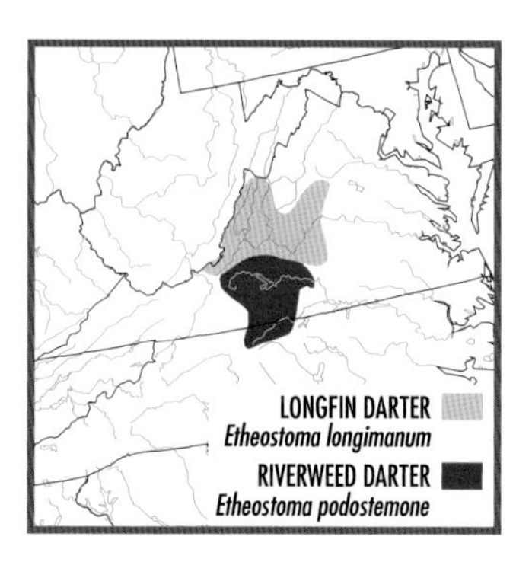

bars on large male. Moderately blunt snout; no premaxillary frenum. Light brown above, 6 dark brown saddles. Dusky teardrop often reduced to spot. Middle orange band, light blue edge and base on 1st dorsal fin (bands faint on female); other fins mostly clear on female, blue and orange on breeding male. Breeding male has blue with orange or gray cheek and opercle. Incomplete lateral line; unscaled or partly scaled cheek; 2 anal spines. Has 39–67 lateral scales. To 2½ in. (6.1 cm). **RANGE:** Mississippi R. basin from w. VA to se. KS, and south to LA; Gulf Slope drainages from Mobile Bay and Pensacola Bay, GA, AL, and FL, to Sabine R., LA. In Clinch and Powell rivers, VA and TN, but replaced in rest of middle and upper Tennessee R. drainage by Blueside Darter, *E. jessiae*. Common; locally abundant. **HABITAT:** Clear sandy and rocky pools of creeks and small to medium, usually fast, rivers. **SIMILAR SPECIES:** (1) See Bluemask Darter, *E. akatulo*, and (2) Blueside Darter, *E. jessiae* (Pl. 55). (3) Johnny Darter, *E. nigrum* (Pl. 45), has Xs and Ws (not squares) *along* lateral line, no red or blue on body or fins, 1 anal spine.

BLUEMASK DARTER *Etheostoma akatulo* **Not shown**

IDENTIFICATION: Similar to Speckled Darter, *E. stigmaeum*, but has mostly or fully scaled cheek, complete lateral line; breeding male has bright blue cheek and opercle, lacks orange and blue on 2d dorsal and anal fins. Has 39–51 lateral scales. To 2¼ in. (5.6 cm). **RANGE:** Upper Caney Fork system (Cumberland R. drainage), TN. Rare; protected as an *endangered species*. **HABITAT:** Rocky pools, runs and riffles of clear creeks and small rivers. **SIMILAR SPECIES:** (1) See Speckled Darter, *E. stigmaeum* (Pl. 55).

BLUESIDE DARTER *Etheostoma jessiae* **Pl. 55**

IDENTIFICATION: Similar to Speckled Darter, *E. stigmaeum*, but has *pointed snout, narrow premaxillary frenum, deep blue bars* on body and fins of male. Has 44–53 lateral scales. To 2¾ in. (6.8 cm). **RANGE:**

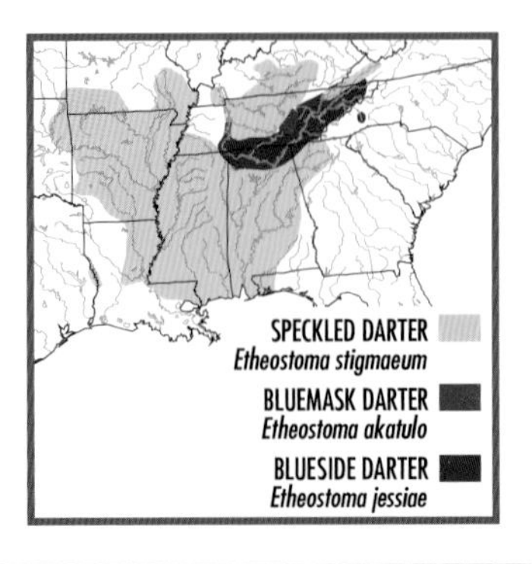

Middle and upper Tennessee R. drainage—above Duck R., TN, VA, NC, GA, AL, and MS. Generally common. **HABITAT:** Rocky pools and adjacent riffles of clear creeks and small fast rivers. **REMARKS:** Often considered a subspecies of Speckled Darter, *E. stigmaeum*. **SIMILAR SPECIES:** (1) See Speckled Darter, *E. stigmaeum* (Pl. 55).

MARYLAND DARTER *Etheostoma sellare* Pl. 45

IDENTIFICATION: *Asymmetrical caudal fin base;* upper half extends farther back than lower half. *Wide, flat head,* closely set eyes; narrowly joined branchiostegal membranes (Fig. 56). Brown to olive above; 4 large dark brown saddles extend down side, often to below lateral line; 3–7 dark brown blotches on lower side; large dusky teardrop; small black caudal spot. Complete lateral line; 43–53 lateral scales. To 3¼ in. (8.4 cm). **RANGE:** Tributaries of lower Susquehanna R., Harford Co., MD. Protected as an *endangered species* but probably extinct. **HABITAT:** Fast rocky riffles of creeks. **SIMILAR SPECIES:** No other darter has asymmetrical caudal fin base.

GREENSIDE DARTER *Etheostoma blennioides* Pl. 48

IDENTIFICATION: *Skin on rear of upper lip fused to skin of snout.* Blunt snout; broadly joined branchiostegal membranes (Fig. 56); no premaxillary frenum; scales on cheek and opercle. Yellow-green; 6–7 square, dark green saddles; brown to dark red spots on upper side; *5–8 green Ws, Us, or bars on side.* Dorsal fins red along base; green on male, clear to dusky on female. Dusky teardrop. Bright green fins and bars on side of breeding male. Complete lateral line; 50–86 lateral scales. Largest *Etheostoma* species—to 6¾ in. (17 cm). **RANGE:** Great Lakes and Mississippi R. basins from NY and MD to e. KS and OK, and from ON south to GA, AL, and AR; Atlantic Slope in Mohawk, Susquehanna, and Potomac drainages, NY to VA. Absent on Former Mississippi Embayment and lowlands of sw. IN and IL. Usually common to abundant; on Atlantic Slope, common only in Potomac R. **HABITAT:** Rocky riffles

MARYLAND DARTER
Etheostoma sellare

GREENSIDE DARTER
Etheostoma blennioides

of creeks and small to medium rivers; shores of large lakes. **REMARKS:** Three subspecies. *E. b. pholidotum* (in Great Lakes, Mohawk, Maumee, Wabash, and Missouri river drainages) has low scale counts (usually 55–65 lateral scales). *E. b. newmanii* (Cumberland, Tennessee, St. Francis, White, Arkansas, and Ouachita river drainages) has high scale counts (usually 66–78 lateral scales). *E. b. blennioides* (Ohio R. basin above confluence of Green R., KY; Potomac R. on Atlantic Slope; upper Genesee R. of Great Lakes basin) has intermediate scale counts (usually 63–70 lateral scales). **SIMILAR SPECIES:** (1) Other darters *lack* fusion of skin on rear of lip to skin on snout. (2) See Rock Darter, *E. rupestre* (Pl. 48).

TUCKASEGEE DARTER *Etheostoma gutselli* Not shown

IDENTIFICATION: Similar to Greenside Darter, *E. blennioides*, but has premaxillary frenum, usually no scales on opercle. To 5 in. (12.5 cm). **RANGE:** Little Tennessee and Pigeon river systems, NC and TN. Uncommon. **HABITAT:** Fast rocky riffles of creeks and small rivers. **REMARKS:** Hybridizes with *E. b. newmanii* in Hiwassee R. drainage; may be subspecies of Greenside Darter. **SIMILAR SPECIES:** (1) See Greenside Darter, *E. blennioides* (Pl. 48).

ROCK DARTER *Etheostoma rupestre* Pl. 48

IDENTIFICATION: Similar to Greenside Darter, *E. blennioides*, but smaller; *lacks* scales on cheek, fusion of skin at rear of upper lip to skin of snout. Has 46–65 lateral scales. To 3¼ in. (8.3 cm). **RANGE:** Mobile Bay drainage, GA, AL, MS, and se. TN. Common, especially in rapids near Fall Line. **HABITAT:** Fast rocky riffles of creeks and small to medium rivers. **SIMILAR SPECIES:** (1) See Greenside Darter, *E. blennioides* (Pl. 48).

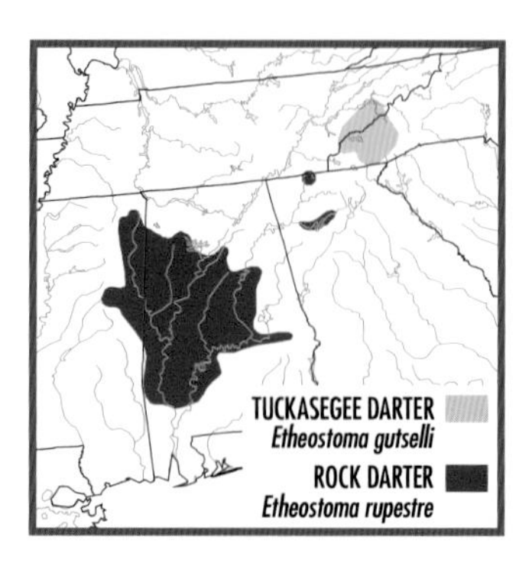

HARLEQUIN DARTER *Etheostoma histrio* Pl. 48

IDENTIFICATION: *Two large dark brown to green spots* on caudal fin base. *Many dark brown or black specks* on yellow belly and underside of head. Yellow to green above, 6–7 dark brown saddles; 7–11 dark brown blotches along side. First dorsal fin clear with dark red edge; dusky green on large male. Black teardrop. Breeding male is emerald green with brown and black mottling. Broadly joined branchiostegal membranes (Fig. 56). Complete lateral line; 45–58 lateral scales. To 3 in. (7.7 cm). **RANGE:** Mostly confined to Former Mississippi Embayment and Gulf Slope drainages from w. KY and se. MO to LA, and from Escambia R., AL and FL, to Neches R., TX; Red and Arkansas drainages west to e. OK and TX. Isolated populations in Wabash R. drainage, IL and IN, and Green R. system, KY. Widely distributed but generally uncommon. **HABITAT:** Sand and gravel runs of small to medium rivers; usually near debris.

SEAGREEN DARTER *Etheostoma thalassinum* Pl. 48

IDENTIFICATION: Light brown, 7 dark brown saddles; about *7 small dark brown blotches* (often Ws) along side. On large male, blue-green bars pass through blotches on side and encircle body near caudal fin. Brown mottling on upper side. *Red edge,* dusky black base on 1st dorsal fin. Blue anal and pelvic fins on breeding male. Black teardrop. *Broadly joined branchiostegal membranes* (Fig. 56). Usually 9–10 dorsal spines. Complete lateral line; 38–50 lateral scales. To 3¼ in. (8 cm). **RANGE:** Santee R. drainage, NC and SC. Common. **HABITAT:** Rocky riffles of creeks and small to medium rivers. **SIMILAR SPECIES:** (1) See Turquoise Darter, *E. inscriptum*. (2) Swannanoa Darter, *E. swannanoa*, (Pl. 48), has *8–9 black, oval to round blotches* on side, usually 12 dorsal spines.

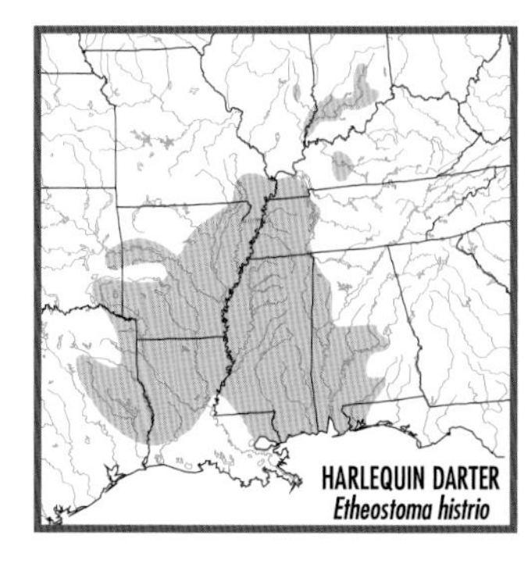

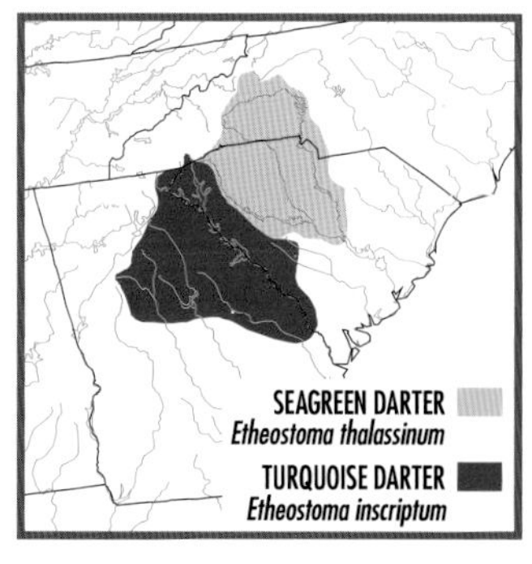

TURQUOISE DARTER *Etheostoma inscriptum* Not shown

IDENTIFICATION: Nearly identical to Seagreen Darter, *E. thalassinum*, but has *6 dark saddles; horizontal rows of small red spots* on side of large male. Has 39–61 lateral scales. To 3¼ in. (8 cm). **RANGE:** Edisto, Savannah, and Altamaha river drainages, NC, SC, and GA. Common. **HABITAT:** Rocky riffles of creeks and small to medium rivers. **SIMILAR SPECIES:** (1) See Seagreen Darter, *E. thalassinum* (Pl. 48).

SWANNANOA DARTER *Etheostoma swannanoa* Pl. 48

IDENTIFICATION: *Rows of small red spots* on side (more prominent on male). Light brown above, 6 black saddles; 8–9 black blotches on side *vertically oval* at front, round at rear. *Orange edge and base* on dusky 1st dorsal fin. Blue-green anal and pelvic fins on breeding male. Dusky to black teardrop. Broadly joined branchiostegal membranes (Fig. 56). Complete lateral line; 46–62 lateral scales; usually 12 dorsal spines. To 3½ in. (9 cm). **RANGE:** Upper Tennessee R. drainage, VA, NC, and TN. Common, especially in French Broad and Little Pigeon river systems. **HABITAT:** Swift rubble riffles of small to medium rivers. **SIMILAR SPECIES:** See Pl. 48. (1) Blenny Darter, *E. blennius*, has 4 large saddles, strongly tapering body. (2) Seagreen Darter, *E. thalassinum*, and (3) Turquoise Darter, *E. inscriptum,* have small dark brown blotches (*not* vertically oval) or Ws along side, usually 9–10 dorsal spines. (4) Greenside Darter, *E. blennioides*, has Ws along side, *green edge* (male) on 1st dorsal fin.

BLENNY DARTER *Etheostoma blennius* Pl. 48

IDENTIFICATION: *Body thick at front,* strongly tapering to *narrow caudal peduncle. Four large dark brown to black saddles* extend down to meet dark blotches along side; white to yellow below. Red edge on 1st dorsal fin. Breeding male has red spots on upper side, bright blue lips, red-purple dorsal fins, blue-black breast and anal and pelvic fins.

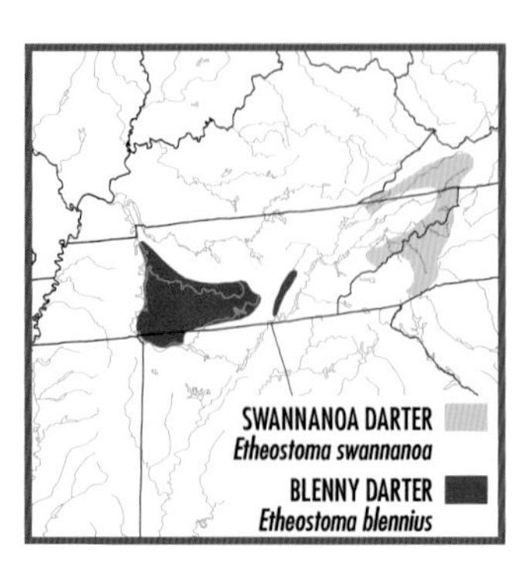

Teardrop usually prominent. Broadly joined branchiostegal membranes (Fig. 56). Complete lateral line; 40–51 lateral scales. To 3¼ in. (8.3 cm). **RANGE:** West- and south-flowing tributaries of Tennessee R. in cen. TN and n. AL (east to Sequatchie R. system). Locally common. **HABITAT:** Fast, rocky riffles of creeks and small to medium rivers. **REMARKS:** Two subspecies. *E. b. sequatchiense*, endemic to Sequatchie R., has 40–44 lateral scales, dark lines on side of large adult; usually has scales on upper opercle. *E. b. blennius* has 42–51 lateral scales, unscaled opercle, no dark lines on side; occupies rest of range except Elk R., s.-cen. TN and n. AL, where intergrades occur. **SIMILAR SPECIES:** See Pl. 48. (1) Swannanoa Darter, *E. swannanoa,* has *6 saddles,* less tapering body. (2) Other darters with large dark saddles are not in Tennessee R. system.

CANDY DARTER *Etheostoma osburni* Pl. 48

IDENTIFICATION: Dark green above, *5 large dark brown saddles; 9–11 green bars* alternate with *orange interspaces* (much brighter on male) on side; yellow to orange below. Orange edge, middle green band on 1st dorsal fin. Bright green and orange breeding male may be most vivid freshwater fish in N. America. Broadly joined branchiostegal membranes (Fig. 56). Unscaled breast. Complete lateral line; *58–70 lateral scales.* To 4 in. (10 cm). **RANGE:** Kanawha R. drainage above Kanawha Falls, WV and VA. Fairly common. **HABITAT:** Fast rubble riffles of small to medium rivers. **SIMILAR SPECIES:** (1) See Kanawha Darter, *E. kanawhae* (Pl. 48). (2) Variegate Darter, *E. variatum* (Pl. 48), has *4 saddles, 48–60 lateral scales,* scales on rear of breast.

KANAWHA DARTER *Etheostoma kanawhae* Pl. 48

IDENTIFICATION: Nearly identical to Candy Darter, *E. osburni*, but has *48–58 lateral scales,* is less brilliantly colored. Has 5–6 large dark brown saddles. Large male has bright orange belly and branchioste-

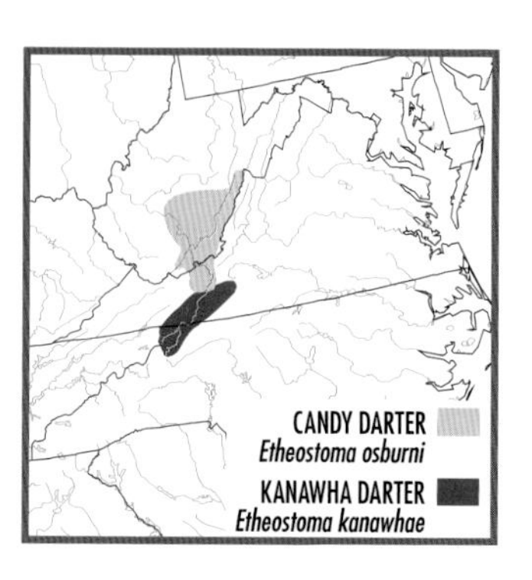

gal membranes. To 3½ in. (8.6 cm). **RANGE:** New R. drainage, VA and NC. Fairly common. **HABITAT:** Fast gravel and rubble riffles of small to medium rivers. **SIMILAR SPECIES:** (1) See Candy Darter, *E. osburni* (Pl. 48).

VARIEGATE DARTER *Etheostoma variatum* Pl. 48

IDENTIFICATION: *Four large brown saddles* angle down and forward to lateral line. *Green and orange bars* on side. Red edge, middle blue band, brown base on 1st dorsal fin. Breeding male is deep blue with orange or red belly, orange spots on side and fins. Broadly joined branchiostegal membranes (Fig. 56). Scales on rear of breast. Large eye (about equal to snout length). Complete lateral line; 48–60 lateral scales. To 4½ in. (11 cm). **RANGE:** Ohio R. basin from sw. NY to VA, e. KY, and s. IN. Only below Kanawha Falls in Kanawha R. system of WV. Common; sometimes abundant in clear streams. **HABITAT:** Swift gravel and rubble riffles of small to medium rivers. **SIMILAR SPECIES:** See Pl. 48. (1) See Missouri Saddled Darter, *E. tetrazonum*. (2) Candy Darter, *E. osburni*, and (3) Kanawha Darter, *E. kanawhae* (both restricted to Kanawha R. system), have *5–6 dark saddles*, unscaled breast.

MISSOURI SADDLED DARTER *Etheostoma tetrazonum* Pl. 48

IDENTIFICATION: Nearly identical to Variegate Darter, *E. variatum*, but has smaller eye, darker saddles; breeding male is blue-green, has orange belly, orange bars on side, red-orange edge on 1st dorsal fin. Has 46–57 lateral scales; usually 12 dorsal spines, 15 pectoral rays. To 3½ in. (9 cm). **RANGE:** Gasconade, Osage, and Moreau river drainages (Missouri R. tributaries), MO. Common. **HABITAT:** Fast gravel and rubble riffles of small to medium rivers. **SIMILAR SPECIES:** See (1) Variegate Darter, *E. variatum* (Pl. 48), and (2) Meramec Saddled Darter, *E. erythrozonum*. (3) Arkansas Saddled Darter, *E. euzonum* (Pl. 48), has 54–71 lateral scales.

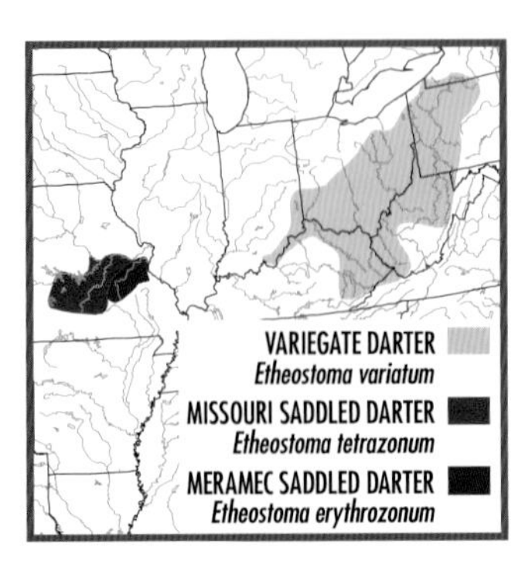

MERAMEC SADDLED DARTER *Etheostoma erythrozonum* **Not shown**

IDENTIFICATION: Nearly identical to Missouri Saddled Darter, *E. tetrazonum*, but usually has 13 dorsal spines, 16 pectoral rays; breeding male *lacks* blue-green body; has *red-orange stripe* along lower side to anal fin, then *red-orange blotches* to caudal fin (Missouri Saddled Darter has bars), 2 rows of red-orange spots on blue anal fin. Has 43–55 lateral scales. To 3¼ in. (8.5 cm). **RANGE:** Meramec R. drainage, MO. Common. **HABITAT:** Fast gravel and rubble riffles of small to medium rivers. **SIMILAR SPECIES:** (1) See Missouri Saddled Darter, *E. tetrazonum* (Pl. 48).

ARKANSAS SADDLED DARTER *Etheostoma euzonum* **Pl. 48**

IDENTIFICATION: Olive above; *4 large dark brown saddles* angle down and forward to lateral line; *green and orange spots* on upper side; yellow lower side. Red-orange edge, middle blue band, dusky green base on 1st dorsal fin. *Large head and eye.* Broadly joined branchiostegal membranes (Fig. 56). Complete lateral line; 54–71 lateral scales. To 4¾ in. (12 cm). **RANGE:** White R. drainage (including Current and Black rivers), MO and AR. Uncommon. **HABITAT:** Deep, fast gravel and rubble riffles of small to medium rivers. **REMARKS:** Two subspecies. *E. e. erizonum*, in Current R. system, has scales on cheek, long snout, relatively small eye. *E. e. euzonum*, occupying rest of range except Black R. system where intergrades occur, lacks scales on cheek, has blunter snout, larger eye. **SIMILAR SPECIES:** (1) Missouri Saddled Darter, *E. tetrazonum* (Pl. 48) and (2) Meramec Saddled Darter, *E. erythrozonum*, have 43–57 lateral scales.

BANDED DARTER *Etheostoma zonale* **Pl. 48**

IDENTIFICATION: Has *9–13 large dark green bars on side* extending *onto belly* and under caudal peduncle to *join* those of other side. Yellow-green, 6 dark saddles; 2 large yellow spots on caudal fin base. Nar-

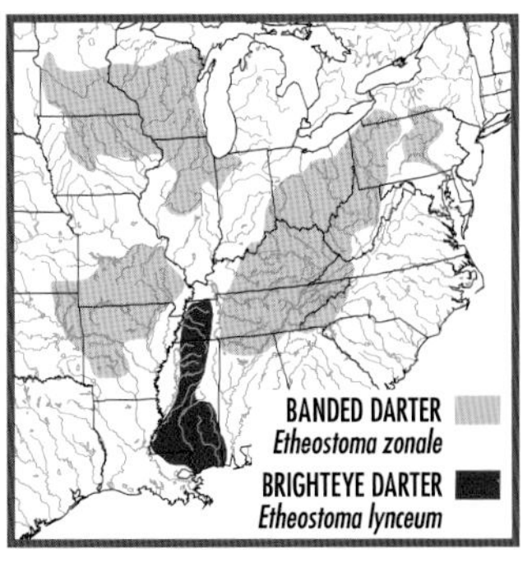

row teardrop, often broken into 2 black spots. Green edge, red base on 1st dorsal fin (colors brighter on male). Wide premaxillary frenum, moderate snout, broadly joined branchiostegal membranes (Fig. 56), 5 (rarely 6) branchiostegal rays. Complete lateral line; 38–63 (usually 45 or more) lateral scales. To 3 in. (7.8 cm). **RANGE:** Three disjunct areas: Ohio R. basin from sw. NY to e. IN and south to n. GA and n. AL; Lake Michigan and Mississippi R. basins from nw. MI to MN, south to nw. IN and cen. IL; Ozark-Ouachita drainages of s. MO, se. KS, AR, and e. OK . Introduced on Atlantic Slope into Susquehanna R., NY, PA, and MD, and headwaters of Savannah R., SC. Common; locally abundant. **HABITAT:** Rocky riffles of creeks and small to medium rivers. **SIMILAR SPECIES:** See Pl. 48. (1) See Brighteye Darter, *E. lynceum*. (2) Harlequin Darter, *E. histrio*, has 2 large dark spots at front of caudal fin. In (3) Greenside Darter, *E. blennioides*, (4) Rock Darter, *E. rupestre*, and (5) Emerald Darter, *E. baileyi*, green bars on side *rarely join* (except occasionally on large male) those of other side; Greenside and Rock darters have *fewer* than 9 bars; Emerald Darter has much blunter snout, usually 8 dark saddles on back.

BRIGHTEYE DARTER *Etheostoma lynceum* **Pl. 48**

IDENTIFICATION: Similar to Banded Darter, *E. zonale*, but green bars on side *darker, with interspaces about as wide as bars* (narrower interspaces in Banded Darter); usually has *fewer than 45 lateral scales.* To 2½ in. (6.4 cm). **RANGE:** Tributaries of Mississippi R. on Former Mississippi Embayment, w. KY, w. TN, MS, and LA. Gulf drainages from Escatawpa R., AL, to Mississippi R., LA. Common. **HABITAT:** Rocky riffles of creeks and small rivers; near debris in sand and gravel runs. **SIMILAR SPECIES:** (1) See Banded Darter, *E. zonale* (Pl. 48).

NEXT 25 SPECIES (snubnose darters through Holiday Darter, *E. brevirostrum*): *Extremely blunt snout; broadly joined branchiostegal membranes* (Fig. 56); *no or narrow premaxillary frenum* (Fig. 57); *8–9 dark saddles.* Most have *bright red spot* at front of 1st dorsal fin. Complete lateral line; usually 10–12 dorsal spines, 10–12 dorsal rays, 6–8 anal rays, 14 pectoral rays; usually *5 branchiostegal rays* (6 in Coosa Darter, *E. coosae*)—most other darters have 6; usually *9 preoperculomandibular pores*—most other darters have 10. Snubnose darters are among the most colorful N. American fishes. All are restricted to se. U.S. in southern tributaries of Ohio R. (Kentucky R., KY, and downstream), eastern tributaries of Mississippi R. south of Ohio R., and Gulf drainages east of Mississippi R. (to Choctawhatchee R., AL and FL). None is on Atlantic Slope, north of Ohio R., or west of Mississippi R.

NEXT 9 SPECIES (through Eastrim Darter, *E. orientale*): Narrow premaxillary frenum, no vomerine teeth.

EMERALD DARTER *Etheostoma baileyi* Pl. 48

IDENTIFICATION: Olive above, 7–10 (usually 8) dark green saddles; *7–11 small emerald green squares* along side (expanded into bars on breeding male); white to yellow below; bright red spot at front, *red edge* on 1st dorsal fin; dusky teardrop. Unscaled breast. Has 45–56 lateral scales. To 2¼ in. (5.6 cm). **RANGE:** Upper Kentucky and upper Cumberland river drainages, e. KY and ne. TN. Locally common. **HABITAT:** Rocky pools and runs, sometimes riffles, of creeks and small to medium rivers. **SIMILAR SPECIES:** (1) Banded Darter, *E. zonale*, and (2) Brighteye Darter, *E. lynceum* (both Pl. 48), have 6 dorsal saddles, *large dark green bars* on side extending onto belly and joining those of other side, more pointed snout.

KENTUCKY SNUBNOSE DARTER *Etheostoma rafinesquei* Pl. 49

IDENTIFICATION: *Crosshatching* (created by dark scale edges), 7–10 *(usually 8) short black bars* on side. First dorsal fin has red spot at front, red edge. Yellow-brown; 7–9 dark saddles, 1st saddle covering most of nape; dusky teardrop. Breeding male has bright red dorsal fins (green base on 1st dorsal fin deepest at front); blue-green head, anal and pelvic fins; orange breast and belly. Rear half of breast usually scaled. Has 37–43 lateral scales. To 2½ in. (6.5 cm). **RANGE:** Upper Green and Gasper river (Barren R. tributary) systems, KY. Locally common. **HABITAT:** Rocky pools, runs, and adjacent riffles of creeks and small rivers. **SIMILAR SPECIES:** (1) Splendid Darter, *E. barrenense* (Pl. 49), only other snubnose darter in Green R. system, *lacks* crosshatching, has black blotches along side fused into stripe, 1st dorsal saddle only on front half of nape, 42–49 lateral scales.

SPLENDID DARTER *Etheostoma barrenense* Pl. 49

IDENTIFICATION: Has *7–10 black blotches on side fused into stripe.* Yellow-brown above; 7–9 black saddles, 1st only on front half of nape; red

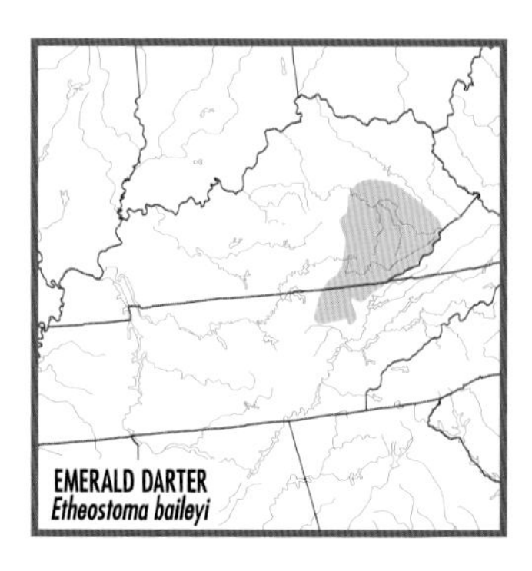

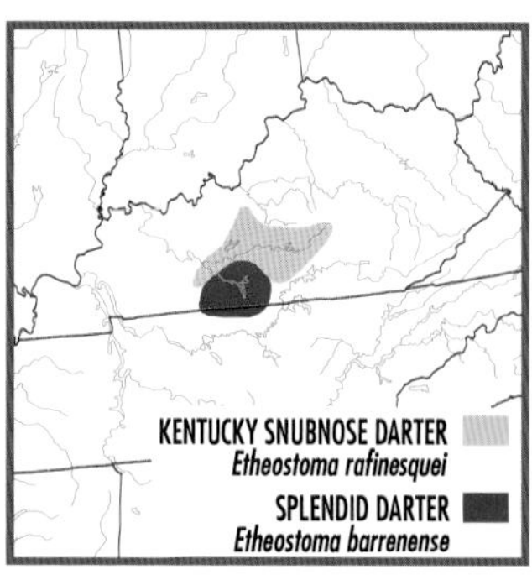

spot at front of 1st dorsal fin; thin, dusky teardrop. Breeding male is *red above and below* black stripe along side; has green head, blue anal and pelvic fins, red edge and green base on 1st dorsal fin, red middle band in 2d dorsal fin. Unscaled breast. Has 42–49 lateral scales. To 2½ in. (6.4 cm). **RANGE:** Barren R. system (except Gasper R.), s.-cen. KY and n.-cen. TN. Common. **HABITAT:** Rocky pools and adjacent riffles of creeks and small rivers. **SIMILAR SPECIES:** (1) Blackside Snubnose Darter, *E. duryi* (Pl. 49), and (2) Saffron Darter, *E. flavum*, lack premaxillary frenum, have orange or yellow (not red) on side. (3) Kentucky Snubnose Darter, *E. rafinesquei* (Pl. 49), has *crosshatching and dark bars* on side, 37–43 lateral scales.

TENNESSEE SNUBNOSE DARTER *Etheostoma simoterum* Pl. 49

IDENTIFICATION: *Many small red spots* (brightest on large male) in short rows on upper side; bright red spot at front, *red edge* on 1st dorsal fin; 2d dorsal fin red with black base. Light green or brown above, 6–10 (usually 8–9) black saddles; 8–9 black blotches fused into *wavy stripe* along side; dusky white to orange below; large black teardrop. Breeding male has black breast, orange belly, mostly red dorsal fins; blue to blue-green snout, anal and pelvic fins. Fully scaled nape, usually unscaled breast. Has 44–54 lateral scales. To 2¾ in. (7.3 cm). **RANGE:** Upper Holston R. system (Tennessee R. drainage), e. TN and w. VA; McClure R. and Russell Fork (Big Sandy R. drainage), w. VA and extreme se. KY. Common. **HABITAT:** Current-swept rocky pools and adjacent riffles of creeks and small to medium rivers. **SIMILAR SPECIES:** (1) Tennessee Snubnose Darter lookalikes: Next 5 species (through Eastrim Darter, *E. orientale*) are most easily distinguished from one another and Tennessee Snubnose Darter as breeding males. It often is necessary to rely on microscopic examination of specimens and geography to identify them. (2) Blackside Snubnose Darter, *E. duryi* (Pl. 49), and (3) Saffron Darter, *E. flavum*, *lack* red spots on upper side, premaxillary frenum.

CUMBERLAND SNUBNOSE DARTER *Etheostoma atripinne* Not shown

IDENTIFICATION: Similar to Tennessee Snubnose Darter, *E. simoterum*, but is mostly *green below* (orange only along lower side); has *olive green egg-shaped blotches* along side, 51–63 lateral scales. Breeding male has *light green to turquoise breast*, bright red spots and black bands on 1st dorsal fin. To 3 in. (7.7 cm). **RANGE:** Tributaries of Cumberland R. system in Nashville Basin, from near mouth of Roaring R. to Mansker Creek (ne. Nashville), TN. Abundant. **HABITAT:** Current-swept rocky pools and adjacent riffles of creeks and small to medium rivers. **SIMILAR SPECIES:** (1) See Tennessee Snubnose Darter, *E. simoterum* (Pl. 49).

TENNESSEE DARTER *Etheostoma tennesseense* Not shown

IDENTIFICATION: Similar to Tennessee Snubnose Darter, *E. simoterum*, but has *olive green egg-shaped blotches* along side. Breeding male has *orange breast*; 1st dorsal fin with bright red spot at front, then red dashes and oblong spots, red wash at rear of fin contiguous with red edge. Has 44–58 lateral scales. To 3 in. (7.4 cm). **RANGE:** Tennessee R. system from w. VA to Hardin Creek (Hardin Co.), w. TN.; absent in upper Holston R. system (North, South, and Middle forks). Also in upper Bluestone R. system (New-Ohio river drainage), w. VA. Common. **HABITAT:** Current-swept rocky pools and adjacent riffles of creeks and small to medium rivers. **SIMILAR SPECIES:** (1) See Tennessee Snubnose Darter, *E. simoterum* (Pl. 49).

DUCK DARTER *Etheostoma planasaxatile* Not shown

IDENTIFICATION: Similar to Tennessee Snubnose Darter, *E. simoterum*, but has *olive green vertical blotches* along side. Breeding male has orange breast, green belly, bright red spots and black vermiculations on 1st dorsal fin. Has 49–60 lateral scales. To 2½ in. (6.4 cm). **RANGE:** Duck R. system, TN. Abundant. **HABITAT:** Current-swept rocky pools and adjacent riffles of creeks and small to medium rivers. **SIMILAR SPECIES:** (1) See Tennessee Snubnose Darter, *E. simoterum* (Pl. 49).

WESTRIM DARTER *Etheostoma occidentale* Not shown

IDENTIFICATION: Similar to Tennessee Snubnose Darter, *E. simoterum*, but has *olive green egg-shaped blotches* along side; breeding male has orange breast and belly, bright red spots and black horizontal bands on 1st dorsal fin. Fully scaled nape. Has 49–60 lateral scales. To 3 in. (7.8 cm). **RANGE:** Tributaries of Cumberland R. from Whites Creek near Nashville, TN, to Little R., KY. Common. **HABITAT:** Current-swept rocky pools and adjacent riffles of creeks and small to medium rivers.

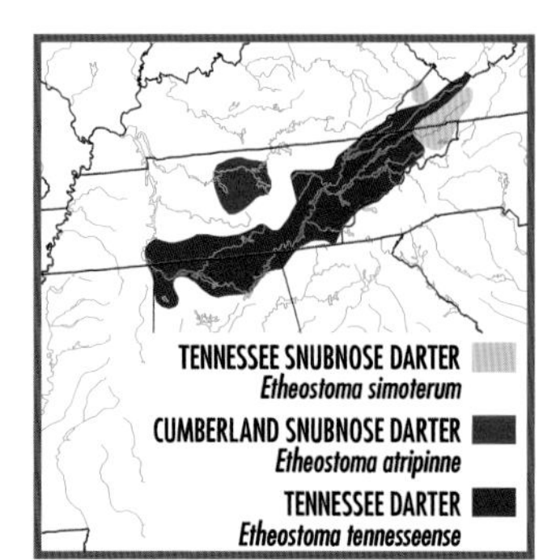

SIMILAR SPECIES: (1) See Tennessee Snubnose Darter, *E. simoterum* (Pl. 49).

EASTRIM DARTER *Etheostoma orientale* **Not shown**

IDENTIFICATION: Similar to Tennessee Snubnose Darter, *E. simoterum*, but has *brown or black egg-shaped blotches* along side (often fused into stripe); breeding male has orange breast and belly, bright red spots and black vermiculations on 1st dorsal fin. Mostly unscaled nape. Has 50–61 lateral scales. To 2½ in. (6.5 cm). **RANGE:** Tributaries of Cumberland R. from Fishing Creek, KY, to just below Obey R., TN. Common. **HABITAT:** Current-swept rocky pools and adjacent riffles of creeks and small to medium rivers. **SIMILAR SPECIES:** (1) See Tennessee Snubnose Darter, *E. simoterum* (Pl. 49).

NEXT 16 SPECIES (through Holiday Darter, *E. brevirostrum*): No premaxillary frenum, usually vomerine teeth.

BLACKSIDE SNUBNOSE DARTER *Etheostoma duryi* **Pl. 49**

IDENTIFICATION: Has *9–10 black blotches along side fused into stripe.* First dorsal fin usually has red spot at front, red edge (faint on female), wavy red and black lines. Light olive-brown to orange above, usually 8 black saddles; green or gray lips; dusky teardrop. Breeding male has *orange upper and lower sides;* 2 red-orange spots on caudal fin base; red dorsal fins. Unscaled breast. Has 38–54 lateral scales. To 2¾ in. (7.2 cm). **RANGE:** Tennessee R. drainage, TN, GA, and AL (absent in extreme upper Tennessee; occurs east to near Knoxville). Common. **HABITAT:** Rocky pools and adjacent riffles of creeks and small rivers. **SIMILAR SPECIES:** (1) See Saffron Darter, *E. flavum.* (2) Tennessee Snubnose Darter, *E. simoterum* (Pl. 49), and lookalike species have red spots on upper side, *little or no fusion* of dark blotches along side, premaxillary frenum. (3) Splendid Darter, *E. barrenense* (Pl. 49), has *bright red side* on large male, premaxillary frenum.

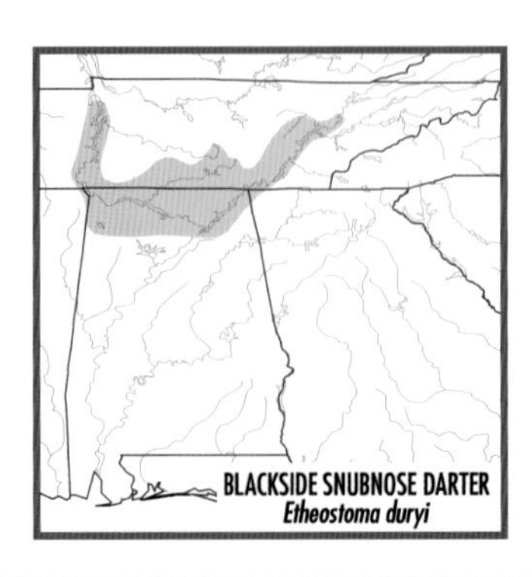

SAFFRON DARTER *Etheostoma flavum* **Not shown**

IDENTIFICATION: Similar to Blackside Snubnose Darter, *E. duryi*, but has *more discrete brown blotches* along side (less fused into stripe). Breeding male has *yellow-gold lips and upper and lower sides*, green caudal peduncle; yellow spots on caudal fin base; *no* red on fins. Has 42–59 lateral scales. To 2¾ in. (7 cm). **RANGE:** Lower Cumberland and lower Tennessee river drainages, KY and TN. Upstream in Cumberland R. drainage to Harpeth R.; upstream in Tennessee R. drainage (on east side of Tennessee R. only) to upper Duck R., upper Buffalo R., and Indian Creek (Wayne Co., TN). Common; locally abundant. **HABITAT:** Rocky pools and adjacent riffles of headwaters, creeks, and small rivers. **SIMILAR SPECIES:** (1) See Blackside Snubnose Darter, *E. duryi* (Pl. 49).

CHERRY DARTER *Etheostoma etnieri* **Pl. 49**

IDENTIFICATION: *Black lines* along upper side of body. First dorsal fin has small red spot at front, thin black bands. Olive above, 7–9 black saddles; 8–10 black blotches along side; white to red below; thin black teardrop. Breeding male has green head and breast; bright red belly, side, and 2d dorsal, anal, and caudal fins. Premaxillary frenum absent or narrow. Unscaled breast. Has 45–57 lateral scales. To 3 in. (7.7 cm). **RANGE:** Upper Caney Fork (of Cumberland R.) system, cen. TN. Locally common. **HABITAT:** Bedrock pools and rocky riffles of creeks and small rivers. **SIMILAR SPECIES:** (1) Other snubnose darters (Pls. 48 & 49) *lack* black lines along upper side.

FIREBELLY DARTER *Etheostoma pyrrhogaster* **Pl. 49**

IDENTIFICATION: First dorsal fin has red spot at front, *red band* near edge (faint on female). *Thin brown stripe* above lateral line *interrupted* by 7–10 black or brown blotches (stripe often obscured on breeding male). Light brown above, 8–9 black saddles; teardrop faint or absent. Breeding male has *wide red band* on dorsal and anal fins; turquoise

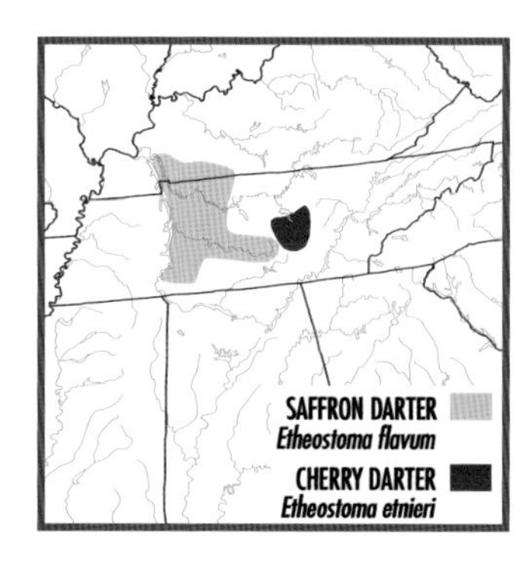

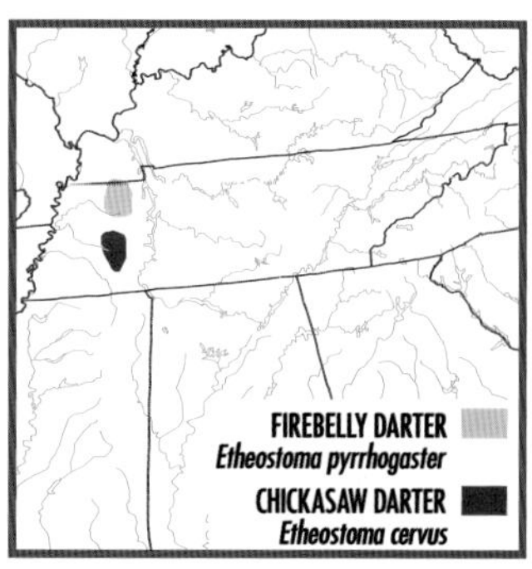

in dorsal, anal, and pelvic fins; *bright red body*; green head; black blotches on front half of side fused into black stripe. Scales (partially embedded) on rear half of breast. Has 39–47, usually 41–44, lateral scales. To 2¾ in. (7 cm). **RANGE:** Obion R. system, w. KY and w. TN. Uncommon. **HABITAT:** Sand- and gravel-bottomed runs and pools of headwaters, creeks, and small rivers. **SIMILAR SPECIES:** (1) See Chickasaw Darter, *E. cervus*. (2) Bandfin Darter, *E. zonistium* (Pl. 49), has *2 thin red bands* through middle of 1st and 2d dorsal fins.

CHICKASAW DARTER *Etheostoma cervus* **Not shown**

IDENTIFICATION: Nearly identical to Firebelly Darter, *E. pyrrhogaster*, but large male *lacks* bright turquoise in dorsal, anal, and pelvic fins (may have pale blue in fins), has straw-colored cheek and opercle, 37–45 (usually 39–41) lateral scales. To 2½ in. (6.2 cm). **RANGE:** Upper Forked Deer R. system, w. TN. Uncommon. **HABITAT:** Sandy runs and pools of headwaters and creeks; usually near woody debris or undercut banks. **SIMILAR SPECIES:** (1) See Firebelly Darter, *E. pyrrhogaster* (Pl. 49).

BANDFIN DARTER *Etheostoma zonistium* **Pl. 49**

IDENTIFICATION: First dorsal fin has red spot at front (see Remarks), blue edge, *2 red bands* (faint on female) through middle. *Thin brown stripe* above lateral line *interrupted* by 7–10 black or brown blotches (stripe often obscured on breeding male). Yellow-brown above, 8 dark saddles; no or thin teardrop. Breeding male has blue edge, 2 bright red bands on 1st dorsal fin; 1–2 red bands on 2d dorsal fin; *bright red body*; green head; blue anal and pelvic fins; black blotches on front half of side fused into black stripe. Scales (partly embedded) on rear half of breast. Has 40–52 lateral scales. To 2¾ in. (7.1 cm). **RANGE:** Tributaries of lower Tennessee R. system, KY, TN, ne. MS, and nw. AL, upstream to Bear Creek system; confined to western tributaries except in Land Between the Lakes, KY, and Hardin Co., TN. Also in Spring Creek (Hatchie R. drainage), w. TN, and Hubbard Creek (extreme upper Sipsey R. system), AL. Common. **HABITAT:** Sand- and gravel-bottomed pools of headwaters, creeks, and small rivers. **REMARKS:** Individuals in upper Bear Creek and Hubbard Creek, AL, lack red spot at front of 1st dorsal fin. **SIMILAR SPECIES:** (1) Firebelly Darter, *E. pyrrhogaster* (Pl. 49), and (2) Chickasaw Darter, *E. cervus*, have *1 red band* near edge of 1st dorsal fin on female, *1 wide red band* on male.

COASTAL DARTER *Etheostoma colorosum* **Pl. 49**

IDENTIFICATION: Blue edge, red middle band, black at base of 1st and 2d dorsal fins (colors faint on female). Red band only on last 3–5 membranes of 1st dorsal fin, *no red spot* at front of fin. Yellow-brown above, 8 dark brown saddles; *thin red-brown stripe* above lateral line interrupted by 7–9 brown round blotches along side; white to orange below; dusky teardrop. Breeding male has *high, arched 1st dorsal fin*; blue edge, bright red band on each dorsal fin; blue head; blue anal, caudal, and pelvic fins; bright orange lower side and caudal peduncle. Unscaled breast. Has 39–50 lateral scales. To 2¾ in. (7 cm). **RANGE:** Gulf Coastal Plain drainages from Choctawhatchee R. to Perdido R., AL and FL. Common. **HABITAT:** Sand- and gravel-bottomed pools of headwaters, creeks, and small rivers. **SIMILAR SPECIES:** See (1) Alabama Darter, *E. ramseyi,* (2) Tallapoosa Darter, *E. tallapoosae* (Pl. 49), (3) Yazoo Darter, *E. raneyi* (Pl. 49), and (4) Tombigbee Darter, *E. lachneri.*

ALABAMA DARTER *Etheostoma ramseyi* Not shown

IDENTIFICATION: Nearly identical to Coastal Darter, *E. colorosum*, but has *bold orange to russet dashes* above lateral line, series of blue-green round blotches along side *connected* to dorsal saddles. Has 40–53 lateral scales. To 2½ in. (6.1 cm). **RANGE:** Alabama R. system, AL; mostly below Fall Line, above Fall Line in Cahaba R. Common. **HABITAT:** Sand- and gravel-bottomed pools of headwaters, creeks, and small rivers. **SIMILAR SPECIES:** See Pl. 49. See (1) Coastal Darter, *E. colorosum,* (2) Tallapoosa Darter, *E. tallapoosae*, (3) Yazoo Darter, *E. raneyi*, and (4) Tombigbee Darter, *E. lachneri.*

TALLAPOOSA DARTER *Etheostoma tallapoosae* Pl. 49

IDENTIFICATION: Similar to Coastal Darter, *E. colorosum*, but *lacks* thin red-brown stripe above lateral line; has *7–10 large, dark brown blotches* along side, *red band* (often fading to brown at front) on 1st dorsal fin *extending to front of fin* (best developed on breeding male). Has 44–57 lateral scales. To 3 in. (7.7 cm). **RANGE:** Above Fall Line in Tallapoosa R. system, GA and AL. Locally common. **HABITAT:** Bedrock pools and rocky riffles of creeks and small rivers. **SIMILAR SPECIES:**

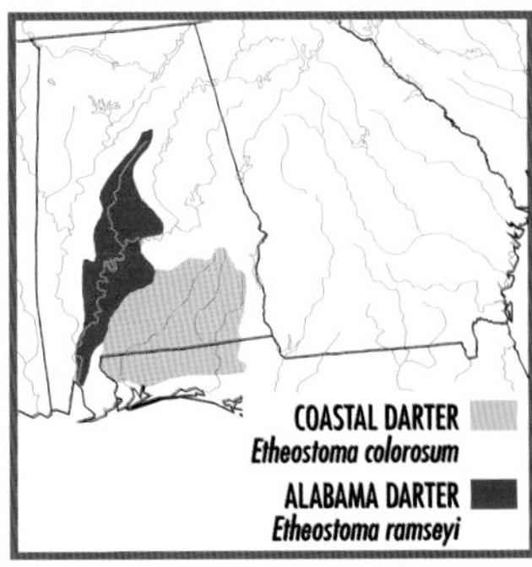

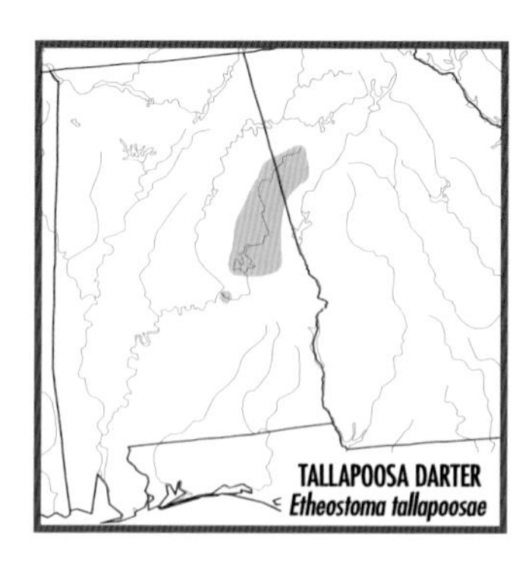

See (1) Coastal Darter, *E. colorosum* (Pl. 49), (2) Alabama Darter, *E. ramseyi*, (3) Yazoo Darter, *E. raneyi* (Pl. 49), and (4) Tombigbee Darter, *E. lachneri.*

YAZOO DARTER *Etheostoma raneyi* Pl. 49

IDENTIFICATION: Similar to Coastal Darter, *E. colorosum*, but has series of *elongated brown blotches* along lateral line, usually fused at front; blotches barely extend below lateral line except on caudal peduncle. Has 41–53 lateral scales. To 2¼ in. (6 cm). **RANGE:** Upper Yazoo R. drainage, MS. Uncommon. **HABITAT:** Sandy pools of headwaters and creeks. **SIMILAR SPECIES:** See (1) Coastal Darter, *E. colorosum* (Pl. 49), (2) Alabama Darter, *E. ramseyi*, (3) Tallapoosa Darter, *E. tallapoosae* (Pl. 49), and (4) Tombigbee Darter, *E. lachneri.*

TOMBIGBEE DARTER *Etheostoma lachneri* Not shown

IDENTIFICATION: Similar to Coastal Darter, *E. colorosum*, but has *orange* between *large green bars* along side; bars best developed on rear half of side. Has 39–51 lateral scales. To 2½ in. (6.2 cm). **RANGE:** Tom-

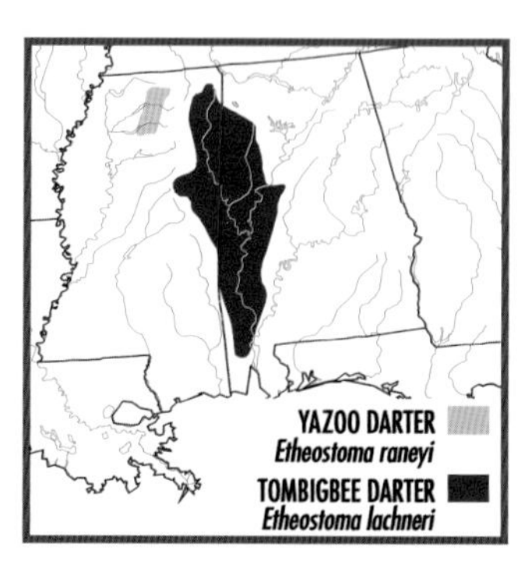

bigbee R. system, AL and MS. Common. **HABITAT:** Sand- and rock-bottomed pools of headwaters, creeks, and small rivers. **SIMILAR SPECIES:** See Pl. 49. See (1) Coastal Darter, *E. colorosum,* (2) Alabama Darter, *E. ramseyi*, (3) Tallapoosa Darter, *E. tallapoosae*, and (4) Yazoo Darter, *E. raneyi.*

WARRIOR DARTER *Etheostoma bellator* Pl. 49

IDENTIFICATION: First dorsal fin has red spot at front, *red band* near edge (faint on female). Yellow-brown above, 8 dark brown saddles; dusky teardrop. Interrupted thin red-brown stripe along side above lateral line, series of black blotches (sometimes confluent) below lateral line. Breeding male has bright red edge on 1st dorsal fin; middle red band on 2d dorsal fin; *bright red-orange narrow stripe* on lower side and caudal peduncle; blue-green head and anal, caudal, and pelvic fins. Unscaled breast. Has 41–55 lateral scales. To 2¾ in. (7 cm). **RANGE:** Above Fall Line in Black Warrior R. system, AL. Locally common. **HABITAT:** Bedrock pools and rocky riffles of creeks and small rivers. **SIMILAR SPECIES:** (1) See Vermilion Darter, *E. chermocki.* (2) Other snubnose darters in Gulf drainages (Pl. 49) have *blue edge* on 1st dorsal fin.

VERMILION DARTER *Etheostoma chermocki* Not shown

IDENTIFICATION: Nearly identical to Warrior Darter, *E. bellator*, except breeding male has *broad* red to orange stripe along lower side—stripe often touching brown blotches on rear half of body; *broader red band* in 2d dorsal fin. Has 43–54 lateral scales. To 2¾ in. (7.2 cm). **RANGE:** Turkey Creek (Black Warrior R. system), Jefferson Co., AL. Uncommon in extremely small range; protected as an *endangered species.* **HABITAT:** Gravelly runs in headwaters and creeks; usually in current near vegetation. **SIMILAR SPECIES:** (1) See Warrior Darter, *E. bellator* (Pl. 49).

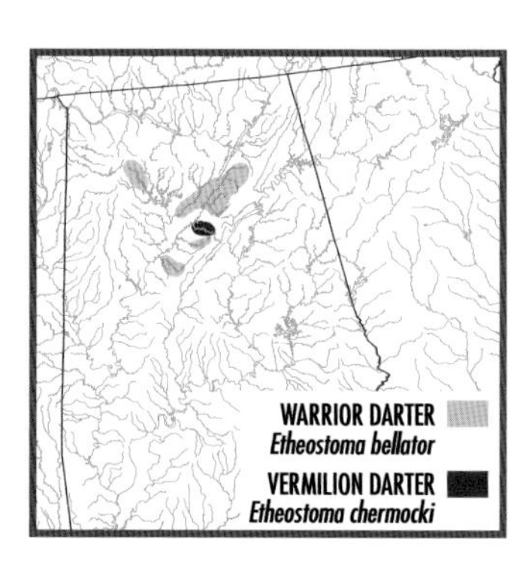

COOSA DARTER *Etheostoma coosae* Pl. 49

IDENTIFICATION: First dorsal fin has bright red spot at front, *1–2 narrow red bands*, blue-green edge (faint on female). Light brown to olive above, 8–9 dark saddles; black teardrop. Breeding male has *8–10 dark blue-brown bars* along side; bright blue edge, 2 red (or red-brown) bands on 1st dorsal fin; red 2d dorsal fin; green head; blue anal and pelvic fins. *Six branchiostegal rays*; unscaled breast. Has 44–58 lateral scales. To 2¾ in. (7.2 cm). **RANGE:** Coosa R. system, GA, AL, and se. TN. Fairly common. **HABITAT:** Rocky pools and adjacent riffles of creeks and small to medium rivers. **SIMILAR SPECIES:** (1) Other snubnose darters (Pls. 48 & 49) have *5 branchiostegal rays*. (2) See Cherokee Darter, *E. scotti*. (3) Holiday Darter, *E. brevirostrum* (Pl. 48), has yellow-white halos around red blotches on lower side, red band on blue anal fin.

CHEROKEE DARTER *Etheostoma scotti* Not shown

IDENTIFICATION: Nearly identical to Coosa Darter, *E. coosae*, but 1st dorsal fin of breeding male is *mostly brick red* (black at base, thin blue edge), lacks bands. Has 5 or 6 branchiostegal rays; 45–58 lateral scales. To 2¾ in. (7.1 cm). **RANGE:** Middle Etowah R. system, n. GA. Uncommon; protected as a *threatened species*. **HABITAT:** Rocky pools and adjacent riffles of creeks and small rivers. **SIMILAR SPECIES:** (1) See Coosa Darter, *E. coosae* (Pl. 49).

HOLIDAY DARTER *Etheostoma brevirostrum* Pl. 48

IDENTIFICATION: Red-brown blotches (appearing as interrupted brown stripe on juvenile) between *8–10 green bars* along side. First dorsal fin has red spot at front, blue edge (faint on female). Yellow-green above, 8 dorsal saddles; black teardrop. Breeding male has green head; bright blue edge, red band on each dorsal fin; short red-brown bars between wide green bars on upper side, *yellow-white halos around red blotches* on lower side, blue pelvic fin, *red band on blue anal fin.*

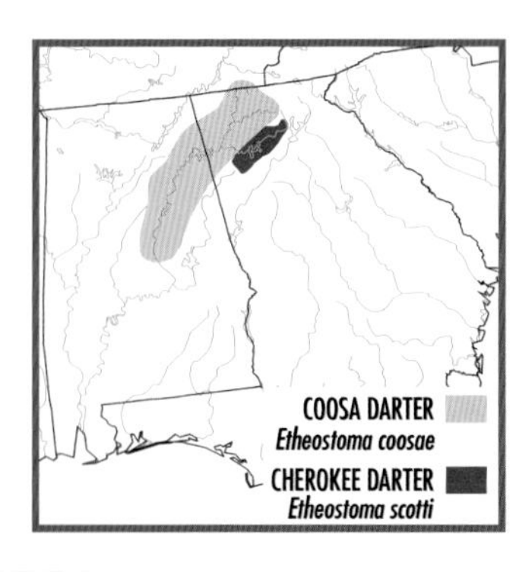

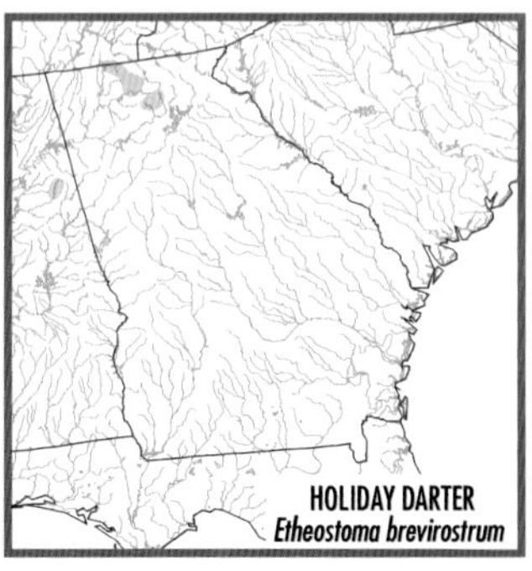

Unscaled breast. Has 42–54 lateral scales. To 2½ in. (6.4 cm). **RANGE:** Upper Coosa R. system, GA, AL, and se. TN. Uncommon. **HABITAT:** Rocky runs and pools, sometimes riffles, of creeks and small to medium rivers. **SIMILAR SPECIES:** (1) Coosa Darter, *E. coosae* (Pl. 49), and (2) Cherokee Darter, *E. scotti,* have *blue-brown bars* along side; *lack* yellow-white halos around red blotches on lower side, red band on blue anal fin.

ASHY DARTER *Etheostoma cinereum* Pl. 55

IDENTIFICATION: *Longitudinal rows of small brown spots* on upper side. *Row of small black rectangles* along side; oblique dusky bars extend down from rectangles. *Red lips* (absent in Tennessee R. system); red wavy lines and red-orange edge on 1st dorsal fin. Preorbital bars form large black V on snout. Breeding male has *greatly enlarged 2d dorsal fin*, red dorsal and pectoral fins, blue anal and pelvic fins. Long, sharp snout. Complete lateral line; 55–61 lateral scales. To 4¾ in. (12 cm). **RANGE:** Cumberland and Tennessee river drainages, VA, KY, TN, GA, and AL. Rare over most of range, extinct in VA, GA, and AL; common locally in a few streams. **HABITAT:** Rocky pools with current in small to medium rivers; usually near vegetation.

GREENBREAST DARTER *Etheostoma jordani* Pl. 50

IDENTIFICATION: Has *8–11 small black blotches* along side just below lateral line; less obvious blotches along upper side. *No* alternating dark and light lines on side of body. Brown above, 8–9 black saddles; white to blue below; dusky yellow fins; dusky teardrop. Dusky to black edge on 2d dorsal, caudal, and anal fins. Breeding male is *bright blue below*; has *small red spots* on side; black spot at front, red edge on 1st dorsal fin; blue edge, red band in 2d dorsal and caudal fins; blue anal and pelvic fins. 44–56 lateral scales, 14–17 transverse scales, 17–23 scales around caudal peduncle. To 3 in. (7.9 cm). **RANGE:** Above and below Fall Line in upper Alabama R. system (Mobile Bay drainage),

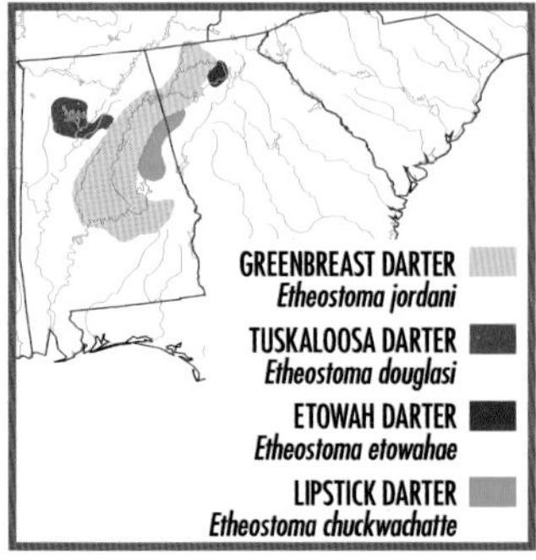

GA, AL, and se. TN. Replaced in upper Etowah R. system above Lake Allatoona, GA, by Etowah Darter, *E. etowahae*, and above Fall Line in Tallapoosa R. system, GA and AL, by Lipstick Darter, *E. chuckwachatte*. Common. **HABITAT:** Fast rocky riffles of creeks and small to medium rivers. **SIMILAR SPECIES:** (1) Greenbreast Darter lookalikes: Next 3 species (through Lipstick Darter, *E. chuckwachatte*) are easily distinguished from one another and Greenbreast Darter only as breeding males. It often is necessary to rely on microscopic examination of specimens and geography to identify them.

TUSKALOOSA DARTER *Etheostoma douglasi* **Not shown**

IDENTIFICATION: Nearly identical to Greenbreast Darter, *E. jordani*, but *lacks scales on opercle*; breeding male lacks red spots on side of body. To 3 in. (7.6 cm). **RANGE:** Upper Black Warrior R. system, AL. Common. **HABITAT:** Rocky riffles of creeks and small to medium rivers. **SIMILAR SPECIES:** (1) See Greenbreast Darter, *E. jordani* (Pl. 50).

ETOWAH DARTER *Etheostoma etowahae* **Not shown**

IDENTIFICATION: Similar to Greenbreast Darter, *E. jordani*, but usually has *11–14 transverse scales, 16–18 scales around caudal peduncle*; breeding male lacks red spots on side of body. To 2½ in. (6.5 cm). **RANGE:** Upper Etowah R. system, GA (above Lake Allatoona). Rare; protected as an *endangered species*. **HABITAT:** Rocky riffles of creeks and small to medium rivers. **SIMILAR SPECIES:** (1) See Greenbreast Darter, *E. jordani* (Pl. 50).

LIPSTICK DARTER *Etheostoma chuckwachatte* **Not shown**

IDENTIFICATION: Similar to Greenbreast Darter, *E. jordani*, but breeding male has *red lips, red band* in anal fin. To 2½ in. (6.8 cm). **RANGE:** Above Fall Line in Tallapoosa R. system, GA and AL. Common. **HABITAT:** Rocky riffles of creeks and small to medium rivers. **SIMILAR SPECIES:** (1) See Greenbreast Darter, *E. jordani* (Pl. 50).

NEXT 14 SPECIES (through Yoke Darter, *E. juliae*): Alternating *dark and light lines* on side of *deep, compressed body.*

REDLINE DARTER *Etheostoma rufilineatum* **Pl. 50**

IDENTIFICATION: *Black dashes* on cheek and opercle. Teardrop broken into *2 black spots. Cream-colored caudal fin base.* Pointed snout. Dark gray above; dark blotches on side. Black edge on 2d dorsal, caudal, and anal fins. Male has red spots on side, orange belly, blue breast, red-orange band on fins; female has brown spots on side, white to dusky blue breast, black spots on fins. Has 43–64 lateral scales. To 3½ in. (8.4 cm). **RANGE:** Cumberland (below Big South Fork) and Ten-

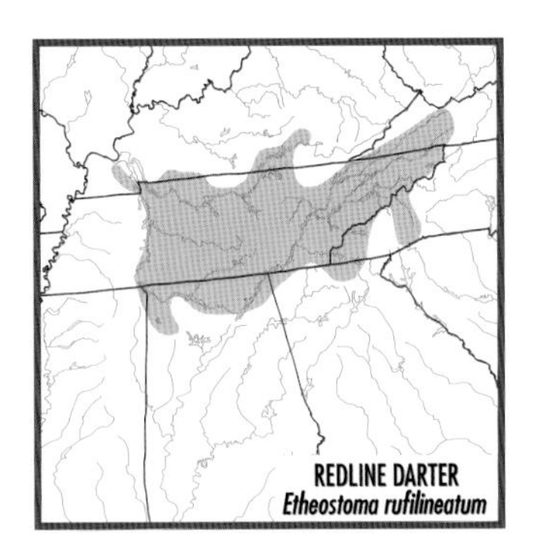

nessee river drainages, VA, KY, NC, TN, GA, AL, and MS. Abundant. **HABITAT:** Clear, fast rocky riffles of creeks and small to medium rivers. **SIMILAR SPECIES:** (1) Similar darters (Pl. 50) *lack* black dashes on cheek, cream-colored caudal fin base.

YELLOWCHEEK DARTER *Etheostoma moorei* Pl. 50

IDENTIFICATION: Dark gray above; black mottling on side; light gray to white below; *blue breast. Green base, red band* on 1st dorsal fin of male; other fins red, black at base. *Black spots* on fins of female. Black edge on dorsal, caudal, and anal fins; black teardrop on dusky cheek. Has 51–60 lateral scales. To 2¾ in. (7.2 cm). **RANGE:** Little Red R. system, n.-cen. AR. Uncommon. **HABITAT:** Fast rocky riffles of small to medium rivers. **SIMILAR SPECIES:** No similar species west of Mississippi R. (1) Bayou Darter, *E. rubrum* (Pl. 50), *lacks* green in fins, has white cheek, red spots on side. (2) Smallscale Darter, *E. microlepidum* (Pl. 50), has red spots on side of male.

NEXT 10 SPECIES (through Orangefin Darter, *E. bellum*): *Discrete, small bright red spots* on side of large male, *small brown spots* on side of female.

BAYOU DARTER *Etheostoma rubrum* Pl. 50

IDENTIFICATION: Dark gray above; large black teardrop on *white cheek. Cream-colored caudal fin base* followed by *2 large black spots*. Black edge on dorsal, caudal, and anal fins. Blue breast. Fins red on male, black spot at front of 1st dorsal fin; black-spotted on female. Has 45–55 lateral scales. To 2¼ in. (5.5 cm). **RANGE:** Bayou Pierre system, sw. MS. Uncommon; protected as a *threatened species*. **HABITAT:** Fast, rocky riffles of creeks and small to medium rivers. **SIMILAR SPECIES:** No similar species within range. (1) Closely related Yellowcheek

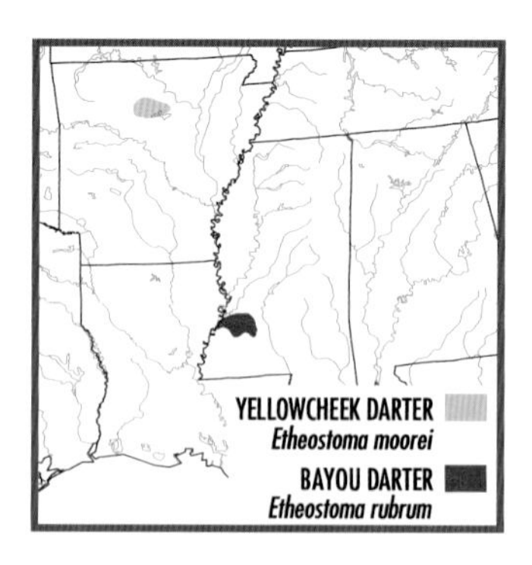

Darter, *E. moorei* (Pl. 50), *lacks* red spots on side; has dusky black cheek, green on 1st dorsal fin of male.

SMALLSCALE DARTER *Etheostoma microlepidum* Pl. 50

IDENTIFICATION: Large male is gray, with *bright green and orange dorsal, caudal, and anal fins*; has *black halos* around red spots on side. Female is brown, with *black spots* on fins. Black edge on 2d dorsal, caudal, and anal fins; black teardrop. Has 55–71 lateral scales. To 2¾ in. (7.2 cm). **RANGE:** Lower Cumberland R. drainage, w. KY and n.-cen. TN. Localized and uncommon. **HABITAT:** Clear, shallow gravel riffles of small rivers. **SIMILAR SPECIES:** (1) Bloodfin Darter, *E. sanguifluum*, *lacks* black edge on fins, green on fins of large male. (2) Redline Darter, *E. rufilineatum* (Pl. 50), has dark dashes on cheek and opercle; *lacks* green on fins.

SPOTTED DARTER *Etheostoma maculatum* Pl. 50

IDENTIFICATION: *Extremely compressed body; narrow pointed snout*; round caudal fin. Gray body; male has *black halos* around red spots on side, blue breast, gray or blue fins. No black edge on 2d dorsal,

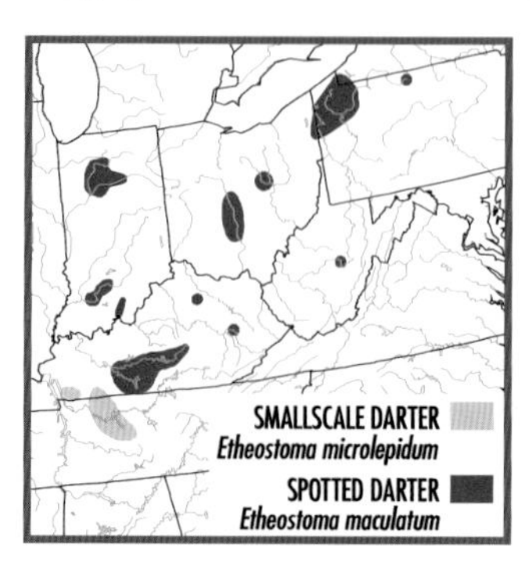

caudal, and anal fins. Female has black mottling on side, *black spots* on fins. Teardrop thin or absent. Has 53–68 lateral scales. To 3½ in. (9 cm). **RANGE:** Ohio R. basin from w. NY and PA to n. IN, and south to WV and KY. Extremely localized and uncommon. **HABITAT:** Fast, rocky riffles of small to medium rivers. **SIMILAR SPECIES:** See Pl. 50: (1) See Bloodfin Darter, *E. sanguifluum*. (2) Wounded Darter, *E. vulneratum*, has bright red spots on 1st dorsal fin; red 2d dorsal and caudal fins; black edge on 2d dorsal, caudal, and anal fins; straight-edged caudal fin. (3) Smallscale Darter, *E. microlepidum*, has green on fins of male. (4) Bluebreast Darter, *E. camurum*, often with Spotted Darter, is *less compressed*, has *blunter snout*, *lacks* black halos around red spots on side of male.

BLOODFIN DARTER *Etheostoma sanguifluum* — Not shown

IDENTIFICATION: Similar to Spotted Darter, *E. maculatum*, but large male has *2 bright red spots* at front, usually 1 at rear, of 1st dorsal fin; red fin membranes. Has 51–66 lateral scales. To 3½ in. (9 cm). **RANGE:** Middle Cumberland R. drainage from Rockcastle R., KY, to Caney Fork, TN. Generally uncommon. **HABITAT:** Fast rocky riffles of small to medium rivers. **SIMILAR SPECIES:** See Pl. 50. See (1) Spotted Darter, *E. maculatum*, (2) Coppercheek Darter, *E. aquali*, and (3) Wounded Darter, *E. vulneratum*.

COPPERCHEEK DARTER *Etheostoma aquali* — Pl. 50

IDENTIFICATION: Nearly identical to Bloodfin Darter, *E. sanguifluum*, but has *wavy copper lines* on cheek and opercle, no teardrop. Has 57–67 lateral scales. To 3¼ in. (8 cm). **RANGE:** Duck R. system (Tennessee R. drainage), w.-cen. TN. Fairly common. **HABITAT:** Clear, fast rocky riffles of small to medium rivers. **SIMILAR SPECIES:** (1) See Bloodfin Darter, *E. sanguifluum*.

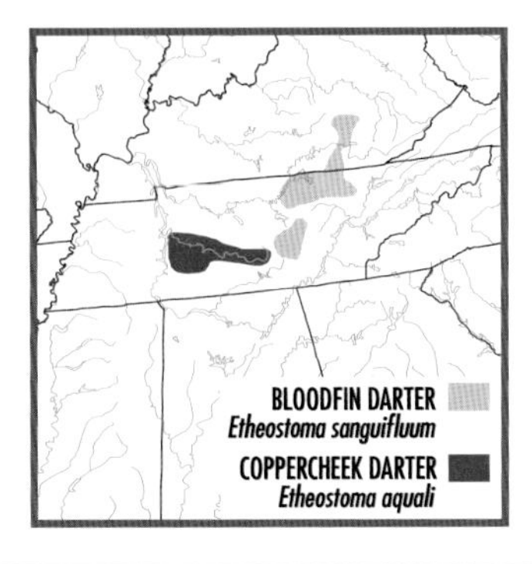

WOUNDED DARTER *Etheostoma vulneratum* Pl. 50

IDENTIFICATION: *Extremely compressed body; narrow pointed snout*; straight-edged caudal fin. Gray body; *black edge* on 2d dorsal, caudal, and anal fins. Male has *black halos* around red spots on side; 2 bright red spots at front, usually 1 at rear, of 1st dorsal fin; red 2d dorsal and caudal fins; blue breast. No red on anal and pelvic fins. Has 51–66 lateral scales. To 3¼ in. (8 cm). **RANGE:** Upper Tennessee R. drainage, VA, NC, and e. TN. Common. **HABITAT:** Fast rocky riffles of small to medium rivers. **SIMILAR SPECIES:** (1) See Boulder Darter, *E. wapiti*. (2) Coppercheek Darter, *E. aquali* (Pl. 50), (3) Bloodfin Darter, *E. sanguifluum*, and (4) Spotted Darter, *E. maculatum* (Pl. 50), *lack* black edge on 2d dorsal, caudal, and anal fins; have rounded caudal fin.

BOULDER DARTER *Etheostoma wapiti* Not shown

IDENTIFICATION: Nearly identical to Wounded Darter, *E. vulneratum*, but *lacks* red on 1st dorsal fin; male has *no red* on body or fins. To 3¼ in. (8.5 cm). **RANGE:** Elk R. and Shoal Creek systems (Tennessee. R. drainage), s. TN and n. AL. Rare in Elk R.; possibly extirpated from Shoal Creek. Protected as an *endangered species*. **HABITAT:** Fast, rocky riffles of small to medium rivers. **SIMILAR SPECIES:** (1) See Wounded Darter, *E. vulneratum* (Pl. 50).

BLUEBREAST DARTER *Etheostoma camurum* Pl. 50

IDENTIFICATION: Male has *bright red spots not* surrounded by black halos on side; red fins, *blue breast*. Female has *brown spots* on side, brown fins, white to light blue breast. Olive green to gray above; light green to white below; black edge on 2d dorsal, caudal, and anal fins. Dusky teardrop, unscaled nape, moderately blunt snout. Has 47–70 lateral scales. To 3¼ in. (8.4 cm). **RANGE:** Ohio R. basin from w. NY to e. IL, and south to Tennessee R. drainage, TN and AL. Locally common but absent from most rivers within range. **HABITAT:** Fast, rocky riffles of small to medium rivers. **SIMILAR SPECIES:** See Pl. 50. (1) Greenfin

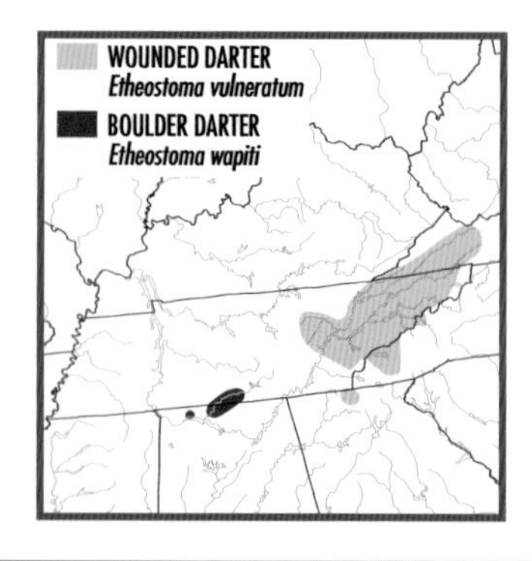

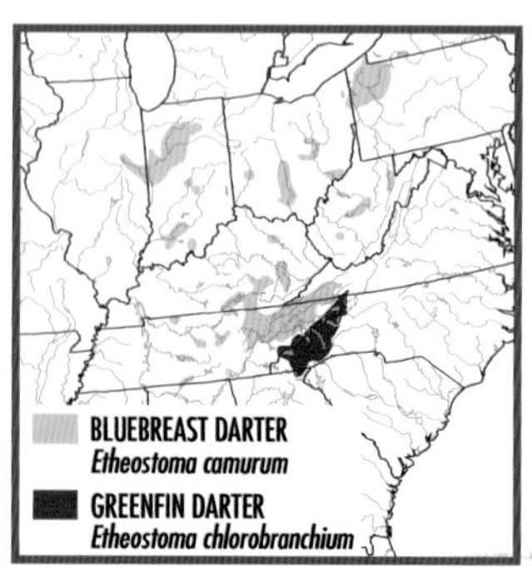

Darter, *E. chlorobranchium*, has green fins, dark teardrop, more pointed snout. (2) Spotted Darter, *E. maculatum*, (3) Bloodfin Darter, *E. sanguifluum*, and (4) Wounded Darter, *E. vulneratum*, have pointed snout, *black halos* around red spots on side. (5) Coppercheek Darter, *E. aquali*, and (6) Redline Darter, *E. rufilineatum*, have pointed snout, distinctive marks on cheek.

GREENFIN DARTER *Etheostoma chlorobranchium* Pl. 50

IDENTIFICATION: Large male is *deep green* with *green fins* (sometimes pink pectoral fins). Female and juvenile are brown above; have blue breast, yellow-brown fins (often with some green), small black and red-brown spots on side. Black edge on dorsal, caudal, and anal fins; black teardrop. Moderately pointed snout. Has 52–72 lateral scales. To 4 in. (10 cm). **RANGE:** Upper Tennessee R. drainage from South Fork Holston R. to Hiwassee R. systems, VA, NC, TN, and GA. Common. **HABITAT:** Very fast, rocky riffles of creeks and small to medium rivers. **SIMILAR SPECIES:** (1) Bluebreast Darter, *E. camurum* (Pl. 50), *lacks* green; has blunter snout, less distinct teardrop.

ORANGEFIN DARTER *Etheostoma bellum* Pl. 50

IDENTIFICATION: *Large black teardrop* (sometimes obscure on large male); black blotches along side; black edge on dorsal, caudal, and anal fins. Male is brown to orange; has blue breast, *orange fins*. Female is brown above, yellow below; has yellow-brown fins. Scales on rear of nape; moderately pointed snout. Has 48–63 lateral scales. To 3½ in. (9 cm). **RANGE:** Upper Green and Barren river systems, KY and n.-cen. TN. Common; locally abundant. **HABITAT:** Fast, rocky riffles of creeks and small to medium rivers. **SIMILAR SPECIES:** (1) Only similar species in Green R. system, Spotted Darter, *E. maculatum* (Pl. 50), is more compressed; has long pointed snout, *small or no teardrop, no* orange on fins, black spots on fins of female.

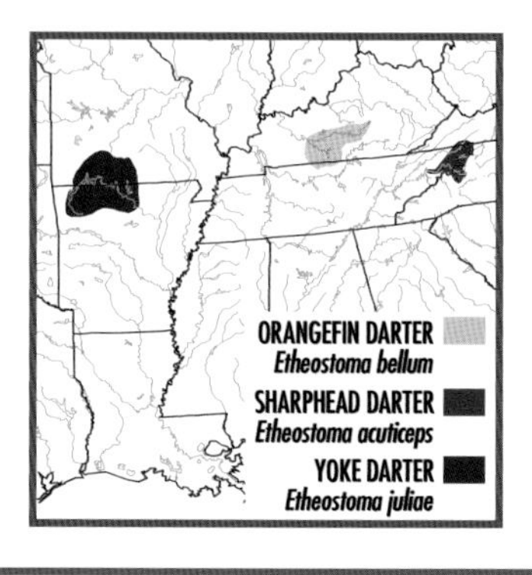

SHARPHEAD DARTER *Etheostoma acuticeps* Pl. 50

IDENTIFICATION: *Compressed body; extremely pointed snout.* Long dusky bars on side. No red on body or fins; no black edge on fins. Female is yellow-brown; no spots on fins. Male is olive to *blue*; has *turquoise fins*, black spot at front of 1st dorsal fin. Teardrop absent or dusky. Unscaled opercle. Complete lateral line; 54-65 lateral scales. To 3¼ in. (8.4 cm). **RANGE:** Holston and Nolichucky river systems (Tennessee R. drainage), w. VA, w. NC, and e. TN. Rare. **HABITAT:** Fast, deep rocky riffles in small to medium rivers. **SIMILAR SPECIES:** See Pl. 50. (1) Wounded Darter, *E. vulneratum*, (2) Redline Darter, *E. rufilineatum*, (3) Bluebreast Darter, *E. camurum*, and (4) Greenfin Darter, *E. chlorobranchium*, have red spots on side of male; black edge on 2d dorsal, caudal, and anal fins; scales on opercle.

YOKE DARTER *Etheostoma juliae* Pl. 50

IDENTIFICATION: *Large black saddle* ("yoke") on nape and down side to pectoral fin; 3 smaller saddles. Brown above; light green below; large *metallic green humeral spot*; wide dusky bars on side; *orange fins*; large teardrop. Pointed snout. Has 50–65 lateral scales. To 3 in. (7.8 cm). **RANGE:** White R. drainage (excluding Black R. system), s. MO and n. AR. Common; locally abundant. **HABITAT:** Clear, fast, rocky riffles of creeks and small to medium rivers. **SIMILAR SPECIES:** (1) Yellowcheek Darter, *E. moorei* (Pl. 50), *lacks* yoke; is gray with red fins on male; female has black-spotted fins. (2) Other close relatives of Yoke Darter (Pl. 50) occur east of Mississippi R.

TIPPECANOE DARTER *Etheostoma tippecanoe* Pl. 50

IDENTIFICATION: *Small*—to 1¾ in. (4.3 cm). Blue-black bars on side, darkest at rear; *last bar large, encircles caudal peduncle*, followed by 2 yellow (female) or orange (male) spots on caudal fin base. Dark saddles on back; 2d saddle begins in front of 1st dorsal fin. Male is orange; has blue breast, dark orange fins. Female is dark brown above, yellow below; has black spots on fins. Moderately pointed snout. No scales on cheek. Incomplete lateral line; 40–65 lateral scales. **RANGE:** Ohio R. basin from w. PA to IN, and south to Cumberland R. drainage, TN. Extremely localized; locally common. **HABITAT:** Shallow gravel riffles of small to medium-sized rivers. **SIMILAR SPECIES:** (1) See Golden Darter, *E. denoncourti*. (2) Other similar darters (Pl. 50) *lack* black bar on caudal peduncle.

GOLDEN DARTER *Etheostoma denoncourti* Not shown

IDENTIFICATION: Similar to Tippecanoe Darter, *E. tippecanoe*, but has scales on cheek behind eye, 2d dorsal saddle under (not in front of) 1st dorsal fin. Has 39–51 lateral scales. To 1¾ in. (4.3 cm). **RANGE:** Tennessee R. drainage, VA and TN. Extremely localized; locally common.

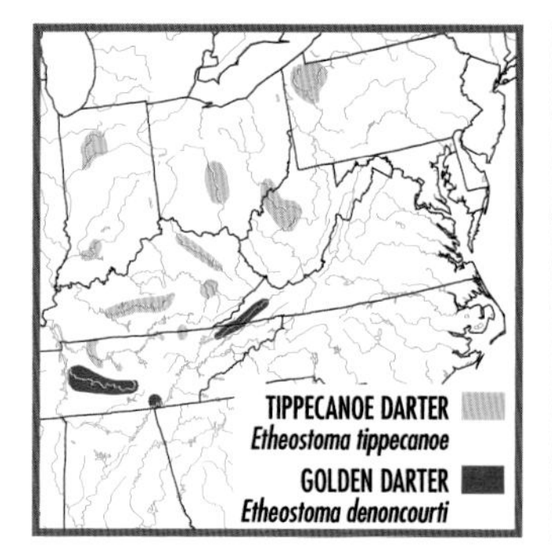

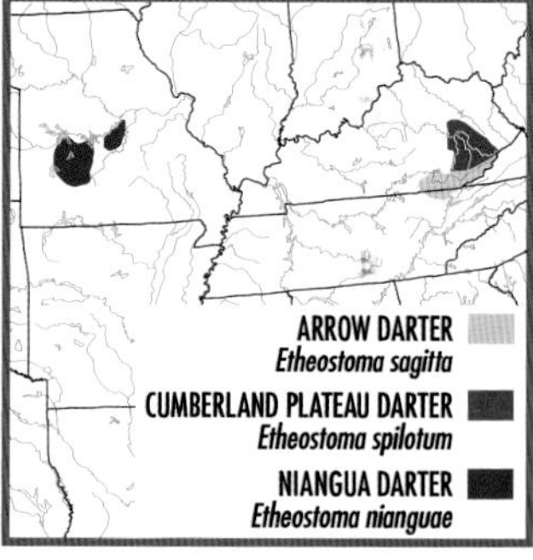

HABITAT: Shallow gravel riffles of small to medium-sized rivers. **SIMILAR SPECIES:** (1) See Tippecanoe Darter, *E. tippecanoe* (Pl. 50).

ARROW DARTER *Etheostoma sagitta* Pl. 51

IDENTIFICATION: Long slender body; *long pointed snout; 2 black spots* on caudal fin base partially fused into short bar. Straw colored to olive above, 6–9 brown saddles; *7–9 large green Us* alternate with orange bars (faint on female) along side; yellow below. Breeding male has bright orange bars on side, red-orange edge on 1st dorsal fin, bands of red-orange spots on 2d dorsal and caudal fins, blue-green anal and pelvic fins. Usually incomplete lateral line; 55–69, usually 63 or more, lateral scales; 49–66, usually more than 52, pored lateral-line scales. To 4¾ in. (12 cm). **RANGE:** Upper Cumberland R. drainage, KY and TN. Uncommon and declining; coal-mine pollution has eliminated many populations. **HABITAT:** Rocky riffles and pools of headwaters, creeks, and small rivers. **SIMILAR SPECIES:** See (1) Cumberland Plateau Darter, *E. spilotum*, and (2) Niangua Darter, *E. nianguae* (Pl. 51).

CUMBERLAND PLATEAU DARTER *Etheostoma spilotum* Not shown

IDENTIFICATION: Nearly identical to Arrow Darter, *E. sagitta*, but has 52–61, *usually 62 or fewer, lateral scales*; 34–54, *usually fewer than 50, pored lateral-line scales.* To 4¾ in. (12 cm). **RANGE:** Upper Kentucky R. drainage, KY. Uncommon and declining; coal-mine pollution has eliminated many populations. **HABITAT:** Rocky riffles and pools of headwaters, creeks, and small rivers. **SIMILAR SPECIES:** (1) See Arrow Darter, *E. sagitta* (Pl. 51).

NIANGUA DARTER *Etheostoma nianguae* Pl. 51

IDENTIFICATION: Similar to Arrow Darter, *E. sagitta*, but has *2 (unfused) jet-black spots* on caudal fin base, 72–82 lateral scales. Breeding male has large bright orange and green bars along side, orange and green bands on dorsal and caudal fins. To 5¼ in. (13 cm). **RANGE:** Osage R. drainage (Missouri R. basin), s.-cen MO. Rare; protected as a *threatened species*. **HABITAT:** Rocky pools and runs of creeks and small to medium rivers. **SIMILAR SPECIES:** (1) See Arrow Darter, *E. sagitta* (Pl. 51).

PINEWOODS DARTER *Etheostoma mariae* Pl. 51

IDENTIFICATION: First dorsal fin with *bright red edge, black spot* at front. *Small black blotches* on lower half of head and breast; thin teardrop broken into spots. Light brown above; *broad dark brown to black stripe* (often broken into wide bars), *yellow lateral line* on side; white to yellow-green below. Dark spots on 2d dorsal, caudal, and anal fins. Broadly joined branchiostegal membranes (Fig. 56). Has 35–39 lateral scales. To 3 in. (7.6 cm). **RANGE:** Little Peedee R. system near Fall Line, NC and SC. Common within small range. **HABITAT:** Gravel riffles and current-swept vegetation in creeks. **SIMILAR SPECIES:** (1) Savannah Darter, *E. fricksium* (Pl. 51), has moderately joined branchiostegal membranes, bright green belly (interrupted by bright orange bars on male); *no* black spot at front of 1st dorsal fin.

SAVANNAH DARTER *Etheostoma fricksium* Pl. 51

IDENTIFICATION: Light brown above, *green below*; bright *orange bars* on belly of male; *broad dark brown to black stripe* on side. Red-brown spots on fins, *red edge* on 1st dorsal fin. *Small black blotches* on lower half of head and breast. Black teardrop, 3–4 black spots on caudal fin base. Moderately joined branchiostegal membranes. Usually complete lateral line; 35–45 lateral scales. To 3 in. (7.4 cm). **RANGE:** Below Fall Line in Edisto, Combahee, Broad, and Savannah river drainages, SC and GA. Common. **HABITAT:** Gravel riffles, gravel and sand runs of creeks and small rivers. Often in vegetation. **SIMILAR SPECIES:** (1) Pinewoods Darter, *E. mariae* (Pl. 51), has broadly joined branchiostegal membranes, black spot at front of 1st dorsal fin, no orange on belly. (2) Christmas Darter, *E. hopkinsi* (Pl. 53), has *middle* red band, dusky green edge on 1st dorsal fin; *no* broad dark brown to black stripe on side.

OKALOOSA DARTER *Etheostoma okaloosae* Pl. 45

IDENTIFICATION: Brown above; *5–8 rows of small dark brown spots* on side, row of larger dark brown dashes just below yellow lateral line; white to yellow below. *Black spots* on lower half of head and breast. Dusky brown fins; thin middle red band on 1st dorsal fin. Pointed snout. Thin, broken teardrop. Complete lateral line; not strongly arched near front. Has 32–37 lateral scales. To 2 in. (5.3 cm). **RANGE:** Choctawhatchee Bay drainage, FL. Common in small area; protected as an *endangered species*. **HABITAT:** Vegetated sandy runs of clear creeks. **SIMILAR SPECIES:** (1) Gulf Darter, *E. swaini* (Pl. 53), has incom-

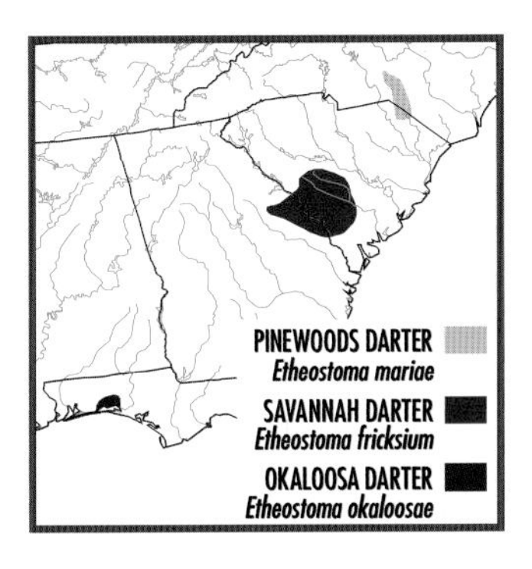

plete lateral line, blunter snout; male has bright blue and orange body and fins. (2) Brown Darter, *E. edwini* (Pl. 55), has lateral line incomplete, arched at front; *no* rows of brown spots on side; bright red spots on body and fins of male.

STIPPLED DARTER *Etheostoma punctulatum* Pl. 51

IDENTIFICATION: Light brown above, 6–8 dark saddles; *dark brown mottling, specks on head and body*, white to *orange* below. *Large black teardrop*. Speckled fins; brown base and edge on 1st dorsal fin, orange band near edge on male. Breeding male has bright orange head and lower half of body (to caudal fin), *wide blue stripe* on rear half of side, orange and blue bands on 1st dorsal fin (clear along base). Pointed snout. Incomplete lateral line; 58–80 lateral scales; usually 31–33 scales around caudal peduncle, 8 infraorbital pores. To 4 in. (10 cm). **RANGE:** Missouri and White river drainages in Ozark Uplands of MO and AR. Isolated population in upper Castor R. (Mississippi R. tributary) of se. MO. Common. **HABITAT:** Rocky pools of headwaters, creeks, and small rivers; often near springs and debris. **SIMILAR SPECIES:** (1) See Sunburst Darter, *E.* species. (2) Arkansas Darter, *E. cragini*, and (3) Paleback Darter, *E. pallididorsum* (both Pl. 51), have strongly bicolored body (upper half brown, lower half white to orange), blunt snout, fewer than 58 lateral scales.

SUNBURST DARTER *Etheostoma* species Not shown

IDENTIFICATION: Similar to Stippled Darter, *E. punctulatum*, but has *4 dark saddles*; *more heavily speckled*, stouter body; usually 28–29 scales around caudal peduncle, 9–10 infraorbital pores. Breeding male has orange *restricted to belly*, no blue stripe on side, dark band along 1st dorsal fin base. To 3 in. (7.7 cm). **RANGE:** Arkansas R. drainage, sw. MO, se. KS, nw. AR, and ne. OK. Common. **HABITAT:** Same as Stippled Darter. **SIMILAR SPECIES:** (1) See Stippled Darter, *E. punctulatum* (Pl. 49).

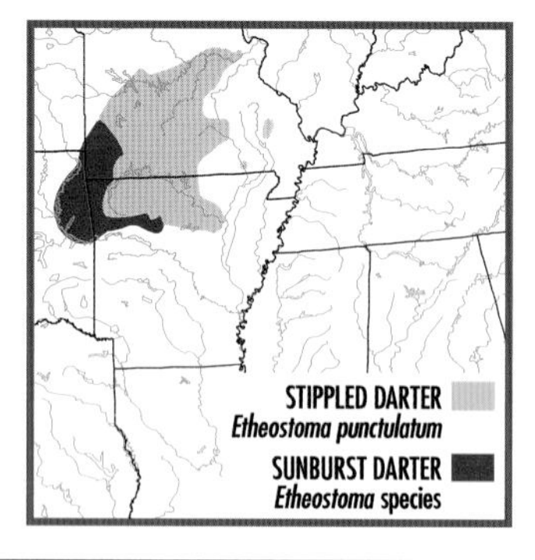

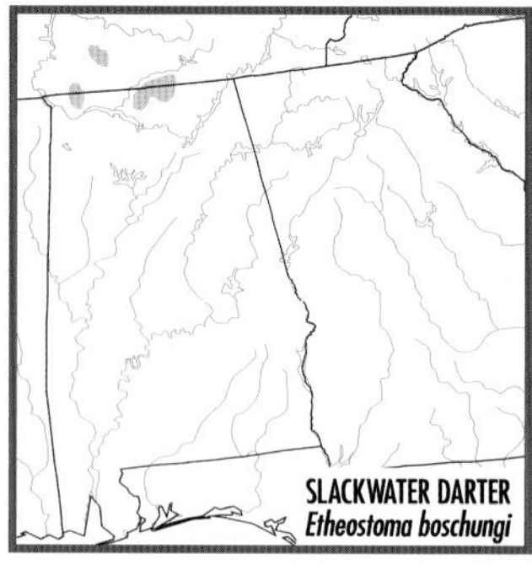

SLACKWATER DARTER *Etheostoma boschungi* Pl. 51

IDENTIFICATION: Brown above, 3 black saddles; *many black specks on back and side*, black blotches on upper side; white to orange (large male) below. *Large black teardrop*. Dusky fins; wide orange edge, blue base on 1st dorsal fin. Large head; blunt snout. Incomplete lateral line (30–44 pores); 43–58 lateral scales. To 3 in. (7.8 cm). **RANGE:** Middle Tennessee R. drainage from Flint R., n. AL, to Buffalo R., s.-cen. TN. Rare; protected as a *threatened species*. **HABITAT:** Gravel-bottomed pools and runs of creeks and small rivers; usually in debris. Spawns in headwaters. **SIMILAR SPECIES:** See Pl. 51. (1) Trispot Darter, *E. trisella*, has *thin teardrop*, 1 anal spine, complete lateral line. (2) Arkansas Darter, *E. cragini*, and (3) Paleback Darter, *E. pallididorsum*, both west of Mississippi R., have fewer than 26 lateral-line pores.

ARKANSAS DARTER *Etheostoma cragini* Pl. 51

IDENTIFICATION: *Strongly bicolored body:* upper half dark brown, lower half white to *orange*. *Black specks* on body and fins; black blotches sometimes on upper side. Blue edge, orange middle band, blue base on 1st dorsal fin of breeding male. Large black teardrop. Large head; blunt snout. Incomplete lateral line (4–25 pores); 42–57 lateral scales. To 2¼ in. (6 cm). **RANGE:** Arkansas R. drainage, sw. MO, nw. AR, KS, OK, and CO. Uncommon. **HABITAT:** Spring-fed vegetated headwaters and creeks, usually over mud. **SIMILAR SPECIES:** (1) See Paleback Darter, *E. pallididorsum* (Pl. 51). (2) Stippled Darter, *E. punctulatum* (Pl. 51), and (3) Sunburst Darter, *E.* species, *lack* strongly bicolored body; have more than 57 lateral scales, longer snout.

PALEBACK DARTER *Etheostoma pallididorsum* Pl. 51

IDENTIFICATION: Nearly identical to Arkansas Darter, *E. cragini*, but is more *slender; has wide, pale olive stripe along back*. Incomplete lateral line (8–19 pores); 43–55 lateral scales. To 2¼ in. (6 cm). **RANGE:** Caddo

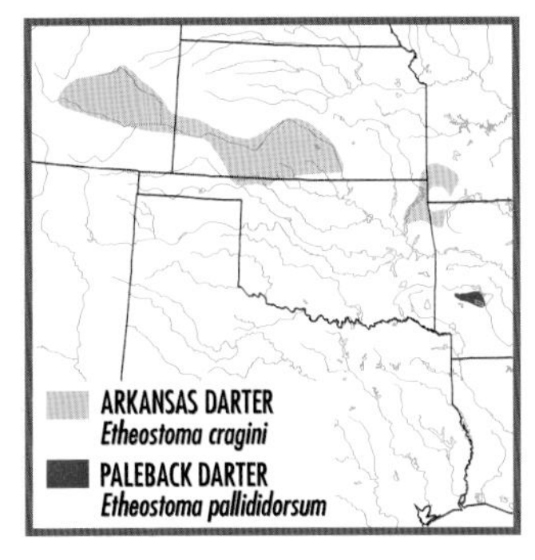

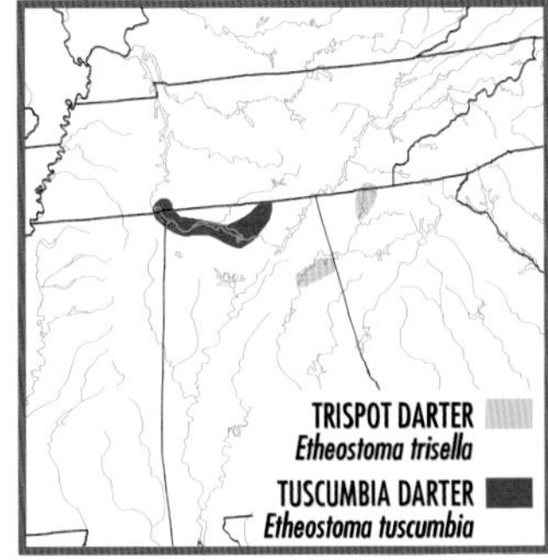

R. and Hallmans Creek (upper Ouachita R. drainage), sw. AR. Fairly common in small range. **HABITAT:** Shallow, rocky pools of headwaters and creeks; vegetated springs. **SIMILAR SPECIES:** (1) See Arkansas Darter, *E. cragini* (Pl. 51).

TRISPOT DARTER *Etheostoma trisella* Pl. 51

IDENTIFICATION: Brown above and on side; *3 dark brown saddles*; white to yellow below. Complete lateral line; *1 anal spine*. Thin teardrop. Breeding male has bright *orange body*, dusky fins, middle red band on 1st dorsal fin. Has 42–52 lateral scales. To 2¼ in. (5.9 cm). **RANGE:** Coosa R. system, GA, AL, and se. TN. Rare. **HABITAT:** Sand and gravel runs of creeks and small rivers. Spawns in headwaters. **SIMILAR SPECIES:** (1) Slackwater Darter, *E. boschungi* (Pl. 51), has large black teardrop, *2 anal spines, incomplete lateral line.*

TUSCUMBIA DARTER *Etheostoma tuscumbia* Pl. 45

IDENTIFICATION: Olive-brown, *gold specks* on back, head, and upper side; 4–6 dark brown saddles; small, dark blotches along side; dark bar on caudal fin base. *Scales on top of head* and often on branchiostegal membranes. *One anal spine*. Long, tubular genital papilla on male. Incomplete lateral line; 37–51 lateral scales. To 2½ in. (6.1 cm). **RANGE:** Springs along Tennessee R. in AL and, formerly, s.-cen. TN. Common in a few springs. **HABITAT:** Large vegetated springs. **SIMILAR SPECIES:** (1) Goldstripe Darter, *E. parvipinne* (Pl. 45), and (2) Trispot Darter, *E. trisella* (Pl. 51), *lack* scales on top of head and on branchiostegal membranes, gold specks; Goldstripe Darter has 2 anal spines; Trispot Darter has complete lateral line.

GOLDSTRIPE DARTER *Etheostoma parvipinne* Pl. 45

IDENTIFICATION: Light gray-brown above, dark brown mottling (sometimes coalesced into short wide bars); light yellow to white below; red

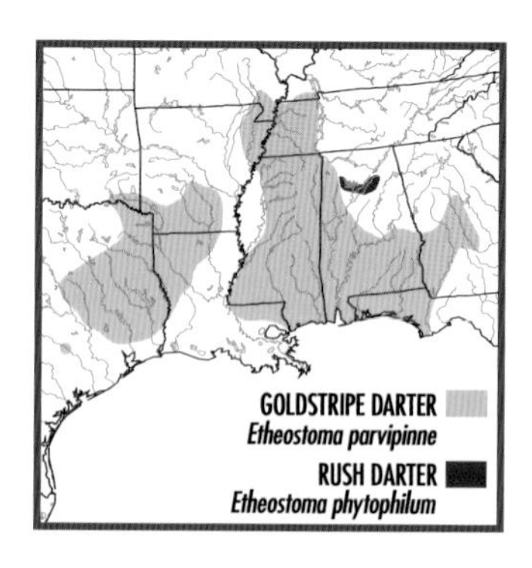

eye; usually incomplete "gold" (yellow) lateral line. *Very short, blunt snout*; small *upturned mouth*. Black teardrop. Large male has black spot at front of 1st dorsal fin, thin stripes on rear half of body. Usually 2 anal spines. Has 40–62 lateral scales. To 2¾ in. (7 cm). **RANGE:** Coastal Plain from Ocmulgee (Atlantic Slope) and Flint (Gulf Slope) river systems, GA, to Colorado R. system, TX; north in Former Mississippi Embayment to w. KY and se. MO. Above Fall Line only in Ocmulgee R. Fairly common. **HABITAT:** Clay- and sand-bottomed runs and pools of vegetated, spring-fed headwaters and creeks. **SIMILAR SPECIES:** (1) See Rush Darter, *E. phytophilum*. (2) Blackfin Darter, *E. nigripinne*, (3) Fringed Darter, *E. crossopterum*, and related darters (Pl. 54) have much larger, nearly *horizontal mouth, more pointed snout*, no black spot at front of 1st dorsal fin. (4) Tuscumbia Darter, *E. tuscumbia* (Pl. 45), has gold specks, scales on top of head, 1 anal spine, no black spot at front of dorsal fin, no thin stripes on side.

RUSH DARTER *Etheostoma phytophilum* — Not shown

IDENTIFICATION: Very similar to Goldstripe Darter, *E. parvipinne*, but usually has *47 or fewer* (vs. usually 48 or more) *lateral scales, 22 or fewer scales* (vs. 23 or more) *around caudal peduncle*, 13 or fewer (vs. 14 or more) transverse scales. To 2¾ in. (7 cm). **RANGE:** Upper Black Warrior R. system (known only from tributaries of Sipsey Fork and Locust Fork), AL. Extremely localized and uncommon. **HABITAT:** Vegetated areas, usually along margins, of spring-fed creeks. **SIMILAR SPECIES:** (1) See Goldstripe Darter, *E. parvipinne* (Pl. 45).

REDFIN DARTER *Etheostoma whipplei* — Pl. 51

IDENTIFICATION: Many small *bright red* (on male) or *yellow* (female) *spots* on side; thin dark lines, many interrupted, on side. Slender, *almost uniformly deep body* from head to caudal peduncle. Mottled body; often dark bars on rear of body, sometimes cut into upper and lower halves by yellow lateral line. Olive above, 8–10 brown saddles; white to orange (on male) below. Faint to black teardrop. Blue edge, middle red band on dorsal, caudal, and anal fins of male and 1st dorsal fin of female (faint). Blue pelvic fins on large male. Incomplete lateral line; 56–76 lateral scales; 26–39 (usually 28–35) scales around caudal peduncle. To 3½ in. (9 cm). **RANGE:** Above Fall Line in White and Arkansas river systems, sw. MO, se. KS, n. AR, and e. OK. Common. **HABITAT:** Rocky pools, sometimes runs and riffles, of headwaters, creeks, and small rivers. **SIMILAR SPECIES:** (1) See Redspot Darter, *E. artesiae*. (2) Orangebelly Darter, *E. radiosum* (Pl. 51), *lacks* red and yellow spots, thin dark lines on side; is *deeper bodied* (as adult).

REDSPOT DARTER *Etheostoma artesiae* Not shown

IDENTIFICATION: Nearly identical to Redfin Darter, *E. whipplei*, but has *45–66 (usually fewer than 59) lateral scales, 18–32 (usually 20–28) scales around caudal peduncle*. To 3 in. (9 cm). **RANGE:** Gulf drainages from Halawakee Creek (Chattahoochee R. drainage), AL, to Neches R., TX; north in Mississippi R. drainage to n. MS and Ouachita R. system, AR. Disjunct population in Bear Creek system (Tennessee R. Drainage), nw. AL and ne. MS. Primarily below Fall Line, but above Fall Line in Mobile Bay drainage, AL. Common. **HABITAT:** Sandy and rocky pools, sometimes runs and riffles, of headwaters, creeks, and small rivers. **SIMILAR SPECIES:** (1) See Redfin Darter, *E. whipplei* (Pl. 51).

ORANGEBELLY DARTER *Etheostoma radiosum* Pl. 51

IDENTIFICATION: Short dark bars (rear ones long) along side, *cut into upper and lower halves by yellow lateral line*. Olive to tan above, 8–10 dark saddles; white to orange below. Black teardrop. Blue edge, middle red band on dorsal, caudal, and anal fins (faint on female). Orange branchiostegal membranes, blue pelvic fins on large male.

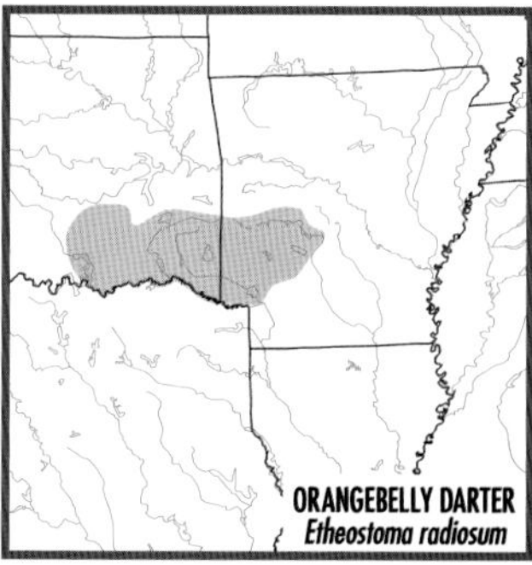

Incomplete lateral line; 47–66 (usually 52–61) lateral scales. To 3½ in. (8.6 cm). **RANGE:** Ouachita and Red river drainages above Fall Line, sw. AR and se. OK; 1 record in Red R. drainage of n. TX (Lamar Co.). Abundant. **HABITAT:** Gravel and rubble riffles and runs of creeks and small to medium rivers. **REMARKS:** Three subspecies. *E. r. cyanorum*, in Blue R., OK, has blunt snout, 39–60 lateral-line pores. *E. r. paludosum*, in Clear Boggy, Kiamichi, and Washita rivers, OK, has row of blue spots on edge of 1st dorsal fin. *E. r. radiosum* occupies rest of range. Orangebelly Darter and Orangethroat darters, *E. spectabile*, hybridize freely in Blue R., OK, forming a "hybrid swarm" (*E. r. cyanorum* x *E. s. pulchellum*) with traits between those of the 2 species. **SIMILAR SPECIES:** (1) Redfin Darter, *E. whipplei* (Pl. 51), and (2) Redspot Darter, *E. artesiae*, are more slender (as adults); have many small red (on male) or yellow (female) spots, thin dark lines on side.

RAINBOW DARTER *Etheostoma caeruleum* Pl. 53

IDENTIFICATION: *Deep body*, deepest under *middle* of 1st dorsal fin. *Dark bars* on side; blue between red on male; dark brown between yellow-white on female. Light brown above; 6–10 dark saddles, 2 or 3 prominent; yellow, green, or red below. Dorsal, caudal, and anal fins *red with blue edge* (faint on female). Large male has red pectoral fin, blue pelvic fin, orange branchiostegal membranes, blue cheek, and, in some populations, rows of red spots on side. Unscaled cheek and breast; uninterrupted infraorbital canal (Fig. 55); usually 13 pectoral rays; incomplete lateral line; 36–57 (usually 41–50) lateral scales. To 3 in. (7.7 cm). **RANGE:** Great Lakes and Mississippi R. basins from s. ON and w. NY to MN, and south to n. AL and AR; isolated populations along lower Mississippi R., including in sw. MS and e. LA. Also in Hudson Bay basin, MN; and upper Potomac R. drainage (Atlantic Slope), WV. Abundant. **HABITAT:** Fast gravel and rubble riffles of

creeks and small to medium rivers. **SIMILAR SPECIES:** See Pls. 52 & 53. (1) Orangethroat Darter, *E. spectabile*, and lookalike species (see Orangethroat Darter), and (2) Redband Darter, *E. luteovinctum*, have body deepest at *nape or at front of* 1st dorsal fin, *no red* in anal fin, infraorbital canal interrupted; Orangethroat Darter usually has 11–12 pectoral rays; Redband Darter has partly scaled cheek and breast, usually 52–59 lateral scales. (3) Mud Darter, *E. asprigene*, has fully scaled cheek, dark blotch at rear of 1st dorsal fin, *no red* on anal fin. (4) Gulf Darter, *E. swaini*, has thin dark lines on upper side, scales on cheek.

MUD DARTER *Etheostoma asprigene* Pl. 53

IDENTIFICATION: *Dark bars* on side (darkest at rear), blue between dull orange on male. *Fully scaled cheek*. First dorsal fin has middle red band, blue edge and base, *large black blotch* at rear (faint on female). Body deepest under middle of 1st dorsal fin. Olive-brown above, 6–10 dark saddles; white to dull orange below. Large dusky teardrop. Middle red band on 2d dorsal fin of male. Incomplete lateral line; 44–54 (usually 48–51) lateral scales. To 2¾ in. (7.1 cm). **RANGE:** Mississippi R. basin lowlands from WI and MN to LA and e. TX (Red R. drainage); on Gulf Slope in Sabine-Neches drainage, LA and TX. Fairly common. **HABITAT:** Sluggish riffles over rocks or debris in small to large rivers; lowland lakes. **SIMILAR SPECIES:** See Pls. 52 & 53. (1) Orangethroat Darter, *E. spectabile*, and lookalike species (see Orangethroat Darter account), (2) Rainbow Darter, *E. caeruleum*, (3) Gulf Darter, *E. swaini*, and (4) Creole Darter, *E. collettei*, *lack* large blotch at rear of 1st dorsal fin, usually are more brightly colored. Body of Orangethroat Darter is deepest at nape or front of 1st dorsal fin; Rainbow Darter has *unscaled cheek*, red in anal fin; Gulf Darter usually has 38–45 lateral scales; Creole Darter has *partly scaled cheek*.

ORANGETHROAT DARTER *Etheostoma spectabile* Pl. 52

IDENTIFICATION: *Arched body, deepest at nape or front of 1st dorsal fin. Thin dark stripes, 6–9 dark bars* (blue between orange on male, brown between yellow-white on female) on side. Olive to light brown above, usually 7–10 dark saddles; white to orange below. Thin black or dusky teardrop. *Blue edge*, orange on dorsal and caudal fins (faint on female). Large male has blue breast and anal fin, blue or black pelvic fin, orange branchiostegal membranes and belly, 2 orange spots on caudal fin base. Interrupted infraorbital canal, *usually 3 posterior infraorbital pores* (Fig. 55); uninterrupted supratemporal canal; incomplete lateral line; partly scaled cheek; 17–20 scales around caudal peduncle; 38–55 (usually 42–50) lateral scales, 17–35 pored; usually 11–12 pectoral rays. To 2¾ in. (7.2 cm). **RANGE:** Lake Erie and Mississippi R. basins from se. MI and OH to e. WY, and south to TN and n. TX; Gulf drainages (Trinity R. to San Antonio R.) of TX, mostly on Edwards Plateau. Abundant. **HABITAT:** Shallow gravel riffles, rocky runs and pools, of headwaters, creeks, and small rivers. **REMARKS:** Three subspecies. *E. s. pulchellum*, from Platte R., NE, to Guadalupe R., TX, has mostly orange 1st dorsal fin, blue-gray breast, prominent dark vertical bars on side. *E. s. squamosum*, in Arkansas R. drainage of sw. MO, se. KS, ne. OK, and nw. AR, has mostly orange 1st dorsal fin, bars darkest on lower side. *E. s. spectabile*, over rest of range, has mostly blue 1st dorsal fin, thin dark stripes on side. Several forms previously recognized as subspecies or variants of Orangethroat Darter now are recognized as species. Additional populations may be named (Pl. 52), including Ozark Darter, with orange belly, red dots on upper side (White R. system, MO and AR); Ihiyo Darter, with white belly, light orange breast with few scales (Caney Fork and nearby tributaries, Cumberland R., TN); Sheltowee Darter, with orange breast and belly, orange and blue rectangles on side (Dix R. system, KY); and Mamequit Darter, with fully scaled blue-gray breast, white belly (lower Cumberland R., KY and TN). **SIMILAR SPECIES:** (1) Orangethroat Darter lookalikes

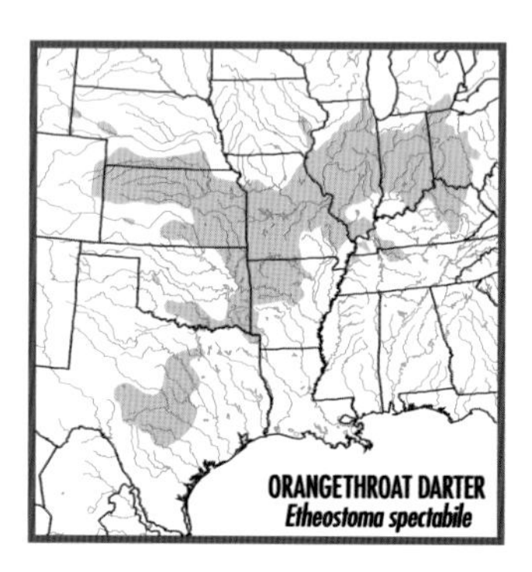

(Pl. 52): Next 7 species (through Headwater Darter, *E. lawrencei*) are most easily distinguished from one another and Orangethroat Darter as breeding males. (2) Redband Darter, *E. luteovinctum* (Pl. 53), has brown stripe along, dark squares below, lateral line; 21–26 scales around narrow caudal peduncle. (3) Rainbow Darter, *E. caeruleum* (Pl. 53), is deepest under *middle* of 1st dorsal fin; has *red* on anal fin of male, uninterrupted infraorbital canal; usually 13 pectoral rays. (4) Greenthroat Darter, *E. lepidum* (Pl. 53), is deepest under *middle* of 1st dorsal fin, has interrupted supratemporal canal (Fig. 55), green branchiostegal membranes and breast. (5) Gulf Darter, *E. swaini*, and (6) Creole Darter, *E. collettei* (both Pl. 53), are deepest under *middle* of 1st dorsal fin, have uninterrupted infraorbital canal, usually 13–14 pectoral rays.

BUFFALO DARTER *Etheostoma bison* Pl. 52

IDENTIFICATION: Similar to Orangethroat Darter, *E. spectabile*, but has *blue belly; dark dashes* on upper side; 36–47 *(usually 37–43) lateral scales*, 14–34 pored. Breeding male has humped nape. To 2¾ in. (6.7 cm). **RANGE:** Lower Duck (Piney R. and downstream) and lower Tennessee R. systems (Indian Creek downstream to Turkey Creek), KY and TN. Common; locally abundant. **HABITAT:** Shallow gravel riffles, rocky runs and pools, of headwaters, creeks, and small rivers. **SIMILAR SPECIES:** (1) See Orangethroat Darter, *E. spectabile,* and related species (Pl. 52).

HIGHLAND RIM DARTER *Etheostoma kantuckeense* Pl. 52

IDENTIFICATION: Similar to Orangethroat Darter, *E. spectabile*, but has *blue-gray breast, belly; faint lines* (series of dots) on side. No scales on breast. Has 37–47 (usually 38–43) lateral scales, 21–33 pored. To 2½ in. (6.4 cm). **RANGE:** Barren R. system, KY and TN. Common. **HABITAT:** Shallow gravel riffles, rocky runs and pools, of headwaters, creeks,

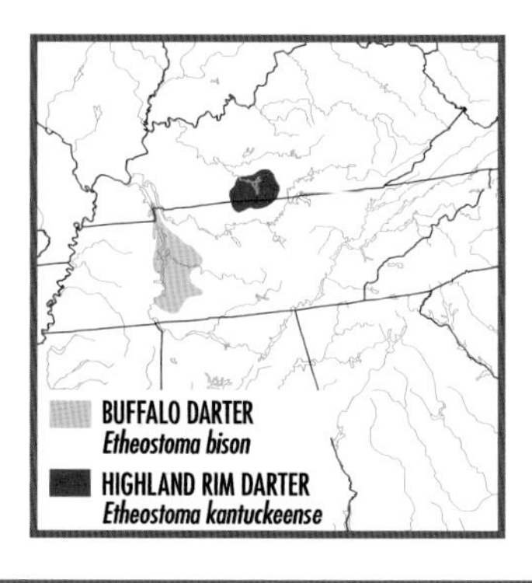

and small rivers. **SIMILAR SPECIES:** (1) See Orangethroat Darter, *E. spectabile,* and related species (Pl. 52).

BROOK DARTER *Etheostoma burri* Pl. 52

IDENTIFICATION: Similar to Orangethroat Darter, *E. spectabile*, but has *red breast, belly; diamond-shaped blue bars* on side. Has 39–50 (usually 43–49) lateral scales, 20–32 pored. To 2½ in. (6.1 cm). **RANGE:** Upper Black R. system downstream to Mississippi R. alluvial plain near Poplar Bluff, se. MO. Locally common. **HABITAT:** Shallow gravel riffles, sometimes rocky runs and pools, of headwaters, creeks, and small rivers. **SIMILAR SPECIES:** (1) See Orangethroat Darter, *E. spectabile,* and related species (Pl. 52).

STRAWBERRY DARTER *Etheostoma fragi* Pl. 52

IDENTIFICATION: Similar to Orangethroat Darter, *E. spectabile*, but has *faint lines, 10–12 oblique turquoise bars* on side; bars *meet those of other side* on belly; *red-orange breast*; *wide orange band* in *middle* of dorsal fin; fully scaled cheek. Has 43–56 (usually 47–53) lateral scales, 21–40 pored. To 2½ in. (6.1 cm). **RANGE:** Strawberry R. system, ne. AR. Abundant. **HABITAT:** Shallow gravel riffles, rocky runs and pools, of headwaters, creeks, and small rivers. **SIMILAR SPECIES:** (1) See Orangethroat Darter, *E. spectabile,* and related species (Pl. 52). (2) Current Darter, *E. uniporum*, has *blue-gray breast and belly, wide blue band* in *lower half* of dorsal fin, *few scales* on cheek.

CURRENT DARTER *Etheostoma uniporum* Pl. 52

IDENTIFICATION: Similar to Orangethroat Darter, *E. spectabile*, but has *faint lines* on side, *8-10 oblique turquoise bars* on side that *meet those of other side* on belly, *blue-gray breast and belly*, wide blue band in *lower half* of dorsal fin, *interrupted supratemporal canal, usually 1 posterior infraorbital pore* (Fig. 55). Few scales on cheek. Has 38–52

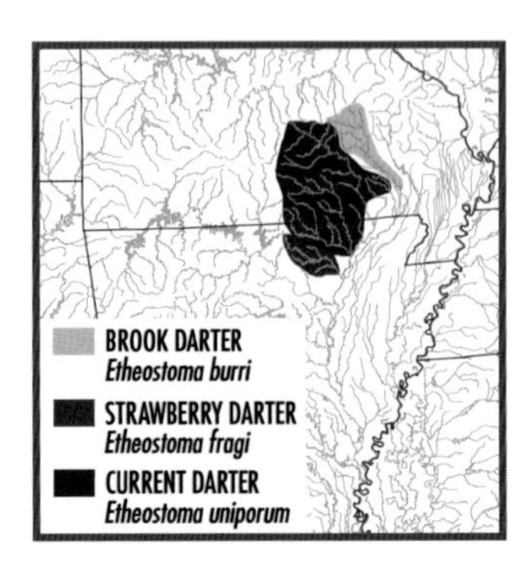

(usually 42–49) lateral scales, 11–32 pored. To 2½ in. (6.1 cm). **RANGE:** Black R. system between Black R., se. MO, and Strawberry R., ne. AR. Abundant. **HABITAT:** Shallow gravel riffles, sometimes rocky runs and pools, of headwaters and creeks. **SIMILAR SPECIES:** (1) See Orange-throat Darter, *E. spectabile,* and related species (Pl. 52). (2) Strawberry Darter, *E. fragi*, has *orange belly, red-orange breast, wide orange band* in *middle* of dorsal fin, fully scaled cheek.

SHAWNEE DARTER *Etheostoma tecumsehi* Pl. 52

IDENTIFICATION: Similar to Orangethroat Darter, *E. spectabile*, but *no dark stripes* on side, breeding male has alternating orange and blue bars prominent along entire side (vs. prominent only on rear half), orange on belly confluent with anterior orange bars (Orangethroat Darter lacks orange bars on front half of body). Has 44–55 (usually 47–49) lateral scales, 27–40 pored. To 2½ in. (6.4 cm). **RANGE:** Upper Pond R. system, w. KY. Locally common. **HABITAT:** Shallow gravel riffles, rocky runs and pools, of headwaters, creeks, and small rivers. **SIMILAR SPECIES:** (1) See Orangethroat Darter, *E. spectabile,* and related species (Pl. 52).

HEADWATER DARTER *Etheostoma lawrencei* Pl. 52

IDENTIFICATION: Nearly identical to Shawnee Darter, *E. tecumsehi*, but usually has 13 dorsal rays (vs. 12), usually 31 or fewer (vs. 32 or more) pored lateral-line scales. To 3 in. (7.4 cm). **RANGE:** Salt R. system, KY; upper Green R. system down to Mud R., KY; Cumberland R. system from Cumberland Falls, KY, to Dillard Creek, TN (including lower tributaries of Caney Fork). Common; locally abundant. **HABITAT:** Shallow gravel riffles, rocky runs and pools, of headwaters, creeks, and small rivers. **SIMILAR SPECIES:** (1) See Shawnee Darter, *E. tecumsehi* (Pl. 52).

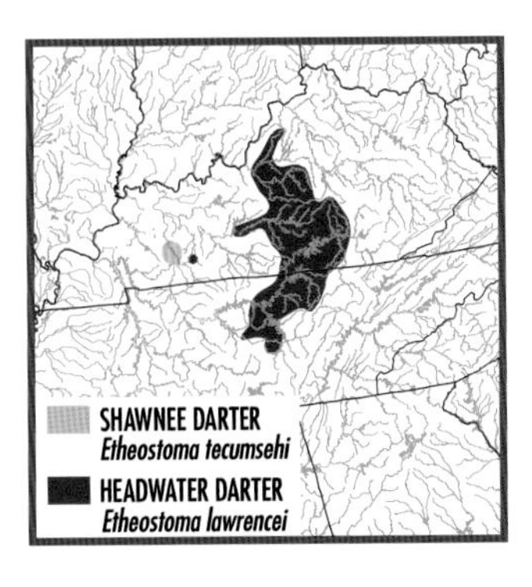

REDBAND DARTER *Etheostoma luteovinctum* Pl. 53

IDENTIFICATION: *Body deepest at front of 1st dorsal fin*, strongly tapering to *narrow caudal peduncle*. Tan above, 7–9 dark saddles; *light brown stripe* along lateral line, *7–9 dark squares* just below lateral line; white to yellow below. Black teardrop. Blue edge, red base on dorsal fins (faint on female). Breeding male has bright blue bars between red-orange on side, black pelvic fin, blue anal fin. Partly scaled cheek and breast; 21–26 scales around caudal peduncle; incomplete lateral line; 50–61 (usually 52–59) lateral scales. To 2¾ in. (6.8 cm). **RANGE:** Stones R. and Collins R. (Cumberland R. drainage), Duck R. and Elk R. (Tennessee R. drainage) systems, cen. TN. Fairly common in Duck R.; rare elsewhere. **HABITAT:** Shallow rocky pools of headwaters, creeks, and small rivers. **SIMILAR SPECIES:** (1) Orangethroat Darter, *E. spectabile,* and lookalike species (see Orangethroat Darter account and Pl. 52), *lack* dark stripe along, and dark squares below, lateral line; have *deeper* caudal peduncle, 17–20 scales around caudal peduncle, usually 37–53 lateral scales. (2) Rainbow Darter, *E. caeruleum* (Pl. 53), is deepest under *middle* of 1st dorsal fin; has unscaled cheek and breast, red in anal fin of male, usually 41–50 lateral scales.

GULF DARTER *Etheostoma swaini* Pl. 53

IDENTIFICATION: *Dark bars* on side (blue-brown between orange on male, brown between yellow-white on female), often obscured by dark mottling on female. Light brown above, about 7 dark saddles; *thin dark lines* (rows of small black spots) on upper side, white to *orange* (large male) below. Three black spots on caudal fin base; dusky to dark teardrop. Blue edge, red base, middle red and blue bands on dorsal fins (faint on female). Blue anal and pelvic fins, blue and red-orange caudal fin on large male. Incomplete lateral line; 35–50 (usually 38–45) lateral scales. Fully to partly scaled cheek; body deepest under middle of 1st dorsal fin; uninterrupted infraorbital and (usually) supratem-

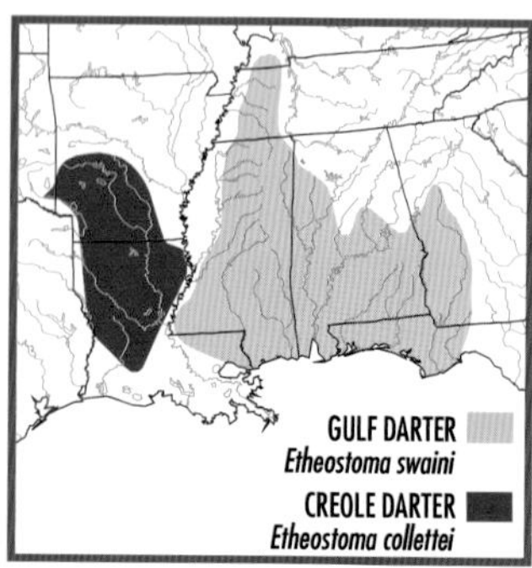

poral canals (Fig. 55). To 3½ in. (9 cm). **RANGE:** Gulf drainages from Ochlockonee R., GA and FL, to Mississippi R., LA, and north on Former Mississippi Embayment (east of Mississippi R. only) to KY; Bear Creek (Tennessee R. drainage), nw. AL and ne. MS. Mostly below Fall Line. Common. **HABITAT:** Shallow rocky riffles, current-swept vegetation in headwaters, creeks, and small to medium rivers. **SIMILAR SPECIES:** See Pl. 53. See (1) Creole Darter, *E. collettei*, and (2) Watercress Darter, *E. nuchale*. (3) Mud Darter, *E. asprigene*, has large black blotch at rear of 1st dorsal fin, usually 48–51 lateral scales; *lacks* dark lines on upper side.

CREOLE DARTER *Etheostoma collettei* Pl. 53

IDENTIFICATION: Similar to Gulf Darter, *E. swaini*, but has *44–60, usually 46–55, lateral scales*; breeding male is more blue, has breeding tubercles on lower body scales (absent on Gulf Darter). To 3 in. (7.4 cm). **RANGE:** Ouachita, Red, Calcasieu, and Sabine river drainages, AR, LA, and OK. Abundant in Ouachita; less common elsewhere. **HABITAT:** Gravel riffles, current-swept vegetation and debris in creeks and small to medium rivers. **SIMILAR SPECIES:** (1) See Gulf Darter, *E. swaini* (Pl. 53). (2) Redspot Darter, *E. artesiae*, and (3) Orangebelly Darter, *E. radiosum* (Pl. 51), *lack* blue and orange bars on side; have red on anal fin of male, 16–23 transverse scales.

WATERCRESS DARTER *Etheostoma nuchale* Pl. 53

IDENTIFICATION: Similar to Gulf Darter, *E. swaini*, but is *smaller, more compressed*; has *12–24 pored lateral-line scales* (Gulf Darter has 28–43), interrupted infraorbital and supratemporal canals (Fig. 55). Has 35–42 lateral scales. To 2¼ in. (5.4 cm). **RANGE:** Springs on Halls and Village creeks (Black Warrior R. system), Jefferson Co., AL. Rare; protected as an *endangered species*. **HABITAT:** Vegetated springs. **SIMILAR SPECIES:** (1) See Gulf Darter, *E. swaini* (Pl. 53).

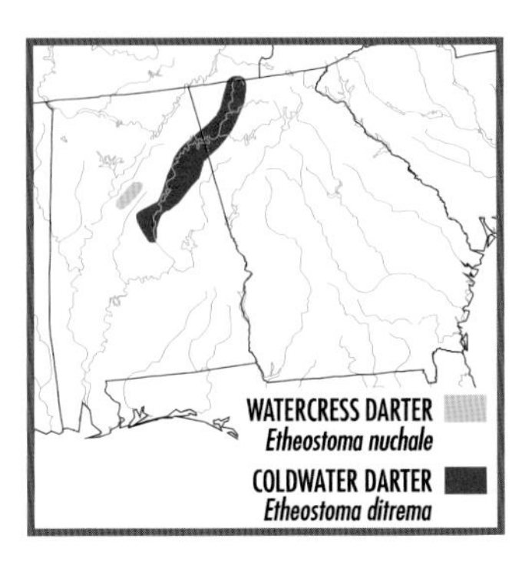

COLDWATER DARTER *Etheostoma ditrema* Pl. 53

IDENTIFICATION: *Dark brown mottling* on yellow back and side; *orange belly* on male. *Three black spots* on caudal fin base. Black teardrop. Blue edge and middle red band on 1st dorsal fin of male, sometimes on 1st dorsal fin of female and 2d dorsal fin of male. Incomplete lateral line; 41–54 (usually 43–50) lateral scales. To 2¼ in. (5.4 cm). **RANGE:** Coosa R. system, GA, AL, and se. TN. Rare and highly localized. **HABITAT:** Vegetated springs and spring runs. **SIMILAR SPECIES:** (1) Gulf Darter, *E. swaini*, and (2) Watercress Darter, *E. nuchale* (both Pl. 53), have *dark bars* and *thin dark lines* on side; large males have blue-brown and orange bars on side.

GREENTHROAT DARTER *Etheostoma lepidum* Pl. 53

IDENTIFICATION: *Red-orange specks or spots* between *long green bars* on side of male; yellow between short brown-black bars on female. Body deepest under middle of 1st dorsal fin. Olive above, dark saddles; white to orange below. Three black spots on caudal fin base; thin black teardrop. Blue-green edge on red 1st dorsal fin; red spots on green 2d dorsal and caudal fins (faint on female). Large male with green branchiostegal membranes and breast; green anal and pelvic fins, sometimes with red bands. Interrupted infraorbital and supratemporal canals (Fig. 55). Incomplete lateral line; 43–67 (usually 48–55) lateral scales. To 2½ in. (6.6 cm). **RANGE:** Colorado, Guadalupe, and Nueces river drainages, TX; Pecos R. system, NM. Common on Edwards Plateau in TX; uncommon in NM. **HABITAT:** Gravel and rubble riffles, especially spring-fed and vegetated riffles, of headwaters, creeks, and small rivers. **SIMILAR SPECIES:** (1) Orangethroat Darter, *E. spectabile* (Pl. 52), is deepest at nape or at front of 1st dorsal fin; has orange branchiostegal membranes and belly, uninterrupted supratemporal canal. (2) Geographically close Rio Grande Darter, *E. grahami* (Pl. 51), is deeper bodied; has many small red (on male) or black (female) spots on side of body, red 1st dorsal fin.

RIO GRANDE DARTER *Etheostoma grahami* Pl. 51

IDENTIFICATION: *Deep bodied. Many small red* (on male) *or black* (female) *spots* on side. *Red 1st dorsal fin* (faint on female). Male has red 2d dorsal, anal, and pelvic fins; yellow caudal and pectoral fins. Olive above, 8–10 dark saddles; often dusky blotches along side; white to yellow below. Faint teardrop. Incomplete lateral line; 40–56 (usually 45–51) lateral scales; interrupted infraorbital and supratemporal canals (Fig. 55). To 2¼ in. (6 cm). **RANGE:** Lower Rio Grande drainage, TX and Mexico. In TX only in Sycamore Creek, Devils R., and lower Pecos R. Common in Devils R. **HABITAT:** Gravel and rubble riffles of creeks and small rivers. **SIMILAR SPECIES:** (1) Greenthroat Darter, *E. lepidum* (Pl. 53), and (2) Orangethroat Darter, *E. spectabile* (Pl. 52), have dark

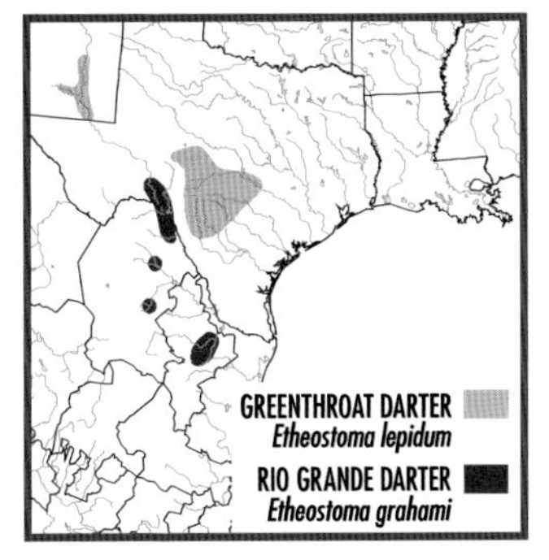

bars on side, blue edge on 1st dorsal fin; Orangethroat Darter *lacks* many small red or black spots on side.

CHRISTMAS DARTER *Etheostoma hopkinsi* Pl. 53

IDENTIFICATION: Has *10–12 dark green bars* on side, separated by *brick red* on large male, yellow on female. Thin green edge, middle red band on 1st dorsal fin. Yellow to green above, 8 dark green saddles; yellow to green below. Three black spots on caudal fin base. Black teardrop. Incomplete lateral line; 39–52 (usually 40–49) lateral scales. To 2½ in. (6.6 cm). **RANGE:** Savannah, Ogeechee, and Altamaha river drainages, SC and GA. Fairly common. **HABITAT:** Rocky riffles of creeks and small to medium rivers. **REMARKS:** Two subspecies. *E. h. binotatum*, in Savannah R. drainage, has unscaled opercle, 2 dark squares on nape. *E. h. hopkinsi*, in Ogeechee and Altamaha river drainages, has fully scaled opercle, no or faint squares on nape. **SIMILAR SPECIES:** (1) Savannah Darter, *E. fricksium* (Pl. 51), has broad dark brown stripe along side, red edge on 1st dorsal fin. (2) Gulf Darter, *E. swaini* (Pl. 53), is deeper bodied, has blue (on male) or brown (female) bars on side, orange belly on male, thin dark lines on upper side.

DIRTY DARTER *Etheostoma olivaceum* Pl. 54

IDENTIFICATION: *Dark brown to black bands* on 2d dorsal and caudal fins (rows of *crescents* on fins of breeding male). *Thin black stripes*, black mottling, sometimes black bars, on dusky (dirty) side. Olive to gray above, white to light brown below. *Long sharp snout*, unscaled opercle, scaled nape, no teardrop. Breeding male black; 2d dorsal fin has black margin; usually 12 rays; 4 branches per ray, 2d and 3d branches equal in length (Fig. 63). Uninterrupted infraorbital canal. Incomplete lateral line; 44–58 lateral scales. To 3¼ in. (8 cm). **RANGE:** Lower Caney Fork system and nearby tributaries of Cumberland R., cen. TN. Common. **HABITAT:** Rocky pools and nearby riffles of headwaters and creeks.

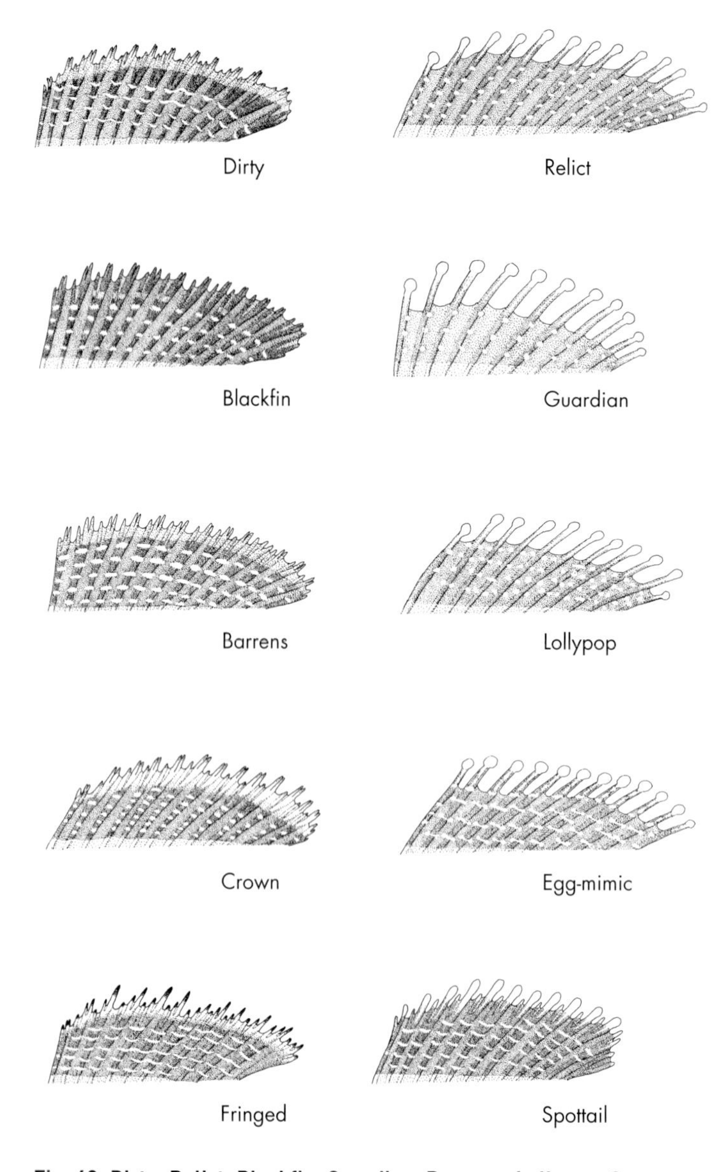

Fig. 63. Dirty, Relict, Blackfin, Guardian, Barrens, Lollypop, Crown, Egg-mimic, Fringed, and Spottail darters—second dorsal fin of breeding male.

Often in very shallow water. **REMARKS:** Dirty Darter and relatives (*next 23 species*—through Striated Darter, *E. striatulum*) are 1 of 2 groups of darters (other is Johnny Darter, *E. nigrum,* and relatives) in which the eggs are laid in a single-layer cluster on underside of a stone and guarded by a male. Females of both groups have a wide, flat genital papilla; papillae of other darters usually are tubular. **SIMILAR SPECIES:** (1) Spottail Darter, *E. squamiceps*, and related species (Pl. 54) have *scales* on cheek and opercle, teardrop, blunter snout; *lack* stripes on side, crescents on caudal fin of breeding male.

SPOTTAIL DARTER *Etheostoma squamiceps* Not shown

IDENTIFICATION: *Three vertically aligned black spots* on caudal fin base; *bold dark brown bands* (alternating black and white bands on breeding male) on 2d dorsal and caudal fins. Brown above, *dark brown mottling* on back and side; black teardrop. Scaled cheek and opercle; *incomplete lateral line*. Breeding male has black head, body, and fins (except pectoral) and during spawning develops wide black and white bars on side. Second dorsal fin of breeding male has *small white knobs* (3 branches per ray; 2d and 3d branches equal in length, adnate, tipped with small white knob; Fig. 63); usually 13 rays; fin membrane extends from base nearly to tips of fin rays. Usually 9 dorsal spines. Interrupted infraorbital canal. Has 38–60 lateral scales. To 3½ in. (8.8 cm). **RANGE:** Lower Ohio R. basin (Green R. drainage, KY and TN, to Bay Creek system, se. IL), including extreme lower Wabash R. drainage, IN and IL; Red R. system (Cumberland R. drainage), KY and TN. Common; locally abundant. **HABITAT:** Rocky pools and adjacent riffles of headwaters, creeks, and small rivers. **SIMILAR SPECIES:** (1) Spottail Darter lookalikes (Pl. 54): Next 8 species (through Relict Darter, *E. chienense*) are most easily distinguished from one another and Spottail Darter as breeding males (Fig. 63). It often is necessary to rely on microscopic examination of specimens and geography to identify them.

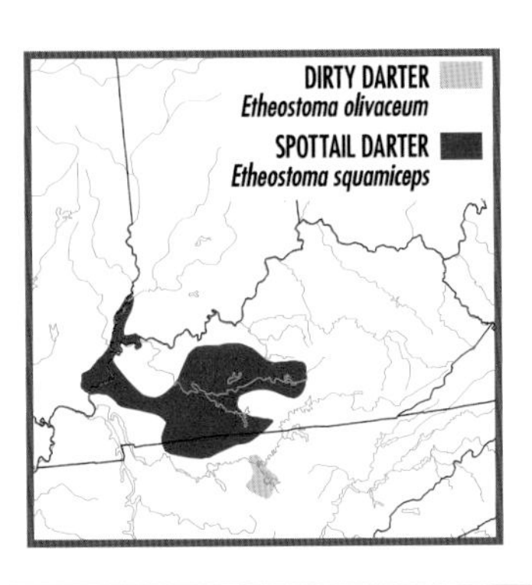

FRINGED DARTER *Etheostoma crossopterum* **Pl. 54**

IDENTIFICATION: Similar to Spottail Darter, *E. squamiceps*, but 2d dorsal fin of breeding male has *white edge* (may be small black tips on rays), *3 non-adnate branches* per ray, *3d branch much longer than 2d* (Fig. 63); usually 12–13 rays; fin membrane extends from base nearly to tips of fin rays. First dorsal fin of large male has dull orange band. Usually 9 dorsal spines. Interrupted infraorbital canal. Has 45–62 lateral scales. To 4 in. (10 cm). **RANGE:** Middle and lower Cumberland R. drainage, KY and TN (mostly below Caney Fork, but isolated records from Obey R. system and Barren Fork Collins R.); middle Duck R. system, including upper Buffalo R., TN; Shoal Creek system (Tennessee R. drainage), TN and AL. Also in Mississippi R. tributaries: Cache R., s. IL, Reelfoot Lake and Bear Creek, w. TN. Common; locally abundant. **HABITAT:** Rocky pools and adjacent riffles of headwaters and creeks. **SIMILAR SPECIES:** (1) See Spottail Darter, *E. squamiceps*, and related species (Fig. 63). (2) Breeding males of Blackfin Darter, *E. nigripinne*, and (3) Dirty Darter, *E. olivaceum* (Pl. 54), have *black edge* on 2d dorsal fin, 3d branch of each dorsal ray *equal* to 2d ray; Blackfin Darter has 8–12 light bands on caudal fin (Fringed Darter has 5–9).

BLACKFIN DARTER *Etheostoma nigripinne* **Not shown**

IDENTIFICATION: Similar to Spottail Darter, *E. squamiceps*, but 2d dorsal fin of breeding male has *black edge* (Fig. 63); 3 branches per ray, 2d and 3d branches equal in length, non-adnate; usually 13 rays; fin membrane extends from base nearly to tips of fin rays. First dorsal fin of large male has thin bright orange band. Usually 8 dorsal spines. Interrupted infraorbital canal. Has 43–53 lateral scales. To 3½ in. (8.8 cm). **RANGE:** Tennessee R. drainage from Paint Rock R. downstream to Duck R. system (in upper and lower Duck R. system), TN, AL, and MS; absent in Cypress Creek and most of Shoal Creek systems. Common; locally abundant. **HABITAT:** Rocky pools and adjacent riffles of headwaters, creeks, and small rivers. **SIMILAR SPECIES:** (1) See Spottail Darter, *E. squamiceps*, and related species (Fig. 63). (2) Barrens Darter, *E. forbesi*, and (3) Crown Darter, *E. corona*, have yellow edge on 2d dorsal fin of breeding male, usually 14–15 dorsal rays.

BARRENS DARTER *Ethesostoma forbesi* **Not shown**

IDENTIFICATION: Nearly identical to Blackfin Darter, *E. nigripinne*, but 2d dorsal fin of breeding male has *bright yellow-gold edge* (Fig. 63); usually 14–15 rays. Usually 8–10 light bands on caudal fin, 4-5 rows of clear to light yellow bars on 2d dorsal fin. Has 41–54 lateral scales. To 3½ in. (8.9 cm). **RANGE:** Barren Fork Collins R. (Caney Fork system), cen. TN. Rare. **HABITAT:** Rocky pools and adjacent riffles of headwaters and creeks. **SIMILAR SPECIES:** (1) See Blackfin Darter, *E. nigripinne*. (2)

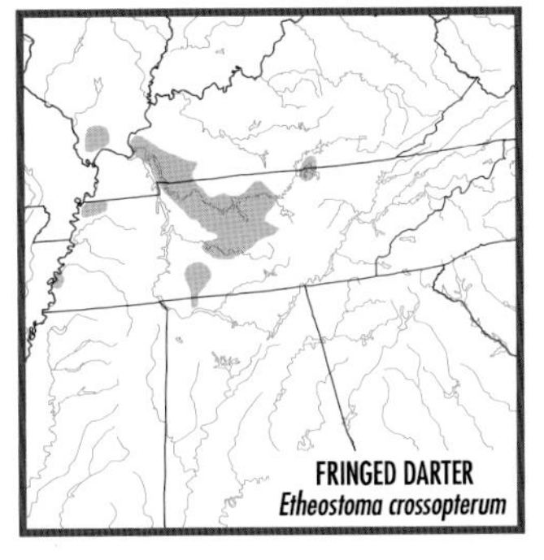

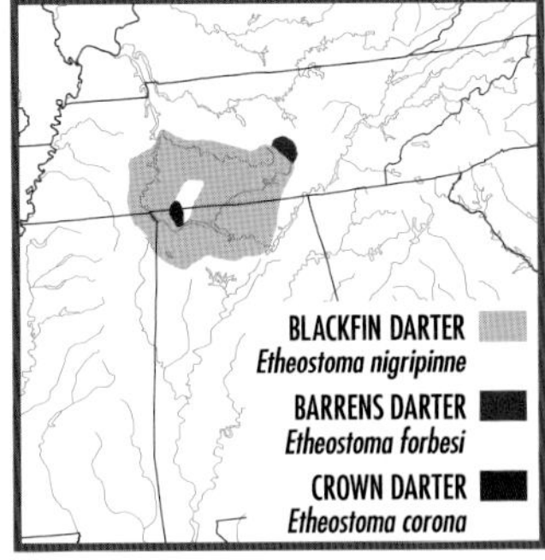

Crown Darter, *E. corona*, usually has 11–12 light bands on caudal fin, 6–7 rows of clear to light yellow bars on 2d dorsal fin.

CROWN DARTER *Etheostoma corona* Not shown

IDENTIFICATION: Nearly identical to Blackfin Darter, *E. nigripinne*, but 2d dorsal fin of breeding male has *bright yellow margin* (Fig. 63); usually 9 dorsal spines, 14–15 rays. Usually 11–12 light bands on caudal fin, 6–7 rows of clear to light yellow bars on 2d dorsal fin. Has 42–56 lateral scales. To 3½ in. (9.2 cm). **RANGE:** Cypress Creek (Tennessee R. drainage), sw. TN and nw. AL. Common; locally abundant. **HABITAT:** Rocky pools and adjacent riffles of headwaters, creeks, and small rivers. **SIMILAR SPECIES:** (1) See Blackfin Darter, *E. nigripinne*. (2) Barrens Darter, *E. forbesi*, usually has 8–10 light bands on caudal fin, 4–5 rows of clear to light yellow bars on 2d dorsal fin.

LOLLYPOP DARTER *Etheostoma neopterum* Pl. 54

IDENTIFICATION: Similar to Spottail Darter, *E. squamiceps*, but 2d dorsal fin of breeding male usually has 11 rays tipped with *large yellow knobs* (Fig. 63); *2 branches per ray, equal in length, adnate*; fin membrane extends about half distance from base to tips of fin rays. Usually 9 dorsal spines. Usually uninterrupted infraorbital canal. Has 41–52 lateral scales. To 3 in. (7.6 cm). **RANGE:** Shoal Creek System (Tennessee R. drainage), s.-cen. TN and nw. AL. Uncommon. **HABITAT:** Rocky and sandy pools of headwaters and creeks. **SIMILAR SPECIES:** (1) See Spottail Darter, *E. squamiceps*, and related species (Fig. 63). (2) See Guardian Darter, *E. oophylax*. (3) Egg-mimic Darter, *E. pseudovulatum*, has interrupted infraorbital canal; breeding male has yellow bars on dorsal rays above fin membrane.

GUARDIAN DARTER *Etheostoma oophylax* Not shown

IDENTIFICATION: Nearly identical to Lollypop Darter, *E. neopterum*, but each row of windows (clear areas) on 2d dorsal fin of breeding male has *2 windows* per interradial membrane (Fig. 63), *8–12 yellow bands* on caudal fin (Lollypop Darter has 1 window per membrane, 5–9 yellow bands on caudal fin). Has 41–59 lateral scales. To 3½ in. (8.9 cm). **RANGE:** Tributaries of lower Tennessee R. in w. KY and w. TN upstream to Decatur and Perry counties, TN, including extreme lower Duck R. system. Common. **HABITAT:** Rocky and sandy pools of headwaters and creeks. **SIMILAR SPECIES:** (1) See Lollypop Darter, *E. neopterum* (Pl. 54).

EGG-MIMIC DARTER *Etheostoma pseudovulatum* Not shown

IDENTIFICATION: Similar to Spottail Darter, *E. squamiceps*, but 2d dorsal fin of breeding male usually has 12 rays tipped with *large yellow knobs; 2 branches per ray, equal in length, adnate* (Fig. 63); fin membrane extends about half distance from base to tips of fin rays; *yellow bars* on dorsal rays above fin membrane. Usually 8 dorsal spines. Usually interrupted infraorbital canal. Has 42–52 lateral scales. To 3 in. (7.6 cm). **RANGE:** Tributaries of Duck R. system, w.-cen. TN. Common in small range. **HABITAT:** Rocky and sandy pools of headwaters and creeks. **SIMILAR SPECIES:** (1) See Spottail Darter, *E. squamiceps*, and related species (Fig. 63). (2) Lollypop Darter, *E. neopterum* (Pl. 54), and (3) Guardian Darter, *E. oophylax*, have uninterrupted infraorbital canal; breeding males *lack* yellow bars on dorsal rays above fin membrane.

RELICT DARTER *Etheostoma chienense* Not shown

IDENTIFICATION: Similar to Spottail Darter, *E. squamiceps*, but 2d dorsal fin of breeding male usually has 12–13 rays; *2 branches per ray, equal in length, adnate; small white knobs* on tips of rays (Fig. 63); fin membrane extends about ⅔ distance from base to tips of fin rays.

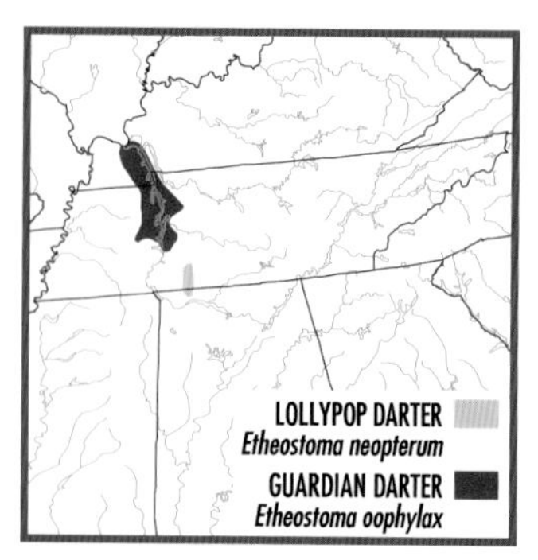

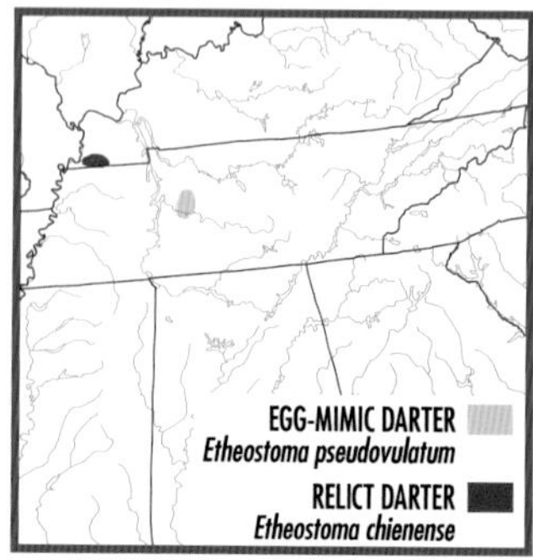

Usually 8–9 dorsal spines. Uninterrupted infraorbital canal. Has 42–54 lateral scales. To 4 in. (10 cm). **RANGE:** Bayou du Chien system, w. KY. Common in small range; protected as an *endangered species.* **HABITAT:** Sandy pools of headwaters and creeks. **SIMILAR SPECIES:** (1) See Spottail Darter, *E. squamiceps*, and related species (Fig. 63). (2) Lollypop Darter, *E. neopterum* (Pl. 54), (3) Guardian Darter, *E. oophylax*, and (4) Egg-mimic Darter, *E. pseudovulatum,* have *large yellow knobs* on 2d dorsal fin of breeding male.

STRIPETAIL DARTER *Etheostoma kennicotti* Pl. 54

IDENTIFICATION: *Black bands* on 2d dorsal, caudal, and often pectoral fins. *Gold knobs* on tips of dorsal spines of adult. Tan to yellow above, 6–7 brown saddles; dark brown blotches on upper side, *larger dark blotches* along side. Teardrop thin or absent; moderate snout; unscaled nape. Widely interrupted infraorbital canal (Fig. 55), 4 front pores, 1 rear pore. Incomplete lateral line; 38–53 lateral scales. Usually 7 dorsal spines, 11–12 dorsal rays, 7 anal rays. To 3¼ in. (8.3 cm). **RANGE:** Tributaries of Ohio R. in s. IL and w. KY; Green R. drainage, KY; upper Cumberland R. drainage—Big South Fork and above, KY and TN; Tennessee R. drainage, KY, TN, GA, AL, and MS. Fairly common but localized. **HABITAT:** Rocky pools of headwaters, creeks, and small rivers. **SIMILAR SPECIES:** See Pl. 54. (1) Fantail Darter, *E. flabellare*, has broadly joined branchiostegal membranes, protruding lower jaw, pointed snout, usually stripes or rows of black spots on side, 2 rear infraorbital canal pores. (2) Spottail Darter, *E. squamiceps*, and related species have scaled nape, *no* gold knobs on dorsal spines, narrowly interrupted infraorbital canal. (3) Teardrop, *E. barbouri*, (4) Slabrock, *E. smithi*, and (5) Striated, *E. striatulum*, darters (Pl. 54), resemble small Stripetail Darters but usually have 8–9 dorsal spines, 13–14 dorsal rays, 9 anal rays.

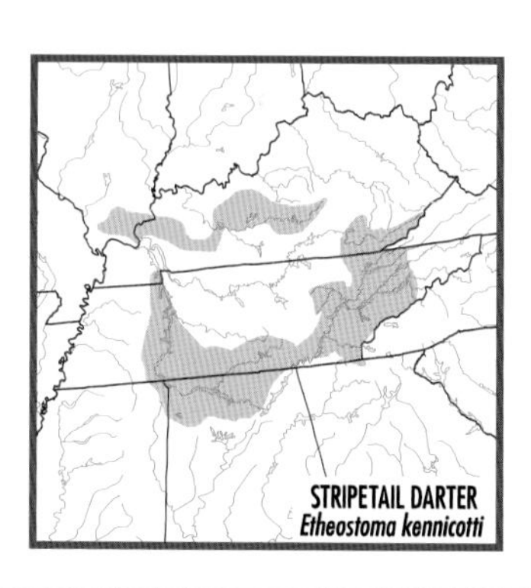

FANTAIL DARTER *Etheostoma flabellare* Pl. 54

IDENTIFICATION: *Black bands* on 2d dorsal and caudal fins. *Gold knobs* on tips of dorsal spines of adult—large on male, small on female. *Broadly joined branchiostegal membranes* (Fig. 56); *protruding lower jaw* (over most of range; see Remarks); pointed snout, unscaled nape. Teardrop thin or absent. Infraorbital canal widely interrupted (Fig. 55), 4 front pores, 2 rear pores. Brown to olive above, yellow to white below. Color highly variable; some populations have thin black stripes on side, others have black bars, mottling, or rows of black spots on side. Breeding male is bright yellow (in parts of KY and TN) to olive brown with black head. Incomplete lateral line; 38–60 (usually 45–55) lateral scales. To 3¼ in. (8.4 cm). **RANGE:** Atlantic, Great Lakes, and Mississippi R. basins from s. QC to MN, and south to SC (Peedee R. system), n. AL, and ne. OK. Abundant. **HABITAT:** Rocky riffles of creeks and small to medium rivers. **REMARKS:** Three subspecies. *E. f. humerale*, in Atlantic drainages from lower Susquehanna R. to Cape Fear R., has terminal mouth, small eye, black anal and pelvic fins on breeding male, usually 10 dark bars along side, 7 dorsal spines, 12 pectoral rays. *E. f.* subspecies (unnamed) in upper Tennessee R. drainage (as far south as Little Tennessee R.), New R., and headwaters of Shavers Fork Cheat R. (Monongahela R. system), has terminal mouth, black anal and pelvic fins on breeding male, usually 8–10 dark bars along side, 8 dorsal spines, 13 pectoral rays. *E. f. flabellare*, elsewhere, has protruding lower jaw, upturned mouth, dusky white anal and pelvic fins on breeding male. **SIMILAR SPECIES:** (1) See Carolina Fantail Darter, *E. brevispinum*. (2) Duskytail Darter, *E. percnurum* (Pl. 54), has large black specks on side of head, large eye (diameter equal to or longer than snout), black edge on pectoral, anal, 2d dorsal, and caudal fins of breeding male. (3) Stripetail Darter, *E. kennicotti* (Pl. 54), has *narrowly* joined branchiostegal membranes, *less* protruding lower jaw; series of black blotches along side, no stripes or rows of small black spots, 1 rear infraorbital canal pore.

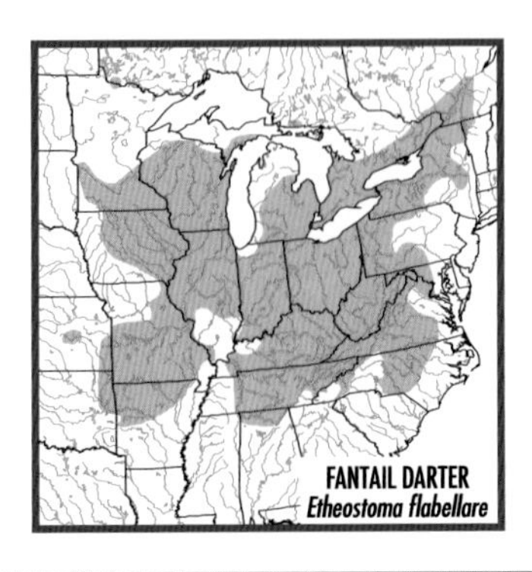

CAROLINA FANTAIL DARTER *Etheostoma brevispinum* Pl. 54

IDENTIFICATION: Similar to Fantail Darter, *E. flabellare*, but has *7 or fewer dark bars on side*; 2 or more bars *wedge-shaped* (rectangular in Fantail Darter) on large male. No dark stripes on side. Breeding male golden tan. Incomplete lateral line; 39–57 (usually 42–50) lateral scales. To 3 in. (8.4 cm). **RANGE:** Above Fall Line in Yadkin (upper Peedee R.), Santee, and Savannah rivers of NC, SC, and VA. Common but localized. **HABITAT:** Rocky riffles of creeks and small to medium rivers. **SIMILAR SPECIES:** (1) See Fantail Darter, *E. flabellare* (Pl. 54).

DUSKYTAIL DARTER *Etheostoma percnurum* Pl. 54

IDENTIFICATION: *Black specks* (largest on juvenile) on side of head; *large eye* (diameter equal to or longer than snout). Broadly joined branchiostegal membranes (Fig. 56); *protruding lower jaw*; unscaled nape. Yellow-brown to olive above, white below. No teardrop. *Gold knobs* on tips of dorsal spines of adult—large on male, small on female. Breeding male has black edge on pectoral, anal, 2d dorsal, and caudal fins; no marbling on 2d dorsal fin, no marbling or bands in middle of caudal fin. Infraorbital canal widely interrupted (Fig. 55), 4 front pores, 2 rear pores. Incomplete lateral line; 38–48 (usually 42–44) lateral scales; 16–31 (usually 20–24) pored lateral-line scales, usually 17–18 branched caudal rays. To 2½ in. (6.4 cm). **RANGE:** Copper Creek (Clinch R. system), VA; formerly South Fork Holston R., TN. Rare; protected as an *endangered species*. **HABITAT:** Gravel, rubble, and slabrock pools and runs of small to medium rivers. **SIMILAR SPECIES:** (1) Marbled Darter, *E. marmorpinnum,* (2) Citico Darter, *E. sitikuense,* and (3) Tuxedo Darter, *E. lemniscatum* (Pl. 54), usually have 15–16 branched caudal rays, black tessellations or bands on middle of caudal fin. Citico and Tuxedo darters usually have more than 25 pored lateral-line scales. Marbled and Citico darters have dusky to black marbling on 2d dorsal fin of breeding male. (4) Fantail Darter,

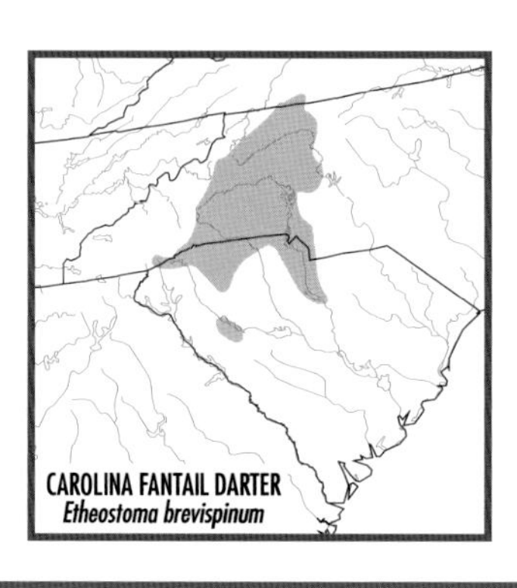

E. flabellare (Pl. 54), has *small specks* on side of head, *smaller eye*, no bold black edge on fins of breeding male.

MARBLED DARTER *Etheostoma marmorpinnum* Not shown

IDENTIFICATION: Nearly identical to Duskytail Darter, *E. percnurum*, but usually has *most of belly covered by scales*, usually 15 branched caudal rays, dusky bands on middle of caudal fin, black marbling on 2d dorsal fin of breeding male. Has 39–47 lateral scales; 19–31 (usually 22–27) pored lateral-line scales. To 2½ in. (6.3 cm). **RANGE:** Little River (Tennessee R. system), TN. Rare. **HABITAT:** Gravel, rubble, and slabrock pools and runs of small to medium rivers. **SIMILAR SPECIES:** (1) Duskytail Darter, *E. percnurum* (Pl. 54), (2) Citico Darter, *E. sitikuense,* and (3) Tuxedo Darter, *E. lemniscatum* (Pl. 54), usually have *less than 30 percent of belly scaled*. Duskytail Darter usually has 17–18 branched caudal rays, no black bands on middle of caudal fin. Duskytail and Tuxedo darters lack black marbling on 2d dorsal fin of breeding male. Citico Darter usually has 30–33 pored lateral-line scales.

CITICO DARTER *Etheostoma sitikuense* Not shown

IDENTIFICATION: Nearly identical to Duskytail Darter, *E. percnurum*, but *usually has 30–33 pored lateral-line scales* (range 27–34), 15–16 branched caudal rays, black tessellations in middle of caudal fin, dusky marbling on 2d dorsal fin of breeding male. Has 40–45 lateral scales. To 2½ in. (6.3 cm). **RANGE:** Citico and Abrams creeks (Tennessee R. system), e. TN. Rare. **HABITAT:** Gravel, rubble, and slabrock pools and runs of creeks. **SIMILAR SPECIES:** (1) Duskytail Darter, *E. percnurum* (Pl. 54), (2) Marbled Darter, *E. marmorpinnum*, and (3) Tuxedo Darter, *E. lemniscatum* (Pl. 54), *usually have 28 or fewer pored lateral-line scales*. Duskytail usually has 17–18 branched caudal rays, lacks black tessellations on middle of caudal fin. Duskytail and Tuxedo darters lack marbling on 2d dorsal fin of breeding male.

TUXEDO DARTER *Etheostoma lemniscatum* Pl. 54

IDENTIFICATION: Similar to Duskytail Darter, *E. percnurum*, but is more slender, has *anal fin origin behind dorsal fin origin* (under 2d or 3d ray); breeding male has *bold black edge* on 2d dorsal and caudal fins, black edge on upper half of pectoral fin, black tessellations on middle of caudal fin, no black marbling on 2d dorsal fin. Usually 15–16 branched caudal rays. Has 42–49 lateral scales; 21–31 (usually 26–28) pored lateral-line scales. To 2½ in. (6.5 cm). **RANGE:** Big South Fork of Cumberland R., KY. Rare. **HABITAT:** Gravel, rubble, and slabrock pools and runs of medium-sized rivers. **SIMILAR SPECIES:** (1) Duskytail Darter, *E. percnurum* (Pl. 54), (2) Marbled Darter, *E. marmorpinnum,* and (3) Citico Darter, *E. sitikuense,* have anal fin origin *under 2d dorsal fin origin*, are deeper bodied; breeding males usually have *dusky edge* on 2d dorsal and caudal fins. Duskytail Darter usually has 17–18 branched caudal rays, usually 20–24 pored lateral-line scales, no black tessellations on caudal fin. Citico Darter usually has 30–33 pored lateral-line scales.

BARCHEEK DARTER *Etheostoma obeyense* Pl. 54

IDENTIFICATION: *Narrow iridescent bar* on cheek, areas in front of and behind teardrop *dusky*. Yellow-brown above; dark brown blotches, sometimes bars, *no stripes or rows of dark spots,* on side. Brown bars on fins; red edge, black spot near front of 1st dorsal fin. Large male has red dorsal, caudal, and anal fins; black paired fins; black edge on anal and caudal fins. Widely interrupted infraorbital canal (Fig. 55). Four front pores, 2 rear pores. Incomplete lateral line (10–26 pores); 39–56 lateral scales. To 3¼ in. (8.4 cm). **RANGE:** Middle Cumberland R. drainage from Big South Fork to Obey R., KY and TN. Fairly common. **HABITAT:** Rocky pools of headwaters, creeks, and small rivers. **SIMILAR SPECIES:** See Pl. 54. Other "Barcheek Darters" (Pl. 54)—next 6 species (through Striated Darter, *E. striatulum*)—are similar to one another and Barcheek Darter but differ in color and scale and fin ray counts. Breeding males are easiest to identify. Two species of this group are never found together. (1) See Slabrock Darter, *E. smithi*. (2) Striped Darter, *E. virgatum*, (3) Corrugated Darter, *E. basilare*, and (4) Stone Darter, *E. derivativum*, have *dark stripes* on side. (5) Teardrop Darter, *E. barbouri*, and (6) Striated Darter, have 3 front and 1 rear infraorbital canal pores, *rows of small dark spots* on side; are smaller—to 2¼ in. (5.5 cm).

SLABROCK DARTER *Etheostoma smithi* Pl. 54

IDENTIFICATION: Similar to Barcheek Darter, *E. obeyense*, but is *smaller*; has *more darkly outlined scales, 3 front and 1 rear* infraorbital canal pores, fewer than 14 lateral-line pores. Has 41–54 lateral scales. To 2½ in. (6.2 cm). **RANGE:** Lower Cumberland (below Caney Fork) and

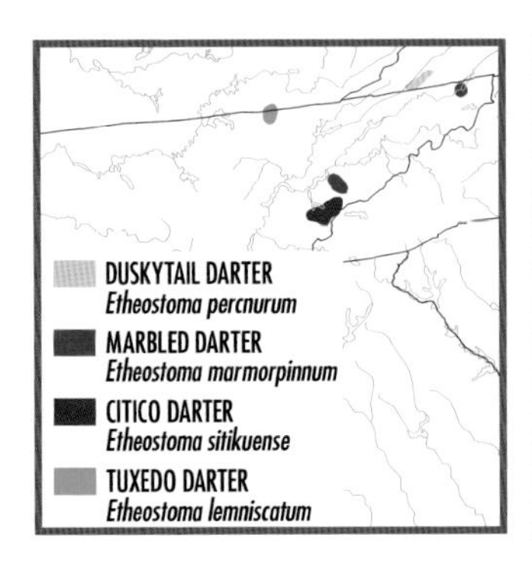

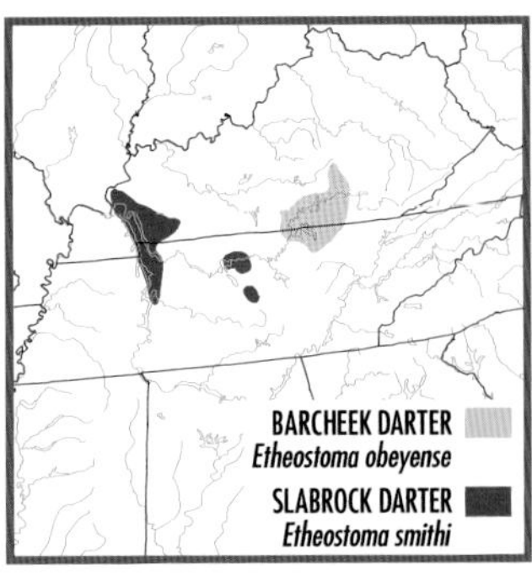

lower Tennessee river (lower Duck R. and downstream) drainages, KY and TN. Fairly common. **HABITAT:** Rocky pools of headwaters, creeks, and small rivers; rocky margins of medium-sized rivers, impoundments. **SIMILAR SPECIES:** (1) See Barcheek Darter, *E. obeyense* (Pl. 54).

STRIPED DARTER *Etheostoma virgatum* Pl. 54

IDENTIFICATION: *Dark brown stripes* on side. *Narrow iridescent bar* on cheek; regions in front of and behind teardrop dusky. Yellow-brown above, 6–8 dark brown saddles; small dark blotches along side; white to yellow below. *Brown bands* on fins; red edge, black spot near front of 1st dorsal fin. Breeding male has red dorsal, caudal, and anal fins; black pelvic fins; *bright white spots* on pectoral fin; dark blue edge on red pectoral and anal fins. Incomplete lateral line. Widely interrupted infraorbital canal (Fig. 55), 4 front pores, 2 rear pores. Usually 9 anal, 13 dorsal rays. Has 41–61 (usually 48 or more) lateral scales. To 3 in. (7.8 cm). **RANGE:** Rockcastle R., Buck Creek, and Beaver Creek systems, e. KY. Common. **HABITAT:** Rocky pools of headwaters, creeks, and small to medium rivers. **SIMILAR SPECIES:** See Pl. 54. See (1) Corrugated Darter, *E. basilare*, and (2) Stone Darter, *E. derivativum*. (3) Barcheek, *E. obeyense*, (4) Teardrop, *E. barbouri*, (5) Slabrock, *E. smithi*, and (6) Striated, *E. striatulum*, *darters lack* dark stripes on side; Teardrop, Slabrock, and Striated darters have 3 front, 1 rear infraorbital canal pores.

CORRUGATED DARTER *Ethesotoma basilare* Not shown

IDENTIFICATION: Similar to Striped Darter, *E. virgatum*, but usually has 10 anal rays, 14 dorsal rays, 41–48 lateral scales; is smaller. Breeding male *lacks* bright white spots on pectoral fin. To 2¾ in. (7.1 cm). **RANGE:** Upper Caney Fork system, cen. TN. Locally common. **HABITAT:** Rocky pools of headwaters, creeks, and small to medium rivers. **SIMILAR SPECIES:** (1) See Striped Darter, *E. virgatum* (Pl. 54).

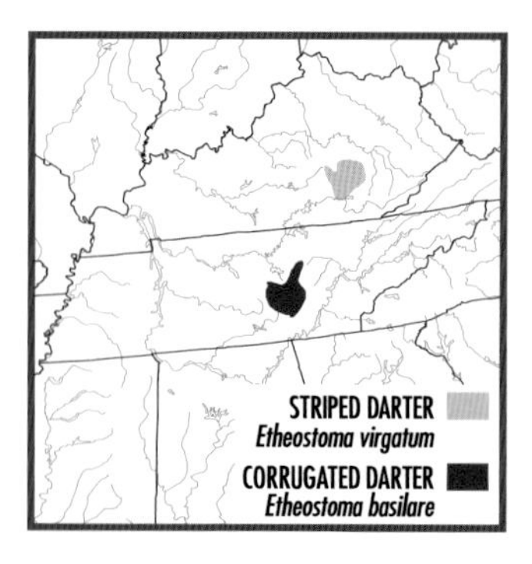

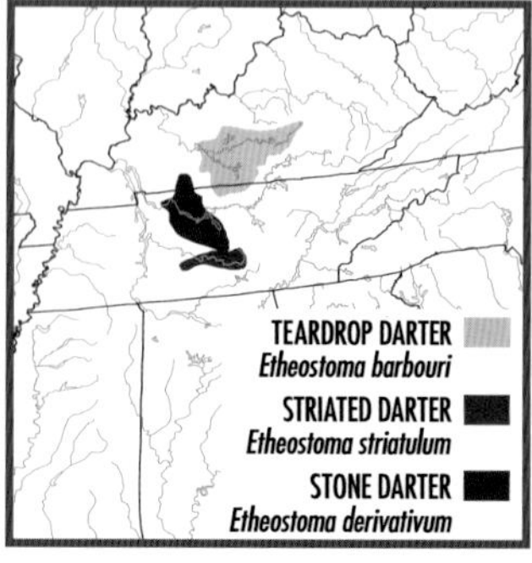

STONE DARTER *Etheostoma derivativum* Not shown

IDENTIFICATION: Similar to Striped Darter, *E. virgatum*, but usually has 15 or fewer pored lateral scales (Striped Darter usually has 13 or more); is smaller. Breeding male *lacks* bright white spots on pectoral fin, has conspicuous *dark blue edge* on 2d dorsal and caudal fins. To 2¾ in. (6.9 cm). **RANGE:** Lower Cumberland R. drainage (West Fork Stones R. to Red R.), KY and TN. Common in TN; rare in KY. **HABITAT:** Rocky pools of headwaters, creeks, and small to medium rivers. **SIMILAR SPECIES:** (1) See Striped Darter, *E. virgatum* (Pl. 54).

TEARDROP DARTER *Etheostoma barbouri* Pl. 54

IDENTIFICATION: *Wide iridescent bar on cheek* (covers most of cheek). *White areas* in front of and behind *large black teardrop*. Yellow-brown above; *rows of small dark brown spots* and blotches on side. Brown bars on fins; red edge, black spot near front of 1st dorsal fin. Large male has red dorsal, caudal, and anal fins; black paired fins; black edge on anal and caudal fins. Widely interrupted infraorbital canal (Fig. 55), 3 front pores, 1 rear pore. Usually 9 dorsal spines. Incomplete lateral line; 40–49 lateral scales. To 2¼ in. (6 cm). **RANGE:** Middle and upper Green R. drainage, KY and TN. Fairly common. **HABITAT:** Rocky pools of headwaters, creeks, and small rivers. **SIMILAR SPECIES:** See Pl. 54. (1) See Striated Darter, *E. striatulum*. (2) Slabrock Darter, *E. smithi*, and (3) Barcheek Darter, *E. obeyense*, have narrow iridescent bar on cheek, *dusky areas* in front of and behind teardrop, *no* rows of dark spots on side; Barcheek Darter has 4 front and 2 rear infraorbital canal pores. (4) Striped Darter, *E. virgatum*, (3) Corrugated Darter, *E. basilare*, and (4) Stone Darter, *E. derivativum*, have *dark stripes* on side, 4 front, 2 rear infraorbital canal pores.

STRIATED DARTER *Etheostoma striatulum* Pl. 54

IDENTIFICATION: Similar to Teardrop Darter, *E. barbouri*, but has *narrow* bar on cheek, *dusky* areas in front of and behind teardrop, darker rows of spots on side. Usually 8 dorsal spines. Has 38–50 lateral scales. To 2¼ in. (5.6 cm). **RANGE:** Duck R. system, cen. TN. Rare. **HABITAT:** Rocky pools of headwaters and creeks. **SIMILAR SPECIES:** (1) See Teardrop Darter, *E. barbouri* (Pl. 54).

LEAST DARTER *Etheostoma microperca* Pl. 55

IDENTIFICATION: *Deep, compressed body. Extremely short (0–3 pores) lateral line*. Olive above, dark green saddles; green blotches along side, rows of dark green spots on upper and lower sides; white to yellow below. *Large teardrop*. Black edge and base, middle red band on 1st dorsal fin of male. *Orange or red anal and pelvic fins, large lateral flap* on pelvic fin of breeding male. Unscaled breast. Usually 2 anal spines; 2–3 infraorbital canal pores (Fig. 55); 30–36 lateral scales. To 1¾ in. (4.4 cm).

RANGE: Great Lakes, Hudson Bay, and Mississippi R. basins from ON to MN south to s. OH, cen. IN, and cen. IL; Ozark-Ouachita drainages of s. MO, se. KS, nw. AR, and e. OK; isolated populations in n. KY near Louisville (extinct), Jefferson Co., MO, and Blue R., OK. Common; sometimes abundant in spring-fed streams. **HABITAT:** Quiet, vegetated lakes, headwaters, creeks, and small rivers. Usually over mud and sand. **SIMILAR SPECIES:** See Pl. 55. (1) Cypress Darter, *E. proeliare*, (2) Fountain Darter, *E. fonticola*, and (3) Iowa Darter, *E. exile*, are more *slender, lack* dark green on body, orange anal and pelvic fins on male; Cypress and Fountain darters have 4 infraorbital canal pores, Iowa Darter has 8.

CYPRESS DARTER *Etheostoma proeliare* Pl. 55

IDENTIFICATION: *Short (0–9 pores), strongly arched lateral line.* Olive above, 6–9 dark brown saddles; *black or brown dashes* along side, spots on upper and lower sides. Thin teardrop. Black edge and base, black spot at front, middle red band on 1st dorsal fin of male. Black anal and pelvic fins, *large lateral flap* on pelvic fin of breeding male. Unscaled breast. Two anal spines; 4 infraorbital canal pores (Fig. 55); 34–38 lateral scales. To 2 in. (4.8 cm). **RANGE:** Mississippi R. basin from s. IL and e. OK to Gulf; Gulf Slope drainages from Choctawhatchee R., FL, to San Jacinto R., TX. Primarily on Coastal Plain. Common. **HABITAT:** Standing or slow-flowing water, usually in vegetation over mud. **SIMILAR SPECIES:** See Pl. 55. (1) See Fountain Darter, *E. fonticola*. (2) Least Darter, *E. microperca*, has 2–3 infraorbital canal pores, orange or red anal and pelvic fins on male, large teardrop; is dark green, deeper bodied. (3) Slough Darter, *E. gracile*, has green bars or squares on side, *13–27 lateral-line pores*, usually 8 infraorbital canal pores.

FOUNTAIN DARTER *Etheostoma fonticola* Pl. 55

IDENTIFICATION: Similar to Cypress Darter, *E. proeliare*, but has *dark brown crosshatching* on upper and lower sides, *1 anal spine*. Has 31–37 lateral scales. To 1¾ in. (4.3 cm). **RANGE:** San Marcos and Comal

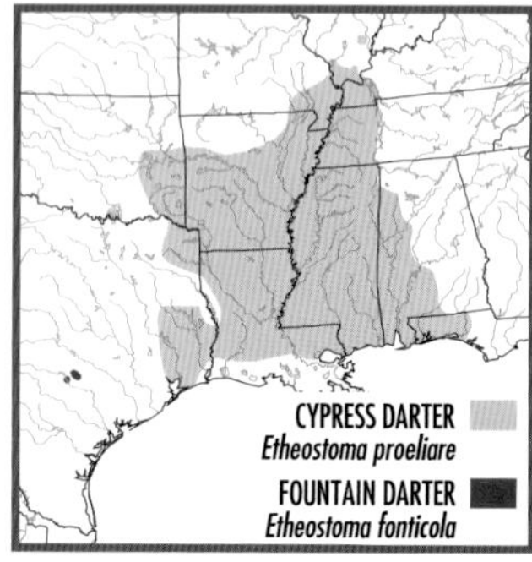

springs and their effluent rivers (Guadalupe R. system), s.-cen. TX. Common in San Marcos Spring; extirpated but reintroduced to Comal Spring. **HABITAT:** Vegetated springs, pools, and runs of effluent rivers. **REMARKS:** Common in San Marcos Spring; protected as an *endangered species*. **SIMILAR SPECIES:** (1) See Cypress Darter, *E. proeliare* (Pl. 55).

SAWCHEEK DARTER *Etheostoma serrifer* Pl. 55

IDENTIFICATION: *Red* around *2 bold black spots* on caudal fin base. *Incomplete yellow lateral line, strongly arched near front,* usually 28–38 pores. Tan above; dark brown mottling, often bars on side; green to white, often with black specks below. Teardrop absent or faint. Clear to dusky fins. Six infraorbital canal pores (Fig. 55); 44–66 lateral scales. Serrated preopercle. To 2¾ in. (6.8 cm). **RANGE:** Atlantic Coastal Plain from Dismal Swamp, s. VA, to Altamaha R. drainage, GA. Common in northern half of range, uncommon in southern half. **HABITAT:** Swamps; lakes; sluggish headwaters, creeks, and small rivers; usually near vegetation. **SIMILAR SPECIES:** (1) Swamp Darter, *E. fusiforme*, and (2) Carolina Darter, *E. collis* (both Pl. 55), *lack* 2 bold black spots at middle of caudal fin base, usually have 28 or fewer lateral-line pores.

SWAMP DARTER *Etheostoma fusiforme* Pl. 55

IDENTIFICATION: *Slender, compressed body.* Green to tan above, small dark saddles; dark green and brown mottling, 10–12 squares on side; white to yellow, *many black and brown specks* below. Thin teardrop; 3 dusky black spots on caudal fin base. Scaled breast. Widely interrupted infraorbital canal, 4–5 pores (Fig. 55). Incomplete lateral line, strongly arched near front; usually 28 or fewer pores; 40–63 (usually 46–56) lateral scales. To 2¼ in. (5.9 cm). **RANGE:** Seaboard Lowlands, Atlantic and Gulf coastal plains from s. ME to LA (Sabine R.) and se. OK (Red R.); Former Mississippi Embayment north to KY and se. MO; isolated population in San Jacinto R. drainage, TX. Introduced into French Broad system, NC. Common to abundant in coastal streams;

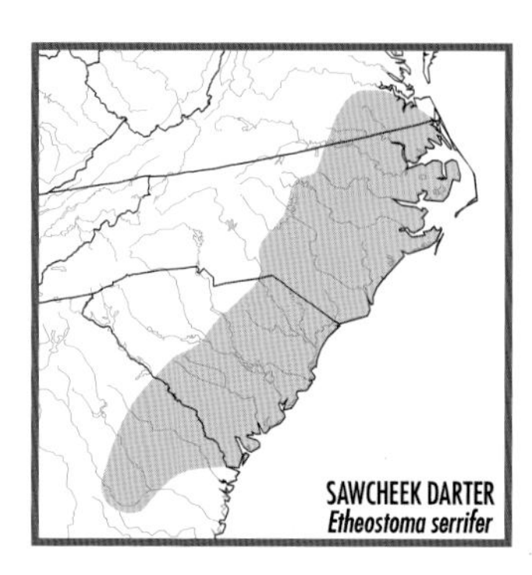

uncommon on Embayment. **HABITAT:** Standing or slow-flowing water over mud, sometimes sand; often in vegetation. **REMARKS:** Two subspecies. *E. f. fusiforme,* ME to Waccamaw R., NC, has 2 rear infraorbital canal pores (Fig. 55), 0–4 interorbital scales. *E. f. barratti,* rest of range, usually has 1 rear infraorbital canal pore, 5 or more interorbital scales. **SIMILAR SPECIES:** See Pl. 55. (1) Slough Darter, *E. gracile*, and (2) Backwater Darter, *E. zonifer,* have bright green bars on side, red band on 1st dorsal fin, unscaled breast. (3) Carolina Darter, *E. collis*, has many small dark brown spots on side, black spot at front of 1st dorsal fin of male, usually 47 or fewer lateral scales.

SLOUGH DARTER *Etheostoma gracile* Pl. 55

IDENTIFICATION: Yellow above, green saddles and wavy lines on back; bright *green bars* on side of male, green squares or mottling on female; yellow to white below. Blue-gray edge and base, *middle red band* on 1st dorsal fin (faint on female). Thin teardrop. Unscaled breast. Uninterrupted infraorbital canal (Fig. 55), usually 8 pores. *Incomplete lateral line, strongly arched near front*; 13–27 pores; 40–55 lateral scales. To 2¼ in. (6 cm). **RANGE:** Mississippi R. basin from cen. IL and ne. MO to LA; extends up Arkansas and Red river drainages to se. KS and e. OK. Gulf Slope drainages from Tombigbee R. (1 record), MS, to Nueces R., TX. Abundant, especially on Coastal Plain. **HABITAT:** Standing or slow-flowing water over mud; often in vegetation. **SIMILAR SPECIES:** See Pl. 55. (1) See Backwater Darter, *E. zonifer*. (2) Swamp Darter, *E. fusiforme*, *lacks* bright green bars on side, red band on 1st dorsal fin; has scaled breast, 4–5 infraorbital canal pores. (3) Cypress Darter, *E. proeliare*, has brown or black dashes, *no green*, on side; 0–9 lateral-line pores; 4 infraorbital canal pores.

BACKWATER DARTER *Etheostoma zonifer* Pl. 55

IDENTIFICATION: Nearly identical to Slough Darter, *E. gracile*, but has *interrupted* infraorbital canal (Fig. 55), usually *6 pores*. Has 41–53 lat-

eral scales. To 1¾ in. (4.4 cm). **RANGE:** Mobile Bay drainage, AL and MS; isolated population in Apalachicola R. drainage, AL and FL. Fairly common. **HABITAT:** Mud-bottomed, often vegetated, pools of sluggish creeks and small rivers. **SIMILAR SPECIES:** (1) See Slough Darter, *E. gracile* (Pl. 55).

CAROLINA DARTER *Etheostoma collis* Pl. 55

IDENTIFICATION: Light brown above; *many small dark brown spots* on side; *brown dashes or blotches* along side; white to yellow below. Black teardrop; black spot at front of 1st dorsal fin of male; 3 black spots on caudal fin base. Incomplete lateral line, strongly arched near front; 5–30 pores. Has 35–49 lateral scales; *1 anal spine*; usually no interorbital pores; 4–5 infraorbital canal pores (Fig. 55). To 2¼ in. (6 cm). **RANGE:** Atlantic Piedmont from Roanoke R. drainage, VA, to Santee R. drainage, SC. Uncommon. **HABITAT:** Muddy and rocky pools and backwaters of sluggish headwaters and creeks. **REMARKS:** Two subspecies. *E. c. lepidinion*, Roanoke and Neuse river drainages, has partly scaled breast, fully scaled nape. *E. c. collis*, Peedee and Santee river drainages, has unscaled breast, unscaled or partly scaled nape. **SIMILAR SPECIES:** (1) Sawcheek Darter, *E. serrifer* (Pl. 55), has 2 bold black spots at middle of caudal fin base, *2 anal spines*. (2) Swamp Darter, *E. fusiforme* (Pl. 55), is more slender, *lacks* small brown spots on side, has *2 anal spines.*

IOWA DARTER *Etheostoma exile* Pl. 55

IDENTIFICATION: *Slender body; long, narrow caudal peduncle.* Tan above; dark brown mottling, often short bars on side. Black teardrop. Blue edge and base, middle red band on 1st dorsal fin (faint on female). Breeding male has orange belly, alternating *blue and brick red bars* on side. Eight infraorbital canal pores (Fig. 55). Incomplete lateral line, 19–34 pores; often arched near front. Has 45–69 lateral scales. To 2¾ in. (7.2 cm). **RANGE:** St. Lawrence-Great Lakes, Hudson Bay, and

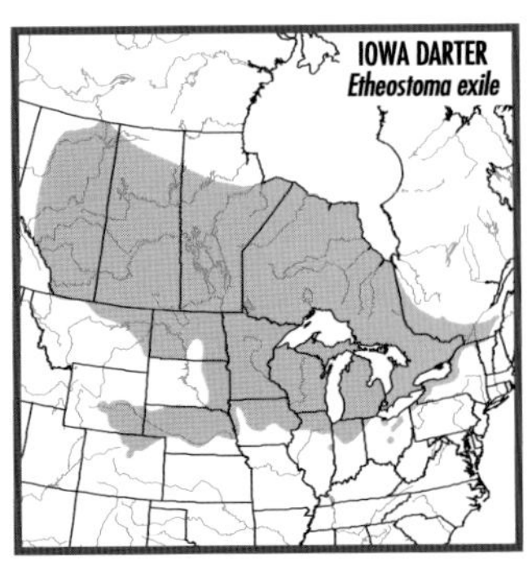

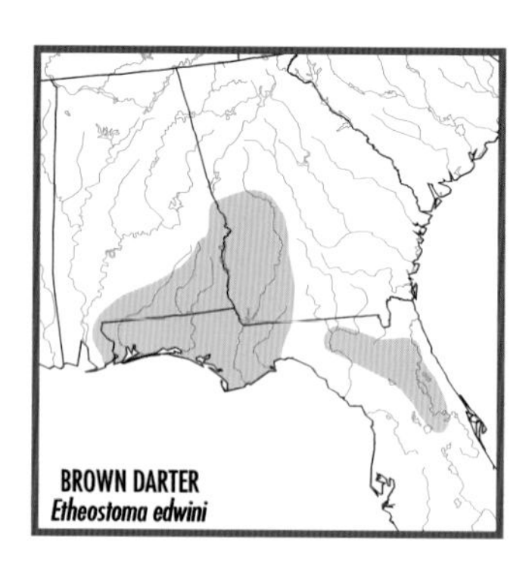

Mississippi R. basins from s. QC to n. AB, and south to OH, IL, and CO. Occurs farthest north and west of any darter. Common. **HABITAT:** Vegetated lakes, pools of headwaters, creeks, and small to medium rivers. **SIMILAR SPECIES:** (1) Least Darter, *E. microperca* (Pl. 55), is green, *deeper bodied*; has 0–3 lateral-line pores, 30–36 lateral scales, 2–3 infraorbital canal pores, pelvic fin flaps on male.

BROWN DARTER *Etheostoma edwini* Pl. 55

IDENTIFICATION: *Bright red spots* on *deep, compressed body* and dorsal, caudal, and anal fins of male (sometimes of female). *Incomplete, yellow lateral line*; may be arched near front. Tan above, brown mottling on side, white to yellow below. Thin teardrop; often black spots on lower half of head. Has 34–42 lateral scales. To 2 in. (5.3 cm). **RANGE:** St. Johns R. drainage, FL, to Perdido R. drainage, AL. Common. **HABITAT:** Sandy runs, especially near vegetation, of creeks and small rivers. **SIMILAR SPECIES:** (1) Swamp Darter, *E. fusiforme* (Pl. 55), is *more slender, lacks* red spots. (2) Okaloosa Darter, *E. okaloosae* (Pl. 45), has complete, *unarched* lateral line; rows of brown spots on side; *no* red spots.

Drums and Croakers: Family Sciaenidae (1 native; 2 introduced)

This widely distributed family has about 210 species. Most drums occupy continental shelves of tropical and temperate oceans. One species is restricted to fresh waters of North America; 2 marine species are established in the Salton Sea in southern California. The name drum (or croaker) refers to the ability to produce sounds using the gas bladder.

Drums have 2 dorsal fins, *the first relatively short* with spines, *the second longer* with rays; 1–2 anal spines; a lateral line *extending to the end of the caudal fin*; thoracic pelvic fins; ctenoid scales; and a lateralis

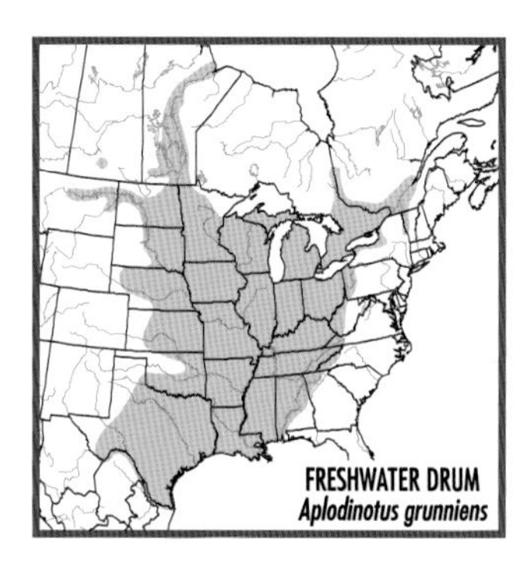

system on the head consisting of large cavernous canals and pores. In most species the body is deep, and highly arched at the origin of the first dorsal fin.

Two drums native to the Gulf of California have been introduced into the Salton Sea in southern California, as game fish. BAIRDIELLA, *Bairdiella icistia*, has a *large, terminal mouth* and rounded caudal fin; is gray above, silver below; and reaches 12 in. (30 cm). ORANGEMOUTH CORVINA, *Cynoscion xanthulus*, has a *troutlike body*, *lower jaw projecting beyond* the upper jaw, rear edge of the caudal fin pointed at the middle; is blue-gray above, with a *yellow caudal fin*; and reaches 36 in. (90 cm).

FRESHWATER DRUM *Aplodinotus grunniens* **Pl. 36**

IDENTIFICATION: Strongly arched body; subterminal mouth. Silver above and on side; dusky fins (except white pelvic fins). Pointed caudal fin. Very long outer pelvic ray. Second dorsal fin about twice as long as 1st dorsal fin; usually 10 dorsal spines, 29–32 rays. To 35 in. (89 cm). **RANGE:** Greatest latitudinal range of any N. American freshwater fish. St. Lawrence-Great Lakes, Hudson Bay, and Mississippi R. basins from QC to n. MB and s. SK, and south to Gulf; Gulf drainages from Mobile Bay, GA and AL, through e. Mexico to Río Usumacinta system, Guatemala. Common. **HABITAT:** Bottom of medium to large rivers and lakes.

Pygmy Sunfishes: Family Elassomatidae (7)

Pygmy sunfishes are *small* (to 1¾ in. [4.7 cm]), have *no lateral line*, no lateralis canal on the mandible, a round caudal fin, *cycloid scales;* 3–5 dorsal spines, and usually 3 anal spines. They have a large eye and protruding lower jaw; are dark olive to light brown with *many black specks* on the head and body; and have rows of black spots on the median fins. Pygmy sunfishes are restricted to the southeastern U.S.

BANDED PYGMY SUNFISH *Elassoma zonatum* Pl. 43

IDENTIFICATION: Has *7–12 dark green to blue-black bars* on side; *1–2 large black spots* (sometimes indistinct) on upper side below dorsal fin origin. *Dark teardrop.* No scales on top of head. Breeding male is black with green-gold flecks, alternating gold and black bars on side, black fins, gold-green bar under eye. Has 4–5 dorsal spines, 9–10 rays; 5–6 anal rays; 28–45 lateral scales; 15–16 pectoral rays. To 1¾ in. (4.7 cm). **RANGE:** Atlantic and Gulf coastal plain drainages from Roanoke R., NC, to St. Johns R., n. FL, and west to Brazos R., TX; north in Former Mississippi Embayment to s. IL and Wabash R. floodplain, s. IN. Rarely above Fall Line. Common. **HABITAT:** Swamps, heavily vegetated sloughs, and small sluggish streams; usually over mud. **SIMILAR SPECIES:** (1) Other pygmy sunfishes (Pl. 43) *lack* large black spot(s) on side, dark teardrop; all but Bluebarred, *E. okatie*, have 3–4 dorsal spines.

SPRING PYGMY SUNFISH *Elassoma alabamae* Pl. 43

IDENTIFICATION: Has *6–8 thin gold or blue bars* along dark brown to black side. *Clear "window"* at rear of 2d dorsal and anal fins of large male. No large black spots on upper side. No scales on head. Usually 3 dorsal spines. Has 10–12 dorsal rays; 6–7 anal rays; 28–30 lateral scales. To 1¼ in. (3 cm). **RANGE:** Springs and spring runs in Tennessee R. drainage in Lauderdale and Limestone counties, AL. Uncommon in small area; several populations extirpated. **HABITAT:** Vegetated, spring-fed pools. **SIMILAR SPECIES:** (1) Other pygmy sunfishes (Pl. 43) *lack* thin gold bars along side, clear window at rear of 2d dorsal and anal fins.

EVERGLADES PYGMY SUNFISH *Elassoma evergladei* Pl. 43

IDENTIFICATION: *Scales* (often embedded) on top of head. Dark lips. No large black spots on upper side. No dark teardrop. Breeding male is shiny black with bright iridescent blue spots on side and below eye.

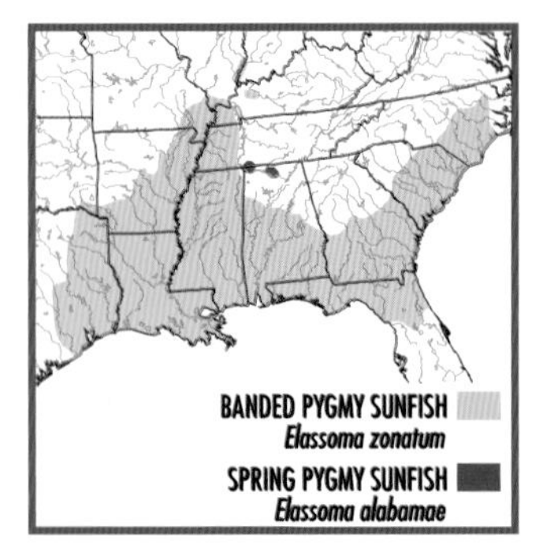

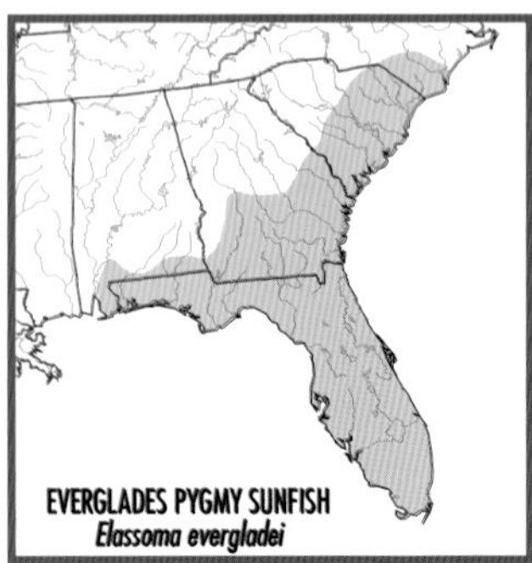

Usually 4 dorsal spines, 8–10 rays; 4–6 anal rays; 23–32 lateral scales. To 1¼ in. (3.4 cm). **RANGE:** Atlantic and Gulf coastal plain drainages from Cape Fear R., NC, to Mobile Bay, AL. Common. **HABITAT:** Swamps; heavily vegetated sloughs and small sluggish streams; usually over mud. **SIMILAR SPECIES:** (1) Other pygmy sunfishes (Pl. 43) *lack* scales on top of head. (2) Okefenokee Pygmy Sunfish, *E. okefenokee*, and (3) Gulf Coast Pygmy Sunfish, *E. gilberti*, have 2 large cream-colored spots on caudal fin base, light lips, 10–13 dorsal rays, 6–9 anal rays; breeding males have *bright blue bars* on side.

OKEFENOKEE PYGMY SUNFISH *Elassoma okefenokee* Pl. 43

IDENTIFICATION: Brown bars (darkest at rear, often broken into vertically aligned blotches) on side; *2 large cream-colored spots* on caudal fin base; front of lips *light-colored* (dark at sides) except in large male. No large black spots on upper side. No dark teardrop. No scales on top of head. Breeding male is shiny black with *bright iridescent blue bars* on side and below eye. Three preopercular canal pores; usually 4 dorsal spines; 10–13 dorsal rays; 7–9 (usually 8) anal rays; 26–33 lateral scales. To 1¼ in. (3.4 cm). **RANGE:** Atlantic Coastal Plain drainages from Altamaha R., GA, to Lake Okeechobee, FL. Also in interior lake basins in n.-cen. FL, and in upper Suwannee, Withlacoochee, and Hillsborough river drainages on Gulf Coast of FL. Common. **HABITAT:** Swamps; heavily vegetated sloughs and small sluggish streams; usually over mud. **SIMILAR SPECIES:** (1) See Gulf Coast Pygmy Sunfish, *E. gilberti*. (2) Everglades Pygmy Sunfish, *E. evergladei* (Pl. 43), *lacks* 2 large cream-colored spots on caudal fin base; has scales on top of head, *dark-colored lips*, 8–10 dorsal rays, 4–6 anal rays. (3) Banded Pygmy Sunfish, *E. zonatum* (Pl. 43), *lacks* 2 large cream-colored spots on caudal fin base; has 1–2 large black spots on side, dark teardrop.

GULF COAST PYGMY SUNFISH *Elassoma gilberti* Not shown

IDENTIFICATION: Nearly identical to Okefenokee Pygmy Sunfish, *E. okefenokee*, but has *4 preopercular canal pores*, usually 7 (often 8) anal rays; 27–32 lateral scales. To 1¼ in. (3.1 cm). **RANGE:** Gulf Coastal Plain drainages of FL and s. GA from Homosassa Springs Run R., cen. FL, to Choctawhatchee Bay, including lower Suwannee R. drainage (replaced by Okefenokee Pygmy Sunfish, *E. okefenokee*, in upper Suwannee). Common. **HABITAT:** Swamps; heavily vegetated sloughs and small sluggish streams; usually over mud. **SIMILAR SPECIES:** (1) See Okefenokee Pygmy Sunfish, *E. okefenokee* (Pl. 43).

BLUEBARRED PYGMY SUNFISH *Elassoma okatie* Pl. 43

IDENTIFICATION: Has 8–14 (usually 10–11) *wide dark bars* along side, about 3 times as wide as light interspaces. No large black spots on upper side. No scales on top of head. Breeding male is black with

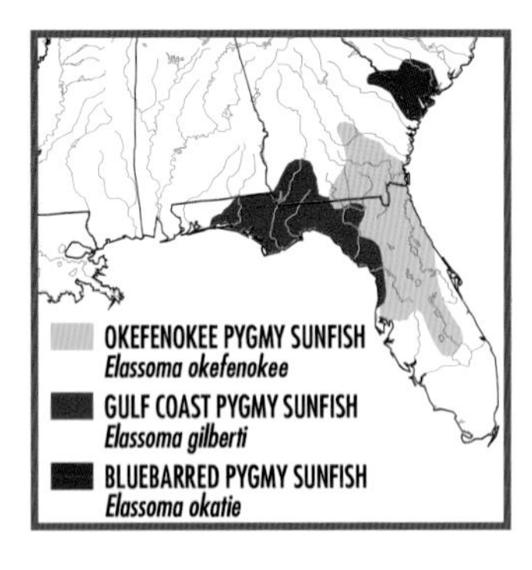

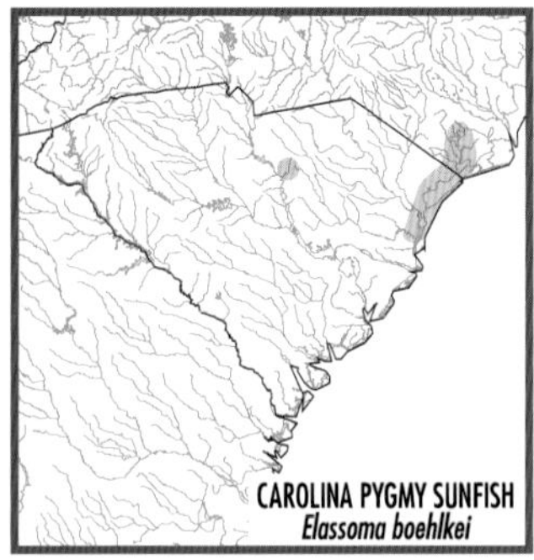

blue-green flecks and alternating blue-green and black bars on side. Usually 5 (often 4) dorsal spines; 8–12 dorsal rays; 4–8 anal rays; 24–30 lateral scales. To 1¼ in. (3.4 cm). **RANGE:** Lower Edisto, New, and Savannah river drainages, SC and GA. Localized and uncommon. **HABITAT:** Heavily vegetated creeks, sloughs, and roadside ditches. **SIMILAR SPECIES:** See Pl. 43. (1) Carolina Pygmy Sunfish, *E. boehlkei,* usually has 12–13 *narrow dark bars* along side, usually 4 dorsal spines. (2) Okefenokee Pygmy Sunfish, *E. okefenokee*, and (3) Gulf Coast Pygmy Sunfish, *E. gilberti*, usually have 4 dorsal spines. (4) Everglades Pygmy Sunfish, *E. evergladei*, has scales on top of head.

CAROLINA PYGMY SUNFISH *Elassoma boehlkei* **Pl. 43**

IDENTIFICATION: Has 10–16 (usually 12–13) *narrow dark bars* along side, about same width as light interspaces. No large black spots on upper side. No scales on top of head. Breeding male is black with blue-green flecks, alternating blue-green and black bars on side. Usually 4 dorsal spines; 8–12 dorsal rays; 4–8 anal rays; 24–30 lateral scales. To 1¼ in. (3.2 cm). **RANGE:** Waccamaw and Santee river drainages, NC and SC. Extremely localized and uncommon. **HABITAT:** Heavily vegetated creeks, sloughs, and roadside ditches. **SIMILAR SPECIES:** See Pl. 43. (1) Bluebarred Pygmy Sunfish, *E. okatie,* usually has 10–11 *wide dark bars* along side, usually 5 dorsal spines. (2) Everglades Pygmy Sunfish, *E. evergladei*, has scales on top of head.

MULLETS: Family Mugilidae (2)

Mullets are moderately elongated silver gray fishes that are *flattened above*, have *2 widely separated dorsal fins*, the first with 4 spines, the second with 1 spine, 7–9 rays, and no obvious lateral line. There are about 74 species worldwide, mostly in coastal marine waters. Two enter fresh waters in our area.

STRIPED MULLET *Mugil cephalus* Pl. 57

IDENTIFICATION: Rounded (in cross section) body in front, compressed at rear; *2 widely spaced dorsal fins, 2d fin with 1 spine and 8 rays.* Small spots on scales form *dusky lateral stripes* along blue-green to silver body. Small terminal mouth; adipose eyelid. Three anal spines, 8 rays. Large scales; 38–42 lateral scales. Has 25 or more rakers on 1st gill arch. To 36 in. (91 cm), but rarely more than 20 in. (50 cm) in fresh water. **RANGE:** Atlantic Coast from NS to s. Mexico; Pacific Coast from s. CA to Chile. Ascends lower reaches of rivers along Atlantic and Gulf coasts, farther upstream in large rivers, including Red R. to cen. OK; Mississippi R. to Missouri R., MO; lower Ohio R., KY. Common. **HABITAT:** Pools and runs of medium to large rivers. **SIMILAR SPECIES:** (1) See Mountain Mullet, *Agonostomus monticola.*

MOUNTAIN MULLET *Agonostomus monticola* Not shown

IDENTIFICATION: Similar to Striped Mullet, *Mugil cephalus*, but has *black spot* at pectoral fin base, another at caudal fin base; yellow median fins; no adipose eyelid. Light brown above, gray on side. Two anal spines, 10 rays. Has 17–20 rakers on 1st gill arch. To 12 in. (30 cm). **RANGE:** Atlantic and Gulf coasts from NC to TX, south to Venezuela and West Indies. Sporadic in streams in FL, LA, and TX. Rare in fresh water. **HABITAT:** Pools and runs of small to medium rivers. **SIMILAR SPECIES:** See (1) Striped Mullet, *Mugil cephalus* (Pl. 57).

CICHLIDS: Family Cichlidae (1 native; 20 introduced)

Cichlids, popular aquarium fishes, are native to Central and South America (1 species extends north to Texas), the West Indies, Africa, Madagascar, the Middle East, and coastal India. Most are freshwater; a few tolerate brackish water. There are about 1560 species; 1 is native to our region, and 20 have been introduced into the U.S. from either the American tropics or Africa.

Some introductions into the U.S. have been (and continue to be) deliberate; others are presumed to have been accidental. Deliberate introductions have been mainly for aquatic vegetation control and aquaculture. Most accidental introductions were from fish farms or release of aquarium pets.

Cichlids have only *1 nostril* on each side, a *2-part lateral line* with the front portion higher on the body than the rear portion, and exceedingly protractile jaws. Most species do not exceed 12 in. (30 cm).

OSCAR *Astronotus ocellatus* Pl. 56

IDENTIFICATION: White or yellow edge on *large, rounded* (fanlike) *2d dorsal, caudal, and anal fins.* Red around *large black spot* on upper caudal fin base; sometimes similar spot at rear of 2d dorsal fin.

Small scales on 2d dorsal and anal fins. Olive blue-green body and fins; large black blotches on body and fins. Young has wavy white and orange bars, spots on black head, body, and fins. Three anal spines. To 16 in. (40 cm). **RANGE:** Native to Ríos Orinoco, Amazon, and La Plata basins, S. America. Established in s. FL. Common. **HABITAT:** Mud- and sand-bottomed canals and standing water bodies.

PEACOCK CICHLID *Cichla ocellaris* Pl. 56

IDENTIFICATION: Elongate body; *large mouth, projecting lower jaw.* Silver halo around *large black spot* on caudal fin. Olive green above; *3 black bars* on side; yellow-white below; white spots on dark gray 2d dorsal fin, upper lobe of caudal fin; other fins gray or black. Large adult has yellow-orange stripe from mouth to caudal fin base; red on anal and pelvic fins, lower half of caudal fin; red iris. Scales on 2d dorsal and anal fins. To 26 in. (66 cm). **REMARKS:** Also referred to as Butterfly Peacock Bass, but not a bass. **RANGE:** Native to Essequibo and other drainages in Guianas. Established in canals and lakes, Miami-Ft. Lauderdale area, se. FL, where common. **HABITAT:** Pools and runs in blackwater rivers in S. America; drainage canals and lakes in FL.

REDSTRIPED EARTHEATER *Geophagus surinamensis* Pl. 56

IDENTIFICATION: *Long snout*; eye high on head; dip between eyes gives fish a "bug-eyed" look. *Black blotch* on side. Lateral line bifurcates on caudal peduncle; straight-edged caudal fin. Olive to green; many iridescent blue-green stripes on side; iridescent powder blue spots on fins (except pectoral). Large male may have rows of red spots on side, red fins (except pectoral), long 1st pelvic ray. Three anal spines. To about 12 in. (30 cm). **RANGE:** Native to Suriname and French Guiana. Possibly established in canals in se. FL. Uncommon. **HABITAT:** Mud- and sand-bottomed canals.

AFRICAN JEWELFISH *Hemichromis letourneuxi* Pl. 56

IDENTIFICATION: *Fairly slender body; rounded caudal fin.* Yellow-green to red-brown body and fins; *large black blotch* on side, smaller blotches on opercular tab and caudal fin base (blotches rarely connected to form stripe). Large individual has iridescent blue spots on *brilliant red body.* Has 13–15 dorsal spines, *3 anal spines.* To 8 in. (20 cm). **RANGE:** Native to rivers and lakes of cen. and w. Africa. Established in s. FL. Common. **HABITAT:** Mud- and sand-bottomed canals, streams, and swampy areas; usually near vegetation.

RIO GRANDE CICHLID *Herichthys cyanoguttatus* Pl. 56

IDENTIFICATION: Dusky to olive above; *4–6 dark blotches* (1st most prominent) along rear half of side, usually confluent with dusky saddles; black blotch on caudal fin base. *Many small white to blue*

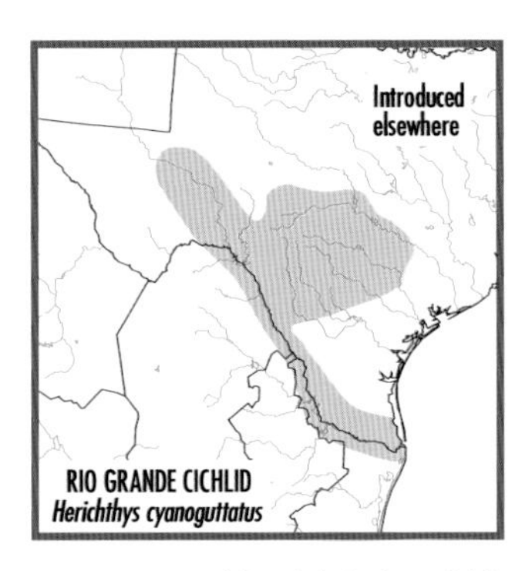

spots on blue-green or gray side. Adult has iridescent blue-green spots or wavy lines on head, body, and fins. Breeding individual has white head, front half of body; black rear half of body. Breeding male has nuchal hump. Has 15–17 dorsal spines, usually 10–12 rays; 5–7 anal spines; 9–10 anal rays. To 12 in. (30 cm). **RANGE:** Our only native cichlid. Originally in Nueces and lower Rio Grande drainages, TX, and south to ne. Mexico. Introduced elsewhere in TX as far north as Brazos R., and in sw. FL. Common. **HABITAT:** Pools and runs of small to large rivers; prefers warm water and vegetation. **SIMILAR SPECIES:** See Pl. 56. (1) Jack Dempsey, *Rocio octofasciata,* usually has 2 dark lines on top of head between eyes, 8–9 anal spines, 17–19 dorsal spines. (2) Convict Cichlid, *Amatitlania nigrofasciata*, has 9–11 anal spines, 7 intense black bars along side. (3) Midas Cichlid, *Amphilophus citrinellus, lacks* iridescent spotting pattern.

JACK DEMPSEY *Rocio octofasciata* Pl. 56

IDENTIFICATION: *Two gray to black lines* between eyes. Olive, gray, or tan above; gray to black connected blotches along side (may form bars); black spot at caudal fin base; white to iridescent blue spots on median fins. Adult usually *deep blue*; has dark bars, *many iridescent white, blue, or green spots* on head, body, and median fins; some have purple or red below, red edge on dorsal fin. Has 17–19 dorsal spines, 9–10 rays; 8–9 anal spines; 7–8 anal rays. To about 10 in. (25 cm). **RANGE:** Native to Atlantic Slope from Río Chachalacas, Veracruz, Mexico, to Río Ulúa, Honduras. Possibly reproducing in canals and ponds in cen. and s. FL. Uncommon. **HABITAT:** Weedy, mud- and sand-bottomed canals, ditches, and ponds. **SIMILAR SPECIES:** See Pl. 56. (1) Rio Grande Cichlid, *Herichthys cyanoguttatus, lacks* 2 dark lines on top of head; usually has 5–7 anal spines, 15–17 dorsal spines. (2) Convict Cichlid, *Amatitlania nigrofasciata*, usually has 9–11 anal spines, 7 intense black bars along side. (3) Midas Cichlid, *Amphilophus citrinellus, lacks* 2 dark lines on top of head, usually has 6–8 anal spines.

CONVICT CICHLID *Amatitlania nigrofasciata* Pl. 56

IDENTIFICATION: Usually *7 black bars* on side extend onto dorsal and anal fins (intensity of bars variable; sometimes 1st 3 appear as blotches); *1st bar Y-shaped*, crosses nape; 7th bar on caudal fin base. *Black spot* on upper half of opercle. Light blue or gray; clear or light blue-gray fins. Large male has intense black bars, black underside of head, long rays at rear of dorsal and anal fins; female has gold yellow lower side. Has 18–19 dorsal spines, 7–9 rays; 9–11 anal spines, 6–8 rays; 12–14 pectoral rays. To 4¾ in. (12 cm). **RANGE:** Native to Cen. America from Guatemala to Panama. Established in springs of Custer Co., ID (aquarium-developed white form), Lincoln and Clark counties, NV, and Grand Teton National Park, WY. Common. **HABITAT:** Warm pools of springs and their effluents. **SIMILAR SPECIES:** (1) Midas Cichlid, *Amphilophus citrinellus* (Pl. 56), *lacks* Y-shaped bar on nape; usually has 6–8 anal spines, 14–18 pectoral rays, 10–13 dorsal rays. (2) Black Acara, *Cichlasoma bimaculatum* (Pl. 56), *lacks* black bars on side; usually has 4 anal spines, 15–16 dorsal spines.

MIDAS CICHLID *Amphilophus citrinellus* Pl. 56

IDENTIFICATION: *Six dusky to black bars* (darkest at middle) on side; *large black blotch* on midside, *smaller black spot* on caudal fin base (spots always prominent; bars vary). Color highly variable (see Remarks). Gray-brown; small white spots on gray fins (except clear pectoral). Large male has nuchal hump, long rays at rear of dorsal and anal fins. Red, orange, yellow, or white variants may be found; some have thick, fleshy lips. Has 15–18 dorsal spines, 10–13 rays; 6–8 anal spines, 5–10 rays; 14–18 pectoral rays. To 12 in. (31 cm). **RANGE:** Native to Atlantic Slope of Nicaragua and Costa Rica. Established in canals in s. FL. Uncommon. **HABITAT:** Rocky canals; crevices used for spawning and protection of young. **REMARKS:** Although most Midas Cichlids are gray with black marks, a few become brightly colored (gold, red, or white), usually gold. Brightly colored individuals are more aggressive and grow faster, but are more vulnerable to predation. **SIMILAR SPECIES:** (1) See Mayan Cichlid, *Cichlasoma urophthalmus*. (2) Convict Cichlid, *Amatitlania nigrofasciata* (Pl. 56), has Y-shaped black bar crossing nape, usually 7–9 dorsal rays, 9–11 anal spines, 12–14 pectoral rays. (3) Black Acara, *Cichlasoma bimaculatum* (Pl. 56), has black stripe on side usually extending onto opercle, 4 anal spines.

MAYAN CICHLID *Cichlasoma urophthalmus* Not shown

IDENTIFICATION: Similar to Midas Cichlid, *Amphilophus citrinellus*, but has 5–7 bold dark green to black bars on side; blue halo around *large black blotch* on upper caudal fin base (spot is about half the depth of caudal peduncle). Olive green above and on side; red-edged dorsal and caudal fins; red on chin, throat, and breast. Yellow iris. Has 5–7 anal spines. To 15 in. (38 cm). **RANGE:** Native to Atlantic Slope from Río Coatzcoalcos, Mexico, to Río Prinzapolka, Nicaragua. Established in s. FL north to Melbourne and Tampa. Uncommon. **HABITAT:** Lakes; freshwater marshes and mangrove swamps. Tolerates wide variety of salinities; breeds in fresh and salt water. **SIMILAR SPECIES:** (1) See Midas Cichlid, *Amphilophus citrinellus* (Pl. 56).

BLACK ACARA *Cichlasoma bimaculatum* Pl. 56

IDENTIFICATION: *Two black blotches* on side, 1 at midbody, 1 on upper caudal fin base; smaller blotches along yellow-gray to dark green side usually extend as *dark stripe onto opercle*; 6–7 olive to dark brown bars on side; black suborbital spot. Adult has blue-gray fins, except yellow pectoral fin; long rays at rear of dorsal and anal fins. Usually 15–16 dorsal spines, 10–11 rays; 4 anal spines, usually 9–10 rays; 13–15 pectoral rays. To 8 in. (20 cm). **RANGE:** Native to Guianas. Established in s. FL south of Lake Okeechobee. Common. **HABITAT:** Canals and bodies of standing water; tolerates low oxygen. **SIMILAR SPECIES:** (1) See Yellowbelly Cichlid, *C. salvini.* (2) Convict Cichlid, *Amatitlania nigrofasciata* (Pl. 56), has 7 intense *black bars* along side, usually 9–11 anal spines, 18–19 dorsal spines. (3) Midas Cichlid, *Amphilophus citrinellus* (Pl. 56), has 6–8 anal spines, usually *no* black stripe on opercle; reaches 12 in. (31 cm).

YELLOWBELLY CICHLID *Cichlasoma salvini* Not shown

IDENTIFICATION: Similar to Black Acara, *C. bimaculatum*, but is *yellow*; has *row of black blotches* (partially connected) along middle of side, 2d row on upper side; orange lower side. Large male is bright yellow, has iridescent blue on dorsal fin. To 7 in. (18 cm). **RANGE:** Native to Atlantic Slope from Río Papaloapan, Mexico, to Guatemala. Established in se. FL. Uncommon. **HABITAT:** Mud-bottomed canals. **SIMILAR SPECIES:** (1) See Black Acara, *C. bimaculatum.*

BANDED CICHLID *Heros severus* Pl. 56

IDENTIFICATION: Deep, compressed body. Usually *5–7 dusky to black bars* on side; bar on caudal fin base extends *onto dorsal and anal fins.* Yellow-olive to dark green head and body. Adult is yellow below; has yellow lower fins. Young has many black spots on purple-tan body; lacks black bars. Has 15–17 dorsal spines, usually 11–13 rays; 6–8 anal spines, usually 11–12 rays. To about 8 in. (20 cm). **RANGE:** Native to Ríos Negro and Orinoco drainages, n. S. America. Possibly established in se. FL. Uncommon. **HABITAT:** Warm spring pools and their effluents.

FIREMOUTH *Thorichthys meeki* Pl. 56

IDENTIFICATION: *Large black blotch* on lower half of gill cover; *bright red or orange underside of head* (brightest on adult). Yellow-olive to gray head and body; 3–5 black blotches along side, 1 on caudal fin base; red edge, rows of blue spots or blotches on clear fins. Large male is dark green with iridescent blue spots on side, brilliant red underside of head; has long rays at rear of dorsal, anal, and caudal fins, iridescent blue spots on red fins (except pectoral). Has 15–17 dorsal spines, 10–13 rays; 8–10 anal spines, 7–9 rays. To 6¾ in. (17 cm). **RANGE:** Native to Atlantic Slope from Río Tonalá, Veracruz, Mexico, to Guatemala. Possibly established in s. FL. Uncommon. **HABITAT:** Mud- and sand-bottomed canals and rocky ponds.

JAGUAR GUAPOTE *Parachromis managuense* Not shown

IDENTIFICATION: *Very large oblique mouth*, rear edge reaching below front of eye; *lower jaw projects* beyond upper jaw. Distinctive lobe on rear of preopercle. Green above; purple tinge on yellow-gold side; *many purple-black spots, blotches on body and fins*; usually row of black squares along side; yellow below. Red iris. Has 17–18 dorsal spines, 10–11 rays; 8–9 anal spines, 8–9 rays. To 25 in. (63 cm). **RANGE:** Native to Atlantic Slope from Honduras to Costa Rica. Established in s. FL. Locally common. **HABITAT:** Mud-bottomed canals.

NEXT 7 SPECIES (through Spotted Tilapia, *Tilapia mariae*): *Three anal spines* (rarely 4); usually straight-edged or slightly forked caudal fin, *black spot* on opercle.

MOZAMBIQUE TILAPIA *Oreochromis mossambicus* Pl. 56

IDENTIFICATION: *Large oblique mouth* reaches under front of eye or beyond. Gray to olive above; 3–4 black spots on dull yellow to gray-green side; yellow below; no dark bars on caudal fin. Large male has *thick blue upper lip*, blue or black body, *white underside of head*, *red edge* on black dorsal and caudal fins, red pectoral fin. Young has 6–8 black bars on silver side. Has 29–33 lateral scales; 15–17 (usually 16) dorsal spines, 10–12 rays; 9–10 anal rays; 14–20 rakers on lower limb of 1st gill arch. To 15 in. (39 cm). **RANGE:** Native to se. Africa. Established in several states, including AZ, CA, CO, FL, ID, and TX; stocked elsewhere for aquaculture. Common. **HABITAT:** Warm, weedy pools of sluggish streams; canals and ponds. **SIMILAR SPECIES:** See (1) Wami Tilapia, *O. urolepis.* (2) Blackchin Tilapia, *Sarotherodon melanotheron* (Pl. 56), has *small mouth*, usually 27–29 lateral scales; large male has *black underside* of head, gold gill cover.

WAMI TILAPIA *Oreochromis urolepis* **Not shown**

IDENTIFICATION: Nearly identical to Mozambique Tilapia, *O. mossambicus*, but has *19–27 rakers* on lower limb of 1st gill arch, *lacks* white underside of head on large male. Has 15–18, usually 17, dorsal spines. To 17 in. (43 cm). **RANGE:** Native to Wami R. basin, Tanzania. Established in irrigation canals and associated drainages in s. CA. Common. **HABITAT:** Warm weedy ditches and canals. **SIMILAR SPECIES:** (1) See Mozambique Tilapia, *O. mossambicus* (Pl. 56).

BLUE TILAPIA *Oreochromis aureus* **Not shown**

IDENTIFICATION: Blue-gray back shading to white belly; pink-red borders on dorsal and caudal fins. Black spot at rear of dorsal fin (lost on adult). Usually *12–15 dorsal rays, 18–26 rakers* on lower limb of 1st gill arch. Large male has bright metallic blue head, pale blue side, *blue-black chin and breast*. To 14½ in. (37 cm). **RANGE:** Native to n. Africa and Jordan Valley, Eurasia. Stocked in U.S. for aquaculture; established in AZ, CA, FL, NC, NV, and TX, possibly in CO, ID, OK, and PA. Common; widespread in peninsular FL. **HABITAT:** Warm ponds and impoundments. Reproduces in both fresh and brackish water. **SIMILAR SPECIES:** (1) See Nile Tilapia, *O. niloticus*.

NILE TILAPIA *Oreochromis niloticus* **Not shown**

IDENTIFICATION: Similar to Blue Tilapia, *O. aureus*, but has *dark bars* on caudal fin. Usually 12–14 dorsal rays, 20–26 rakers on lower limb of 1st gill arch. Large male has pink on lower head, side, and pectoral fins; often dark bars on side. To 25 in. (63 cm). **RANGE:** Native to n. and cen. Africa. Established in Pascagoula R. and Biloxi Bay drainages, s. MS, possibly in peninsular FL. Uncommon. **HABITAT:** Warm ponds and impoundments. **SIMILAR SPECIES:** (1) See Blue Tilapia, *O. aureus*.

BLACKCHIN TILAPIA *Sarotherodon melanotheron* **Pl. 56**

IDENTIFICATION: *Small mouth* does not reach front of eye. Orange or gold yellow on back and upper side, pale blue below. Large male has gold gill cover, *black underside* of head, black edge on median fins, often dark bars on side. Usually 27–29 lateral scales; 15–16 dorsal spines, 10–12 dorsal rays, 8–10 anal rays, 12–19 rakers on lower limb of 1st gill arch. To 10 in. (26 cm). **RANGE:** Native to estuaries and river deltas in w. Africa. Established in peninsular FL. Common. **HABITAT:** Lower reaches of streams; estuaries. **SIMILAR SPECIES:** (1) Mozambique Tilapia, *Oreochromis mossambicus* (Pl. 56), has *large mouth*, usually 29–33 lateral scales; large male has *white underside* of head. (2) Blue Tilapia, *O. aureus,* has *metallic blue head* on large male, 12–15 dorsal rays, 18–26 rakers on lower limb of 1st gill arch.

REDBELLY TILAPIA *Tilapia zillii* **Not shown**

IDENTIFICATION: *Large, nearly horizontal mouth.* Head wider than body. Silver gray to dark olive above; 6–7 faint dark bars on light olive to yellow-brown side (often with metallic green or red sheen); yellow to brown belly and fins. Black blotch on upper edge of opercle; many yellow spots around black spot on 2d dorsal fin. Adult has 6–8 black bars on side, *blood red* (brightest on male) underside of head, belly, and caudal peduncle; blue-green vermiculations on blue-black head. Has 14–16 dorsal spines, 10–13 rays; 7–10 anal rays. To 12½ in. (32 cm). **RANGE:** Native to n. and cen. Africa, Middle East. Established in AZ, CA, NC, SC, and TX. Common. **HABITAT:** Irrigation canals, ponds, and springs where stocked for aquatic weed control or as a food fish. **SIMILAR SPECIES:** (1) See Spotted Tilapia, *T. mariae* (Pl. 56).

SPOTTED TILAPIA *Tilapia mariae* **Pl. 56**

IDENTIFICATION: Similar to Redbelly Tilapia, *T. zillii*, but *lacks* blood red below, has 6–9 black blotches or bars on side that *continue onto dorsal fin*. Large individual has faint bars on side, *bold black blotches* along midside; white edge on dorsal and caudal fins. Has 15–16 dorsal spines, 12–15 rays; 10–12 anal rays. To 13 in. (33 cm). **RANGE:** Native to Atlantic Slope of w.-cen. Africa. Established in s. FL north to Melbourne and in Rogers Spring, Clark Co., NV. Common. **HABITAT:** Mud- to sand-bottomed canals; warm springs. **REMARKS:** A similar species from w. Africa, Hornet Tilapia, *T. buttikoferi*, with bold black and white bars on adult, black edge on caudal fin of juvenile, may be established in s. FL. **SIMILAR SPECIES:** (1) See Redbelly Tilapia, *T. zillii*.

SURFPERCHES: Family Embiotocidae (1)

Most surfperches are marine and occur inshore, in the surf zone, in kelp, and in tidepools. One species, Tule Perch, *Hysterocarpus traskii*, lives in fresh water in northern California. The other 23 species of this family occur off Japan and Korea and along the Pacific Coast of North America. Surfperches have a compressed, elliptical to oblong perchlike body; a continuous (unnotched) dorsal fin with usually 9–11 spines, 19–28 rays; a *scaled ridge* along the dorsal fin base; 3 anal spines; 15–35 anal rays; a forked caudal fin; and *cycloid scales*. Most are silver and barred or striped. All are livebearers.

TULE PERCH *Hysterocarpus traskii* **Fig. 64**

IDENTIFICATION: Deep body, terminal mouth, forked caudal fin. *Ridge of scales* at base of dorsal fin. Dark blue or purple above, white to yellow below. Three color phases: wide dark bars, narrow dark bars, or no bars on side. Adult may have large hump on nape. Has 20–26 anal

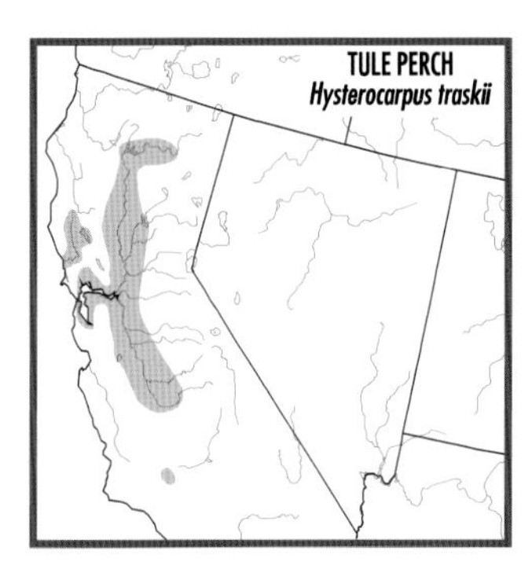

rays. To 5¾ in. (15 cm). **RANGE:** Clear Lake; Russian, Sacramento-San Joaquin, and Pajaro-Salinas river drainages, CA. Common in north; severely decimated in Pajaro-Salinas and San Joaquin river drainages. **HABITAT:** Mud- to gravel-bottomed pools and runs of small to large, low-elevation rivers; lakes. Usually near emergent aquatic plants or overhanging banks. Rare in brackish water. **REMARKS:** Three subspecies: *H. t. lagunae*, in Clear Lake; *H. t. pomo*, in Russian R.; *H. t. traskii,* in rest of range.

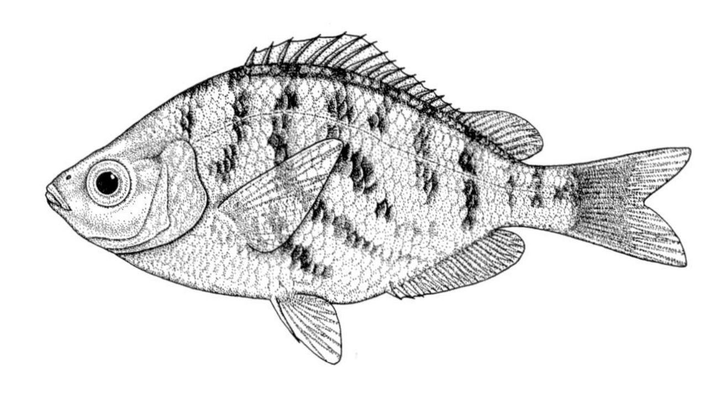

Fig. 64. Tule Perch.

SNAKEHEADS: Family Channidae (2 introduced)

These air-breathing freshwater fishes have a long cylindrical body, tubular anterior nostrils, long dorsal and anal fins, thoracic pelvic fins, and large scales on top of the head (most species). The large mouth has a protruding lower jaw, usually with canine teeth. Snakeheads are native to Asia and tropical Africa; about 30 species are recognized. They are sold in pet trade and fish markets in the U.S. Several species recently have been reported from the U.S. (records in 25 states); 2 are established.

BULLSEYE SNAKEHEAD *Channa marulius* **Not shown**

IDENTIFICATION: Long, cylindrical body; rounded caudal fin. *Long dorsal (45–55 rays) and anal (28–36 rays) fins.* Scales on top of head. Large terminal mouth; lower jaw with *7–18 canine teeth. Pale-edged ocellus* on upper caudal fin base (usually absent in adult over 16 in. [40 cm]). Body light with black blotches along side to dark purple-black with white specks; dark fins. Has 60–70 lateral scales. To 4 ft. (1.3 m). **RANGE:** Native to s. Asia. Established in lakes and canals in Broward Co., FL. Uncommon, but increasing. **HABITAT:** Clear or tannic-stained pools or lakes over sand or rock bottoms, often near aquatic vegetation. **SIMILAR SPECIES:** (1) See Northern Snakehead, *C. argus.* (2) Bowfin, *Amia calva* (Pl. 3), has abdominal pelvic fins, *short anal fin* (9–10 rays), gular plate, *no scales on top of head.*

NORTHERN SNAKEHEAD *Channa argus* **Not shown**

IDENTIFICATION: Similar to Bullseye Snakehead, *C. marulius*, but *lacks* ocellus on caudal fin, has dark stripes from eye to edge of gill cover. Color variable; body brown with dark brown blotches along side and on back; dark fins. To 33 in. (85 cm). **RANGE:** Native to China and Korea. Established in Potomac R. drainage, MD. Possibly established in lower White R. system, AR. **HABITAT:** Pools over sand or rocks, often near vegetation. **SIMILAR SPECIES:** (1) See Bullseye Snakehead, *C. marulius.*

SLEEPERS: Family Eleotridae (1)

Sleepers are goby-like, bottom-dwelling fishes with no lateral line and separate dorsal fins, but with *6 branchiostegal rays* and *pelvic fins not fused* to one another. They occur in tropical, subtropical, and temperate regions in marine, brackish, and fresh waters. About 170 species are recognized; 1 occurs in fresh water in our area.

BIGMOUTH SLEEPER *Gobiomorus dormitor* **Not shown**

IDENTIFICATION: Slender body; *wide, flat head;* lower jaw projects in front of upper jaw. Dark brown or olive above, often mottled; *black stripe* along side to caudal fin; dark lines on cheek. Has 55–65 lateral scales. To 2 ft. (0.6 m). **RANGE:** Atlantic and Gulf coasts from s. FL and s. TX to Brazil; Caribbean islands. Enters fresh water in s. FL; occasional in lower Rio Grande, TX. Common. **HABITAT:** Over sand streams; usually in current near cover. **SIMILAR SPECIES:** (1) Gobies have 5 branchiostegal rays; most have pelvic fins *fused* together.

GOBIES: Family Gobiidae (5 native; 4 introduced)

Gobies comprise one of the largest families of fishes with about 2000 species worldwide. They have 5 branchiostegal rays, no lateral line, and 2 dorsal fins—the first with spines and separated from the second with 0–1 spine and 9–25 rays. Most are bottom-dwellers, have *pelvic fins fused,* and are under 4 in. (10 cm). Gobies are abundant in the tropics, less so in temperate regions. Most live in shallow to moderately deep salt and brackish water, but some are in fresh water. There are 5 species native to coastal waters of the U.S.; 2 Asian species established in California; and 2 species established in the Great Lakes, probably through the release of ballast water from ships.

RIVER GOBY *Awaous banana* Not shown

IDENTIFICATION: *Long conical snout; small eyes* high on head; thick lips on large mouth, *upper lip extends in front of lower lip;* rounded caudal fin. *Skin flaps on shoulder girdle* project into gill chamber (lift gill cover). Small, dark blotches and vermiculations on yellow-tan body; 61–69 lateral scales. To 12 in. (30 cm). **RANGE:** Atlantic and Gulf coasts from SC to n. S. America (apparently absent from AL, MS, LA, and TX). Uncommon in U.S.; occasional in fresh water in FL. **HABITAT:** Over sand in flowing pools and runs of streams. **SIMILAR SPECIES:** No other goby in fresh waters of U.S. has long conical snout, rounded caudal fin.

DARTER GOBY *Ctenogobius boleosoma* Pl. 57

IDENTIFICATION: Long, slender body; long, pointed caudal fin. *No or few scales on nape.* Has 4 or 5 brown spots or bars on tan side, bars under 2d dorsal fin may be joined to form Vs; *large black spot* on caudal fin base; black spot at top of gill cover. Small brown spots on dusky dorsal and anal fins; orange edge on dorsal, caudal, and pectoral fins of male. Has 29–35 lateral scales, 6 spines in 1st dorsal fin, 11 dorsal rays, 12 anal rays, 16 pectoral rays. To 3 in. (7.5 cm). **RANGE:** Atlantic and Gulf coasts from NJ (rarely n. of NC) to Brazil, including Bermuda, Bahamas, and West Indies. Common. **HABITAT:** Nearshore muddy and grassy areas. Usually in estuaries; enters fresh water. **SIMILAR SPECIES:** (1) See Freshwater Goby, *C. shufeldti* (Pl. 57).

FRESHWATER GOBY *Ctenogobius shufeldti* Pl. 57

IDENTIFICATION: Similar to Darter Goby, *C. boleosoma*, but has *scales on nape*, 12 dorsal rays, 13 anal rays, 17 pectoral rays. Has 28–40 lateral scales. To 3¼ in. (8 cm). **RANGE:** Atlantic and Gulf coasts from NC to TX. Apparently absent from west coast of peninsular FL. Common. **HABITAT:** Nearshore muddy and grassy areas. Usually in bays and estuaries; enters fresh water. **SIMILAR SPECIES:** (1) See Darter Goby, *C. boleosoma* (Pl. 57).

SLASHCHEEK GOBY *Ctenogobius pseudofasciatus* **Not shown**

IDENTIFICATION: *Dark brown bar* on lower part of cheek (preopercle to above corner of jaw). Long, pointed caudal fin. Light brown to straw above, 5 dark blotches on side; orange edge on 1st dorsal fin. Elongated 3d spine in 1st dorsal fin of male. Six spines in 1st dorsal fin; 1 spine, 11 rays in 2d dorsal fin, 12 anal fin rays; 29–34 lateral scales. To 2¼ in. (5.7 cm). **RANGE:** Caribbean Coast of Cen. America and n. S. America. Highly disjunct population in e.-cen. FL in Loxahatchee R., St. Lucie R., and Sebastian Creek and associated freshwater canals. Highly localized. **HABITAT:** Over sand in flowing pools and runs of streams.

LONGJAW MUDSUCKER *Gillichthys mirabilis* **Not shown**

IDENTIFICATION: Huge mouth; *upper jaw bones* nearly reach gill opening in adult (reaches back of eye in juvenile). Olive, brown, or blue-black above, dark mottling or bars on side; yellow below. Small black blotch at rear of 1st dorsal fin. Has 4–8 spines in low 1st dorsal fin; 10–17 rays in 2d dorsal fin; 60–100 lateral scales. To 8¼ in. (21 cm). **RANGE:** Along coast from Tomales Bay to Baja California, Gulf of California, and Salton Sea, CA. Abundant. **HABITAT:** Muddy tidal sloughs; intertidal creeks.

YELLOWFIN GOBY *Acanthogobius flavimanus* **Not shown**

IDENTIFICATION: Has 5–8 dusky blotches along light brown side; dark spots on dorsal fins; *dusky zigzag bars* on upper half of caudal fin. First dorsal fin with 8 spines, *taller* than 2d dorsal fin with 14 rays. Large ctenoid scales; 55–65 lateral scales. To 10¾ in. (27 cm). **RANGE:** Native to shallow coasts of Japan, China, and Korea. Established (in 1963) and abundant in San Francisco Bay and Sacramento R. to Knights Landing, CA. Occasional in Delta-Mendota Canal, California Aqueduct, San Luis Reservoir, and Los Angeles Harbor; may be found along coast as far south as San Diego Bay. **HABITAT:** Shallow, soft-bottomed areas.

SHIMOFURI GOBY *Tridentiger bifasciatus* **Not shown**

IDENTIFICATION: *Blunt, flattened head*; large mouth, jaws extend to back of eye; thick caudal peduncle. Mottled overall or with dark brown stripe along side; white specks on head. *White bar* at base of pectoral fin; *orange edge* on 2d dorsal and anal fins; tiny rows of spots on caudal fin. Has 6–7 spines in 1st dorsal fin; 1 spine, 11–14 rays in 2d dorsal fin; 54–60 lateral scales. To 4¼ in. (11 cm). **RANGE:** Native to estuaries of Japan and mainland along Sea of Japan. First discovered in Suisan Marsh, CA, in 1985; now abundant in upper San Francisco Estuary, Pyramid Reservoir, and Piru Creek, CA; expected in any water body connected to California Aqueduct. **HABITAT:** Shallow pools with rocky or vegetated substrate; spawns in estuaries and fresh water.

ROUND GOBY *Neogobius melanostomus* **Not shown**

IDENTIFICATION: *Large head*, eye high on head. *Bold black spot* on rear of 1st dorsal fin. Black specks and mottling on gray body; no dark bars on fins; *small dusky spot* at base of upper pectoral rays. Breeding male black overall. Short tubular anterior nostrils not reaching upper lip. Has 18–19 pectoral rays; usually 48–55 lateral scales. To 12 in. (30 cm) in native range, 7 in. (18 cm) in our area. **RANGE:** Native to Black and Caspian sea basins, Eurasia. Found in St. Clair R., MI-ON, in 1990; now throughout Great Lakes and tributaries in MI, OH, IN, and IL; Illinois R. (Mississippi R. basin), IL. Abundant in Lakes Michigan, Huron, and Erie; rare in Ontario and Superior. **HABITAT:** Rocky and vegetated lake shores and areas of large rivers; to 70 ft. (20 m) deep. **SIMILAR SPECIES:** (1) Only other goby in Great Lakes is Tubenose Goby, *Proterorhinus semilunaris*, which has long barbel-like nostrils overhanging upper lip, *no* black spot on rear of 1st dorsal fin.

TUBENOSE GOBY *Proterorhinus semilunaris* **Not shown**

IDENTIFICATION: *Long barbel-like nostrils* overhang upper lip. Mottled brown overall; dark bars on fins; no conspicuous black spot on rear of 1st dorsal fin; no dark spot at base of upper pectoral rays. Has 14–16 pectoral rays; usually 42–47 lateral scales. To 5 in. (13 cm). **RANGE:** Native to Eurasia, including tributaries and estuaries of Black, Caspian, Aegean, and Aral seas. Established in St. Clair R. (1990) and Lake St. Clair, MI, west shore of Lake Erie, MI, probably in n. OH. Uncommon. **HABITAT:** Rocky and weedy areas of streams and lakes. **REMARKS:** Considered endangered within its native range. **SIMILAR SPECIES:** (1) Only other goby in Great Lakes is Round Goby, *Neogobius melanostomus*, which has *short* tubular nostrils; black spot on rear of 1st dorsal fin; no dark bars on fins.

SAND FLOUNDERS: Family Paralichthyidae (1)

Sand flounders have an extremely *flattened body* with long dorsal and anal fins covering nearly all of the upper and lower edges of the body, the *origin of the dorsal fin over or slightly in front of the eyes, and the eyes on the left side of the head*. Most are marine; few live in fresh water. They number 112 species, in the Atlantic, Pacific, and Indian oceans.

SOUTHERN FLOUNDER *Paralichthys lethostigma* Pl. 57

IDENTIFICATION: Lateral line *strongly arched* over pectoral fin. Oblong body, pointed snout. Color varies to match substrate; typically dark brown spots, mottling on light brown body. Has 85–100 lateral scale rows; 80–95 dorsal fin rays; 63–74 anal fin rays. To 30 in. (76 cm). **RANGE:** Atlantic and Gulf coasts from Albemarle Sound, NC, to n. Mexico. Uncommon in fresh water. **HABITAT:** Near shore, usually over mud; ascends large rivers in FL, AL, and TX. **SIMILAR SPECIES:** (1) Hogchoker, *Trinectes maculatus* (Pl. 57), has *origin of dorsal fin near mouth, eyes on right side of head, straight lateral line*, no pectoral fins, bluntly rounded snout.

Righteye Flounders: Family Pleuronectidae (1)

Righteye flounders have an extremely *flattened body* with long dorsal and anal fins covering nearly all of the upper and lower edges of the body, the *origin of the dorsal fin over or slightly in front of the eyes, the eyes usually on the right side of the head*. There are 101 species worldwide; most are marine (occasionally brackish, rarely in fresh water).

STARRY FLOUNDER *Platichthys stellatus* Not shown

IDENTIFICATION: *Large dark bars alternate with yellow to orange bars* on median fins. *Rough, star-shaped tubercles* on brown to nearly black upper side; sometimes dark blotches on white lower side. Oblong body, pointed snout. Eyes on right or left side of head. Has 68–78 lateral scales, 52–64 dorsal fin rays, 38–47 anal fin rays. To 3 ft. (1 m). **RANGE:** Pacific and Arctic coasts from Santa Barbara, CA, to Japan and Bathurst Inlet, Canada. Common; sporadic in fresh water. **HABITAT:** Mostly near shore, often in estuaries, over sand and mud. **SIMILAR SPECIES:** Only flatfish regularly in fresh waters along Pacific and Arctic coasts of N. America.

American Soles: Family Achiridae (1)

American soles have an extremely *flattened body* with long dorsal and anal fins covering nearly all of the upper and lower edges of the body, the *origin of the dorsal fin far forward—near the mouth—and the eyes on the right side of the head*. Thirty-three species occur in marine and fresh waters from the U.S. to Argentina.

HOGCHOKER *Trinectes maculatus* Pl. 57

IDENTIFICATION: Oblong body, *bluntly rounded snout. No pectoral fins.* Lateral line nearly straight; 7–11 black bars, often dark blotches, on dusky or dark brown upper side, white below. Has 65–77 lateral scales, 48–57 dorsal rays, 38–42 anal rays. To 8 in. (20 cm). **RANGE:** Atlantic and Gulf coasts from MA to Venezuela; freshwater streams, often far inland. Common. **HABITAT:** Sandy runs and pools of streams; less often over mud. Moves between estuaries and fresh water seasonally and with age. **SIMILAR SPECIES:** (1) Southern Flounder, *Paralichthys lethostigma* (Pl. 57), has *pointed snout, pectoral fins, origin of dorsal fin over eyes, eyes on left side of head*, lateral line arched over pectoral fin.

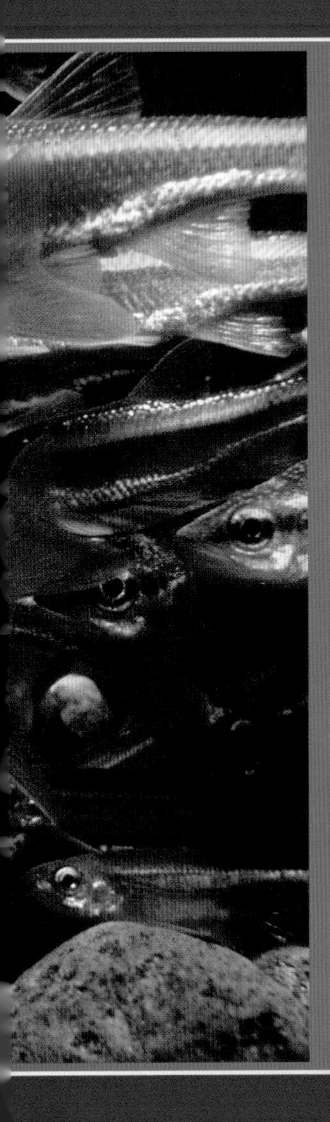

ACKNOWLEDGMENTS
GLOSSARY
REFERENCES
PHOTO CREDITS
INDEX

ACKNOWLEDGMENTS

Many colleagues contributed information and otherwise helped in the preparation of this guide. We are extremely grateful for their help; the finished product is much more accurate and useful than it would have been without their assistance. We thank Craig W. Ronto for Figures 6 and 52 and Jason Bourque for Figure 25.

Contributors deserving special thanks are those who reviewed portions of the manuscript or provided information on North American fishes, including William Eschmeyer, Nicholas Mandrak, James Williams, Robert Jenkins, Robert Robins, Michael Retzer, Alfred Thomson, Mark Sabaj, Jonathan Armbruster, David Etnier, Henry Bart, Carter Gilbert, Richard Mayden, David Eisenhour, Dean Hendrickson, Henry Robison, David Neely, Kyle Piller, Matt Thomas, Steven Norris, Carol Johnston, Rebecca Blanton, Patrick Ceas, Jean Porterfield, Andrés López, Nicholas Lang, Thomas Near, Frank Veraldi, Michael Hardman, Jason Knouft, Gabriela Hogue, Philip Harris, Gerald Smith, Robert Cashner, Andrew Simons, Anna George, Rebecca Fuller, Bernard Kuhajda, William Loftus, Brant Fisher, Erling Holm, Steve Phelps, William Voiers, William Smith-Vaniz, Brady Porter, Bruce Collette, Bruce Bauer, Franklin Snelson, Byron Freeman, Mary Freeman, Carl Ferraris, John Friel, John Lundberg, Christopher Taylor, Kevin Cummings, Christine Mayer, Molly Phillips, Michael Compton, Christopher Scharpf, Donovan German, Whit Bronaugh, Tim Berra, Thomas Buchanan, Richard Franz, Douglas Carney, Maynard Raasch, Edward Murdy, Edward Wiley, Howard Jelks, Leo Nico, Steve Walsh, Rudolf Arndt, James Albert, Will Crampton, Jeremy Wright, Robert Edwards, Gordon Weddle, Jason Seitz, Robert Wood, Jay Stauffer, Jeffrey Stewart, Joseph Nelson, Christine Thacker, Morgan Raley, Nelson Rios, Jeff Koppelman, John Bruner, Pamela Schofield, Philip Willink, Stuart Welsh, Dan Cincotta, Samantha Hilber, Justin Havird, J. R. Shute, Bradley

Ennis, Steve Powers, Tamra Mendelson, Pam Fuller, Mark Pyron, Thomas Simon, Rex Strange, Fred Rohde, Melvin Warren, George Burgess, John Switzer, Frank Schwartz, Carl Bond, Miles Coburn, Anthony Echelle, Ray Birdsong, David Lindquist, Jamie Thomerson, James Grady, William LeGrande, William Taylor, F. William Beamish, Philip Cochran, Clark Hubbs, Alex Peden, Robert Behnke, Robert Miller, Walter Courtenay, Reeve Bailey, Philip Smith, Thomas Baugh, Jacques Bergeron, Carl Bond, Noel Burkhead, Paul Monaco, Peter Moyle, Jerry Johnson, David Propst, John Rinne, William Roston, Wayne Starnes, Harold Tyus, Ginny Adams, Reid Adams, Frank Cross, William Pflieger, Edward Menhinick, Neil Douglas, Timothy Bonner, Bobby Whiteside, Dean Fletcher, Stephen Ross, Mollie Cashner, Ralph Yerger, Camm Swift, James Rogers, Steve Kelsch, Herbert Boschung, John Ramsey, Richard Wydoski, Richard Whitney, J. D. McPhail, David Lee, Steve Platania, Thomas Turner, Alexandra Snyder, W. L. Minckley, C. Lavett Smith, Maurice Mettee, Patrick O'Neil, Karsten Hartel, George Becker, John Lyons, Edwin Cooper, David Swofford, Bret Albanese, Molly Phillips, and Weerapongse Tangitjaroen.

Completion of this revision was supported in part by the All Catfish Species Inventory award from the U.S. National Science Foundation (DEB-0315963).

Some terms below have broader or alternate meanings but are defined as applied to fishes in this guide.

Abdomen. Belly; lower surface of a fish between pelvic fins and anus (Fig. 1).

Adipose eyelid. Translucent tissue that partially covers eyeball in some fishes (Fig. 8).

Adipose fin. Small fleshy fin, usually without spines or rays, on back between dorsal fin and caudal fin (see Pl. 26).

Adnate. Joined together congenitally.

Allopatric. Occurring in different geographic areas. See Sympatric.

Ammocoete. Blind larva of lamprey (Fig. 4).

Anadromous. Moving from ocean into fresh water to spawn, as in salmons and shads.

Anal fin. Median fin located on undersurface, usually just behind anus (Fig. 1).

Axillary process. Fleshy flap (actually modified scale), usually narrow and projecting to rear, just above pectoral or pelvic fins.

Backwater. Quiet pool on side of stream channel. See Pool.

Band. Bar or stripe on fin. See Bar, Stripe.

Bar. (1) Vertical band of color. See Band, Stripe. (2) Ridge of sand or gravel in stream or along shore, formed by water currents.

Barbel. Fleshy projection (sometimes whiskerlike) on head (usually near mouth) (Fig. 16). Barbels are sensitive to touch and taste.

Basioccipital process. Extension of bone at lower rear edge of skull (Fig. 21).

Benthic. Living on or near the bottom.

Blotch. Irregularly shaped color mark. See Speck, Spot.

Branchiostegal membranes. Membranes connecting branchiostegal rays.

Branchiostegal ray. Raylike bony support of branchiostegal membranes.

Breast. Chest; lower surface of body between gill openings and pelvic fins (Fig. 1).

Bridle. Color mark across snout that suggests a bridle (Fig. 62).

Canine. Pointed, conical tooth, usually larger than surrounding teeth.

Carnivorous. Meat eating; feeding on animals.

Cartilage. Material that forms skeleton of young fishes and which persists in adults of some species (notably sharks and rays) but is largely converted to bone in most fishes.

Catadromous. Moving from fresh water to ocean to spawn, as in American Eel.

Caudal fin. Median fin at rear of body; "tail" (Fig. 1).

Caudal fulcrum. V-shaped, spinelike scale at front of caudal fin.

Caudal peduncle. Rear, usually slender, part of body between base of last dorsal- and anal-fin rays and caudal-fin base (Fig. 1).

Caudal spot. Spot at base (origin) of caudal fin.

Channel. Main course of a stream.

Coastal Plain. Plain extending inland along Atlantic (Atlantic Coastal Plain) and Gulf of Mexico (Gulf Coastal Plain) coasts; extends north to southern Illinois on Former Mississippi Embayment.

Compressed. Flattened from side to side.

Concave. Bowed or curved inward.

Convex. Bowed or curved outward.

Ctenoid. Type of scale with toothed rear edge, making scale rough to touch.

Cusp. Principal projecting point of tooth.

Cycloid. Type of scale with smooth rear edge, making scale smooth to touch.

Deciduous. In reference to scales, loosely attached to body and easily shed when fish is handled.

Decurved. Curved downward.

Depressed. Flattened from top to bottom.

Disc. See Oral disc.

Dorsal. Upper part of fish; above.

Dorsal fin. Median rayed fin on back, often notched or subdivided; sometimes fully divided into two fins (Fig. 1).

Ear flap. Fleshy or bony extension on rear edge of opercle, as on sunfishes.

Embedded. Covered by skin (usually refers to scales).

Endemic. Restricted to particular drainage, lake, etc.

Endorheic. Referring to surface drainage not reaching sea.

Epural. Elongate detached bone above urostyle and behind last neural spine supporting caudal fin rays.

Extirpated. Not extant; eliminated by human activities as opposed to extinct from natural causes.

Falcate. Deeply indented or sickle-shaped, e.g., edge of fin.

Fall Line. Boundary between Coastal Plain and Piedmont, typically marked by waterfalls or large rapids on rivers.

Fin base. Part of fin attached to body.

Forked. In reference to caudal fin, used when rear edge is distinctly indented.

Former Mississippi Embayment. See Mississippi Embayment.

Frenum. Fleshy bridge or connection, as between snout and upper lip (Fig. 57).

Fusiform. Cylindrical and tapering at both ends (usually refers to body shape).

Ganoid. Type of scale covered with hard enamel (e.g., on gars).

Gas bladder. Sac located between spinal column and gut cavity; also called air or swim bladder.

Genital papilla. Small, fleshy projection at genital pore (immediately behind anus) in some fishes.

Genus (plural: genera). Taxonomic category including one species or a closely related (i.e., all species sharing a common ancestor) group of species.

Gill. Breathing organ in fishes, including highly vascularized filaments used to extract oxygen from water.

Gill arch. Bony or cartilaginous support to which gill filaments and gill rakers are attached (Fig. 2).

Gill chamber. Cavity where gills are located.

Gill cover. Bony flap covering outside of gill chamber (Fig. 2).

Gill filament. Feathery projection on rear of gill arch; for exchange of respiratory gases.

Gill opening. Opening at rear of head, from gill chamber to outside (most fishes have one on each side); called gill slits in sharks and rays.

Gill raker. Toothlike projection on front of gill arch; often used to trap food items between gill arches (Fig. 2).

Gonopodium. Front rays of anal fin of male livebearers, modified to serve as an intromittent organ.

Gular plate. Bony plate on throat.

Habitat. Place where a fish lives; usually defined in terms of substrate, current, and stream size.

Halo. Circle of color around spot of another color.

Head length. Distance from tip of snout, lip, or chin—whichever is farthest forward—to rear edge of gill cover (Fig. 1).

Herbivorous. Vegetarian; feeding on plants.

Hermaphrodite. Having both male and female gonads in one individual.

Herringbone lines. Pattern of parallel slanting lines (e.g., as caused by ribs of a herring or blood vessels on upper side of a minnow).

Heterocercal. Type of caudal fin in which vertebral column extends into upper lobe, which is usually longer (e.g., on sturgeons; Pl. 2).

Homocercal. Type of caudal fin in which all principal rays of fin attach to hypural plate (modified last vertebra). This type of caudal fin is usually symmetrical (e.g., on shads; Pl. 4).

Humeral spot. Large spot at upper edge of pectoral fin base (often on "humeral scale").

Ichthyologist. Person who studies fishes.

Infraorbital. Below eye; e.g., infraorbital canal or pores (Fig. 55).

Intergrade. Individual with characters intermediate between those of two subspecies and found in "zone of intergradation."

Interorbital. Between eyes (orbits).

Interpelvic width. Straight-line distance between pelvic fin bases.

Invertebrate. Animal lacking backbone (e.g., insect, crayfish, worm).

Isthmus. Triangular, frontmost part of underside of body; largely separated from head, in most bony fishes, by gill openings.

Jugular. In throat area; e.g., in reference to location of pelvic fins.

Juvenile. Young; usually small version of adult.

Keel. Shelflike fleshy or bony ridge.

Lacrimal. Bone between eye and nostril (Fig. 50).

Larva (plural: larvae). Newborn; developmental stage of a fish before it becomes a juvenile.

Lateral. On the side.

Lateralis system. Sensory system consisting of a series of pores and canals on head, body, and sometimes caudal fin of a fish; used to detect water movements.

Lateral line. Canal along body, usually single and located roughly at midside, but sometimes branched and variously placed. Rearward extension of sensory canal system (lateralis system) on head; contains sense organs that detect pressure change (Fig. 1).

Lateral-scale count. Number of scales along lateral line, if present and complete, or along midside if lateral line is absent or incomplete (Fig. 1).

Lateral scales. Row of scales along midside (usually along lateral line) from rear edge of gill cover to base of caudal fin (Fig. 1). Often called lateral-line scales. See Lateral-scale count.

Leptocephalus. Transparent, ribbonlike larva of eels and tarpons.

Littoral. Occurring at or in immediate vicinity of shoreline.

Mandible. Lower jaw.

Maxilla. Rear bone of two bones forming upper jaw.

Medial. In middle plane or axis of body.

Median fins. Unpaired fins located on median plane of body; dorsal, caudal, and anal fins (Fig. 1).

Melanophore. Cell containing melanin, a dark brown or black pigment. When contracted, these cells appear as pepperlike dots; when expanded, large areas of fish may become dark.

Midwater. In or near middle water stratum, as opposed to at surface or on bottom (see Benthic).

Mississippi Embayment. Low plain from southern Illinois to Gulf of Mexico; submerged under sea through much of its history, now traversed by Mississippi River. Also called Former Mississippi Embayment.

Monotypic. Referring to a genus or family containing only one species.

Myomeres. Body segments. In lampreys, trunk myomeres extend from first segment after last gill pore to, and including, segment before anus.

Nape. Part of back immediately behind head; in spiny-rayed fishes, part between head and point where first (spiny) dorsal fin begins (Fig. 1).

Neoteny. Retention of juvenile features in adult.

Nipple. Small projection resembling a teat.

Nocturnal. Active at night.

Nostril. Nasal opening (fishes usually have two on each side).

Nuchal. Pertaining to nape (e.g., nuchal hump).

Ocellus (plural: ocelli). Eyelike spot; usually dark, bordered by ring of light pigment.

Omnivorous. Feeding on both plants and animals.

Opercle. Uppermost and largest of bones forming gill cover (Fig. 1).

Oral disc. Fleshy circular structure surrounding mouth of lamprey (Fig. 4).

Orbital. Related to eye.

Origin. Point where fin begins—point at which most anterior ray is inserted.

Oxbow. Lake formed in abandoned channel of stream meander after stream has cut through land at narrow point in meander.

Paedomorphism. Retention of juvenile characteristics in adult.

Paired fins. Collectively, pectoral and pelvic fins (Fig. 1).

Palatine. One of pair of bones on roof of mouth, one on each side, between jaw and midline; often used to describe teeth on this bone.

Palatine teeth. Teeth on palatine bones.

Papilla (plural: papillae). Small nipplelike projection of tissue.

Papillose. Having papillae.

Parr marks. Dark elliptical bars on side of young fish; usually absent on adult.

Pectoral fin. One of pair of fins (one on each side) attached to shoulder girdle, just behind head (Fig. 1).

Peduncle. See Caudal peduncle.

Pelagic. Living in open water away from bottom.

Pelvic fin. One of pair of fins on lower part of body (Fig. 1); position ranges from on belly just in front of anal fin (abdominal), to under pectoral fin (thoracic), to isthmus (jugular).

Peritoneum. Lining of abdominal cavity. May be pigmented and visible externally.

Pharyngeal. Of, or near, pharynx.

Piedmont. Hilly, upland region between Coastal Plain and Appalachian Mountains, stretching from southeastern New York to central Alabama.

Placoid. Type of toothlike scale; of dentin with pointed backward projection of enamel.

Plankton. Small plants (phytoplankton) and animals (zooplankton); mostly free-floating.

Plica (plural: plicae). Small fold of skin.

Plicate. Having folds of skin.

Pool. Quiet, often relatively deep, segment of stream. See Backwater, Riffle, Run.

Pore. Tiny opening in skin; usually involved with sensory perception.

Pored scale. Scale with pore; e.g., lateral-line scales.

Predorsal scales. Row of scales along middle of back between head and dorsal fin.

Premaxilla. One of pair of bones at front of upper jaw.

Premaxillary frenum. Bridge of flesh connecting upper lip and snout (Fig. 57).

Preopercle. Bone at rear of cheek and in front of gill cover, often separated from gill cover by groove (Fig. 1).

Preopercular spine. Any spine along rear or lower edges of preopercle (Fig. 53).

Preoperculomandibular canal (or pores). Canal (or pores) extending along rear edge of preopercle and onto mandible (Fig. 55).

Preorbital. In front of eye; e.g., preorbital bar.

Protrusible. Referring to mouth with upper lip not attached to snout and which may be extended far forward to catch prey.

Punctate lateral line. Pattern of rows of black spots above and below lateral line (Fig. 29).

Pyloric caecum (plural: caeca). Fingerlike tube at junction of stomach and intestine.

Ray. Flexible, segmented fin ray; often branched (Fig. 1). Also used to refer to bony element that supports and spreads branchiostegal membranes. See Branchiostegal ray.

Reticulate. Color markings in chainlike pattern or network.

Riffle. Fast-flowing, shallow segment of stream where surface of water is broken over rocks or debris. See Pool, Run.

Rudiment. Small, incompletely developed fin ray or gill raker (Figs. 2 & 3).

Run. Transitional segment of stream between riffle and pool, with moderate current and depth. See Pool, Riffle.

Saddle. Color mark, more or less rectangular, on back.

Scalation. Arrangement of scales; squamation.

Scute. Enlarged scale, often with one or more bony projections (Fig. 58).

Seaboard Lowlands. Low coastal border of New England.

Sea-run. Anadromous; moving from ocean into fresh water to spawn.

Serrae. Sawtooth-like notches along an edge.

Serrate. Having serrae.

Shoal. Shallow, usually gravel-bottomed, area of stream, often submerged area of bar.

Snout. Portion of head in front of eyes and above mouth (Fig. 1).

Soft ray. See Ray.

Speck. Very small blotch. See Blotch, Spot.

Spine. (1) Bony projection, usually on head. (2) Hard, unsegmented, and unbranched ray in fin; sometimes called a spinous ray (Fig. 1).

Spiracle. Opening (behind eye) of separate duct or canal leading to gill chamber in sharks, rays, and certain ancient bony fishes. (Not the gill opening.)

Spiral valve. Fold of tissue spiraling through intestine.

Spot. Circular color mark. See Blotch, Speck.

Squamation. Arrangement of scales.

Standard length. Straight-line distance from tip of snout, lip, or chin—whichever is farthest forward—to rear end of vertebral column (end of hypural plate; locate by lifting caudal fin and noting crease at caudal fin base). Used as standard measure of length of fish (Fig. 1).

Stripe. Horizontal band of color. See Band, Bar.

Submandibular pores. Pores along underside of mandible (lower jaw).

Submarginal. In reference to fins, area along, but not including, edge of fin.

Suborbital. Below eye.

Subspecies. Geographically diagnosable population of a species. Subspecies name consists of three parts: genus, species, subspecies, e.g., *Percina caprodes fulvitaenia*.

Subterminal. In reference to position of mouth, used when mouth opens below foremost point of head. See Terminal, Upturned.

Supratemporal canal (or pores). Canal (or pores) across back of head (Fig. 55).

Sympatric. Occurring in same geographic area. See Allopatric and Syntopic.

Symphyseal knob. Bony protuberance at junction of two bones (e.g., halves of lower jaw).

Syntopic. Occurring at same place (as opposed to sympatric—only in same general area). Two species may be sympatric without being syntopic.

Teardrop. Vertical color mark under eye. Suborbital bar.

Terminal. In reference to position of mouth, used when mouth opens at foremost point of head, upper and lower jaws being equally far forward. See Subterminal, Upturned.

Territorial. Defending a particular area.

Thoracic. On breast (e.g., in reference to pelvic fin).

Transverse scales. Row of scales from anal fin origin to dorsal fin (or middle of back) (Fig. 1).

Truncate. Straight-edged, as opposed to pointed or forked.

Tubercle. Small, white, hard (keratinized) protuberances on skin; usually seasonal in occurrence and only on breeding males (Fig. 27).

Tubercular ridge. Ridge of keratinized tissue along scales or fin rays.

Tubercular spot. White area where tubercle will develop.

Upturned. In reference to position of mouth, used when mouth opens above foremost point of head. See Subterminal, Terminal.

Ventral. Lower part of fish; below.

Vermiculation. Color pattern of wavy (wormlike) lines.

Vertebrate. Animal with backbone (e.g., fish, frog, squirrel).

Viscera. Internal body organs.

Vomerine. Pertaining to vomer, a median bone in front of roof of mouth; often used to describe teeth on this bone.

REFERENCES

The information used to prepare a field guide is drawn primarily from technical papers. The papers are not cited here because of established field guide practice and space limitations. Listed below are general references that provide information and serve as sources for technical literature.

Berra, T. M. 2001. *Freshwater Fish Distribution*. San Diego: Academic Press.

Eschmeyer, W. N., ed. 2010. *Catalog of Fishes*. Available at www.calacademy.org/research/ichthyology/catalog/fishcatsearch.html.

Eschmeyer, W. N., and E. S. Herald. 1983. *A Field Guide to Pacific Coast Fishes of North America*. Boston: Houghton Mifflin.

Lee, D. S., C. R. Gilbert, C. H. Hocutt, R. E. Jenkins, D. E. McAllister, and J. R. Stauffer, Jr., eds. 1980 et seq. *Atlas of North American Freshwater Fishes*. Raleigh: North Carolina State Museum of Natural History.

Nelson, J. S. 2006. *Fishes of the World*. 4th ed. New York: John Wiley & Sons.

Nelson, J. S., E. J. Crossman, H. Espinosa-Perez, L. T. Findley, C. R. Gilbert, R. N. Lea, and J. D. Williams. 2004. *Common and Scientific Names of Fishes from the United States, Canada, and Mexico*. 6th ed. Bethesda, MD: American Fisheries Society Special Publication 29.

Page, L. M. 1983. *Handbook of Darters*. Neptune City, NJ: T.F.H. Publications.

Robins, C. R., and G. C. Ray. 1986. *A Field Guide to Atlantic Coast Fishes of North America*. Boston: Houghton Mifflin.

David Snyder: 133
Lawrence Page: 134
Matthew Thomas: vi, vii, ix, 119, 625, 627, 635, 636, 637
William Roston: ii–iii, 622–623
Jeremy Monroe/Freshwaters Illustrated: xx–xxi, 116–117
Jim Berry: v

INDEX